D1703651

The Claisen Rearrangement

Edited by
Martin Hiersemann and
Udo Nubbemeyer

1807–2007 Knowledge for Generations

Each generation has its unique needs and aspirations. When Charles Wiley first opened his small printing shop in lower Manhattan in 1807, it was a generation of boundless potential searching for an identity. And we were there, helping to define a new American literary tradition. Over half a century later, in the midst of the Second Industrial Revolution, it was a generation focused on building the future. Once again, we were there, supplying the critical scientific, technical, and engineering knowledge that helped frame the world. Throughout the 20th Century, and into the new millennium, nations began to reach out beyond their own borders and a new international community was born. Wiley was there, expanding its operations around the world to enable a global exchange of ideas, opinions, and know-how.

For 200 years, Wiley has been an integral part of each generations journey, enabling the flow of information and understanding necessary to meet their needs and fulfill their aspirations. Today, bold new technologies are changing the way we live and learn. Wiley will be there, providing you the must-have knowledge you need to imagine new worlds, new possibilities, and new opportunities.

Generations come and go, but you can always count on Wiley to provide you the knowledge you need, when and where you need it!

William J. Pesce
President and Chief Executive Officer

Peter Booth Wiley
Chairman of the Board

The Claisen Rearrangement

Methods and Applications

Edited by
Martin Hiersemann and Udo Nubbemeyer

WILEY-VCH Verlag GmbH & Co. KGaA

Editors:

Priv.-Doz. Dr. Martin Hiersemann
Institute of Organic Chemistry
University of Dortmund
Otto-Hahn-Strasse 6
44227 Dortmund
Germany

Prof. Dr. Udo Nubbemeyer
Institute of Organic Chemistry
University of Mainz
Duesbergweg 10–14
55099 Mainz
Germany

Library of Congress Card No.: applied for
British Library Cataloguing-in-Publication Data
A catalogue record for this book is available from the British Library

Bibliographic information published by The Deutsche Nationalbibliothek
The Deutsche Nationalbibliothek lists this publication in the Deutsche Nationalbibliografie; detailed bibliographic data is available in the Internet at http://dnb.d-nb.de

Printed in the Federal Republic of Germany
Printed on acid-free paper

Typesetting Kühn & Weyh, Satz und Medien, Freiburg
Printing betz-druck GmbH, Darmstadt
Bookbinding Litges & Dopf Buchbinderei GmbH, Heppenheim
Wiley Bicentennial Logo Richard J. Pacifico

ISBN 978-3-527-30825-5

Contents

The Claisen Rearrangement. Edited by M. Hiersemann and U. Nubbemeyer

ISBN: 978-3-527-30825-5

Preface

Historically, the thermal rearrangement of aromatic and aliphatic allyl vinyl ether was first published in 1912 by Ludwig Claisen. The neat carbon variant of this 3,3 sigmatropic bond reorganization, the Cope rearrangement, was reported in 1940, 38 years later. Thus, the Cope reaction should have been termed as a 3-carba-Claisen rearrangement. However, the reverse is found within the literature: the Claisen rearrangement is termed as a 3-oxa-Cope rearrangement. Consequentily, the hetero Claisen reactions are found as 3-hetero-Cope conversions displaying heteroatoms such as nitrogen and sulfur in position 3 of the rearrangement framework. Carrying out a keyword supported literature search, this inconsistent use of synonyms describing one and the same process should be strongly considered. Within our book, we will use the historically exact name *Claisen* rearrangement. Considering the widespread applications of the Claisen rearrangement, we should keep in mind that Mother Nature has been utilizing the aliphatic version already for a much longer period of time: the enzyme-catalyzed rearrangement of chorismate into prephenate also follows the same mechanism.

Although nowadays almost anybody seems to know something about the Claisen rearrangement, the exact nature of the transition state and the way substituents and solvents influence the rate and the selectivity of the reaction can be very difficult to elucidate. However, for the vast majority of applications, qualitative guidelines are sufficient to predict and/or explain the course of a Claisen rearrangement. One of the main conclusions from this book is that there isn't *the* Claisen rearrangement but a truly amazing number of mechanistically related variations of it that have been and are being developed. In this context, the first *Claisen book* presents a platform concerning basics and the state of the art.

From the breathtaking number of applications in target-oriented synthesis it becomes evident that the Claisen rearrangement (and its variants) is one of the most powerful stereoselective carbon-carbon-bond forming reactions. The efficiency of the reaction clearly profits from its atom economy. However, to be honest, access to the actual substrate for the rearrangement may prove costly. A particular strength is the predictability of the stereochemical course of the rearrangement based on the knowledge of the geometry of the cyclic transition state.

Still, even after more than 90 years of development, optimization and application of Claisen rearrangements, there is still plenty of room for further research.

The Claisen Rearrangement. Edited by M. Hiersemann and U. Nubbemeyer

ISBN: 978-3-527-30825-5

With this in mind, we intended to provide interested researchers with a useful guide to the scope and limitations of this versatile rearrangement. To realize this task, we had to rely on various specialists who were originally contacted in the beginning of 2003 and, indeed, many of them agreed to contribute to the *Claisen book*. We are deeply indebted to all the authors who spent their limited time resources to compile a truly outstanding collection of facts concerning the various Claisen rearrangements. This book will certainly serve as a reference for many years to come.

Dortmund and Mainz *Martin Hiersemann and Udo Nubbemeyer*

List of Contributors

Katsuhiro Akiyama
Department of Applied Chemistry
Tokyo Institute of Technology
2-12-1S1-29 Ookayama, Meguro-ku
Tokyo 152-8552
Japan

Carole Alayrac
Organisch-Chemisches Institut der
Westfälischen Wilhelms-Universität
Münster
Corrensstrasse 40
48149 Münster
Germany

Stefan N. Gradl
Department of Chemistry
University of California at Berkeley
628 Latimer Hall
Berkeley, CA 94720
USA

Hong Guo
Department of Biochemistry,
Cellular and Molecular Biology
University of Tennessee
M407 Walters Life Sciences Building
Knoxville, TN 37996-0840
USA

Mark A. Hatcher
Department of Chemistry
The Johns Hopkins University
3400 North Charles Street
Baltimore, MD 21218
USA

Martin Hiersemann
Institute of Organic Chemistry
University of Dortmund
Otto-Hahn-Strasse 6
44227 Dortmund
Germany

Hayato Ichikawa
Osaka University of Pharmaceutical
Sciences
4-20-1 Nasahara
Takatsuki, Osaka, 569-1094
Japan

Hisanaka Ito
Department of Pharmacy and
Life Science
Tokyo University
1432-1 Horinouchi, Hachioji
Tokyo 192-0392
Japan

The Claisen Rearrangement. Edited by M. Hiersemann and U. Nubbemeyer

ISBN: 978-3-527-30825-5

Uli Kazmaier
Institute of Organic Chemistry
University of the Saarland
Building C 4.2
P.O. Box 15 11 50
66041 Saarbrücken
Germany

Mukund G. Kulkarni
Department of Chemistry
University of Pune
Pune 411 007
India

Yves Langlois
Laboratoire de Synthèse des
Substances Naturelles
Université de Paris-Sud
Bâtiment 410
91405 Orsay
France

Keiji Maruoka
Department of Chemistry
Kyoto University
Sakyo, Kyoto 606-8502
Japan

Christopher M. McFarland
Department of Chemistry and
Biochemistry
University of Arkansas
Fayetteville, AR 72701
USA

Matthias C. McIntosh
Department of Chemistry and
Biochemistry
University of Arkansas
Fayetteville, AR 72701
USA

Patrick Metzner
Laboratoire de Chimie Moléculaire et
Thioorganique (UMR CNRS 6507)
ENSICAEN-Université de Caen
6 Boulevard du Maréchal Juin
14050 Caen
France

Koichi Mikami
Department of Applied Chemistry
Tokyo Institute of Technology
2-12-1S1-29 Ookayama, Meguro-ku
Tokyo 152-8552
Japan

Udo Nubbemeyer
Institute of Organic Chemistry
University of Mainz
Duesbergweg 10–14
55128 Mainz
Germany

Stéphane Perrio
Laboratoire de Chimie Moléculaire et
Thioorganique (UMR CNRS 6507)
ENSICAEN-Université de Caen
6 Boulevard du Maréchal Juin
14050 Caen
France

Gary H. Posner
Department of Chemistry
The Johns Hopkins University
3400 North Charles Street
Baltimore, MD 21218
USA

Niny Rao
Department of Biochemistry,
Cellular and Molecular Biology
University of Tennessee
M407 Walters Life Sciences Building
Knoxville, TN 37996-0840
USA

Vincent Reboul
Laboratoire de Chimie Moléculaire et
Thioorganique (UMR CNRS 6507)
ENSICAEN-Université de Caen
6 Boulevard du Maréchal Juin
14050 Caen
France

Julia Rehbein
Institute of Organic Chemistry
University of Dortmund
Otto-Hahn-Strasse 6
44227 Dortmund
Germany

Takeo Taguchi
Department of Pharmacy and
Life Science
Tokyo University
1432-1 Horinouchi, Hachioji
Tokyo 192-0392
Japan

Dirk Trauner
Department of Chemistry
University of California at Berkeley
602 Latimer Hall
Berkeley, CA 94720
USA

1
Chorismate-Mutase-Catalyzed Claisen Rearrangement

Hong Guo and Niny Rao

1.1
Introduction

Chorismic acid is the key branch point intermediate in the biosynthesis of aromatic amino acids in microorganisms and plants (Scheme 1.1a) [1]. In the branch that leads to the production of tyrosine and phenylalanine, chorismate mutase (CM, chorismate-pyruvate mutase, EC 5.4.99.5) is a key enzyme that catalyzes the isomerization of chorismate to prephenate (Scheme 1.1b) with a rate enhancement of about 10^6–10^7-fold. This reaction is one of few pericyclic processes in biology and provides a rare opportunity for understanding how Nature promotes such unusual transformations. The biological importance of the conversion from chorismate to prephenate and the synthetic value of the Claisen rearrangement have led to extensive experimental investigations [2–43].

In addition, the reaction catalyzed by chorismate mutase is a paradigm for the study of enzyme mechanism and has been a subject of extensive computational investigations [44, 47–83]. One of the main reasons for the current focus on the mechanism of this enzyme is the fact that the reaction is a straightforward unimolecular rearrangement of the substrate with no chemical transformations in the enzyme or the solvent during the reaction. This eliminates many of the problems that arise for other cases and may help to settle some of the long-standing issues concerning the origin of the catalysis [84].

Experimental results for the CM-catalyzed and uncatalyzed reaction, as well as structural information for chorismate mutase, have been extensively discussed in two previous reviews [2, 3]. There has been a rapid growth of literature in computational studies of chorismate mutase in the last few years. In this chapter, we shall begin by summarizing some key experimental data related to the Claisen rearrangement along with existing structural information for chorismate mutase. We will then review the results of computational studies of chorismate mutase and discuss different proposals that have been suggested for the mechanism of the CM-catalyzed reaction.

The Claisen Rearrangement. Edited by M. Hiersemann and U. Nubbemeyer

ISBN: 978-3-527-30825-5

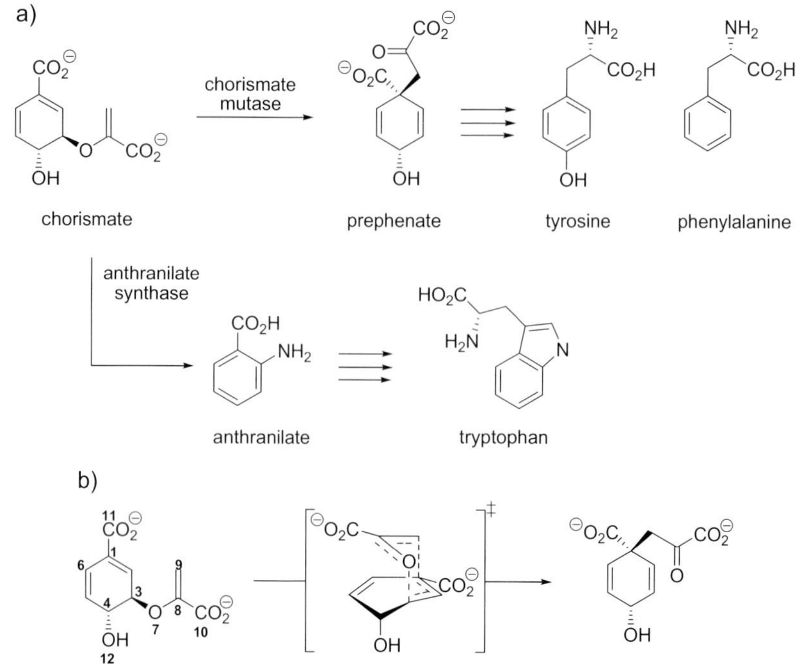

Scheme 1.1

1.2
Experimental Studies

1.2.1
Substrate Binding

Knowles and coworkers [13, 14] demonstrated that the rearrangement of chorismate to prephenate proceeds through the same transition state (**1**, TS in Scheme 1.2) in solution and at the enzyme active site. The atoms of the [3,3]-pericyclic region in this TS are arranged in a "chair-like" configuration. The result of Knowles and coworkers has led to the suggestion that the bond breaking and making process starts from a chair-like pseudodiaxial conformer of chorismate (**2**, CHAIR in Scheme 1.2), where C_1 and C_9 are positioned to form the carbon-carbon bond, as required for the Claisen rearrangement. Thus, one straightforward way for chorismate mutase to catalyze the rearrangement is to bind the CHAIR conformer preferentially from solution and then catalyze its chemical transformation at the active site. A requirement for such a mechanism is a sufficiently large population of the CHAIR conformer in solution. To determine the population of CHAIR in solution, Copley and Knowles [15] measured the temperature variation of the ^1H coupling constants for the protons in the ring of chorismate. It was shown that although the dominant conformer(s) is a pseudodiequatorial conformation (see **3** of Scheme 1.2 for a schematic diagram), a pseudodiaxial conformer(s) exists at a reasonable level (~12%) in solution. Copley and Knowles [15] assumed that the pseudodiaxial conformer they observed in the NMR experiment was the CHAIR conformer and concluded that the enzyme could bind this reactive conformer directly from solution and catalyze its chemical transformation at the active site. But a later study of the transferred nuclear Overhauser effects for chorismate by Hilvert and his coworkers [17] failed to find evidence for the existence of CHAIR in solution. Recent molecular dynamics (MD) simulations [82, 83] suggested that the NMR data could correspond to other pseudodiaxial conformer(s) rather than CHAIR (see below).

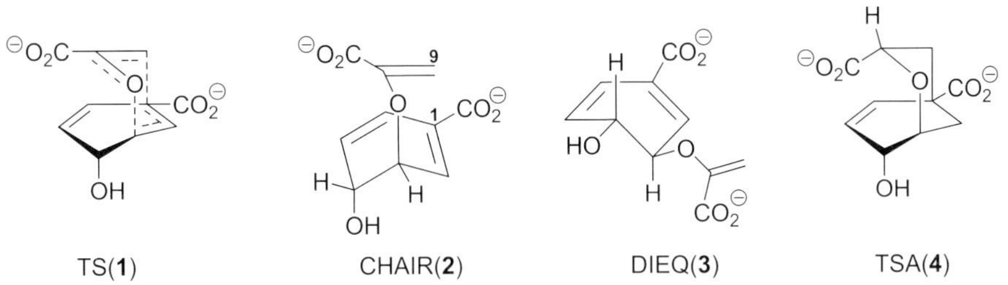

TS(**1**) CHAIR(**2**) DIEQ(**3**) TSA(**4**)

Scheme 1.2

1.2.2 Substrate Structural Requirements for Catalysis

The structural features of the substrate required for binding and catalysis by *Escherichia coli* chorimate mutase (P-protein EcCM) and *Bacillus subtillus* chorismate mutase (BsCM) have been studied [22, 23]. Besides the allyl vinyl ether, the two carboxylic acid groups in chorismic acid were found to be very important for the catalysis. For instance, experimental studies [22] showed that ester **5** (see Scheme 1.3) was not a substrate or inhibitor for EcCM, suggesting that the presence of the sidechain carboxyl group is crucial for the binding and catalysis. EcCM and BsCM were also unable to catalyze the rearrangement of **6** (which lacks the ring carboxylic acid group) [22, 23], even though **6** proved to be a weak to modest competitive inhibitor (K_i of **6** is 0.4 mM and 0.5 mM for EcCM and BsCM, repectively; for chorismate K_m is 0.32 mM and 0.28 mM, respectively). Thus, the existence of the ring carboxyl group is also essential for the catalysis, but may not

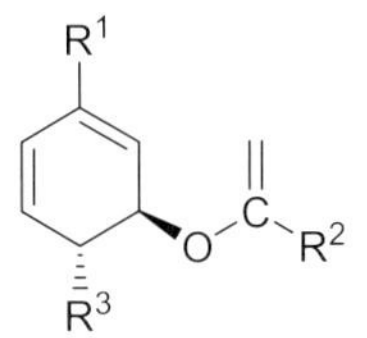

5: $R_1 = CO_2H$; $R_2 = CO_2Me$; $R_3 = OH$
6: $R_1 = H$; $R_2 = CO_2H$; $R_3 = OH$
7: $R_1 = CO_2H$; $R_2 = CO_2H$; $R_3 = OMe$

Scheme 1.3

be required for the binding. Analog **7** was a reasonable substrate for EcCM (K_m = 1.9 mM and k_{cat} = 0.56 s^{-1}) with a rate acceleration (k_{cat}/k_{uncat}) of 2×10^4 by the enzyme; for chorismic acid $k_{cat}/k_{uncat} = 2 \times 10^6$. Thus, the free hydroxyl group at C_4 may not be required for the catalysis by EcCM, but it is not clear whether this is also the case for BsCM (see below).

1.2.3
X-ray Structures of Chorismate Mutase

A number of X-ray structures for chorismate mutase are available. The structures of BsCM and *Saccharomyces cerevisiae* (yeast) CM complexed with an *endo*-oxabicyclic transition state analog inhibitor (**4**, TSA in Scheme 1.2) [19] have been determined by Lipscomb and coworkers [6, 7, 12, 34]; the structures without TSA bound were also obtained for BsCM and yeast CM as well as for some of their mutants [8, 10, 12, 31]. The X-ray structures for the monofunctional amino-terminal chorismate mutase domain engineered from the P-protein (EcCM) and a less active catalytic antibody 1F7 complexed with TSA have been determined by Lee et al. [4] and Haynes et al. [33], respectively. Both EcCM and yeast CM are homodimers, whereas BsCM is a homotrimer. It has been demonstrated that the dimer of EcCM can be superimposed onto a monomer of yeast CM [4, 11, 34], indicating a common evolutionary origin of the two CMs with an ancestral protein that was structurally closer to EcCM than to yeast CM [34]. Moreover, there was a possible gene duplication event in the evolution of yeast CM [34], allowing the formation of the regulatory domain for this enzyme. The structure of BsCM, which consists mainly of β-sheets, is different from the almost all-helical structures of EcCM and yeast CM.

Scheme 1.4 shows the schematic diagrams for the active site structures of EcCM [4], yeast CM [12, 34], BsCM [6, 7] and catalytic antibody 1F7 [33]. The active site of BsCM is somewhat open and more solvent accessible than the more buried catalytic packets in EcCM and yeast CM. As is evident from Scheme 1.4a and b, most of the active site residues in EcCM and yeast CM are conserved. For instance, in the both cases the guanidinium groups of two Arg residues (Arg28 and Arg11′ in EcCM and Arg16 and Arg157 in yeast CM, respectively) form salt bridges with the carboxylate groups of the inhibitor. Lys39 (Lys168) in EcCM (yeast CM) is in hydrogen bond distances to the sidechain carboxylate group and the ether oxygen of TSA. A major difference between the two active sites is that the other residue interacting with the ether oxygen is Gln in EcCM (Gln88), but is Glu in yeast CM (Glu246). It has been shown that Glu246 has to be protonated for functionality of yeast CM. The replacement of Glu246 by Gln changes the pH optimum for the activity from a narrow to a broad pH range, even though the kinetic parameters are not significantly affected by the mutation (e.g., the effect on k_{cat}/K_m is less than 10-fold) [28]. Consistent with these observations on yeast CM, the replacement of Gln88 in EcCM by Glu leads to loss of activity of 700-fold at pH 7.5, but the activity of the Gln88Glu mutant can be reduced almost 10^3-fold by simply lowering the pH to 4.9 [27] (see Table 1.1 and the next section for more details on the effects of mutations).

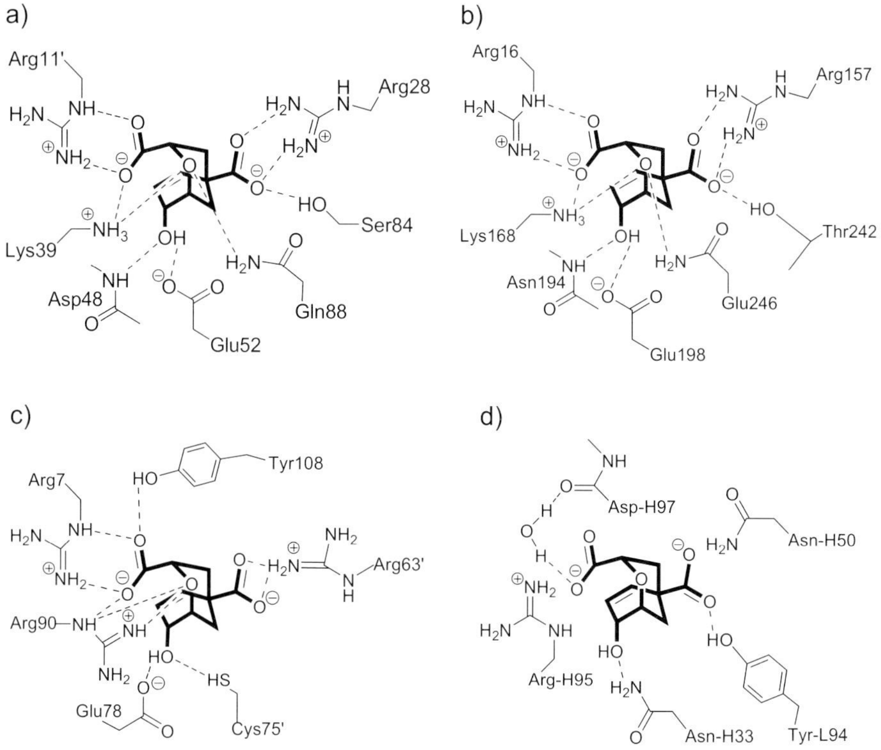

Scheme 1.4 The active sites of the CM complexes, (a) EcCM; (b) yeast CM; (c) BsCM; (d) catalytic antibody 1F7.

Scheme 1.4c shows that the active site of BsCM also consists of highly charged residues. Arg7 forms a similar salt bridge with the sidechain carboxylate group of TSA as Arg11′ in the EcCM complex. Arg90 interacts with the both ether oxygen and sidechain carboxylate group. Arg63 was not visible in the electron-density map in an earlier X-ray structure determination [7]. But a more recent X-ray structure [8] of higher resolution (1.3 Å) without TSA bound showed that Arg63 is turned inward toward the active site and may therefore interact with the ring carboxylate group of TSA. Another interaction that exists in all the three CMs is the hydrogen bond between the C_4-hydroxyl group of TSA and a Glu residue (Glu52 in EcCM, Glu198 in yeast CM and Glu78 in BsCM). This Glu residue appears to play a more important role for the reaction catalyzed by BsCM than by EcCM (see below). Comparison of the active site structures of EcCM, BsCM and yeast CM with that of the catalytic antibody (1F7) (Scheme 1.4d) shows that the enzymes provide many more hydrogen bonding and electrostatic interactions to the functional groups of TSA than does the antibody. The lack of the multiple interactions is believed to be responsible for the observed 10^4-times lower activity of the antibody relative to that of the natural chorismate mutase [33].

1.2.4 Effects of Mutations

For the uncatalyzed Claisen rearrangement k_{uncat} is about 10^{-5} s^{-1} [20, 31], and the k_{cat} value for the CM-catalyzed reaction is approximately 46–72 s^{-1} (Table 1.1). Thus, the enzyme is able to accelerate the rate of the reaction by 10^6 to 10^7-fold. To identify the key residues that play an important role in the catalysis, a number of active site mutants were generated and characterized for EcCM [27, 35], yeast CM [28] and BsCM [25, 29, 31, 36] and the effects of mutations on the activity have been determined.

For EcCM, Arg28, Arg11′ and Lys39 are involved in the direct interactions with the two carboxylate groups as well as the ether oxygen of TSA in the X-ray structure (Scheme 1.4a). Table 1.1 shows that these positively charged residues play a very important role in the catalysis [27]. For instance, the k_{cat}/K_m values for the Arg28Lys and Arg11Lys are approximately 10^3 lower than wild-type, whereas the values for Lys39Ala and Lys39Arg are about 10^4 lower. Similar observations were made for the related yeast CM, where the Arg157Ala, Arg16Ala and Lys168Ala mutants showed no detectable chorismate mutase activity [28]. The hydrogen bond between Gln88 (Glu246 in yeast CM) and the ether oxygen was also found to be very important. For instance, the replacement of Gln88 by Ala leads to a reduction of the activity by 10^4-fold. For the Glu52 mutants, the order of activity is Glu52 > Gln52 > Asp52 > Ala52. Glu52 interacts with the C_4-hydroxyl group in the X-ray structure. The higher activity of Glu52Gln than Glu52Asp seems to indicate that the existence of a carboxylate group in the vicinity of the C_4-hydroxyl may not be necessary. This seems to be consistent with the earlier discussions of substrate structural requirements for the catalysis where it was shown that the free hydroxyl group at C_4 may not be required in the case of the EcCM-catalyzed reaction (see above).

The kinetic parameters for BsCM mutants are also available [25, 29, 31, 36] and listed in Table 1.1. Arg7, which forms a similar interaction with TSA as Arg11′ in EcCM, was found to be very important. For instance, the replacement of Arg7 by Ala leads to an approximately 5×10^5-fold reduction in k_{cat}/K_m. Arg90, which interacts with the both ether oxygen and sidechain carboxylate group (Scheme 1.4c), is also crucial for the catalysis. For instance, the k_{cat} and k_{cat}/K_m values for Arg90Gly are more than five orders of magnitude lower than those of the wild-type enzyme. Moreover, the importance of the positive charge on Arg90 was demonstrated by Hilvert and coworkers [36] who showed that there is a significant reduction of the activity (> 10^4-fold in k_{cat}) when Arg90 was replaced by citrulline, an isosteric but neutral arginine analog. Interestingly, the double mutants Cys88Lys/Arg90Ser and Cys88Ser/Arg90Lys restore a factor of more than 10^3 in k_{cat} compared to Arg90Gly [31]. Another important residue for the catalysis is Glu78. Glu78 is in a similar location as Glu52 in EcCM. Table 1.1 shows that the k_{cat}/K_m values for Glu78Ala and Glu78Gln are about 10^4 lower than wild-type. By contrast, the activity of Glu78Asp is only 30-fold lower. This seems to suggest that the existence of a carboxylate group in the vicinity of the C_4-hydroxyl is more important for the

Table 1.1 Kinetic constants for EcCM and BsCM mutants.

Enzyme	Mutant	k_{cat} (s^{-1})	K_m (μM)	k_{cat}/K_m ($M^{-1}s^{-1}$)	K_i for 4 (μM)	Reference
EcCM	Wild type	72	296	2.4×10^5	3.66	[27]
	R11A		>2000	26		[27]
	R11K		>2000	230		[27]
	R28A		>2000	170		[27]
	R28K		>2000	230		[27]
	K39A		>2000	4.3		[27]
	K39R		>2000	1.9		[27]
	E52A	0.49	4580	110	218	[27]
	E52D	3.1	1440	2.2×10^3	78.4	[27]
	E52Q	24	1080	2.3×10^4	26.8	[27]
	Q88A		>2000	12		[27]
	Q88E		>2000	361		[27]
	Q88E (pH 4.9)	72	296	2.4×10^5	3.66	[27]
BsCM	Wild type	46	67	6.9×10^5	3	[36]
	R90G	2.7×10^{-4}	150	31		[31]
	R90K			31		[25]
	R90A			<1		[25]
	R7K			717		[25]
	R7A			1		[25]
	C75D/E78A			1.66×10^3		[25]
	E78D	35.7	1297	2.75×10^4	43.6	[25]
	E78Q			75		[25]
	E78A			33		[25]
	C88K/R90K	0.29	4300	67	>>1000	[31]
	C88K/R90S	0.32	1900	170	1100	[31]
	R90Cit	0.0026	270	230	6.8	[36]

BsCM-catalyzed reaction than for the EcCM-catalyzed reaction. Consistent with this suggestion, the activity of the Glu78Ala mutant is rescued 50-fold by replacing C75 (which is also near the C_4-hydroxyl group) with Asp in double mutant Glu78Ala/C75Asp [25]. The studies of substrate structural requirements for the catalysis (see above) showed that **6**, which lacks the ring carboxylate group, is not a substrate for EcCM [22] and BsCM [23]. For EcCM, the residue that interacts with the ring carboxylate group is Arg 28, and the replacement of Arg28 by another residue leads to a significant reduction of the activity. However, for BsCM the corresponding residue has not been clearly identified. A recent X-ray structure [8] for BsCM suggested that Arg63 may interact with the ring carboxylate group, but the mutagenesis study for the Arg63 mutants has not been available.

1.2.5 Activation Parameters

The activation parameters for the CM-catalyzed and uncatalyzed Claisen rearrangement are listed in Table 1.2 [20, 21, 26, 42]. For the uncatalyzed reaction, the activation barrier ($\Delta G^{\ddagger}$) is 24.5 kcal/mol. Chorismate mutase is able to reduce the activation barrier by 7–10 kcal/mol. Table 1.2 shows that the rate acceleration is due to a reduction in the entropy of activation to near zero and a decrease in the enthalpy of activation by about 5 kcal/mol; the only exception is the BsCM-catalyzed reaction for which there is a significant unfavorable $\Delta S^{\ddagger}$. However, the reliability of these data has been called into question [44], and it was suggested [44] that both the substrate binding and product leaving are expected to show large solvent compensation effects involving $\Delta H^{\ddagger}$ and $\Delta S^{\ddagger}$ [45, 46].

Table 1.2 Activation parameter for the catalyzed and uncatalyzed reaction[a].

Enzyme	$\Delta H^{\ddagger}$ (kcal·mol^{-1})	$\Delta S^{\ddagger}$ (e.u.)	$\Delta G^{\ddagger}$ (kcal·mol^{-1})	$\Delta\Delta G^{\ddagger}$ (kcal·mol^{-1})
BsCM	12.7	−9.1	15.4	−8.9
EcCM	16.3	−3.0	17.2	−7.3
K. pneumoniae	15.9	−1.1	16.2	−8.3
S. aureofaciens	14.5	−1.6	15.0	−9.5
Uncatalyzed	20.5	−12.9	24.5	–

a) All entries are as cited in Refs. [21] and [26]. $\Delta G^{\ddagger}$ calculated at 25 °C.

1.3 Catalytic Mechanism of Chorismate Mutase

The exact mechanism of the action by chorismate mutase is still not clear, in spite of extensive experimental and theoretical investigations. Several suggestions have been proposed concerning the origin of the catalysis. They include: (a) the stabilization of transition state by the enzyme, presumably through electrostatic interactions from the active site residues; (b) the promotion of substrate conformational transition to generate the reactive CHAIR conformer at the active site (see Scheme 1.2); (c) the increase of populations of near attack conformers (NACs); and (d) strain effects and conformational compression. These proposals will be discussed below.

1.3.1 Stabilization of Transition State by Active Site Residues

The X-ray structures of the CM complexes [4, 6, 7, 12, 34] showed that the enzymes consist of highly charged active site residues and form extensive hydrogen bonding interactions with the inhibitor (Scheme 1.4a–c). Several residues that interact directly with the inhibitor were proved to be crucial for the catalysis from mutagenesis studies (see above). For instance, each enzyme forms two hydrogen bonds with the ether oxygen of TSA. Replacement of the corresponding residue(s) involved the interaction(s) (e.g., Gln88→Ala or Lys39→Ala mutation in EcCM or Arg90→Ala mutation in BsCM) and led to a significant reduction of the catalytic efficiency (Table 1.1). Hilvert and his coworkers [36] further showed that in the case of BsCM the existence of positive charge on Arg90 is very important, as the replacement of Arg90 by a neutral arginine analog (citrulline) causes $>10^4$-fold decrease in k_{cat}. These observations have led to the suggestion that one possible origin of the catalysis is electrostatic interactions from the active site residues that stabilize the developing charges in the transition state (TS) and lower the activation barrier.

Several computational studies [49, 58, 67, 71, 75, 80] of the BsCM catalysis support the suggestion that chorismate mutase works by stabilization of transition state through electrostatic interactions from the active site residues. For instance, Lyne et al. [49], Lee et al. [67] and Ranaghan et al. [76] performed hybrid quantum mechanical/molecular mechanical (QM/MM) studies of BsCM. In a hybrid QM/MM approach, the flexibility of a quantum mechanical description for a small number of atoms at the active site or in solution (e.g., the atoms in chorismate) is combined with the efficiency of an empirical force field representation for the bulk of the solvated system. So both the chemical events of bond breaking and making (e.g., the change from chorismate to prephenate) as well as the effects of the environment (e.g., the reduction of activation barrier by the enzyme) could be described. Lyne et al. [49], Lee et al. [67] and Ranaghan et al. [76] used a perturbation approach [84] based on fixed BsCM-CHAIR and BsCM-TS structures to study the contribution of each residue to the lowering of the reaction barrier in going from the CHAIR conformer to TS. In this perturbation approach, the charges on the atoms of the residue under consideration are turned off. As a result, the reaction barrier would increase (decrease) if that residue stabilizes (destabilizes) the transition state. They showed that, consistent with the experimental observations mentioned above, the deletion of Arg90 led to the increase of the reaction barrier by 3.1–9.6 kcal/mol. Other important residues identified from the perturbation approach include Arg7 and Glu78, and the deletions of these two residues led to increases of the barrier by 1.2–4.0 and 1.6–3.5 kcal/mol, respectively. The results of these computational studies therefore support the suggestion that the enzyme works by electrostatic stabilization of transition state from the active site residues. Interestingly, it was found from one of the studies [67] that the deletion of Asp118 decreased the reaction barrier by as much as 5 kcal/mol (corresponding a rate acceleration of more than 10^3-fold), whereas the mutagenesis study showed that the rate is actually lower by a factor of 2. The perturbation approach has only been used for BsCM. It would be of interest to use this approach to study the EcCM and yeast CM catalysis, where a neutral polar residue (Gln88 in EcCM and Glu246 in yeast CM) interacting with the ether oxygen plays a crucial role in the catalysis.

It should be pointed out that different conclusions may sometime be reached from different calculations. For instance, Worthington et al. [54] showed that the electrostatic stabilization of Arg90 was not stronger for the transition state than for the reactant based on their calculations using the method of effective fragment potentials. The suggestion that the enzyme works by electrostatic stabilization of the transition state is also supported by other computational studies, where the activation barriers in solution and in the enzyme are compared. It was found that there is a significant reduction of the barrier (about 10 kcal/mol) in going from solution to the enzyme active site [58, 71, 75, 80]. However, an important factor that needs to be taken into account in this type of studies is that the stable conformations of chorismate in solution and enzyme are different. Such conformational differences may have significant energetic consequences on the activation barriers for the Claisen rearrangement in the different environments. A question that needs to be addressed is whether the reduction of the activation barrier in going from solution to the enzyme active site obtained from computational studies actually comes from the lowering of the TS energy or from some other origin. We will discuss the stabilization of transition state again in connection with other proposals.

1.3.2
Substrate Conformational Transition and the Role of Active Site Residues

Knowles and coworkers [13, 14] demonstrated that the transition state (**1** in Scheme 1.2) for the Claisen rearrangement of chorismate has their atoms in the [3,3]-pericyclic region arranged in a chair-like configuration. The pseudodiaxial CHAIR conformer (**2** in Scheme 1.2) is the substrate conformation that can reach this chair-like transition state directly. The bond breaking and making process may therefore start from CHAIR during the rearrangement. One way for CM to catalyze the reaction is to bind this CHAIR conformer from solution and lower the activation barrier for its chemical transformation at the active site. A requirement for this mechanism is a sufficiently large population of CHAIR in solution. Copley and Knowles [15] performed a NMR study and showed that a pseudodiaxial conformer(s) exists at a reasonable level in water. They assumed that this pseudodiaxial conformer observed in the NMR experiment was CHAIR and suggested that enzyme could bind this reactive conformer directly. Many later discussions of the CM catalysis have been based on this assumption. However, there are different possible pseudodiaxial conformers for chorismate, and these conformers cannot simply be distinguished by the coupling constants measured in the NMR experiment [15]. Moreover, a later study of the transferred nuclear Overhauser effects for chorismate by Hilvert and his coworkers [17] failed to support the original suggestion the stable pseudodiaxial conformer in solution is CHAIR. Consistent with this later experimental study, high level *ab initio* calculations [47, 48, 50, 68, 70, 82] showed that the reactive CHAIR conformer is considerably less stable than other pseudodiaxial and pseudodiequatorial conformers (see Table 1.3). For instance, the energy difference between CHAIR and the most stable pseudodiequatorial conformer of chorismate can be as much as 18 kcal/mol. It is interesting to note from Table 1.3 that while the results from the SCC-DFTB semiempirical method are rather close to the *ab initio* results, the AM1 method can overestimate the stability of the CHAIR conformer by as much as 10 kcal/mol in the gas phase. This suggests that care must be excised in choosing proper computational methods for the study of the Claisen rearrangement of chorismate.

To explore the possible solution conformers, Guo et al. [82] performed QM/MM molecular dynamics simulations using the CHARMM program [86]; the CHAIR

Table 1.3 Energies and structural parameters for some chorismate conformers[a)].

Conformer	ΔE[b)] (kcal·mol^{-1})	R_1 (Å)	τ_1 (degrees)	τ_2 (degrees)	τ (degrees)	δ (degrees)
CHAIR	17.9(13.8)[6.5]	3.8	–133	91	73	21
DIEQ$_1$	2.9(3.5)[0.3]	4.4	–61	169	79	–54
DIEQ$_2$	0.0(0.0)[0.0]	5.3	–75	161	–106	–43
DIAX[c)]		–4.9	160	71	–58	45
Ex-DIAX	12.3	5.2	–160	71	–121	45

a) Except where otherwise noted, ΔE values and the structural parameters are based on B3LYP/6–31G* calculations from Ref. [82]. The energy of DIEQ$_2$ is taken as the zero. $R_1 = R$ ($C_1...C_9$), $\tau_1 = \tau$ (O_7–C_3–C_4–O_{12}), $\tau_2 = \tau$ (H–C_3–C_4–H), $\tau = \tau$ (C_8–O_7–C_3–C_2), and $\delta = 1/2\,(|\tau_1| - |\tau_2|)$ (see Scheme 1.5).

b) B3LYP/6–31G* (SCC-DFTB) [AM1].

c) Calculated from the energy minimization at HF/6–31G* level with eight bridging waters between the two carboxylate groups of chorismate.

conformer was used as the initial conformation for chorismate. Chorismate was described by QM (SCC-DFTB) [86], and the rest of the system (explicit aqueous solvent) was treated by MM [87]. The quantum mechanical description of the substrate is advantageous because it does not require specific MM parameters to be determined and provides a more realistic treatment of the fluctuations of the covalent bond distances. Although CHAIR is a high energy, local minimum in the gas phase, it is not stable in solution. Instead, it is rapidly (within 10–20 ps) converted to another pseudodiaxial conformer, called DIAX (see Scheme 1.5). Chorismate spends a significant portion of the time in the DIAX conformation during the remainder of the simulations, and the CHAIR conformer was not observed. The values of the dihedral angles describing the ring [i.e., τ_1 (O_7–C_3–C_4–O_{12}) and τ_2 (H–C_3–C_4–H)] are similar for DIAX and CHAIR. DIAX is distinguished from CHAIR by τ(C_8–O_7–C_3–C_2); i.e., τ is about 70° in CHAIR, whereas it is about −30° to −70° in DIAX. Scheme 1.5 shows that DIAX has a sidechain carboxylate group, rather than the sidechain methylene group, over the C_1 atom and is therefore an inactive conformation. However, it may well be the one observed by Copley and Knowles [15] in their solution NMR studies; its structure is consistent with the NMR measurement of Hilvert and coworkers [17]. The results obtained in the solution simulations indicate that the original proposal in which the enzyme preferentially binds the CHAIR conformer is not tenable, because its concentration in solution seems to be too low from the simulations study. Instead, a likely possibility is that DIAX or other conformers, which are relatively stable in solution, are ones bound by the enzyme.

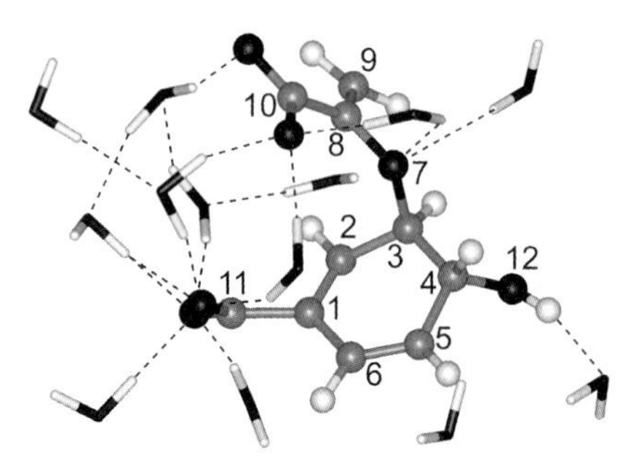

Scheme 1.5 A snapshot of chorismate in solution from the molecular dynamics simulations. The conformation is DIAX with the sidechain carboxylate group over the C_1 atom.

The dynamics of CHAIR and DIAX in the active sites of yeast CM [82], BsCM [83] and EcCM (unpublished results) were studied using the QM/MM molecular dynamics simulations, which take into account the motions of the substrate as well as the enzyme. Similar to the solution study, Chorismate was described by QM (SCC-DFTB) [85], and the rest of the system (CM) was treated by MM with the CHARMM force field [88]. It was shown that, in contrast to the motion of CHAIR in solution (see above), no conformational transition occurs in the active sites of the enzymes [82]; the substrate remains in the neighborhood of CHAIR during the MD simulations. Moreover, the important interactions with the active site residues observed in the X-ray structures are retained. The stability of CHAIR in the active sites of EcCM and BsCM was confirmed by Hur and Bruice [60–64] from molecular dynamics simulations with a special set of parameters for the substrate (i.e., without the QM description). We will discuss the results of Hur and Bruice later.

For the DIAX conformer, it was shown from the QM/MM MD simulations [82–83] that when the inactive conformer is bound to any of the CM active sites, it is rapidly converted to the reactive CHAIR conformer with the important interactions observed in the X-ray structures recovered. Schemes 1.6 and 1.7 show the active site structures before and after the conformational transition for the cases of EcCM and BsCM, respectively; for the structures of the yeast CM complex, (see Ref. [82]). The initial orientation of DIAX in the active sites is such that the sidechain carboxylate group (and the ether oxygen) forms the observed interactions with the active residues. For instance, in the case of BsCM, Arg7 was allowed to form a salt bridge with the sidechain carboxylate group of the substrate in the initial docking configuration, whereas Arg90 formed two hydrogen bonds with the sidechain carboxylate and the ether oxygen, respectively (see Scheme 1.7a). Such initial orientations of the substrate in the active sites are consistent with the results of the experimental studies [22, 23]. Indeed, it was shown that the substrate analogs (e.g., **5** in Scheme 1.3) without the sidechain carboxylate group are not inhibitors or substrates for chorismate mutase [22, 23] (see Section 1.2.2 above), whereas the analogs without the ring carboxylate group can be modest competitive inhibitors. This indicates that the interactions involving the sidechain

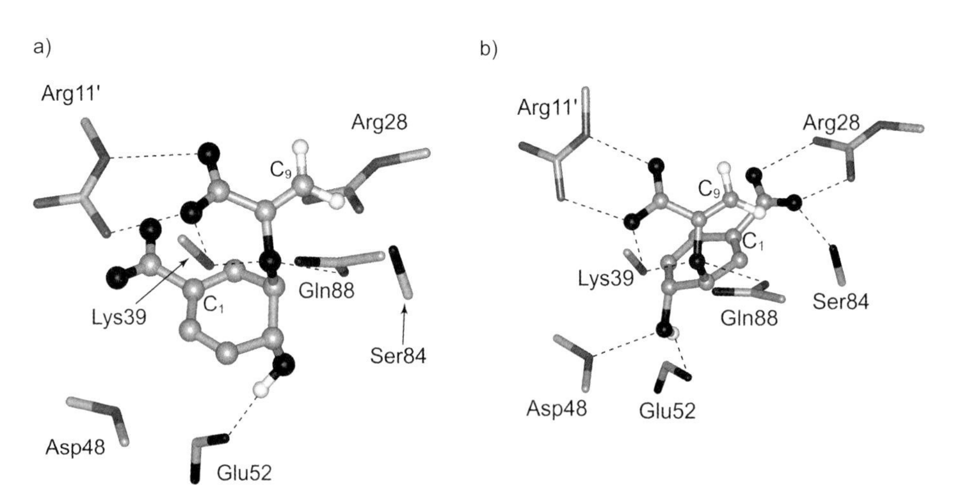

Scheme 1.6 The active site structures before and after the substrate conformational transition for the EcCM-catalyzed reaction. (a) The structure before the transition. The substrate is in the inactive DIAX conformation. (b) The structure after the transition. The substrate is in the CHAIR conformation.

carboxylate group play a very important role for the binding process [2]. As is evident from Schemes 1.6 and 1.7, the ring of chorismate undergoes a significant clockwise rotation in the active sites of EcCM and BsCM during the MD simulations, and such rotation generates the reactive CHAIR conformation from the solution DIAX conformation.

For EcCM, Arg28, Ser84 and the backbone amide group of Asp48 seem to provide the main driving forces for the rotation of the ring of the substrate (Scheme 1.6). Glu52 already forms the hydrogen bond with the C_4-hydroxyl group before the rotation and is unlikely to play a role for the conformational transition. By contrast, Glu78 in BsCM does not form the hydrogen bond with the C_4-hydroxyl group of DIAX, and this hydrogen bond is formed only when the substrate changes to CHAIR (Scheme 1.7b). Thus, unlike the case involving Glu52 in EcCM, the negatively charged field from Glu78 and the potential hydrogen bonding interaction with the C_4-hydroxyl group may make a significant contribution to the rotation of the ring during the conformational transition. It is interesting to note that the results of the MD simulations seem to be consistent with the mutagenesis studies (Table 1.1), where it was shown that the existence of the carboxylate group in the vicinity of the C_4-hydroxyl is more important for the BsCM-catalyzed reaction than for the EcCM-catalyzed reaction. Another residue that might also provide the driving force for the conformational transition is Arg63′, although the mutagenesis study for the Arg63 mutants has not been available. Conformational transitions from another inactive pseudodiequatorial conformer ($DIEQ_1$) to CHAIR were also observed in the active sites [60, 82]. These results from the QM/MM MD simulations suggest that one contribution of the enzyme is to bind the

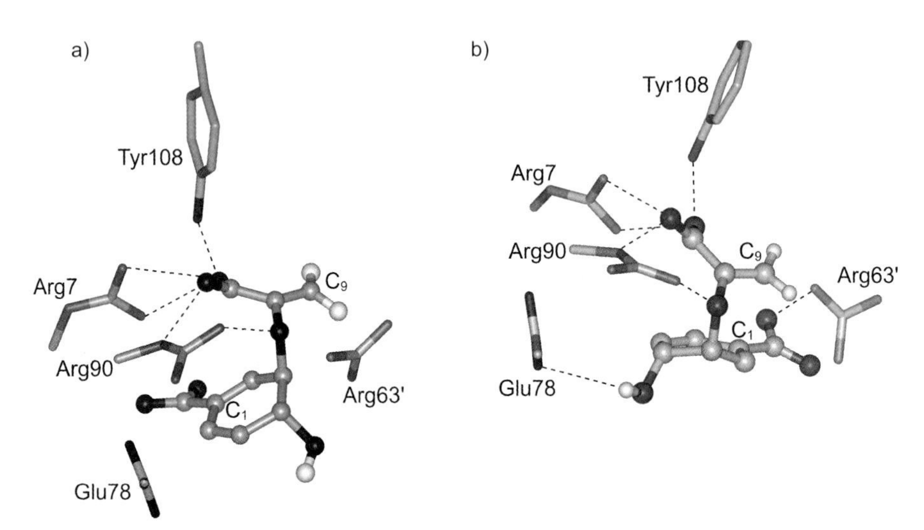

Scheme 1.7 The active site structures before and after the substrate conformational transition for the BsCM-catalyzed reaction. (a) The structure before the transition. The substrate is in the DIAX conformation. (b) The structure after the transition. The substrate is in the CHAIR conformation.

more prevalent nonreactive conformers and transform them into the active form in a step before the chemical reaction. This suggestion seems to be supported by the studies of substrate structural requirements for catalysis (Section 1.2.2). Indeed, it was shown that EcCM and BsCM are unable to catalyze the rearrangement of **6** (which lacks the ring carboxyl group) [22, 23], even though **6** is an inhibitor for the both enzymes. The deletion of the ring carboxylate group would remove some of the important driving forces required for the conformational transition (e.g., the interaction between Arg28 and the ring carboxylate) and destroy the ability of the enzyme to generate the reactive CHAIR conformation in the active site, leading a reduction of the catalytic efficiency.

To examine the free energy involved in the stabilization of the CHAIR form in the active site and the role of the active site residues, the QM/MM molecular dynamics simulations and umbrella sampling simulations were performed for the wild-type BsCM as well as mutant enzymes [83]. This study showed that Arg90 is essential for stabilizing the reactive substrate conformation (CHAIR) in the active site, as both the Arg90Ala and Arg90Gly mutations destroy the ability of the wild-type enzyme to stabilize CHAIR and stabilize the inactive DIAX conformation instead. CHAIR was found to be about 10 kcal/mol more stable than DIAX in the wild-type BsCM, and it becomes 5 kcal/mol less stable in Arg90Ala. The results of the MD and free energy simulations therefore suggested an additional role for Arg90. Its existence appears to be essential for generating and stabilizing the CHAIR conformation of the substrate in the active site. The extra energy cost (~5 kcal/mol) necessary for the formation of CHAIR in Arg90Ala and Arg90Gly obtained from the simulations would contribute to the loss of activity of the mutants [31]. Thus, the stabilization of the chair-like conformations of chorismate and the transition state [25] seems to require a proper balance of different interactions at the active site. The removal of the single key residue, such as Arg90, can destroy such delicate balance of the interactions, leading to a dramatic reduction in the catalytic efficiency.

The mutagenesis study [31] showed that the double mutant Cys88Lys/Arg90Ser and Cys88Ser/Arg90Lys can restore a factor of more than 10^3 in k_{cat} (see above). This restoration of the catalytic efficiency may be explained by a similar argument. That is, the relatively high activities for Cys88Lys/Arg90Ser and Cys88Ser/Arg90Lys could be related to the more balanced interactions generated by the double mutations that are able to stabilize a substrate conformation close to CHAIR. Indeed, the CHAIR conformer was found to be more stable in the double mutant Cys88Lys/Arg90Ser than Arg90Ala from the free energy simulations [83]. In the Cys88Lys/Arg90Ser mutant, the substrate is somewhat distorted, and additional energy (circa 2 kcal/mol) is necessary to generate the optimal reactive CHAIR conformation. This factor, as well as the lack of a stable hydrogen-bonding interaction with the ether oxygen atom, would make the double mutants less efficient than the wild-type enzyme (k_{cat} values are about 10^2-fold less than that of wild-type).

1.3.3
Contribution of the Near Attack Conformers (NACs)

The concept of the near attack conformers [89, 90] has also been applied to the mechanism of the CM-catalyzed reaction by Hur and Bruice [60–64]. The near attack conformers (NACs) are special substrate conformations in which the bond-forming atoms are at the van der Waals distance and at an angle near the one in the transition state [60–64]. One convenience of the NAC approach in understanding enzymes is that quantum mechanical calculations may not be necessary for the description of chemical process of bond breaking and making. Therefore, one only needs to focus on the geometrical arrangements of the atoms in the *substrate* that are involved in the bond formation. However, it has been pointed out [80, 91] that one intrinsic problem with this concept is that there does not seem to be a unique way of defining NAC. This is in contrast with the concept of the transition state. A transition state is a well-defined stationary point on the surface and satisfies a variational criterion [91]. According to the results based on the NAC approach, chorismate mutase speeds up the Claisen rearrangement by binding its substrate primarily in its NAC form, rather than by lowering the transition-state barrier [61–62]. However, as Garcia-Viloca et al. [91] pointed out, the essential NAC effect is likely to be that "the activation free energy barrier is reduced by the substrate conformational change induced by the enzyme" (see above).

Different definitions for NCA have been used in the study of the Claisen rearrangement of chorismate [60–64, 77–80]. Some of the inconsistencies in the results as well as in the interpretations might be related to the use of the different definitions (see below). In the earlier studies [60, 62–64], Hur and Bruice used two geometrical parameters to define NAC, and the parameters include the distance between C_1 and C_9 (R_1) and the attacking angle θ_1 (see Scheme 1.8a). A conformation of chorismate was considered to be NAC if $R_1 \leq 3.7$ Å and $\theta_1 \leq 40°$ [60]. In some other investigations [77–78], only the condition on R_1 (i.e., $R_1 \leq 3.7$ Å) was used in the definition of NAC. In a more recent study, Hur and Bruice [61] included an additional parameter, θ_2, in their definition of NAC (see Scheme 1.8b) and suggested that the three conditions for being NAC are: (1) $R_1 \leq 3.7$ Å, (2) $\theta_1 \leq 28.2°$ and (3) $\theta_2 \leq 38.2°$. It should be mentioned that the NACs defined by the single parameter (R_1), the two parameters (R_1 and θ_1) or even three parameters (R_1, θ_1 and θ_2) may not correspond to the reactive CHAIR conformation (see Scheme 1.2). As we mentioned earlier, the CHAIR conformer of the substrate can change to the chair-like transition state directly with the C_1 and C_9 carbon atoms well positioned to form the carbon-carbon bond. The conformers that satisfy the NAC conditions may not posses these properties. To demonstrate this point, Scheme 1.9a and 1.9b show two typical conformations of chorismate in solution that were observed during the QM/MM MD simulations using AM1 method [93] (Guo et al., unpublished results; see also Ref. [61]). The conformation in Scheme 1.9a satisfies the condition on R_1 (i.e., $R_1 = 3.55$ Å. Also, θ_1 is only 2.6° greater than 40°). Nevertheless, the conformer is inactive because the π orbital of C_9 is not positioned to form the carbon-carbon bond with C_1. The conformation in Scheme 1.9b

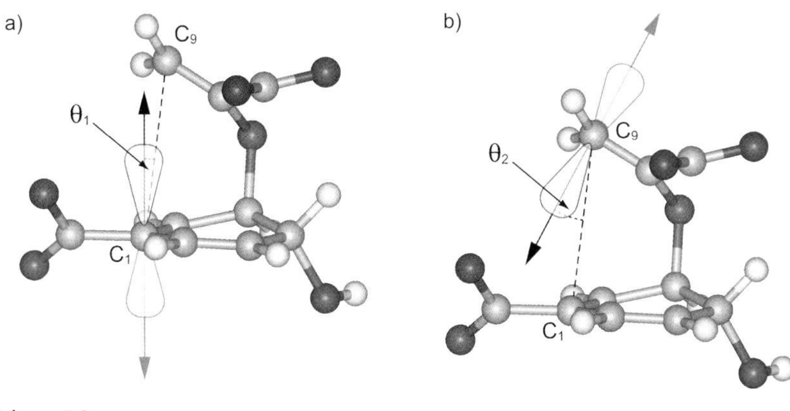

Scheme 1.8

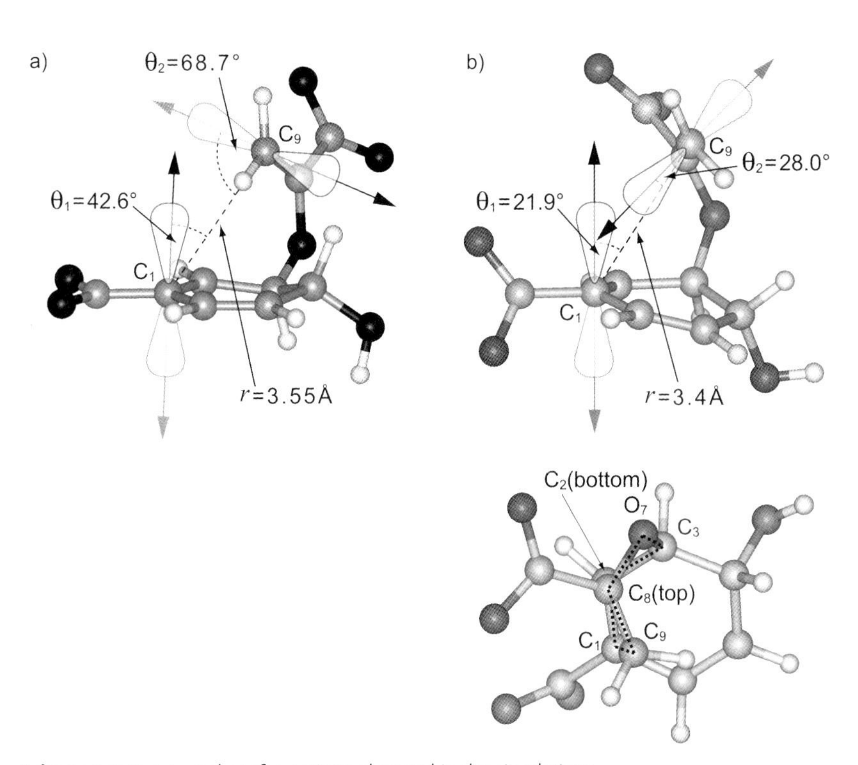

Scheme 1.9 Two typical conformations observed in the simulations of chorismate in solution using the AM1 method for chorismate. (a) An inactive conformation that satisfies $R_1 \leq 3.7$ Å. (b) A "boat-like" NAC that satisfies all the three conditions for NAC proposed in Ref. [61].

satisfies all the three conditions proposed by Hur and Bruice (i.e., $R_1 \leq 3.7$ Å, $\theta_1 \leq 28.2°$ and $\theta_2 \leq 38.2°$). However, this conformer will not lead to the chair-like transition state. As we mentioned earlier, it has been demonstrated by Knowles and coworkers [13, 14] that the Claisen rearrangement of chorismate to prephenate proceeds through the chair-like transition state in solution and enzyme active site. In fact, the NAC in Scheme 1.9b will lead to a "boat-like" transition state in which the atoms in the [3,3]-pericyclic region are arranged in a boat-like configuration. Because the reaction does not proceed through the boat-like transition state either in solution or in the enzyme active site, the formation of the boat-like NACs may not be of great interest. The key point in the NAC concept is that the geometrical parameters for the atoms associated with the bond formation are close to those in the transition state. For the Claisen rearrangement, these geometrical parameters are similar in the chair-like and boat-like transition states. Therefore, the use of these parameters alone is not sufficient to distinguish the boat-like NAC from the chair-like NAC (which is essentially the reactive CHAIR conformation).

Hur and Bruice [61–63] considered the formation of the transition state as two different stages, namely, the formation of NAC with a free energy of formation of ΔG°_N and the formation of the transition state from NAC with a free energy of $\Delta G^\ddagger_{TS}$:

$$\Delta G^\ddagger = \Delta G^\circ_N + \Delta G^\ddagger_{TS} \tag{1.1}$$

Here $\Delta G^\ddagger$ is the activation energy for the Claisen rearrangement. Thus, important insights concerning the origin of the catalysis may be obtained by examining how $\Delta G^\ddagger$ changes with ΔG°_N and $\Delta G^\ddagger_{TS}$ under different environments [61]. ΔG°_N can be estimated from the probability of occurring in molecular dynamics simulations [63] or from free energy simulations, whereas $\Delta G^\ddagger_{TS}$ may be obtained from QM/MM calculations. Hur and Bruice [61] performed molecular dynamics simulations using the CHARMM program with a special (unpublished) set of parameters for chorismate. They estimated ΔG°_N in different environments (solution, 1F7 catalytic antibody, the R90Cit mutant of BsCM, the E52A mutant of EcCM, wild-type BsCM, and wild-type EcCM) and plotted the experimental $\Delta G^\ddagger$ values as a function of ΔG°_N. It was found that $\Delta G^\ddagger$ changes with ΔG°_N linearly with a slope of 1.1. Moreover, the $\Delta G^\ddagger_{TS}$ values obtained by subtracting ΔG°_N from the experimental $\Delta G^\ddagger$ values are essentially a constant (see Table 1.4) in the different environments, as one would expect based on the linear relationship between $\Delta G^\ddagger$ and ΔG°_N. These data support the proposal that the relative rate of the Claisen rearrangement overwhelmingly depends on the efficiency of formation of NAC [61]. Although the results of Hur and Bruice are of considerable interest, there are some significant inconsistencies with other computational studies. Several investigations have suggested that the rate acceleration by chorismate mutase is mainly due to the electrostatic stabilization of the transition state at the enzyme active site [49, 58, 67, 71, 75, 80] and that the efficiency of formation of NAC should not be considered as the major origin for the catalysis [77–78, 80]. For instance, Štrajbl et al. [80] performed empirical valence bond (EVB) and binding free energy calcu-

Table 1.4 Relationship between the free energies in different environments[a].

Environment	$\Delta G^{\ddagger}$	–	ΔG°_{N}	=	$\Delta G^{\ddagger}_{TS}$
Solution	24.2		8.1		16.1
Wild type EcCM	15.2		0.1		15.1
Wild type BsCM	15.4		0.3		15.1
E52A (EcCM)	18.2		1.3		16.9
R90Cit (BsCM)	21.2		4.1		17.1
1F7	21.3		5.5		15.8

a) From Ref. [61] and in $kcal \cdot mol^{-1}$.

lations. They showed that the major catalysis effect of chorismate mutase is electrostatic transition state stabilization (TSS). Although Štrajbl et al. also obtained the NAC effect of about 5 kcal/mol, they suggested that this effect was not the reason for the catalysis but the result of the TSS.

There are different possible reasons for the inconsistencies. For instance, different definitions for NAC have been used in different investigations that could result in discrepancies in the free energy of formation for NAC. For instance, the NACs defined only by $R_1 \leq 3.7$ Å would lead to a smaller free energy of conformation for NAC than that defined by the three conditions. This is because more conformations (e.g., the conformation in Scheme 1.9a) would be counted as NACs. In the analysis of the origin of enzyme catalysis based on the NAC concept, the effect of the enzyme on the activation barrier is divided into the two different contributions (see above). An underestimate (overestimate) of the free energy for the NAC formation would result in an overestimate (underestimate) of $\Delta G^{\ddagger}_{TS}$, leading to a different conclusion concerning the importance of the NAC formation. Thus, a consistent definition of NAC should be very important for the comparison of different simulations results. Another possible reason for the inconsistencies would be the use of different computational approaches and force field parameters. For instance, the AM1 method has been widely used in the study of the Claisen rearrangement of chorimate in solution and enzyme. However, the gas phase calculations showed that the AM1 method overestimates the stability of the CHAIR conformer significantly (see Table 1.3). Such deficiencies in the AM1 method would affect the accuracy of the simulation results as well as the conclusion concerning the origin of the catalysis.

1.3.4
Strain Effects and Conformational Compression

Strains and conformational compression of the substrate at the enzyme active site were also proposed to explain the rate acceleration by chorismate mutase [66, 70,

78]. For instance, Jorgensen and his coworkers [78] suggested, based on the results of their calculations and analysis, that the stabilization of the transition state in chorismate mutase plays only a secondary role and that the rate enhancement is primarily due to conformational compress of the substrate. However, the analysis of Jorgensen and coworkers has been called into question [80], and it was shown that the results might be interpreted differently [80].

1.4 Conclusion

The discussions given in this chapter have shown that although chorismate mutase has been a subject of extensive experimental and theoretical investigations, there are still considerable uncertainties concerning how the Claisen rearrangement from chorismate to prephenate is actually catalyzed by the enzyme. The computational investigations have led different possibilities, but experimental studies with modern techniques are necessary to identify the most likely mechanism of the CM catalysis.

Acknowledgment

The work described herein was supported by the Center of Excellence for Structural Biology, University of Tennessee. We thank Prof. Martin Karplus for insightful discussions.

List of Abbreviations

BsCM	*Bacillus subtillus* chorismate mutase
CHAIR	chair-like configuration
CM	chorismate-pyruvate mutase, EC 5.4.99.5
DIAX	pseudodiaxial conformer
DIEQ	inactive pseudodiequatorial conformer
EcCM	*Escherichia coli* chorismate mutase from P-protein
EVB	empirical valence bond
MD	molecular dynamics
MM	molecular mechanics
NAC	near attack conformers
SCC-DFTB	self-consistent charge density functional tight binding
TS	transition state
TSA	transition state analog

References

1 Haslam, E. *Shikimic Acid Metabolism and Metabolites*, John Wiley & Sons, New York, 1993.
2 Lee, A. Y.; Stewart, J. D.; Clardy, J.; Ganem, B. *Chem. Biol.* **1995**, *2*, 195–203.
3 Ganem, B. *Angew. Chem. Int. Ed.* **1996**, *35*, 936–945.
4 Lee, A. Y.; Karplus, P. A.; Ganem, B.; Clardy, J. *J. Am. Chem. Soc.* **1995**, *117*, 3627–3628.
5 Stewart, J.; Wilson, D. B.; Ganem, B. *J. Am. Chem. Soc.* **1990**, *112*, 4582–4584.
6 Chook, Y. M.; Ke, H. M.; Lipscomb, W. N. *Proc. Nat. Acad. Sci. USA*. **1993**, *90*, 8600–8603.
7 Chook, Y. M.; Gray, J. V.; Ke, H. M.; Lipscomb, W. N. *J. Mol. Biol.* **1994**, *240*, 476–500.
8 Ladner, J. E.; Reddy, P.; Davis, A.; Tordova, M.; Howard, A. J.; Gilliland, G. L. *Acta Crystallogr. Sect. D Biol. Crystallogr.* **2000**, *56*, 673–683.
9 Hilvert, D.; Carpenter, S. H.; Nared, K. D.; Auditor, M. T. M. *Proc. Nat. Acad. Sci. USA* **1988**, *85*, 4953–4955.
10 Xue, Y. F.; Lipscomb, W. N.; Graf, R.; Schnappauf, G.; Braus, G. *Proc. Nat. Acad. Sci. USA* **1994**, *91*, 10814–10818.
11 Xue, Y. F.; Lipscomb, W. N. *J. Mol. Biol.* **1994**, *241*, 273–274.
12 Strater, N.; Hakansson, K.; Schnappauf, G.; Braus, G.; Lipscomb, W. N. *Proc. Nat. Acad. Sci. USA* **1996**, *93*, 3330–3334.
13 Sogo, S. G.; Widlanski, T. S.; Hoare, J. H.; Grimshaw, C. E.; Berchtold, G. A.; Knowles, J. R. *J. Am. Chem. Soc.* **1984**, *106*, 2701–2703.
14 Copley, S. D.; Knowles, J. R. *J. Am. Chem. Soc.* **1985**, *107*, 5306–5308.
15 Copley, S. D.; Knowles, J. R. *J. Am. Chem. Soc.* **1987**, *109*, 5008–5013.
16 Andrews, P. R.; Haddon, R. C. *Aust. J. Chem.* **1979**, *32*, 1921–1929.
17 Campbell, A. P.; Tarasow, T. M.; Massefski, W.; Wright, P. E.; Hilvert, D. *Proc. Nat. Acad. Sci. USA* **1993**, *90*, 8663–8667.
18 Clarke, T.; Stewart, J. D.; Ganem, B. *Tetrahedron* **1990**, *46*, 731–748.
19 Bartlett, P. A.; Nakagawa, Y.; Johnson, C. R.; Reich, S. H.; Luis, A. *J. Org. Chem.* **1988**, *53*, 3195–3210.
20 Andrews, P. R.; Smith, G. D.; Young, I. G. *Biochemistry* **1973**, *12*, 3492–3498.
21 Galopin, C. C.; Zhang, S.; Wilson, D. B.; Ganem, B. *Tetrahedron Lett.* **1996**, *37*, 8675–8678.
22 Pawlak, J. L.; Padykula, R. E.; Kronis, J. D.; Aleksejczyk, R. A.; Berchtold, G. A. *J. Am. Chem. Soc.* **1989**, *111*, 3374–3381.
23 Galopin, C. C.; Ganem, B. *Bioorg. Med. Chem. Lett.* **1997**, *7*, 2885–2886.
24 Crespo, A.; Scherlis, D. A.; Marti, M. A.; Ordejon, P.; Roitberg, A. E.; Estrin, D. A. *J. Phys. Chem. B* **2003**, *107*, 13728–13736.
25 Cload, S. T.; Liu, D. R.; Pastor, R. M.; Schultz, P. G. *J. Am. Chem. Soc.* **1996**, *118*, 1787–1788.
26 Kast, P.; AsifUllah, M.; Hilvert, D. *Tetrahedron Lett.* **1996**, *37*, 2691–2694.
27 Liu, D. R.; Cload, S. T.; Pastor, R. M.; Schultz, P. G. *J. Am. Chem. Soc.* **1996**, *118*, 1789–1790.
28 Schnappauf, G.; Strater, N.; Lipscomb, W. N.; Braus, G. H. *Proc. Nat. Acad. Sci. USA* **1997**, *94*, 8491–8496.
29 Kast, P.; Hartgerink, J. D.; AsifUllah, M.; Hilvert, D. *J. Am. Chem. Soc.* **1996**, *118*, 3069–3070.
30 Kast, P.; AsifUllah, M.; Jiang, N.; Hilvert, D. *Proc. Nat. Acad. Sci. USA* **1996**, *93*, 5043–5048.
31 Kast, P.; Grisostomi, C.; Chen, I. A.; Li, S. L.; Krengel, U.; Xue, Y. F.; Hilvert, D. *J. Biol. Chem.* 2000, *275*, 36832–36838.
32 Ife, R.; Ball, L. F.; Lowe, P.; Haslam, E. *J. Chem. Soc. Perkin Trans. I* **1976**, 1776–1783.
33 Haynes, M. R.; Stura, E. A.; Hilvert, D.; Wilson, I. A. *Science* **1994**, *263*, 646–652.
34 Strater, N.; Hakansson, K.; Schnappauf, G.; Braus, G.; Lipscomb, W. N. *Structure* **1997**, *5*, 1437–1452.
35 Zhang, S.; Kongsaeree, P.; Clardy, J.; Wilson, D. B.; Ganem, B. *Bioorg. Med. Chem.* **1996**, *4*, 1015–1020.
36 Kienhofer, A.; Kast, P.; Hilvert, D. *J. Am. Chem. Soc.* **2003**, *125*, 3206–3207.

37 Zhang, S.; Pohnert, G.; Kongsaeree, P.; Wilson, D. B.; Clardy, J.; Ganem, B. *J. Biol. Chem.* **1998**, *273*, 6248–6253.
38 Gustin, D. J.; Mattei, P.; Kast, P.; Wiest, O.; Lee, L.; Cleland, W. W.; Hilvert, D. *J. Am. Chem. Soc.* **1999**, *121*, 1756–1757.
39 Lee, A.L.; Zhang S.; Kongsaeree; Clardy, J.; Ganem, B.; Erickson, J. W.; Xie, D. *Biochemistry* **1998**, *37*, 9052–9057.
40 Addadi, L.; Jaffe, E. K.; Knowles, J. R. *Biochemistry* **1983**, *22*, 4494–4501.
41 Andrews, P. R.; Cain, E. N.; Rizzardo, E.; Smith, G. D. *Biochemistry* **1977**, *16*, 4848–4852.
42 Gorisch, H. *Biochemistry* **1978**, *17*, 3700–3705.
43 Gamper, M.; Hilvert, D; Kast, P. *Biochemistry*, **2000**, *39*, 14087–14094.
44 Ma, J. P.; Zheng, X. F.; Schnappauf, G.; Braus, G.; Karplus, M.; Lipscomb, W. N. *Proc. Nat. Acad. Sci. USA* **1998**, *95*, 14640–14645.
45 Sturtevant, J. M. *Proc. Nat. Acad. Sci. USA* 1977, 74, 2236–2240.
46 Naghibi, H.; Tamura, A.; Sturtevant, J. M. *Proc. Nat. Acad. Sci. U.S.A.* **1995**, *92*, 5597–5599.
47 Wiest, O.; Houk, K. N. *J. Am. Chem. Soc.* **1995**, *117*, 11628–11639.
48 Wiest, O.; Houk, K. N. *J. Org. Chem.* **1994**, *59*, 7582–7584.
49 Lyne, P. D.; Mulholland, A. J.; Richards, W. G. *J. Am. Chem. Soc.* **1995**, *117*, 11345–11350.
50 Davidson, M. M.; Gould, I. R.; Hillier, I. H. *J. Chem. Soc., Chem. Commun.* **1995**, 63–64.
51 Davidson, M. M.; Gould, I. R.; Hillier, I. H. *J. Chem. Soc., Perkin Trans. 2* **1996**, 525–532.
52 Davidson, M. M.; Guest, J. M.; Craw, J. S.; Hillier, I. H.; Vincent, M. A. *J. Chem. Soc., Perkin Trans. 2* **1997**, 1395–1400.
53 Worthington, S. E.; Krauss, M. *J. Phys. Chem. B* **2001**, *105*, 7096–7098.
54 Worthington, S. E.; Roitberg, A. E.; Krauss, M. *J. Phys. Chem. B* **2001**, *105*, 7087–7095.
55 Worthington, S. E.; Roitberg, A. E.; Krauss, M. *Int. J. Quantum Chem.* **2003**, *94*, 287–292.
56 Hall, R. J.; Hindle, S. A.; Burton, N. A.; Hillier, I. H. *J. Comput. Chem.* **2000**, *21*, 1433–1441.
57 Carlson, H. A.; Jorgensen, W. L. *J. Am. Chem. Soc.* **1996**, *118*, 8475–8484.
58 Crespo, A.; Scherlis, D. A.; Marti, M. A.; Ordejon, P.; Roitberg, A. E.; Estrin, D. A. *J. Phys. Chem. B* **2003**, *107*, 13728–13736.
59 Guimaraes, C. R. W.; Repasky, M. P.; Chandrasekhar, J.; Tirado–Rives, J.; Jorgensen, W. L. *J. Am. Chem. Soc.* **2003**, *125*, 6892–6899.
60 Hur, S.; Bruice, T. C. *Proc. Nat. Acad. Sci. USA* **2002**, *99*, 1176–1181.
61 Hur, S.; Bruice, T. C. *Proc. Nat. Acad. Sci. USA* **2003**, *100*, 12015–12020.
62 Hur, S.; Bruice, T. C. *J. Am. Chem. Soc.* **2003**, *125*, 10540–10542.
63 Hur, S.; Bruice, T. C. *J. Am. Chem. Soc.* **2003**, *125*, 5964–5972.
64 Hur, S.; Bruice, T. C. *J. Am. Chem. Soc.* **2003**, *125*, 1472–1473.
65 Kast, P.; Tewari, Y. B.; Wiest, O.; Hilvert, D.; Houk, K. N.; Goldberg, R. N. *J. Phys. Chem. B* **1997**, *101*, 10976–10982.
66 Khanjin, N. A.; Snyder, J. P.; Menger, F. M. *J. Am. Chem. Soc.* **1999**, *121*, 11831–11846.
67 Lee, Y. S.; Worthington, S. E.; Krauss, M.; Brooks, B. R. *J. Phys. Chem. B* **2002**, *106*, 12059–12065.
68 Madurga, S.; Vilaseca, E. *Phys. Chem. Chem. Phys.* **2001**, *3*, 3548–3554.
69 Madurga, S.; Vilaseca, E. *J. Phys. Chem. A* **2002**, *106*, 11822–11830.
70 Marti, S.; Andres, J.; Moliner, V.; Silla, E.; Tunon, I.; Bertran, J. *J. Phys. Chem. B* **2000**, *104*, 11308–11315.
71 Marti, S.; Andres, J.; Moliner, V.; Silla, E.; Tunon, I.; Bertran, J. *Theor. Chem. Acc.* **2001**, *105*, 207–212.
72 Marti, S.; Andres, J.; Moliner, V.; Silla, E.; Tunon, I.; Bertran, J. *Chem. Eur. J.* **2003**, *9*, 984–991.
73 Marti, S.; Andres, J.; Moliner, V.; Silla, E.; Tunon, I.; Bertran, J. *Theochem. J. Mol. Struct.* **2003**, *632*, 197–206.
74 Marti, S.; Andres, J.; Moliner, V.; Silla, E.; Tunon, I.; Bertran, J. *J. Am. Chem. Soc.* **2004**, *126*, 311–319.
75 Marti, S.; Andres, J.; Moliner, V.; Silla, E.; Tunon, I.; Bertran, J.; Field, M. J. *J. Am. Chem. Soc.* **2001**, *123*, 1709–1712.
76 Ranaghan, K. E.; Ridder, L.; Szefczyk, B.; Sokalski, W. A.; Hermann, J. C.; Mulholland, A. J. *Mol. Phys.* **2003**, *101*, 2695–2714.

77 Repasky, M. P.; Guimaraes, C. R. W.; Chandrasekhar, J.; Tirado-Rives, J.; Jorgensen, W. L. *J. Am. Chem. Soc.* **2003**, *125*, 6663–6672.

78 Guimaraes, C. R. W.; Repasky, M. P.; Chandrasekhar, J.; Tirado-Rives, J.; Jorgensen, W. L. *J. Am. Chem. Soc.* **2003**, *125*, 6892–6899.

79 Roca, M.; Marti, S.; Andres, J.; Moliner, V.; Tunon, M.; Bertran, J.; Williams, A. H. *J. Am. Chem. Soc.* **2003**, *125*, 7726–7737.

80 Strajbl, M.; Shurki, A.; Kato, M.; Warshel, A. *J. Am. Chem. Soc.* **2003**, *125*, 10228–10237.

81 Woodcock, H. L.; Hodoscek, M.; Sherwood, P.; Lee, Y. S.; Schaefer, H. F.; Brooks, B. R. *Theor. Chem. Acc.* **2003**, *109*, 140–148.

82 Guo, H.; Cui, Q.; Lipscomb, W. N.; Karplus, M. *Proc. Nat. Acad. Sci. USA* **2001**, *98*, 9032–9037.

83 Guo, H.; Cui, Q.; Lipscomb, W. N.; Karplus, M. *Angew. Chem. Int. Ed.* **2003**, *42*, 1508–1511.

84 (a) Schowen, R. L. *Proc. Nat. Acad. Sci. USA* **2003**, *100*, 11931–11932. (b) *Chem. Eng. News* **2004**, *82*, 35–39.

85 Brooks, B.; Bruccoleri, R. E.; Olafson, B. D.; States, D. J.; Swaminathan, S.; Karplus, M. *J. Comput. Chem.* **1983**, *4*, 187.

86 (a) Elstner, M.; Porezag, D.; Jungnickel, G.; Elsner, J.; Haugk, M.; Frauenheim, T.; Suhai, S.; Seigert, G. *Phys. Rev. B* **2001**, *58*, 7260–7268. (b) Cui, Q.; Elstner, M.; Kaxiras, E.; Frauenheim, T.; Karplus, M. *J. Phys. Chem. B* **2001**, *105*, 569–585.

87 (a) Jorgensen, W. L. *J. Am. Chem. Soc.* **1981**, *103*, 335. (b) Neria, E.; Fischer, S.; Karplus, M. *J. Chem. Phys.* **1996**, *105*, 1902–1921.

88 MacKerell, A. D. Jr.; Bashford, D.; Bellott, M.; Dunbrack, R. L.; Evanseck, J. D.; Field, M. J.; Fischer, S.; Gao, J.; Guo, H.; Ha, S.; Joseph-McCarthy, D.; Kuchnir, L.; Kuczera, K.; Lau, F. T. K.; Mattos, C.; Michnick, S.; Ngo, T.; Nguyen, D. T.; Prodhom, B.; Reiher, W. E.; Roux, B.; Schlenkrich, M.; Smith, J. C.; Stote, R.; Straub, J.; Watanabe, M.; Wiorkiewicz-Kuczera, J.; Yin, D.; Karplus, M. *J. Phys. Chem. B* **1998**, *102*, 3586–3616.

89 Bruice, T. C. *Acc. Chem. Res.* **2002**, *35*, 139–148.

90 Bruice, T. C.; Benkovic, S. J. *Biochemistry* **2000**, *39*, 6267.

91 Garcia-Viloca, M.; Gao, J.; Karplus, M.; Truhlar, D. G. *Science* **2004**, *303*, 186–195.

92 Dewar, M. J. S.; Zoebisch, E. G.; Healy, E. F.; Steward, J. J. P. *J. Am. Chem. Soc.* **1985**, *107*, 3902–3909.

2
Chiral-Metal-Complex-Catalyzed Aliphatic Claisen Rearrangement

Koichi Mikami and Katsuhiro Akiyama

2.1
Introduction

Since its discovery in 1912 [1], the Claisen rearrangement, namely thermal [3,3]-sigmatropic rearrangement of an allyl vinyl ether, has received great attention as one of the most powerful methods for stereoselective carbon–carbon bond formation. Metal catalysis has been introduced to significantly accelerate the Claisen rearrangement even at room temperature. A literature survey on metal catalysis of Claisen rearrangements is available in excellent reviews by Lutz, Overman, Frauenrath and Hiersemann [2].

Therefore, we focus our attention only on the recent developments of chiral metal catalysis rather than acceleration or promotion by an equimolar amount of metal complexes. Excellent contributions based on main-group metals by more than one molar amount of Al^{III} and B^{III} [3] complexes in particular will be described in this chapter just as an introduction to metal-catalyzed Claisen rearrangement.

The Claisen rearrangement is a concerted, thermally-allowed [3,3]-sigmatropic rearrangement that takes place through highly ordered six-membered transition states. The stereochemical course of the Claisen rearrangement can be generally predicted on the basis of conformational analysis in chair-like rather than boat-like transition states. However, metal catalysis sometimes results in a change in the stereoselectivity generally predicted from the chair-like transition states, depending on the binding modes of late transition metals with diene portions. Chiral metal complexes have been further exploited for asymmetric catalysis. However, metal-chelation-controlled Claisen rearrangements, aza-Claisen rearrangements, and imidate and acetate rearrangements are beyond the scope of this chapter.

The Claisen Rearrangement. Edited by M. Hiersemann and U. Nubbemeyer

ISBN: 978-3-527-30825-5

2.2 Binding Modes of Main-group and Late Transition Metals

Binding modes of metal complexes with Claisen systems, allyl vinyl ethers, are classified into two patterns (Scheme 2.1). The first involves binding with ether oxygen using hard main-group metals and enantioselectively using such a chiral metal complex stems from the differentiation of two enantiomeric chair-like transition states by the selective coordination of two enantiotopic lone pairs on oxygen. However, carbonyl products are generally more coordinative to these oxophilic metals. This is the reason why one molar amount or more of main-group metals are generally required for the complete progress of the Claisen rearrangements of allyl vinyl ether systems [4].

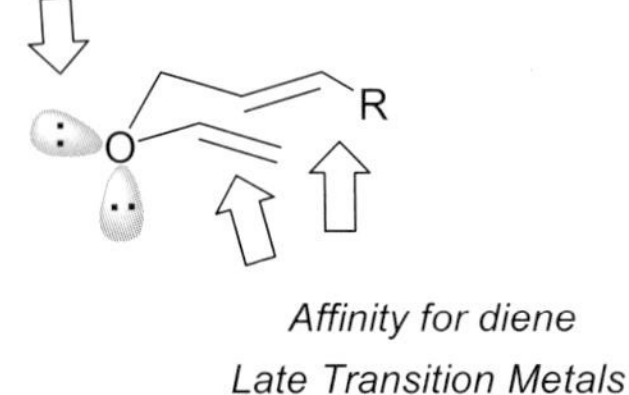

Scheme 2.1 Binding modes of main-group and late transition metals.

In sharp contrast, the late transition metals are more coordinative to soft carbon–carbon multiple bonds rather than hard oxygen. These binding modes are further classified into mono- and bi-dentate coordinations, depending on the ligands on the metal catalysts, or the substituent pattern in the Claisen diene systems and solvents employed. Bi-dentate coordination of the Claisen substrate is advantageous over the weak mono-dentate coordination of the Claisen rearrangement product, γ,δ-unsaturated carbonyl compounds, to release the metal complex allowing the catalytic cycle. Furthermore, enantiodiscrimination by chiral late transition metal complexes is based on the discrimination of two enantiotopic diene faces in the enantiomeric six-membered transition states.

2.3 Aluminum(III)-promoted Claisen Rearrangement

The Lewis acidity of organoaluminum reagents (R_2AlX) is maximized in the monomeric structure and increased by the electron-withdrawing substituent. The order of the monomeric aluminum Lewis acidity is as follows: $R_2AlNR'_2 < R_2AlOR' < R_2AlOAr < R_2AlOCOR'$. However, aluminum reagents are generally

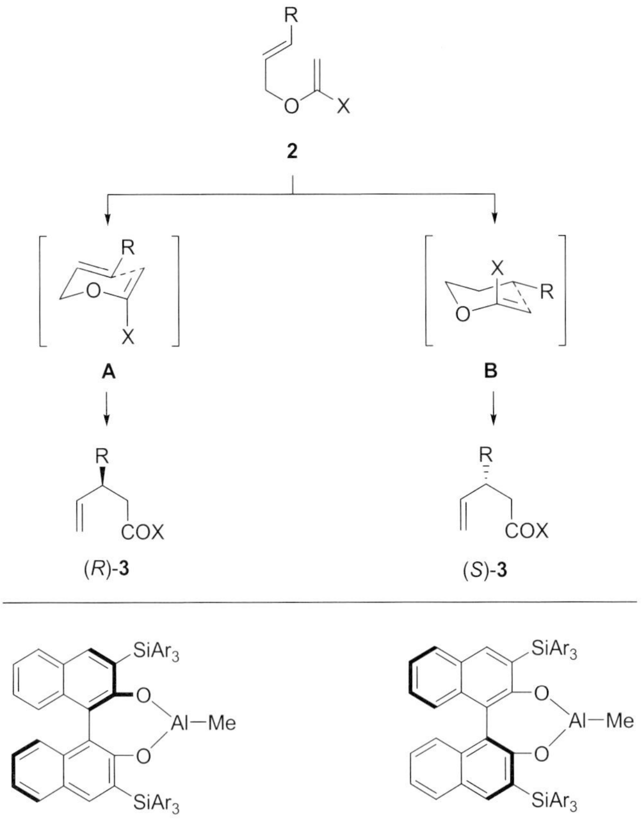

Scheme 2.2 Al^{III}-promoted asymmetric Claisen rearrangement.

stabilized as the dimeric complexes in solution. In this case, the Lewis acidity is in the order: $R_2AlNR'_2 < R_2AlOR' < R_2AlOCOR' < R_2AlOAr$. Therefore, R_2AlOAr is expected to be the most Lewis acidic in solution. Bulky ligands can be used for aluminum reagents to suppress the formation of the dimeric complexes, and hence increase the Lewis acidity.

Yamamoto and Maruoka have reported the first aluminum-promoted asymmetric Claisen rearrangement [5]. They used a chiral 3,3′-bis(triarylsilyl)-substituted binaphthol (BINOL) aluminum reagent **1** to promote the enantioselective coordination to one of the two enantiomeric oxygen lone pairs of **2**. Enantioselective discrimination is based on two chair-like transition states **A** and **B** to give two enantiomeric products **3**, respectively (Scheme 2.2).

In this case, the reaction proceeds between –78 and 0 °C (Table 2.1). The best solvent is CH_2Cl_2. Use of other solvents such as toluene, ether, or THF decreases the rate of the rearrangement. The bulky substituent X (X = $SiMe_3$, $GeMe_3$) is nec-

Tab. 2.1 3,3′-Bis(triarylsilyl) BINOL-Al complex-catalyzed asymmetric Claisen rearrangement.[a)]

Entry	(*R*)-1	Two stage reaction condition (°C, h)	Yield [%]	*ee* [%]
		2a → **3a** (Ph, SiMe$_3$)		
1	$Ar_3 = tBuMe_2$	–40, 0.1; –20, 24	22	14 (*S*)
2	Ar = Ph	–40, 0.1; –20, 8	86	80 (*S*)
3	$Ar_3 = tBuPh_2$	–40, 0.1; –20, 3	99	88 (*S*)
4[b)]	Ar = Ph	–40, 0.1; –20, 5	85	80 (*R*)
		2b → **3b** (Ph, SiMe$_2$Ph)		
5	Ar = Ph	–78, 0.1; –40, 16	65	85 (*S*)
6	$Ar_3 = tBuPh_2$	–78, 0.1; –40, 8	76	90 (*S*)
		2c → **3c** (Ph, SiMe$_3$)		
7	Ar = Ph	–20, 0.1; 0, 2	77	67 (*S*)
		2a → **3a** (Ph, SiMe$_3$)		
8	Ar = Ph	–78, 0.1; –40, 15	73	91 (*S*)
9	$Ar_3 = tBuPh_2$	–78, 0.1; –40, 16	68	93 (*S*)

a) The rearrangement was carried out with 1.1–2 equivalents of (*R*)-**1** in CH_2Cl_2.
b) Use of (*S*)-**1** as a catalyst.

essary to obtain high enantioselectivity. The triarylsilyl ($SiAr_3$) group is required as the 3,3′-disubustituent of the BINOL ligand. The 3,3′-diphenyl substituent is not effective. Among various trialkylsilyl substituents for the BINOL–Al complexes for substrate **2a** (R = Ph, X = $SiMe_3$), the bulky *tert*-butyldiphenylsilyl group gives the highest enantioselectivity (entry 3 vs. 1 and 2). The introduction of an alkyl group at the *para* position of the triphenylsilyl group results in decrease in selectivity. Use of the (*Z*)-configuration of the substrate in the allyl group **2c** provides the same absolute configuration of (*S*)-product **3c** (entry 7). In the (*Z*)-substrate, the reaction proceeds via boat transition states because of steric repulsion between R group and Si group in the chair transition state [6].

The transition state models in asymmetric Claisen rearrangement are shown in Fig. 2.1. The space filling models **A** and **B** of allyl vinyl ether **2a** (R = Ph; X = $SiMe_3$) exemplify the two enantiomeric chair-like transition state models. The orientation of the vinyl ether double bond of the substrate is the most important for achieving the highest enantioselectivity. The conformation of **A** is a good match for the molecular cleft of chiral aluminum reagents to give the (*S*)-configured product **3a**. In contrast, the conformation of **B** is disfavored in approaching the cleft of the aluminum reagent because of the hindrance of the methylene substituent.

They have also reported a chiral ATPH **4** [7] aluminum tris(2,6-diphenylphenoxide) analogue, namely ATBN **5a** and ATBN-F **5b**, for asymmetric Claisen re-

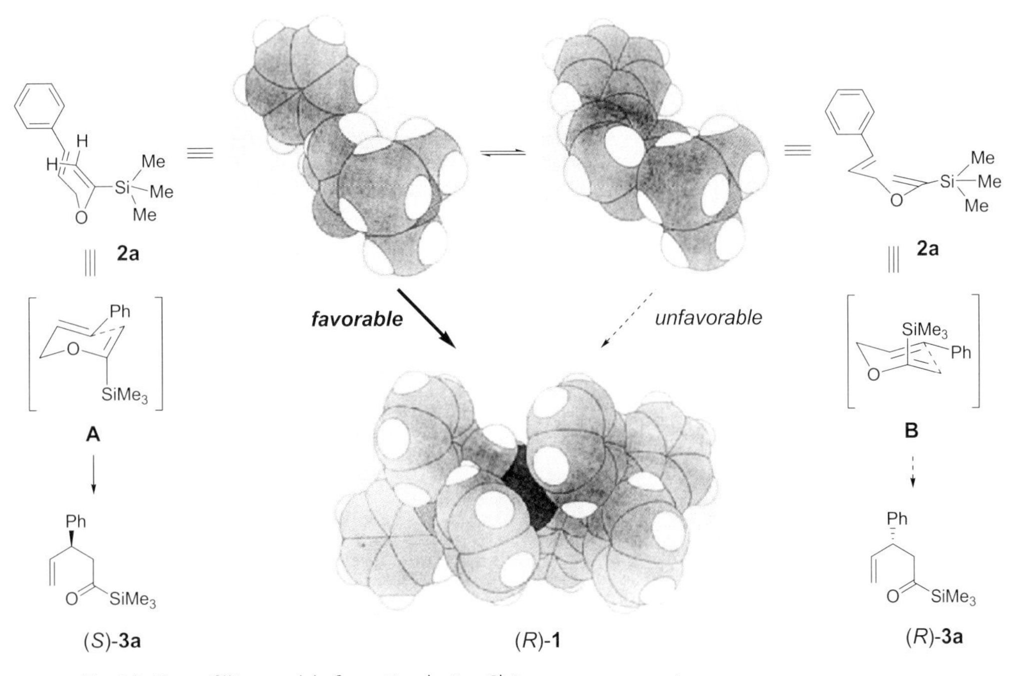

Fig. 2.1 Space-filling model of enantioselective Claisen rearrangement.

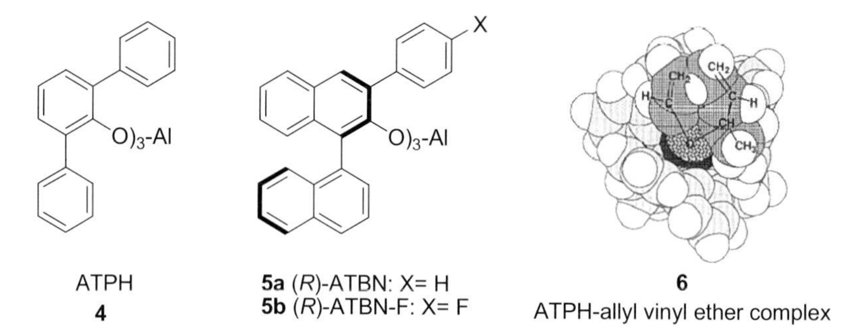

Fig. 2.2 ATPH, ATBN and space filling model of ATPH-allyl vinyl.

arrangement (Fig. 2.2) [8]. ATPH serves as a tight reaction pocket for allyl vinyl ethers and coordinates to the six-membered chair-like transition state **6** (Fig. 2.2).

The reaction proceeds at –78°C in toluene (Table 2.2). In this system, the substituent at C-2 position of the substrate is not necessary. In all substrates, the enantioselectivity of ATBN-F **5b** is higher than ATBN **5a**.

In the transition state for the asymmetric Claisen rearrangement of allyl vinyl ether **7c** with (*R*)-ATBN **5a**, the one conformation of the ether substrate is the matched system for the C_3-symmetric molecular cleft of the chiral aluminum reagent **9**, which gives the (*S*)-product (Fig. 2.3). Increased enantioselectivity using

Tab. 2.2 (*R*)-ATBN- or (*R*)-ATBN-F-promoted asymmetric Claisen rearrangement.

R, O, **7** → (*R*)-ATBN **5a** or (*R*)-ATBN-F **5b**, toluene, -78 °C → R, O, **8**

Entry	Substrate	Al^{III}-reagent	Yield [%]	*ee* [%]
1	R = *t*-Bu **7a**	(*R*)-ATBN	63	63 (*S*)
2		(*R*)-ATBN-F	70	91 (*S*)
3	R = c-C_6H_{11} **7b**	(*R*)-ATBN	78	61 (*S*)
4		(*R*)-ATBN-F	85	86 (*S*)
5	R = Ph **7c**	(*R*)-ATBN	93	61 (*S*)
6		(*R*)-ATBN-F	97	76 (*S*)
7	R = $SiMe_3$ **7d**	(*R*)-ATBN-F	78	92

The reaction was carried out with 1.1–2 equivalents of (*R*)-ATBN or (*R*)-ATBN-F for 10–40 h.

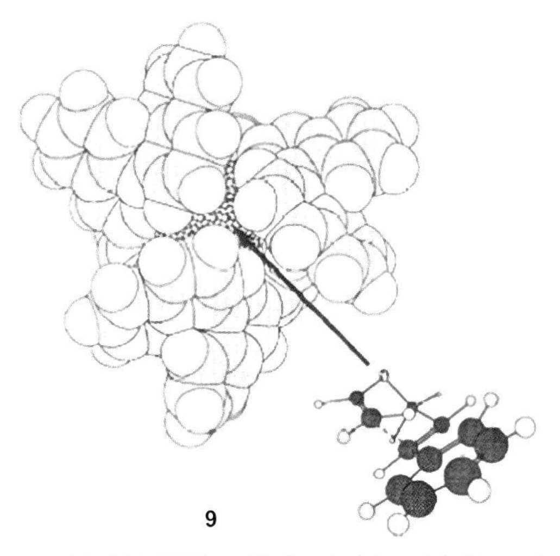

Fig. 2.3 Space-filling model of (R)-ATBN, and ball and stick model of *trans*-cinnamyl vinyl ether.

(*R*)-ATBN-F **5b** is explained by the effective shielding of one side of the aluminum center with three *para* fluorophenyl groups, which prevent substrate approach from the back side in **9** (Fig. 2.3).

Recently, Maruoka reported a new chiral bimetallic organoaluminum reagent **10** for asymmetric Claisen rearrangement (Scheme 2.3) [9]. This reagent activates the allyl vinyl ether to promote rearrangement with double coordination to oxygen.

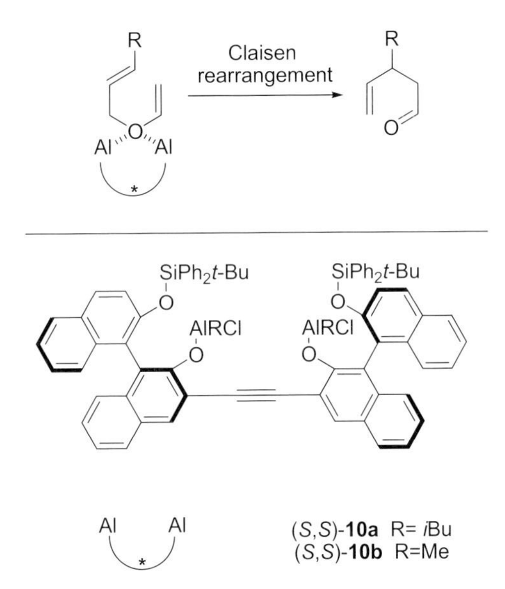

Scheme 2.3 Chiral bimetallic organoaluminium-promoted asymmetric Claisen rearrangement.

In this reaction, PPh_3 is used as an additive to avoid fragmentation of the substrate (Table 2.3). Generally, (*S*,*S*)-**10a** and (*S*,*S*)-**10b** seem to be complementary to each other in terms of reactivity and selectivity. Higher enantioselectivity is obtained with more sterically bulky substituents R on the substrate.

Tab. 2.3 Chiral bimetallic organo-aluminium-catalyzed asymmetric Claisen rearrangement.

(*S*,*S*)-**10** (1.1 eq), PPh_3 (2.2 eq), CH_2Cl_2: **7** → **8**

Entry	Substrate	Al^{III}-reagent	Two stage reaction condition (°C, h)	Yield [%]	*ee* [%]
1	R = *t*-Bu **7a**	**10a**	–78, 1.5; –45, 4	75	80 (*S*)
2		**10b**	–78, 1.5; –45, 4	70	85 (*S*)
3	R = c-C_6H_{11} **7b**	**10a**	–78, 0.1; –45, 4	82	62 (*S*)
4		**10b**	–78, 0.1; –45, 4	66	71 (*S*)
5	R = Ph **7c**	**10a**	–78, 0.1; –45, 1	92	51 (*S*)
6		**10b**	–78, 0.1; –45, 1	50	57 (*S*)
7	R = $SiMe_3$ **7d**	**10b**	–78, 0.1; –45, 2 –35, 1; –23, 1	63	83

2.4
Copper(II)-catalyzed Claisen Rearrangement

Copper complexes can generally exist in three oxidation states: Cu^0, Cu^I, and Cu^{II}. The Lewis acidity of copper depends on their oxidation state and the counter ions. The Cu^{II} oxidation state is generally the most Lewis acidic [10].

Recently, Hiersemann reported the first catalytic enantioselective Claisen rearrangement (Scheme 2.4) [11]. The 2-alkoxycarbonyl-substituted allyl vinyl ethers **11** are reactive under the Lewis acid catalysis. Therefore, the Claisen rearrangements proceed catalytically [12]. Usually the Lewis-acid-catalyzed Claisen rearrangement does not proceed catalytically because of a higher affinity of the carbonyl product for the Lewis acids than the ether substrate. But this 2-alkoxycarbonyl-substituted substrate **11** can coordinate to metals in a bidentate fashion. This 2-alkoxycarbonyl substrate has higher affinity for Lewis acidic Cu complexes than the simple ether substrate. In this system, chiral copper (II) bisoxazoline {Cu^{II} (box)} complex **13** is effective for the enantioselective Claisen rearrangement.

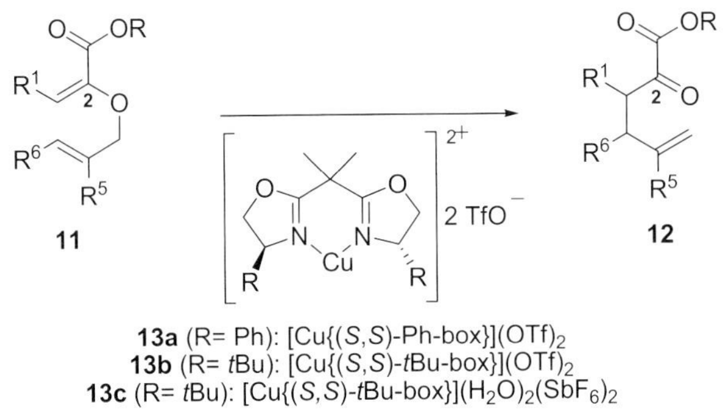

13a (R= Ph): [Cu{(*S*,*S*)-Ph-box}](OTf)$_2$
13b (R= *t*Bu): [Cu{(*S*,*S*)-*t*Bu-box}](OTf)$_2$
13c (R= *t*Bu): [Cu{(*S*,*S*)-*t*Bu-box}](H_2O)$_2$(SbF_6)$_2$

Scheme 2.4 Cu^{II}-catalyzed enantioselective Claisen rearrangement using 2-alkoxycarbonyl substrate.

The [Cu^{II}(box)](OTf)$_2$ (**13**)-catalyzed Claisen rearrangement of allyl vinyl ethers, in which only one center of chirality is formed, has been investigated (Table 2.4).

The absolute configuration of the product is determined by the configuration of the vinyl ether double bond and absolute configuration of the catalyst. The catalyst **13b** (R=*t*-Bu) displays lower reactivity than **13a** (R=Ph). The combination of MS4A improves the catalyst activity and enantioselectivity. The same configuration of the rearrangement products are obtained when the Cu^{II} catalysts, (*S,S*)-**13b** (R=*t*-Bu) and (*R,R*)-**13a** (R=Ph) are employed. This effect is also described for the hetero-Diels–Alder reaction by using Cu^{II}(box) catalysts because of changing coordination mode of copper [13]. In this reaction, alkyl or alkenyl groups can be used as the substituents on a vinyl group. The methyl ester substrate also gives high enantioselectivity. This reaction requires substrates substituted at the C_6 position to provide high enantioselectivitiy.

Tab. 2.4 Cu^{II}-catalyzed enantioselective Claisen rearrangement.

11a → Cu^{II}-catalyst, CH_2Cl_2, r.t. → (*R*)-**12a** + (*S*)-**12a**

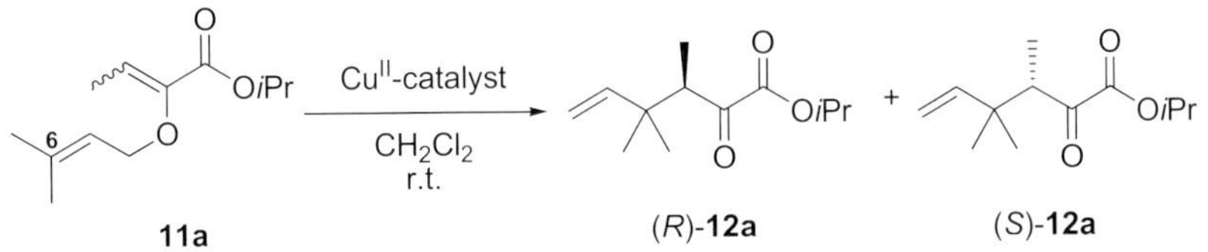

Entry	Substrate	Z:E	Catalyst	*t* [h]	Yield [%]	*ee* [%]
1	(*E*)	4:96	(*S,S*)-**13a**, 5 mol%	1	99	82 (*S*)
2	(*Z*)	96:4	(*S,S*)-**13a**, 5 mol%	1	100	82 (*R*)
3	(*Z*)	96:4	(*R,R*)-**13a**, 5 mol%	1	100	82 (*S*)
4	(*Z*)	96:4	(*S,S*)-**13b**, 10 mol%	24	47	88 (*S*)
5	(*Z*)	96:4	(*S,S*)-**13b**, 10 mol% + MS4A	24	99	88 (*S*)

The diastereoselective construction of two stereogenic centers is further examined using **11b** (Table 2.5). Allyl vinyl ethers with (*Z*)-configured allyl ether double bonds rearrange with very high diastereo- and enantioselectivities. The configuration of the product can be determined by that of the vinyl ether double bond as in the case of thermal rearrangement. (*Z*,*Z*)- and (*E*,*Z*)-substrates provide *syn*- and *anti*-products, respectively (entry 1, 2). This observation suggests a chair-like transition state when substrates have a (*Z*)-configured allyl ether double bond. In contrast, an (*E*)-configured allyl ether double bond does not rearrange with complete diastereoselectivity (entry 3, 4).

Tab. 2.5 [Cu(*S*,*S*)-Ph-box]$(OTf)_2$-catalyzed Claisen rearrangement.

5 mol% **13a**, CH_2Cl_2, r.t.

11b → (3*S*,4*R*)-*syn* **12b**; (3*R*,4*R*)-*anti* **12b**; (3*R*,4*S*)-*syn* **12b**; (3*S*,4*S*)-*anti* **12b**

Entry	Substrate	*t* [h]	Yield [%]	*syn:anti*	*ee* [%]
1	1*E*, 5*Z*	38	99	3:97	88 (3*S*, 4*S*)
2	1*Z*, 5*Z*	38	98	99:1	84 (3*R*, 4*S*)
3	1*E*, 5*E*	12	100	86:14	82 (3*S*, 4*R*)
4	1*Z*, 5*E*	4	100	28:72	72 (3*R*, 4*R*)

The enantioselectivity of [Cu(*S*,*S*)-*t*Bu-box]$(OTf)_2$ (**13b**)-catalyzed Claisen rearrangement is explained as follows (Fig. 2.4). The alkoxycarbonyl and ether oxygens coordinate in a bidentate fashion to the Cu^{II}(box) complexes. The square planer geometry around the copper(II) cation has been proposed and a chair-like transition-state model is suggested. The allylic ether moiety should approach the vinyl ether moiety from the opposite direction of the *t*-Bu substituents on the box-ligand. The Cu^{II}(box) catalyst differentiates between two enantiomeric chair-like transition state by selective coordination of enantiotopic lone pairs on oxygen to form (*S*,*S*,*pro*-*S*)-14a.

A more Lewis acidic catalyst [Cu^{II}(*t*Bu-box)]$(H_2O)_2(SbF_6)_2$ (**13c**) affords higher enantioselectivity [14]. In this system, the highly polarized transition state **14b** is proposed (Scheme 2.5).

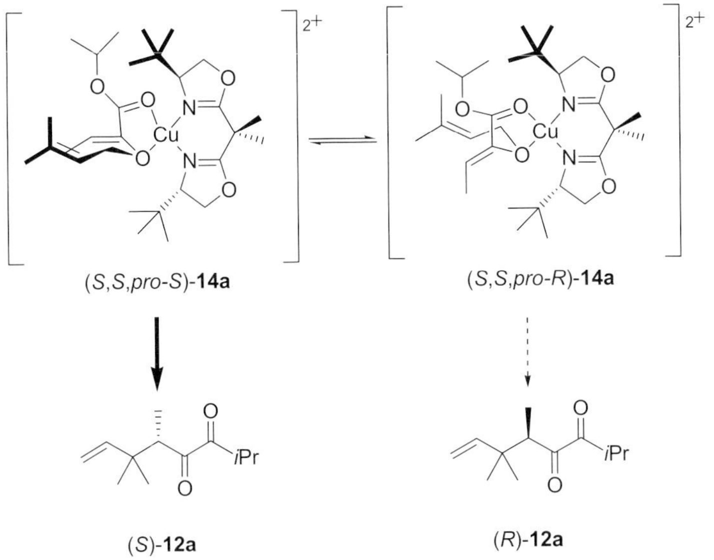

O O N Cu O N O 2+ ⇌ O O N Cu O O N O 2+

(S,S,pro-S)-**14a** (S,S,pro-R)-**14a**

O iPr O O iPr O

(S)-**12a** (R)-**12a**

Fig. 2.4 Proposed stereochemical course of the Cu^{II}(box)-catalyzed Claisen rearrangement.

O OiPr O

11a

5 mol%
$[Cu\{(S,S)\text{-}t\text{Bu-box}\}](H_2O)_2(SbF_6)_2$ (**13c**)

CH_2Cl_2, r.t., 2 h

tBu O O N Cu N O δ⁻ O tBu R1Z R1E Oi-Pr δ⁺ ‡

14b

O OiPr O

(R)-**12a**

O OiPr O

(S)-**12a**

(E)-**11a**: 94 %, 99 % *ee* (S)
(Z)-**11a**: 100 %, 99 % *ee* (R)

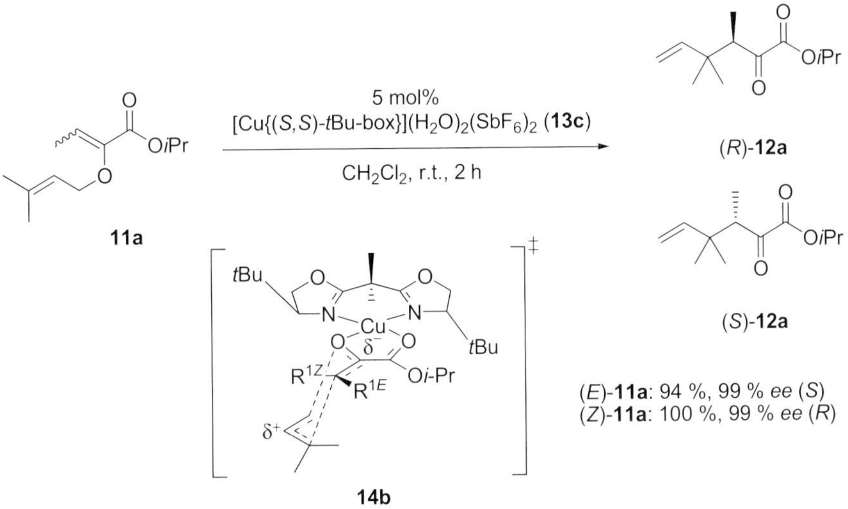

Scheme 2.5 Cu^{II}-catalyzed enantioselective Claisen rearrangement.

The 2-oxazoline-substituted substrate **15** has been examined but the selectivity is moderate (Table 2.6) [15]. The diastereoselectivity depends on the configuration of the vinyl ether double bond (entry 1, 3). In this case, the diastereoselectivity is mainly dictated by substrate configuration (entry 1, 2).

Tab. 2.6 Cu^{II}-catalyzed diastereoselective Claisen rearrangement using 2-oxazoline-substituted substrate.

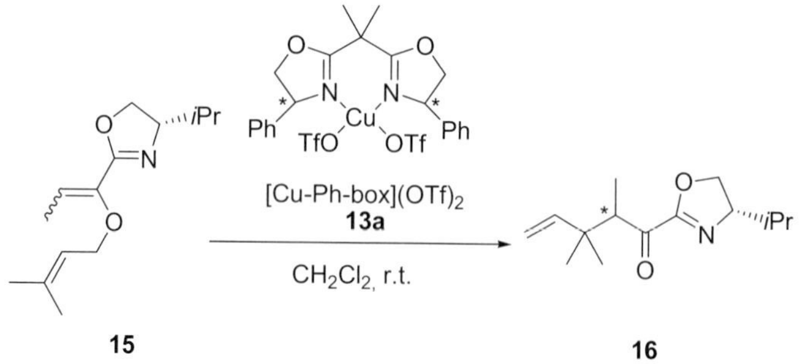

[Cu-Ph-box](OTf)$_2$ **13a**, CH_2Cl_2, r.t.; **15** → **16**

Entry	Substrate	Catalyst	*t* [h]	Yield [%]	*ee* [%]
1	(*Z*)-15	R,R	2	95	64 (*R*)
2		S,S	2	93	62 (*R*)
3	(*E*)-15	R,R	3	88	36 (*S*)

The copper(II) bis(oxazoline) complex [16] also catalyzes the asymmetric carbonyl-ene reaction [17]. Hiersemann has reported the asymmetric domino Claisen rearrangement and intra-molecular carbonyl-ene reaction (Scheme 2.6) [18].

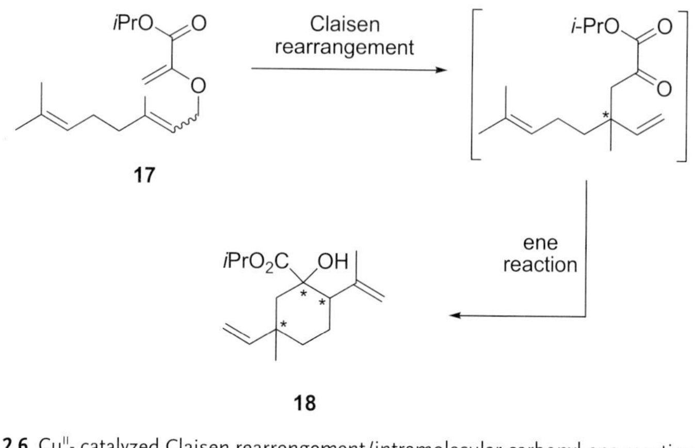

Scheme 2.6 Cu^{II}- catalyzed Claisen rearrengement/intramolecular carbonyl-ene reaction.

In this reaction, two enantiomerically pure diastereomers **18** out of eight possible stereoisomers are obtained (Table 2.7).

The following asymmetric "ene" reaction proceeds with complete diastereoselectivity to afford exclusively two diastereomers in which the hydroxyl and isopropenyl groups are in a *cis* arrangement (Scheme 2.7).

Tab. 2.7 Cu^{II}-catalyzed domino Claisen rearrangement and intramolecular carbonyl-ene reaction.

10 mol% [Cu(*S*,*S*)-Ph-box](OTf)$_2$ **13a**, CH_2Cl_2, r.t., 12 h

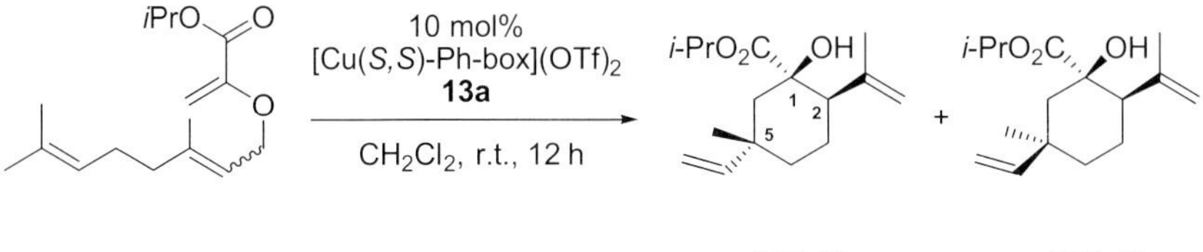

17 → (5*S*)-**18** + (5*R*)-**18**

Entry	Substrate	Yield [%]	(5S)-18:(5R)-18	*ee* [%]
1	*E*	98	89:11	98
2	*Z*	93	19:81	99

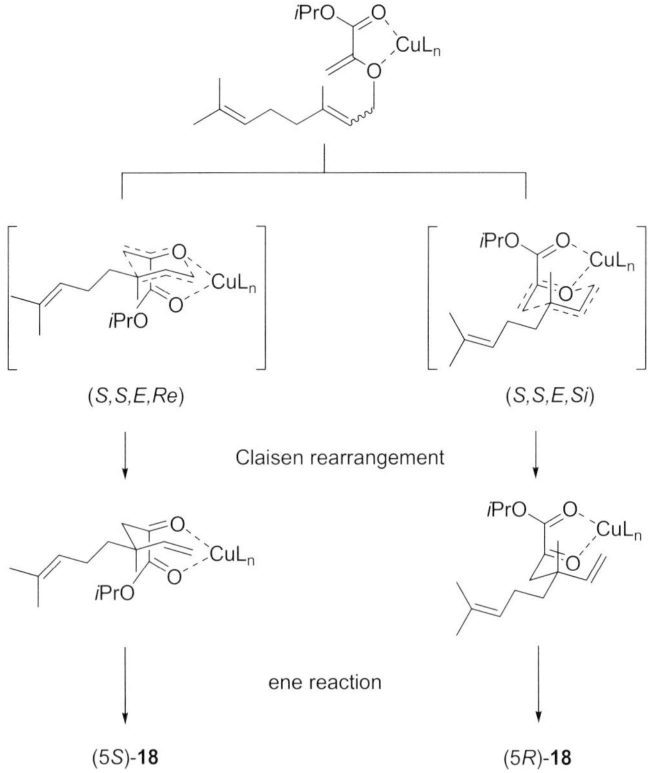

(*S*,*S*,*E*,*Re*) (*S*,*S*,*E*,*Si*)

Claisen rearrangement

ene reaction

(5*S*)-**18** (5*R*)-**18**

Scheme 2.7 Stereochemical course of the sequential Claisen rearrangement intramolecular carbonyl-ene reaction.

2.5 Palladium(II)-catalyzed Claisen Rearrangement

Late transition-metal-catalyzed asymmetric Claisen rearrangement takes place in a different mode from that of Lewis-acid-catalyzed Claisen rearrangement. Late transition metal catalysis is based on affinity for the Claisen diene system. Among late transition metals, palladium complexes are the most useful and effective for the Claisen rearrangement.

Van der Baan and Bickelhaupt (1986), as well as Mikami and Nakai (starting from 1985), have reported $PdCl_2(CH_3CN)_2$-catalyzed Claisen rearrangements (Scheme 2.8) [19]. In this system, interesting stereochemical features are observed [19, 20]. Both the (*E*)- and (*Z*)-configuration of allyl ethers **19** selectively gives *anti*-**20**. This stereoconvergent phenomenon is quite different in nature from that of conventional thermal rearrangement.

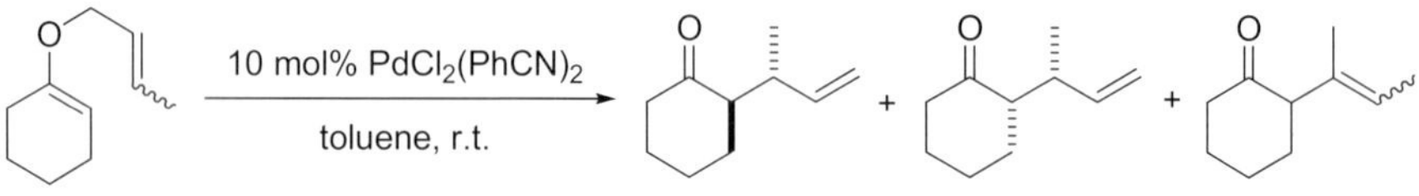

10 mol% $PdCl_2(PhCN)_2$, toluene, r.t.

19	*anti*-**20**	*syn*-**20**	**21**
(*E*)-19 55%	90	10	
(*Z*)-19 79%	77	9	14

Scheme 2.8 $PdCl_2(PhCN)_2$-catalyzed Claisen rearrangement.

The intriguing diastereoselectivity is rationalized as follows (Scheme 2.9). In substrate (*E*)-**19**, the reaction proceeds via a boat transition state (**A**) where palladium coordinates to the Claisen substrate in bidentate fashion to give *anti*-**20**. In substrate (*Z*)-**19**, the boat transition state (**D**) is disfavored due to large steric repulsion between Pd and *cis*-methyl group. Therefore the chair transition state (**C**) is favored over the boat transition state (**D**) to eventually give *anti*-**20**.

The extension of palladium catalysts to asymmetric catalysis has been examined for this system (Scheme 2.10) [21]. There are two types of palladium catalyst complexes: cationic and neutral. Cationic $[Pd(II)\text{-}BINAP](SbF_6)_2$ complex **22** and DABNTf-Pd complex **23** [22] have both been examined.

Under cationic conditions, the product is not obtained because of its high Lewis acidity. In contrast, using neutral complexes, the reactions proceed and give (*R*,*R*)-*anti*-**20** with good enantioselectivity, moderate diastereoselectivity and low yield. This result suggests that sulfonamide ligand DABNTf [23] is effective for the asymmetric Claisen rearrangement.

Among several chiral trifluoromethanesulfonylamide ligands, the DABNTf ligand **26** gives the highest enantioselectivity (Table 2.8). The best solvent is acetonitrile and the best temperature is 40–60°C to give the product in moderate yield.

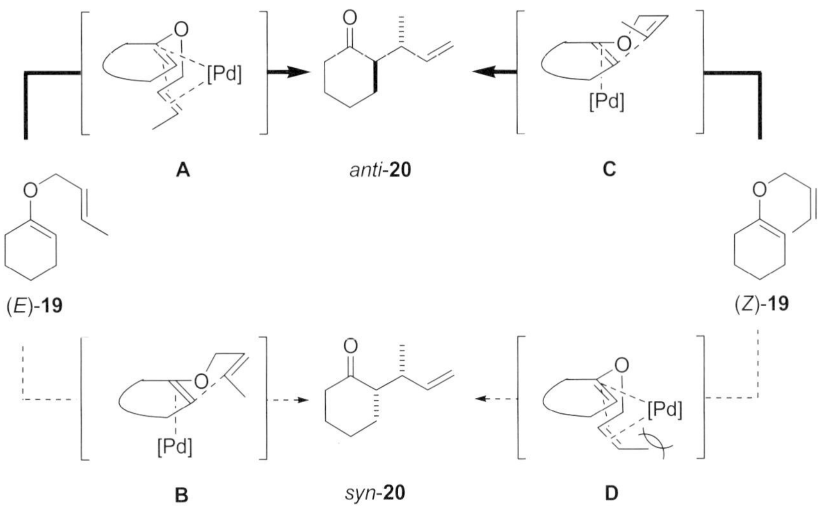

[Pd]
O
A
anti-**20**
C
(E)-**19**
(Z)-**19**
B
syn-**20**
D

Scheme 2.9 The proposed mechanism of Pd^{II}-catalyzed Claisen rearrangement.

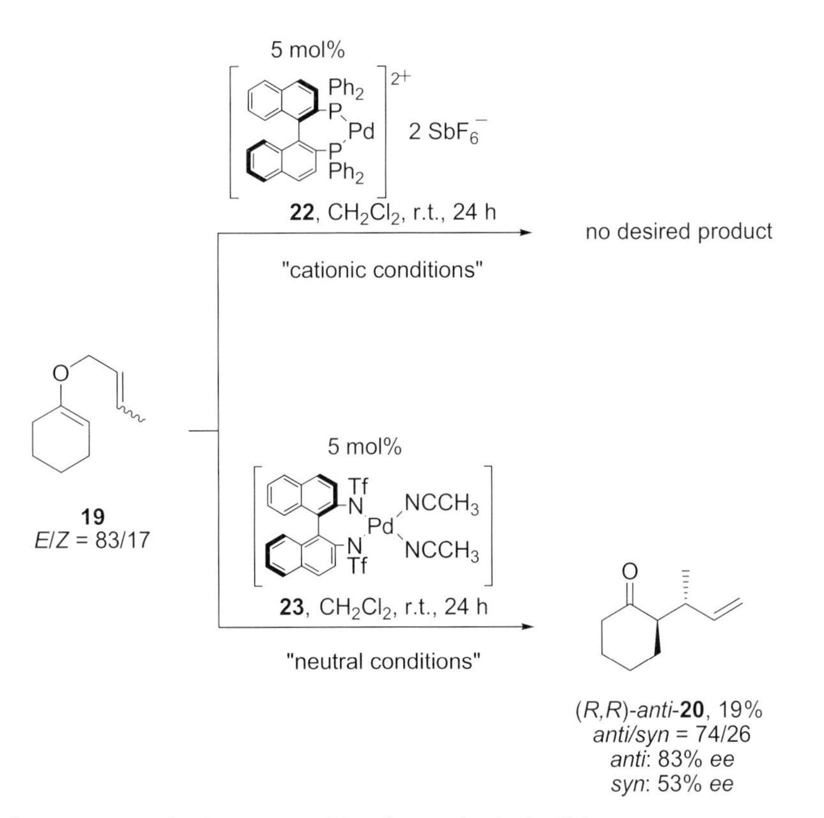

Scheme 2.10 Neutral and cationic conditions for enantioselective Claisen rearrangement.

Tab. 2.8 Pd^{II}-catalyzed enantioselective Claisen rearrangement.

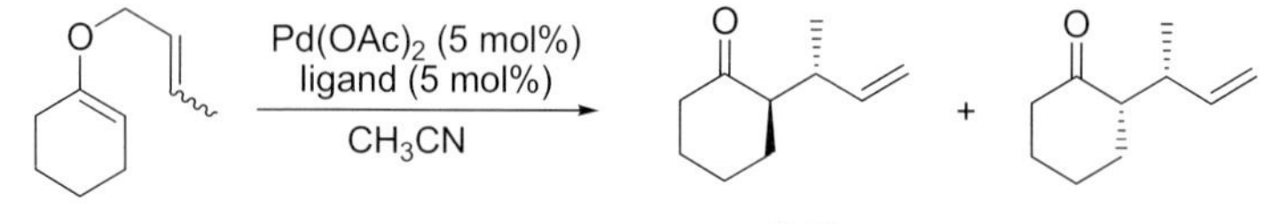

19, E/Z = 83/17 → Pd(OAc)$_2$ (5 mol%), ligand (5 mol%), CH$_3$CN → anti-20 + syn-20

Entry	Ligand	T [°C]	Yield [%][a]	*anti/syn*	*ee* [%]
1	(*R,R*)-DPENTf **24**	*r.t.*	6	80/20	28/21
2	(*R,R*)-DACyTf **25**	*r.t.*	29	94/6	63/51
3	(*R*)-DABNTf **26**	*r.t.*	24	85/15	84/50
4		*40*	42	85/15	83/57
5		*60*	54	82/18	81/53
6		*80*	69	78/22	77/32

a) NMR yield

Chiral ligands:

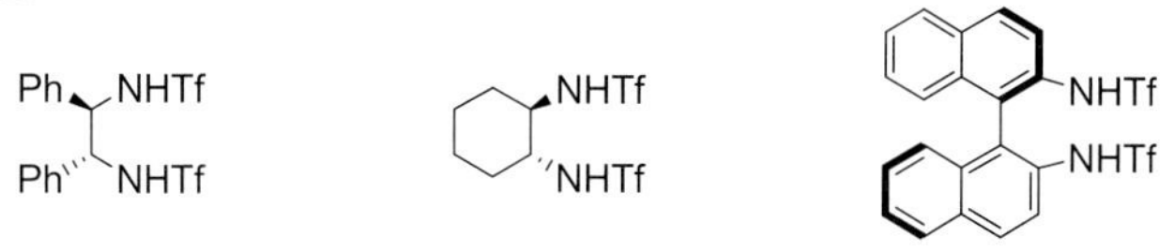

(*R,R*)-DPENTf, **24** (*R,R*)-DACyTf, **25** (*R*)-DABNTf, **26**

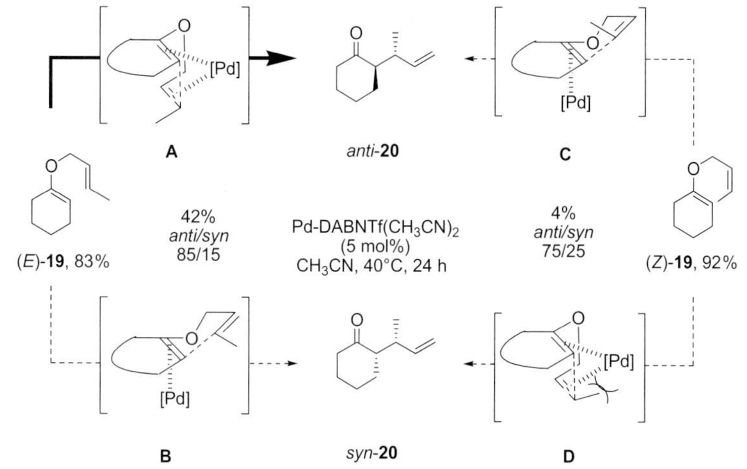

(*E*)-**19**, 83% — 42%, *anti/syn* 85/15 — Pd-DABNTf(CH$_3$CN)$_2$ (5 mol%), CH$_3$CN, 40°C, 24 h — 4%, *anti/syn* 75/25 — (*Z*)-**19**, 92%

A, B, C, D; *anti*-**20**, *syn*-**20**

Scheme 2.11 Neutral and cationic conditions for enantioselective Claisen rearrangement.

In order to consider the reaction mechanism, the effect of configuration of allyl group (*E*)- and (*Z*)-**19** is examined (Scheme 2.11). This asymmetric reaction mechanism could be the same as that in $PdCl_2(CH_3CN)_2$ catalysis. In this asymmetric reaction, (*E*)-**19** gave the *anti*-**20** product. The reaction should proceed via a six-membered boat transition state (**A**) via the bidentate coordination to the Pd catalyst. In contrast, (*Z*)-**19** does not give the product, because the DABNTf ligand is very bulky to destabilize the transition state (**C**).

Based on the X-ray structure of (*R*)-DABNTf-Pd-(*S*)-BINAP complex [22], the sense of enantioselectivity is predicted as follows (Fig. 2.5). Due to the steric repulsion of *trans*-methyl group between the Claisen substrate and CF_3–SO_2N in DABNTf, a six-membered boat transition state (*R,R*)-**19** is favored over the enantiomeric (*S,S*)-**19**. Therefore, (*R,R*)-*anti*-**20** would be obtained.

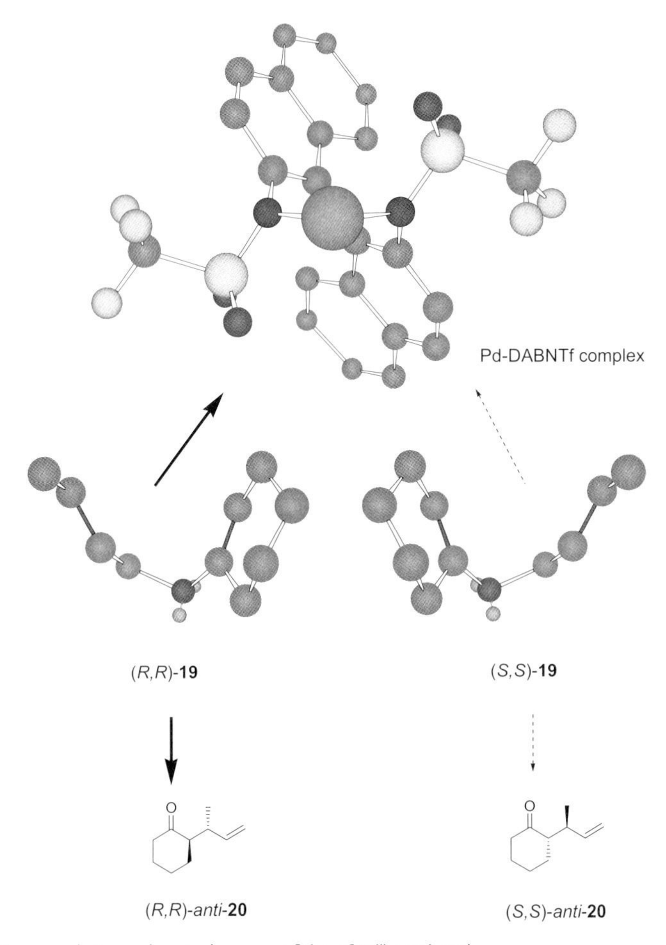

Fig. 2.5 Proposed stereochemical course of the of Pd^{II}-catalyzed asymmetric Claisen rearrangement.

References

1 L. Claisen, *Berichte* **1912**, *45*, 3157.

2 Excellent reviews: (a) R. P. Lutz *Chem. Rev.* **1984**, *84*, 205. (b) L. E.Overman *Angew. Chem. Int. Ed. Engl.* **1984**, *23*, 579; *Angew. Chem.* **1984**, *96*, 565. (c) H. Frauenrath, in *Stereoselective Synthesis* **1996** G. Helmchen, R. W. Hoffmann, J. Mulzer; E. Schaumann (eds.) Georg-Thieme Verlag, Stuttgart, vol. 6, p. 3301. (d) M. Hiersemann, L. Abraham, *Eur. J. Org. Chem.* **2002**, 1461.

3 E. J. Corey, D. H. Lee, *J. Am. Chem. Soc.* **1991**, *113*, 4026.

4 Examples of achiral Lewis acid catalyzed Claisen rearrangement: (a) B. M. Trost, F. D. Toste, *J. Am Chem. Soc.* **1998**, *120*, 815. (b) B. M. Trost, F. D. Toste, *J. Am. Chem. Soc.* **2000**, *122*, 3785. (c) G. Koch, P. Janser, G. Kottirsch, E. Romero-Giron, *Tetrahedron Lett.* **2002**, *43*, 4837. (d) G. Koch, G. Kottirsch, B. Wietfeld, E. Kusters *Org. Proc. Res. Dev.* **2002**, *6*, 652.

5 (a) K. Maruoka, H. Banno, H. Yamamoto, *J. Am. Chem. Soc.* **1990**, *112*, 7791. (b) K. Maruoka, H. Banno, H. Yamamoto, *Tetrahedron Asymmetry*, **1991**, *2*, 647. (c) K. Takai, I. Mori, K. Oshima, H. Nozaki, *Tetrahedron Lett.* **1981**, *22*, 3985. (d) A. Itoh, S. Ozawa, K. Oshima, H. Nozaki, *Bull. Chem. Soc. Jpn*, **1981**, *54*, 274. (e) K. Maruoka, K. Nonoshita, H. Banno, H. Yamamoto, *J. Am. Chem. Soc.* **1988**, *110*, 7922. (f) K. Maruoka, H. Banno, K. Nonoshita, H. Yamamoto, *Tetrahedron Lett.* **1989**, *30*, 1265. (g) K. Nonoshita, H. Banno, K. Maruoka, H. Yamamoto, *J. Am. Chem. Soc.* **1990**, *112*, 316. (h) K. Maruoka, J. Sato, H. Banno, H. Yamamoto, *Tetrahedron Lett.* **1990**, *31*, 377. (i) H. Yamamoto, K. Maruoka, *Pure Appl. Chem.* **1990**, *62*, 2063. (j) K. Nonoshita, K. Maruoka, H. Yamamoto, *Bull. Chem. Soc. Jpn.* **1992**, *65*, 541.

6 K. Maruoka, H. Yamamoto, *Synlett* **1991**, 793.

7 (a) K. Maruoka, H. Imoto, S. Saito, H. Yamamoto, *J. Am. Chem. Soc.* **1994**, *116*, 4131. (b) S. Saito, M. Shiozawa, T. Nagahara, M. Nakadai, H. Yamamoto, *J. Am. Chem. Soc.* **2000**, *122*, 7847. (c) S. Sano, T. Nagahara, M. Shiozawa, M. Nakadai, H. Yamamoto, *J. Am. Chem. Soc.* **2003**, *125*, 6200.

8 K. Maruoka, S. Saito, H. Yamamoto, *J. Am. Chem. Soc.* **1995**, *117*, 1165.

9 E. Tayama, A. Saito, T. Ooi, K. Maruoka, *Tetrahedron* **2002**, 8307.

10 M. P. Sibi, G. R. Cook, in *Lewis Acids in Organic Synthesis* **2000** H. Yamamoto (eds.) Wiley, vol. 2, p. 543.

11 (a) L. Abraham, R. Czerwonka, M. Hiersemann. *Angew. Chem. Int. Ed.* **2001**, *40*, 4700. (b) M. Hiersemann, L. Abraham, *Eur. J. Org. Chem.* **2002**, 1461.

12 (a) M. Hiersemann, *Synthesis*, **2000**, *9*, 1279. (b) M. Hiersemann, L. Abraham, *Org. Lett.* **2001**, *3*, 49.

13 (a) M. Johannsen, K.A. Jørgensen, *J. Org. Chem.* **1995**, *60*, 5757. (b) D. A. Evans, J. S. Johnson, C. S. Burgey, K. R. Campos, *Tetrahedron Lett.* **1999**, *40*, 2879. (c) K. A. Jørgensen, *Angew. Chem. Int. Ed.* **2000**, *39*, 3558.

14 L. Abraham, M. Korner, M. Hiersemann, *Tetrahedron Lett.* **2004**, *45*, 3647.

15 H. Helmboldt, M. Hiersemann, *Tetrahedron Lett.* **2003**, *59*, 4031.

16 D. A. Evans, S. W. Tregay, C. S. Burgey, N. A. Paras, T. Vojkovsky, *J. Am. Chem. Soc.* **2000**, *122*, 7936.

17 For reviews on carbonyl-ene reactions: (a) K. Mikami, M. Shimizu, *Chem. Rev.* **1992**, *92*, 1021. (b) B. B. Snider in *Comprehensive Organic Synthesis* **1991** B. M. Trost, I. Fleming (eds.) Pergamon, London, vols. 2 and 5. (c) K. Mikami, M. Shimizu, T. Nakai, *J. Synth. Org. Chem. Jpn.* **1990**, *48*, 292.

18 S. Kaden, M. Hiersemann, *Synlett*, **2002**, *12*, 1995.

19 (a) J. L. van der Baan, F. Bickelhaupt, *Tetrahedron Lett.* **1986**, *27*, 6267. (b) K. Mikami, K. Takahashi, T. Nakai, *Tetrahedron Lett.* **1987**, *28*, 5879. (c) K. Mikami, K. Takahashi, T. Nakai, *J. Am. Chem. Soc.* **1990**, *112*, 4035.

20 K. Mikami, K. Takahashi, T. Nakai, T. Uchimaru, *J. Am. Chem. Soc.* **1994**, *116*, 10948. M. Sugiura, T. Nakai, *Chem. Lett.* **1995**, 697.

21 K. Akiyama, K. Mikami, *Tetrahedron Lett.* **2004**, *45*, 7217.

22 K. Mikami, Y. Yusa, K. Aikawa, M. Hatano, *Chirality* **2003**, *15*, 105.

23 (a) K. Mikami, Y. Motoyama, M. Terada, *Inorganica Chemica Acta* **1994**, *222*, 71. (b) M. Terada, Y. Motoyama, K. Mikami, *Tetrahedron Lett.* **1994**, *35*, 6693. (c) K. Mikami, O. Kotera, Y. Motoyama, H. Sakaguchi, *Synlett* **1995**, 975.

3
Aliphatic and Aromatic Claisen Rearrangement

3.1
Aliphatic Claisen Rearrangement

Hayato Ichikawa and Keiji Maruoka

3.1.1
Introduction

The aliphatic Claisen rearrangement is applied not only in early steps, but also in key steps in total synthesis. The rearrangement proceeds only by heating an allyl vinyl ether at 150–200 °C without any reagents to produce a γ,δ-unsaturated aldehyde, which is a useful synthetic intermediate, with extremely high regio- and stereoselectivity. The great number of synthetic efforts on record that employ a Claisen rearrangement in a key C–C bond forming step attests to the power of the reaction. In fact, a number of reviews have been prepared on an aliphatic Claisen rearrangement [1–5]. It is noteworthy that Ziegler's review [2] is the most comprehensive with historical overviews, mechanistic aspects and synthetic applications of the aliphatic rearrangement up to 1988. Accordingly, this chapter mainly comprises thermal simple aliphatic Claisen rearrangement research since 1988. On the other hand, the aliphatic Claisen rearrangement gives numerous carbonyl compounds, which depend on the type of allyl vinyl ethers. In the beginning of this chapter, the synthetic methods of allyl vinyl ethers will be described. Then, the reaction mechanism, stereoselectivity and application for organic synthesis are discussed according to the type of allyl vinyl ethers. This chapter also describes tandem reactions including an aliphatic Claisen rearrangement and the carbanion-accelerated Claisen rearrangement.

The Claisen Rearrangement. Edited by M. Hiersemann and U. Nubbemeyer

ISBN: 978-3-527-30825-5

3.1.2
Synthesis of Allyl Vinyl Ethers

3.1.2.1 Hg-Catalyzed Synthesis

Allyl vinyl ethers are usually obtained by either mercury salts or acid-catalyzed vinyl ether exchange with allylic alcohols, or acid-catalyzed vinylation of allylic alcohols with acetals. However, yields in these reactions are often low and care must be used when using mercury [1]. Mercuric salts of weak carboxylic acids, for example, acetate, have been reported to be particularly effective in catalyzing the transfer of vinyl groups from vinyl ethers to alcohols (Eq. 3.1.1) [6].

OH + O nBu — 8 mol% $Hg(OAc)_2$, azeotrope → O 75 %

Equation 3.1.1

3.1.2.2 From Ammonium Betaine

The use of mercuric salts has, in principle, been unacceptable due to environmental problems. Buchi and Vogel developed a modification of the Claisen rearrangement for *primary* and *secondary* allylic alcohols that does not require catalysis by mercuric salts or acids, and gives aldehydes directly [7]. Furthermore, sealed tubes or other high-pressure vessels are not necessary. The betaine is shown to combine with alkoxides to give the corresponding *E*-3-alkoxyacrylic acids (Eq. 3.1.2) [8]. These crude products are heated with a trace of hydroquinone to give the corresponding Claisen rearrangement products with concomitant decarboxylation.

CO_2H — Me_3N → $Me_3\overset{+}{N}$ CO_2^- **1** — HO Ph, NaH, THF → O CO_2H Ph — 160-180 °C, - CO_2 → OHC Ph

Equation 3.1.2

3.1.2.3 Acid-Catalyzed Synthesis

Using *p*-toluenensulfonic acid as an acid catalyst, allylic alcohols and carbonyl compounds are converted to allyl vinyl ethers. A mixture of 2-hydroxy-3-butenoates with an equivalent amount of carbonyl compounds or derivative acetals are refluxed in the presence of a catalytic amount of *p*-toluenensulfonic acid (Eq. 3.1.3) [9]. The corresponding 6-oxo-2-hexenoate **2** is produced as a mixture of *E*/*Z*-configured diastereomers.

TsOH, toluene

2, 53 %, *E*:*Z*= 2:1

Equation 3.1.3

3.1.2.4 Wittig Olefination

The route for Wittig olefination of ketones and aldehydes to furnish the corresponding allyl vinyl ethers is one promising method for the vinyl ether synthesis [10]. The reaction of allyl chloromethyl ether with triphenylphosphine in benzene gave allyloxymethylenetriphenylphosphonium chloride **3** in nearly quantitative yield. Treating an equimolar mixture of the phosphonium salts **3** and carbonyl compounds with potassium *tert*-butoxide in *tert*-butanol afforded the corresponding allyl vinyl ether in good yield (Eq. 3.1.4). Reaction of an enolizable acyclic aliphatic ketone and aldehyde furnished the corresponding allyl vinyl ether in good yields (Eq. 3.1.5).

*t*BuOK, THF
0 °C, 30 min

3 91 %

Equation 3.1.4

*t*BuOK, THF
0 °C, 30 min

81 %

Equation 3.1.5

3.1.2.5 Sulfoxide Elimination

Tertiary allylic alcohols can be vinylated by a Michael-type addition to a vinyl sulfoxide **4**, followed by elimination of PhSOH (Eq. 3.1.6) [11]. A solution of the sulfoxide **5** was heated in the presence of both sodium bicarbonate and α-pinene, which is utilized to scavenge released sulfinic acid and minimizing unwanted side reactions, to give the bicyclic alcohol **6** in 45% yield.

For the synthesis of the intermediate of the sesquiterpene (±)-ceratopicanol, the *secondary* allylic alcohol **7** was converted to a new alcohol **8** via one-pot etherification, Claisen rearrangement and reduction with $LiAlH_4$ (Eq. 3.1.7) [12]. This trans-

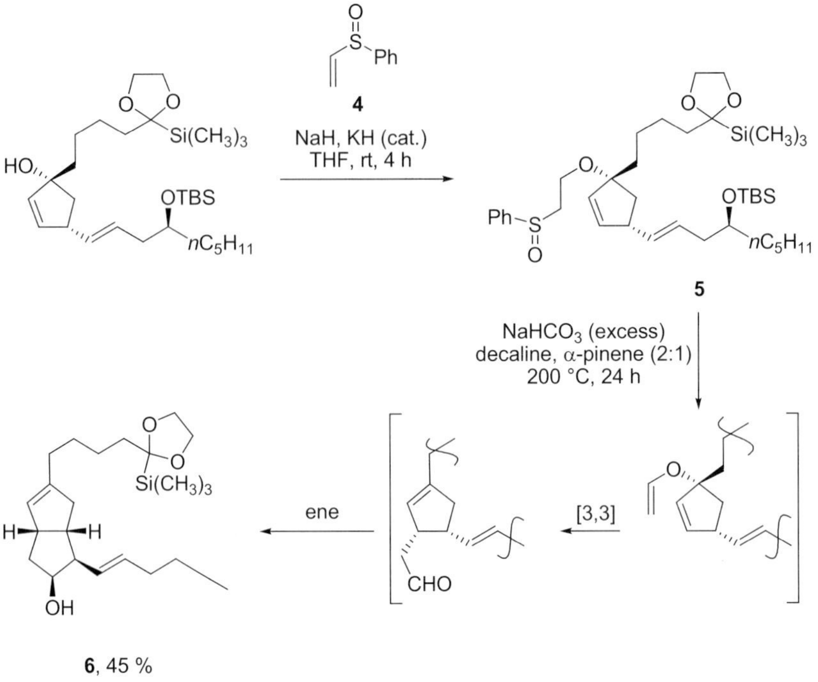

Equation 3.1.6

formation was best done (73% overall yield) by treating the *secondary* allylic alcohol with phenyl vinyl sulfoxide and heating the product. The *in situ* allyl vinyl ether generated under these conditions rearranged to the aldehyde, which was reduced without purification.

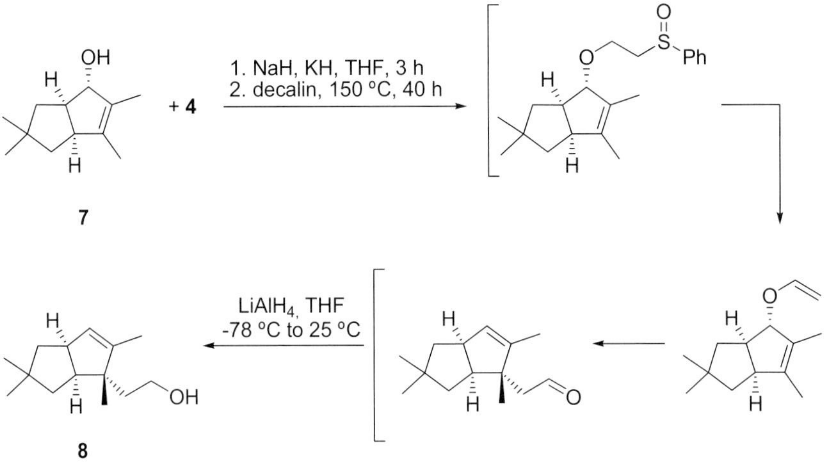

Equation 3.1.7

3.1.2.6 Selenoxide Elimination

Elimination of selenoxides is also able to make the vinyl group by pyrolysis of selenoxides *in situ* (Eq. 3.1.8). The oxidation of the selenoxide with sodium periodate took place concurrent to selenoxide elimination and Claisen rearrangement [13]. However, the rearrangement of **9** is difficult because the examples studied earlier were either free of the side chain or carried it into the chair-like transition state in an equatorial disposition. The large pendant chain of **9** must be projected axially as depicted in the transition state **A** to lead to the rearrangement of **9**. Although the energetic costs associated with the desired conversion to cyclooctenone were certain to be higher, the manner in which this factor had to be dealt with experimentally was approached empirically.

$NaIO_4$, NaHCO, CH_3OH, H_2O then Et_3N, $EtOCH{=}CH_2$, DMF 220 °C, sealed tube

Equation 3.1.8

The elimination of the selenoxide through heating in DMF, a polar aprotic solvent, was examined. As before, triethylamine was also added to neutralize the electrophilic PhSeOH liberated during the generation of **9** and to guard against undesired processes induced from that direction.

3.1.2.7 From Ketal

Ketals are converted to allyl vinyl ethers after Wittig olefination [14]. Wittig reactions permit the incorporation of various substituents into the allylic terminal of the ether. Enol ethers have been conveniently prepared by cleavage of acetals with various Lewis acids. Cleavage of the ketals with the Lewis acid, for example, triethylsilyl triflate, in the presence of diisopropylethylamine in refluxing 1,2-dichloroethane afforded the enol ethers (Eq. 3.1.9). The resulting enol ethers were then heated to effect the Claisen rearrangement without isolation.

nPr PPh3 TESOTf, iPr2NEt ClCH2CH2Cl nPr TESO 70 °C, 5 h TESO nPr O 61 %

Equation 3.1.9

3.1.2.8 **From Allene**

The allyl vinyl ethers arise from the addition of allylic alcohols to allenes [15]. The procedure is to use a catalytic amount of NaH and alcohol as the solvent for **10** and about one equivalent of NaH for the product **11**, respectively (Eq. 3.1.10) [16]. In the observations of Stirling [17] and others [18], β,γ-unsaturated sulfones are more stable than their α,β-isomers by about 2.5 kcal/mol. For unsubstituted allyl vinyl ethers, the α,β-isomer **11** appears to be more stable than the β,γ-isomer **10** contrary to the trend described in Steele's report [19], but consistent with Stirling's observations with methoxide [17]. This presumably derives from a synergistic interaction between the strong donor (RO) and acceptor (arylSO_2) groups. Consequently, the kinetic adduct **10** can only be obtained in good yield in protic media, which suppresses isomerization. The relative proportion of the isomer **10** could be increased in these cases by operating at lower temperature and using *tert*-butanol as a substitute for the allylic alcohols.

PhO2S + HO 7 mol% NaH rt, 10 min PhO2S O **10**, 84 % 88 mol% NaH rt, 10 min PhO2S O **11**, 52 %

Equation 3.1.10

3.1.2.9 Ir-Catalyzed Synthesis

Miyaura has reported the iridium-catalyzed isomerization of unsymmetrical diallyl ethers to (*E*)-allyl vinyl ethers [20]. The iridium catalyst **12** active for the isomerization of diallyl ethers was prepared *in situ* by bubbling hydrogen into a red solution of $[Ir(cod)(PPh_2Me)_2]PF_6$ by the method reported by Felkin (Eq. 3.1.11) [21]. The cinnamyl ethers were treated with an iridium catalyst to furnish the corresponding vinyl ethers (Eq. 3.1.12). The utility of this catalyst was demonstrated in the stereoselective synthesis of allyl vinyl ethers. 2,3,3-Trimethyl-4-pentenal was obtained in 51% yield when the isomerization of the diallyl ether to give the allyl vinyl ether was followed by heating to proceed the Claisen rearrangement (Eq. 3.1.13).

$$[Ir(cod)(PPh_2Me)_2]PF_6 \xrightarrow{H_2} [IrH_2(solv)(PPh_2Me)_2]PF_6$$

12

Equation 3.1.11

3 mol% **12**, THF, rt, 1 h

60 %, *E*:*Z*= 99:1

Equation 3.1.12

3 mol% **12**, THF, rt, 1 h

180 °C, 5 h

65 %, *E*:*Z*= 99:1

51 %

Equation 3.1.13

3.1.2.10 Cu-Catalyzed Synthesis

The Ullmann-type coupling of aliphatic allylic alcohols with vinyl iodides are also able to produce allyl vinyl ethers. The reaction was effected using 10 mol% of CuI, 20 mol% of **13** as a ligand and Cs_2CO_3 for the coupling of allylic alcohol with vinyl iodide to afford 68% of the allyl vinyl ether and 4% of the Claisen rearrangement products (Eq. 3.1.14). These procedures are applied to the diastereoselective generation of a compound with a *tertiary* center vicinal to a *quaternary* one, as well as one with two adjacent *quaternary* stereogenic centers.

The idea enabled a domino process by combining copper-catalyzed C–O bond formation and thermal Claisen rearrangement. Subjecting allylic alcohol and vinyl iodide to the reaction conditions led to the clean formation of the desired rearrangement product in 55% yield with high stereochemical purity (Eq. 3.1.15). The preparation of two adjacent *quarternary* stereocenters from geraniols is successfully prepared using the aforementioned reaction conditions (Eq. 3.1.16).

*n*Oct, OH + I, *n*Oct

10 mol% CuI, 20 mol% **13**
150 mol% Cs_2CO_3
toluene, air, 80 °C, 24 h

13

*n*Oct, *n*Oct, O
68 %
+
O, *n*Oct, *n*Oct
4 %

Equation 3.1.14

OH + I, nC_5H_{11}

10 mol% CuI, 20 mol% **13**
300 mol% Cs_2CO_3
o-xylene, 120 °C, 2 d

O, nC_5H_{11}
55 %, d.r.= 92:8

Equation 3.1.15

OH + I, nC_5H_{11}

10 mol% CuI, 20 mol% **13**
300 mol% Cs_2CO_3
o-xylene, 120 °C, 2 d

O, nC_5H_{11}
77 %, d.r.= 92:8

Equation 3.1.16

3.1.2.11 Tebbe Reagent

The Tebbe reagent is a powerful tool for transformation of esters to vinyl ethers [23]. Tebbe methylenation of a glycal ester can be followed by the thermal rearrangement, thereby allowing stereoselective access to *C*-glycosides [24]. The Tebbe reaction proceeds smoothly to yield the corresponding enol ethers. The thermal rearrangement of enol ethers, which occurred upon heating to 180 °C in either benzonitrile or tributylamine, then produced the corresponding *β*-*C*-glycosides in good to excellent yield (Eq. 3.1.17). It is noteworthy that the rearrangement includes the Boc-protected amino ester, since Boc-protected amino acids may be suitable substrates for this reaction sequence, thereby allowing access to *C*-glycosyl amino acids.

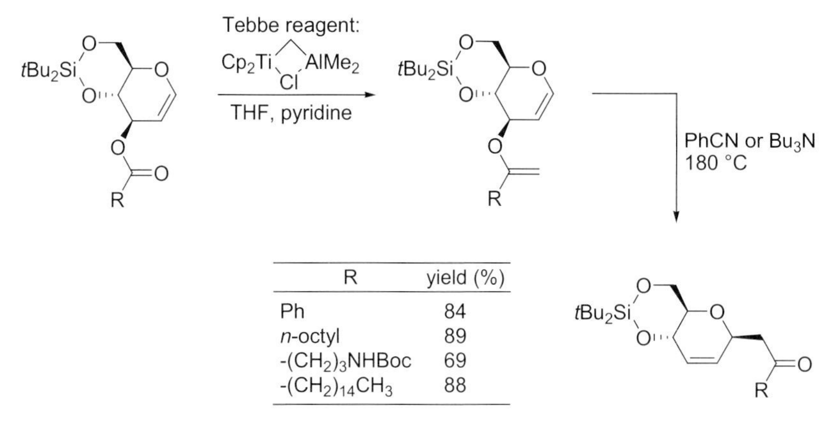

R	yield (%)
Ph	84
n-octyl	89
$-(CH_2)_3NHBoc$	69
$-(CH_2)_{14}CH_3$	88

Equation 3.1.17

3.1.3 Acyclic Aliphatic Claisen Rearrangement

3.1.3.1 Transition State of Aliphatic Claisen Rearrangement

The Claisen rearrangement is a suprafacial, concerted, nonsynchronous pericyclic process that is considered occasionally as an intramolecular S_N2' alkylation (Eq. 3.1.18) [2]. When the Claisen rearrangement can produce enantiotopic faces at both terminals of allyl vinyl ether, the rearrangement can proceed through two pairs of chiral transition states to prepare two racemic diastereomers bearing newly created stereocenters of the products **15** and **16** (Eq. 3.1.19). Achiral allyl vinyl ether **14** can provide two enantiomeric chair-like transition states chair 1 and 2, both of which lead to the racemic, diastereomeric aldehyde **15**. Similarly, the enantiomeric boat-like transition states boat 1 and 2 provide racemic diastereomer **16**. The two transition states are essentially different in energy and the ratio **15**/**16** reflects the transition state geometry.

Claisen rearrangement

Equation 3.1.18

Schmid has examined the rate and stereochemistry of rearrangement of four crotyl propenyl ethers **14a**–**14d** [25, 26]. As a result, the *E*,*E* isomer **14a** is found to rearrange an order of magnitude faster than the *Z*,*Z* isomer **14b**. The *E*,*E* and *Z*,*Z* isomers rearrange through a chair-like transition state to give the *syn* diastereomers **15a** and **15b** as the major product. Moreover, the *Z*,*E* and *E*,*Z* isomers give the *anti* diastereomers **15c** and **15d** as the predominant stereoisomer. Since all

14

a, (*E*,*E*): $R^1 = R^3 = Me$, $R^2 = R^4 = H$
b, (*Z*,*Z*): $R^1 = R^3 = H$, $R^2 = R^4 = Me$
c, (*Z*,*E*): $R^2 = R^3 = Me$, $R^1 = R^4 = H$
d, (*E*,*Z*): $R^2 = R^3 = H$, $R^1 = R^4 = Me$

chair 1

chair 2

boat 1

boat 2

15 (+ *ent*-**15**)

16 (+ ent-**16**)

Equation 3.1.19

four double bond isomers, **14a–14d**, proceed through the chair-like transition state, a change in the configuration of a single double bond leads to the formation of a diastereomer. Indeed, any pair change in olefin geometry for a given transition state, or single change of olefin configuration and change in transition state, results in the formation of the enantiomer of the same diastereomer [27].

3.1.3.2 Secondary Allylic Ethers

The aliphatic Claisen rearrangement of *secondary* allylic ethers has been recognized to provide *E* double bonds (Eq. 3.1.20) [28, 29]. The Claisen rearrangement of C4 and C5 substituted vinyl ether provided a 90:10 ratio of (*E*)- to (*Z*)-unsaturated aldehydes (Table 3.1.1, entry 1). An increase in the steric bulk of the C5 substituent produced higher stereoselectivity (entry 3).

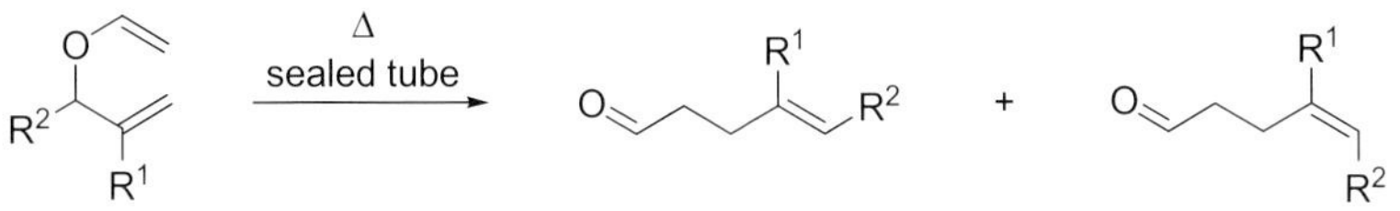

Equation 3.1.20

Tab. 3.1.1 The *E*:*Z* ratio of the Claisen rearrangement of *secondary* allyl vinyl ethers (Eq. 3.1.20) [29].

Entry	R^1	R^2	*T* (°C)	*E*:*Z*
1	Me	Et	110	90:10
2	Me	Et	205	86:14
3	Me	*i*-Pr	110	93:7
4	Et	Et	110	90:10

These results indicated that the *E*/*Z* isomer ratios in the Claisen rearrangements corresponded to the free-energy change for the conversion of a substituent from the equatorial to the axial position of a cyclohexane. Furthermore, using the well-established chair model for the transition state (Eq. 3.1.21), Faulkner has expected that the axial substituent R^3 would introduce a relatively large 1,3-diaxial interaction with substituent R^2 and thus increase the stereoselectivity of the Claisen rearrangement [27].

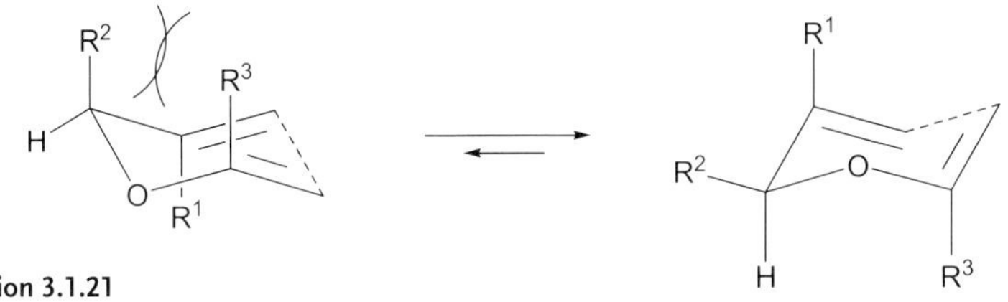

Equation 3.1.21

With this manner, the Claisen rearrangement of vinyl ethers of *secondary* allylic alcohols provides *E*-olefinic aldehyde and takes advantage of *E*-selectivity for total synthesis of many natural products.

The synthesis of the C1–C13 fragment of amphidinolide B, which is a 26-membered macrolide, is proceeded via Claisen rearrangement of the ether as a key step in an expected procedure, leading to the aldehyde in excellent yield (Eq. 3.1.22) [30].

amphidinolide B

200 °C, decaline

97 %

Equation 3.1.22

The optically active (20*R*)-de-*AB*-cholesta-8(14),22-dien-9-one (**19**) and (20*S*)-de-*AB*-isocholesta-8(14),22-dien-9-one (**20**), which are synthetic precursors for metabolites of vitamin D, were prepared by Suzuki and Kametani's group [31]. The Claisen rearrangement provided the separable aldehydes **17** and **18** (Eq. 3.1.23). Following decarbonylation of **17** with $(Ph_3P)_3RhCl$ produced **19** and the same treatment of the C20 epimer **18** gave **20**.

150 °C collidine

17: R^1= CH_2CHO, R^2 = H
18: R^1= H, R^2= CH_2CHO

$(Ph_3P)_3RhCl$, benzene, reflux

19: R^1= Me, R^2= H
20: R^1= H, R^2= Me

Equation 3.1.23

3.1.3.3 Substituted Vinyl Ethers

Recently, a Wittig olefination/Ir-catalyzed etherification sequence was developed as a method for selective preparation of substituted vinyl ether (see Section 3.1.2.9). However, no convenient syntheses are available for the selective preparation of simple propenyl ethers. Accordingly, Hiersemann has investigated the rearrangement of allyl vinyl ethers bearing ester at the C2 position (Eq. 3.1.24) [32]. The Claisen rearrangement of the (*Z*,*E*)-**21** and (*E*,*E*)-**21** mixture afforded the *α*-keto esters as an *anti* and *syn* mixture. As shown in Section 3.1.3.1, the configuration of the rearrangement products is based on the established chair-like transition state. Therefore, the thermal rearrangement of (*Z*,*E*)- and (*E*,*E*)-**21** should lead to *anti*- and *syn*-*α*-keto esters, respectively.

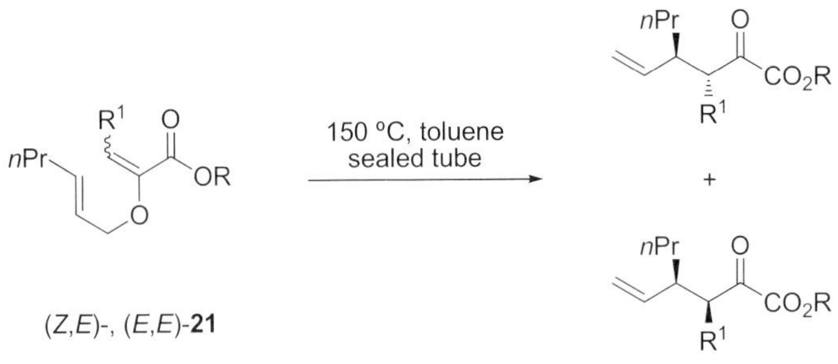

Equation 3.1.24

In the synthesis of (±)-fluorobotryodiplodin, the thermal rearrangement of vinyl substituted fluorinated methyl ether with the (*Z*,*Z*) geometry proceeded in a highly stereoselective manner, giving a ketone with *syn* geometry in 92% yield (Eq. 3.1.25) [33].

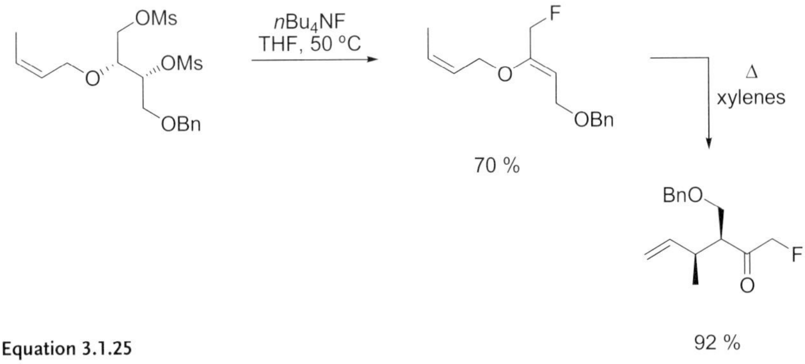

Equation 3.1.25

3.1.3.4 Allyl Allenyl Ethers

Allenes undergo the Claisen rearrangement at temperatures as low as 70 °C [34]. The rearrangement of **22** yields only the *Z*-isomer of **23** (Eq. 3.1.26). Examination of molecular models reveals that the chair-like transition state **C** leading to the *E*-isomer of **23** suffers destabilizing steric interactions from the bulky *tert*-butyl group (Eq. 3.1.27). These interactions are absent from the diastereomeric transition state **B** that affords the *Z*-isomer.

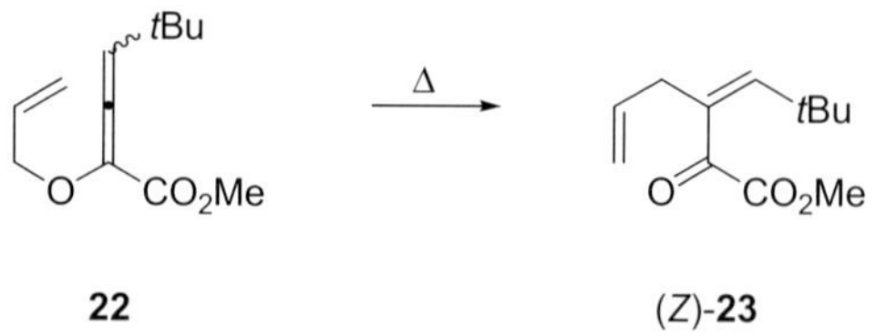

Equation 3.1.26

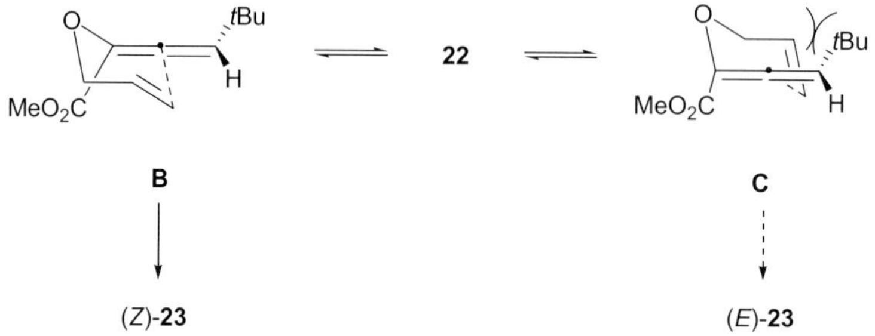

Equation 3.1.27

3.1.3.5 Disubstituted Vinyl Ether

The Claisen rearrangement of disubstituted allyl vinyl ether **24** was expected to proceed via to a chair-like transition state **D** because of favorable less-hindered geometry (Eq. 3.1.28) [35]. The importance of the favorable steric control elements found in **D** to the excellent stereochemical complementary can best be emphasized by comparison with the rearrangement of **26** bearing the *E* allyl ether under the same reaction conditions as that of **24** (Eq. 3.1.29). In this case, 25% of the aldehydic product mixture arises by way of boat-like transition state **E** because it is rather less crowded than **F**.

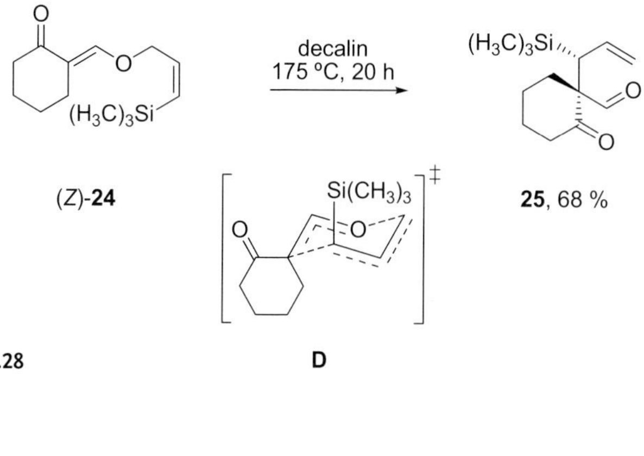

Equation 3.1.28

(H$_3$C)$_3$Si, O, E, ‡, 25, 25 %, decalin, 175 °C, 20 h, O, O, Si(CH$_3$)$_3$, (*E*)-26, O, O, Si(CH$_3$)$_3$, F, ‡, (H$_3$C)$_3$Si, O, O

Equation 3.1.29

3.1.3.6 Water-Promoted Claisen Rearrangement

The solvent effects of the Claisen rearrangement have been investigated and reviewed by Gajewski [36, 37]. Hydrophobic acceleration was observed in the Claisen rearrangement. For instance, when the rearrangement of **29** bearing acetylated sugar was performed in toluene and compared with that of allyl vinyl ether **27**, the the rearrangement was achieved after 13 days in toluene in the case of **29**, but after only 1 h in water in the case of **27** (Eqs. 3.1.30, 3.1.31) [38]. The Claisen rearrangement of **27** was conducted in water at pH 12 in the presence of $NaBH_4$ to furnish the alcohol **31** as a diastereomeric mixture (Eq. 3.1.32). The Claisen rearrangement of **29** by heating gave the aldehyde **30** in 92% yield as a mixture of diastereomers ($R/S = 60/40$) (Eq. 3.1.33). The facial selectivity of the rearrangement of **29** induced by the sugar moiety was due to a preferred attack of the *Re* face of the allyl vinyl ether both in toluene and water, to avoid the 1,3-*syn* diaxial interaction as depicted in Scheme 3.1.1. It is interesting to note that the glucose moiety induced the same chirality as in the Diels–Alder reaction [39, 40].

OH, HO, O, HO, OH, O, 27, 80 °C, H$_2$O, 1 h, OH, HO, O, HO, OH, CHO, 28

Equation 3.1.30

OAc, AcO, O, AcO, OAc, O, 29, 80 °C, toluene, 13 d, OAc, AcO, O, AcO, OAc, CHO, 30

Equation 3.1.31

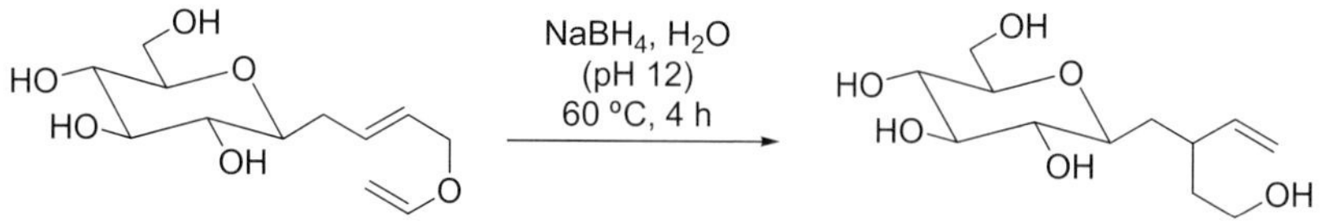

NaBH$_4$, H$_2$O (pH 12) 60 °C, 4 h

27 → 31

Equation 3.1.32

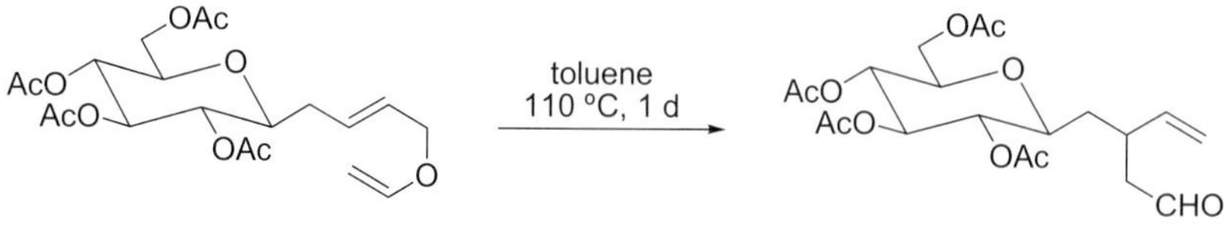

toluene 110 °C, 1 d

29 → **30**, 92 %, *R*:*S*= 60:40

Equation 3.1.33

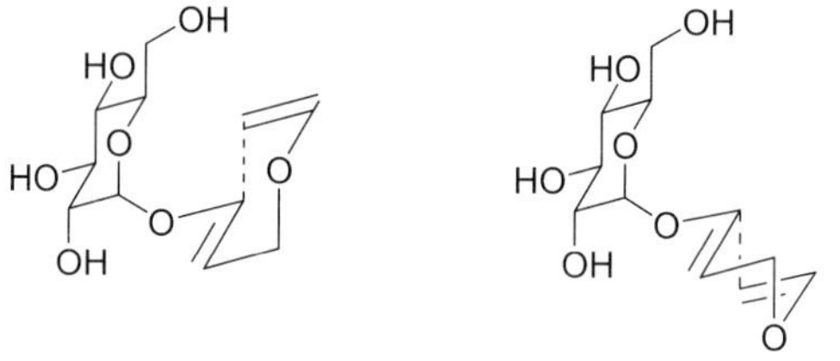

Scheme 3.1.1 *Si* *Re*

3.1.3.7 Diastereoselective Rearrangement Using Chiral Sulfoxide Groups

It is known that allyl vinyl ethers with sulfinyl groups rearrange to unsaturated carbonyl compounds [41]. Recently, a highly diastereoselective Claisen rearrangement of allyl vinyl ethers with sulfinyl auxiliary was reported [42]. Diastereomer **32** gave aldehydes **34** containing (*Z*)-alkene as single diastereomers accompanying decarboxylation (Eq. 3.1.34, Table 3.1.2). In contrast, diastereomer **36** bearing an opposite stereoisomer at C4 gave a mixture of rearrangement products **38** and **39** (Eq. 3.1.35, Table 3.1.3).

These results may be tentatively rationalized in terms of diastereomeric transition states derived from conformers, for which an *s-cis* conformation around the C–S bond is proposed (Scheme 3.1.2). In the case of 5-*E* allyl ether, **32** displays a reinforcing relationship of stereocontrolling elements with **G** accounting for the observed selectivity, since **I** would be destabilized by an axial R_2 and bulky aryl group pointing toward the incoming vinyl residue. For nonreinforcing diastereomer **36**, the energy difference between **H** and **J** should be smaller than for **G** and **I**, with **H** being more stable.

Tab. 3.1.2 Sulfinyl-mediated Claisen rearrangement (Eq. 3.1.34).

Entry	Substrate	*T* (°C)	*t* (min)	*Z*:*E* [a)]	Yield (%)
1	(+)-**32a**	130	60	100:0, (–)-**34a**	79
2	**32b**	134	180	100:0, **34b**	78
3	**32c**	130	75	100:0, **34c**	79
4	**33a**	120	180	8:92, **34a**, **35a**	74
5	**33d**	110	420	0:100, **35d**	77

a) *Z*/*E* ratios include two byproducts obtained in some cases [42].

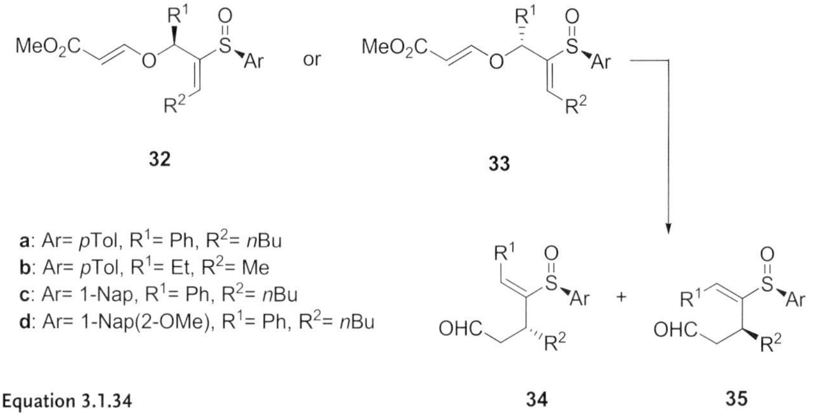

32
33
a: Ar= pTol, R1= Ph, R2= nBu
b: Ar= pTol, R1= Et, R2= Me
c: Ar= 1-Nap, R1= Ph, R2= nBu
d: Ar= 1-Nap(2-OMe), R1= Ph, R2= nBu
34
35

Equation 3.1.34

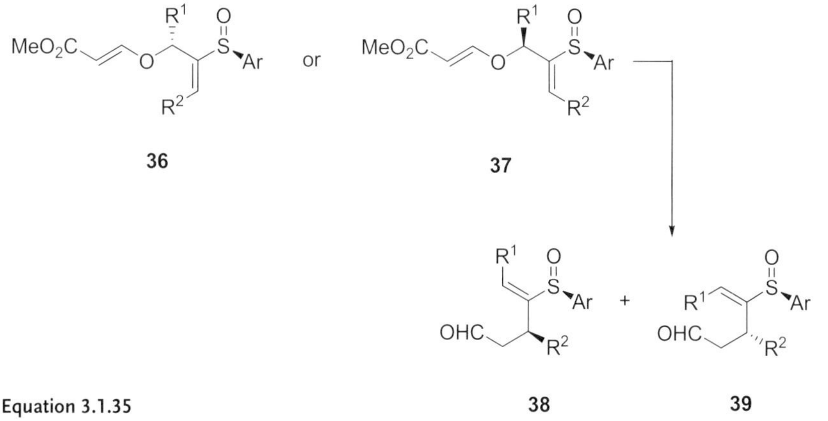

36
37
38
39

Equation 3.1.35

Tab. 3.1.3 Sulfinyl-mediated Claisen rearrangement (Eq. 3.1.35).

Entry	Substrate	*T* (°C)	*t* (min)	*Z*:*E*[a]	Yield (%)
1	(+)-**36a**	134	180	73:27, (–)-**38a**, (+)-**39a**	73
2	**36b**	134	180	74:26, **38b**, **39b**	76
3	**36c**	138	420	64:36, **38c**, **39c**	71
4	**37a**	126	150	24:76, **38a**, **39a**	74
5	**37d**	110	240	0:100, **39d**	77

a) *Z*/*E* ratios include two byproducts obtained in some cases [42].

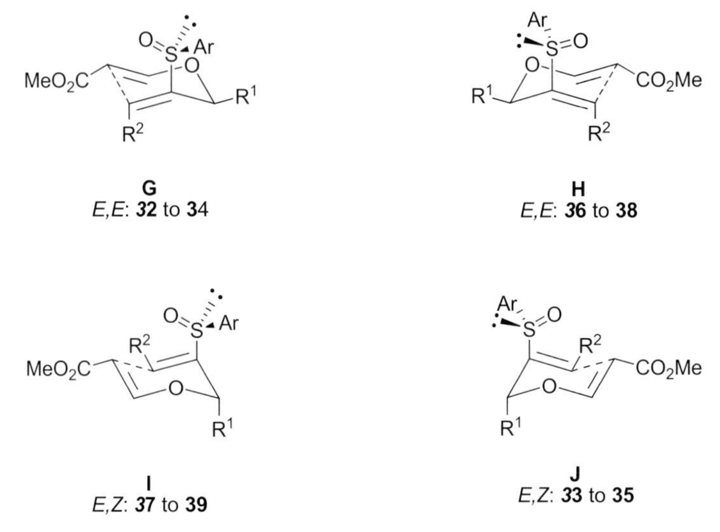

Scheme 3.1.2

3.1.4 Claisen Rearrangement of Cyclic Allyl Vinyl Ethers

3.1.4.1 Ring Expansion Claisen Rearrangement

The Claisen rearrangement of cyclic allyl vinyl ethers provides corresponding ring expansion products. This method is widely used for total syntheses.

In particular, Paquette has reported that the treatment of esters with the Tebbe reagent and subsequent thermal rearrangement of the resulting allyl vinyl ethers gave carbonyl compounds.

For example, bicyclic allyl vinyl ether **40** rearranged to a mixture of eight-membered bicyclic compounds **41** and **42** rich in the α-4-methyl isomer (15:1)

(Eq. 3.1.36) [43]. This rearrangement exhibited several interesting features. If **40** was not well purified such that organometallic residues from the Tebbe reagent remained, **41** and **42** were formed more rapidly and their ratio was closely 1:1. It should also be noted that the yield was modestly increased to about 60%. Apparently, the organometallic impurities serve as Lewis acid promoters to accelerate the [3,3] sigmatropic process. Consequently, it appears that the rearrangement of **41** proceeded smoothly via a chair-like transition state **K**. However, prior equilibration of (*Z*)- to (*E*)-propenyl would allow for competitive isomerization via **L** to **42**. The relative importance of the boat form, an alternative transition state to **42**, is not observed.

180 °C

40 **41**, 36 % + **42**

K **L**

Equation 3.1.36

Treatment of the ester **43** with the Tebbe reagent gave rise to **44** as a 10:1 isomeric mixture via the smooth rearrangement (Eq. 3.1.37) [44]. The exceptionally low temperature and short reaction times required for this conversion suggest that strain relief within the bridged diquinane components operates at the rate-determining transition states and provides a useful driving force for facilitating the electronic reorganization.

From among the two geometries available to **43**, the choices are to orient the cyclopentenyl unit as in **M** or **N**. In the first instance, since a chair-like conformation shall develop, a *Z* double bond will be generated, and H-5 will be projected to the α-face. On the other hand, another conformation **N**, which requires a boat-like geometry, results in the evolution of an *E* double bond, and fixes H-5 in a β-orientation. In addition, **N** suffers from steric compression between the exocyclic methylene group and the silyl substituent. On the basis of these many factors, transition state **M** should be kinetically favored.

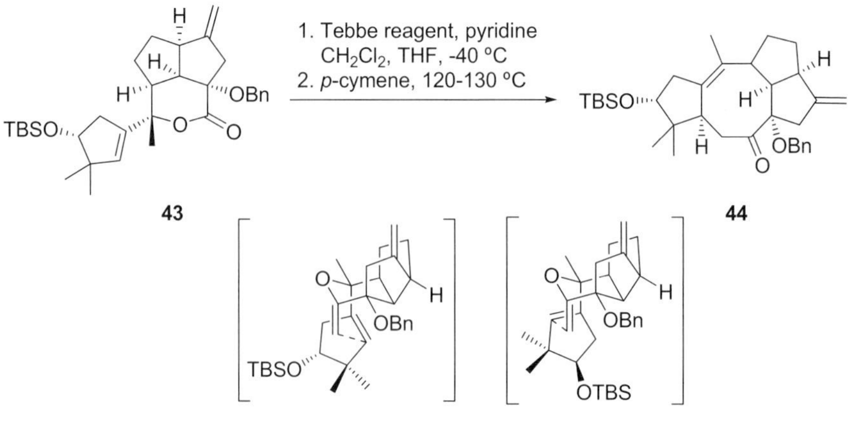

1. Tebbe reagent, pyridine CH_2Cl_2, THF, -40 °C
2. *p*-cymene, 120-130 °C

43 **44** **M** **N**

Equation 3.1.37

Paquette also investigated the transition state of the ring expansion The Claisen rearrangement in the synthesis of (–)-sclerophytin A [45]. In this rearrangement **46** was transformed into **47** more rapidly than was **45**. In fact, **46** rearranged quantitatively into **47** in just 3 h upon being heated in neat conditions at 55 °C. By comparison, **45** required higher temperatures for a longer period. Even under these

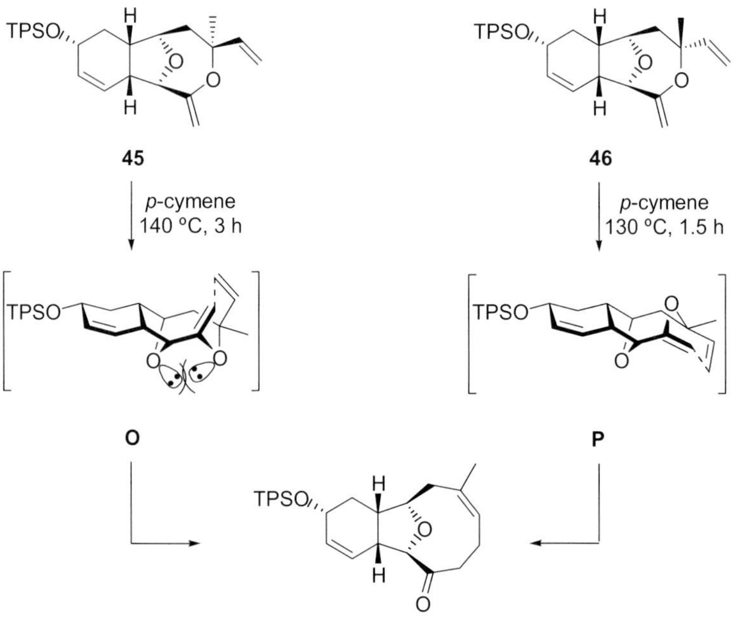

45 *p*-cymene 140 °C, 3 h

46 *p*-cymene 130 °C, 1.5 h

O **P** **47**

Scheme 3.1.3

circumstances, the conversion of **47** was incomplete, and some decomposition was observed. Finally, a compromise position was taken to reconcile this kinetics imbalance. The interesting rate difference can be rationalized by comparing the two chair-like transition states **O** and **P** that are presumably involved. Although the adoption of theses chair geometries guarantees that both enol ethers will give rise to product having a *Z* double bond, the sterically and electronically disfavored interactions that are incurred in **O** but not in **P** elevate the energy barrier with observable consequences.

Bi- and tricyclic fused heterocycles, including allyl vinyl ether moieties, undergo thermal rearrangement in refluxing toluene, xylenes or mesitylene (Eq. 3.1.38) [48]. The transformation proceeds very efficiently to give the expected bridged ring system as sole products in very good yields (Table 3.1.4).

MeO_2C R^1 R^2 CO_2Me $\xrightarrow{\Delta}$ R^1 R^2 CO_2Me MeO_2C

Equation 3.1.38

Tab. 3.1.4 Preparation of Bicyclo[4.n.1]alkanones (Eq. 3.1.38).

Entry	n	R_1	R_2	Solvent	*t* (h)	Yield (%)
1	1	H	H	xylenes	48	88
2	1	Me	H	xylenes	27	80
3	1	Me	Me	toluene	7	96
4	2	Me	H	mesitylene	50	90
5	2	Me	Me	xylenes	17	100

The allylic ether bearing selenoxide **48** was heated to effect sequential elimination of PhSeOH and Claisen rearrangement (Eq. 3.1.39) [49]. Since the latter sigmatropic process is usually *Z*-selective in its ring expansion mode, a kinetic bias for vinyl ether conformer **Q** was anticipated because this transition state pathway is uniquely capable of leading to the *cis* double bond isomer **49**. The stereoselectivity for an orientation of the *secondary* methyl also follows from adoption of the chair-like topography given by **Q**. Although **49** was produced efficiently, care had to be exercised to avoid migration of the double bond conjugation with the lactone carbonyl.

Equation 3.1.39

Numerous examples have established that the Claisen ring expansion is controlled by an overriding thermodynamic preference for setting a *Z* double bond in the 4-cyclooctenone ring via a chair-like transition state if at all possible.

An inseparable mixture of **51** and **54** (3:2 ratio) was methylenated with dimethyltitanocene reagent to furnish enol ethers **52** and **55** in 90% yield (Scheme 3.1.4). Thermal rearrangement of the mixture in a sealed tube produced the desired cyclooctenone **53** and the isomeric 6–5 system **58**. The Claisen rearrangement of **52** proceeded with high stereoselectivity. The formation of **58** is a result of the partial isomerization of **52** to **57** followed by an alternative Claisen rearrangement. Another product obtained was the enol ether **56**, formed by the isomerization of **55**. It was too hindered to perform the Claisen rearrangement of **55** and **56**. Molecular models indicated that the chair-chair transition state **R** should be more favored over the chair-boat transition state [50].

Thermal rearrangement of the dioxolane **59** was carried out in a degassed sealed tube without solvent (Eq. 3.1.40). Although the rearrangement seems to proceed by the concerted Claisen rearrangement, mechanisms involving ionic or biradical intermediates would be also possible. However, formation of **60** could not be suppressed by ethylene glycol, BHT or copper powder [51]. These results support the [3,3] sigmatropic process.

Equation 3.1.40

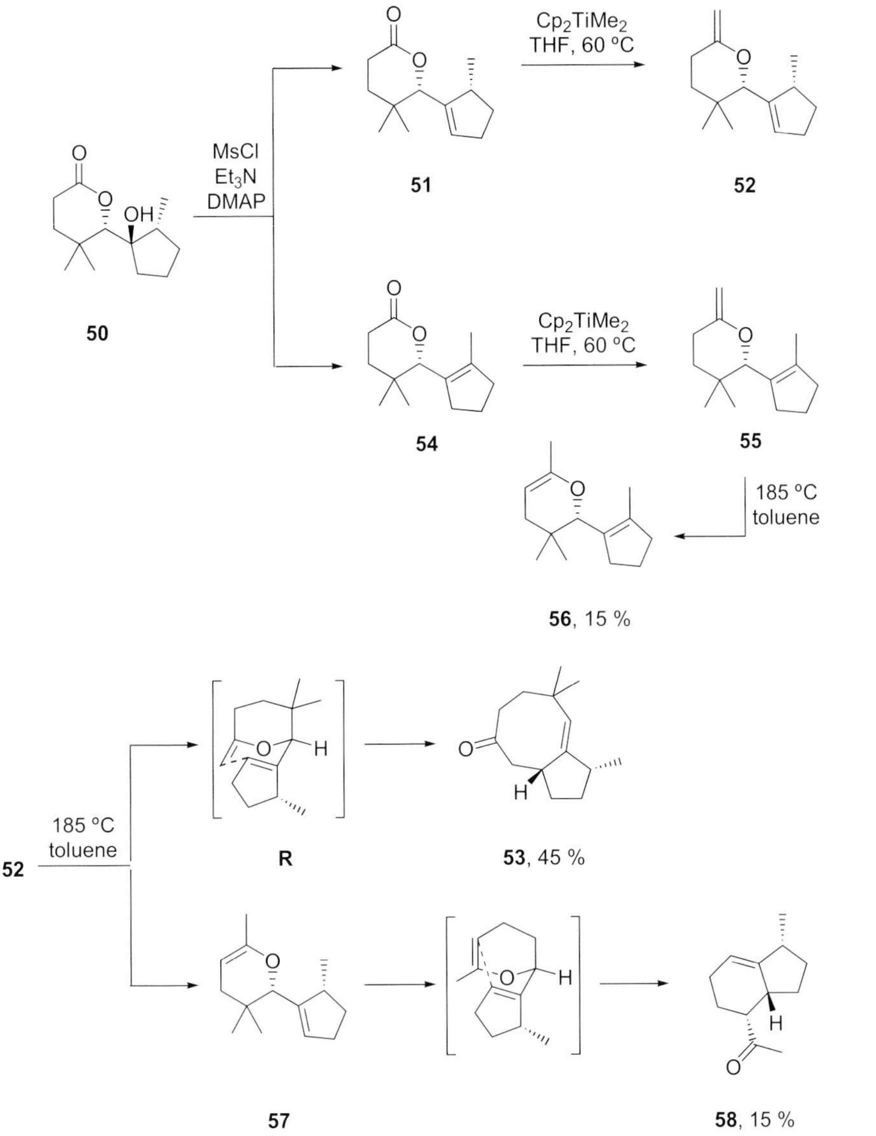

MsCl
Et_3N
DMAP

Cp_2TiMe_2
THF, 60 °C

50

51

52

54

55

185 °C
toluene

56, 15 %

52

185 °C
toluene

R

53, 45 %

57

58, 15 %

Scheme 3.1.4

3.1.4.2 Cyclohexene Synthesis

Pseudo-sugars, in which the ring oxygen atom of a sugar is replaced by a methylene group, were prepared from normal sugars [52]. Allyl vinyl ether **61**, which was prepared from D-glucose was converted to chiral carbocycle **62** via Claisen rearrangement (Eq. 3.1.41). Reduction of **62** with $NaBH_4$ gave functionalized chiral alcohol **63**, which was utilized to prepare the four pseudo-sugars, namely, pseudo-α-D-glucopyranose, pseudo-α-D-mannnopyranose, pseudo-β-D-glucopyranose and pseudo-β-D-mannopyranose.

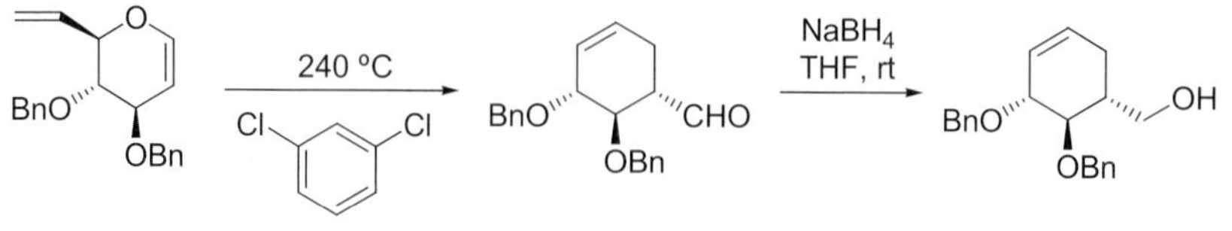

240 °C; NaBH4, THF, rt

61 **62**, 84 % **63**

Equation 3.1.41

The dihydropyranethylene **64**, which a synthetic intermediate of pancratistatin that exhibits a high level of *in vitro* and *in vivo* cancer cell growth inhibitory activity and antiviral activity was rearranged to the *cis*-disubstituted cyclohexene **65** as a single isomer (Eq. 3.1.42) [53]. This rearrangement proceeds through a boat-like transition state of the cyclohexene [54].

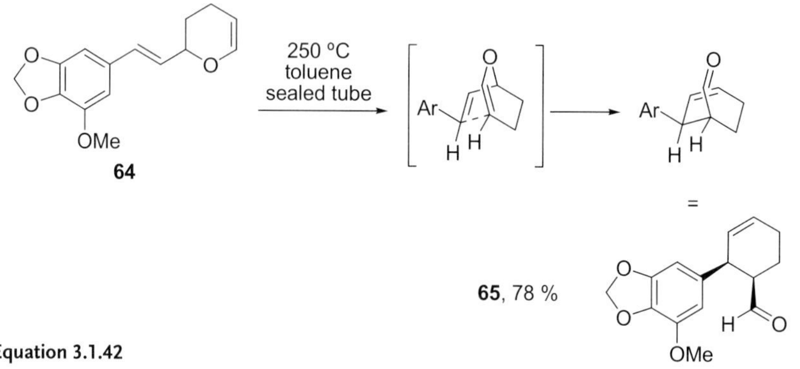

250 °C, toluene, sealed tube

64 **65**, 78 %

Equation 3.1.42

3.1.5 Cyclic Vinyl Ethers

Recently, Ley found that a highly diastereoselective, microwave-induced Claisen rearrangement of an appropriately substituted propargylic enol ether allows the formation of the sterically congested C8–C14 bond of azadirachtin [55].

azadirachtin

reflux, xylene

Equation 3.1.43

Thermal rearrangement of the simple allyl ether was successful in preliminary studies (Eq. 3.1.43) [56]. Unfortunately, repeated attempts to induce the Claisen rearrangement of **66** resulted in nearly complete recovery of the starting material (Eq. 3.1.44, Table 3.1.5, entry 1). With these results at hand, the microwave irradiation was attempted to induce the desired rearrangement. As a result, a small increase in yield was observed when **66** was reacted under the conditions given in Table 3.1.5, entry 2. The microwave irradiation of a solution of **68** after only 15 min resulted in a 71% yield of the desired allene **69** as a single diastereoisomer. After extensive optimization studies, it was shown that microwave irradiation with 15 consecutive 60-s pulses allowed the formation of the **69** in 88% yield.

66, R= TES
68, R= H

67, R= TES
69, R= H

Equation 3.1.44

Tab. 3.1.5 Claisen rearrangement studies using microwave irradiation (Eq. 3.1.44).

Entry	Substrate	T (°C)	Conditions	Yield (%)
1	**66**	180	DCB, 48 h	7
2	**66**	180	DCB, MWI, 1 h	25
3	**68**	180	DCB, MWI, 15 min	71
4	**68**	180	DCB, MWI, 15 min	88[a)]

a) Employed 60-s microwave pulses instead of continuous irradiation (DCB = 1,2-dichlorobenzene; MWI = microwave irradiation).

3.1.6
Cyclic Allyl Ethers

The thermal isomerization of **70** required heating at high temperature to proceed efficiently (Eq. 3.1.45) [57]. It appeared that transition state **T** was adopted preferentially to **S** in order to prevent the development of a 1,3-diaxial interaction in the preliminary expectation. If this kinetic ordering were adopted, the new C–C bond would be installed *cis* to the butyronitrile side chain in the desired manner. Actually, **70**, however, isomerizes preferentially via **S** to set an axial bond and deliver **71** and **72** in a 2.7:1 ratio.

200 °C, xylene
sealed tube

70 **71** **72**

71:**72** = 2.7:1

S **T**

Equation 3.1.45

The rearrangement has occurred stereospecifically from the desired α-face of the double bond to provide optically active **73**, presumably via a chair transition state (Eq. 3.1.46) [58]. This rearrangement has significant implications for the enantiospecific synthesis of the macroline/sarpagine/ajmaline alkaloids, because intermediate **76** has been functionalized at C_{15} with the absolute configuration common to all three families of alkaloids.

145 °C, benzene
sealed tube

Equation 3.1.46 **73**

3.1.7
Tandem Reactions Including Aliphatic Claisen Rearrangement

3.1.7.1 Vinylation/Claisen Rearrangement

A number of vinylation-Claisen rearrangements have been reported as a tandem reaction. Since no additional reagents are required in the thermal Claisen rearrangement, the allyl vinyl ethers can rearrange without isolation. No convenient introduction methods of basically reactive vinyl groups are available for the selective preparation of functionalized vinyl ethers [11, 13, 14, 22, 24].

Fukuyama has found that a one-pot transformation involving vinylation and the Claisen rearrangement was quite effectively performed in *n*-butyl vinyl ether at reflux temperature [59]. A mixture of allylic alcohols, *n*-butyl vinyl ether and catalytic $Hg(OAc)_2$ and NaOAc was stirred at reflux to prepare unsaturated aldehydes (Eq. 3.1.47).

10 mol% $Hg(OAc)_2$
10 mol% NaOAc
500 mol%
reflux

79 %

Equation 3.1.47

The reaction of *secondary* allylic alcohols bearing *trans*-1,2-disubstituted olefins proceeded smoothly to furnish the desired aldehydes in good yields (Table 3.1.6, entry 1). Geraniols, which form *quarternary* carbon centers, rearranged in moderate yield because of low reactivity of vinyl ethers formation. And the one-pot rearrangement of styrene derivatives proceeded in low yield due to preparation of deconjugated products (entries 2–4).

Tab. 3.1.6 One-pot Claisen rearrangement using *n*-butyl vinyl ether (Eq. 3.1.49).

Entry	Substrate	Condition[a)]	*t* (h)	Product	Yield (%)
1	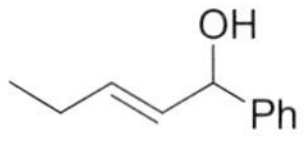 OH, Ph	A	10	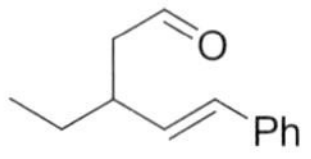 O, Ph	79
2	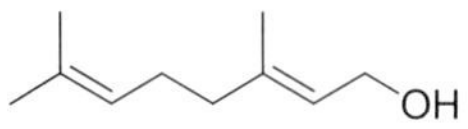 OH	A	36	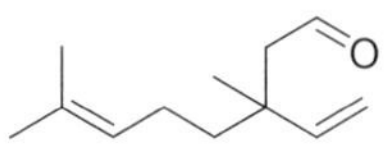 O	61
		B	13		67
3	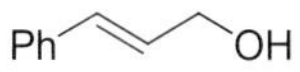 Ph, OH	A	40	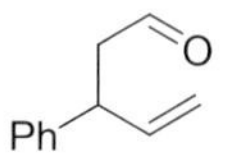 O, Ph	47
		B	4		52
4	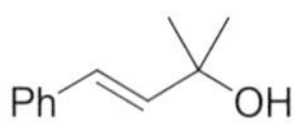 Ph, OH	B	8	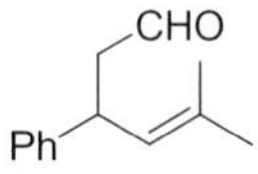 CHO, Ph	53

a) Condition A; $Hg(OAc)_2$ (0.1 equiv), NaOAc (0.1 equiv), *n*-butyl vinyl ether (5 equiv), reflux. Condition B; $Hg(OAc)_2$ (0.1 equiv), NaOAc (0.1 equiv), *n*-butyl vinyl ether (5 equiv), 180 °C in a sealed tube.

In developing a general method for preparing 16-ene C,D-ring with natural C20-(*S*) stereochemistry of 1*α*,25-dihydroxyvitamin D_3, Hatcher and Posner envisaged various allylic vinyl ethers undergoing Claisen rearrangement on the *α*-face of the D-ring [60]. This rearrangement proceeds via a chair-like transition state giving the natural absolute stereochemistry at C20 (Eq. 3.1.48). In order to achieve these rearrangements, the reaction of allylic alcohol **74** with ethyl vinyl ether and $Hg(OAc)_2$ afforded the desired enol ether, which underwent thermal Claisen rearrangement *in situ* to give aldehyde **75** (Eq. 3.1.49). Vinyl ethers of the cyclic *secondary* allylic alcohol provide cyclic *Z*-olefinic aldehydes via Claisen rearrangement because the ring of these cyclic ethers is too small to lead to *E*-olefinic aldehydes.

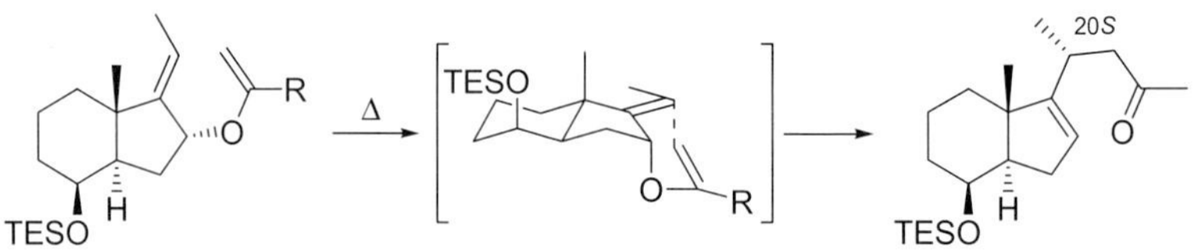

TESO, H, O, R — Δ → [TESO, O, R] → 20*S*, O, TESO, H

Equation 3.1.48

$EtOCH=CH_2$
$Hg(OAc)_2$, 120 °C

74 75

Equation 3.1.49

The chiral aldehyde **76** was employed as the starting material of intramolecular ene-reaction to prepare chiral bicyclo [4.3.1] decanes [61]. The one-pot Claisen rearrangement of carveol with ethyl vinyl ether in the presence of a catalytic amount of $Hg(OAc)_2$ furnished the aldehyde **76** stereospecifically (Eq. 3.1.50). For the synthesis of stereoselective side chain introduced aldehydes as the starting materials of intramolecular ene-reaction to prepare chiral bicyclo [3.3.1] decanes [61], a one-pot Claisen rearrangement of carveol with ethyl vinyl ether in the presence of a catalytic amount of $Hg(OAc)_2$ stereospecifically furnished the aldehyde **76** (Eq. 3.1.50).

$EtOCH=CH_2$
$Hg(OAc)_2$, 180 °C

Equation 3.1.50

76, 84 %

3.1.7.2 Allylation/Claisen Rearrangement

Attempts to perform angular allylation of the enolate of 77 with LDA and allyl bromide resulted in complete recovery of the corresponding enone, probably due to steric hindrance imposed by bis-neopentyl position [62]. On the other hand, treatment of enone **77** with potassium hydride and allyl bromide furnished unstable *O*-allylated dienol ether **78** (Eq. 3.1.51). The resulting solution was heated to cause the Claisen rearrangement, and the desired allylenone **79** was produced as a single isomer in this one-pot operation. The stereoselective introduction of the allyl group from the β-face of the enone **77** was finally proved by transformation of **79** into cyclomyltaylane-5α-ol. The present stereoselective introduction of the allyl unit at angular position is explained by the fact that the Claisen rearrangement proceeded from the less sterically congested conformer **U** by assuming a chair-like transition state.

The aldehyde **80** was converted to allyl enol ether **81** and, without isolation, **81** was heated to afford the aldehyde **82** (Eq. 3.1.52) [63].

KH, $H_2C{=}CH_2CH_2Br$
THF, 0 °C, 2 h

toluene
reflux, 12 h

77 **78** **79**, 96 %

8 steps

U

more favorable than

cyclomyltaylane-5α-ol

Equation 3.1.51

NaH, $H_2C{=}CHCH_2Br$
18-crown-6, THF, reflux

185 °C, 7 h

80 **81** **82**, 79 %

Equation 3.1.52

3.1.7.3 **Anionic Cyclization/Claisen Rearrangement**

Ovaska has developed a tandem anionic cyclization-Claisen rearrangement sequence as a highly stereoselective route to a tricyclic ring system [64–66].

This rearrangement sequence was used to effect a facile, one-pot conversion of an appropriately substituted 4-alkyn-1-ol to the tetracyclic carbon core structure of phorbol. Thus, **83** was sequentially converted to the fused 5–7–6–3 ring system under normal conditions for the tandem 5-exo-dig cyclization-Claisen rearrangement by exposure to catalytic MeLi, and the cyclic ketone **84** was obtained as a single stereoisomer in 76% isolated yield (Eq. 3.1.53).

10 mol% MeLi, Ph_2O
185 °C, 1 h

83

84, 76 %

Equation 3.1.53

3.1.7.4 Claisen Rearrangement/Ene Reaction

The combination of pericyclic reactions constitutes a powerful synthetic strategy to make multiple C–C bonds in an efficient manner.

In the total synthesis of (–)-furodysinin, allylic vinyl ether **85**, which was prepared from (+)-limonene oxide, was heated at high temperature to proceed the tandem Claisen rearrangement and an intramolecular ene reaction (Eq. 3.1.54) [67]. The route was actually designed to take advantage of the stereoselectivity of the Claisen rearrangement to establish the aldehyde side chain and the stereochemistry of the bicyclic system. At lower temperatures and shorter reaction times, the intermediary aldehyde could be isolated.

200 °C, ene

85 **86**

Equation 3.1.54

Mikami and Nakai have investigated one-pot Claisen-ene sequence to achieve the asymmetric total synthesis of (+)-9(11)-dehydroestron methyl ether [68]. Thus, the allylic alcohol and cyclic enol ether in the presence of 2,6-dimethylphenol (10 mol%) was heated in a sealed tube (Eq. 3.1.55). The tandem Claisen-ene product **91** was isolated in 76% yield after hydrolysis. The key feature of the present strategy is the successful development of the asymmetric tandem Claisen-ene sequence for the double carbocyclization of D and C rings that allows for the relatively short construction of the estrogen framework in a highly stereocontrolled fashion.

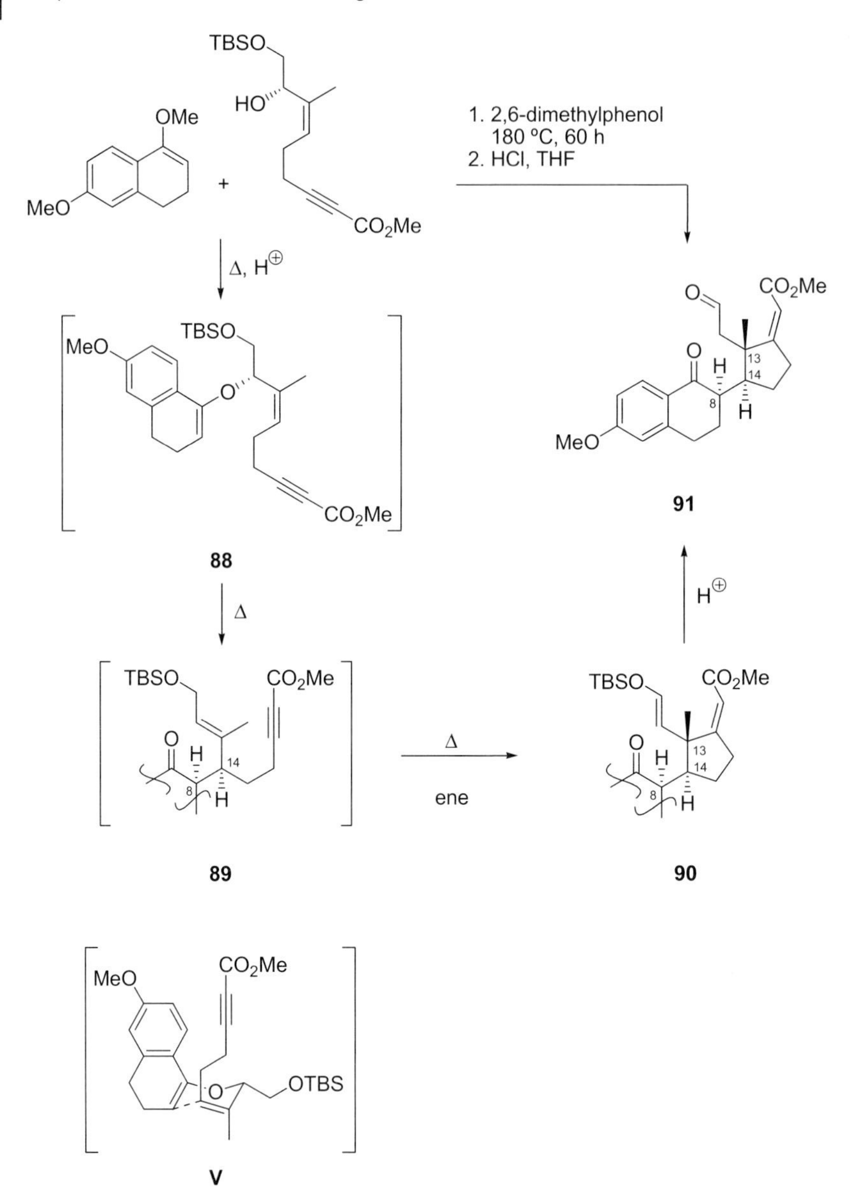

Equation 3.1.55

Particularly noteworthy is the stereochemical course of the asymmetric tandem sequence. The Claisen rearrangement of the cyclic enol ether **88** proceeds through the chair-like transition state **V**. Thus, the (*S*,*Z*)-configuration of the alcohol **87** is completely transmitted to the (14*S*) configuration in the Claisen product, along with a high 8,14-*syn* diastereoselectivity. The subsequent ene-cyclization of **89** proceeds via a bicyclic-endo transition state with the A,B ring at the sterically favorable pseudoequatorial position to eventually establish the 13,14-*trans* configuration in the ene product **90**, along with the formation of the silyl enol ether side chain.

3.1.7.5 Claisen Rearrangement/Conia-Type Oxa-Ene Reaction

The allyl vinyl ether **93**, which was obtained via Wittig olefination between α-hydroxy allyl ester **92** and cumulated phosphorus ylide, was thermally converted to *Z*- and *E*-isomers **96** by heating the solutions [69, 70].

The plausible mechanism for this formal [2,3]-rearrangement is illustrated in Eq. 3.1.56. It seems that Claisen rearrangement of **93** produces the 3-allyltetronic acid **94**, which is not isolable but undergoes a consecutive Conia-type oxa-ene reaction to give the spirocyclopropane. Then, this intermediate can open the cyclopropane ring using the H atom from any of the three surrounding methyl groups, leading either backwards to the normal Claisen products **94** or to the "abnormal" product **95**. The latter avoids the reverse reaction via a [1,3]-H shift followed by a keto-enol tautomerisation to eventually yield *E*/*Z* isomeric mixtures of **96**.

Equation 3.1.56

3.1.7.6 **Oxy-Cope/Ene/Claisen Rearrangement**

A mixture of allyl ether **97** and DBU was heated in a CEM microwave to give lactol **98** (Eq. 3.1.58). It is recognized that this reaction is a new synthetic method based on tandem oxy-Cope/transannular ene/Claisen reaction sequences that can generate up to four contiguous stereogenic centers including two quaternary carbons at C5 and C9 [71].

DBU, toluene
220 °C
CEM microwave
oxy-Cope
ene
Claisen
97
98

Equation 3.1.57

The high diastereoselectivity of this triple tandem reaction can be explained by the mechanism depicted in Scheme 3.1.5. At first glance, the thermal oxy-Cope rearrangement of **99** leads to enol **100**, which tautomerizes *in situ* to produce the ketone **101**. This ketone can adopt two chair-like conformations at the transition state, **W** and **X**. A close examination of the transition state **Z** reveals a psuedo-1,3-diaxial *O*-allyl-ring methylene interaction. Therefore, this favors the transition state **W** over **X** as the reactive conformer to provide the enol **102**. Finally, the Claisen rearrangement proceeds *anti* to the bridgehead alcohol at C5 to afford the desired bicyclic product **103**.

3.1.8
The Carbanion-Accelerated Claisen Rearrangement

3.1.8.1 **Sulfonyl-Stabilized Anions**

Denmark has developed the carbanion accelerated Claisen rearrangement [15, 72–78]. It is known that the nature of the transition state of Claisen rearrangement is dependent on the substituents present [79]. The effects of substituents have been studied in detail by Carpenter [80]. At positions C1, C2, and C4 both donor and acceptor substituents accelerate the reaction with respect to hydrogen (Eq. 3.1.58). Furthermore, the synthetic versatility of the Claisen rearrangement primarily resides in the variety of the substituents X that determine the carbonyl derivative

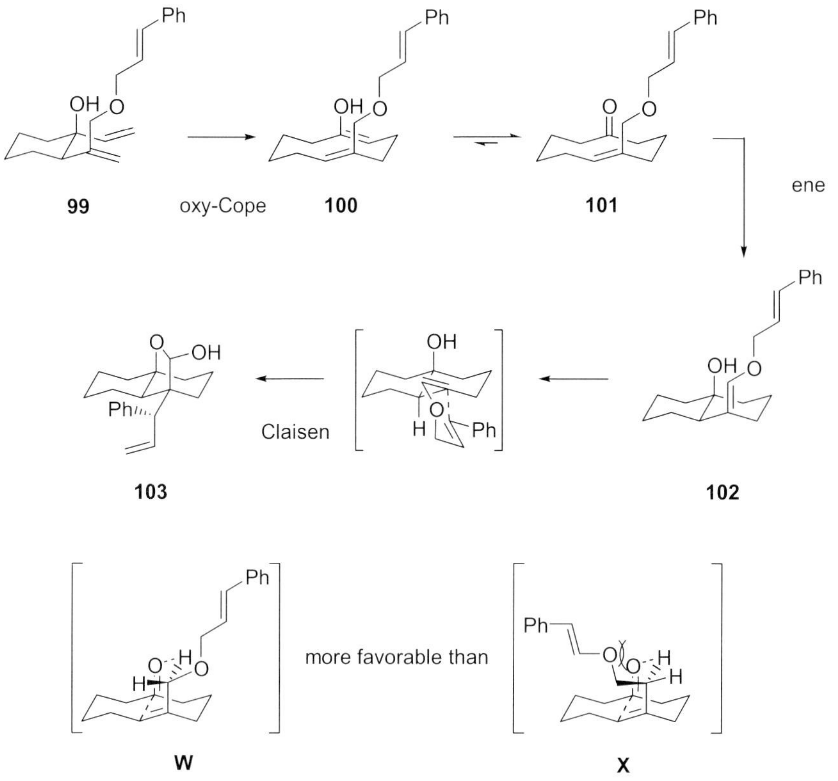

Ph OH O 99 oxy-Cope 100 101 ene 102 Claisen 103 W more favorable than X

Scheme 3.1.5

compounds. In the above consideration, Denmark and coworkers noted that a carbanionic π-donor at C2 was able to accelerate the Claisen rearrangement. In addition, an anion-stabilizing group was determined to introduce an aryl sulfonyl group.

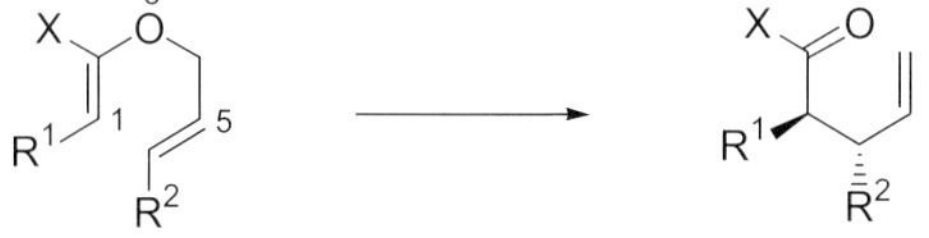

Equation 3.1.58

Unsubstituted **104** and the tautomeric sulfone **105** reacted with potassium hydride and the crown ether in HMPA/THF to afford the rearrangement product **106** in 78% and 76% yield, respectively (Eq. 3.1.59). The position of the double bond had no effect on the yield or regiochemical outcome of the reaction. Substrates disubstituted at C1 rearranged much faster than the unsubstituted **104** (Eq. 3.1.60). It is expected that the anionic charge should not only provide accelerating potential but should also contribute to the exothermicity of the reaction. The

PhSO2 104 PhSO2 105 KH, HMPA 50 °C, 4 h PhSO2 106, 76-78 %

Equation 3.1.59

PhSO2 KH, DMSO 20 °C, 15 min PhSO2 91 %

Equation 3.1.60

PhSO2 base, DMSO PhSO2

base	mol%	T [°C]	t [h]	*anti:syn*	yield [%]
KH	210	20	6.75	86:14	7
NaH	220	50	3	94:6	64
KH, LiCl	260, 1500	50	1.5	95:5	85

Equation 3.1.61

result would be a reduction in both the kinetic and thermodynamic barriers to rearrangement.

When the allyl vinyl ether has at least a substituent at terminal allyl moiety, the addition of LiCl improves the yield of rearrangement products (Eq. 3.1.61).

3.1.8.2 Phosphine Oxide and Phosphonate-Stabilized Anions

It is found that the sulfur-based groups are problematic in preparation and rearrangement of the functionalized allyl vinyl ethers. It became apparent that the phosphorus-based groups as alternative anion-stabilizing moieties, mainly because of the ease of preparation of allenylphosphorus precursors and a well-established chemistry of phosphorus-stabilized allyl anions [77].

At similar levels of substitution, the phosphonate rearranged 10–20 times faster than the corresponding sulfone (Eq. 3.1.62). The rearrangements of C1 monosub-

KH, LiCl
DMSO, 20 °C

X	t [min]	yield [%]	*syn*:*anti*
p-TolSO$_2$	240	73	97:3
(tBuO)$_2$P(=O)	15 min	93	96:4

Equation 3.1.62

210 mol% KH
DMSO, 20 °C, 45 min

74 %

Equation 3.1.63

stituted allyl vinyl ether was complete in a very short time at 20 °C with exclusive γ-regioselectivity (Eq. 3.1.63). It is found that LiCl plays a significant role in the yield of the rearrangement of phosphonates (Eq. 3.1.64, Table 3.1.7, entries 1 and 2). The high diastereoselectivity observed in the entries 3 and 4 is expected in both thermal and anionic variants. However, the *anti* diastereomer was not directly available by anionic rearrangement of the (*Z*)-2-butenyl ether (Table 3.1.7, entry 5). The failure to observe rearrangement may be due to the bulky *tert*-butyl phosphonate moiety. Considering steric and chelation effects, the preferred orientation of the phosphonate places a bulky *tert*-butoxy group over the ring strongly disfavoring either the chair or boat transition structure compared to competing, nonproductive processes (Scheme 3.1.6).

base
solvent
condition

R^1= Me, R^2= H
or
R^1= H, R^2= Me

syn

anti

Equation 3.1.64

Tab. 3.1.7 Carbanion-accelerated and thermal Claisen rearrangements of the phosphonates (Eq. 3.1.64).

Entry	R_1	R_2	KH/LiCl (equiv)	Solvent	T (°C)	t (h)	Yield (%)
1	H	H	1.9/0	DMSO	20	0.25	46
2	H	H	1.9/3.2	DMSO	20	0.25	84
3	Me	H	0/0	THF	100	10.0	90 (93:7)
4	Me	H	2.1/5.1	DMSO	20	0.25	93 (96:4)
5	H	Me	0/0	THF	100	8.0	19 (14:86)

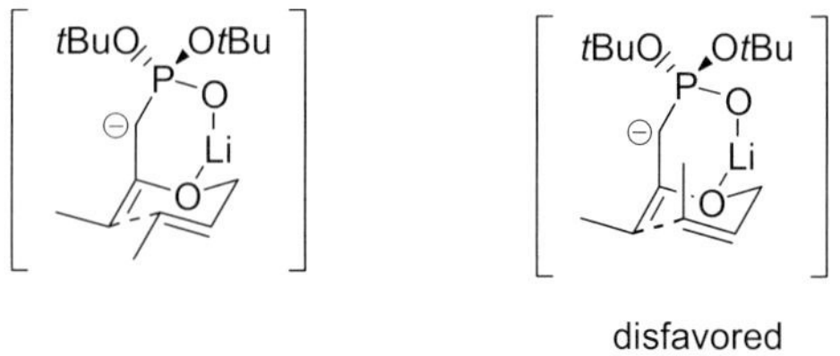

Scheme 3.1.6

3.1.8.3 Phosphonamide-Stabilized Anions

Denmark has also investigated the carbanion accelerated Claisen rearrangement of allyl vinyl ethers with cyclic phosphonamides as carbanion-stabilizing groups [78], because phosphonamide allyl anion is well established and has potential for chiral modification.

The simple C_3 monomethyl substrate with *N,N'*-dibenzyldiazaphospholidine as the stabilizing group was rearranged under anion-accelerated and thermal conditions (Eq. 3.1.65, Table 3.1.8). Unfortunately, Li^+-**107**$^-$ rearranged in THF with very disappointing diastereoselectivity (entries 1 and 2). However, the diastereoselectivity of the anion accelerated Claisen rearrangement by KH improved in the opposite direction (entry 3). Remarkably, thermal rearrangement proceeded with modest selectivity, favoring the same product (entry 5 vs. entries 1 and 2).

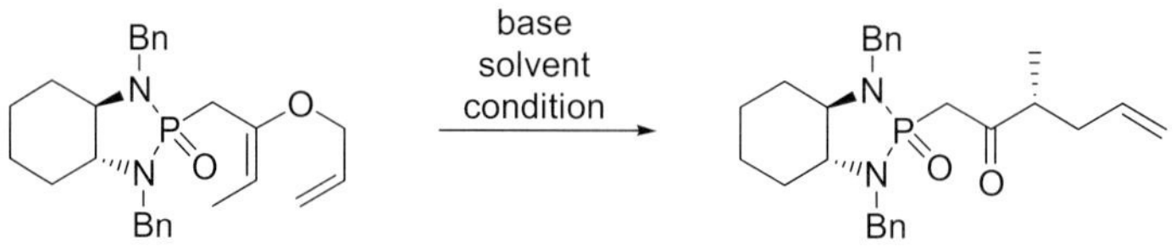

Equation 3.1.65

Tab. 3.1.8 Carbanion-accelerated and thermal Claisen rearrangements of the chiral diazaphospholidine (Eq. 3.1.65).

Entry	Base (equiv)	Solvent	T (°C)	t (h)	ds	Yield (%)
1	*n*-BuLi (1.9)	THF	0	1	58:42	85
2	LDA (1.9)	THF	0	0.75	58:42	64
3	KH (2.3)	DMSO	20	2.0	31:69	79
4	KH (2.1)/LiCl (8.0)	DMSO	20	2.0	48:52	78
5	none	toluene	110	2.0	63:38	81

3.1.9 Conclusion

In past 15 years, the aliphatic Claisen rearrangement has been employed for constructing quantities of complicated carbon skeletons in view of its high stereoselectivity. The simple reaction, requiring only heat, can often be a part of a tandem reaction. This rearrangement has also attracted attention from "green" chemistry, because it is a nonmetallic process. Although the rearrangement was discovered about 100 years ago, its advantages will keep brightening the future.

References

1 Wipf, P. In *Comprehensive Organic Synthesis* **1991** Trost, B.; Fleming, I. (eds.) Pergamon Press, New York, Vol. 5, p. 827.
2 Ziegler, F. E. *Chem. Rev.* **1988**, *88*, 1423.
3 Lutz R. P. *Chem. Rev.* **1984**, *84*, 206.
4 Nowicki, J. *Molecules* **2000**, *5*, 1033. (See also: http://www.mdpi.org/molecules/papers/50801033.pdf).
5 Nubbemeyer, U. *Synthesis* **2003**, 961.
6 Watanabe, W. H.; Conlon, L. E. *J. Am. Chem. Soc.* **1957**, *79*, 2828.
7 Buchi, G.; Vogel, D. E. *J. Org. Chem.* **1983**, *48*, 5406.
8 Vogel, D. E.; Buchi, G. In *Org. Synth. Coll. Vol. 8* **1993** Freeman, J. P. (ed.) John Wiley & Sons, New York, p. 536.
9 Freiría, M.; Whitehead, A. J.; Motherwell, W. B. *Synlett* **2003**, 805.
10 Kulkarni, M. G.; Pendharkar, D. S.; Rasne, R. M. *Tetrahedron Lett.* **1997**, *38*, 1459.
11 Mandai, T.; Matsumoto, S.; Kohama, M.; Kawada, M.; Tsuji, J. Saito, S.; Moriwake, T. *J. Org. Chem.* **1990**, *55*, 5671.
12 Clive, D. L. J.; Magnuson, S. R. *Tetrahedron Lett.* **1995**, *36*, 15.
13 Wang, T.; Pinard, E.; Paquette, L. A. *J. Am. Chem. Soc.* **1996**, *118*, 1309.
14 Hoye, R. C.; Rajapakse, H. A. *Synth. Commun.* **1997**, *27*, 663.
15 Denmark, S. E.; Harmata, M. A. *J. Am. Chem. Soc.* **1982**, *104*, 4972.
16 Denmark, S. E.; Harmata, M. A.; White, K. S. *J. Org. Chem.* **1987**, *52*, 4031.
17 Stirling, C. J. M. *J. Chem. Soc. C* **1964**, 5863.
18 O'Connor, D. E.; Lyness, W. I. *J. Am. Chem. Soc.* **1964**, *86*, 3840.
19 Mackle, H.; McNally, D. V.; Steele, W. V. *Trans. Faraday Soc.* **1969**, 2060.

20 Yamamoto, Y.; Fujikawa, R.; Miyaura, N. *Synth. Commun.* **2000** *30*, 2383.
21 Baudry, D.; Ephritikhine, M.; Felkin, H. *J. Chem. Soc. Chem. Commun.* **1978**, 694.
22 Nordmann, G.; Buchwald, S. L. *J. Am. Chem. Soc.* **2003**, *125*, 4978.
23 Tebbe, F. N.; Parshall, G. W.; Reddy, G. S. *J. Am. Chem. Soc.* **1978**, *100*, 3611.
24 Godage, H. Y.; Fairbanks, A. J. *Tetrahedron Lett.* **2000**, *41*, 7589.
25 Vittorelli, P.; Winkler, T.; Hansen, H. J.; Schmid, H. *Helv. Chim. Acta.* **1968**, *51*, 1457.
26 Hansen, H. J.; Schmid, H. *Tetrahedron* **1974**, *30*, 1959.
27 Perrin, C. L.; Faulkner, D. J. *Tetrahedron Lett.* **1969**, *10*, 2783.
28 Brannock, K. C. *J. Am. Chem. Soc.* **1959**, *81*, 3379.
29 Faulkner, D. J.; Petersen, M. R. *Tetrahedron Lett.* **1969**, *10*, 3243.
30 Ohi, K.; Shima, K.; Hamada, K.; Saito, Y.; Yamada, N.; Ohba, S.; Nishiyama, S. *Bull. Chem. Soc. Jpn.* **1998**, *71*, 2433.
31 Suzuki, T.; Sato, E.; Unno, K.; Kametani, T. *J. Chem. Soc., Chem. Commun.* **1988**, 724.
32 Hiersemann, M. *Synlett* **1999**, 1823.
33 Nakahara, Y.; Shimizu, M.; Yoshioka, H. *Tetrahedron Lett.* **1988**, *29*, 2325.
34 Sleeman, M. J.; Meehan, G. V. *Tetrahedron Lett.* **1989**, *30*, 3345.
35 Paquette, L. A.; Ladouceur, G. *J. Org. Chem.* **1989**, *54*, 4278.
36 Gajewski, J. J. *Acc. Chem. Res.* **1997**, *30*, 219 (and references cited therein).
37 Lindstroem, U. M. *Chem. Rev.* **2002**, *102*, 2751.
38 Lubineau, A.; Auge, J.; Bellanger, N.; Caillebourdin, S. *Tetrahedron Lett.* **1990**, *31*, 4117.
39 Lubineau, A.; Queneau, Y. *Tetrahedorn Lett.* **1985**, *26*, 2653.
40 Lubineau, A.; Queneau, Y. *Tetrahedorn* **1989**, *45*, 6697.
41 Cookson, R. C.; Gopalan, R. *J. Chem. Soc., Chem. Commun.* **1978**, 608.
42 Fernández de la Pradilla, R.; Montero, C.; Tortosa, M. *Org. Lett.* **2002**, *4*, 2373.
43 Paquette, L. A.; Kang, H. *J. Am. Chem. Soc.* **1991**, *113*, 2610.
44 Borrelly, S.; Paquette, L. A. *J. Am. Chem. Soc.* **1996**, *118*, 727.
45 Bernardelli, P.; Moradei, O. M.; Friedrich, D.; Yang, J.; Gallou, F.; Dyck, B. P.; Doskotch, R. W.; Lange, T.; Paquette, L. A. *J. Am. Chem. Soc.* **2001**, *123*, 9021.
46 Hikota, M.; Sakurai, Y.; Horita, K.; Yonemitsu, O. *Tetrahedron Lett.* **1990**, *31*, 6367.
47 Hikota, M.; Tone, H.; Horita, K.; Yonemitsu, O. *Tetrahedron* **1990**, *46*, 4613.
48 Lvoisier, T.; Charonnet, E.; Rodriguez, J. *Synthesis* **1997**, 1258.
49 Ezquerra, J.; He, W.; Paquette, L. A. *Tetrahedron Lett.* **1990**, *31*, 6979.
50 Petasis, N. A.; Patane, M. A. *Tetrahedron Lett.* **1990**, *31*, 6799.
51 Sugiyama, J.; Tanikawa, K.; Okada, T.; Noguchi, K.; Ueda, M.; Endo, T. *Tetrahedron Lett.* **1994**, *35*, 3111.
52 Sudha, A. V. R. L.; Nagarajan, M. *Chem. Commun.* **1998**, 925.
53 Kim, S.; Ko, H.; Kim, E.; Kim, D. *Org. Lett.* **2002**, *4*, 1343.
54 Büchi,G.; Powell, J. E. *J. Am. Chem. Soc.* **1970**, *92*, 3126.
55 Reville, T.; Gobbi, L. B.; Gray, B. L.; Ley, S. V.; Scott, J. S. *Org. Lett.* **2002**, *4*, 3847.
56 Ley, S. V.; Gutteridge, C. E.; Pape, A. R.; Spilling, C. D.; Zumbrunn, C. *Synlett* **1999**, 1295.
57 Paquette, L. A.; Wang, T.; Philippo, C. M. G.; Wang, S. *J. Am. Chem. Soc.* **1994**, *116*, 3367.
58 Zhang, L. H.; Cook, J. M. *J. Am. Chem. Soc.* **1990**, *112*, 4088.
59 Tokuyama, H.; Makido, T.; Ueda, T.; Fukuyama, T. *Synth. Commun.* **2002**, *32*, 869.
60 Hatcher, M. A.; Posner, G. H. *Tetrahedron Lett.* **2002**, *43*, 5009.
61 Srikrishna, A.; Dinesh, C.; Anebouselvy, K. *Tetrahedron Lett.* **1999**, *40*, 1031.
62 Hagiwara, H.; Sakai, H.; Uchiyama, T.; Ito, Y.; Morita, N.; Hoshi, T.; Suzuki, T.; Ando, M. *J. Chem. Soc., Perkin Trans. 1* **2002**, 583.
63 Tsunoda, T.; Amaike, M.; Tambunan, U. S. F.; Fujise, Y.; Ito, S.; Kodama, M. *Tetrahedron Lett.* **1987**, *28*, 2537.
64 Ovaska, T. V.; Roark, J. L.; Shoemaker, C. M. *Tetrahedron Lett.* **1998**, *39*, 5705.

65 Ovaska,T. V.; Roses, J. B. *Org. Lett.* **2000**, *2*, 2361.
66 Ovaska, T. V.; Reisman, S. E.; Flynn, M. A. *Org. Lett.* **2001**, *3*, 115.
67 Ho, T.; Chein, R. *Chem. Commun.* **1996**, 1147.
68 Mikami, K.; Takahashi, K.; Nakai, T. *J. Am. Chem. Soc.* **1990**, *112*, 4035.
69 Schobert, R.; Siegfried, S.; Gordon, G.; Nieuwenhuyzen, M.; Allenmark, S. *Eur. J. Org. Chem.* **2001**, 1951.
70 Schobert, R.; Siegfried, S.; Gordon, G.; Mulholland, D.; Nieuwenhuyzen, M. *Tetrahedron Lett.* **2001**, *42*, 4561.
71 Barriault, L.; Denissova, I. *Org. Lett.* **2002**, *4*, 1371.
72 Denmark, S. E.; Harmata, M. A. *J. Org. Chem.* **1983**, *48*, 3369.
73 Denmark, S. E.; Harmata, M. A. *Tetrahedron Lett.* **1984**, *25*, 1543.
74 Denmark, S. E.; Marlin, J. E. *J. Org. Chem.* **1987**, *52*, 5742.
75 Denmark, S. E.; Rajendra, G.; Marlin, J. E. *Tetrahedron Lett.* **1989**, *30*, 2469.
76 Denmark, S. E.; Harmata, M. A.; White, K. S. *J. Am. Chem. Soc.* **1989**, *111*, 8878.
77 Denmark, S. E.; Marlin, J. E. *J. Org. Chem.* **1991**, *56*, 1003.
78 Denmark, S. E.; Stadler, H.; Dorow, R. L.; Kim, J. *J. Org. Chem.* **1991**, *56*, 5063.
79 Gajewski, J. J.; Gilbert, K. E. *J. Org. Chem.* **1984**, *49*, 11.
80 Carpenter, B. K. *Tetrahedron* **1978**, *34*, 1877.

3.2 Aromatic Claisen Rearrangement

Hisanaka Ito and Takeo Taguchi

3.2.1 Introduction

The [3,3]-sigmatropic rearrangement of aryl allyl ethers is called the aromatic Claisen rearrangement [1]. The thermal [3,3]-sigmatropic rearrangement of allyl vinyl ethers and allyl aryl ethers was first described by Claisen in 1912 [2]. Following the mechanistic proposal of this rearrangement by Claisen [3], a variety of factors related to this reaction have been extensively examined using numerous substrates and this reaction has been developed as a very attractive tool for the preparation of *ortho*-substituted phenol derivatives. Synthetic utilization of this reaction can also be found in a further elaboration of *ortho*-allylated products to poly-functionalized phenol derivatives, which are important intermediates for the synthesis of various functional molecules and biologically active compounds. Normally, the aromatic Claisen rearrangement requires high reaction temperature (180–225 °C), which often leads to the formation of several byproducts through undesired side reactions. There are two major side reactions in the aromatic Claisen rearrangement: one is the rearrangement of the allylic moiety at the *para* position through the subsequent Cope rearrangement of the dienone intermediate and the other is the abnormal Claisen rearrangement. In the former case, the use of a *para*-substituted substrate prevents the Cope rearrangement and thus affords the desired *ortho*-substituted phenol derivatives. A number of catalytic conditions have also been developed to obtain the desired *ortho*-rearranged product under mild conditions. In this section, the reaction mechanism and reaction conditions of the aromatic Claisen rearrangement and its application to the synthesis of functional molecules and biologically active compounds are discussed.

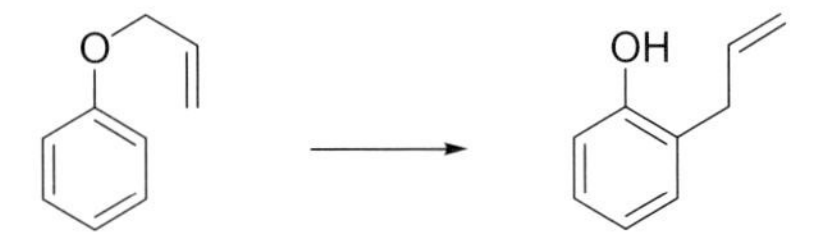

Scheme 3.2.1

3.2.2 Mechanism

3.2.2.1 *Ortho* and *Para* Rearrangement

Continuous efforts for the elucidation of reaction mechanism of the aromatic Claisen rearrangement have been made since the reaction was discovered [4]. The reaction is considered to be a concerted pericyclic pathway. The first step is the [3,3]-sigmatropic rearrangement (Claisen rearrangement) resulting in the bond formation between γ-position of the allyl moiety and the *ortho* position of the aromatic ring to generate *ortho* dienone intermediate **2**. When the *ortho* carbon bearing the allyl group does not have an additional substituent, facile rearomatization of the dienone intermediate due to a rapid enolization results in the formation of *ortho* allyl phenol **3** as a product. If the *ortho* carbon of the dienone intermediate has a substituent, subsequent [3,3]-sigmatropic rearrangement (Cope rearrangement) takes place leading to *para* allylated phenol derivative **5**, in which the α-carbon of the allyl group in the allyl aryl ether attaches to the aromatic ring. In some cases, even without a substituent on the *ortho* position, rearrangement to *para* position competitively occurs. Therefore, existence of a substituent at the *para* position is important to prevent the production of a *para*-rearranged product. Recently, the importance of the *para*-rearrangement product has been increasing for the preparation of functional molecules containing polyaromatic rings such as calixarene and related derivatives.

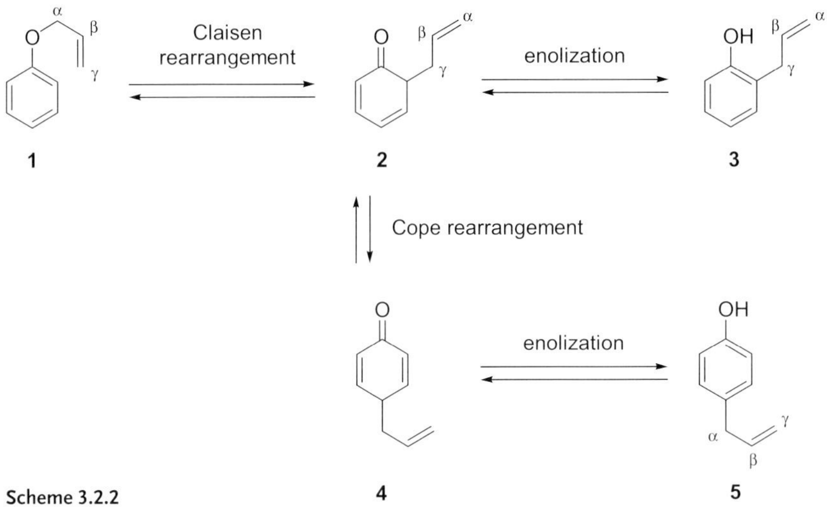

Scheme 3.2.2

3.2.2.2 **Transition State**

Several approaches to elucidate the transition state of the Claisen rearrangement have been discussed. In the case of the aliphatic Claisen rearrangement, the stereochemical outcome of a chiral substrate provides important information suggesting that the chair-like transition state is favorable [5]. In the case of ally aryl ether **6** having substituents at both α- and γ-position of allyl moiety, one of the newly formed two chiral centers in the *ortho* dienone intermediate changes to nonchiral carbon by the enolization leading to more stable aromatic structure **8**. Although few reports deal with the intramolecular chirality transfer based on the absolute stereochemistry of the newly formed chiral center and the geometry of the carbon–carbon double bond, it is suggested that the aromatic Claisen rear-

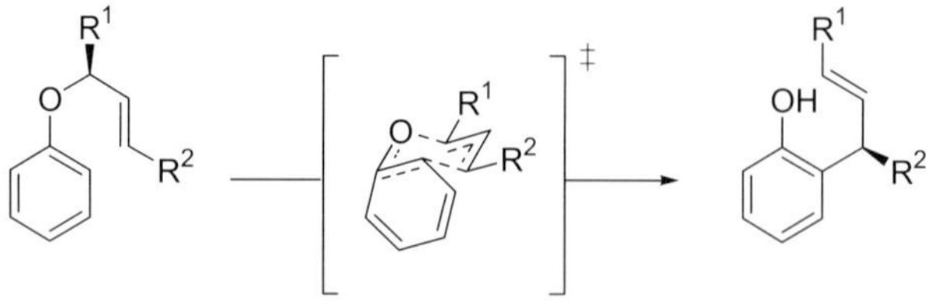

6 7 8

Scheme 3.2.3

rangement also proceeds through a six-membered ring chair-like transition state **7** as shown below [6].

The kinetic isotope effect and theoretical calculation of transition state model of the aromatic Claisen rearrangement were also reported. For example, Singleton et al. investigated the nature of the transition state of the Claisen rearrangement by a combined experimental and theoretical study [7]. It was shown that the experimental isotope effects coincide well with the predicted kinetic isotope effects. Detail of the theoretical studies is discussed in Chapter 11.

3.2.2.3 **Abnormal Claisen Rearrangement**

One of the most common side reactions is the so-called abnormal Claisen rearrangement observed with the allyl aryl ether substrate, having alkyl substituents at α- or γ- position on the allyl moiety [8]. The abnormal Claisen rearrangement was first reported by Lauer and Filbert in 1936 [9] and soon after Hund et al. also

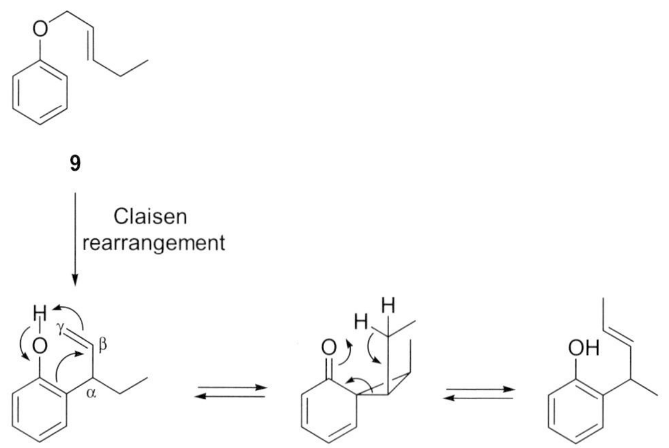

10 normal product

11

12 abnormal product

Scheme 3.2.4

observed the production of the abnormal product under thermal conditions [10]. The mechanism of the abnormal Claisen rearrangement was proposed by Marvell et al. [11] in which phenolic hydroxyl group and an α-alkyl substituent on the *ortho* allylic moiety in the normally rearranged product **10** participate in this reaction. Thus, the reaction pathway involves intramolecular hydrogen transfer from the phenolic hydroxyl group to the γ-position of allylic substituent through the formation of spiro[2,5]octadienone derivative **11** followed by cleavage of the originally existing cyclopropane bond leading to reproduction of phenolic compound **12**. This was verified by labeling experiments [12].

3.2.3
Substrate and Substituent Effect

3.2.3.1 Preparation of Substrate

The most common preparative method to prepare the aryl allyl ether is the Williamson's ether synthesis [1a,b]. Typically, aryl allyl ethers can be obtained from phenol derivatives and allylic halide under basic conditions (K_2CO_3) in refluxing acetone. This method is convenient for the preparation of simple allyl aryl ethers. However, some side reactions such as a competitive *C*-allylation (S_N2' type reaction) often accompany the formation of undesired byproducts. Mitsunobu reaction of phenol derivatives with allylic alcohols instead of allylic halides can be used under mild conditions [13]. In particular, when the allyl halide is unstable, this procedure is effective instead of the Williamson's ether synthesis. This method is also useful for the preparation of chiral allyl aryl ether from chiral allylic alcohol with inversion at the chiral center. Palladium catalyzed *O*-allylation of phenols is also applicable, but sometimes a lack of site-selectivity with unsymmetrical allylic carbonate [14] may be a problematic issue.

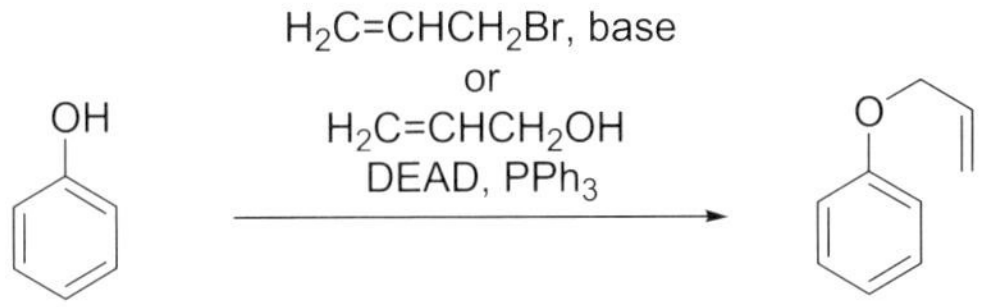

OH

$H_2C{=}CHCH_2Br$, base
or
$H_2C{=}CHCH_2OH$
DEAD, PPh_3

O

Scheme 3.2.5

3.2.3.2 Aryl Unit

The electronic nature of the substituent on the aryl unit actually brings about influences to the rate and the direction of the rearrangement, but it is a minor effect. Other experimental conditions (solvent, reaction temperature, and reaction time) are more important for determining the rate and the isomer ratio (*ortho* vs. *para*) of the rearranged product. In some cases these observations are interpreted in terms of possible polar mechanisms for the aromatic Claisen rearrangement [15].

The effect of the *meta* substituent on the direction of the rearrangement was also reported [16]. Although the regioselectivity of the rearrangement is not so high, it roughly depends on the electronic nature of the substituent. Thus, rearrangement to the 6 position is favorable with the substrate having an electron-donating group (X = OMe), while electron-withdrawing groups (X = COPh, CN) promotes the rearrangement to the hindered 2 position. The theoretical study of the regioselectivity with *meta*-substituted aryl ether was also reported [17].

160 °C

X =	
OMe	31 : 69
CH_3	62 : 38
Br	71 : 29
Cl	66 : 34
COPh	79 : 21
CN	69 : 31

Scheme 3.2.6

A number of Claisen rearrangements of hetero-aromatic compounds have been reported, which are also an important means for the preparation of substituted heterocycles [18]. Under thermal conditions, the reaction of 2-allyloxypyridine **13** gave a mixture of *N*-allylated and *C*-allylated products **14, 15** [19]. A transition metal catalyst is effective for the selective formation of *N*-allylated product **14** under mild conditions [20].

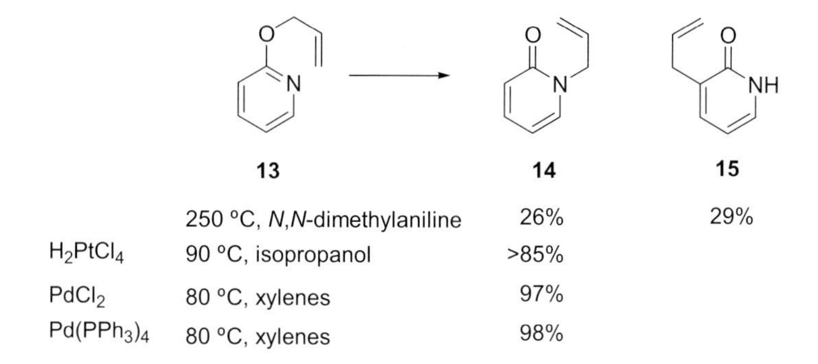

Scheme 3.2.7

3.2.3.3 Allyl and Propargyl Unit

In general, the substituent on the allyl moiety exerts a slight effect on the reaction rate [21]. Goering et al. reported the relative rates for the *ortho* rearrangement of allyl and methylated allyl phenyl ethers [15b] in diphenyl ether. Except for the α-methylated substrate **16**, which markedly enhances the rate, the relative rates of β- and γ-methylated substrates **17, 18** are similar to that of the parent allyl compound **1**.

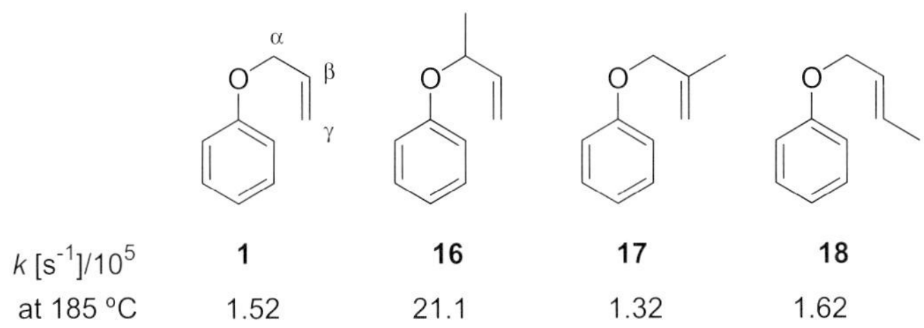

k [s^{-1}]/10^5 at 185 °C	1	16	17	18
	1.52	21.1	1.32	1.62

Scheme 3.2.8

The effect of geometry of the allyl group on the rate of Claisen rearrangement was examined by the use of both *E* and *Z* isomers of crotyl and cinnamyl derivatives [22]. The reaction was carried out around at 200 °C in diphenyl ether. The results indicated that the *E* isomer is slightly more reactive than the *Z* isomer.

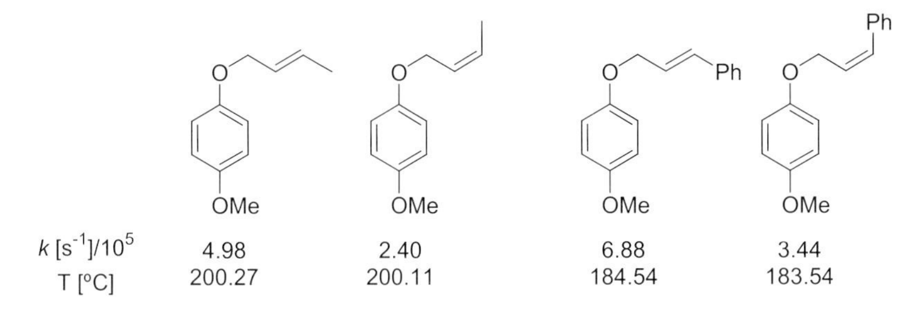

Scheme 3.2.9

Propargyl aryl ether is also an attractive substrate of the Claisen rearrangement for the construction of heterocycles such as chromene derivatives.

The substrate can be prepared by standard etherification reaction of a phenol derivative with a propargylic halide [1b]. Further carbon–carbon bond forming reactions of the terminal acetylenic carbon via metal acetylide provides a variety of substituted propargyl aryl ethers. Direct preparation using substituted propargyl ether is also possible.

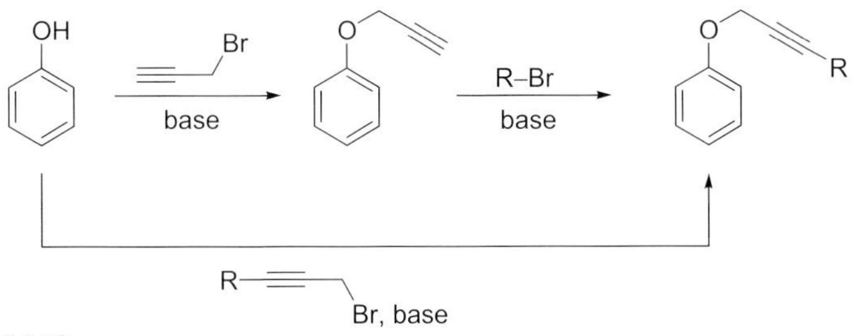

Scheme 3.2.10

Rearrangement of propargyl aryl ether **19** smoothly proceeds under thermal conditions to afford chromene derivatives **22** [23, 24]. The mechanism involves the formation of *ortho* allenyl phenol **20** followed by a 1,5-hydrogen shift and electrocyclic ring closure sequence via **21**. For example, the Claisen rearrangement of propargyl aryl ether **23** prepared by the Mitsunobu reaction smoothly took place at 180 °C to give a flav-3-ene derivative **24** in excellent yield [25].

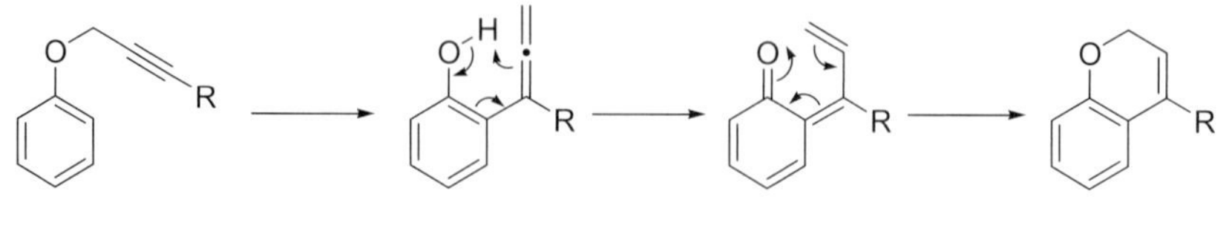

Scheme 3.2.11

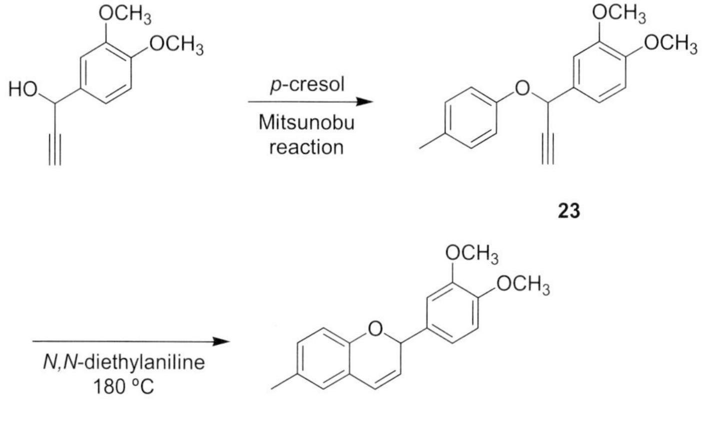

Scheme 3.2.12

3.2.4 Reaction Conditions

The Claisen rearrangement is generally considered a concerted process. Although a typical pericyclic reaction, such as the Cope rearrangement, is usually insensitive to solvent and polarity of the medium, the rate of the Claisen rearrangement is clearly influenced by these factors. In this section, the effects of these factors are discussed.

3.2.4.1 **Thermal Conditions**

As mentioned above, the aromatic Claisen rearrangement of allyl aryl ether requires high temperature. The range of reaction temperature is normally 180–225 °C. Under such somewhat drastic conditions, undesired side reactions often occur competitively. Furthermore, in several cases, achievement of high regio- and stereoselectivity is also very difficult. To prevent the production of undesired product and to increase the selectivity, reaction conditions have been well examined about the solvent and catalyst.

3.2.4.2 **Solvent Effect**

The aromatic Claisen rearrangement proceeds at around 200 °C. Although the reaction is the first order in diphenyl ether as the solvent, it was observed that the reaction rate increases with the production of phenol derivatives [26]. Independently, addition of a phenolic compound results in the enhancement of the reaction rate. These observations indicate that a phenol derivative formed since the reaction product works as an autocatalyst.

Although solvent effect on the reaction rate had already been shown by White [15a] and Goering [15b] in 1958, later (in 1970) White reported the rate constants of the rearrangement of allyl *p*-tolyl ether at 170 °C using 17 solvents of different polarity [27]. The selected results are shown below. In protic solvent, the reaction is faster than that in aprotic solvent and the reaction in *p*-chlorophenol proceeds 300 times faster than that under neat conditions.

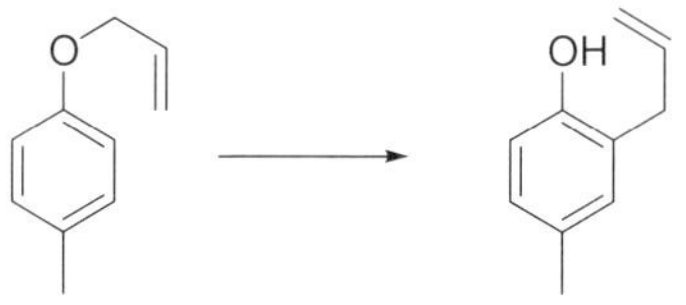

solvent	rate constant k [s^{-1}]/10^6 at 170 °C
none (gas phase)	1.01
n-butyl ether	3.27
n-desylamine	4.52
2-octanol	9.65
1-decanol	11.1
octanoic acid	24.2
2-aminoethanol	27.1
ethylene glycol	67.3
p-cresol	73.1
phenol	103
28.5% ethanol–water	107
p-chlorophenol	303

Scheme 3.2.13

3.2.4.3 Brønsted Acid Catalyst

As mentioned in the solvent effect section, protic solvents accelerate the Claisen rearrangement. In particular the reaction rate enhances by increasing the acidity of the protic solvent used and trifluoroacetic acid (pK_a=0.2) is one of the most effective catalysts. It was reported that in trifluoroacetic acid the Claisen rearrangement of allyl phenyl ether proceeds at room temperature to afford the *ortho*-rearrangement product **3** and 2-methylcoumaran **25** along with a small amount of phenol and unreacted starting ether [28, 29].

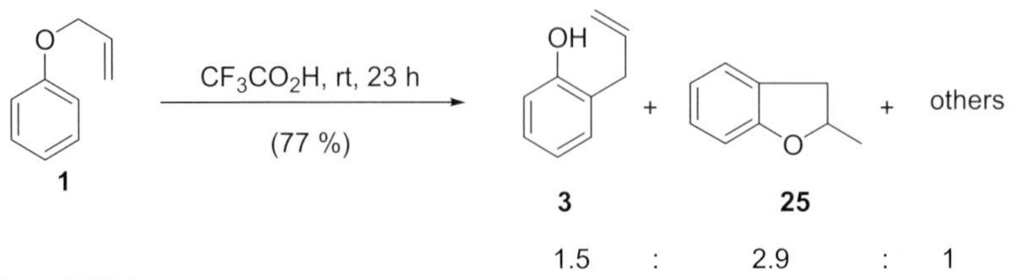

CF_3CO_2H, rt, 23 h (77 %)

1 → **3** + **25** + others

1.5 : 2.9 : 1

Scheme 3.2.14

3.2.4.4 Lewis Acid Catalyst

The first example of Lewis-acid catalyzed aromatic Claisen rearrangement was reported in 1941 [30], that is, the BF_3–AcOH-catalyzed reaction, which proceeded under relatively mild conditions (80 °C). Since this discovery of acceleration by Lewis acid, numerous examples such as BCl_3 [31], BF_3 [32], Et_2AlCl [33], $ZnCl_2$ [34], and $TiCl_4$ [35] have been reported. For details regarding catalysts for the Claisen rearrangement an excellent review article by Lutz [1h] is recommended.

Usually, more than a stoichiometric amount of these catalysts is required for the reaction to efficiently proceed. Among these Lewis acids, diethylaluminum chloride is an excellent catalyst and 2 molar equivalents of diethylaluminum chloride efficiently accelerated the rearrangement of allyl phenyl ether **1** at room temperature to give the *ortho*-rearranged product **3** in 93% yield [33].

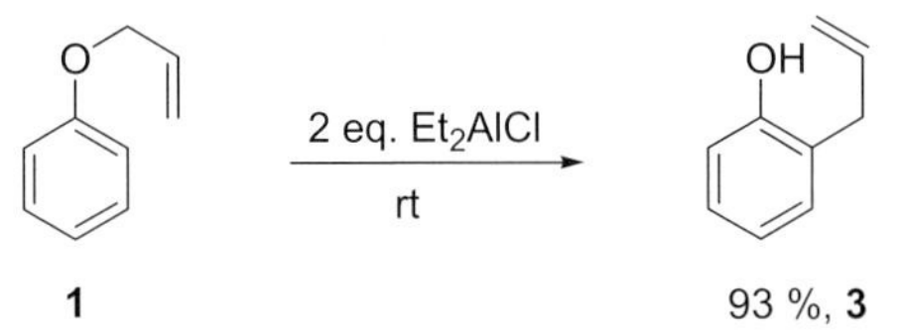

2 eq. Et_2AlCl, rt

1 → 93 %, **3**

Scheme 3.2.15

Although by using these catalysts the Claisen rearrangement proceeds at relatively low temperature within a short period of time, a complex mixture of several products often results due to the charge-accelerated mechanism along with various competing reactions. The Trost group developed the practical aromatic Claisen rearrangement catalyzed by 10 mol% $Eu(fod)_3$ and they demonstrated a complete intramolecular chirality transfer with an optically pure substrate (*vide infra*) [64].

Recently, Wipf et al. reported that the aromatic Claisen rearrangement is accelerated in the presence of trimethylaluminum and water or alumoxanes [36]. Although 4 molar equivalents of trimethylaluminum and the addition of one molar equivalent of water or MAO were required, the reaction smoothly proceeded under mild conditions to give *ortho* allyl phenol in excellent yield. They proposed that the addition of water provides a transient strong Lewis acid, which activates the substrate by complexation to the ethereal oxygen atom.

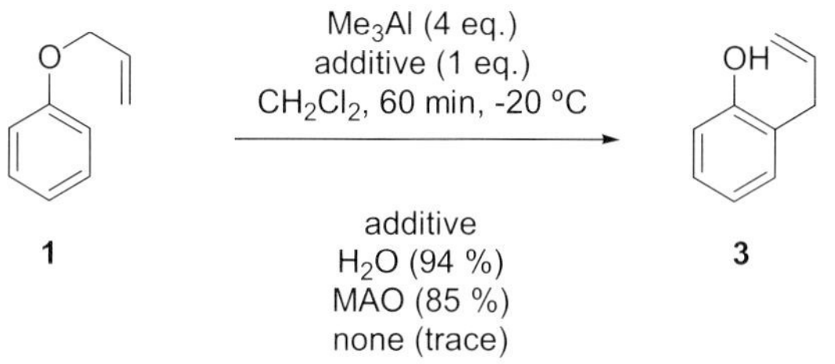

Me_3Al (4 eq.)
additive (1 eq.)
CH_2Cl_2, 60 min, -20 °C

1 → 3

additive
H_2O (94 %)
MAO (85 %)
none (trace)

Scheme 3.2.16

They also examined *ortho* and *para* selectivity by the use of *ortho*-mono-substituted allyl phenyl ether **26** [36]. Under thermal conditions, the *ortho*-rearranged product **27** was predominantly obtained. On the other hand, an increase in the ratio of the *para*-rearranged product **28** was observed in the presence of an aluminum–water catalyst system.

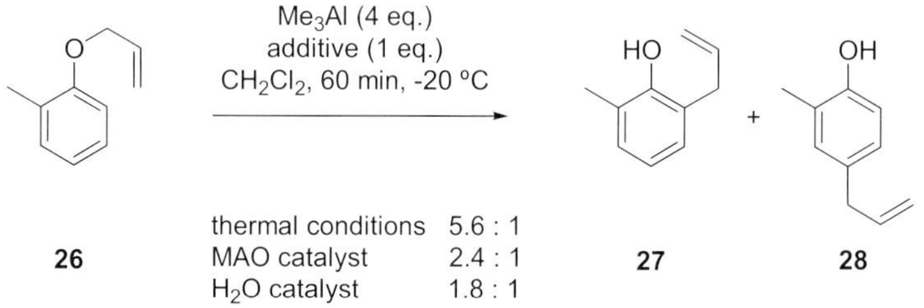

Me_3Al (4 eq.)
additive (1 eq.)
CH_2Cl_2, 60 min, -20 °C

26 → 27 + 28

thermal conditions 5.6 : 1
MAO catalyst 2.4 : 1
H_2O catalyst 1.8 : 1

Scheme 3.2.17

Yamamoto et al. reported unusual behavior for the aromatic Claisen rearrangement by the use of bulky aluminum catalyst [37]. Under the thermal conditions, the allyl 2,4-dimethylphenyl derivative **29** exclusively afforded the *ortho*-rearrangement product **30** in high yield, because one of the *ortho* positions and the *para* position of the substrate are occupied by methyl substituents. On the other hand, treatment of this substrate **29** with a bulky aluminum catalyst **35** gave the *para*-rearranged cyclohexadienone **34**. This result can be explained by considering the coordination of the aluminum catalyst to the allyl aryl ether to form complex **32**, which is sterically more favorable than complex **31**. Thus, the initial Claisen rearrangement occurs at the 2 position to give the geminally disubstituted intermediate **33**, which subsequently converts to the *para*-rearranged product **34** through the Cope rearrangement.

heat

29

30

$(ArO)_2AlCH_3$ (**35**)

31

32

Claisen

33

Cope

34

35
bulky aluminum catalyst

Scheme 3.2.18

3.2.4.5 **Base Catalyst**

A significant catalytic effect using a base catalyst was reported for the reaction of catechol mono-allyl ethers [38]. The rearrangement of catechol mono-allyl ether **36** under thermal conditions (160–200 °C) proceeded to give a mixture of two isomers, *ortho*- and *para*-rearranged products **37** and **38**. The reaction was accelerated by the addition of an equimolar amount of NaOEt proceeding at 78 °C. The isomer ratio of the products was different from that obtained under thermal conditions. The authors proposed a different mechanism for the base-promoted reaction, which involves the [2,3]-Wittig rearrangement and the following Cope rearrangement.

Katsuki et al. reported the Claisen rearrangement of aryl cinnamyl ether **39** for the preparation of a chiral salen complex [39]. They noted that the addition of an equimolar amount of $CaCO_3$ is effective to minimize the *para* rearrangement.

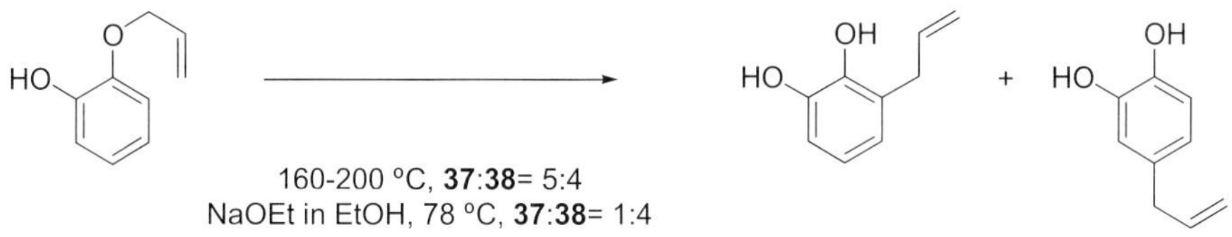

160-200 °C, **37**:**38**= 5:4
NaOEt in EtOH, 78 °C, **37**:**38**= 1:4

36 **37** **38**

Scheme 3.2.19

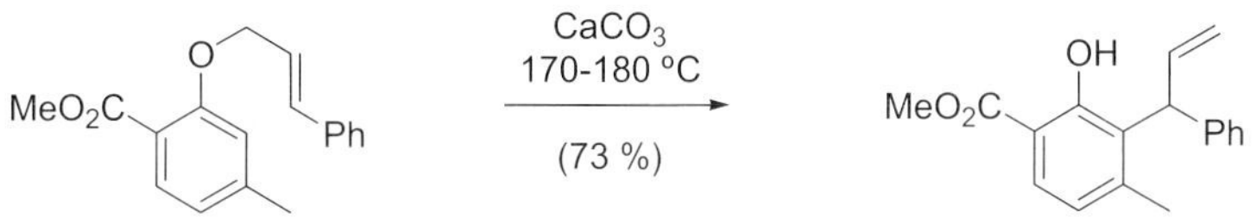

$CaCO_3$
170-180 °C
(73 %)

39 **40**

Scheme 3.2.20

3.2.4.6 **Transition Metal Catalyst**

Transition-metal-catalyzed processes often work nicely for the aromatic Claisen rearrangement. Mechanistic aspects might be different from those under usual thermal conditions. Platinum complexes catalyzed the reaction of allyl 2-naphthyl ether **41** to afford 1-allyl-2-naphthol **42** regioselectively in excellent yield [40]. Molybdenum hexacarbonyl also catalyzed the one-pot conversion of allyl aryl ethers to coumaran derivatives such as **43** from **17** [41].

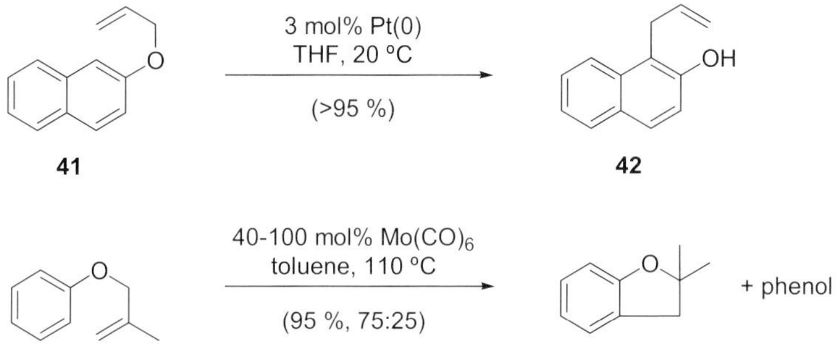

3 mol% Pt(0)
THF, 20 °C
(>95 %)

41 **42**

40-100 mol% $Mo(CO)_6$
toluene, 110 °C
(95 %, 75:25)
+ phenol

17 **43**

Scheme 3.2.21

3.2.4.7 **Other Conditions**

To prevent the formation of abnormal Claisen product, Kishi et al. demonstrated trapping the resulting phenolic hydroxyl group by acetylation by carrying out the reaction in acetic anhydride in the presence of potassium or sodium acetate under thermal conditions [42]. For example, 3-methyl-2-butenyl 1-naphthyl ether **44** was converted to the acetate form of the normal Claisen-rearrangement product **45** in 76% yield. On the other hand, by switching to butyric anhydride as a trapping agent [43, 44] the abnormal Claisen product **46** along with the *para*-rearranged product **47** was obtained in good yield.

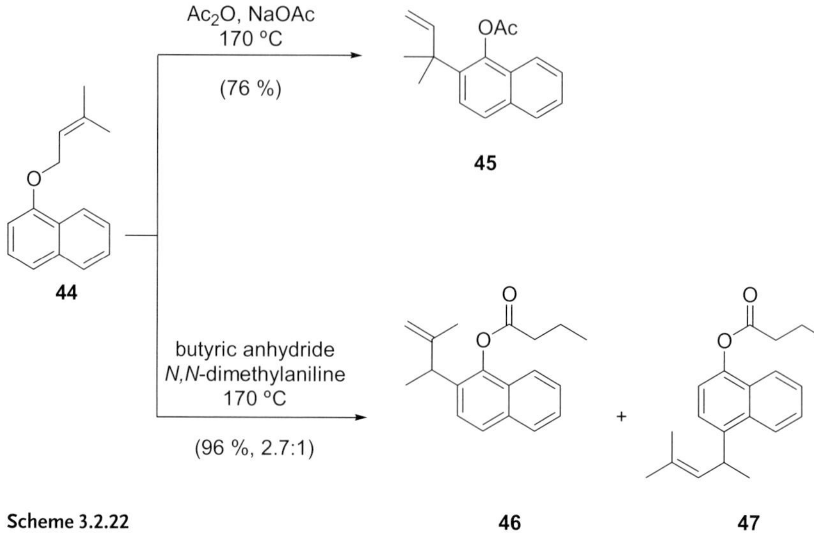

Ac_2O, NaOAc
170 °C
(76 %)

OAc

45

44

butyric anhydride
N,N-dimethylaniline
170 °C
(96 %, 2.7:1)

46 + **47**

Scheme 3.2.22

An alternative efficient method to prevent the abnormal Claisen rearrangement by using a silylating agent such as 1,1,1,3,3,3-hexamethyldisilazane or *N,O*-bis-(trimethylsilyl)acetamide to trap the normal Claisen-rearrangement product has also been reported [45].

Acceleration of the aromatic Claisen rearrangement under microwave irradiation conditions [46] and the photo-Claisen rearrangement has also been reported [47].

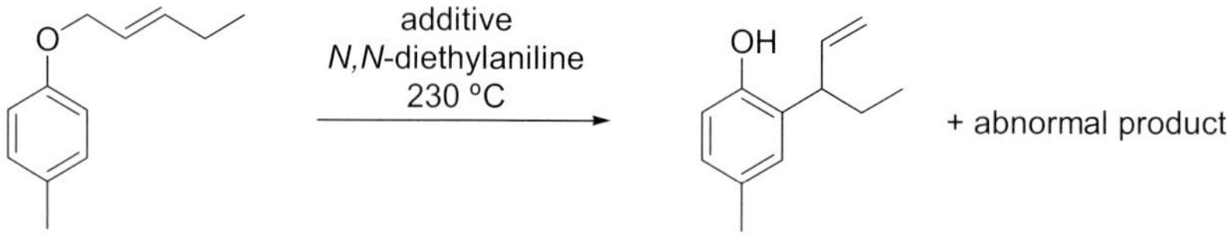

additive
N,N-diethylaniline
230 °C

OH

+ abnormal product

additive	yield	normal : abnormal
none	63 %	58 : 42
10 eq. 1,1,1,3,3,3-hexamethyldisilazane	70 %	>99 : <1
5 eq. *N,O*-bis(trimethylsilyl)acetamide	81 %	>99 : <1

Scheme 3.2.23

3.2.5
Thio-, Amino-, and Related Claisen Rearrangement

The thio-Claisen rearrangement, a sulfur analog of the simple Claisen rearrangement was first reported in 1930 by Hurd et al. [48, 49]. In the 1960s, details of this reaction, especially using simple allyl phenyl sulfide, were examined by Kwart et al. They reported that the thermal reaction of allyl phenyl sulfide **48** gave a mixture of a thiacoumaran derivative **50** along with 1-propenyl phenyl sulfide **51** as a byproduct [50]. In the case of the thio-Claisen rearrangement, the migration of carbon–carbon double bond gives rise to vinyl sulfide derivatives such as **51** and causes a decrease in the yield of the rearrangement product under thermal conditions. The *ortho* allylated thiophenol derivatives, e.g., **49**, more easily cyclizes than the similar phenol derivatives.

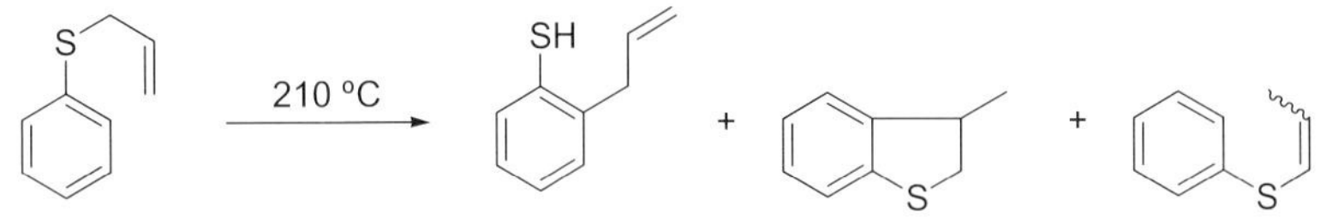

Scheme 3.2.24

The thermal rearrangement of allyl phenyl sulfide **48** in quinoline at 230 °C gave a 1:1 mixture of 2-methylthiacoumaran **50** and thiachroman **52** [51, 52].

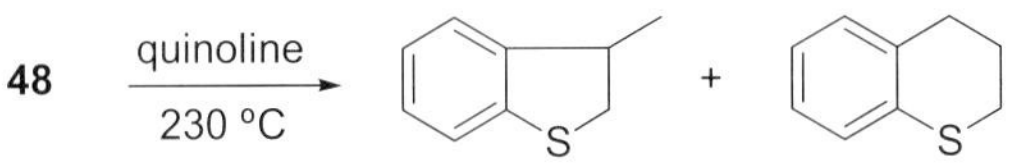

Scheme 3.2.25

A similar tendency was also found in the thermal reaction of propargyl phenyl sulfide. Thus, in the presence of cyclopentadiene on heating propargyl phenyl sulfide **53** in quinoline at 200 °C, thiachromen **56** and the Diels–Alder adduct **55** formed from allenyl sulfide **54** and cyclopentadiene were obtained [53]. In the absence of cyclopentadiene, a mixture of thiachromen **56** and thiacoumaren **57** was obtained.

Heteroaromatic allyl sulfide is also a good substrate for the thio-Claisen rearrangement. One example is shown below [54].

The nitrogen analog of the simple Claisen rearrangement with *N*-allylaniline and related derivatives is called the amino-Claisen rearrangement (or aza-Claisen rearrangement) [55]. Under thermal conditions, more drastic conditions (310–340 °C) compared with that for the oxygen substrates is required to obtain the *ortho* allyl aniline **59** [56]. *N*-Allyl aniline derivatives are, however, not stable at such

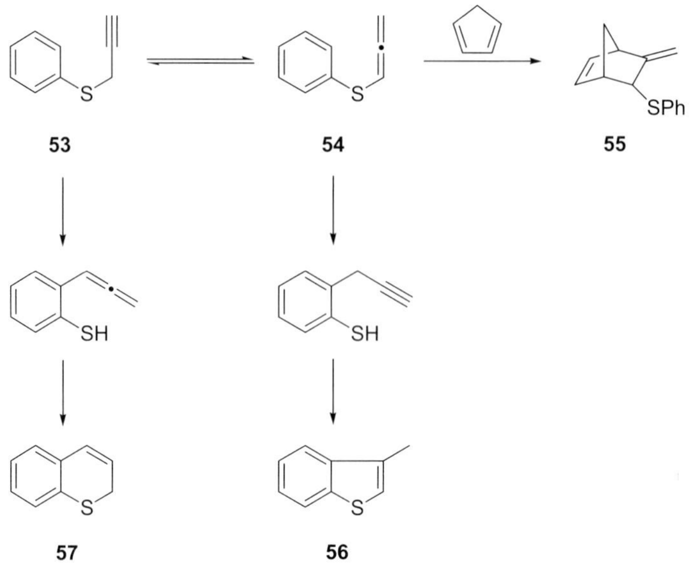

53 54 55 SPh SH SH 57 56

Scheme 3.2.26

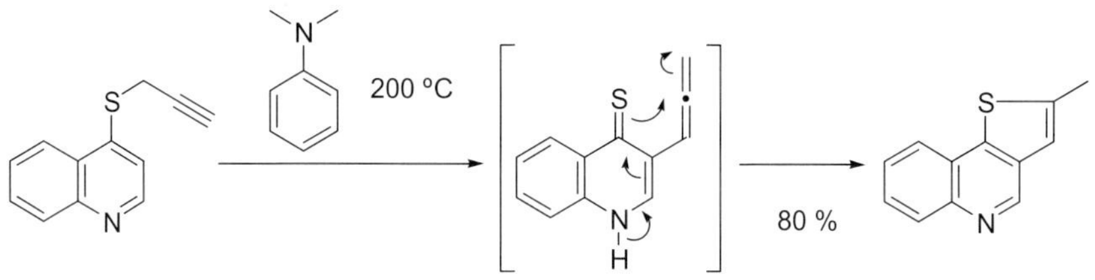

200 °C 80 %

Scheme 3.2.27

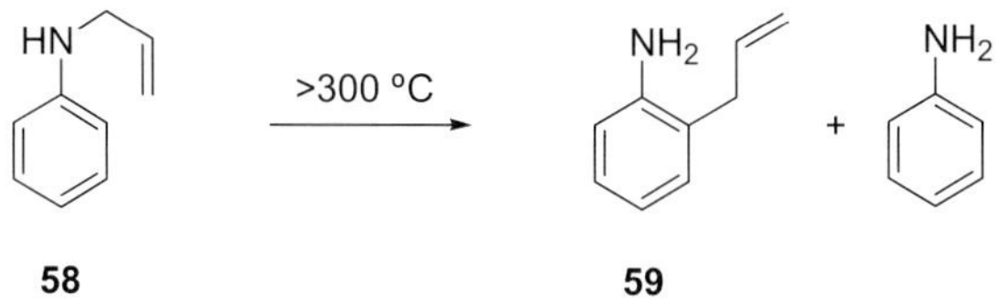

>300 °C 58 59

Scheme 3.2.28

a high temperature and a significant amount of deallylated aniline is formed by thermal decomposition. Thus, the thermal reaction of simple allyl aniline at 175 °C gave no rearrangement product, but just aniline [57].

The reaction is accelerated in the presence of a Brønsted acid or Lewis acid. For example, in the presence of 2 N sulfuric acid, the reaction of *N*-crotylaniline **60** proceeds at 60 °C to give the *ortho*-rearrangement product **61** in 70% yield along with a trace amount of aniline [56].

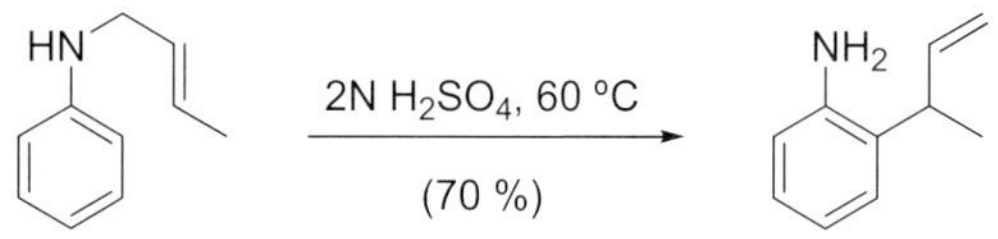

HN ... 2N H_2SO_4, 60 °C (70 %) → NH_2 ...

60 61

Scheme 3.2.29

Similarly, a Brønsted-acid-promoted rearrangement using 10 mol% *p*-toluene-sulfonic acid in acetonitrile/water (10:1) at 65 °C was also reported [58].

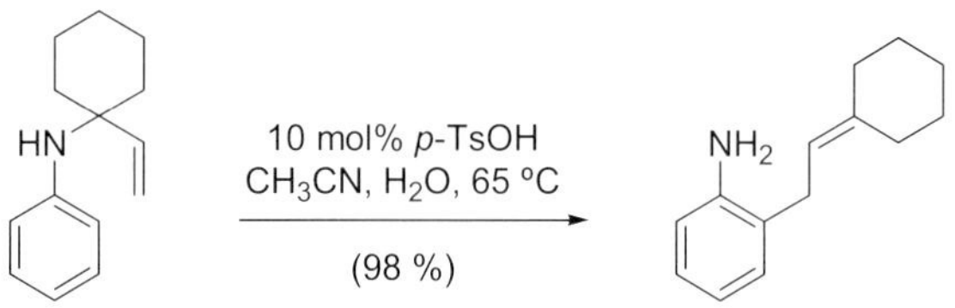

HN ... 10 mol% *p*-TsOH, CH_3CN, H_2O, 65 °C (98 %) → NH_2 ...

62 63

Scheme 3.2.30

A Lewis acid such as $BF_3 \cdot OEt_2$ also effects the amino-Claisen rearrangement as reported by Stille et al. [59].

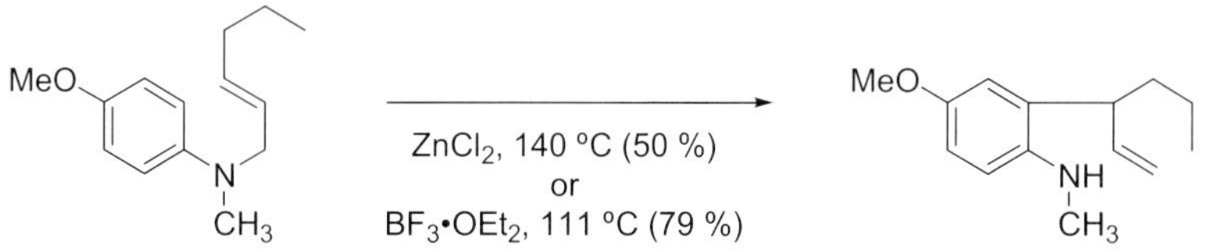

MeO ... N CH_3 → $ZnCl_2$, 140 °C (50 %) or $BF_3 \cdot OEt_2$, 111 °C (79 %) → MeO ... NH CH_3

64 65

Scheme 3.2.31

Scheiner reported a high yield synthesis of nitrogen-containing seven-membered ring **67** through the amino-Claisen rearrangement of a 2-vinylaziridine derivative **66** under relatively mild conditions [60].

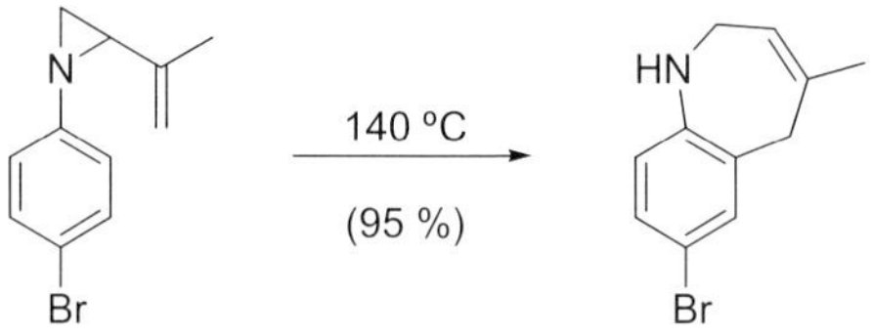

N ... Br → 140 °C (95 %) → HN ... Br

66 67

Scheme 3.2.32

It is possible to incorporate other heteroatoms instead of the oxygen atom. For example, an iodonio-Claisen rearrangement of an *in situ* generated allenyliodonio intermediate **68** was reported by Ochiai et al. [61].

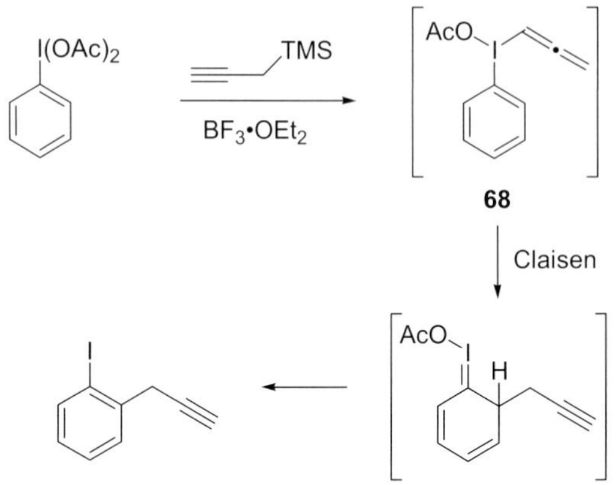

Scheme 3.2.33

3.2.6
Asymmetric Synthesis

3.2.6.1 Intramolecular Chirality Transfer

Goering et al. reported asymmetric aromatic Claisen rearrangement by intramolecular chirality transfer related to the study on elucidation of reaction mechanism [62]. The thermal rearrangement of the *E* isomer of (*R*)-but-3-en-2-yl phenyl ether **69** proceeded at 200 °C to give a mixture of *E* isomer **70** and *Z* isomer **71** in a ratio of 82:18, although the optical purity of each product was not shown.

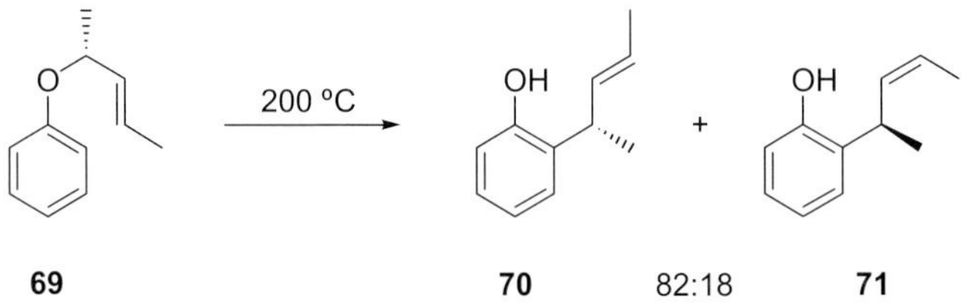

Scheme 3.2.34

A similar example of intramolecular chirality transfer rearrangement employing thermal conditions and a Lewis acid (BCl_3) as the catalysis conditions was also reported [31a].

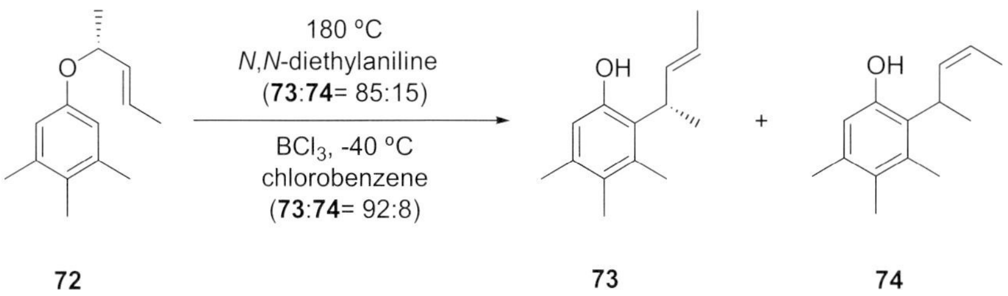

180 °C
N,*N*-diethylaniline
(**73**:**74**= 85:15)

BCl_3, -40 °C
chlorobenzene
(**73**:**74**= 92:8)

72 **73** **74**

Scheme 3.2.35

One of the successful examples of an intramolecular chirality transfer reaction was reported by Takano et al. [63], which is a key step for the total synthesis of (+)-latifine, isolated as the first example of tetrahydroisoquinoline alkaloid possessing a 4-phenyl-5,6-dioxygenated substitution pattern. The rearrangement of the chiral substrate **75** prepared from D-mannitol proceeded under the thermal conditions (*N*,*N*-dimethylaniline, reflux) to give the product **76** in 76% yield. Formation of the *E*-configuration was explained by the chair-like transition state **77**.

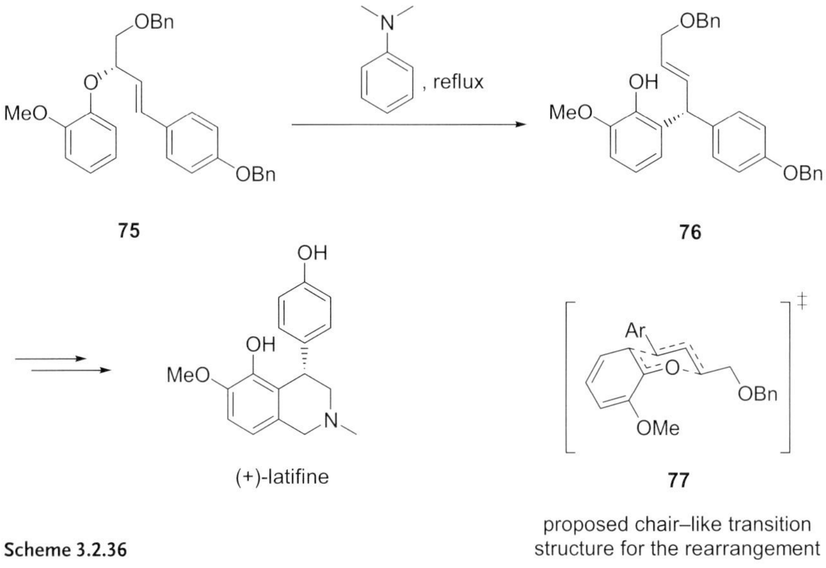

75 **76**

(+)-latifine **77**

proposed chair–like transition structure for the rearrangement

Scheme 3.2.36

Recently, some elegant work was reported on the preparation of chiral *ortho*-substituted phenol derivatives through intramolecular chirality transfer by Trost et al. [64]. Chiral substrate **78** was prepared in excellent enantiomeric excess from phenol and racemic allylic carbonate through asymmetric *O*-allylation with dynamic kinetic asymmetric transformation. They showed that a europium(III) tris(6,6,7,7,8,8,8-heptafluoro-2,2-dimethyl-3,5-octanedionate) $Eu(fod)_3$-catalyzed rearrangement proceeds at 50 °C to give product **79** with complete chirality transfer.

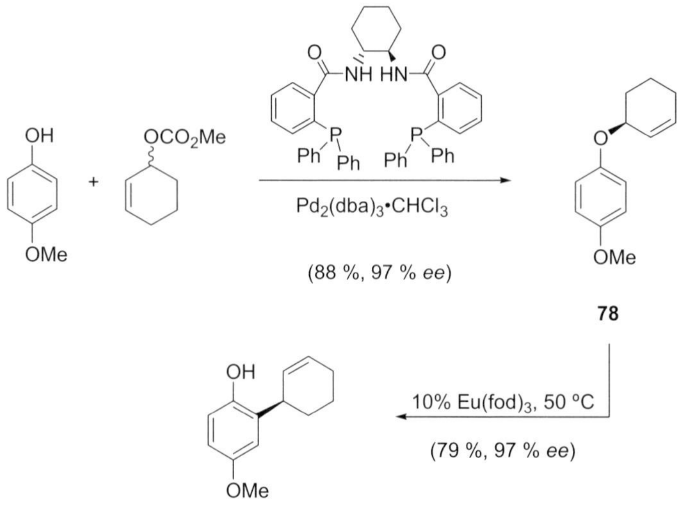

OH
OMe
OCO2Me
+
NH HN
O
O
P
P
Ph
Ph
Ph
Ph
Pd2(dba)3•CHCl3
(88 %, 97 % ee)
O
OMe
78
10% Eu(fod)3, 50 °C
(79 %, 97 % ee)
OH
OMe

Scheme 3.2.37 **79**

3.2.6.2 **Enantioselective Rearrangement**

Enantioselective aromatic Claisen rearrangement was reported from our group in 1997 [65]. The Claisen rearrangement of catechol mono allyl ether derivative **80** was catalyzed by chiral bis-sulfonamide-boron reagent **81**. Although a stoichiometric amount of chiral reagent is required for this reaction, this is only one successful example of the enantioselective aromatic Claisen rearrangement.

O
OH
Ph
Ph
Ts–N
N–Ts
B
Br
81, Et3N
(89 %, 94 % ee)
OH
OH

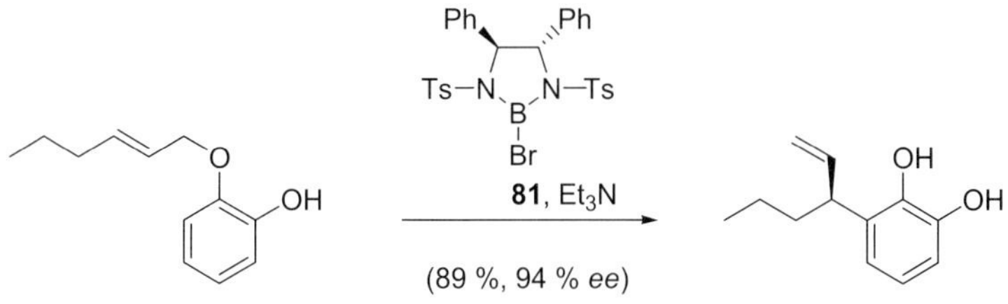

80 **82**

Scheme 3.2.38

3.2.7
Synthetic Applications

Aromatic Claisen rearrangement is an attractive means for the construction of some useful functional groups. In this section, applications of the aromatic Claisen rearrangement to the syntheses of complex molecules are described.

3.2.7.1 Consecutive Cyclization

Aromatic Claisen rearrangement is a powerful tool for the introduction of substituents into coumaran and chroman derivatives [66].

The *β*-methyl substituent on the allyl moiety in the substrate of the aromatic Claisen rearrangement accelerates the formation of the coumaran derivative, often observed as a byproduct arising from the initially formed *ortho* allyl phenol. This sequential cyclization is strongly dependent on the nature of the solvent used [67]. As shown in the reaction of methallyl phenyl ether **17**, the formation of coumaran **43** was accelerated in phenolic solvent. On the other hand, *N,N*-dimethylaniline dramatically depressed the isomerization of double bond and cyclization and it was the best solvent to obtain the *ortho* allyl phenol derivative **83**.

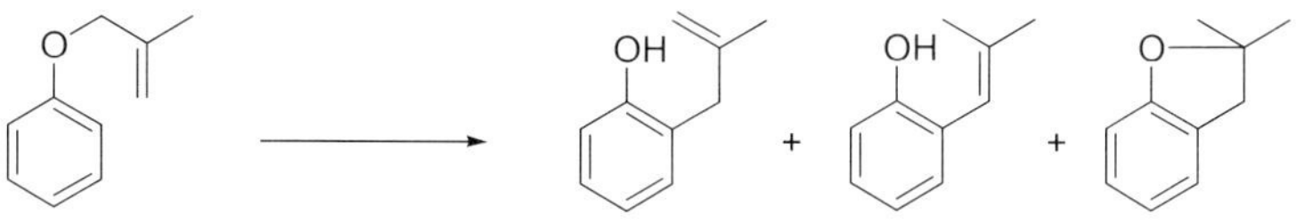

17 → **83** + **84** + **43**

solvent (T, t)	83	84	43
none (205-216 °C, 3.3 h)	4.4	1	2.3
2,6-xylenol (198-199 °C, 3.5 h)	1	1.7	12.2
N,N-dimethylaniline (199-205 °C, 4.8 h)	90	2	1

Scheme 3.2.39

A Lewis-acid-catalyzed reaction often facilitates not only the rearrangement reaction but also the consecutive cyclization as shown below [34a, 68]. Furthermore, additional substituent(s) on the allyl moiety of the substrates effects this consecutive cyclization, giving rise to coumaran derivatives in good yield.

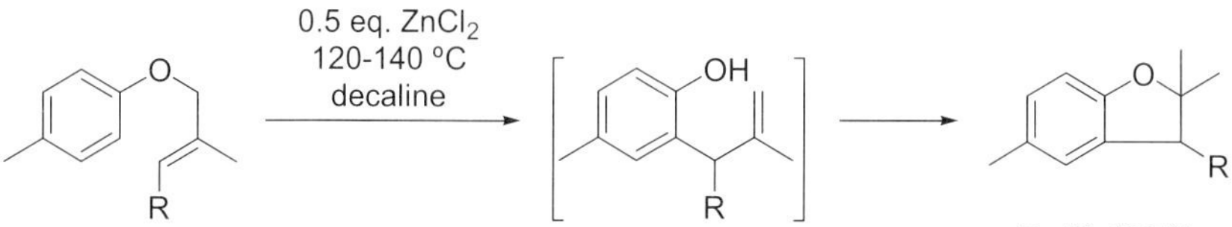

0.5 eq. $ZnCl_2$, 120-140 °C, decaline

R= H, 100 %
R= Me, 44 %

Scheme 3.2.40

The Claisen rearrangement of *γ*-phenoxy-*α*,*β*-enoate **85**, although high temperature was required for the rearrangement reaction, accompanied the following lactonization to give benzofuranone derivative **86** in good yield [69].

In situ generation of acetal-type allylic moiety **87** by mixing 2-methoxy-1,3-butadiene and phenol and the following Claisen rearrangement gave 2-methoxychroman **88** [70].

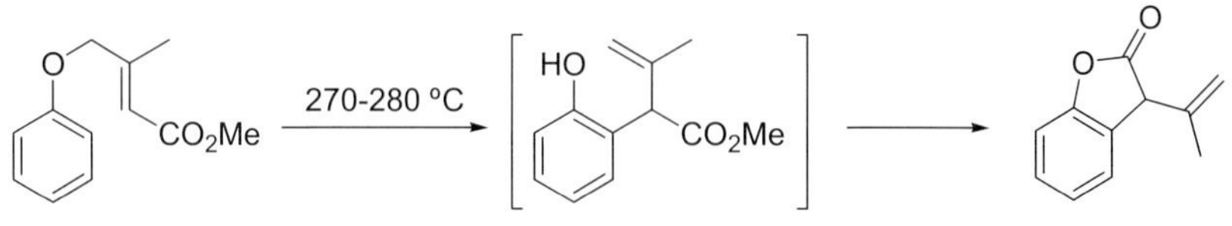
CO2Me
270-280 °C
HO
CO2Me
O
O

85 **86**, 74 %

Scheme 3.2.41

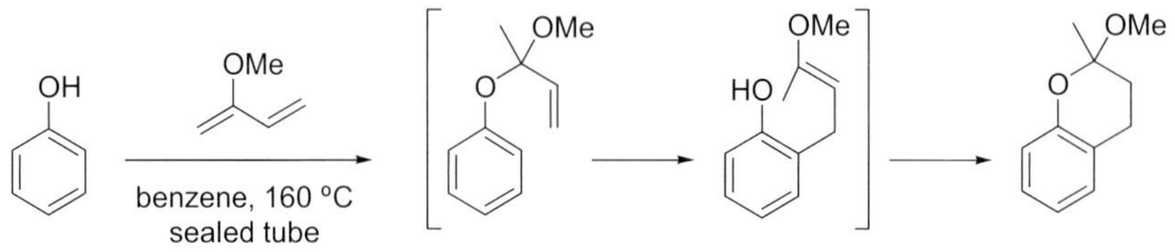
OH
OMe
benzene, 160 °C
sealed tube
OMe
O
OMe
HO
OMe
O

87 **88**, 82 %

Scheme 3.2.42

As one of the variants of allyl ether moiety, 2,3-butadienyl ether is of interest [71]. While the rearrangement proceeded under almost the same conditions as that for simple (typical) allyl ether, the product distribution of cyclized product and uncyclized diene derivative depended on the reaction temperature and time. Thus, the Claisen rearrangement of the allenyl derivative **89** occurs at around 200 °C within 20 min to give the uncyclized diene **90**, while at higher temperature (reflux in diethylene glycol (bp 245 °C) for 30 min), exclusive formation of the cyclized chromen **91** was observed.

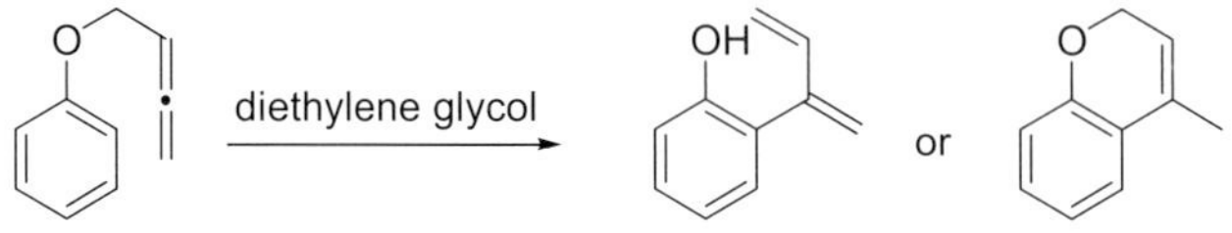
O
diethylene glycol
OH
or
O

89 **90** **91**

200-210 °C, 20 min: 66 % of **90**
reflux, 40 min: 80 % of **91**

Scheme 3.2.43

3.2.7.2 Tandem Reaction

Tandem Claisen and Cope rearrangement is the most common tandem reaction in the aromatic Claisen rearrangement [72]. *Para* rearrangement is one of these reactions.

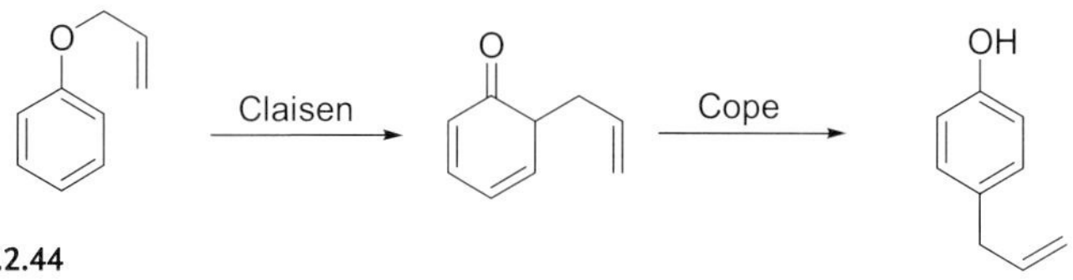

Claisen
Cope

Scheme 3.2.44

Followed by the aromatic Claisen rearrangement of allyl *ortho* vinylaryl ether **92**, the Cope rearrangement of 1,5-hexadiene unit constructed at the *ortho* position proceeded to give *ortho*-1,4-pentadienylphenol derivatives **93**, prior to the *para* rearrangement [73].

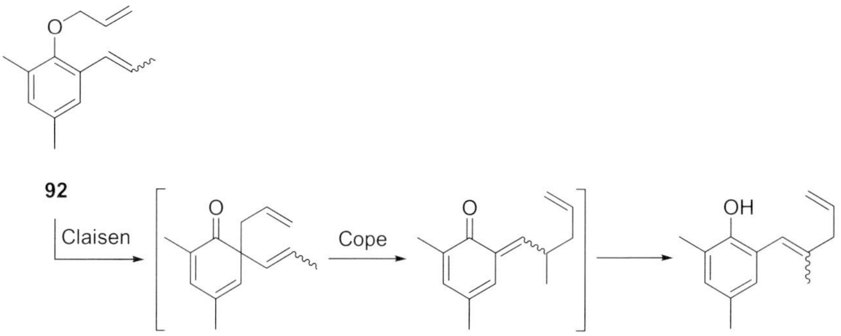

92
Claisen
Cope
93

Scheme 3.2.45

Two types of tandem Claisen rearrangement/Diels–Alder reaction, in which the position of the diene moiety is different are shown here. One is the use of a cyclohexadienone intermediate formed by the Claisen rearrangement as a diene moiety and the other is the use of the substrates in which the diene moiety pre-exists.

The biomimetic synthesis of morellin and related compounds is an example of the former case, and was first reported by Scheinmann and his coworkers [74]. Although the yield was not reported, the synthesis of 4-oxatricyclo[4.3.1.0]decan-2-one skeleton **95** through the tandem Claisen rearrangement/Diels–Alder reaction was achieved.

Recently, the Nicolaou group reported the total synthesis of 1-*O*-methylforbesione having similar structural feature to morellin including 4-oxatricyclo[4.3.1.0]-decan-2-one. Although the tandem reaction gave some structural isomers, desired compound **97** was obtained as a major isomer [75].

Theodorakis et al. also reported a model study for the synthesis of 4-oxatricyclo[4.3.1.0]decan-2-one core by employing acrylate as the dienophile resulting in an increase in the selectivity and yield of the product **99** [76]. They also achieved a total synthesis of seco-lateriflorone, desoxymorellin and forbesione through a tandem Claisen rearrangement/Diels–Alder reaction as a key step [77].

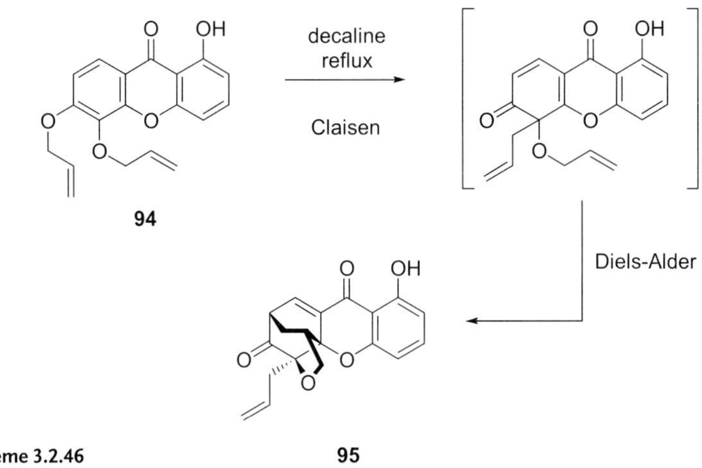

O
OH
decaline
reflux
Claisen
Diels-Alder
94
95

Scheme 3.2.46

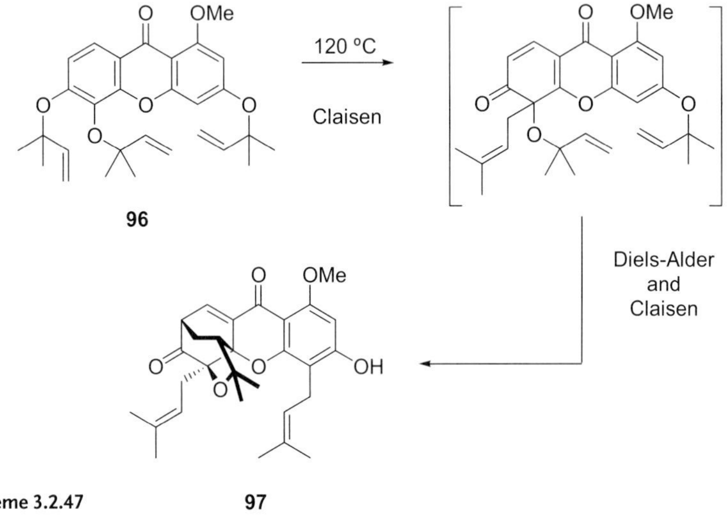

OMe
120 ºC
Claisen
Diels-Alder
and
Claisen
OH
96
97

Scheme 3.2.47

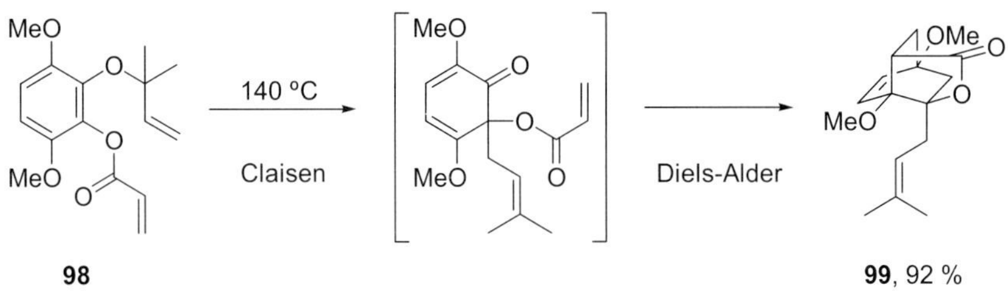

MeO
140 ºC
Claisen
Diels-Alder
OMe
98
99, 92 %

Scheme 3.2.48

As in the latter case, in which pre-existing diene and dienophile parts are located on the side chains of the parent aromatic ring, Kraus et al. reported a high yield and regioselective Claisen rearrangement followed by the intramolecular Diels–Alder reaction leading to anthracyclines **101** from the hydroquinone mono allyl ether derivative **100** [78].

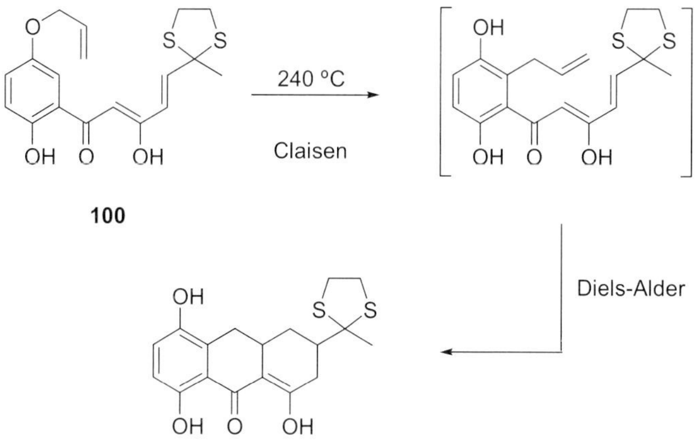

240 °C
Claisen
100
Diels-Alder
101, 85 %

Scheme 3.2.49

Wipf et al. reported tandem aromatic Claisen rearrangement/catalytic asymmetric carboalumination using a chiral zirconocene catalyst [36].

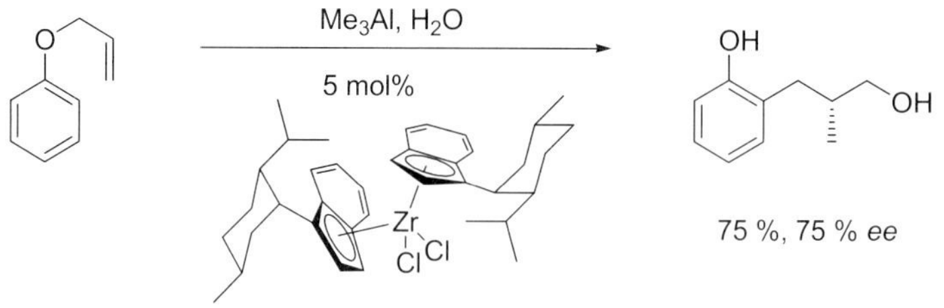

Me_3Al, H_2O
5 mol%
75 %, 75 % *ee*

Scheme 3.2.50

3.2.7.3 **Functional Molecule**

Hiratani et al. reported the synthesis of calixarene analogs **103** by the successive Claisen rearrangement of the 1,1-bis((aryloxy)methyl)ethylene unit giving rise to bis(hydroxyaryl) derivatives [79].

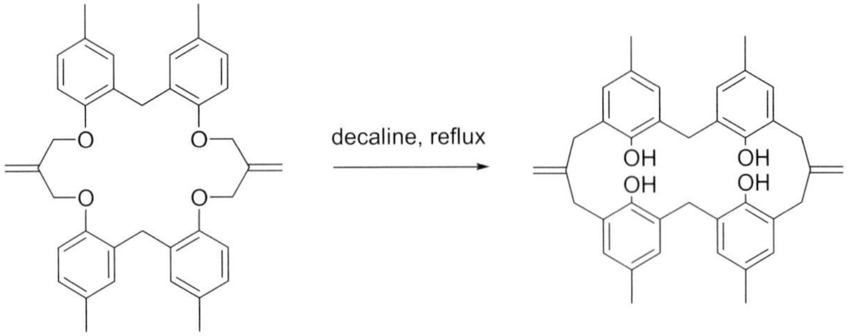

decaline, reflux

102

103

Scheme 3.2.51

3.2.7.4 Natural Products and Biologically Active Compounds

Since the aromatic Claisen rearrangement is a useful synthetic tool for the regioselective introduction of carbon functional group at the *ortho* position of phenol derivatives, a number of applications of this reaction to the synthesis of natural products and biologically active compounds have been reported.

The Eli Lilly group reported the synthesis of zatosetron, a potent, selective, and long-acting $5HT_3$ receptor antagonist, through the thermal aromatic Claisen rearrangement and the following acid-catalyzed cyclization to form a coumaran skeleton **106** as a key step [80].

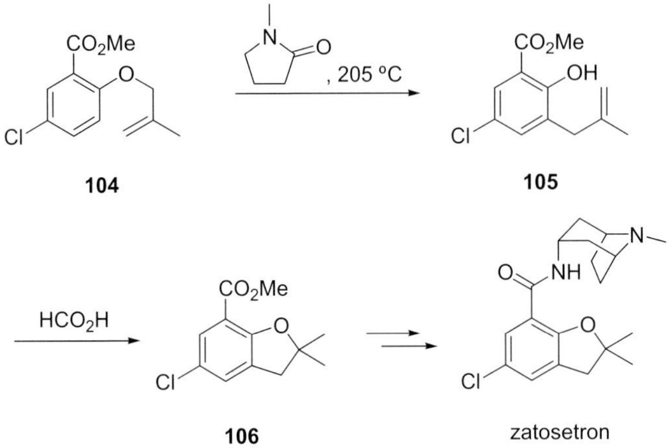

, 205 °C

104

105

HCO_2H

106

zatosetron

Scheme 3.2.52

Very recently, the enantioselective total synthesis of hexahydropyrrolo[2,3-*b*]-indole alkaloids, (–)-pseudophynaminol, through tandem olefination, isomerization and asymmetric Claisen rearrangement was reported [81]. Using a 3-ketodihydroindole derivative **107**, the reaction smoothly proceeded under extremely mild conditions (–78 °C to rt) to give the chiral 3,3-disubstituted 2-oxoindole **108** without loss of the enantiomeric purity of the starting material **107**, for which a chair-like transition state model **109** was proposed. They also achieved the total synthesis of hexahydropyrrolo[2,3-*b*]indole alkaloids, flustramine C, by employing the similar strategy [82].

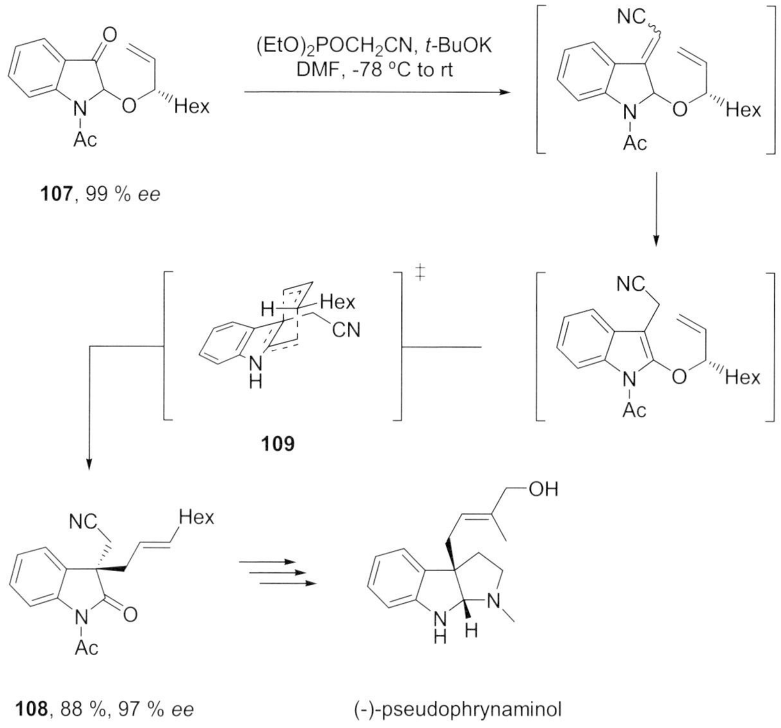

(EtO)$_2$POCH$_2$CN, *t*-BuOK
DMF, -78 °C to rt

107, 99 % *ee*

109

108, 88 %, 97 % *ee*

(-)-pseudophrynaminol

Scheme 3.2.53

Büchi et al. reported the total synthesis of atrovenetin via a base-promoted site-selective Claisen rearrangement followed by the formation of a benzofuran framework **111** [83].

Danishefsky et al. developed a stereospecific route for the synthesis of deoxy analog of mitomycin via an aromatic Claisen rearrangement [84]. The allyl aryl ether **112** was prepared from allylic alcohols and phenol derivative via the Mitsunobu reaction. The aromatic Claisen rearrangement of 1,3-disubstituted allylic ether **112** proceeded under thermal conditions (*N*,*N*-dimethylaniline reflux) to afford the *ortho* rearrangement product **113** in 80% yield.

In the total synthesis of illicinone and tricycloillicinone, which are members of an intriguing class of "small molecule" neurotrophic factors, Pettus et al. employed an ingeniously controlled aromatic Claisen rearrangement twice [85]. In the first Claisen rearrangement, the *para*-rearranged product **116** was predom-

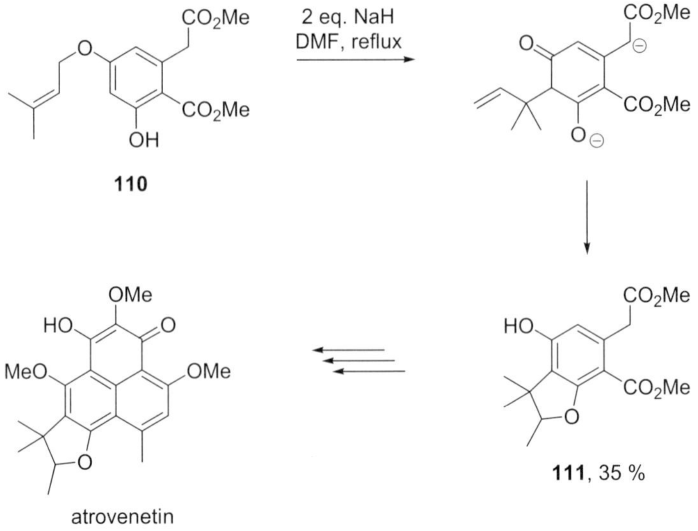

Scheme 3.2.54

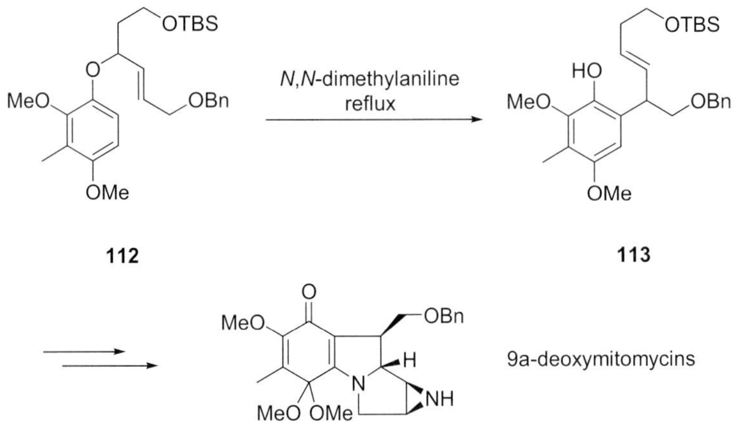

Scheme 3.2.55

inantly obtained when the allyl moiety does not have an additional substituent on the starting substrate **114** (R = H). On the other hand, introduction of benzenesulfonyl group into the allylic moiety resulted in a dramatic change of regioselectivity to afford the desired *ortho*-rearranged product **117** in excellent yield.

The second Claisen rearrangement of the 1,1-dimethylallyl ether **118**, leading to the *ortho*-prenylated compound **119**, proceeded under relatively mild conditions (toluene, 100 °C) in quantitative yield.

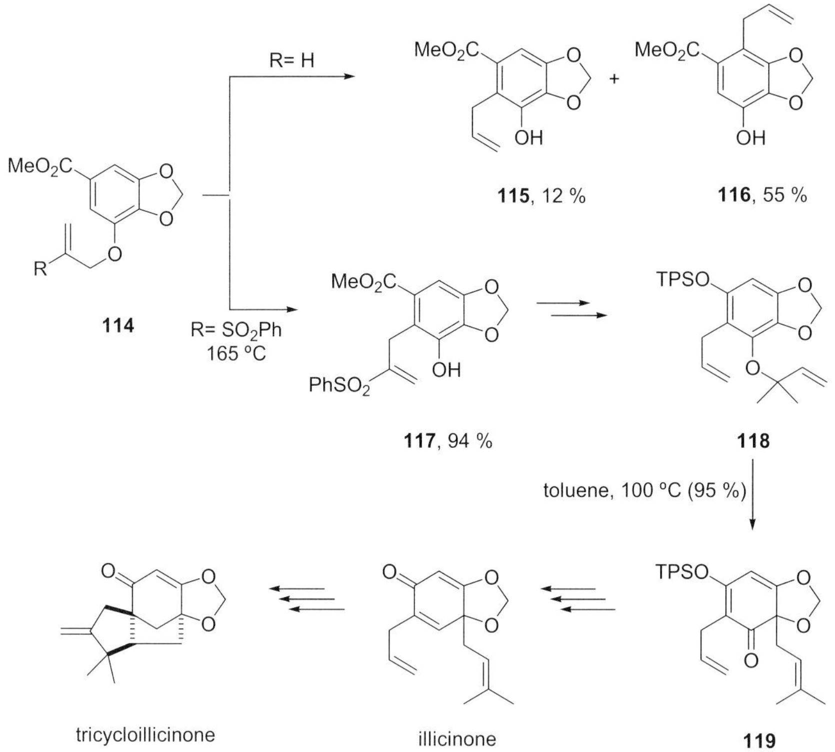

Scheme 3.2.56

References

1 For reviews of the Claisen rearrangement, including the aromatic version, see: a) Tarbell, D. S. *Org. React. (N.Y.)* **1944**, *2*, 1–48. b) Rhoads, S. J.; Raulins, N. R. *Org. React. (N.Y.)* **1975**, *22*, 1–252. c) Hansen, von H.-J.; Schmid, H. *Chimia* **1970**, *24*, 89–124. d) Hansen, von H.-J.; Schmid, H. *Chemistry in Britain* **1969**, *5*, 111–116. e) Gajewski, J. J. in: Grieco, P. A. (ed.) *Organic Synthesis in Water* **1998**, Blackie, London, pp. 82–101. f) Bennett, G. B. *Synthesis* **1977**, 589–606. g) Jefferson, A.; Scheinmann, F. *Q. Rev.* **1968**, *22*, 391–421. h) Lutz, R. P. *Chem. Rev.* **1984**, *84*, 206–247.

2 a) Claisen, L. *Ber. Dtsch. Chem. Ges.* **1912**, *45*, 3157–3166. b) Claisen, L.; Eisleb, O. *Liebigs Ann. Chem.* **1913**, *401*, 21–119.

3 a) Claisen, L.; Tietze, E. *Chem. Ber.* **1925**, *58*, 275–281. b) Claisen, L.; Tietze, E. *Liebigs Ann. Chem.* **1926**, *449*, 81–101.

4 For a review of the mechanism of the Claisen rearrangement, see: Ganem, B. *Angew. Chem. Int. Ed. Engl.* **1996**, *35*, 936–945.

5 For a review, see: Wipf, P. in *Comprehensive Organic Synthesis* **1991** Trost, B. M.; Fleming, I.; Paquette, L. A. (eds.) **1991**, Pergamon, Oxford, vol. 5, pp. 827–873.

6 a) Marvell, E. N.; Stephenson, J. L. *J. Org. Chem.* **1960**, *25*, 676–677. b) Burgstahler, A. W. *J. Am. Chem. Soc.* **1960**, *82*, 4681–4685. c) Alexander, E. R.; Kluiber, R. W. *J. Am. Chem. Soc.* **1951**, *73*, 4304–4306. d) Hart, H. *J. Am. Chem. Soc.* **1954**, *76*, 4033–4035. e) Marvell, E. N.; Stephenson, J. L.; Ong, J. *J. Am. Chem. Soc.* **1965**, *87*, 1267–1274.

7 Meyer, M. P.; DelMonte, A. J.; Singleton, D. A. *J. Am. Chem. Soc.* **1999**, *121*, 10865–10874 and references cited therein.

8 For reviews of the abnormal Claisen rearrangement, see: a) Hansen, H. J. *Mechanism of Molecular Migrations* **1971**, *3*, 177–236. See also Refs. 1b and 1f.

9 Lauer, W. M.; Filbert, W. F. *J. Am. Chem. Soc.* **1936**, *58*, 1388–1392.

10 Hurd, C. D.; Pollack, M. A. *J. Org. Chem.* **1938**, *3*, 550–569.

11 Marvell, E. N.; Anderson, D. R.; Ong, J. *J. Org. Chem.* **1962**, *27*, 1109–1110.

12 a) Habich, von A.; Barner, R.; Roberts, R. M.; Schmid, H. *Helv. Chim. Acta* **1962**, *45*, 1943–1950. b) Lauer, W. M.; Johnson, T. A. *J. Org. Chem.* **1963**, *28*, 2913–2914.

13 a) Kozikowski, A. P.; Ma, D.; Du, L.; Lewin, N. E.; Blumberg, P. M. *J. Am. Chem. Soc.* **1995**, *117*, 6666–6672. b) Chang, S.; Grubbs, R. H. *J. Org. Chem.* **1998**, *63*, 864–866. c) Tsai, Y.-F.; Peddinti, R. K.; Liao, C.-C. *Chem. Commun.* **2000**, 475–476.

14 Cacchi, S.; Fabrizi, G.; Moro, L. *Synlett* **1998**, 741–745.

15 For the effects of the substituent on the para position, see: a) White, W. N.; Gwynn, D.; Schlitt, R.; Girard, C.; Fife, W. *J. Am. Chem. Soc.* **1958**, *80*, 3271–3277. b) Goering, H. L.; Jacobson, R. R. *J. Am. Chem. Soc.* **1958**, *80*, 3277–3285.

16 White, W. N.; Slater, C. D. *J. Org. Chem.* **1961**, *26*, 3631–3638.

17 Gozzo, F. C.; Fernandes, S. A.; Rodrigues, D. C.; Eberlin, M. N.; Marsaioli, A. J. *J. Org. Chem.* **2003**, *68*, 5493–5499.

18 For reviews of the aromatic Claisen rearrangement of heterocycles, see: a) Makisumi, Y. *J. Synth. Org. Chem. Jpn.* **1969**, *27*, 593–608. b) Moody, C. J. *Adv. Heterocycle Chem.* **1987**, *42*, 203–244. c) Thyagarajan, B. S. *Adv. Heterocyclic Chem.* **1967**, *8*, 143–163.

19 a) Moffett, R. B. *J. Org. Chem.* **1963**, *28*, 2885–2886. b) Dinan, F. J.; Tieckelmann, H. *J. Org. Chem.* **1964**, *29*, 892–895.

20 a) Stewart, H. F.; Seibert, R. P. *J. Org. Chem.* **1968**, *33*, 4560–4561. b) Itami, K.; Yamazaki, D.; Yoshida, J. *Org. Lett.* **2003**, *5*, 2161–2164.

21 White, W. N.; Norcross, B. E. *J. Am. Chem. Soc.* **1961**, *83*, 3265–3269.

22 a) Huestis, L. D.; Andrews, L. J. *J. Am. Chem. Soc.* **1961**, *83*, 1963–1968. b) White, W. N.; Norcross, B. E. *J. Am. Chem. Soc.* **1961**, *83*, 1968–1974.

23 a) Hurd, C. D.; Cohen, F. L. *J. Am. Chem. Soc.* **1931**, *53*, 1068–1077. b) Iwai, I.; Ide, J. *Chem. Pharm. Bull.* **1963**, *11*, 1042–1049.

24 a) Harfenist, M.; Thom. E. *J. Org. Chem.* **1972**, *37*, 841–848. b) Sarcevic, N.; Zsindely, J.; Schmid, H. *Helv. Chim. Acta* **1973**, *56*, 1457–1476. c) Anderson, W. K.; LaVoie, E. J. *J. Org. Chem.* **1973**, *38*, 3832–3835.

25 Subramanian, R. S.; Balasubramanian, K. K. *Tetrahedron Lett.* **1988**, *29*, 6797–6800.

26 Kincaid, J. F.; Tarbell, D. S. *J. Am. Chem. Soc.* **1939**, *61*, 3085–3089.

27 White, W. N.; Wolfarth, E. F. *J. Org. Chem.* **1970**, *35*, 2196–2199.

28 a) Svanholm, U.; Parker, V. D. *J. Chem. Soc., Chem. Commun.* **1972**, 645–646. b) Svanholm, U.; Parker, V. D. *J. Chem. Soc., Perkin Trans. II* **1974**, 169–173.

29 Widmer, U.; Hansen, H.-J.; Schmid, H. *Helv. Chim. Acta* **1973**, *56*, 2644–2648.

30 Bryusova, L. Y.; Ioffe, M. L. *Zh. Obshch. Khim.* **1941**, *11*, 722–728.

31 For example: a) Borgulya, von J.; Madeja, R.; Fahrni, P.; Hansen, H.-J.; Schmid, H.; Barner, R. *Helv. Chim. Acta* **1973**, *56*, 14–75. b) Fahrni, von P.; Habich, A.; Schmid, H. *Helv. Chim. Acta* **1960**, *43*, 448–452.

32 Maruyama, K.; Nagai, N.; Naruta Y. *J. Org. Chem.* **1986**, *51*, 5083–5092.

33 a) Sonnenberg, F. M. *J. Org. Chem.* **1970**, *35*, 3166–3167. b) Bender, D. R.; Kanne, D.; Frazier, J. D.; Rapoport, H. *J. Org. Chem.* **1983**, *48*, 2709–2719.

34 a) Bunina-Krivorukova, L. I.; Rossinskii, A. P.; Bal'yan, Kh.V. *Zh. Org. Khim.* **1978**, *14*, 586–593. b) Aleksandrova, E. K.; Bunina-Krivorukova, L. I.; Moshimskaya, A. V.; Bal'yan, Kh. V. *Zh. Org. Khim.* **1974**, *10*, 1039–1046.

35 Narasaka, K.; Bald, E.; Mukaiyama, T. *Chem. Lett.* **1975**, 1041–1044.

36 a) Wipf, P.; Ribe, S. *Org. Lett.* **2001**, *3*, 1503–1505. b) Wipf, P.; Rodriguez, S. *Adv. Synth. Catal.* **2002**, *344*, 434–440.

37 Maruoka, K.; Sato, J.; Banno, H.; Yamamoto, H. *Tetrahedron Lett.* **1990**, *31*, 377–380.

38 Ollis, W. D.; Somanathan, R.; Sutherland, I. O. *J. Chem. Soc., Chem. Commun.* **1974**, 494–496.

39 a) Irie, R.; Noda, K.; Ito, Y.; Katsuki, T. *Tetrahedron Lett.* **1991**, *32*, 1055–1058. b) Irie, R.; Noda, K.; Ito, Y.; Matsumoto, N.; Katsuki, T. *Tetrahedron: Asymmetry* **1991**, *2*, 481–494.

40 Balavoine, G.; Bram, G.; Guibe, F. *Nouv. J. Chim.* **1978**, *2*, 207–209.

41 Bernard, A. M.; Cocco, M. T.; Onnis, V.; Piras, P. P. *Synthesis* **1997**, 41–43.

42 Karanewsky, D. S.; Kishi, Y. *J. Org. Chem.* **1976**, *41*, 3026–3027.

43 Jefferson, A.; Scheinmann, F. *J. Chem. Soc. C* **1969**, 243–245.

44 Murray, R. D. H.; Ballantyne, M. M. *Tetrahedron* **1970**, *26*, 4667–4671.

45 Fukuyama, T.; Li, T.; Peng, G. *Tetrahedron Lett.* **1994**, *35*, 2145–2148.

46 Bagnell, L.; Cablewski, T.; Strauss, C. R.; Trainor, R. W. *J. Org. Chem.* **1996**, *61*, 7355–7359 and references cited therein.

47 Pincock, A. L.; Pincock, J. A.; Stefanova, R. *J. Am. Chem. Soc.* **2002**, *124*, 9768–9778 and references cited therein.

48 Hurd, C. D.; Greengard, H. *J. Am. Chem. Soc.* **1930**, *52*, 3356–3358.

49 For reviews of the thio-Claisen rearrangement, see: a) Majumdar, K. C.; Ghosh, S.; Ghosh, M. *Tetrahedron* **2003**, *59*, 7251–7271. b) Morin, L.; Lebaud, J.; Paquer, D.; Chaussin, R.; Barillier, D. *Phosphorus and Sulfur* **1979**, *7*, 69–80.

50 Kwart, H.; Hackett, C. M. *J. Am. Chem. Soc.* **1962**, *84*, 1754–1755.

51 Meyers, C. Y.; Rivaldi, C.; Banoli, L. *J. Org. Chem.* **1963**, *28*, 2440–2442.

52 Kwart, H.; Evans, E. R. *J. Org. Chem.* **1966**, *31*, 413–419.

53 Kwart, H.; George, T. J. *J. Chem. Soc., Chem. Commun.* **1970**, 433–434.

54 Makisumi, J.; Murabayashi, A. *Tetrahedron Lett.* **1969**, 1971–1974.

55 For review of the amino-Claisen rearrangement, see: Majumdar, K. C.; Bhattacharyya, T. *J. Indian Chem. Soc.* **2002**, *79*, 112–121.

56 Jolidon, S.; Hansen, H.-J. *Helv. Chim. Acta* **1977**, *60*, 978–1032.

57 Carnahan, F. L.; Hurd, C. D. *J. Am. Chem. Soc.* **1930**, *52*, 4586–4595.

58 Cooper, M. A.; Lucas, M. A.; Taylor, J. M.; Ward, A. D.; Williamson, N. M. *Synthesis* **2001**, 621–625.

59 Beholz, L. G.; Stille, J. R. *J. Org. Chem.* **1993**, *58*, 5095–5100.

60 Scheiner, P. *J. Org. Chem.* **1967**, *32*, 2628–2630.

61 Ochiai, M.; Ito, T.; Takaoka, Y.; Masaki, Y. *J. Am. Chem. Soc.* **1991**, *113*, 1319–1323.

62 Goering, H. L.; Kimoto, W. I. *J. Am. Chem. Soc.* **1965**, *87*, 1748–1753.

63 Takano, S.; Akiyama, M.; Ogasawara, K. *J. Chem. Soc., Perkin Trans. 1* **1985**, 2447–2453.

64 Trost, B. M.; Toste, F. D. *J. Am. Chem. Soc.* **1998**, *120*, 815–816.

65 Ito, H.; Sato, A.; Taguchi, T. *Tetrahedron Lett.* **1997**, *38*, 4815–4818.

66 For a review of the Claisen rearrangement of coumarins for the introduction of allylic substituent on the coumarins, see: Majumdar, K. C.; Das, U. *J. Indian Chem. Soc.* **1997**, *74*, 884–890.

67 Shulgin, A. T.; Baker, A. W. *J. Org. Chem.* **1963**, *28*, 2468–2469.

68 Kim, K. M.; Kim, H. R.; Ryu, E. K. *Heterocycles* **1993**, *36*, 497–505.

69 Padmanathan, T.; Sultanbawa, M. U. S. *J. Chem. Soc.* **1963**, 4210–4218.

70 Dolby, L. J.; Elliger, C. A.; Esfandiari, S.; Marshall, K. S. *J. Org. Chem.* **1968**, *33*, 4508–4511.

71 Balasubramanian, T.; Balasubramanian, K. K. *J. Chem. Soc., Chem. Commun.* **1992**, 1760–1761.

72 For a review, see: Ziegler, F. E. in *Comprehensive Organic Synthesis* **1991** Trost B. M.; Fleming, I.; Paquette, L. A. (eds.) Pergamon, Oxford, vol. 5, pp. 875–898.

73 The first example of this tandem reaction was reported by Claisen (see Ref. 3a). See also: a) Schmid, K.; Fahrni, P.;

Schmid, H. *Helv. Chim. Acta* **1956**, *39*, 708–721. b) Lauer, W. M.; Wujciak, D. W. *J. Am. Chem. Soc.* **1956**, *78*, 5601–5606. c) Mülly, von M.; Zsindely, J.; Schmid, H. *Helv. Chim. Acta* **1975**, *58*, 610–640.

74 Quillinan, A. J.; Scheinmann, F. *Chem. Commun.* **1971**, 966–967.

75 Nicolaou, K. C.; Li, J. *Angew. Chem. Int. Ed.* **2001**, *40*, 4264–4268.

76 Tisdale, E. J.; Chowdhury, C.; Vong, B. G.; Li, H.; Theodorakis, E. A. *Org. Lett.* **2002**, *4*, 909–912.

77 a) Tisdale, E. J.; Vong, B. G.; Li, H.; Kim, S. H.; Chowdhury, C.; Theodorakis, E. A. *Tetrahedron* **2003**, *59*, 6873–6887. b) Tisdale, E. J.; Slobodov, I.; Theodorakis, E, A. *Org. Biomol. Chem.* **2003**, *1*, 4418–4422.

78 a) Kraus, G. A.; Fulton, B. S. *Tetrahedron* **1984**, *40*, 4777–4780. b) Kraus, G. A.; Fulton, B. S. *J. Org. Chem.* **1985**, *50*, 1782–1784. c) Kraus, G. A.; Woo, S. H. *J. Org. Chem.* **1987**, *52*, 4841–4846. d) Kraus, G. A.; Liras, S.; Man, T. O.; Molina, M. T. *J. Org. Chem.* **1989**, *54*, 3137–3139.

79 a) Hiratani, K.; Kasuga, K.; Goto, M.; Uzawa, H. *J. Am. Chem. Soc.* **1997**, *119*, 12677–12678. b) Hiratani, K.; Nagawa, Y.; Tokuhisa, H.; Koyama, E. *J. Syn. Org. Chem. Jpn.* **2003**, *61*, 111–122. c) Hiratani, K.; Takahashi, T.; Kasuga, K.; Sugihara, H.; Fujiwara, K.; Ohashi, K. *Tetrahedron Lett.* **1995**, *36*, 5567–5570.

80 Robertson, D. W.; Lacefield, W. B.; Bloomquist, W.; Pfeifer, W.; Simon, R. L.; Cohen, M. L. *J. Med. Chem.* **1992**, *35*, 310–319.

81 Kawasaki, T.; Ogawa, A.; Takashima, Y.; Sakamoto, M. *Tetrahedron Lett.* **2003**, *44*, 1591–1593.

82 Kawasaki, T.; Terashima, R.; Sakaguchi, K.; Sekiguchi, H.; Sakamoto, M. *Tetrahedron Lett.* **1996**, *37*, 7525–7528.

83 Büchi, G.; Leung, J. C. *J. Org. Chem.* **1986**, *51*, 4813–4818.

84 Danishefsky S.; Berman, E.; Ciufolini, M.; Etheredge, S. J.; Segmuller, B. E. *J. Am. Chem. Soc.* **1985**, *107*, 3891–3898.

85 Pettus, T. R. R.; Chen, X.-T.; Danishefsky, S. J. *J. Am. Chem. Soc.* **1998**, *120*, 12684–12685.

4
The Ireland–Claisen Rearrangement (1972–2004)

Christopher M. McFarland and Matthias C. McIntosh

4.1
Introduction

Since its introduction in the 1972, the Ireland–Claisen rearrangement has become widely used in the synthesis of a diverse range of natural products and other targets (Scheme 4.1) [1, 2]. The popularity of the reaction is due to several factors: (1) the ease of preparation of the allylic ester reactants; (2) the ability to control the *E*/*Z* geometry of the ester enolate and hence the relative stereochemistry between C2 and C3 of the pentenoic acid product; (3) the frequently high chirality transfer between the allylic stereocenter of the allyl ketene acetal (C5, pentenoic acid numbering) and the newly formed stereocenter(s) at C2 and/or C3 of the pentenoic acid; and (4) the generally high level of alkene stereocontrol. By comparison to other Claisen rearrangement methods such as the parent Claisen and the Johnson and Eschenmoser variants, the Ireland protocol also has the advantages that (5) the reaction often proceeds at substantially lower temperatures; (6) the reaction is performed under basic rather than acidic conditions; and (7) only a 1:1 stoichiometry is needed between the allylic alcohol and the carbonyl components.

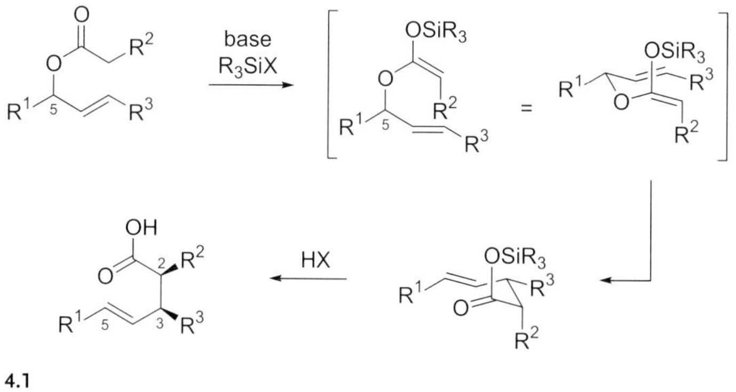

Scheme 4.1

The Claisen Rearrangement. Edited by M. Hiersemann and U. Nubbemeyer

ISBN: 978-3-527-30825-5

More recently, several variations of the Claisen rearrangement have been reported that also possess some of the features of the Ireland variant, including the amide Claisen, *S,S*- and *N,S*-ketene acetal Claisen and the zwitterionic Claisen rearrangements [2]. This review will focus on the Ireland variant of the Claisen rearrangement in which a silyl ketene acetal is an intermediate, although mention will be made of boron ketene acetals where appropriate.

The purpose of this chapter is to review the Ireland–Claisen rearrangement literature beginning with Ireland's seminal 1972 report through to December 2004. This review is not intended as an exhaustive review of the Ireland–Claisen literature. Instead, we will generally highlight those reports that describe the first example of a notable variation of the rearrangement as well as representative applications to natural products synthesis.

After some preliminary discussion of history and nomenclature, the review will briefly cover issues of rearrangement temperature, substituent effects and catalysis. Transition state structure as determined by isotope effects combined with theoretical studies will then be described. Stereochemical aspects of the reaction will be examined in detail with accompanying examples. The remainder of the chapter will describe the various methods of ketene acetal formation, structural variety in the allylic ester substrates and applications to natural product synthesis. The presentation of the material is somewhat arbitrary, and indeed most examples could well have been placed under more than one sub-heading.

4.2 History

The first report of an ester enolate Claisen rearrangement appeared in 1937. Tseou and Wang reported the formation in low yield of pent-4-enoic acid upon attempted acetoacetic ester condensation of allyl acetate using Na metal [3]. Over the next 35 years there were scattered reports of ester enolate Claisen rearrangements that employed Na metal, NaH or Et_2NMgBr to generate the ester enolates from the corresponding allylic esters. These reactions were often plagued by low yields and high reaction temperatures. (For a complete survey of ester enolate Claisen rearrangements leading up to the Ireland variant, see [2 i]).

In 1972, Ireland and Mueller reported the transformation that has come to be known as the Ireland–Claisen rearrangement (Scheme 4.2) [1]. Use of a lithium dialkylamide base allowed for efficient low temperature enolization of the allylic ester. They found that silylation of the ester enolate suppressed side reactions such as decomposition via the ketene pathway and Claisen-type condensations. Although this first reported Ireland–Claisen rearrangement was presumably diastereoselective (*vide infra*, Section 4.6.1), the stereochemistry of the alkyl groups was not an issue in its application to the synthesis of dihydrojasmone.

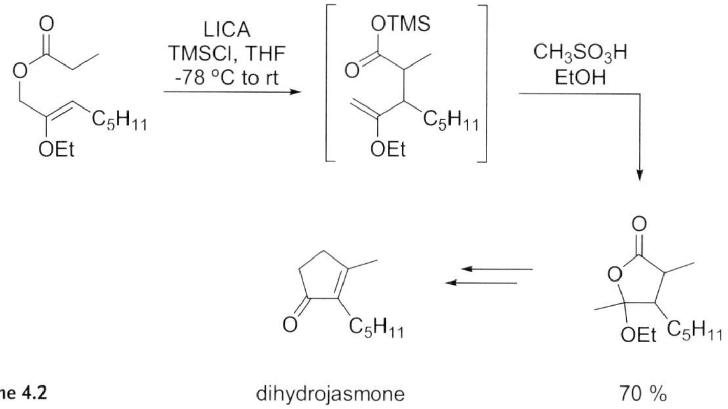

LICA
TMSCl, THF
-78 °C to rt

CH_3SO_3H
EtOH

dihydrojasmone

70 %

Scheme 4.2

4.3 Numbering and Nomenclature

The numbering of the silyl ketene acetal reactant and the corresponding pentenoic acid product differs and may lead to confusion when following the fate of the sigmatropic atoms over the course of the rearrangement. By convention, the numbering of the allyl silyl ketene acetal begins with atom 1 as the terminal carbon of the ketene acetal irrespective of additional substitution at that carbon (Scheme 4.3). The oxygen atom is numbered sequentially with the carbon atoms. The pentenoic acid product is numbered according to IUPAC convention. In the text the type of numbering will be explicitly indicated where appropriate.

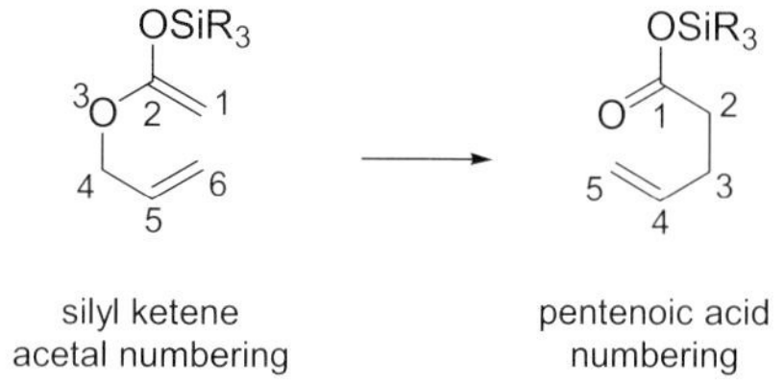

silyl ketene acetal numbering

pentenoic acid numbering

Scheme 4.3

A key aspect of the stereochemistry of the rearrangement results from the stereochemistry of the silyl ketene acetal. The silyl ketene acetals are designated as *Z*– when the higher priority substituent at C1 is *cis* to the $OSiR_3$ group and *E*– when it is *trans* (Scheme 4.4). For simplicity, the geometries of the corresponding enolates and ketene acetals other than silyl will designated as *Z*–(*O*)–M– or *E*–(*O*)–M–, where M = metal, etc.

Z *E* *Z*-(O)-M- *E*-(O)-M-

Scheme 4.4

The *syn* and *anti* stereochemical designations of the pentenoic acid products are given based on the extended chain conformation in which the C1/C2/C3/C4 dihedral angle is 180° (Scheme 4.5). However, the closed conformation in which the C1/C2/C3/C4 dihedral angle is 0° is often more useful in showing how the product stereochemistry derives from the transition state of the rearrangement.

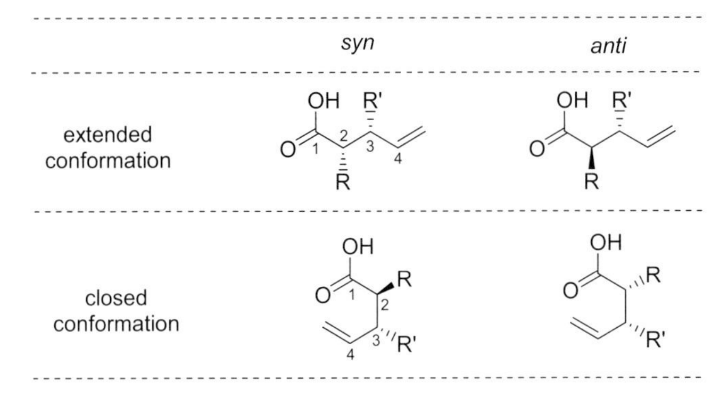

Scheme 4.5

4.4 Rearrangement Temperature, Substituent Effects and Catalysis

4.4.1 Rearrangement Temperature

One of the advantages of the Ireland–Claisen rearrangement relative to other variations is that the rearrangement frequently occurs at lower temperature. There is a net rate acceleration of *ca.* 10^6 relative to allyl vinyl ether itself [1]. Acyclic silyl ketene acetals often rearrange at reasonable rates at ambient temperature (Eq. 1, Scheme 4.6), although moderate heating is often employed to accelerate the rearrangement (Eq. 2) [4]. Allyl silyl ketene acetals derived from cycloalkenes often require heating, although the temperature depends upon the substitution and stereochemistry of the cycloalkenyl and silyl ketene acetal (Eqs. 3 and 4) [1, 5]. Typically the silyl ketene acetal is not isolated and the rearrangement occurs *in situ*. However, in cases where vigorous heating is required, the intermediate acetal may be isolated prior to rearrangement (Eq. 3) [5].

LICA, TMSCl
THF, -78 °C

OTMS

-78 °C to rt
then
NaOH, H_2O

HO

75 %

TBSO

LDA, TBSCl, -78 °C
23 vol-% HMPA, THF

OTBS

TBSO

-78 °C to reflux
then
NaOH, H_2O

TBSO

CO_2H

80 %

OBn

LDA, TBSCl
23 vol-% HMPA, THF

OBn

OTBS

xylenes
190 °C, 18 h
then NaOH
H_2O, THF

HO_2C

OBn

64 %

LDA, TBSCl, -78 °C
23 vol-% DMPU, THF

OTBS

-78 °C to reflux, 6 h
then NaOH, H_2O

HO_2C

H

60 %

Scheme 4.6

4.4.2
Substituent Effects

Ireland noted in an early report that the rate of the rearrangement increased significantly with alkyl substitution at C1, C4 or C6 (Scheme 4.7) [1]. This is consistent with extensive computational and experimental studies of the parent allyl vinyl ether [6], although the corresponding calculations of 2-OR-ketene acetals have not been reported.

ketene acetal	$t_{1/2}$, min	ketene acetal	$t_{1/2}$, min
OTMS, O	210 ± 30	OTMS, O	5 ± 1
OTMS, O	150 ± 30	OTMS, O	<<1
OTMS, O, C_5H_{11}	6 ± 1		

Scheme 4.7

Wilcox and Babston later studied the effect of alkyl substitution at C5 for the silyl ketene acetals derived from acetate and propionate esters (Scheme 4.8) [7]. They noted a decrease in rate of the 5-methyl substituted silyl ketene acetals relative to the parent alkene. They also observed a general increase in the rate of rear-

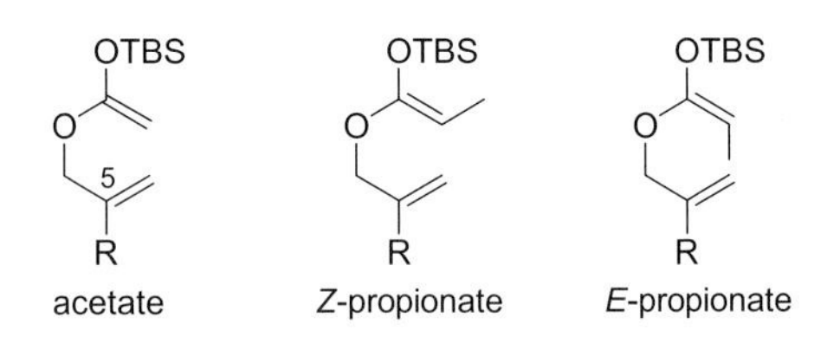

entry	R	$t_{1/2}$ (min) acetate	Z-propionate	E-propionate
1	H	107	4.7	3.8
2	Me	251	12.3	38
3	Et	199	9.4	33.1
4	n-Pr	181	8.1	25.1
5	i-Pr	151	6.8	41.7
6	CH_2SiMe_3	33.6	1.4	3.6

Scheme 4.8

rangement as the 5-alkyl substituent increased in size from methyl to neopentyl for the parent acetate and the *Z*-silyl ketene acetal of the propionate ester. The *E*-silyl ketene acetal showed a similar trend, except for R = *i*Pr (entry 5), which was anomalously slow. These observations have not been addressed computationally.

4.4.3 Catalysis

4.4.3.1 Pd(II) Catalysis

Although there are reports of Pd(II), Hg(II) and Cu(I) catalysis of Claisen rearrangements of allyl vinyl ethers, allyl phenyl ethers and allyl ketene *S,N*-acetals [2], there is only one report of transition metal catalysis of the Ireland–Claisen rearrangement. Yamazaki et al. reported the Pd(II)-catalyzed Ireland–Claisen rearrangement of an allyl silyl ketene acetal (Scheme 4.9) [8]. The rearrangement presumably proceeds via a σ-Pd(II) intermediate by analogy to mechanism proposed for allyl vinyl ethers [9], allylic esters and allyl imidates [10]. In the absence of the $PdCl_2(NCPh)_2$ complex, larger amounts of products resulting from protodesilylation of the silyl ketene acetal were obtained.

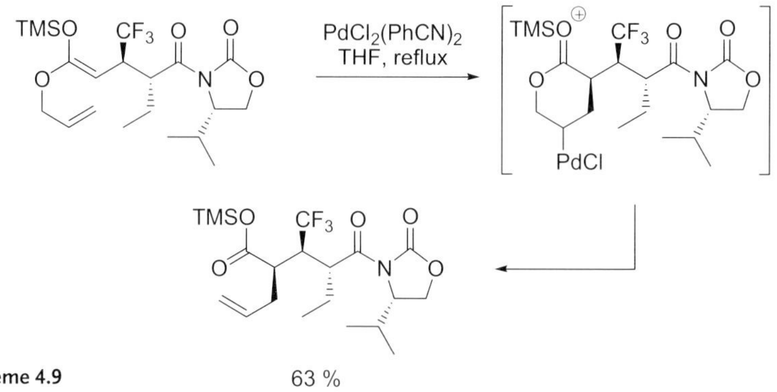

TMSO CF$_3$ O O N O
$PdCl_2(PhCN)_2$
THF, reflux
TMSO$^{\oplus}$ CF$_3$ O O N O PdCl
TMSO CF$_3$ O O N O

Scheme 4.9

63 %

4.4.3.2 Lewis Acid Catalysis

Koch et al. recently reported that Ireland–Claisen rearrangements of allyl silyl ketene acetals derived from allyl phenyl acetates proceeded at lower temperature and with higher diastereoselectivity when catalyzed by Lewis acids such as $TiCl_4$, $ZnCl_2$ and $SnCl_4$ (Scheme 4.10) [11]. Optimal yield and diastereoselectivities were observed when LHMDS and TMSCl were employed to generate the silyl ketene acetal and 2% $TiCl_4$ was used as catalyst. In the absence of catalyst it was necessary to heat the reaction under reflux in THF to obtain the rearrangement products, although the yield and diastereoselectivity were lower. If one assumes a kinetic enolization to the *E*-(*O*)-Li enolate, then silylation would yield the *E*-silyl ketene acetal. Rearrangement via the expected chair transition state would afford the observed *anti* isomer as the major product.

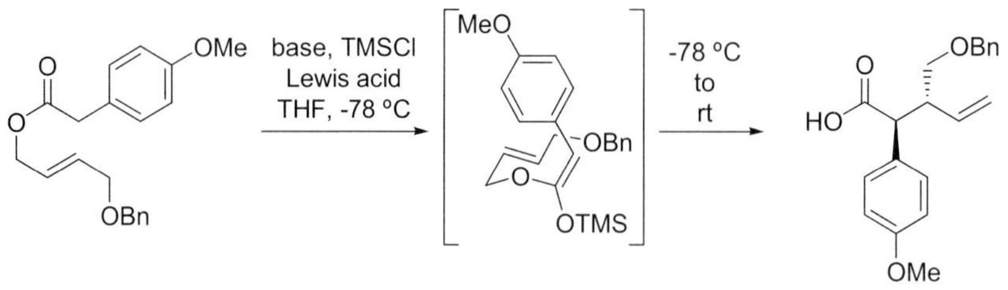

base	Lewis acid	mol%	yield (%)	*anti:syn*
LDA	-	-	75	2:1
LHMDS	$SnCl_4$	5	95	12:1

Scheme 4.10

Although the authors did not specifically address the issue of silyl ketene acetal geometry, the stereoselectivity of the silyl ketene acetal in these cases is noteworthy. Ireland had previously shown that phenyl acetic esters gave low *E*-stereoselectivity upon treatment with LDA and TBSCl (*vide infra*, Section 4.6.1.1). Either LHMDS/TMSCl gives higher *E*-selectivity in the case of phenyl acetic esters or the Lewis acid may be playing a role in both ketene acetal isomerization as well as rearrangement.

4.4.3.3 **Phosphine Catalysis**

Inanaga et al. reported the first example of a nucleophile-catalyzed Ireland–Claisen rearrangement (Scheme 4.11) [12]. Conjugate addition of tricyclohexylphosphine to allyl acrylates generated intermediate phosphonium allyl silyl ketene acetals, which underwent rearrangement to yield α-substituted acrylates after elimination of the phosphine.

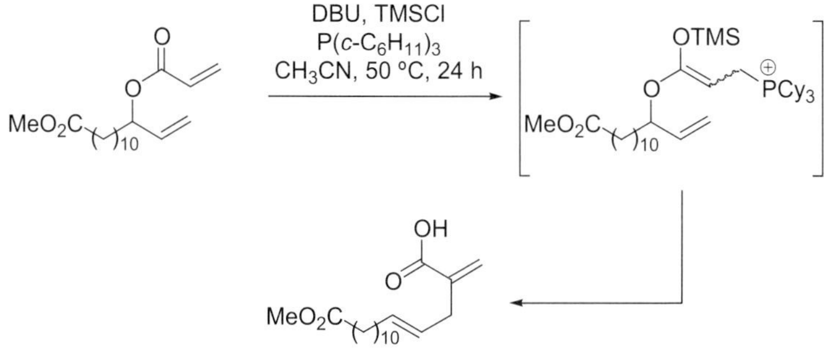

Scheme 4.11

Thomas and Smith adapted the Inanaga procedure in an approach to galbonolide B (Scheme 4.12) [13]. Tricyclohexylphosphine catalysis afforded the desired acrylate in almost quantitative yield.

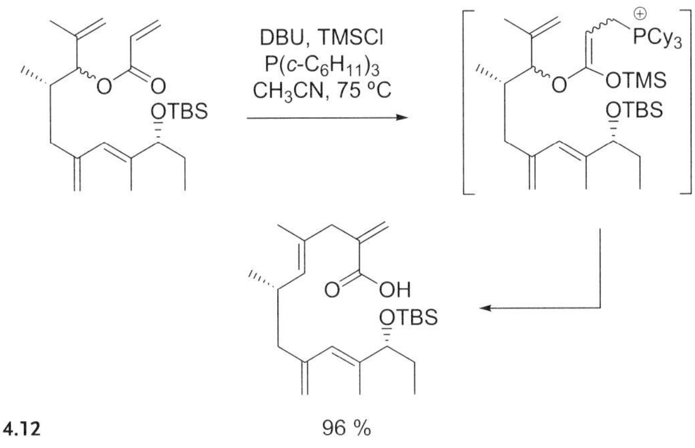
OTBS
DBU, TMSCl
P(c-C6H11)3
CH3CN, 75 °C
PCy3
OTMS
OTBS
OH
OTBS
96 %

Scheme 4.12

4.5 Transition State Structure

This section will describe experimental and theoretical studies of the structure of the transition state with a focus on the extent of bond formation and cleavage. A discussion of chair versus boat transition states is left to the section on stereochemistry.

4.5.1 Isotope Effect Studies

4.5.1.1 Deuterium Isotope Effects

Gajewski and Emrani examined the deuterium isotope effects in the rearrangement of the (D_2)C1 and (D_2)C4 TMS ketene acetals derived from allyl acetate (Scheme 4.13) [14]. They determined that the C4–O bond was 85% broken and the C1–C6 bond was 29% formed in the transition state. They concluded that the

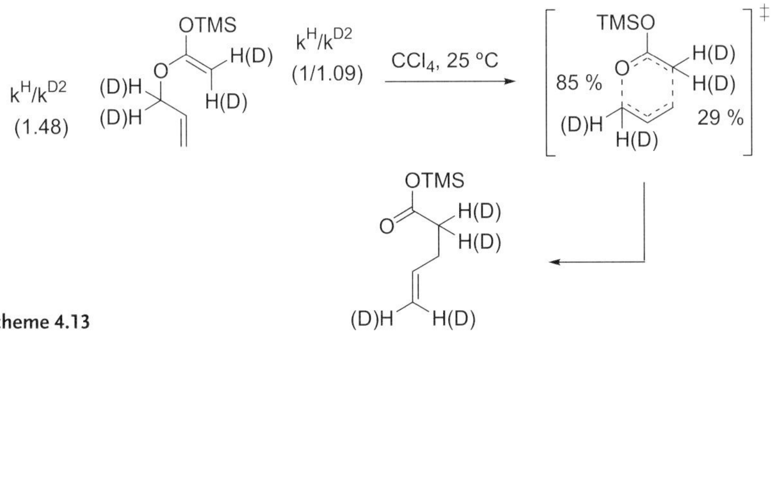
OTMS
H(D)
k^H/k^D2
(1.48)
(D)H
(D)H
H(D)
k^H/k^D2
(1/1.09)
CCl4, 25 °C
TMSO
H(D)
H(D)
85 %
29 %
(D)H
H(D)
OTMS
H(D)
H(D)
(D)H
H(D)

Scheme 4.13

effect of the C2-trimethylsilyloxy substituent was to alter the transition state toward a more oxallyl-allyl radical pair relative to the parent allyl vinyl ether due to a stabilization of the π bond of the oxallyl moiety by the C2 oxygen substituent.

4.5.1.2 ^{14}C Isotope Effects

Saunders, Shine and coworkers studied the ^{14}C kinetic isotope effects of the allyl vinyl ether and 2-OTMS allyl vinyl ether (Scheme 4.14) [15]. The calculated KIEs using the BEBOVIB program gave results similar to those of Gajewski and Emrani, with 80% bond breaking for the C4–O bond and 20% bond formation for the C1–C6 bond.

^{14}C KIE's	
C1	1.0164
C2	1.0241
C4	1.1048
C6	1.0174

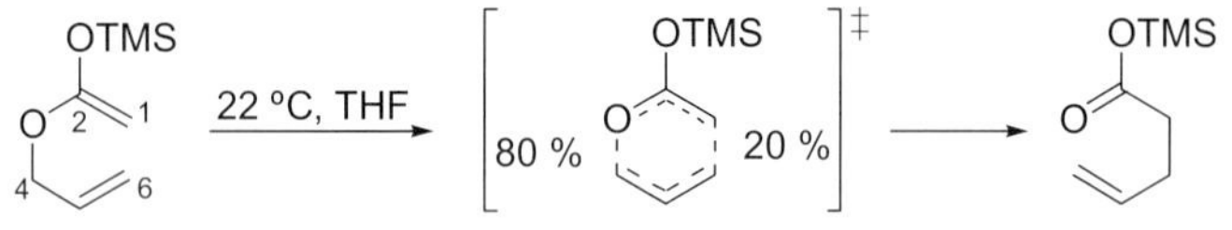

Scheme 4.14

4.5.2 Theoretical Studies

Extensive theoretical studies have been reported for the parent Claisen rearrangement and for allyl vinyl ethers related to the chorismate to prephenate Claisen rearrangement catalyzed by chorismate mutase [16]. There has been, by comparison, much less study of the Ireland–Claisen rearrangement.

4.5.2.1 Calculated vs. Experimental Isotope Effects and Transition State Structure

4.5.2.1.1 2-OH Allyl Vinyl Ether

Houk and Yoo investigated the effect of OH substituents on the Claisen rearrangement of allyl vinyl ether, including C2–OH, which is analogous to the Ireland–Claisen rearrangement [6c]. They found that CASSCF/6–31G* calculations showed the best agreement between experimental and calculated deuterium isotope effects, although the CASSCF/6–31G* transition state structure overestimated the degree of bond breaking relative to B3LYP/6–31G* and HF/6–31G*.

The bond lengths in the optimized transition structures at the three levels of theory are tabulated (Table 4.1). B3LYP/6–31G* gave the best agreement with activation energy relative to the experimentally observed activation every of the rearrangement of 2-OTMS allyl vinyl ether. The calculated relative activation energies for the 2-OH versus the parent allyl vinyl ether were most accurate at the RHF/6–31G* level.

Tab. 4.1 Calculated and experimental transition state structures and energies of C2–OH allyl vinyl ether.

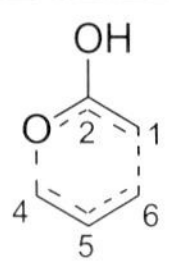

	RHF/6–31G* (Å)	B3LYP/6–31G* (Å)	MCSCF/6–31G* (Å)	Experimental
C1–C2	1.368	1.375	1.378	
C2–O	1.261	1.290	1.264	
O–C4	1.878	1.841	2.023	
C4–C5	1.397	1.413	1.408	
C5–C6	1.370	1.375	1.372	
C1–C6	2.334	2.418	2.605	
E_a (kcal/mol)	39.8	20.1	32.6	23.8
ΔE_a (kcal/mol)	9.1	6.7	9.9	8

4.5.2.1.2 Allyl Vinyl Ether

Singleton et al. later examined natural abundance ^{2}D, ^{13}C and ^{17}O isotope effects for the Claisen rearrangement of allyl vinyl ether [17]. They found that calculated kinetic isotope effects are most accurately predicted using MP4/6–31G*. They concluded that the MP4/6–31G* transition state structure therefore most accurately resembled the true structure of the transition state. They also proposed that the transition state structure is best thought of as allyl/oxallyl fragments bound by cyclic delocalization, rather than as a structure intermediate between an allyl/oxallyl radical pair and a 1,4-diyl. Although these results pertain to the Claisen rearrangement of allyl vinyl ether, similar conclusions may result from a study of the Ireland variant.

4.5.2.2 Cyclohexenyl Allyl Methyl Ketene Acetals

Houk et al. investigated the origins of the boat and chair preferences in the Ireland–Claisen rearrangements of cyclohexenyl esters [18]. Calculations carried out on the O-methyl ketene acetals using B3LYP/6–31G* were consistent with the experimentally observed levels of stereoselectivities in the rearrangements of a series of substituted cyclohexenyl O-trimethylsilyl ketene acetals. They found that the stereochemistry of the silyl ketene acetal and substitution on the cyclohexene ring both played significant roles in determining whether the rearrangement occurred via a chair-like or boat-like transition state. Although the details of this study will not be summarized here, these and earlier studies described above suggest that calculations using B3LYP/6–31G* level of theory may give sufficiently accurate results to predict the stereochemical outcome of Ireland–Claisen rearrangements when the result is not *a priori* obvious.

4.6 Stereochemical Aspects

One of the many attractive features of the Ireland–Claisen rearrangement lies in its ability to reliably transfer stereochemistry to one or two newly formed sp^3 stereocenters (C2 and/or C3 of the pentenoic acid) as well as to the resulting C4–C5 alkene. It has been found that the stereochemical outcome of the rearrangement may be governed by a variety of influences, some easily rationalized and others more subtle.

4.6.1 Simple Diastereoselection: Chair vs. Boat Transition States

A key aspect of stereoselectivity in the Ireland–Claisen rearrangement is the preferential formation of *syn* (erythro) or *anti* (threo) pentenoic acids from appropriately substituted allyl silyl ketene acetals. The stereochemical outcome of the reaction is determined by two features: (1) the geometry of the silyl ketene acetal and allylic alkene, and (2) whether the rearrangement proceeds via a chair-like or boat-like transition state.

4.6.1.1 Enolate and Silyl Ketene Acetal Geometry

In 1975 Ireland reported that ester enolates of propionates and related esters could be stereoselectively generated to give either the *Z*-(*O*)-Li-enolate or the *E*-(*O*)-Li-enolate (because the stereochemical designation of the enolates reverse upon silylation due to changes in substituent priority, the enolate geometries are referred to with respect to the enolate oxygen) (Scheme 4.15) [19]. When THF was used as solvent, the *E*-(*O*)-Li-enolate was trapped with TBSCl to give the *E*-silyl ketene acetal. When a 23% HMPA/THF mixture was used, the *Z*-(*O*)-Li-enolate was trapped as the *Z*-silyl ketene acetal. The *E*-selectivities were generally high (91:9 to 97:3) with the exception of phenyl acetates. (Enolization of esters where R = OR or NR_2 will be covered in Section 4.8.1).

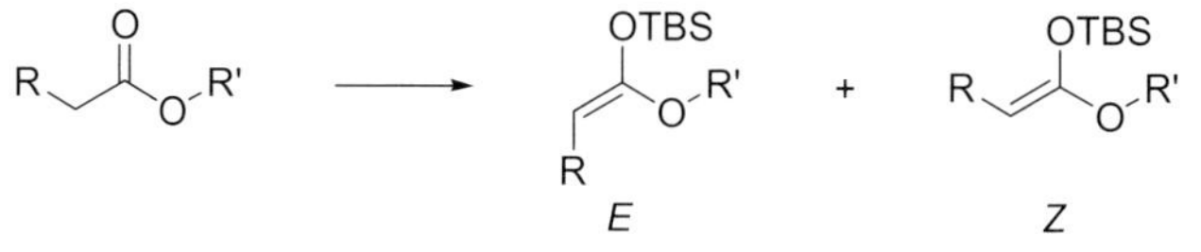

R	R'	*E*/*Z* (THF)	*E*/*Z* (23 vol% HMPA, THF)
Et	Me	91/9	16/84
t-Bu	Me	97/3	9/91
Et	*t*-Bu	95/5	23/77
Ph	Me	29/71	5/95

Scheme 4.15

4.6.1.2 Acyclic Allyl Silyl Ketene Acetals

Ireland demonstrated that the *anti* 2,3-dimethyl pentenoic acid isomer could be obtained either by rearrangement of the *E*-silyl ketene acetal of *E*-crotyl propionate or with the *Z*-silyl ketene acetal of the *Z*-crotyl propionate at comparable levels of diastereoselectivity (Scheme 4.16) [1]. The analogous results for the *syn* pentenoic acid can be obtained using the *Z*-silyl ketene acetal of *E*-crotyl propionate or with the *E*-silyl ketene acetal with *Z*-crotyl propionate. The diastereoselectivities varied from 5:1 to 8:1. Since the allyl silyl ketene acetals are achiral, the products are of course racemic.

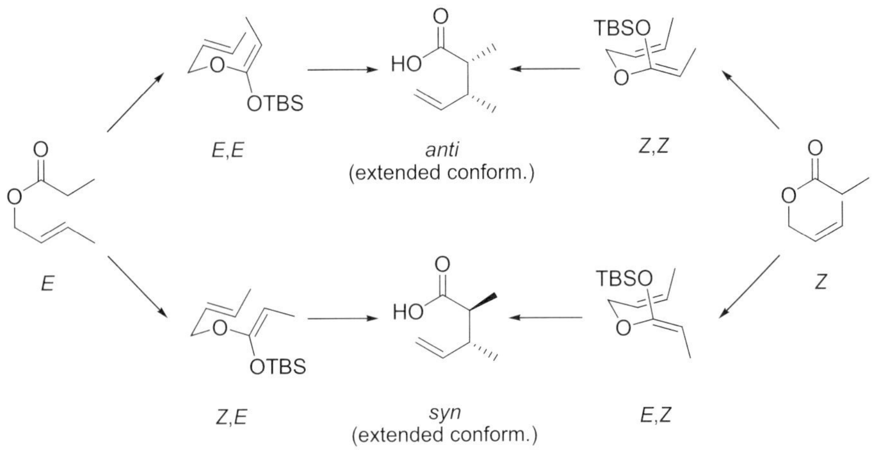

Scheme 4.16

The stereochemical outcomes are consistent with preferential rearrangement via a chair-like transition state. The calculated energy difference between the chair and boat transition states for 2-OCH_3 allyl vinyl ether at the B3LYP/6–31G* level of theory was reported to be 2.3 kcal/mol [18].

4.6.2 Diastereoface Differentiation: Cyclic Allyl Silyl Ketene Acetals

In contrast to the acyclic allyl silyl ketene acetals, cyclic substrates may preferentially rearrange through either the chair or boat transitions states. In 1981 Bartlett and Pizzo reported that treatment of cyclohexenyl propionates under either set of conditions reported by Ireland resulted in the formation of the same major isomer (Scheme 4.17) [20]. They concluded that the *E*-silyl ketene acetal rearranged preferentially via a chair-like transition state, while the *Z*-silyl ketene acetal rearranged via a boat-like transition state. These conclusions were recently supported computationally by Houk et al., who reported a 1.0 kcal/mol preference for the boat transition state for the *Z*-geometry in the analogous OMe ketene acetal and a 1.4 kcal/mol preference for the chair transition state for the *E*-geometry in the OMe ketene acetal [18].

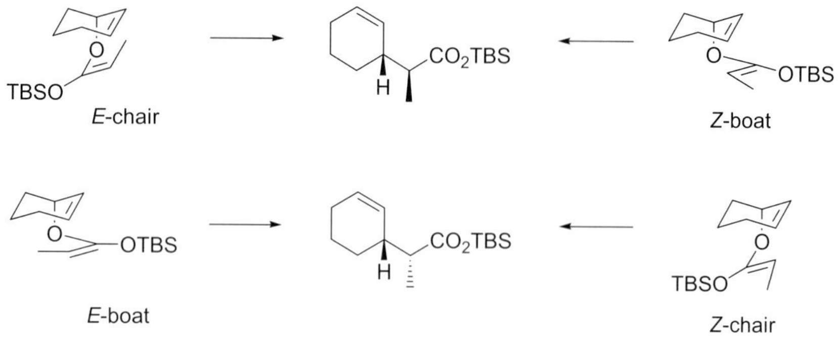

E-chair Z-boat

E-boat Z-chair

E-silyl ketene acetal: chair/boat= 85/15
Z-silyl ketene acetal: chair/boat= 25/75

Scheme 4.17

Ireland and others have found that the preference for a chair-like versus boat-like transition state for cyclic allyl silyl ketene acetals is dependent upon ring size, ring constitution (e.g., cyclohexene versus dihydropyran) and ring substituent stereochemistry. In the case of dihydropyran- and dihydrofuran-derived allylic esters, Ireland has suggested that the transition state is shifted to a more product-like geometry relative to the carbocyclic systems (Scheme 4.18) [21]. In these cases the boat transition state is favored irrespective of the silyl ketene acetal geometry. He proposed that the effect of the pyran oxygen was to increase the degree of C4–O bond cleavage in the transition state and hence diminish the steric interactions between the C1 methyl group and the pyran ring. He also proposed that the greater degree of bond cleavage results in a more polarized transition state. The boat transition state was argued to better stabilize the greater dipolar character due to improved overlap between the two allyl moieties. These proposals have not been evaluated theoretically.

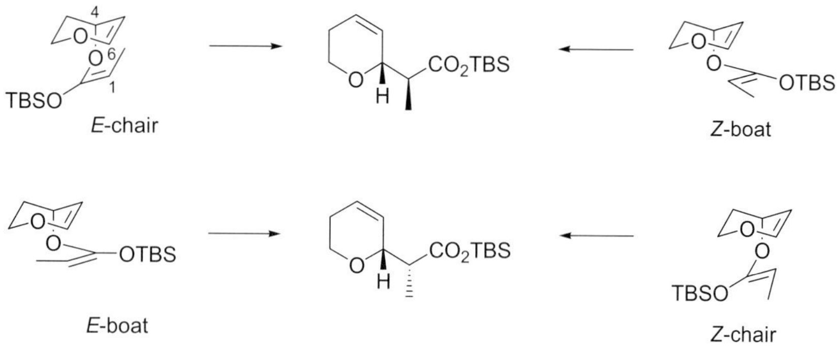

E-chair Z-boat

E-boat Z-chair

E-silyl ketene acetal: chair/boat= 29/71
Z-silyl ketene acetal: chair/boat= 14/86

Scheme 4.18

4.6.3
Alkene Stereochemistry

The rearrangement of allylic esters derived from 1° alcohols affords terminal alkenes. If the alcohol precursor is 2° or 3°, there is the possibility of two alkene isomers being formed. Because of the strong preference for the chair-like transition state in acyclic systems, the stereochemistry of the alkene is highly predictable in the case of 2° carbinol derived esters. The rearrangement occurs so as to place the larger C4 substituent in a pseudo-equatorial position. This is illustrated with the rearrangement reported by Katzenellenbogen and Christy in 1974 in a synthesis of a butterfly pheromone in which the *E*-alkene is selectively formed in the rearrangement (Scheme 4.19) [22].

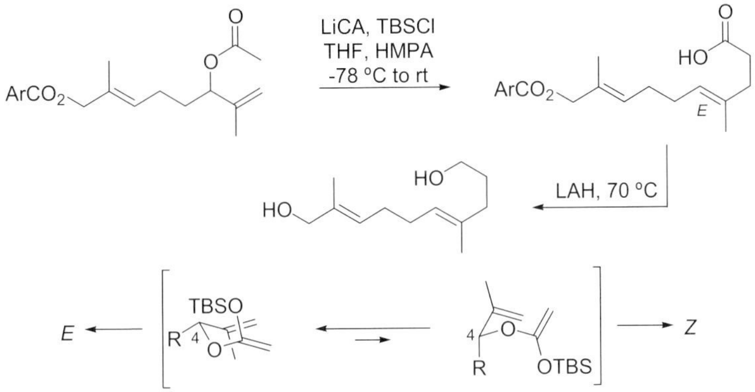

LiCA, TBSCl
THF, HMPA
-78 °C to rt

LAH, 70 °C

Scheme 4.19

4.6.4
Chirality Transfer

4.6.4.1 Allylic Esters Possessing One Stereocenter: Absolute Stereocontrol

Combining the concepts illustrated in Schemes 14 and 17 leads to the conclusion that use of a chiral non-racemic 2° alcohol will enable the transfer of chirality from the carbinol center to the newly formed stereocenter(s) at C2 and/or C3 of the pentenoic acid product. The first example of this type of rearrangement was reported by Ireland in which he employed a C4 TBS crotyl propionate as a chiral 1° alcohol equivalent (Scheme 4.20) [23]. This example also illustrates how the 1,2-shift of the alkene can alter functional group reactivity, i.e., the allyl silane in the reactant is converted to a vinyl silane in the product. The *ee*'s of the pentenoic acid products were not reported.

Panek et al. have made extensive use of this feature of the Ireland–Claisen rearrangement (Scheme 4.21) [24]. Treatment of the (*R*)-*O*-methyl glycolate with LDA and TMSCl gave the acid in excellent yield and diastereoselectivity. The resulting allyl silanes were used as nucleophiles in diastereoselective $S_E{}'$ additions.

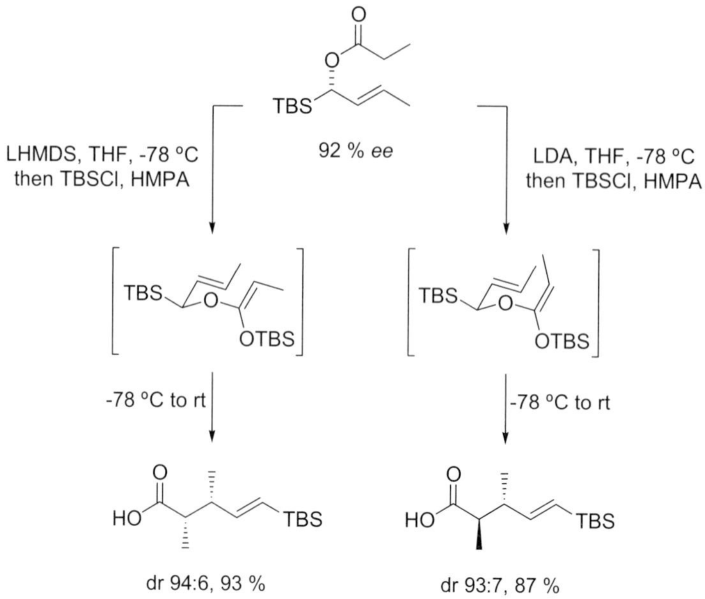

Scheme 4.20

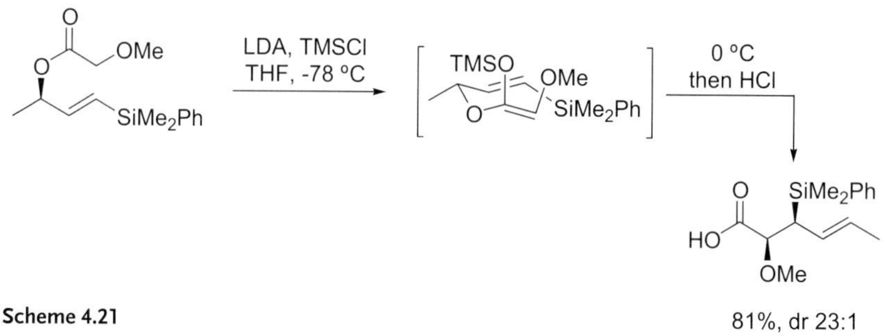

Scheme 4.21

4.6.4.2 **Allylic Esters Possessing Multiple Stereocenters: Relative Stereocontrol**

If the allylic ester possesses stereocenters in addition to the carbinol stereocenter, the Ireland–Claisen rearrangement may be used to control the relative configuration between the stereocenters. Transfer of chirality in the Claisen rearrangement occurs from the carbinol center in a 1,3-sense for the β-carboxyl stereocenter (indicated with an asterisk, Eq. 1, Scheme 4.22) or a 1,4-sense for the α-carboxyl stereocenter (indicated with an asterisk, Eq. 2). If the reaction to generate the carbinol stereocenter is coupled with a subsequent Claisen rearrangement, a net 1,n-stereocontrol will result.

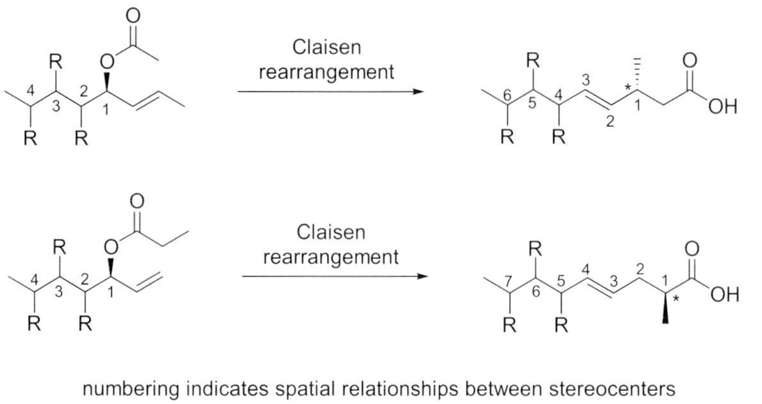

Claisen rearrangement

Claisen rearrangement

numbering indicates spatial relationships between stereocenters

Scheme 4.22

4.6.4.2.1 **1,4- and 1,5-Asymmetric Induction**

A wide variety of methods exist for the asymmetric synthesis of alcohols with 1,2- and 1,3-stereocontrol relative to a second stereocenter (cf. Scheme 4.22), such as aldol additions, substituted allyl metal additions to aldehydes, vinyl metal additions to α-chiral aldehydes, dihydroxylations, etc. When coupled with a subsequent Claisen rearrangement, a net 1,4- and 1,5-asymmetric induction can be achieved. A representative example will be shown; others can be seen in subsequent sections.

In 1988 Heathcock reported the Johnson, Eschenmoser and Ireland–Claisen rearrangements of ketene acetals derived from chiral non-racemic allylic alcohols (Scheme 4.23) [25]. The alcohols were themselves derived from Evans aldol additions. While the Johnson and Eschenmoser rearrangements were used to illus-

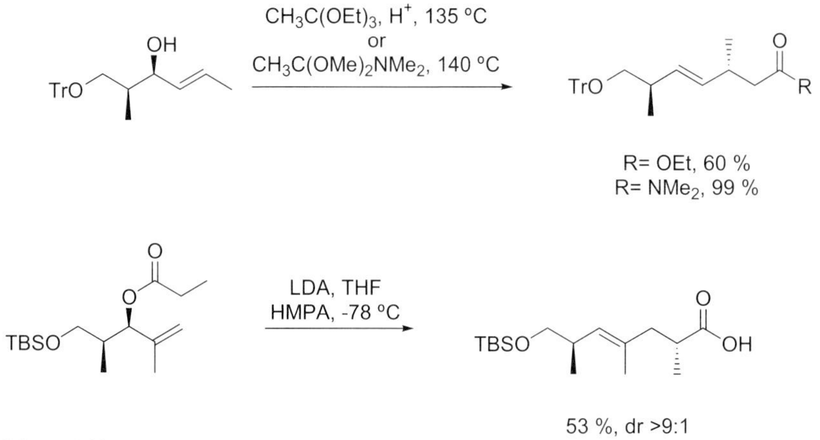

$CH_3C(OEt)_3$, H^+, 135 °C or $CH_3C(OMe)_2NMe_2$, 140 °C

R= OEt, 60 %
R= NMe_2, 99 %

LDA, THF HMPA, -78 °C

53 %, dr >9:1

Scheme 4.23

trate the 1,4-asymmetric induction (Eq. 1), the Ireland variant was the method of choice for 1,5-asymmetric induction because of the ability to control the silyl ketene acetal stereochemistry (Eq. 2).

4.6.4.2.2 1,6- and 1,7-Asymmetric Induction

To achieve a net 1,6- or 1,7-asymmetric induction, it would be necessary to first attain 1,3- or 1,4-asymmetric induction in the synthesis of the carbinol precursor to the Claisen rearrangement (cf. Scheme 4.22). High levels of 1,3- or 1,4-asymmetric induction are often observed in additions of metal nucleophiles to cyclohexenones. McIntosh et al. have reported high levels of 1,6- and 1,7-asymmetric induction using additions to cyclohexenones followed by Ireland–Claisen rearrangement (Scheme 4.24) [26]. The 3° *bis*-allylic esters were prepared by axially selective vinyl MgBr addition to the corresponding cyclohexenones with good (10:1, Eq. 2) to excellent (>20:1, Eq. 1) diastereoselectivity. Ireland–Claisen rearrangement then afforded the pentenoic acids with high levels of 1,6- and 1,7-asymmetric induction. A notable feature of carbonyl additions to cyclohexenones is that high levels of 1,3- and 1,4-asymmetric induction are easily obtained. By contrast, typically only modest levels of 1,2-asymmetric induction in acyclic systems when the stereogenic center is not substituted with a coordinating heteroatom, while 1,3- and 1,4-asymmetric induction is not generally possible.

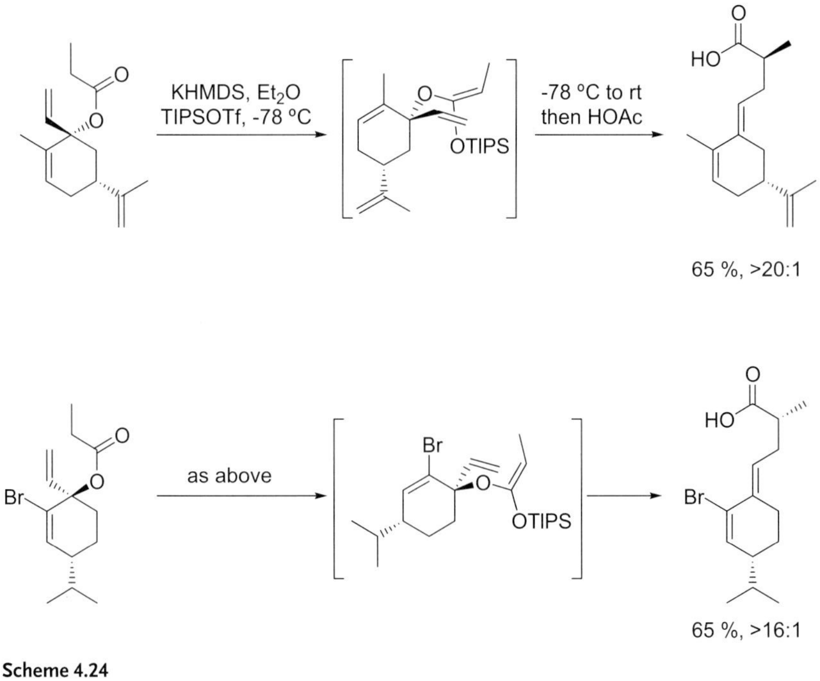

KHMDS, Et_2O
TIPSOTf, -78 °C

OTIPS

-78 °C to rt
then HOAc

HO

65 %, >20:1

Br

as above

OTIPS

65 %, >16:1

Scheme 4.24

4.6.5
Influence of Remote Stereocenters

The C4 stereocenter dominates the stereochemical outcome of the Ireland–Claisen rearrangements in appropriately substituted allyl silyl ketene acetals. For substrates lacking a C4 stereocenter, other more remote stereocenters may play a significant role in the stereochemical outcome. There are comparatively few cases of remote stereocontrol, so most examples will be described in this section. The stereocenters will be discussed based on their location relative to the allyl silyl ketene acetal. Carbons *α* to the C1 carbon will be designated C1′, alpha to C4, C4′ and so on (Fig. 4.1).

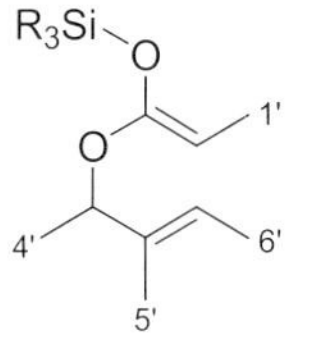

Figure 4.1

4.6.5.1 C1′ Stereocenters

4.6.5.1.1 C1′ CF_3 Substituent

In the Pd(II) catalyzed Ireland–Claisen rearrangement reported by Yamazaki et al. mentioned previously (cf. Scheme 4.9) a pronounced stereochemical effect was noted due to the C1′ CF_3 group (Scheme 4.25) [8]. The rearrangement apparently occurred via the *Z*-silyl ketene acetal and exhibited high 1,2-asymmetric induction. The lowest energy conformation about the C1–C1′ bond should have the

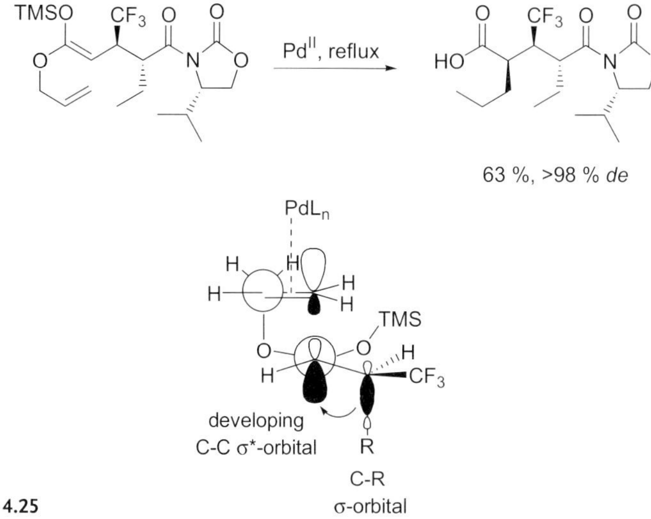

Scheme 4.25

geometry shown, with the allylic (C1′) hydrogen eclipsed with the *cis* oxygen [27]. Preferential attack of the allyl group *syn* to the CF_3 group would afford the observed stereoisomer. The authors suggested that the facial selectivity of the rearrangement was due to the Cieplak effect [28], with the new bond being formed *anti* to the more electron rich C–C σ bond. This would allow for donation of electron density from the C–R σ bond to the developing C1–C6 σ^* bond (allyl ketene acetal numbering). Replacement of the CF_3 group with a CH_3 group resulted in significantly lower diastereoselectivity.

4.6.5.1.2 C1′ CH_3 Substituent

Gilbert et al. [29] recently found that Ireland–Claisen rearrangement of a prenyl cyclopentyl ester with a CH_3 substituent at C1′ gave high facial selectivity (Scheme 4.26). The facial selectivity was due to transition states in which the prenyl group approached *anti* to the C1′ CH_3 group. The CH_3 group is presumably disposed pseudo-axially with respect to the cyclopentane ring due to allylic strain with the ketene acetal oxygens. Rearrangement of both *E*- and *Z*-silyl ketene acetals through either chair or boat transition states would afford the same diastereomer. When *Z*- or *E*-crotyl esters were employed, the same stereoselectivity was obtained at the newly formed 4° center (C2, pentenoic acid numbering), but the diastereoselectivity at C3 was low, presumably due to a lack of stereoselectivity in the formation of the silyl ketene acetal.

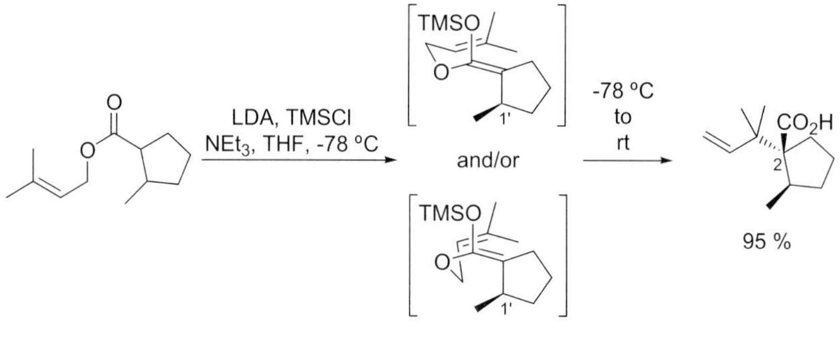

LDA, TMSCl
NEt_3, THF, -78 °C
TMSO
O
1'
and/or
TMSO
O
1'
+ *Z*-silyl ketene acetals
-78 °C
to
rt
CO_2H
2
95 %

Scheme 4.26

4.6.5.1.3 C1′ CO_2Bn Substituent

Martin et al. observed modest to excellent diastereoselectivity for substituted succinate esters (Scheme 4.27) [30]. A similar Cieplak-type effect as in the Yamazaki report (cf. Scheme 4.25) could be argued in these cases as well. Allylic strain would orient the C1′ stereocenter as shown with the C–H bond eclipsed with the ether oxygen. Attack of the allylic ether would then occur *anti* to the more electron rich cyclopentyl C–C bond. Alternatively, a simple allylic strain argument could account for the observed results [27]. Attack of the allylic alkene would occur *syn* to the smaller CO_2Bn group.

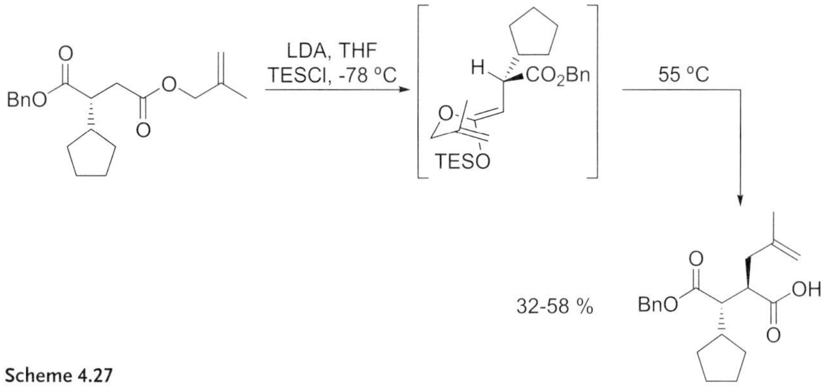

Scheme 4.27

4.6.5.1.4 C1′ NR_2 Substituents

Knight et al. found that the Ireland–Claisen rearrangement of C1′ NR_2-substituted allyl silyl ketene acetals gave poor to excellent diastereoselectivity depending on whether the C1′ stereocenter was contained within a 5- or 6-membered ring (Scheme 4.28) [31]. A piperidine derivative (n= 1) gave 94:6 diastereoselectivity, while a pyrrolidine derivative (n= 0) gave only 1.3:1. They rationalized the stereochemical outcome in the piperidinyl case by assuming a *Z*-silyl ketene acetal based on earlier precedent [46]. The silyl ketene acetal was also presumed to adopt an axial disposition relative to the piperidine ring to avoid allylic strain with the Boc group. If the C1′ hydrogen is oriented approximately co-planar to the OTMS group, as would be expected by allylic strain, then approach of the allyl group over the sp^2 hybridized nitrogen of the carbamate should be less sterically crowded than approach over the ring.

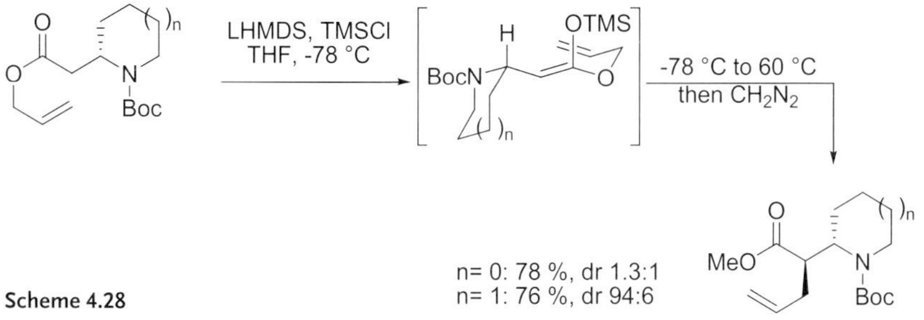

Scheme 4.28

4.6.5.1.5 C1′ OR Substituents

In 1984 Fujisawa et al. observed a directing effect due to a C1′ oxygen stereocenter (Scheme 4.29) [32]. Tandem deprotonation and silylation of the β-hydroxy allylic ester presumably gave the intermediate silyloxy silyl ketene acetal. The *Z*-configuration of the silyl ketene acetal was a consequence of chelation of the intermediate dianion by the Li cation. The authors postulated that the allylic alkene preferentially approaches the silyl ketene acetal *anti* to the TMSO group. The C1,C1′ bond presumably adopted a conformation with the C1′ hydrogen eclipsed with the OTMS group of the ketene acetal to minimize allylic strain. The substrate-induced diastereoselectivity was modest (21:79 for the *E*-crotyl ester, 28:72 for the *Z*-crotyl ester) as was the simple (*syn*/*anti*) diastereoselectivity (25:75 for the *E*-crotyl ester, 29:71 for the *Z*-crotyl ester for the major products).

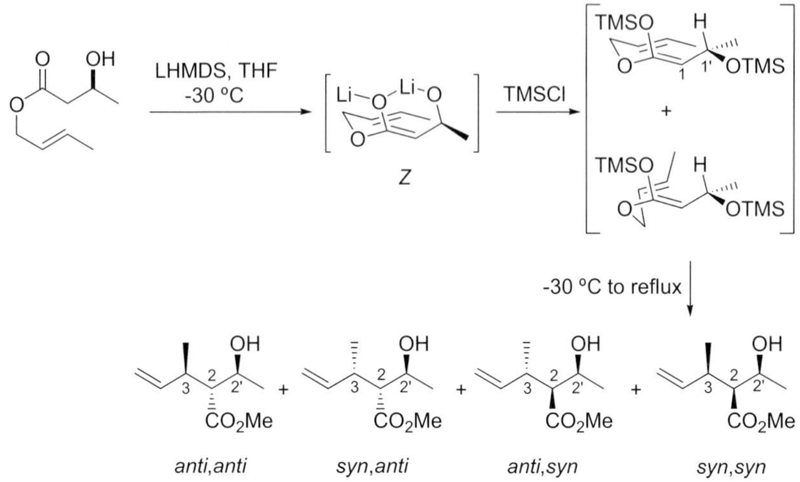

LHMDS, THF, -30 °C; TMSCl; -30 °C to reflux

anti,anti + *syn,anti* + *anti,syn* + *syn,syn*

E-crotonate: 37 %, dr 3:18:72:7
substrate-induced diastereoselectivity: C2/C2′-*anti* : C2/C2′-*syn*= 21:79
simple (*syn*/*anti*) diastereoselectivity: C2/C3-*syn* : C2/C3-*anti*= 25:75
Z-crotonate (not depicted): 24 %, dr 25:3:4:68
substrate-induced diastereoselectivity: C2/C2′-*anti* : C2/C2′-*syn*= 28:72
simple (*syn*/*anti*) diastereoselectivity: C2/C3-*syn* : C2/C3-*anti*= 71:29

Scheme 4.29

The diastereofacial selectivity is inconsistent with a Cieplak model, since the crotyl group would have to approach from the side *syn* to the C1′ methyl. The low C2, C3 *syn*/*anti* selectivity could be due to competitive rearrangement via a boat transition state, although it is not obvious why the boat transition state should be of similar energy in this case.

Gilbert et al. also observed a difference in diastereoselectivity as a function of ring size in an approach to the trichodienes (Scheme 4.30) [33]. Enolization of the C1′ methoxy ester at low temperature preferentially afforded the *Z*-silyl ketene acetal as a result of chelation between the enolate and the OMe group. Ireland–Claisen rearrangement occurred via a chair transition state with the allylic alkene approaching *syn* to the C1′ OMe group. This somewhat surprising result was rationalized as a consequence of competing steric interactions. Allylic strain disfavors the diequatorial cyclohexane conformer, but 1,3-diaxial strain disfavors the diaxial conformer. The authors proposed that approach of the allylic alkene occurred in an equatorial fashion to the diaxial conformer. Axial approach of the allylic alkene to either diequatorial conformer would be disfavored by diaxial interactions between the allyl group and the axial hydrogens. Equatorial approach to the diequatorial conformer would result in an unfavorable eclipsing interaction between the C1′ OMe group and the forming carboxyl group. By contrast, equatorial approach to the diaxial conformer would avoid both of those interactions.

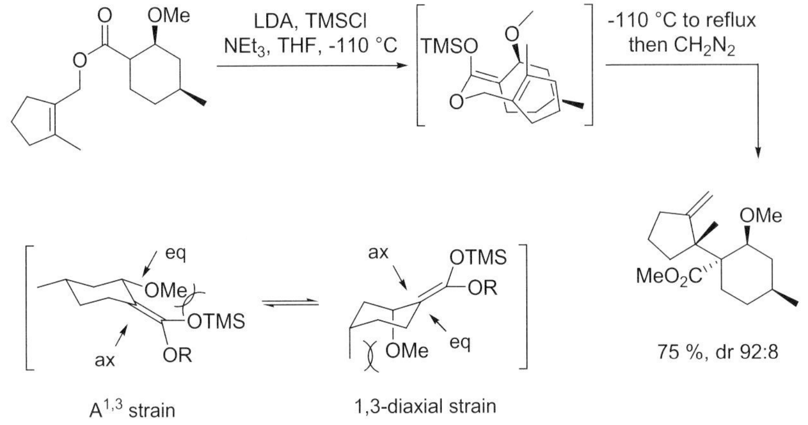

Scheme 4.30

In contrast to the systems described above, rearrangement of the analogous cyclopentyl substrate resulted in an almost complete lack of facial selectivity (Scheme 4.31).

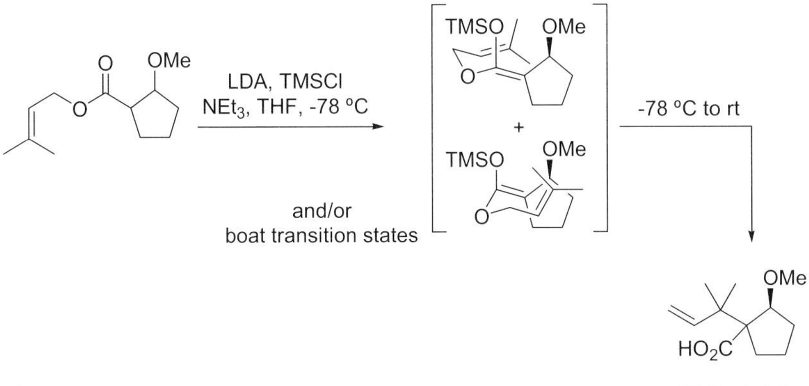

Scheme 4.31

4.6.5.1.6 C1′ SiR_3 Substituents

Fleming and Betson reported that Ireland–Claisen rearrangement of either the *E*- or *Z*-silyl ketene acetals bearing a stereocenter at C1′ possessing a TBDPS group gave moderate (R = *i*-Pr) to excellent (R = Me, Ph) 1,2-asymmetric induction (Scheme 4.32) [34]. The transition states were proposed to adopt a conformation with the C1′ hydro-

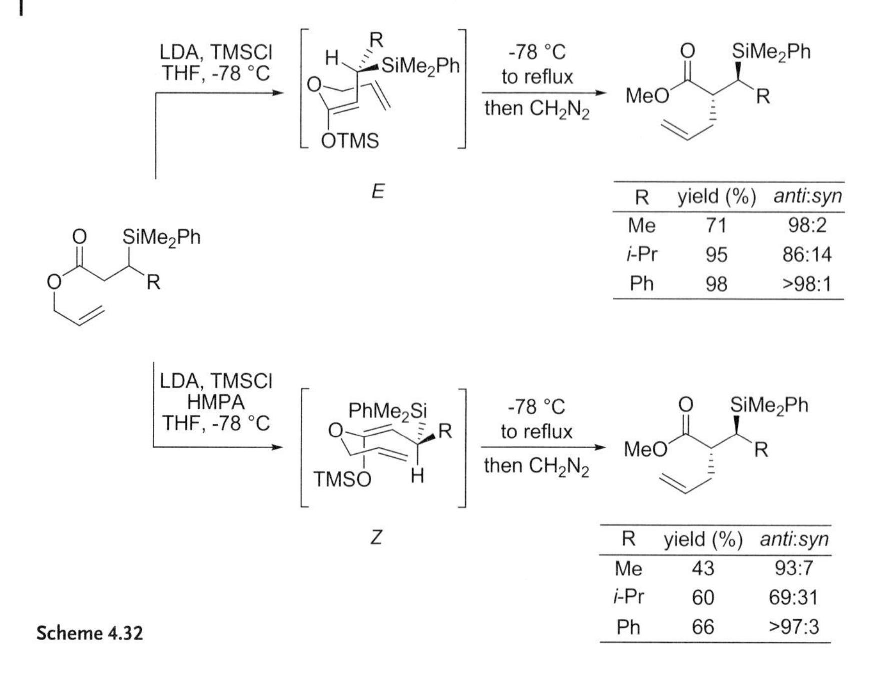

R	yield (%)	*anti*:*syn*
Me	71	98:2
i-Pr	95	86:14
Ph	98	>98:1

R	yield (%)	*anti*:*syn*
Me	43	93:7
i-Pr	60	69:31
Ph	66	>97:3

Scheme 4.32

gen eclipsed with either the alkyl ether oxygen (*E*-isomer) or the silyl ether oxygen (*Z*-isomer) so as to minimize allylic strain. The authors compared the results to those obtained by Yamazaki, in which the allylic alkene approached *anti* to the more electron rich C–Si σ bond. However, it seems equally plausible that allylic strain could be the controlling factor, with the allyl group approaching *syn* to the smaller R group.

4.6.5.2 **C5′ Stereocenters**

Ishizaki et al. found that use of a triisopropylsilyl (TIPS) protecting group gave optimal facial selectivity in the Ireland–Claisen rearrangement of C5′ hydroxyalkyl substituted allylic alkene, with the silyl ketene acetal attacking *anti* to the O-silyl group (Scheme 4.33) [35]. The rearrangement presumably proceeded via a chair-

LDA
TBSCl
HMPA
-78°C-rt

SiR_3	yield (%)	*trans*:*cis*
TIPS	61	10.5:1
TBS	64	3.7:1
$SiPh_3$	51	2:1

Scheme 4.33

like transition state, although in this case both chair and boat transition states would yield the same product. Assuming the $OSiR_3$ group is pseudo-axial to avoid $A^{1,2}$ strain with the acetoxymethyl group, the Ireland–Claisen rearrangement occurred via equatorial attack. This is the opposite facial selectivity to that previously observed by Ireland in a sterically unbiased system (cf. Section 4.6.5.4.2).

4.6.5.3 C6′ Stereocenters

4.6.5.3.1 C6′ NR_2 Substituents

In 1993 Mulzer and Shanyoor reported the Ireland–Claisen rearrangement of an allyl silyl ketene acetal derived from an allyl glycolate that possessed a C6′ *N*-Boc substituted stereocenter (Scheme 4.34) [36]. The rearrangement occurred with complete diastereoselectivity and high yield. The stereochemical outcome was rationalized using a frontier molecular orbital analysis proposed by Kahn and Hehre [37]. The electron deficient allylic alkene orients itself so as to place the lower energy *C–N σ** orbital approximately parallel to the *π** orbital of the allylic alkene, thus lowering the energy of the combined orbitals (LUMO). The analysis parallels the Felkin–Anh analysis of nucleophilic additions to *α*-chiral aldehydes [38]. It is noteworthy that allylic strain would contribute little to stabilizing the conformer shown above, since the alkene bears only a proton substituent at C5 [27].

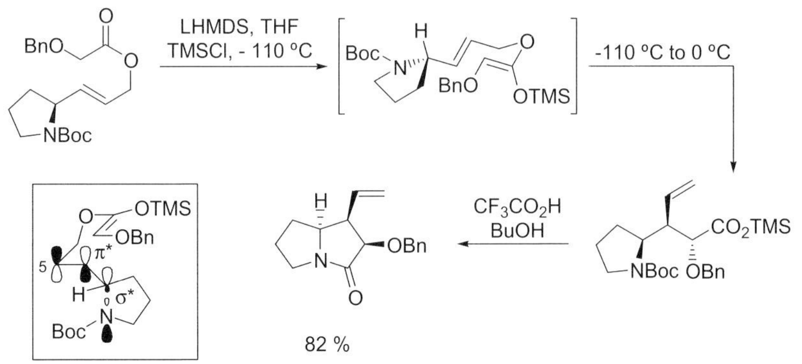

BnO
NBoc
LHMDS, THF
TMSCl, - 110 °C
Boc
H
O
BnO
OTMS
-110 °C to 0 °C
OTMS
OBn
π*
5
H
σ*
Boc
N
H
N
OBn
O
82 %
CF_3CO_2H
BuOH
CO_2TMS
NBoc OBn

Scheme 4.34

Also in 1993, Hauske and Julin reported a similar Ireland–Claisen rearrangement of an acyclic C6′ carbamate (Scheme 4.35) [39]. The authors examined three different silyl ketene acetals in the rearrangement, although no experimental details were provided. All three examples apparently proceeded with complete facial selectivity with respect to the allylic alkene to afford the *syn* stereochemistry between the allyl group and the NHBoc group in the conformation shown. The same rationale for facial selectivity can be applied as for Mulzer's results in the previous scheme. The reason for the low C2,C3 *syn*/*anti* diastereoselectivity in the propionate example was not addressed. A lack of control of enolate geometry or post-rearrangement epimerization are both possible.

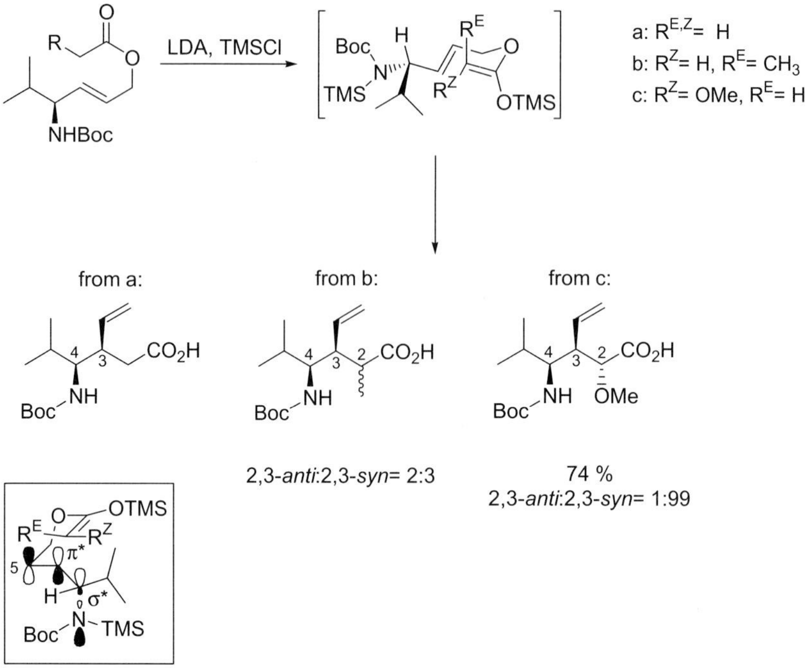

LDA, TMSCl
Boc
H
N
TMS
R^E
R^Z
O
OTMS
a: R^{E,Z}= H
b: R^Z= H, R^E= CH3
c: R^Z= OMe, R^E= H
NHBoc
from a:
from b:
from c:
CO2H
NH
Boc
OMe
2,3-anti:2,3-syn= 2:3
74 %
2,3-anti:2,3-syn= 1:99
π*
σ*

Scheme 4.35

4.6.5.3.2 C6′ OR Substituents

The earliest examples of Ireland–Claisen rearrangements of allyl silyl ketene acetals bearing a stereocenter at C6′ were reported by Cha and Lewis in 1984 (Scheme 4.36) [40]. In contrast to the nitrogen C6′ substituents, oxygen substituents exhibited considerably less facial bias. Rearrangements of the acetate esters of either the *E* or *Z* alkenes gave only 1.3:1 and 1.4:1 C3,C6′ *anti*:*syn* ratios, respectively.

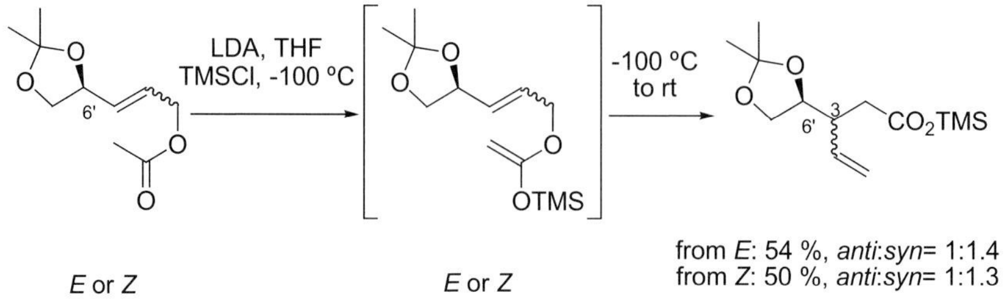

LDA, THF
TMSCl, -100 °C
-100 °C
to rt
OTMS
CO2TMS
E or Z
E or Z
from E: 54 %, anti:syn= 1:1.4
from Z: 50 %, anti:syn= 1:1.3

Scheme 4.36

The optimal facial selectivity was obtained when *O*-Me glycolate esters were employed using the *Z* allylic alkene (Scheme 4.37). All four possible isomers were obtained in a 0.5:0.3:9:1 ratio. A 7.3:1 ratio of C3,C4 *syn* isomers were obtained

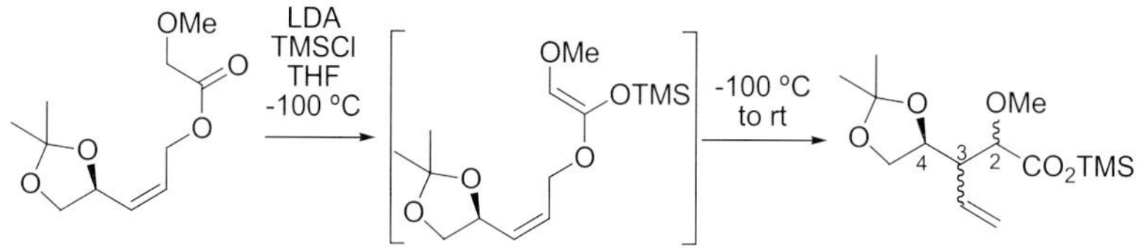

48 %
2,3-*syn*-3,4-*syn* : 2,3-*syn*-3,4-*anti*= 9:1
2,3-*syn*-3,4-*syn* : 2,3-*anti*-3,4-*syn*= 12.5:1

Scheme 4.37

relative to the C3,C4 *anti* isomers. A 12.5:1 ratio was obtained for the two *syn* C2,C3 isomers relative to the two *anti* C2,C3 isomers.

The higher diastereofacial selectivity obtained with the glycolate relative to the acetate ester is consistent with the Kahn and Hehre analysis (cf. Scheme 4.34 and 4.35). The more electron rich methoxy silyl ketene acetal might be expected to undergo rearrangement at a lower temperature than the unsubstituted silyl ketene acetal and hence a higher diastereofacial selectivity is observed. From an FMO perspective, the effect of the OMe group on the silyl ketene acetal is to raise the HOMO energy relative to the acetate-derived silyl ketene acetal. The better HOMO/LUMO energy match would lead to a lowering of the activation energy and thus reaction temperature.

4.6.5.3.3 **C6′ SiR_3 Substituents**

Fleming and Betson found that rearrangement of C6′ silyl substituted allyl silyl ketene acetals proceeded with high levels of diastereoselectivity when the allylic alkene was *Z* configured, but low level when the alkene was *E* configured (Scheme 4.38) [34]. These results are consistent with the higher allylic strain

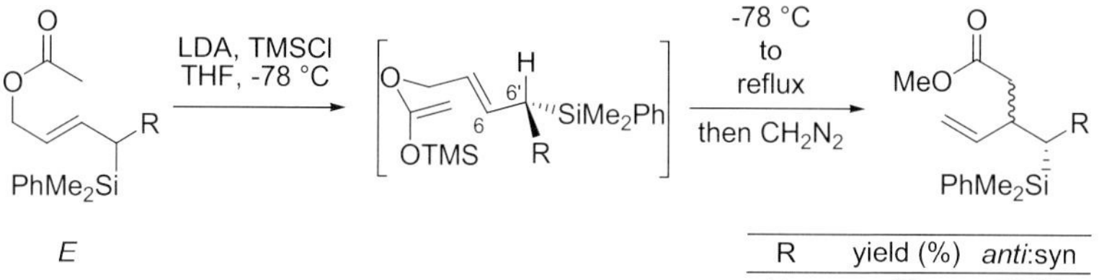

E

R	yield (%)	*anti*:syn
Me	68	38:62
i-Pr	29	52:48

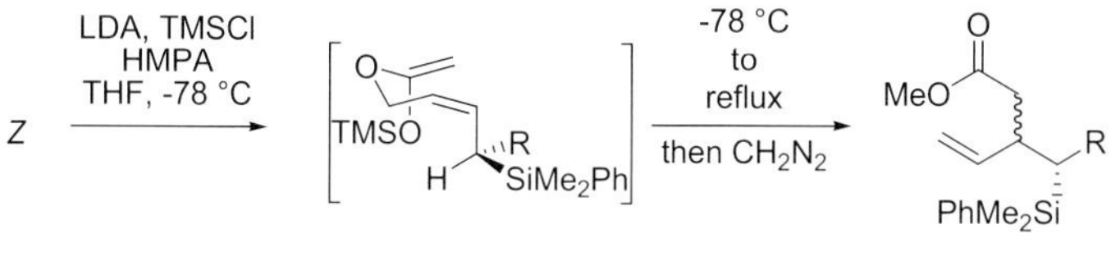

Z

R	yield (%)	*anti*:syn
Me	53	93:7
i-Pr	44	98:2

Scheme 4.38

inherent in the *Z* diastereomer. Rotation about the C6, C6′ bond in the *E* isomer results in a significantly lower energy penalty than in the *Z* isomer. Approach of the silyl ketene acetal occurs preferentially *anti* to the C–Si bond.

These two results taken together suggest a solely sterically driven transition state preference. In Eq. 1, the lack of controlling allylic strain results in low diastereoselectivity, while in Eq. 2, the more restricted conformation results in attack opposite to the bulky $SiMe_2Ph$ group. If electronic factors were predominant, the reaction in Eq. (1) should also afford higher diastereoselectivity (cf. Scheme 4.35).

4.6.5.4 Other Remote Stereocenters

4.6.5.4.1 C1″ Substituent

Djerassi and Shu observed no asymmetric induction in the Ireland–Claisen rearrangement of a steroid derived crotyl ester possessing an all-carbon stereocenter in the C1″ position (Scheme 4.39) [41]. In the case of the *Z*–silyl ketene acetal, all four possible isomers were obtained in a 7:7:1:1 ratio, with the major products being the expected *syn* isomers.

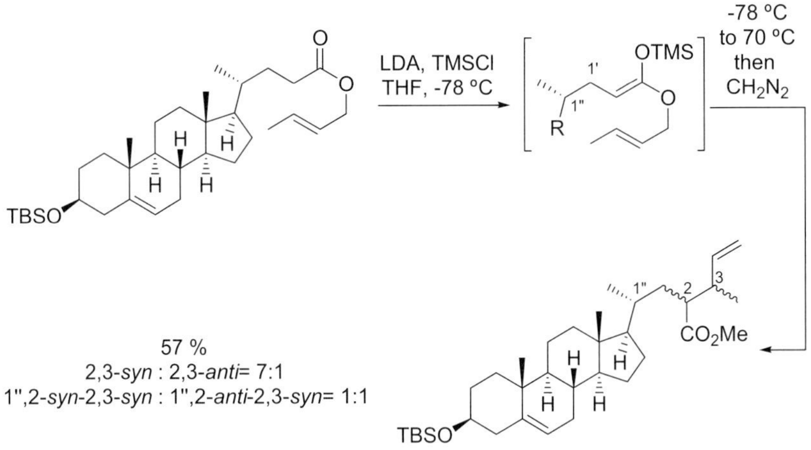

LDA, TMSCl
THF, -78 °C

-78 °C
to 70 °C
then
CH_2N_2

57 %
2,3-*syn* : 2,3-*anti*= 7:1
1″,2-*syn*-2,3-*syn* : 1″,2-*anti*-2,3-*syn*= 1:1

Scheme 4.39

4.6.5.4.2 Axial vs. Equatorial Attack

In 1983 Ireland and Varney examined the propensity for axial versus equatorial attack in the rearrangement of cyclohexenyl acetates (Scheme 4.40) [42]. They observed a *ca.* 10:1 preference for axial attack of the silyl ketene acetal to selectively afford the higher energy 1,4-*cis* isomer. Although either chair or boat transition states would afford the observed product, the authors argued that the chair transition state is less hindered for the acetate-derived silyl ketene acetal. They also studied the rearrangements of the propionate-derived *E*- and *Z*-silyl ketene acetals. While the axial attack was still favored, the axial/equatorial ratio decreased to

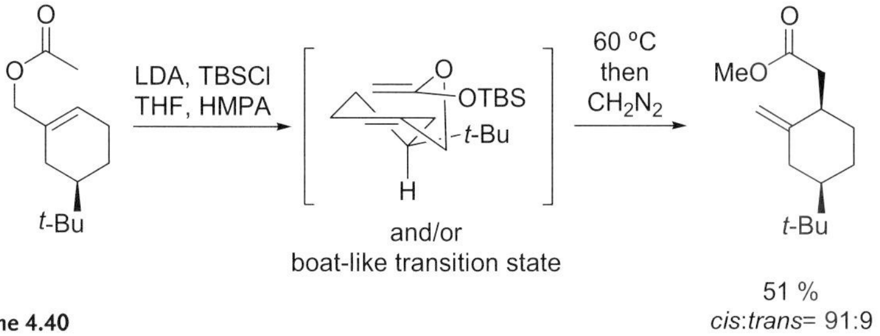

Scheme 4.40

71:29 for the *E*- and 87:13 for the *Z*-silyl ketene acetals. The *syn/anti* diastereomeric ratio was 1:1 for both axial rearrangement products.

By contrast, Deslongchamps et al. observed selective equatorial attack of the *Z*-silyl ketene acetal in a spiroketal (Scheme 4.41). Axial attack would be presumably inhibited by 1,3-diaxial interactions between the silyl ketene acetal and the axial oxygen substituent [43].

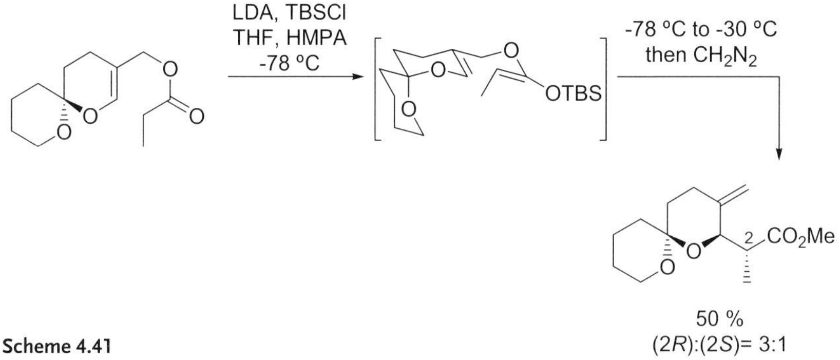

Scheme 4.41

4.6.6 Chiral Auxiliary Mediated Asymmetric Ireland–Claisen Rearrangements

4.6.6.1 Chiral Glycolates

Kallmerten and Gould found that rearrangement of allyl glycolates in which the glycolate oxygen was substituted with a chiral phenethyl group induced modest selectivities at C2 and or C3 of the pentenoic acid products (Scheme 4.42) [44]. Treatment of the crotyl glycolate using the (R)-phenethyl ether afforded a 3:1 mixture of *syn* pentenoic esters in good yield.

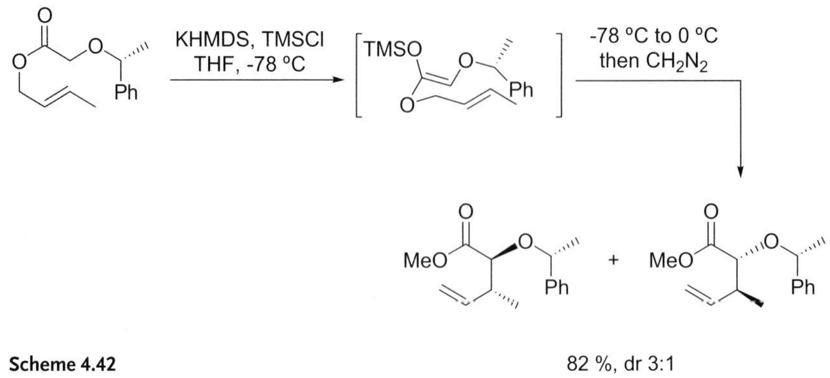
KHMDS, TMSCl
THF, -78 °C
TMSO
Ph
-78 °C to 0 °C
then CH_2N_2
MeO
Ph
+
MeO
Ph
82 %, dr 3:1

Scheme 4.42

4.6.6.2 **Chiral Glycinates**

Saigo et al. reported a related approach to absolute asymmetric induction by use of an aminoindanol auxiliary (Scheme 4.43) [45]. Treatment of the allyl glycinate with NaHMDS and the unusual silylating agent $HSiMe_2Cl$ gave 92:8 ratio of C2 stereoisomers. The authors proposed that an *E*-silyl ketene acetal was formed and adopted the conformation shown to minimize dipolar interactions. Approach of the allylic alkene from the more sterically accessible face of the oxazolidine would then yield the observed product. The authors did not indicate how the auxiliary would be removed.

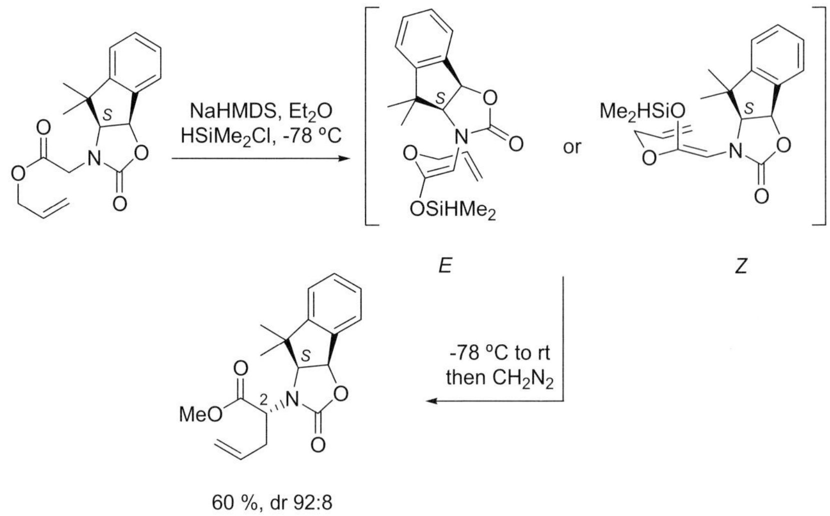
NaHMDS, Et_2O
$HSiMe_2Cl$, -78 °C
$OSiHMe_2$
or
Me_2HSiO
E
Z
-78 °C to rt
then CH_2N_2
MeO
60 %, dr 92:8

Scheme 4.43

Previous precedent with related carbamates would suggest, however, that chelation would favor a *Z*-(*O*)-Li-enolate and hence a *Z*-silyl ketene acetal (cf. Scheme 4.28) [31, 46]. Adoption of the less sterically congested conformation shown would also lead to the observed product.

4.6.6.3 Chiral Boron Ketene Acetals

To date the most general chiral auxiliary mediated asymmetric Ireland–Claisen rearrangement is that of Corey et al. (Scheme 4.44) [47]. They found that treatment of crotyl propionates and related esters afforded good yields, diastereoselectivities and enantioselectivities of the pentenoic acid products. The rearrangements also occurred at significantly lower temperature than the silyl ketene acetals. A key advantage of the chemistry is that the chiral auxiliary attachment, Ireland–Claisen rearrangement, and auxiliary removal all occur in one pot.

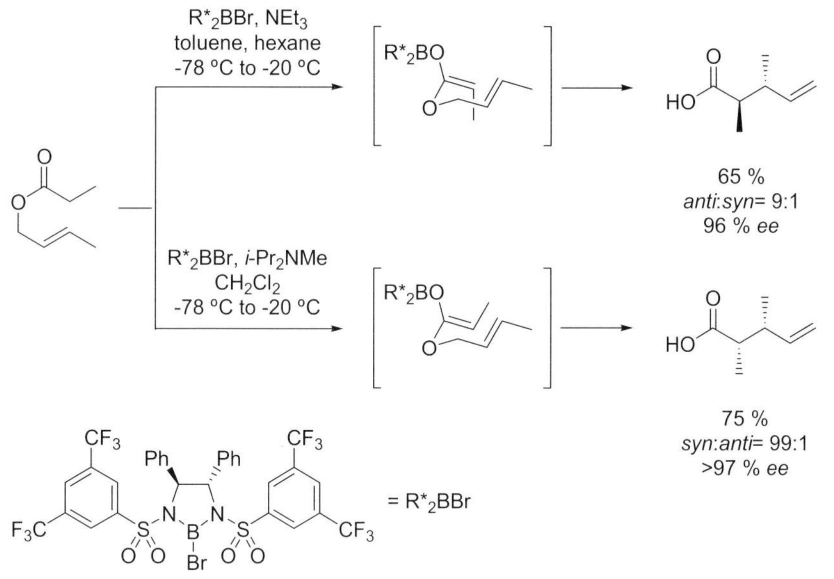

Scheme 4.44

4.7 Methods of Ketene Acetal Formation

The most frequently employed method of formation of silyl ketene acetals is deprotonation of an allylic ester by a strong base, most commonly LDA, LHMDS or KHMDS. A variety of other strategies have been developed for generation of the silyl ketene acetal intermediate.

4.7.1
Chemoselective Deprotonations

Since the Ireland–Claisen rearrangement typically begins with deprotonation of an allylic ester, the scope of the reaction is potentially limited by the presence of other acidic protons in the molecule. Several examples of selective deprotonation of esters in the presence of other carbon acids have been reported.

4.7.1.1 Ester vs. Ketone

In 1983, Burke et al. reported the enolization of an allylic glycolate in the presence of the unprotected cyclopentenone (Scheme 4.45) [48]. Only slightly over one equivalent of base was necessary to effect the enolization at −100 °C, indicating that concomitant enolization of the enone was not competitive. The stereochemical outcome of the rearrangement of the *Z*-silyl ketene acetal is consistent with the expected chair-like transition state.

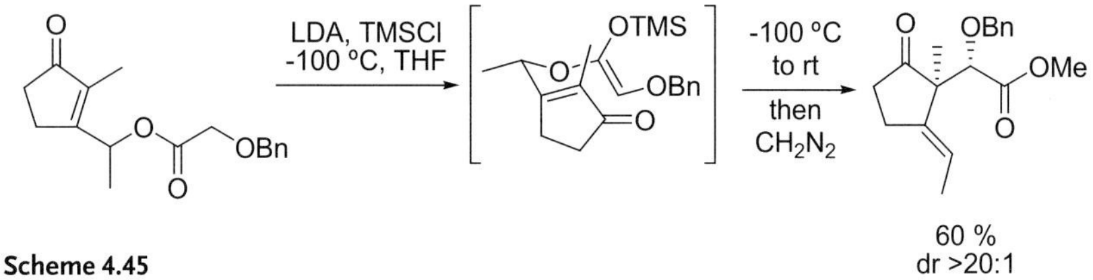

LDA, TMSCl
-100 °C, THF

OTMS
OBn

-100 °C
to rt
then
CH_2N_2

OBn
OMe

60 %
dr >20:1

Scheme 4.45

Paterson et al. reported several examples of enolizations of propionate esters in the presence of ketones bearing two different enolizable hydrogens using an internal quench protocol (Scheme 4.46) [49]. The conformation of the highly congested ketone presumably inhibited abstraction of the *α*-keto protons. However, concentrations of LDA greater than 15 mM resulted in 10–20% of elimination products resulting from ketone enolization.

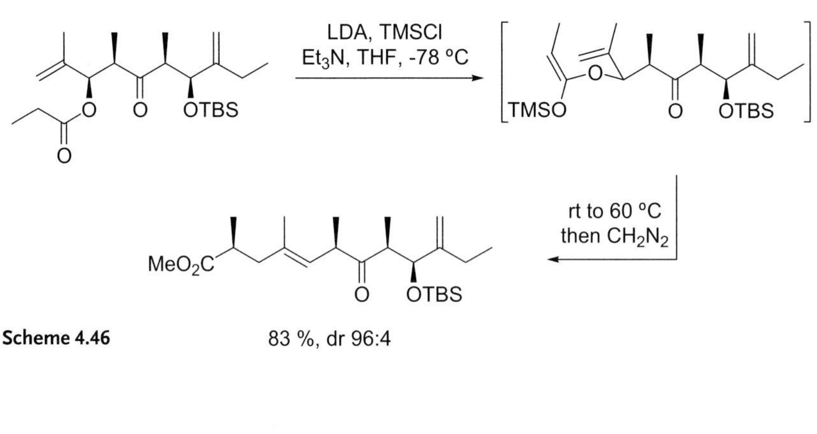

LDA, TMSCl
Et_3N, THF, -78 °C

OTBS

TMSO
OTBS

rt to 60 °C
then CH_2N_2

MeO_2C
OTBS

Scheme 4.46

83 %, dr 96:4

4.7.1.2 Ester vs. Butenolide

In 1992 Burke et al. reported the selective enolization of a glycolate in the presence of a butenolide (Scheme 4.47) [50]. The rearrangement of the *Z*-silyl ketene acetal occurred via a chair-like transition state to afford the *exo* methylene butyrolactone. The pentenoic acid products were transformed into isoavenaciolide and related targets.

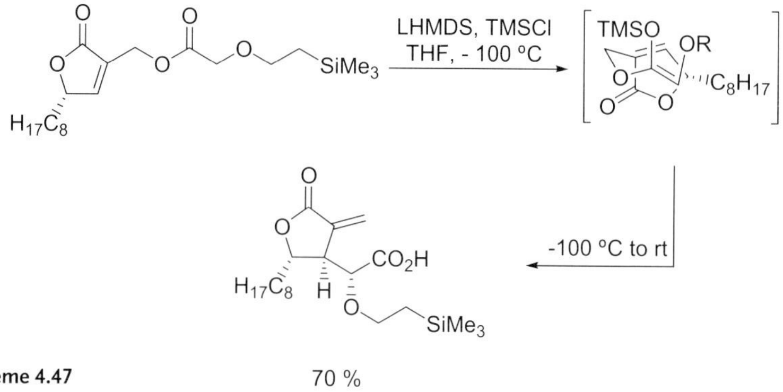
LHMDS, TMSCl
THF, - 100 °C
TMSO
OR
C_8H_{17}
$H_{17}C_8$
$SiMe_3$
-100 °C to rt
CO_2H
H
70 %

Scheme 4.47

4.7.1.3 Ester vs. Branched Ester

Martin et al. have reported several examples of selective enolization of an allyl ester in the presence of a vicinal branched ester (Scheme 4.48) [30]. Modest to excellent diastereoselectivities were observed depending upon the branched ester substituent (cf. Scheme 4.27).

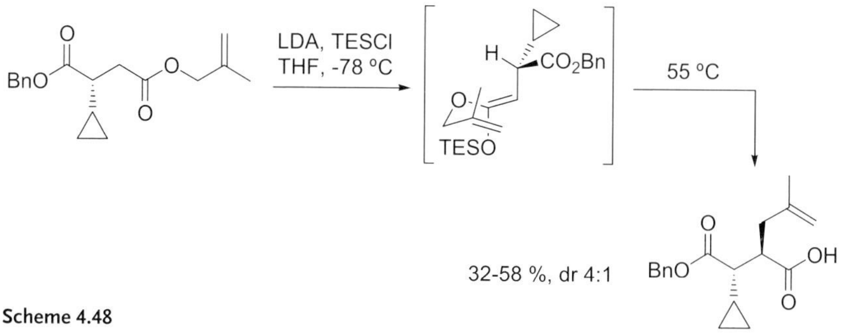
BnO
LDA, TESCl
THF, -78 °C
H
CO_2Bn
TESO
55 °C
32-58 %, dr 4:1
OH

Scheme 4.48

4.7.2
γ-Deprotonations of Allyl Acrylates

Mestres et al. reported that γ-deprotonation of allyl acrylates gave good selectivities for the rearrangement products resulting from *Z*-selective enolization (Scheme 4.49) [51]. The optimal yields were obtained using LTMP as the base. *Syn*/*anti* selectivities were *ca.* 8:1 for the *E*-allylic alkenes and 1:9 for the *Z*-allylic alkenes.

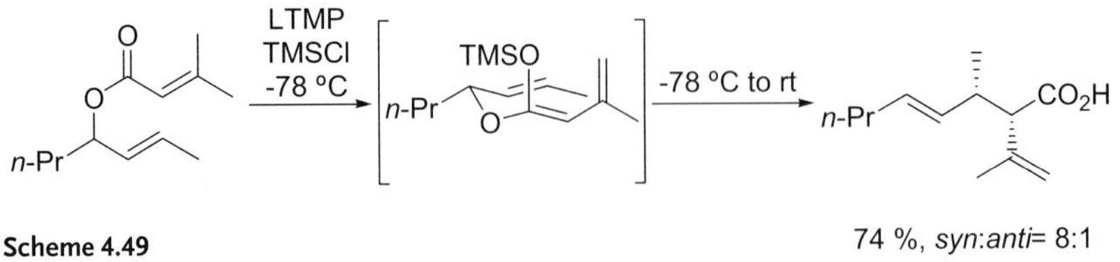

Scheme 4.49

74 %, *syn*:*anti*= 8:1

4.7.3
Silyl Triflates and Tertiary Amine Bases

In 1984 Nakai et al. reported that TMSOTf and NEt_3 could effectively form silyl ketene acetals for trifluoromethylated allylic esters (Scheme 4.50) [52]. The pentenoic acid products were formed in a 72:28 ratio. The authors did not assign the structure of the major isomer, but it seems likely that the *Z*-silyl ketene acetal was formed preferentially based on later results (*vide infra*), which should lead to the *syn* isomer as the major product. While esters derived from 2° alcohols also gave good yields of rearrangement products, 3° esters afforded low yield, apparently due to elimination to the diene.

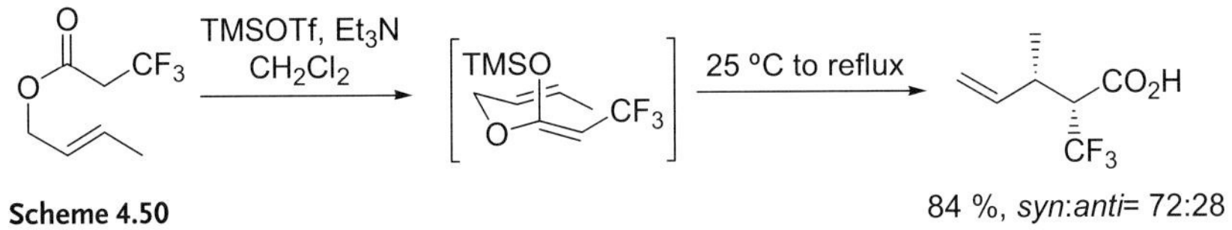

Scheme 4.50

84 %, *syn*:*anti*= 72:28

The Nakai conditions in some cases can lead to greatly increased stereoselectivities relative to the Ireland conditions. Welch et al. found that use of silyl triflates could lead to significantly improved diastereoselectivities and/or yields in the rearrangement of allyl fluoroacetates (Scheme 4.51) [53]. Rearrangement of the *Z*-crotyl fluoroacetate under the Nakai conditions proceeded via the *Z*-silyl ketene acetal and afforded excellent yield and high diastereoselectivity for the *anti* isomer. Treat-

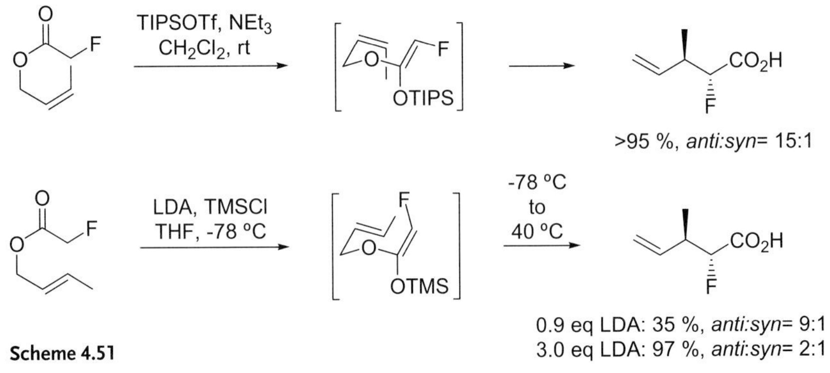

>95 %, *anti*:*syn*= 15:1

Scheme 4.51

0.9 eq LDA: 35 %, *anti*:*syn*= 9:1
3.0 eq LDA: 97 %, *anti*:*syn*= 2:1

ment of the *E*-crotyl fluoroacetate under the Ireland conditions proceeded via the *E*-silyl ketene acetal and gave either high yield or high diastereoselectivity for the *anti* isomer depending upon the equivalents of LDA added.

Nakai et al. later reported a study of the effect of the size of the trialkylsilyl triflate and the tertiary amine base on the diastereoselectivity of the Ireland-Claisen rearrangement (Scheme 4.52) [54]. The best diastereoselectivity (92:8 *syn:anti*) was obtained when a bulky amine ($[c\text{-}C_6H_{11}]_2NMe$) and a bulky silylating agent (TBSOTf) were used. Under these conditions the rearrangements occurred principally via the *Z*-silyl ketene acetal.

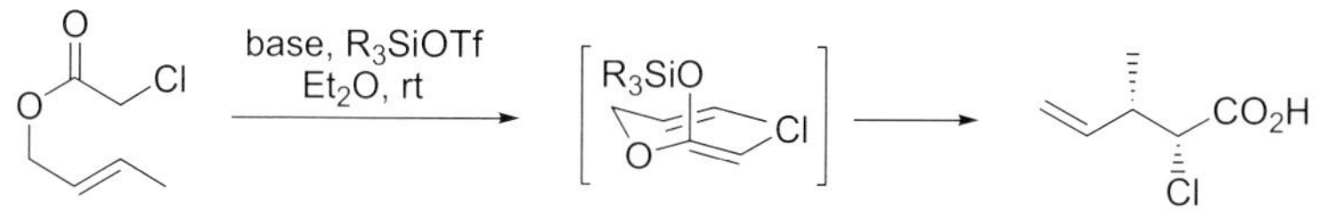

base, R_3SiOTf
Et_2O, rt

R_3Si	base	*syn:anti*
TMS	NEt_3	70:30
TBS	$NMe(c\text{-}C_6H_{11})_2$	92:8

Scheme 4.52

Porco and Hu recently described the first solid phase Ireland–Claisen rearrangement on a polystyrene resin (Scheme 4.53) [55]. Treatment of allylic esters with a polymer supported silyl triflate yielded the desired pentenoic acids after cleavage from the support.

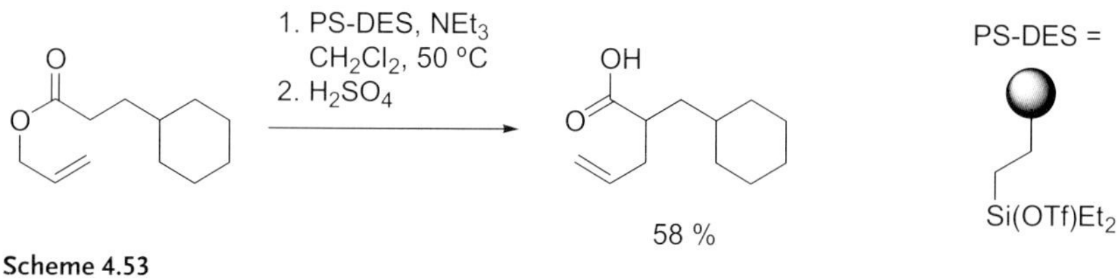

1. PS-DES, NEt_3
CH_2Cl_2, 50 °C
2. H_2SO_4

58 %

PS-DES = Si(OTf)Et_2

Scheme 4.53

4.7.4
N,O-Bis(trimethylsilyl)acetamide and CuOTf

Ito et al. reported a novel silylation method for the Ireland–Claisen rearrangement of allyl isocyanates (Scheme 4.54) [56]. Treatment of the allyl isocyanates with *N,O-bis*(trimethylsilyl)acetamide (BSA) and 3 mol% CuOTf afforded generally good to excellent yields of the rearranged products. The authors state that the CuOTf accelerated the rearrangement, but it seems more likely that it accelerated the silylation of the ester. The authors reported that the silyl ketene acetals prepared using the Ireland conditions underwent rearrangement at room temperature, whereas treatment of the same substrates with BSA at 50 °C resulted in only a trace of rearranged product. It is also possible that both the silylation and rearrangement are catalyzed by the CuOTf [9].

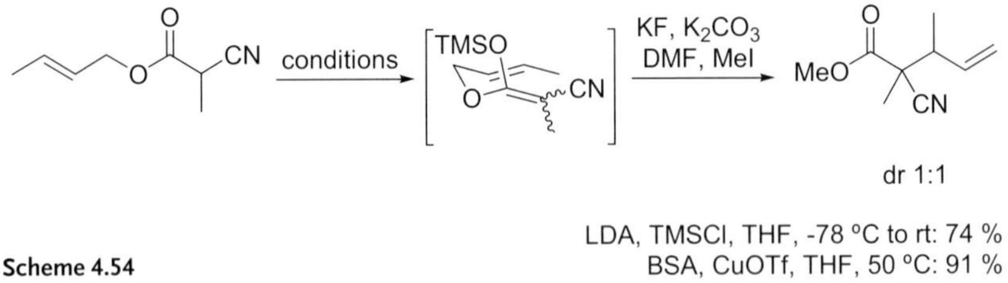

CN
conditions
TMSO
CN
KF, K2CO3
DMF, MeI
MeO
CN
dr 1:1
LDA, TMSCl, THF, -78 °C to rt: 74 %
BSA, CuOTf, THF, 50 °C: 91 %

Scheme 4.54

4.7.5
1,4-Additions

Several groups have employed 1,4-additions of nucleophiles, radicals or silanes to allyl acrylates. The intermediate silyl ketene acetals then underwent Ireland–Claisen rearrangement (Scheme 4.55).

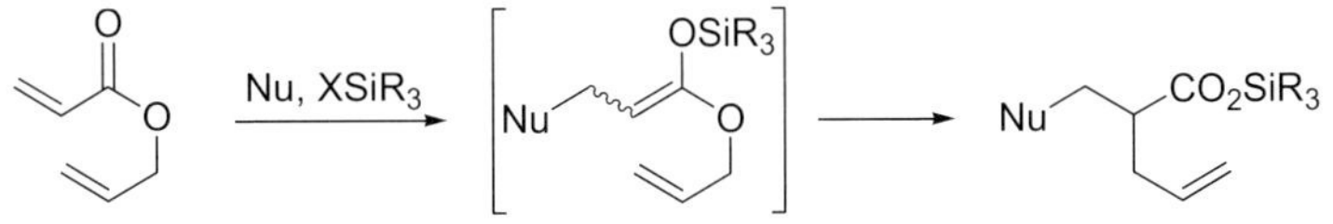

Nu, XSiR3
OSiR3
Nu
Nu
CO2SiR3

Scheme 4.55

4.7.5.1 By Alkyl Cu Reagents

Kuwajima and Aoki first reported a 1,4-addition/Ireland–Claisen rearrangement sequence (Scheme 4.56) [57]. Cu-catalyzed addition of MeMgBr to a series of allyl acrylates yielded the desired pentenoic acids, albeit in modest diastereoselectivity. The low diastereoselectivity almost certainly reflected a lack of control of enolate geometry.

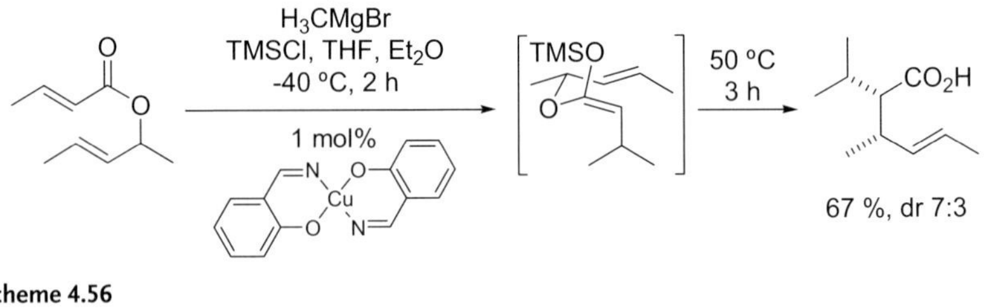

H3CMgBr
TMSCl, THF, Et2O
-40 °C, 2 h
1 mol%
TMSO
50 °C
3 h
CO2H
67 %, dr 7:3

Scheme 4.56

Olsson et al. found that the diastereoselectivity of the pentenoic acid formation was influenced by the nature of the Cu nucleophile (Scheme 4.57) [58]. Use of MeCuLiI(LiI) and TMSI yielded principally the *anti*-pentenoic acid, whereas addition of $Me_2CuLi(LiI)$ and TMSCl yielded the *syn*-pentenoic acid as the major diastereomer.

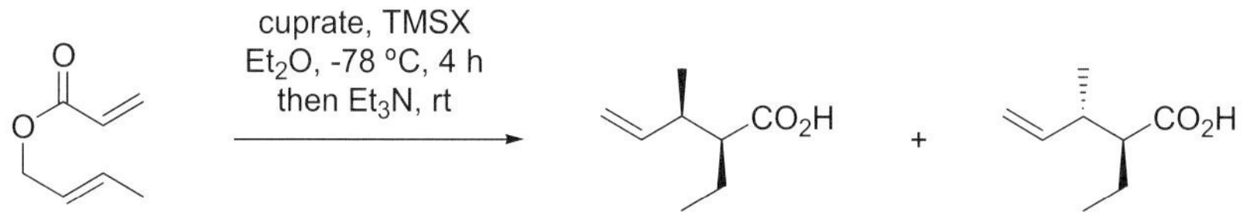

cuprate, TMSX
Et_2O, -78 °C, 4 h
then Et_3N, rt

cuprate	X	yield (%)	*syn:anti*
MeCu(LiI)	I	68	17:83
Me_2CuLi(LiI)	Cl	93	82:18

Scheme 4.57

4.7.5.2 **By Alkyl Radicals**

Takai et al. reported that the $PbCl_2$-catalyzed addition of alkyl radicals to allyl acrylates yielded pentenoic acids in good yields although essentially no diastereoselectivity was observed (Scheme 4.58) [59]. As with the examples cited above, the low diastereoselectivity reflected a lack of control of enolate geometry. The conjugate addition was thought to occur via Mn-mediated reduction of the alkyl iodide to the alkyl radical. The radical underwent conjugate addition to the allyl acrylate to afford the α-carboxyl radical. Further reduction of the carboxyl radical was followed by *in situ* trapping as the silyl ketene acetal.

Mn^0, $PbCl_2$
i-PrI, TMSCl
NMI, THF, DMF

Mn^0

83 %, dr 1:1

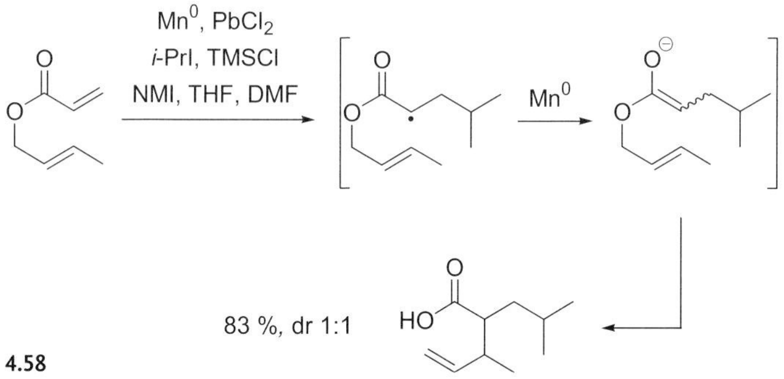

Scheme 4.58

4.7.5.3 **By Enolates**

Yamazaki et al. employed the Evans oxazolidinone enolate in diastereoselective Michael additions to β-CF_3 acrylates to afford intermediate allyl silyl ketene acetals [8]. The products were isolated as *ca.* 2:1 mixtures of pentenoic acids and Michael addition adducts (Scheme 4.59). The rearrangement of the silyl ketene acetal was catalyzed by $PdCl_2(CH_3CN)_2$. The rearrangement apparently occurred via the *Z*-silyl ketene acetal and exhibited high 1,2-asymmetric induction. Aspects of stereochemical control and Pd catalysis have been discussed previously (cf. Scheme 4.25).

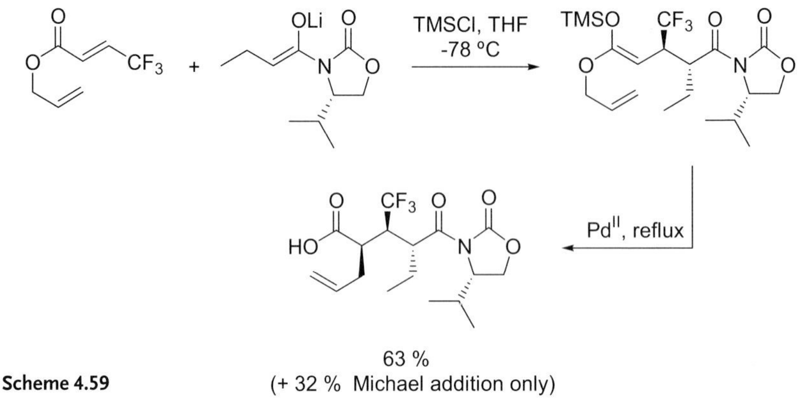

TMSCl, THF -78 °C

Pd^{II}, reflux

63 %
(+ 32 % Michael addition only)

Scheme 4.59

4.7.5.4 By Silanes

In 2002 Miller and Morken reported a Rh(I)-catalyzed conjugate addition of silanes to allyl acrylates to afford Ireland–Claisen rearrangement products (Scheme 4.60) [60]. The optimal conditions employed $[(cod)RhCl]_2$ as the catalyst, MeDuPhos as ligand and $HSiMe_2Cl$ as the silylating agent in benzene at ambient temperature. Diastereoselectivities were generally high (11–25:1). In order to obtain the observed *anti* product as the major isomer, the *E*-silyl ketene acetal would have to be the intermediate. While it is clear that the Rh(I) complex catalyzes the hydrosilylation, the reaction times suggest that Rh(I) does not serve as a catalyst for the Ireland–Claisen rearrangement itself (cf. Scheme 4.7).

The example shown illustrates the potential for application to large scale synthesis. A catalyst loading of 0.5 mol% Rh and DuPhos was sufficient to convert 10 g of allyl acrylate to the pentenoic acid in good yield and stereoselectivity.

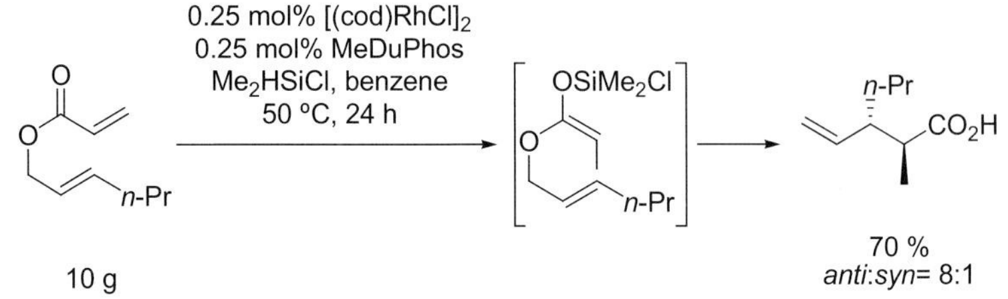

0.25 mol% $[(cod)RhCl]_2$
0.25 mol% MeDuPhos
Me_2HSiCl, benzene
50 °C, 24 h

10 g

70 %
anti:*syn*= 8:1

Scheme 4.60

4.7.6
Electrochemical Reduction

Troll and Wiedemann dimerized allyl acrylates to the dimeric silyl ketene acetals via electrochemical reduction (Scheme 4.61) [61]. The diallyl adipic diacids were isolated as 1:1 mixtures of diastereomers.

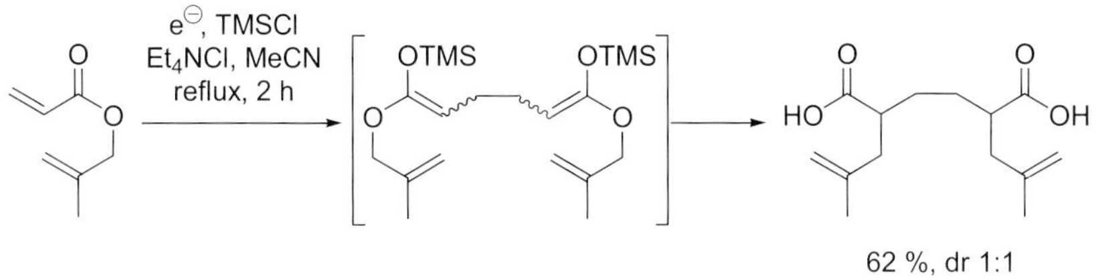

Scheme 4.61

4.7.7 Diels–Alder Cycloaddition

Neier et al. made use of the Diels–Alder reaction to generate an allyl silyl ketene acetal *in situ* (Scheme 4.62) [62]. Diels–Alder cycloaddition of a silyl ketene acetal derived from a dienol propionate yielded an intermediate allyl silyl ketene acetal which underwent *in situ* Ireland–Claisen rearrangement. The pentenoic acid products were formed in 60% yield as a 47:29:24 mixture of diastereomers, from which the major diastereomer shown below was isolated.

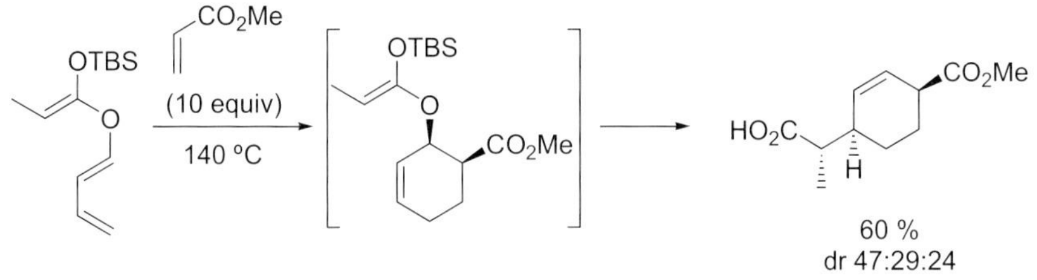

Scheme 4.62

4.7.8 Brook Rearrangement

In their total synthesis of (+)-ophiobolin in 1989, Kishi et al. found that treatment of a cyclopentenyl ester under the typical Ireland conditions gave principally *C*-silylated ester [63]. Heating of a *C*-silyl ester (prepared by acylation using a *C*-silyl acyl chloride) at 230 °C resulted in a 1,3-Brook rearrangement followed by an Ireland–Claisen rearrangement to give the desired product as a 6:1 ratio of isomers at C2 of the pentenoic acid (Scheme 4.63). The major product could have arisen through either a chair transition state of the *Z*-silyl ketene acetal or a boat transition state of the *E*-silyl ketene acetal.

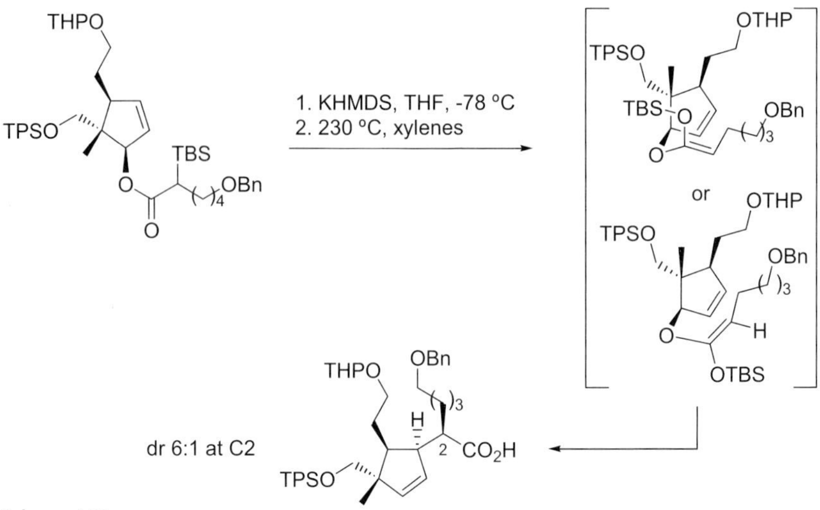

Scheme 4.63

4.7.9
Boron Ketene Acetals

In contrast to silyl ketene acetals, boron ketene acetals have had considerably less use in Ireland–Claisen rearrangements. This is somewhat surprising given the extensive use of boron enolates in aldol and related reactions. As mentioned previously, Corey et al. found that rearrangements of boron ketene acetals may occur at significantly lower temperature than the silyl counterparts (cf. Scheme 4.44) [47]. Corey also demonstrated that the *Z*-(*O*)-*B*-ketene acetal may be obtained in at least as high selectivity as the corresponding silyl ketene acetal. A potential disadvantage of the boron ketene acetals which has not yet been examined is their compatibility with Lewis acid sensitive substrates.

Oh et al. obtained high diastereoselectivity in the Ireland–Claisen rearrangement of allyl mandelates by using $(Ipc)_2BOTf$ to give a 94:6 diastereoselectivity favoring the *syn* diastereomer, presumably via the chelated boron ketene acetal (Scheme 4.64) [64]. Interestingly, although the boron was substituted with chiral non-racemic substituents, the enantioselectivity of the rearrangement was very

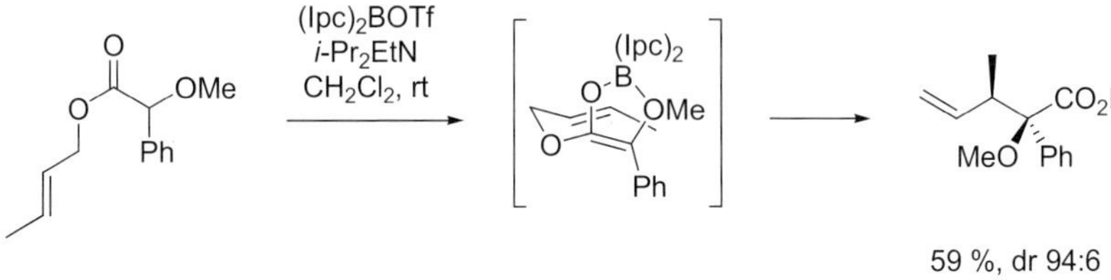

Scheme 4.64

low (< 10%). The mandelate boron ketene acetals underwent rearrangement at considerably lower temperatures than the silyl ketene acetals derived from lactates.

4.7.10
Post-Rearrangement Enolization

In Ireland and Mueller's seminal report in 1972, they noted that 2–6% of *C*-silylated pentenoic acid was invariably isolated along with the desired product when the pentenoic acid product retained an enolizable proton (Scheme 4.65) [1a]. The *C*-silylated product was likely due to enolization of the product silyl ester and subsequent *C*-silylation. An alternative pathway, which has also been observed (*vide infra*, Scheme 4.104) would involve initial *C*-silylation followed by *O*-silylation and rearrangement.

LICA (1.05 eq), TMSCl, THF, -78 °C to rt; LICA, TMSCl; 78 %; 2-6 %

Scheme 4.65

Zhabinskii et al. made use of this phenomenon to *C*-silylate the pentenoic acid product *in situ* (Scheme 4.66) [65]. The *C*-silylated acid was reduced to the alcohol and converted to the skipped diene using a Peterson olefination.

LDA, TMSCl (excess), THF, -70 °C to reflux; LAH; MsCl, pyridine

Scheme 4.66

With the exception of the above example, post-rearrangement epimerization is generally an undesired side reaction. McIntosh et al. found that Ireland–Claisen rearrangement of some *bis*-allylic esters using excess KHMDS and TIPSOTf led to anomalously low and variable diastereoselectivities (Scheme 4.67) [26]. Based on the Ireland report, they hypothesized that post-rearrangement enolization was occurring. They found that high and reproducible stereoselectivities were obtained when the solution of silyl ketene acetal, residual base and silylating agent was treated with HOAc at –78 °C. While excess KHMDS and presumably TIPSOTf were consumed by reaction with HOAc, the silyl ketene acetal remained unaffected.

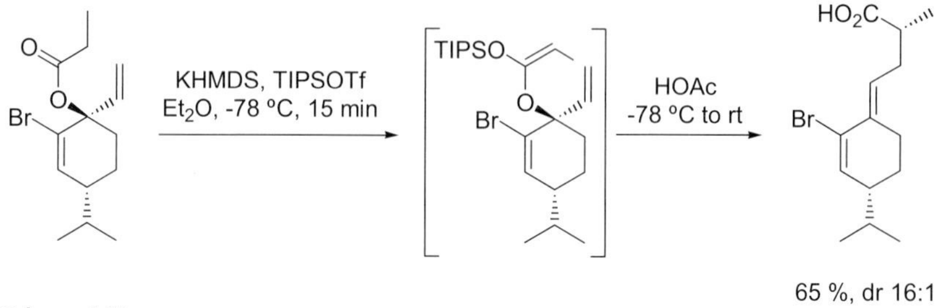

KHMDS, TIPSOTf
Et_2O, -78 °C, 15 min

HOAc
-78 °C to rt

65 %, dr 16:1

Scheme 4.67

The present authors have noted in compiling this review that there are several examples of unusually low diastereoselectivity in systems that would be expected to give good to high levels of stereoselectivity. It seems likely that, at least in some cases, post-rearrangement epimerization is taking place.

4.8 Structural Variations in Allylic Esters

4.8.1 Allylic Esters with α-Heteroatoms

4.8.1.1 Glycolates

Bartlett et al. reported the first example of an Ireland–Claisen rearrangement of an allylic glycolate in 1982 (Scheme 4.68) [66]. Treatment of crotyl glycolate with 2.25 equivalents of LICA and TMSCl in THF gave 50% yield of the 2-hydroxy-pentenoic acids with essentially no stereoselectivity.

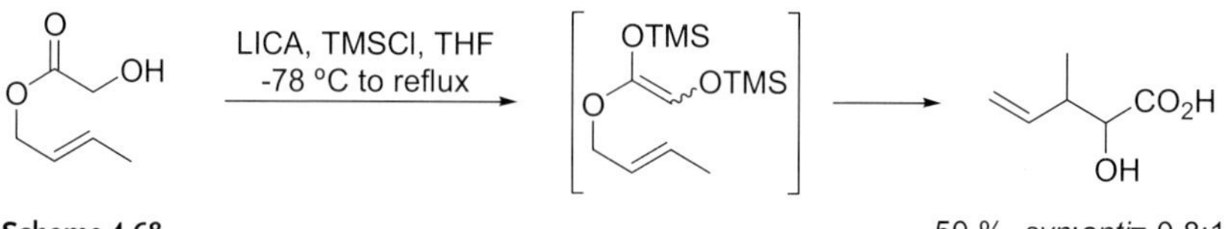

LICA, TMSCl, THF
-78 °C to reflux

50 %, *syn*:*anti*= 0.8:1

Scheme 4.68

In 1983 the Burke [48], Kallmerten [67] and Fujisawa [68] groups reported Ireland–Claisen rearrangements of *O*-alkyl glycolates generally proceeded with good to excellent stereoselectivity (Scheme 4.69). The major products are consistent with *Z*-selective enolization of the glycolate. Burke noted that enolization at temperatures higher than –100 °C resulted in significant amounts of *C*-silylation (entry 1). Kallmerten employed an inverse addition of the LDA to afford comparable yields and selectivity (entry 2). Fujisawa used the unprotected glycolate to give the analogous pentenoic acid in comparable yield and stereoselectivity to the protected glycolates (entry 3). Use of allylic *Z*-alkenes gave *anti* pentenoic acids with equivalent levels of diastereoselectivity.

The difference in yield and stereoselectivity between the Fujisawa and Bartlett examples is noteworthy (cf. Scheme 4.68 and entry 1, Scheme 4.69), considering that the only differences are the base (LICA vs. LHMDS) and the reaction temperature (THF reflux vs. ambient temperature). The low diastereoselectivity in the Bartlett case might be due to post-rearrangement enolization.

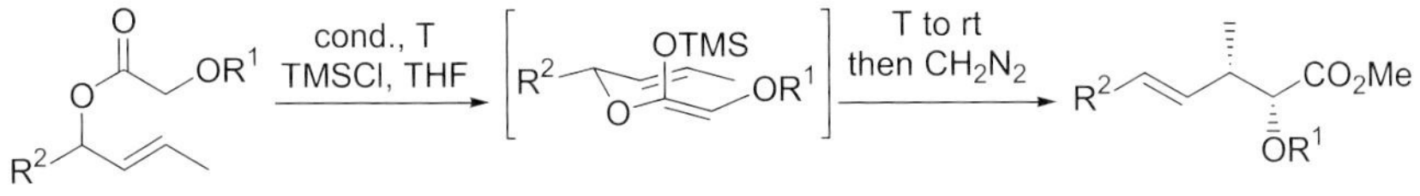

cond., T
TMSCl, THF
OTMS
T to rt
then CH_2N_2
CO_2Me
OR^1
R^2

entry	R^1	R^2	cond.	T (°C)	yield (%)	*syn*:*anti*
1	Bn	H	LDA, NEt_3	-100	77	91:9
2	Me	Me	LDA, inverse addn	-78	73	95:5
3	H*	H	LHMDS	-78	84	92:8

* TMS in transition state

Scheme 4.69

Yamamoto and Hattori found that the *E*-glycolate enolate could be selectively formed when the glycolate oxygen is protected with a bulky silyl group and LTMD is used as the base with TMSCl as an internal quench (Scheme 4.70) [69]. The resulting *syn* diastereomers were obtained in high yields and diastereoselectivities. Use of LHMDS as the base followed by addition of TBSCl reversed the stereochemical outcome to give the *anti* pentenoic acids, also with high yield and diastereoselectivity.

OTBS
E:Z= 2:98
LTMP, TMSCl
THF, -100 °C
TMSO
OTBS
-100 °C to rt
2. CH_2N_2
3. Bu_4NF
CO_2Me
OH
86 %, syn:anti= 95:5
LHMDS, TBSCl
THF, HMPA
-100 °C
TBSO
OTBS
-100 °C to rt
2. CH_2N_2
3. Bu_4NF
CO_2Me
OH
80 %, syn:anti= 3:97

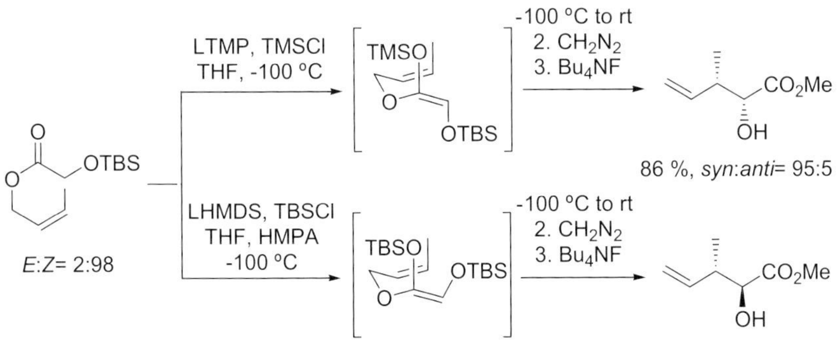

Scheme 4.70

Burke et al. employed the glycolate Claisen rearrangement in the synthesis of the C12–C26 subunit of rhizoxin (Scheme 4.71) [70]. Ireland–Claisen rearrangement of the glycolate gave the *syn* diastereomer and the *E*-alkene. The C15 oxygen was installed via bromolactonization to give a 1:1 mixture of pentanolide and butanolide. The undesired pentanolide was recycled to the acid by reduction with Zn.

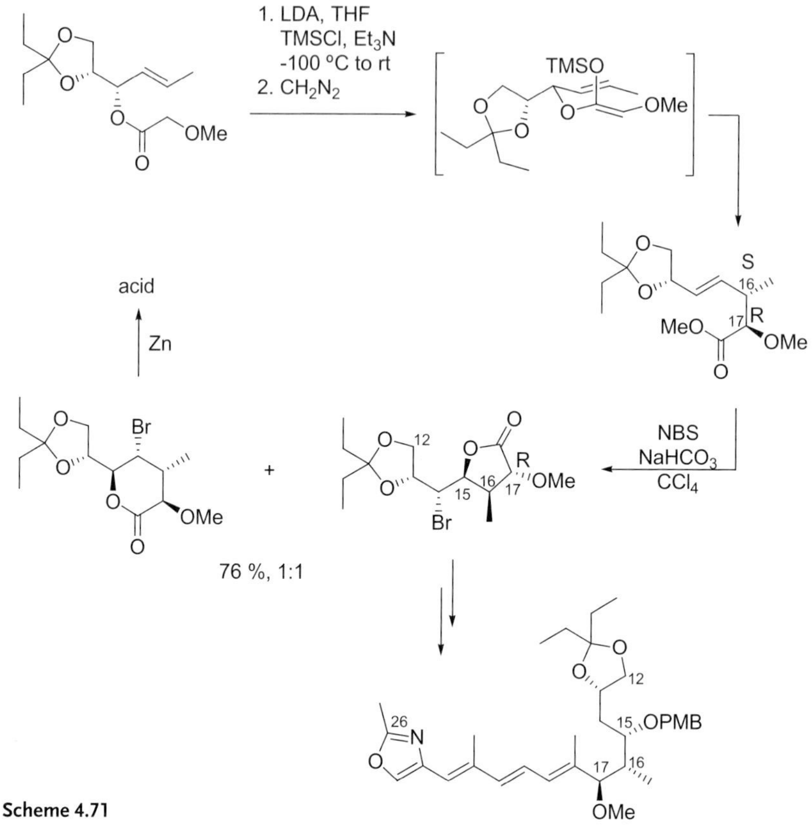

1. LDA, THF, TMSCl, Et_3N, -100 °C to rt
2. CH_2N_2

NBS, $NaHCO_3$, CCl_4

76 %, 1:1

Zn → acid

Scheme 4.71

In studies directed toward spongistatin I, Heathcock et al. noted a low diastereoselectivity in the Ireland–Claisen rearrangement of PMB-protected glycal-based allylic glycolates (Scheme 4.72) [71]. Based on earlier studies by Kallmerten and Gould on an asymmetric Ireland–Claisen rearrangement (cf. Scheme 4.42) [44], Heathcock was able to improve the diastereoselectivity by using enantiomerically pure *α*-methylbenzyl glycolates. While the achiral PMB ester yielded only a 2:1 ratio of *syn* and *anti* acids, the matched (*S*)-*α*-methylbenzyl-protected glycolate increased the selectivity to 6:1.

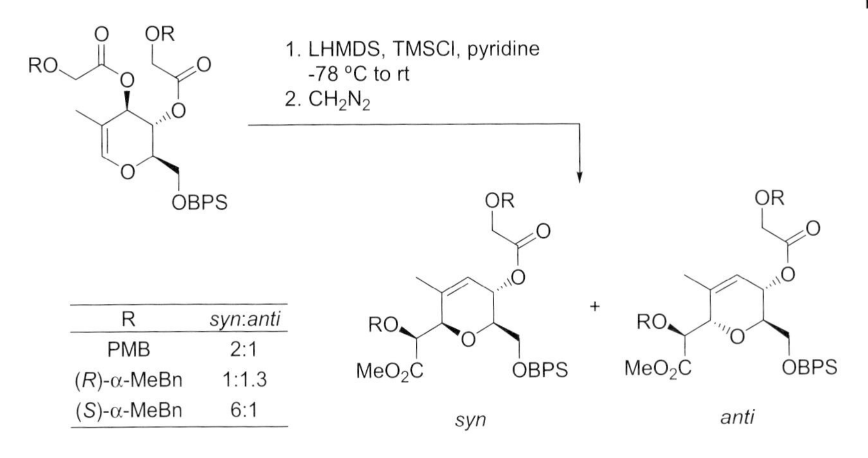

R	*syn*:*anti*
PMB	2:1
(*R*)-α-MeBn	1:1.3
(*S*)-α-MeBn	6:1

Scheme 4.72

Because the Ireland–Claisen rearrangement retains an alkene in the product, the pentenoic acids and derivatives thereof can become substrates for alkene metathesis to generate carbocycles and heterocycles. In 1998 Piscopio et al. reported the rearrangement of *O*-allyl glycolates to afford *O*-allyl pentenoic acids (Scheme 4.73) [72]. The pentenoate ester products were treated with either the Grubbs first generation Ru catalyst or the Schrock–Hoveyda Mo based catalyst to afford high yields of the corresponding pyrans. Several other oxygen, nitrogen and sulfur heterocycles as well as carbocycles were prepared using analogous methods.

1. LDA, THF, TMSCl, -78 °C to rt; 2. BnOH, EDCI, DMAP, CH_2Cl_2 — $SnBu_3$ — CO_2Bn, $SnBu_3$: 98 %, *syn*:*anti*= 10:1 — *i*-Pr, CF_3, Mo, Ph; benzene, rt — CO_2Bn, $SnBu_3$: 91 %

Scheme 4.73

Burke et al. simultaneously published a similar approach to related pyrans (Scheme 4.74) [73]. The alkene metathesis reactions required high catalyst loading (20 mol%) and long reaction times with the Grubbs first generation catalyst.

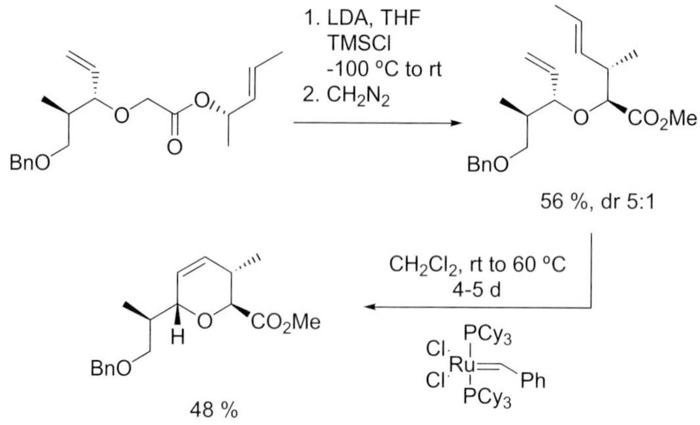
1. LDA, THF
TMSCl
-100 °C to rt
2. CH_2N_2
BnO
O
CO_2Me
56 %, dr 5:1
CH_2Cl_2, rt to 60 °C
4-5 d
PCy_3
Cl
Ru
Ph
H
48 %

Scheme 4.74

4.8.1.2 Lactates

In contrast to the extensive use of glycolate esters, lactate, mandelate and other higher α-oxygenated esters have been used much less frequently. Early work by Bartlett et al. showed that the Ireland–Claisen rearrangement could be used to prepare allylated lactic acids in good yield, but the diastereoselectivity was disappointingly low (Scheme 4.75) [20].

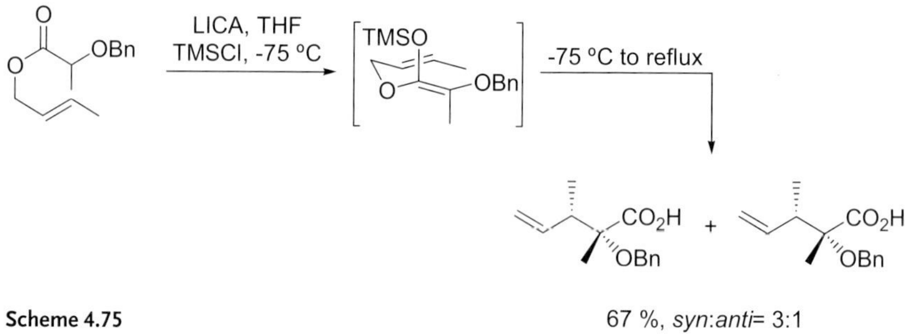
OBn
LICA, THF
TMSCl, -75 °C
TMSO
-75 °C to reflux
CO_2H
+
67 %, *syn*:*anti*= 3:1

Scheme 4.75

Takano et al. used the Ireland–Claisen rearrangement of an allyl lactate in the total synthesis of calcitriol lactone (Scheme 4.76) [74]. The rearrangement proceeded with a 6.7:1 diastereoselectivity via *O*-silylation of the intermediate Li-chelated *Z*-enolate. Takano noted that the corresponding benzyl ether gave significantly lower de (70%) than the PMB ether. This is presumably due to the PMB ether's greater Lewis basicity and hence its greater propensity to coordinate a lithium cation.

Scheme 4.76

4.8.1.3 **Mandelates**

Bartlett et al. found that substantially higher diastereoselectivity was obtained in the Ireland–Claisen rearrangement of crotyl mandelate when TMSCl was omitted (Scheme 4.77) [20]. Oh et al. obtained comparable yields and stereoselectivities using boron ketene acetals (cf. Scheme 4.64).

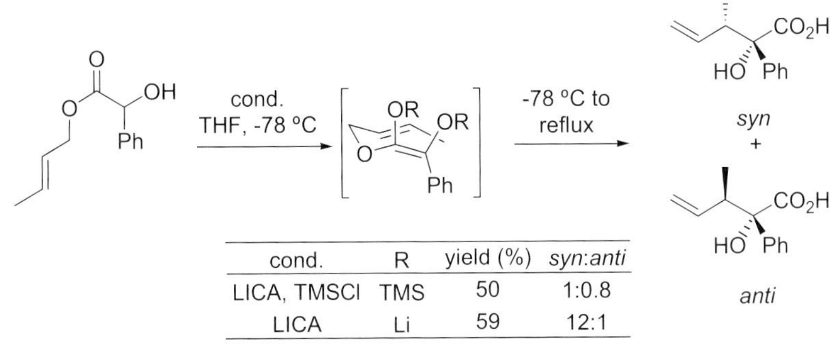

cond.	R	yield (%)	*syn*:*anti*
LICA, TMSCl	TMS	50	1:0.8
LICA	Li	59	12:1

Scheme 4.77

4.8.1.4 **Other Higher Esters**

Recently Langlois et al. showed that high (95:5) diastereoselectivity could be achieved in the Ireland–Claisen rearrangement of an allyl 2-OPMB-hexenoate ester by using KHMDS and TMSCl in toluene at –78 °C (Scheme 4.78) [75]. The diastereoselectivity was substantially lower when LDA was used as the base. The pentenoic ester product was further elaborated to the aldehyde shown using a ring closing metathesis (RCM) reaction to close the carbocyclic ring (*vide infra*).

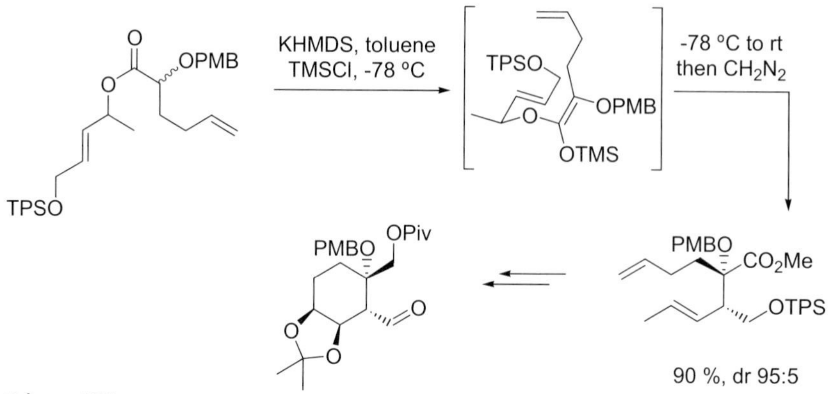

OPMB
TPSO
KHMDS, toluene
TMSCl, -78 °C
TPSO
OPMB
OTMS
-78 °C to rt
then CH2N2
PMBO
CO2Me
OTPS
90 %, dr 95:5
PMBO
OPiv

Scheme 4.78

The aldehyde was projected to serve as an intermediate in the synthesis of fumagillin and ovalicin.

4.8.1.5 Glycinates and Other Higher Esters

Bartlett and Barstow reported the first synthesis of α-amino acids using the Ireland–Claisen rearrangement (Scheme 4.79) [20]. Only in the case of the crotyl glycinate was a silylating agent employed. All other cases used the enolate without silylation. The enolate Claisen rearrangement of glycinates and higher esters is reviewed by Kazmaier in this monograph.

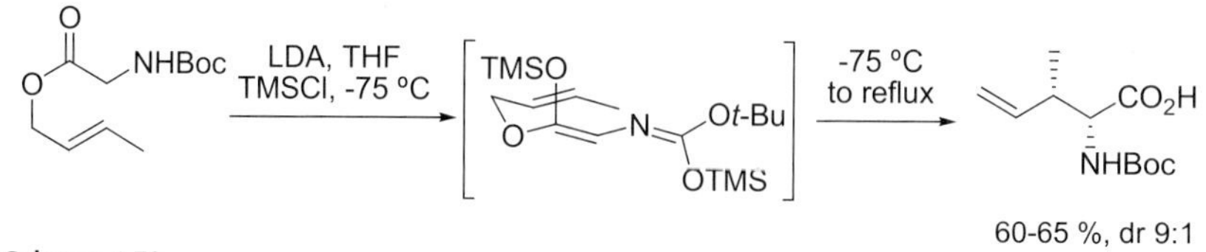

NHBoc
LDA, THF
TMSCl, -75 °C
TMSO
Ot-Bu
OTMS
-75 °C
to reflux
CO2H
NHBoc
60-65 %, dr 9:1

Scheme 4.79

Barrett et al. employed an Ireland–Claisen/alkene metathesis strategy in the synthesis of bicyclic β-lactams using glycinates in which the nitrogen atom was contained with a β-lactam ring (Scheme 4.80) [76]. The rearrangement occurred in excellent yield albeit with no diastereoselectivity at C2 of the pentenoic acid. Ring closing metathesis proceeded in quantitative yield using 20 mol% of the Grubbs first generation catalyst.

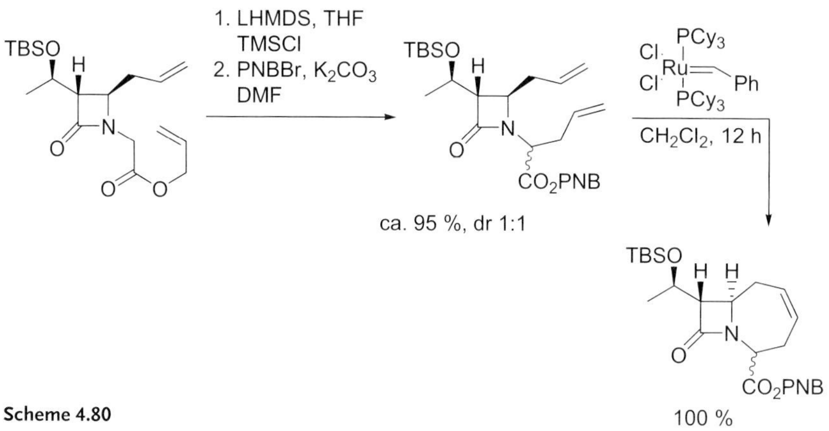
TBSO
H
1. LHMDS, THF
TMSCl
2. PNBBr, K2CO3
DMF
CO2PNB
ca. 95 %, dr 1:1
PCy3
Cl
Ru
Ph
CH2Cl2, 12 h
100 %

Scheme 4.80

4.8.2 Allyl Silanes and Stannanes

The 1,2-shift of the alkene inherent in the Claisen rearrangement can alter the reactivity of adjacent functional groups. Panek et al. have made extensive use of the Claisen rearrangement of esters of vinyl silanes to afford allyl silane products which were used in subsequent S_E' reactions (Scheme 4.81) (cf. Scheme 4.21) [24].

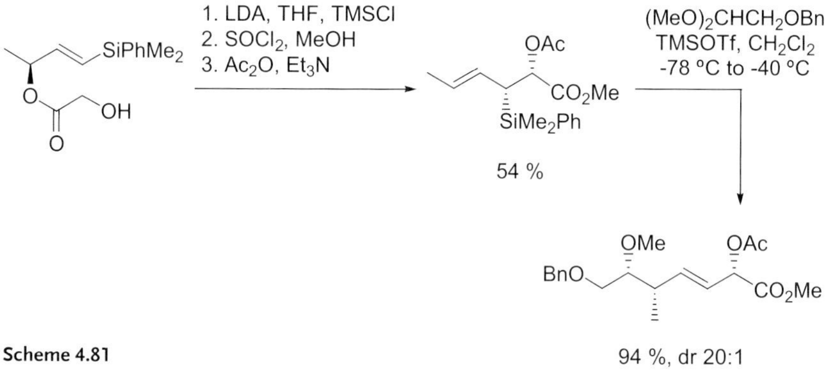
SiPhMe2
OH
1. LDA, THF, TMSCl
2. SOCl2, MeOH
3. Ac2O, Et3N
OAc
CO2Me
SiMe2Ph
54 %
(MeO)2CHCH2OBn
TMSOTf, CH2Cl2
-78 °C to -40 °C
OMe
BnO
94 %, dr 20:1

Scheme 4.81

Kocienski et al. have used a vinyl silane as a masked vinyl alcohol in the synthesis of the C26 to C32 fragment of rapamycin (Scheme 4.82) [77]. Ireland–Claisen rearrangement of the vinylsilane glycolate ester via the *Z*-silyl ketene acetal yielded the corresponding anti pentenoic acid with high diastereoselectivity. Oxidative cleavage of the furyl substituent and Tamao oxidation then afforded the allylic alcohol.

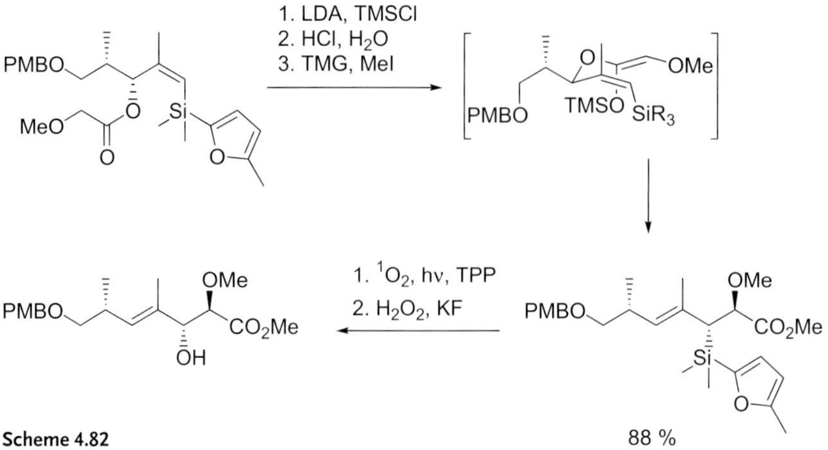
1. LDA, TMSCl
2. HCl, H2O
3. TMG, MeI
PMBO
MeO
O
Si
OMe
TMSO
SiR3
1. 1O2, hv, TPP
2. H2O2, KF
CO2Me
OH
88 %

Scheme 4.82

Parsons et al. employed an allyl silane Ireland–Claisen rearrangement as a route to 1,3-dienes (Scheme 4.83) [78]. Rearrangement of the allyl silane was followed by acid induced fragmentation of the *β*-epoxy allyl silane to give the diene with high alkene stereoselectivity. The asymmetric induction at C2 of the pentenoic acid was surprisingly low for this type of Claisen rearrangement.

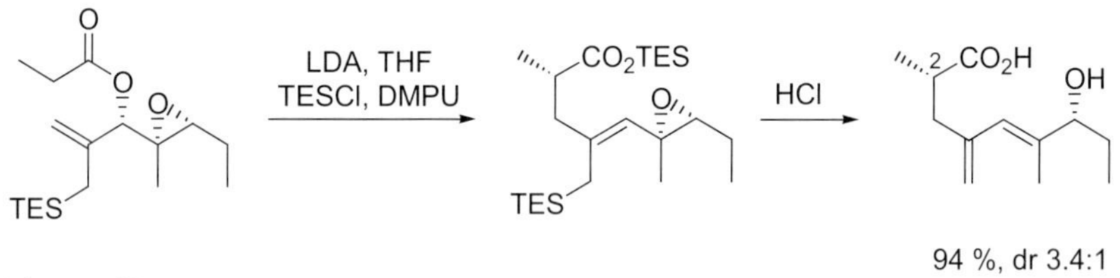
LDA, THF
TESCl, DMPU
CO2TES
TES
HCl
2
CO2H
OH
94 %, dr 3.4:1

Scheme 4.83

Fujisawa et al. used a glycolate variant of the Ireland–Claisen rearrangement to generate the expected *syn* product with high diastereoselectivity (Scheme 4.84) [79]. Peterson olefination of the *β*-hydroxysilane gave either the (2*E*,4*E*)- or (2*Z*,4*E*)-dienes when basic or Lewis acidic conditions were used, respectively.

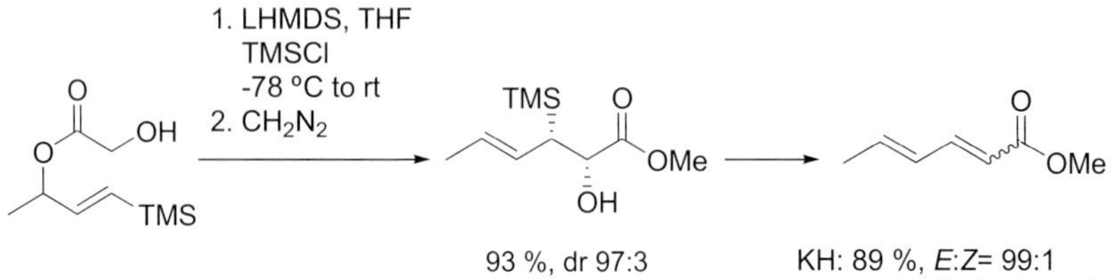
1. LHMDS, THF
TMSCl
-78 °C to rt
2. CH2N2
OH
TMS
OMe
93 %, dr 97:3
KH: 89 %, E:Z= 99:1
F3B•OEt2: 65 %, E:Z= 3:97

Scheme 4.84

Allyl stannanes have also been prepared via Claisen rearrangement (Scheme 4.85). Ritter showed that the rearrangement of the vinyl stannane glycolate proceeded through the expected *Z*-silyl ketene acetal to give the *syn* pentenoic acid in excellent diastereoselectivity [80].

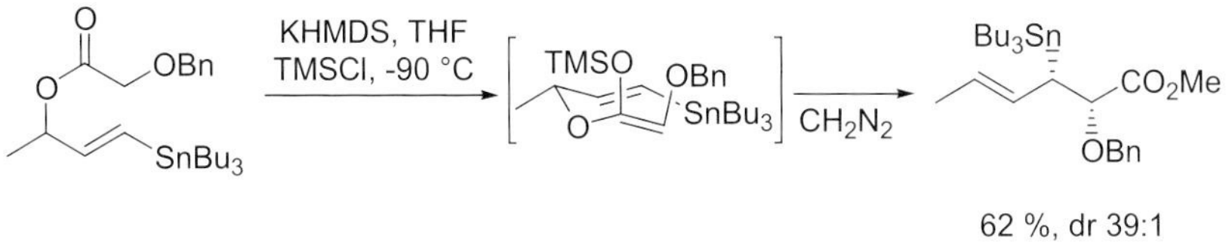
KHMDS, THF
TMSCl, -90 °C
TMSO
OBn
SnBu3
CH2N2
Bu3Sn
CO2Me
OBn
62 %, dr 39:1

Scheme 4.85

4.8.3
Glycals

Ireland recognized early on the potential for the application of the Claisen rearrangement to the synthesis of *C*-glycosides and related oxacycles. In 1979 he reported that the silyl ketene acetals of acetoxy glycals gave the rearranged glycoside in *ca*. 50–80% yields (Scheme 4.86) [23].

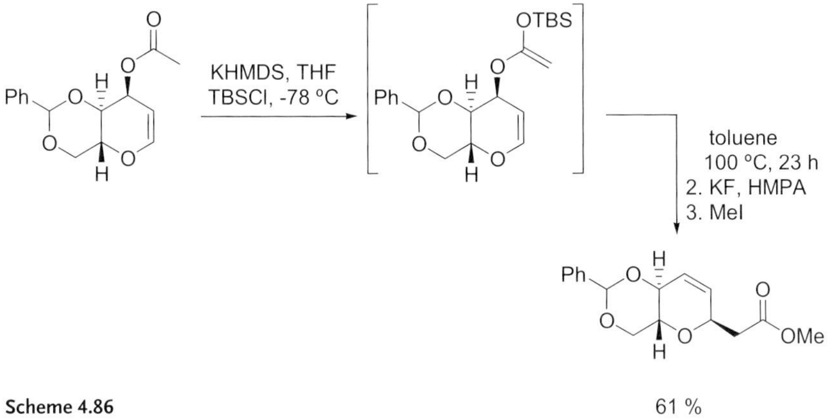
Ph
KHMDS, THF
TBSCl, -78 °C
OTBS
toluene
100 °C, 23 h
2. KF, HMPA
3. MeI
OMe
61 %

Scheme 4.86

Pyranyl propionate esters afforded moderate levels of diastereoselectivity in the rearrangement (Scheme 4.87). Use of THF as solvent gave an 20:80 ratio of *syn* and *anti* adducts, while use of 23% HMPA/THF afforded a 83:17 *syn*:*anti* ratio. In both cases the reactions proceeded preferentially via a boat transition state as was observed in simpler substrates (cf. Scheme 4.18). Furanyl esters gave comparable levels of diastereoselectivity.

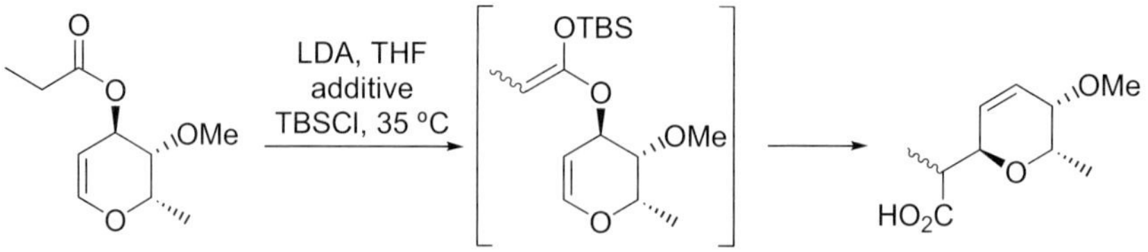

LDA, THF
additive
TBSCl, 35 °C
OTBS
OMe
HO2C

additive	silyl ketene acetale	yield (%)	*anti:syn*
-	*E*	75	80:20
23 vol% HMPA	*Z*	74	17:83

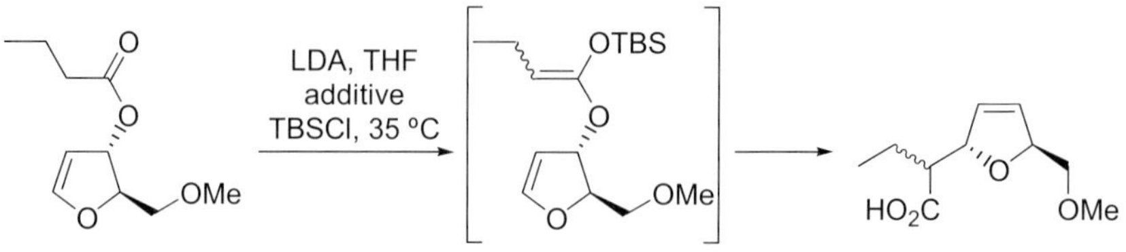

LDA, THF
additive
TBSCl, 35 °C
OTBS
OMe
HO2C

additive	silyl ketene acetale	yield (%)	*anti:syn*
-	*E*	75	79:21
23 vol% HMPA	*Z*	62	9:91

Scheme 4.87

Langlois et al. reported the Ireland–Claisen rearrangement of methylidene enol ethers derived from sugars to yield novel *C*-glycosides (Scheme 4.88) [81]. Although the yields were high, the diastereoselectivities were generally very modest. The authors did not comment on the reasons for the low level of C2-stereocontrol.

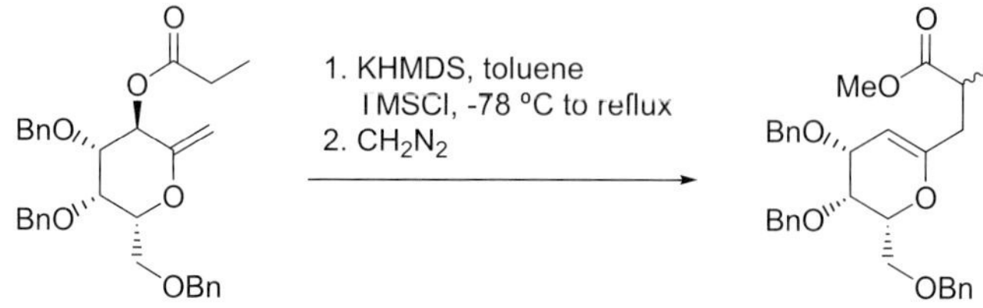

1. KHMDS, toluene
TMSCl, -78 °C to reflux
2. CH2N2
BnO
OBn
MeO

Scheme 4.88

91 %, dr 2:1

4.8.4
Allyl Lactones

A number of groups have employed the Ireland–Claisen rearrangement of allylic lactones to form a variety of carbocyclic and heterocyclic rings. This section will be divided into those lactones that possess exocyclic allylic alkenes and those with endocyclic allylic alkenes. The former give no change in ring size, while the latter result in a 4-atom ring contraction.

4.8.4.1 Lactones with Exocyclic Allylic Alkenes

4.8.4.1.1 Carbocycle Synthesis

In 1980 Danishefsky et al. reported the first examples of Ireland–Claisen rearrangements of vinyl lactones to afford carbocyclic carboxylic acids (Scheme 4.89) [82]. Treatment of 6-vinyl pentanolides under standard Ireland conditions and heating of the reaction mixture to 105 °C afforded the rearranged acids in generally good yields. The products necessarily arose from a boat transition state, since the chair transition states would be exceptionally strained.

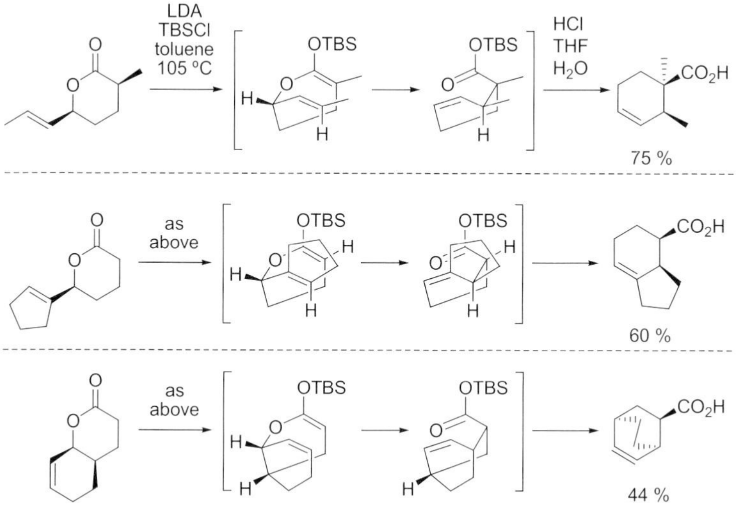

Scheme 4.89

Fukumoto et al. found that a furan alkene could be employed as the allylic alkene in analogous Ireland–Claisen rearrangements (Scheme 4.90) [83]. Treatment of a stereoisomeric mixture of furyl decalones under conditions similar to Danishefsky's afforded the furyl decalins after rearomatization of the furan.

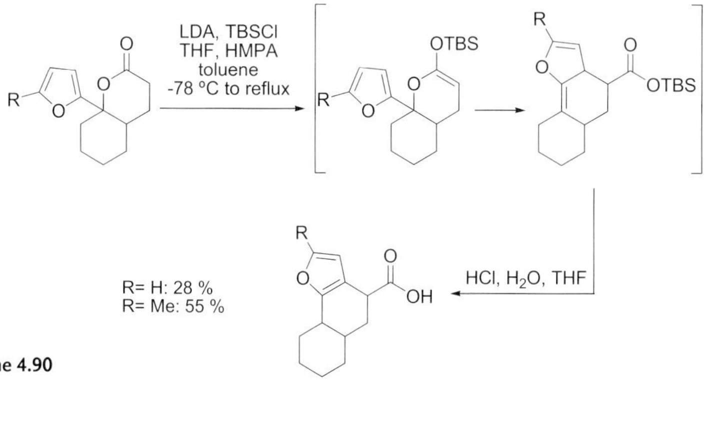

Scheme 4.90

In an approach to chlorothricolide, Ireland and Varney formed the C15–C16 bond using an Ireland–Claisen rearrangement of a macrocyclic allylic lactone (Scheme 4.91) [84]. The stereoselectivity at the *α*-carboxylic acid stereocenter was not an issue, since the carboxyl group was later excised under radical conditions. The alkene was formed solely as the *E*-isomer.

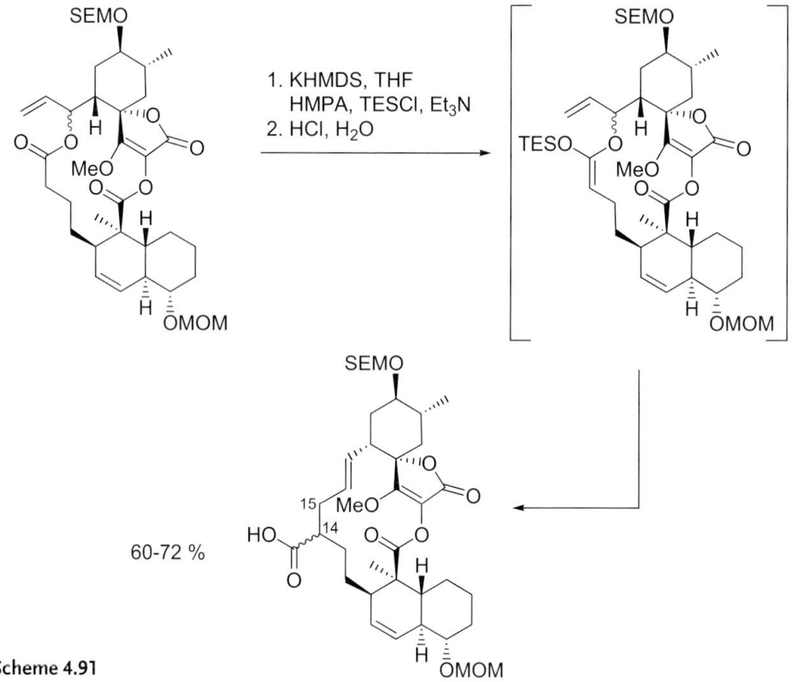

SEMO
H
MeO
OMOM
1. KHMDS, THF
HMPA, TESCl, Et3N
2. HCl, H2O
TESO
60-72 %
HO
15
14

Scheme 4.91

4.8.4.1.2 Heterocycle Synthesis

Burke et al. have employed the Ireland–Claisen rearrangement of 6-vinyl-dioxanones in the synthesis of substituted pyrans (Scheme 4.92) [85]. Treatment of the

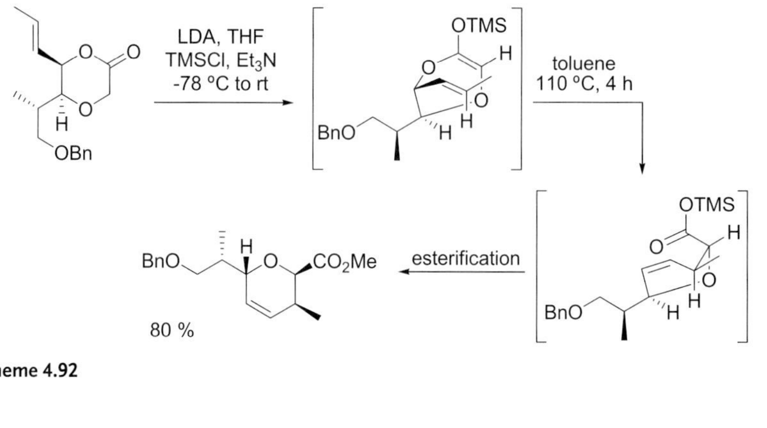

OBn
LDA, THF
TMSCl, Et3N
-78 °C to rt
OTMS
BnO
toluene
110 °C, 4 h
esterification
CO2Me
80 %

Scheme 4.92

dioxanone with LDA and TMSCl, followed by heating to 110 °C afforded good yield of the substituted pyran. The rearrangements necessarily proceeded via boat transition states to afford excellent diastereoselectivities. The rearrangements formed the basis for approaches to the pyran moiety of indanomycin and to the C7–C13 subunit of erythronolide B.

Angle et al. used the Ireland–Claisen rearrangement of 6-vinyl morpholinones to synthesize enantiopure pipecolic acids using alanine as a starting material (Scheme 4.93) [86]. The silyl ketene acetal was generated using the Nakai conditions (TIPSOTf, NEt_3). Heating at toluene reflux afforded the pipecolic acid in good yield. As with the previously described lactones, the rearrangement proceeded through a boat transition state.

TIPSOTf, Et_3N $CDCl_3$, 75 °C; OTIPS; CO_2TIPS; CH_3; 78 %

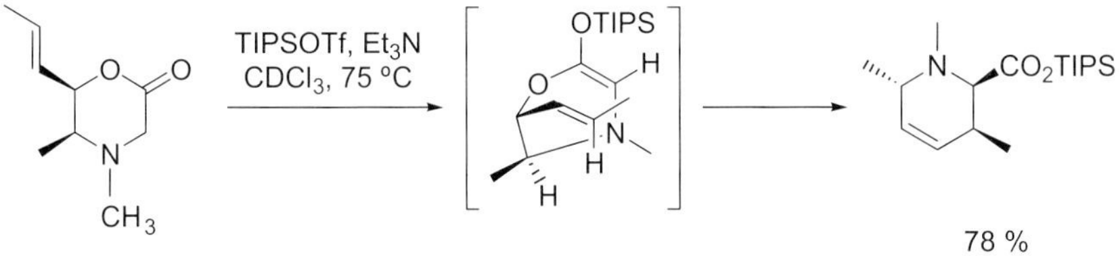

Scheme 4.93

4.8.4.2 Lactones with Endocyclic Allylic Alkenes

4.8.4.2.1 Carbocycle Synthesis

In 1982 Funk et al. reported the Ireland–Claisen rearrangement of medium and large ring lactones bearing endocyclic allylic alkenes to afford carbocycles (Scheme 4.94) [87]. The rearrangements resulted in a net 4-atom ring contraction

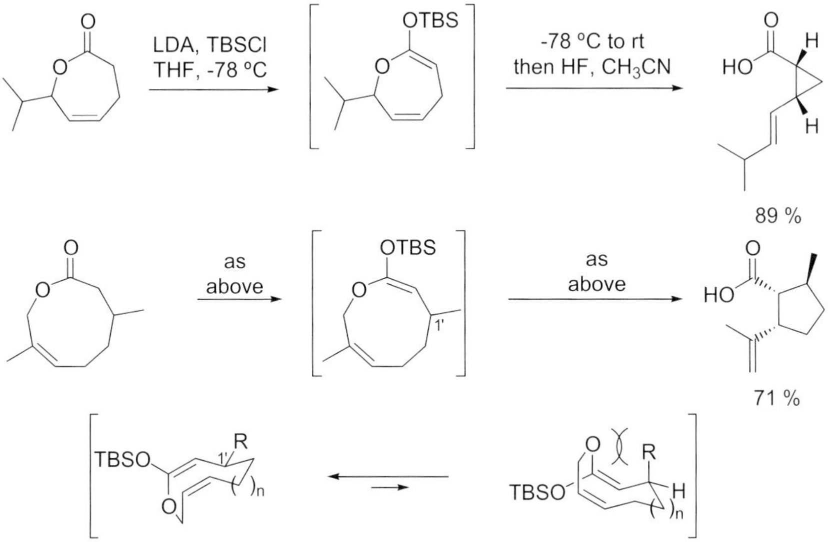

LDA, TBSCl THF, -78 °C; OTBS; -78 °C to rt then HF, CH_3CN; HO; 89 %

as above; OTBS; as above; HO; 71 %

TBSO; R; n; TBSO; R; H; n

Scheme 4.94

relative to the starting lactone, with C2 and C3 of the pentenoic acid being the adjacent atoms in the carbocyclic ring. For lactone rings containing 12 or fewer atoms and a *Z*-allylic alkene, the products invariably resulted in the carboxylic acid and alkene groups being in a *cis* relationship.

The authors reasoned that the boat transition states were the only energetically accessible ones. For lactones bearing a C1′ stereocenter, the transition state placing the C1′ substituent in a pseudo-equatorial position would be preferred to avoid allylic strain present in the alternative transition state. The method led to concise syntheses of the monoterpenoids chrysanthemic acid and isodihydronepatalactone.

For a 15-membered lactones and a 12-membered lactone possessing an *E*-allylic alkene, both *cis* and *trans* isomers were formed (Scheme 4.95). In these cases both chair and boat transition states were accessible.

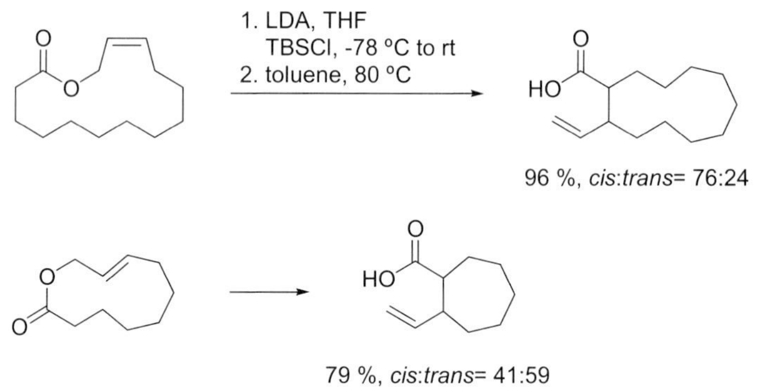

1. LDA, THF
TBSCl, -78 °C to rt
2. toluene, 80 °C

96 %, *cis*:*trans*= 76:24

79 %, *cis*:*trans*= 41:59

Scheme 4.95

Also in 1982, Knight and Cameron reported Ireland–Claisen rearrangements of lactones possessing both endocyclic and exocyclic allylic alkenes [88]. They also found low *cis*/*trans* selectivity in the rearrangement of 14- and 15-membered ring lactones.

Magriotis et al. used the Ireland–Claisen rearrangement of macrolactones to prepare enediynes capable of undergoing the Bergman cyclization (Scheme 4.96) [89]. The authors found that both *E*- and *Z*-silyl ketene acetals were accessible with ca. 10:1 stereoselectivity in the absence or presence of HMPA, respectively. Isolation of the silyl ketene acetals and heating in a sealed tube at 140 °C in the presence of 1,4-cyclohexadiene as a hydrogen atom donor afforded initially the Claisen rearrangement product. The resulting enediyne underwent spontaneous Bergman cyclization to give the tetrahydronaphthalene in 45% overall yield from the lactone. The 7:1 ratio of isomers at C2 and C3 of the pentenoic acid moiety indicated that the Ireland–Claisen rearrangement occurred through a boat transition state.

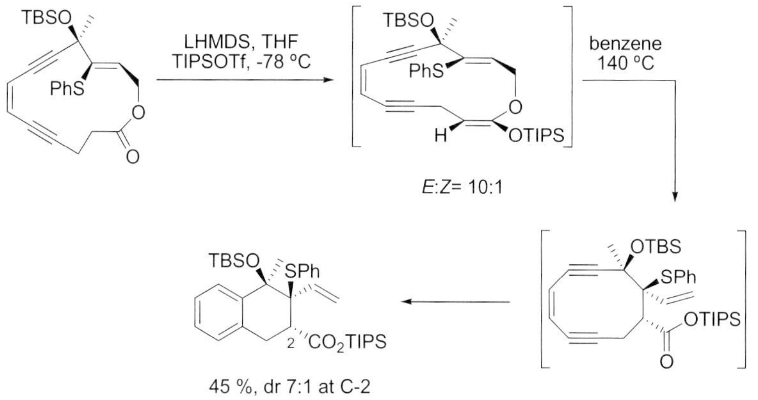

Scheme 4.96

In a subsequent study Magriotis et al. reported an interesting ring contraction variant of the Ireland–Claisen rearrangement in studies directed toward the endiyne antibiotics [90]. They found that the bending of the alkyne that is induced upon its complexation with $Co_2(CO)_6$ resulted in a facile rearrangement of an otherwise unreactive allyl silyl ketene acetal (Scheme 4.97). *In situ* decomplexation of the metal gave the desired carbocycle. In this case the *Z*-silyl ketene acetal was formed exclusively and the rearrangement occurred via a chair transition state. The high diastereoselectivity of the rearrangement was argued to be due to transannular interactions and the antiperiplanar effect, in which the 3° allylic oxygen substituent was disposed *anti* to the forming *C–C* bond (cf. Scheme 4.34).

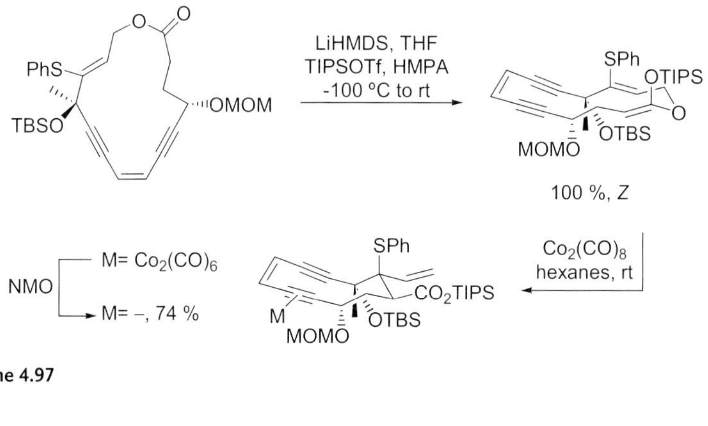

Scheme 4.97

Roush and Works developed an Ireland–Claisen rearrangement of a 16-membered tetraenyl lactone that afforded a 12-membered ring carbocycle possessing a diene and dienophile (Scheme 4.98) [91]. The resulting product underwent spontaneous Diels–Alder cycloaddition to give a tricyclic product in moderate yield. Calculations suggested that both *E*- and *Z*-silyl ketene acetals would lead to the same *cis* configuration in the Claisen rearrangement product.

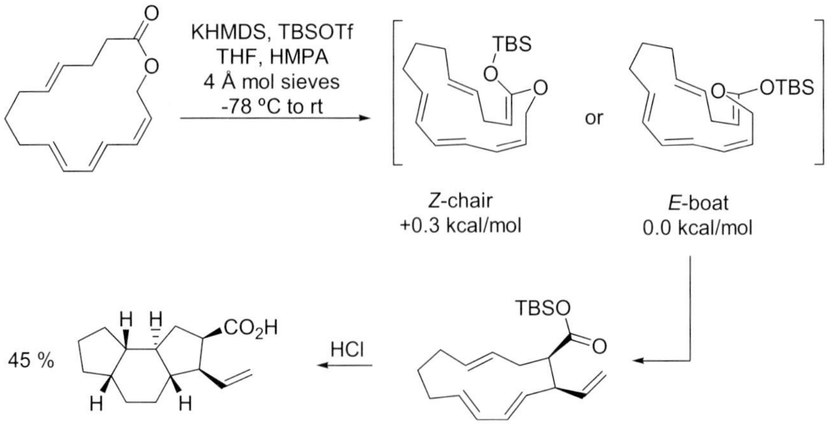

KHMDS, TBSOTf
THF, HMPA
4 Å mol sieves
-78 °C to rt
TBS
O
O
or
O
OTBS
Z-chair
+0.3 kcal/mol
E-boat
0.0 kcal/mol
TBSO
O
HCl
45 %
H H
CO2H
H
H

Scheme 4.98

4.8.4.2.2 **Heterocycle Synthesis**

Funk et al. reported an early example of heterocycle synthesis via a ring contraction of a azalactone via an Ireland–Claisen rearrangement (Scheme 4.99) [87]. Since the lactone was 10-membered, the rearrangement occurred exclusively via the boat transition state to afford the *cis* piperidine diastereomer. The product was elaborated in several steps to a protected version of meroquinene, an intermediate in cinchona alkaloid syntheses.

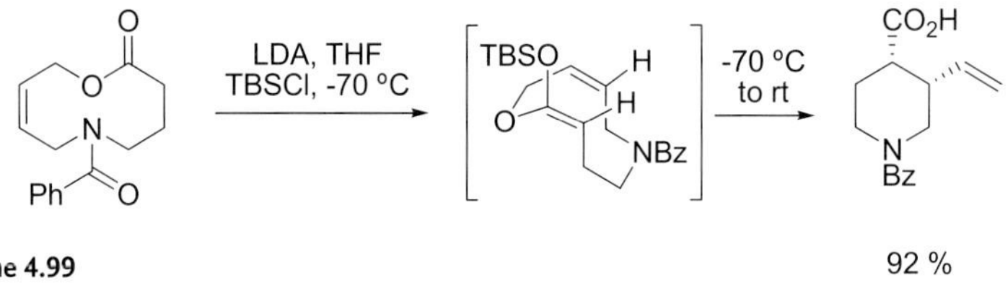

O
O
N
Ph
O
LDA, THF
TBSCl, -70 °C
TBSO
H
H
O
NBz
-70 °C
to rt
CO2H
N
Bz
92 %

Scheme 4.99

Knight et al. employed a related rearrangement in the synthesis of (–)-*α*-kainic acid (Scheme 4.100). The lactone starting material is derived from *L*-aspartic acid [92].

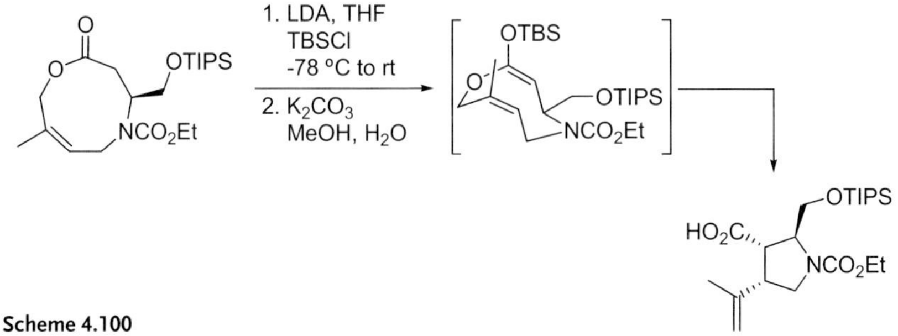

O
O
OTIPS
NCO2Et
1. LDA, THF
TBSCl
-78 °C to rt
2. K2CO3
MeOH, H2O
OTBS
O
OTIPS
NCO2Et
OTIPS
HO2C
NCO2Et

Scheme 4.100

4.8.5 Tertiary Alcohol-Derived Allylic Esters

Compared to 1° and 2° alcohols, 3° alcohol derived allylic esters have received comparatively little attention as substrates in the Ireland–Claisen rearrangement. This may to some extent reflect difficulty in the synthesis of the requisite 3° ester. In addition, stereocontrol in the formation of the resultant alkene may be problematic. The difference in size of the two allylic substituents is in general large for 2° allylic stereocenters since one substituent is always H (e.g., $R^1 = H$, $R^2 \neq H$, Scheme 4.101). However, for 3° allylic esters, a smaller size difference will lead to low alkene stereoselectivity. In addition, if there is a lack of differentiation between the two chair transition states, the C2 and C3 stereocenters will be formed in both antipodal forms.

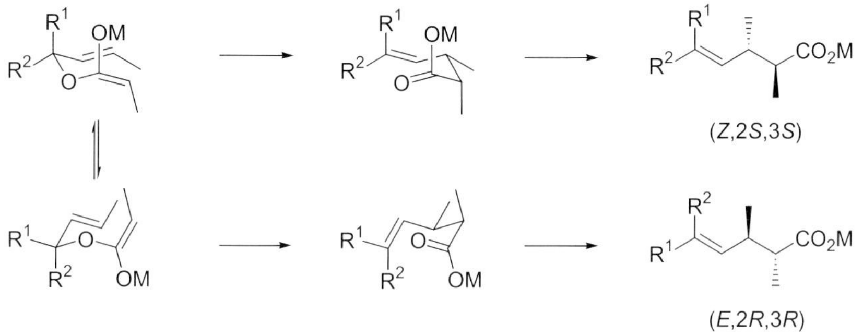

R1 OM R2 O → R1 OM R2 O → R1 R2 CO2M (Z,2S,3S)
R1 O R2 OM → R1 O R2 OM → R2 R1 CO2M (E,2R,3R)

Scheme 4.101

In 1977 Still and Schneider used an Ireland–Claisen rearrangement of a 3° allylic ester in the synthesis of (±)-frullanolide (Scheme 4.102) [93]. Rearrangement to the *β*-pyrrolidinomethyl ester was followed by Cope elimination to the *exo*-methylene lactone. Stereocontrol of the alkene was of course not an issue in this case since the alkene was confined within a ring.

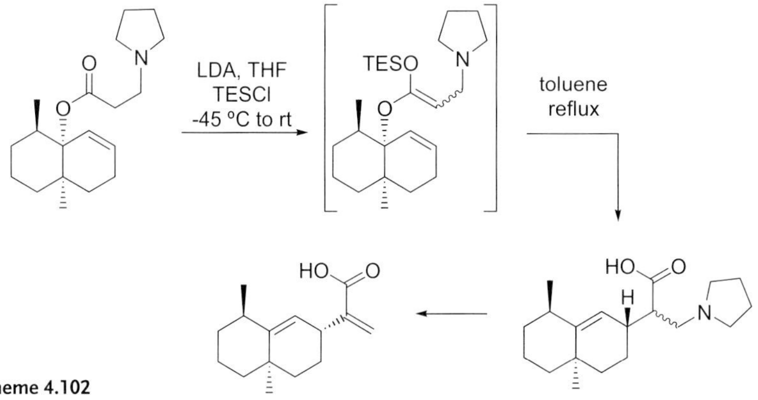

O N O → LDA, THF TESCl -45 °C to rt → [TESO N O] → toluene reflux → HO O H N ← HO O

Scheme 4.102

Johnson et al found that a tetrasubstituted fluoro alkene could be prepared with modest alkene stereoselectivity by rearrangement of a tertiary acetate ester (Scheme 4.103) [94]. Surprisingly, the major product was the E-isomer. This implies that the transition state with the nominally larger CH_2CH_2R group in the pseudo-axial position is favored. The corresponding Johnson-Claisen gave a 1:1 *E*:*Z* mixture.

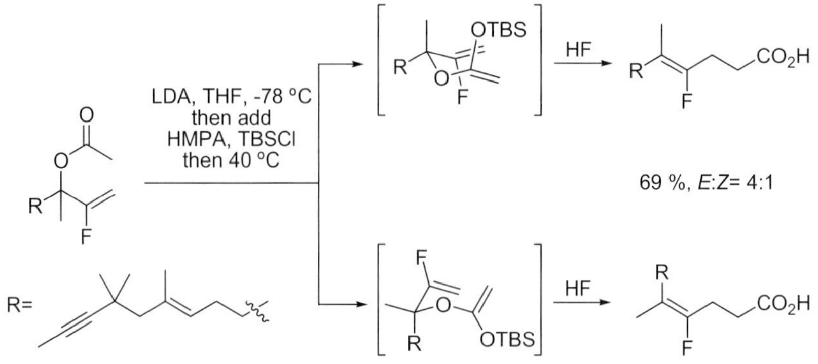

LDA, THF, -78 °C then add HMPA, TBSCl then 40 °C

HF

69 %, *E*:*Z*= 4:1

R=

Scheme 4.103

Overman et al. used the Ireland–Claisen rearrangement of a 3° allylic ester in the synthesis of (+)-pumiliotoxin A (Eq. 1) (Scheme 4.104) [95]. Rearrangement of the *Z*-silyl ketene acetal gave the *E*-alkene as the sole alkene stereoisomer and with 7:1 diastereoselectivity. The stereochemical outcome is consistent with the larger benzyloxypropyl substituent occupying the pseudo-equatorial position of the chair rearrangement transition state.

Interestingly, in model studies directed toward ambruticin, Davidson et al. found that the stereoselectivity of the Ireland–Claisen rearrangement depended on which diastereomer of the ester was used (Eq. 2) (Scheme 4.104) [96]. While the (*R**,*S**) diastereomer shown afforded 15:1 stereoselectivity with respect to the alkene, the (*R**,*R**) stereoisomer gave only a 3:1 selectivity, although in both cases the larger pyranyl substituent preferentially occupied the pseudo-equatorial position of the Ireland–Claisen rearrangement transition state. By contrast the stereochemically analogous acyclic (*S*,*S*) diastereomer that Overman employed gave high alkene stereoselectivity (cf. Scheme 4.103). Apparently the conformation of the pyran ring may play a significant role in determining the stereochemical outcome.

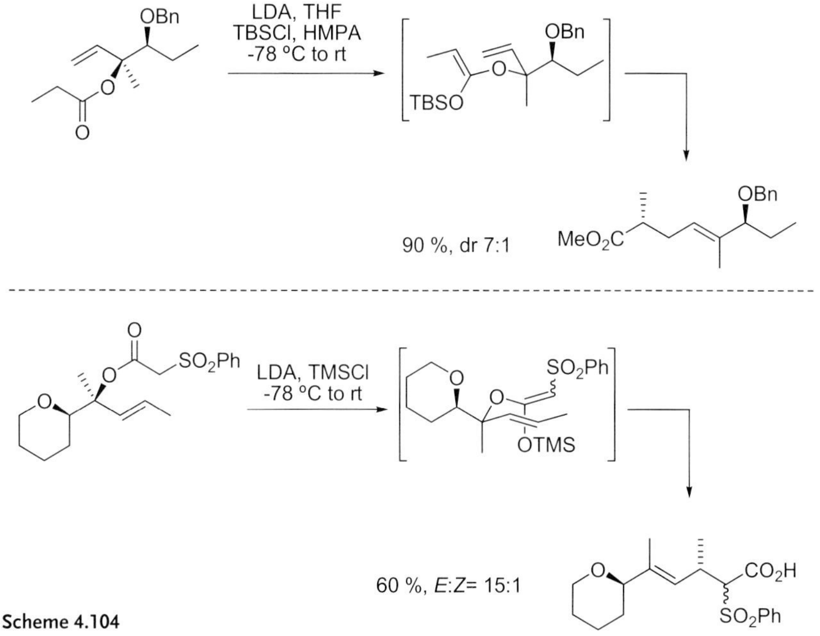
OBn
LDA, THF
TBSCl, HMPA
-78 °C to rt
O
O
TBSO
O
OBn
90 %, dr 7:1
MeO2C
OBn
O
SO2Ph
O
O
LDA, TMSCl
-78 °C to rt
O
O
SO2Ph
OTMS
60 %, E:Z= 15:1
O
CO2H
SO2Ph

Scheme 4.104

Bienz et al. have reported the Ireland–Claisen rearrangement of 3° allylic esters in which one of the allylic substituents is silicon (Scheme 4.105) [97]. The larger silicon substituent occupies the pseudo-equatorial position of the chair to yield the *E*-vinylsilane product. Interestingly, the products apparently result from initial *C*-silylation of the propionate ester followed by a second *Z*-selective enolization of the resulting silyl propionate. The structure of the pentenoic acid product was verified by X-ray crystallographic analysis.

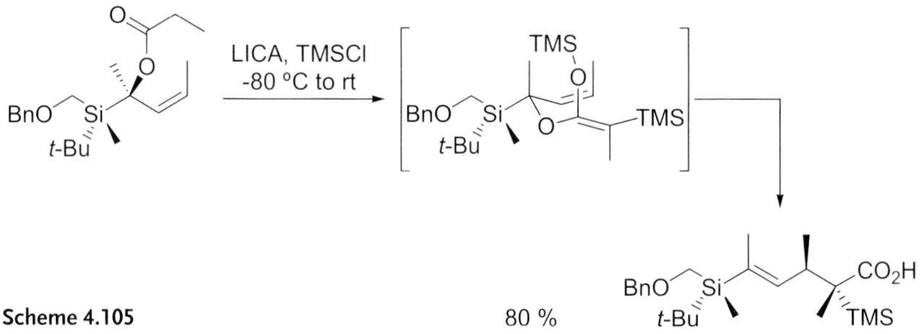
O
O
BnO
Si
t-Bu
LICA, TMSCl
-80 °C to rt
TMS
O
BnO
Si
t-Bu
O
TMS
80 %
BnO
Si
t-Bu
CO2H
TMS

Scheme 4.105

4.8.6
bis-Allylic Esters

The Claisen rearrangement of *bis*-allyl vinyl ethers and related systems has been studied by several groups [98]. Modest to high levels of regioselectivity were obtained in the parent Claisen as well as Johnson, Eschenmoser and Ireland variants of the reaction. Parker and Farmar applied the ester enolate variant of the Ireland–Claisen rearrangement of a *bis*-allylic ester to the synthesis of biflora-4,10(19),15-triene, using the rearrangement to install the 1,3-diene component for a subsequent intramolecular Diels–Alder reaction [99].

In 1990 Tanis et al. reported a surprising dependence of the regioselectivity of the rearrangement of a *bis*-allylic ester (Scheme 4.106) [100]. Treatment of the furyl cyclopentenyl propionate with a series of silyl chlorides resulted in selective rearrangement via the cyclopentene for TBSCl, but via the furan for TIPSCl.

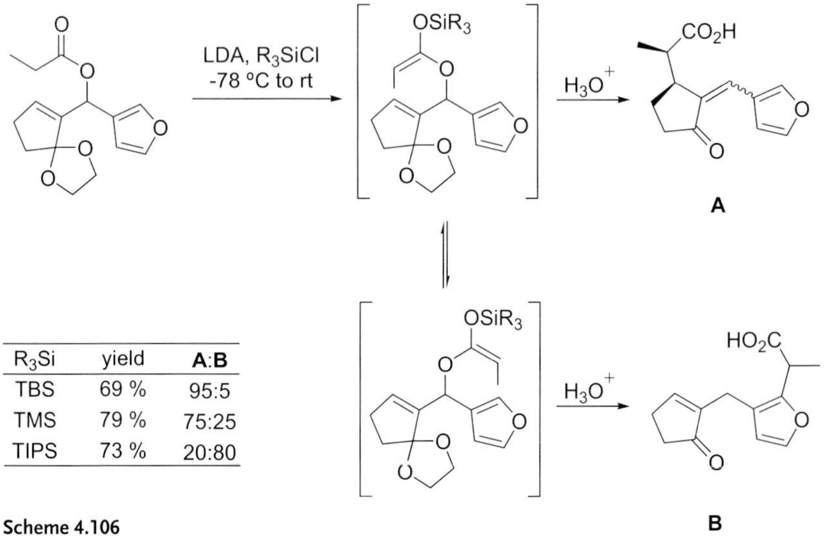

R_3Si	yield	**A**:**B**
TBS	69 %	95:5
TMS	79 %	75:25
TIPS	73 %	20:80

Scheme 4.106

McIntosh et al. have reported a regioselective Ireland–Claisen rearrangement of *bis*-allylic esters derived from cycloalkenones, in which the carbinol carbon is 3° and contained within the cycloalkene ring (Scheme 4.107) [26]. All previous examples of Claisen rearrangements of *bis*-allylic systems employed substrates in which the carbinol carbon was acyclic (e.g., Scheme 4.106). The rearrangement proceeded with very high alkene stereoselectivity for substrates bearing substituents at either carbon adjacent to the carbinol carbon. The rearrangements presumably occurred via chair-like transition states with the larger 3° allylic substituent disposed in a pseudo-equatorial position. Optimal selectivity in this unique 4° to 4° stereochemical relay of the O-benzyl lactate shown were obtained using an adaptation of the Langlois conditions (cf. Scheme 4.78). In contrast to Langlois' studies,

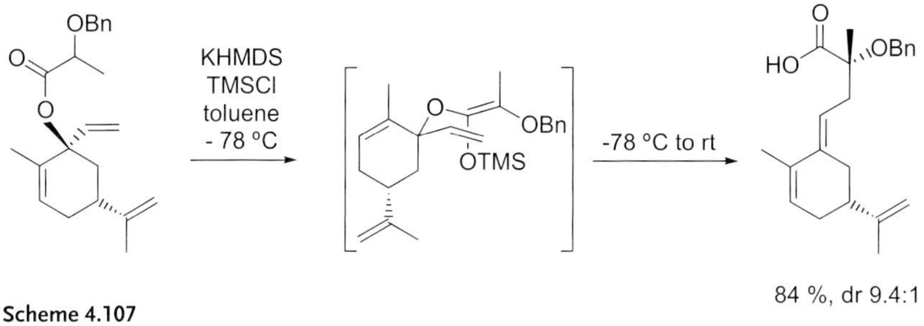
OBn
O
O
KHMDS
TMSCl
toluene
- 78 °C
O
OBn
OTMS
-78 °C to rt
O
HO
OBn
84 %, dr 9.4:1

Scheme 4.107

the authors found that the stereoselectivity was significantly improved by delaying the addition of TMSCl 30–90 min after addition of the KHMDS to the ester at –78 °C.

4.8.7 Fe-Diene Complexes

Roush and Works reported a novel diastereofacially selective Ireland–Claisen rearrangement of Fe-complexed trienic allylic esters (Scheme 4.108) [101]. Although the rearrangement proceeded with excellent facial selectivity, with the allylic alkene attacking *anti* to the Fe-complex in the conformer shown, only modest *syn*/*anti* selectivity was obtained.

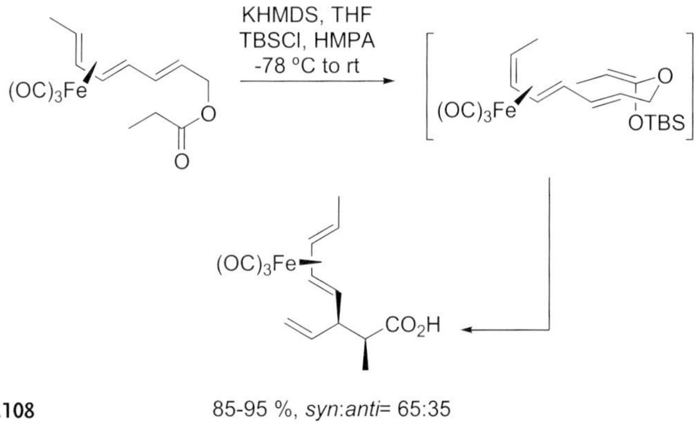
(OC)3Fe
O
O
KHMDS, THF
TBSCl, HMPA
-78 °C to rt
(OC)3Fe
O
OTBS
(OC)3Fe
CO2H
85-95 %, syn:anti= 65:35

Scheme 4.108

4.8.8 Hindered Esters

Magnus and Westwood have reported an Ireland–Claisen rearrangement approach to the taxol skeleton (Scheme 4.109) [102]. The rearrangement presumably occurred via the highly congested *Z*-silyl ketene acetal to install the C2 and (neopentyl) C3 stereocenters with high diastereoselectivity. The unusually high

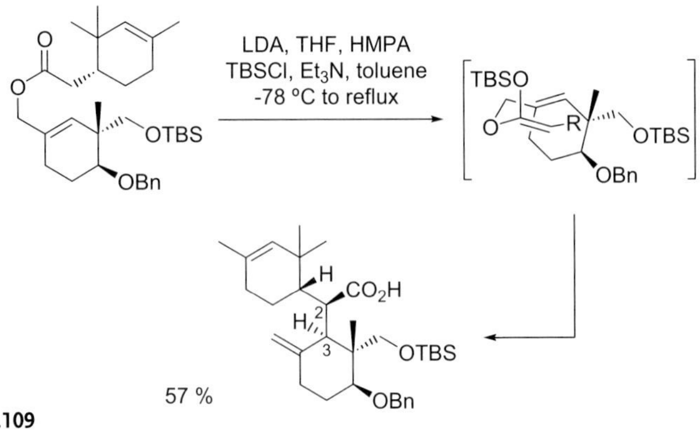

LDA, THF, HMPA
TBSCl, Et_3N, toluene
-78 °C to reflux

57 %

Scheme 4.109

temperature of the rearrangement was presumably necessary to overcome steric hindrance in the transition state.

4.9
Applications to Natural Product Synthesis

This section is devoted to illustrating the manifold applications of the Ireland–Claisen rearrangement in natural products synthesis. The examples were generally chosen because they illustrate the first example of a particular variant of the Ireland–Claisen rearrangement to natural products synthesis.

4.9.1
Prostanoids

In 1976 Ireland et al. reported the application of the rearrangement to the synthesis of the prostanoid skeleton (Scheme 4.110). The stereocenters of the cyclopentene ring were installed via equilibration rather than by the rearrangement itself [103].

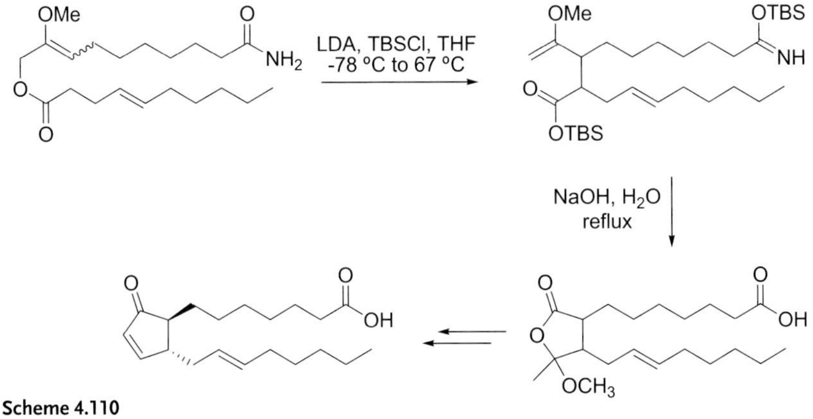

LDA, TBSCl, THF
-78 °C to 67 °C

NaOH, H_2O
reflux

Scheme 4.110

4.9.2 Nonactic Acid

Ireland recognized early on the potential for the Claisen rearrangement in the synthesis of furan and pyran containing natural products. In 1980 Ireland and Vevert published the syntheses of both (–)- and (+)-nonactic acid using the Claisen rearrangement to establish the C2,C3 stereocenters (Scheme 4.111). Rearrangement of the *E*-silyl ketene acetal via the boat transition state afforded the methyl ester with an 86:14 diastereoselectivity [104].

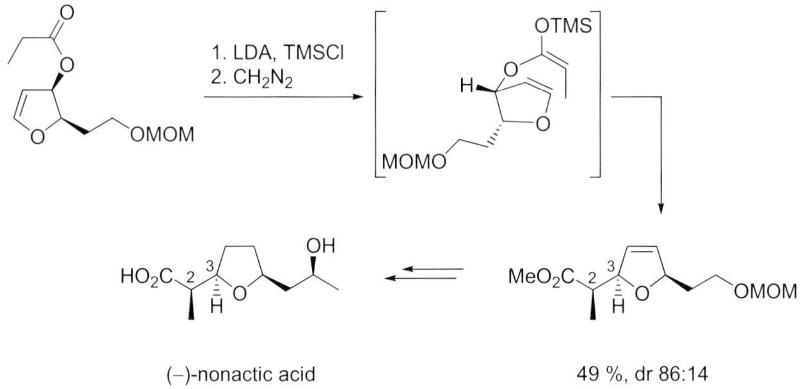

1. LDA, TMSCl
2. CH_2N_2

(–)-nonactic acid

49 %, dr 86:14

Scheme 4.111

4.9.3 Lasalocid A

The Ireland group next applied the rearrangement to the synthesis of increasingly complex polyether antibiotics (Scheme 4.112). In 1980 Ireland et al. reported a formal synthesis of lasolocid A using the rearrangement twice to install both the 3° and 4° stereocenters *α* to the furan oxygen [105]. The first rearrangement was similar to that used in the nonactic acid synthesis (cf. Scheme 4.111). The resulting dihydrofuran was converted to the furanyl pyran. Treatment of the complex ester under the usual reaction conditions afforded a *ca*. 3:1 ratio of isomers favoring the desired *α*-carboxyl 4° center on the furan ring. The product was further elaborated to the ketone shown, which has previously been converted to lasalocid A.

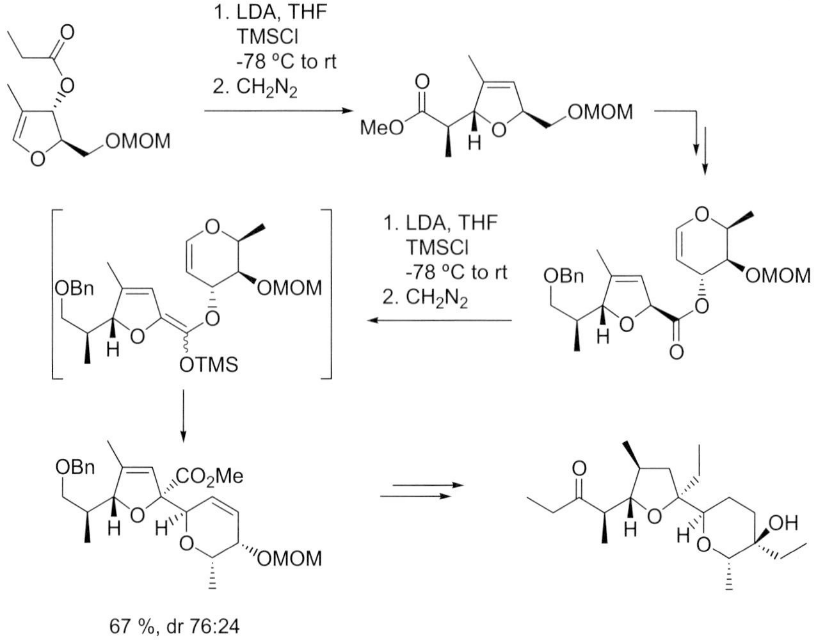

Scheme 4.112

4.9.4
Tirandamycic Acid

In 1981 Ireland et al. [106] reported the total synthesis of tirandamycic acid (Scheme 4.113). Rearrangement of the *Z*-silyl ketene acetal via a boat transition state afforded a 81:19 ratio of diastereomers. Similar rearrangements were employed by Ireland in the synthesis of the structurally related (+) streptolic acid [107] and by Kishi et al. in the synthesis of the C27–C38 fragment of the halichondrins [108].

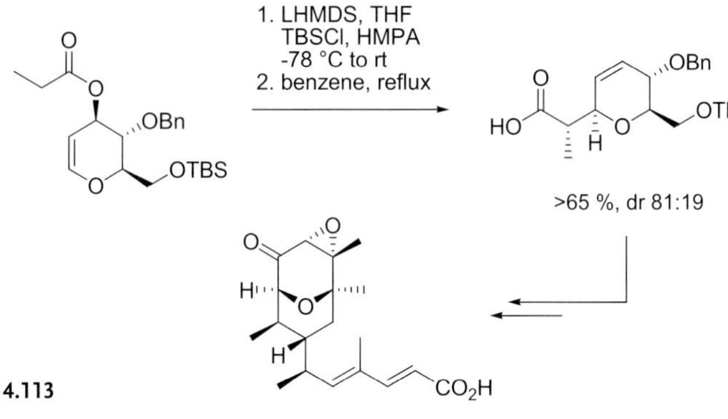

Scheme 4.113

4.9.5 Monensin A

The most complex application of the Ireland–Claisen rearrangement is Ireland's synthesis of monensin A (Fig. 4.2). Claisen rearrangements were ultimately used to establish the C4–C5, C12–C13, and C16–C17 bonds.

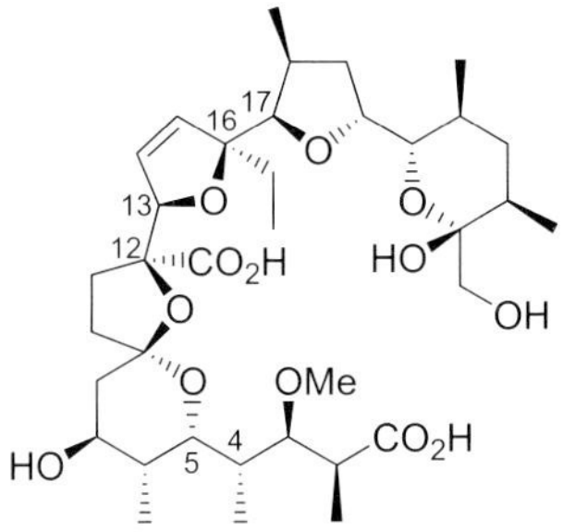

Figure 4.2 Structure of monensin

In early studies toward monensin, Ireland and Norbeck reported an approach to the monensin spiroketal that employed a particularly interesting variant of the rearrangement [109]. The allylic ester they used possessed a leaving group in β-position to the enolate (Scheme 4.114). Although at –78 °C, enolization afforded only products of β-elimination, treating the ester at –100 °C with LDA and TMSCl in 10% HMPA/THF afforded a ca. 80% yield of rearrangement products after warming the reaction mixture to ambient temperature, albeit with little diastereoselectivity.

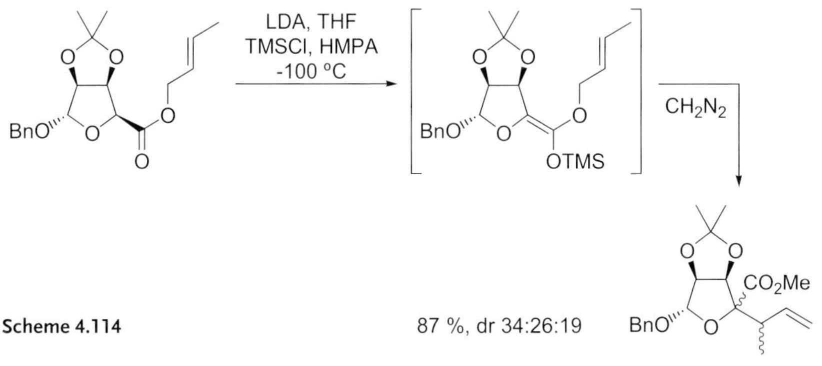

Scheme 4.114

In 1993 Ireland et al. reported the total synthesis of monensin A using a modified approach (Scheme 4.115) [110]. Rearrangement of the pyranyl propionate as in the tirandamycic acid synthesis established the C4–C5 bond with 7:1 diastereoselectivity (Eq. 1). The challenging C16 4° center was installed in high yield, albeit with no stereoselectivity (Eq. 2). Finally, the C12 4° stereocenter was installed in 51% overall yield, but with the desired isomer as the minor product (Eq. 3).

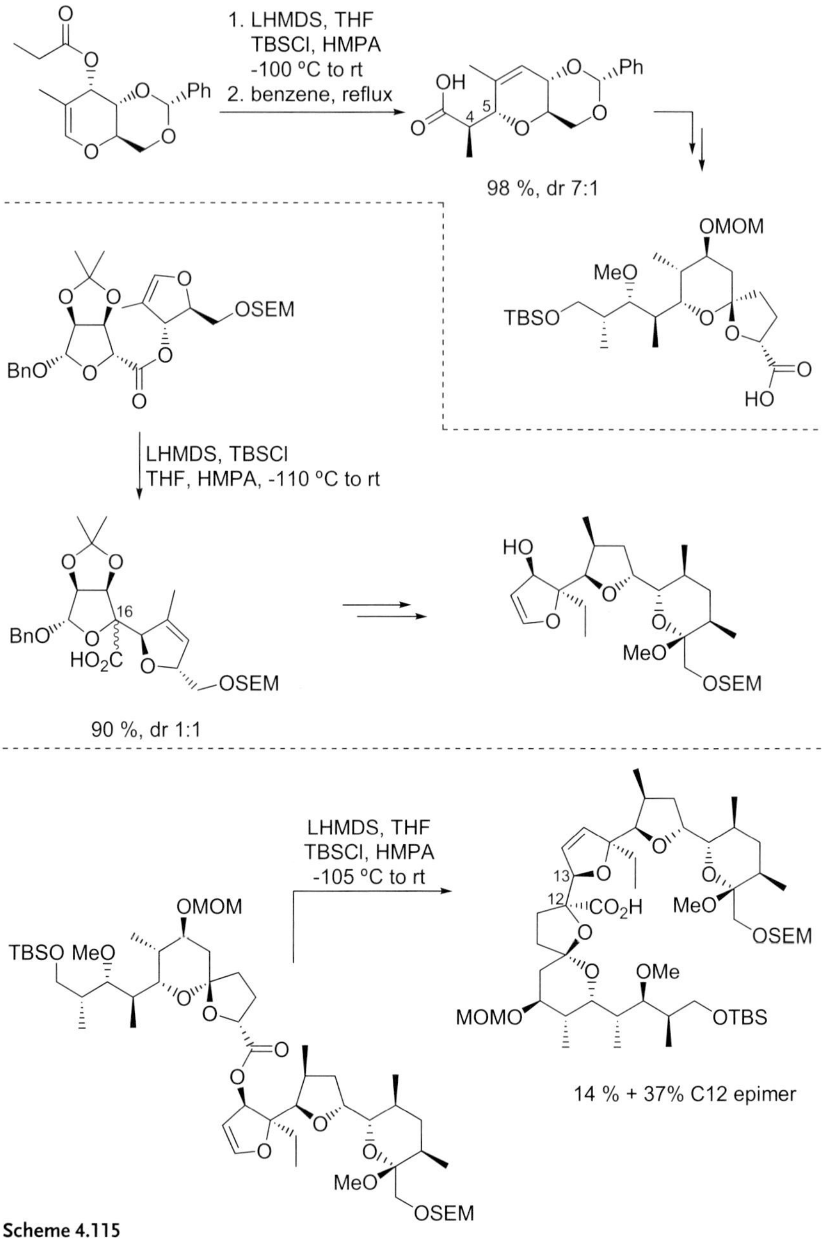

1. LHMDS, THF
TBSCl, HMPA
-100 °C to rt
2. benzene, reflux
OH
Ph
98 %, dr 7:1
OMOM
MeO
TBSO
HO
OSEM
BnO
LHMDS, TBSCl
THF, HMPA, -110 °C to rt
16
HO_2C
HO
MeO
90 %, dr 1:1
LHMDS, THF
TBSCl, HMPA
-105 °C to rt
13
12
CO_2H
OMe
MOMO
OTBS
14 % + 37% C12 epimer

Scheme 4.115

4.9.6
Sphydofuran

Rizzacasa and Di Florio made use of the discovery of Ireland et al. described above (cf. Scheme 4.114) in their synthesis of sphydofuran (Scheme 4.116) [111]. Ireland–Claisen rearrangement of the β-benzyloxy ester under the Ireland conditions afforded a good yield and diastereoselectivity for the desired allylated ester, which was further elaborated to sphydofuran. The allylation preferentially occurred *anti* to the β-benzyloxy group. Rizzacasa and McVinish used an analogous rearrangement in synthetic studies directed toward the squalestatins and zaragozic acids [112].

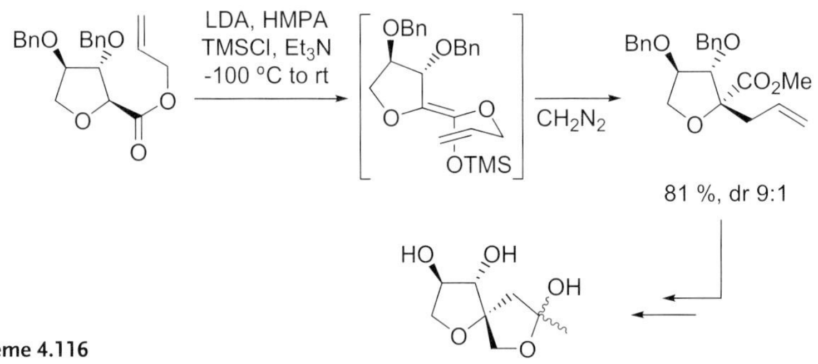

Scheme 4.116

4.9.7
Calcimycin

In 1982 Grieco et al. reported the total synthesis of calcimycin (Scheme 4.117) [113]. They took advantage of the stereospecific feature of the Ireland–Claisen rearrangement by converting two epimeric alcohols to the same rearrangement product. For the β-propionate, they employed the Ireland conditions for generating the *E*-silyl ketene acetal and for the α-propionate, the *Z*-silyl ketene acetal. Both transformations proceeded in high yield.

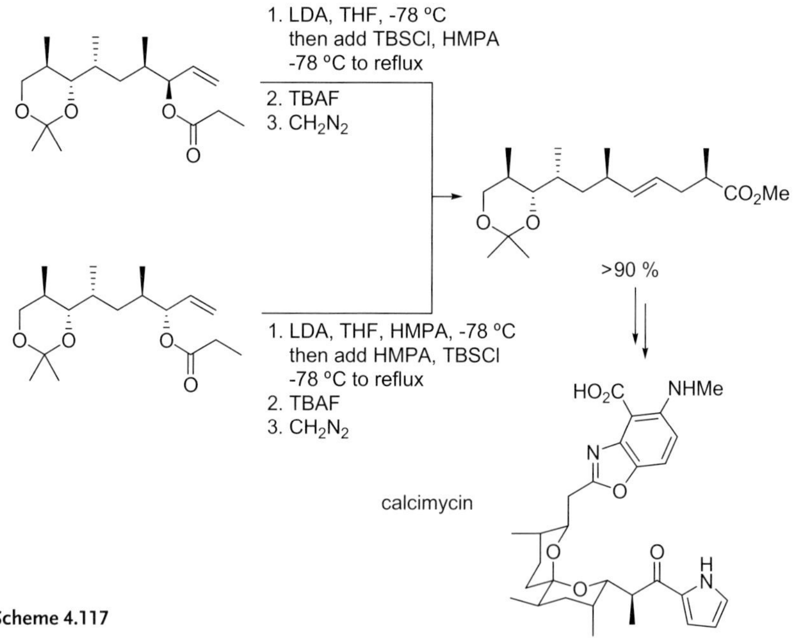

1. LDA, THF, -78 °C then add TBSCl, HMPA -78 °C to reflux
2. TBAF
3. CH_2N_2

CO_2Me

>90 %

1. LDA, THF, HMPA, -78 °C then add HMPA, TBSCl -78 °C to reflux
2. TBAF
3. CH_2N_2

HO_2C NHMe

calcimycin

Scheme 4.117

4.9.8 Ceroplasteric Acid

Rigby et al. employed the Ireland–Claisen rearrangement in an approach to the ophiobolane ceroplasteric acid (Scheme 4.118) [114]. The product was isolated as a 4:1 mixture of diastereomers at C2 (pentenoic acid numbering). The major product is consistent with rearrangement occurring via a chair transition state for an *E*-silyl ketene acetal or a boat transition state for a *Z*-silyl ketene acetal.

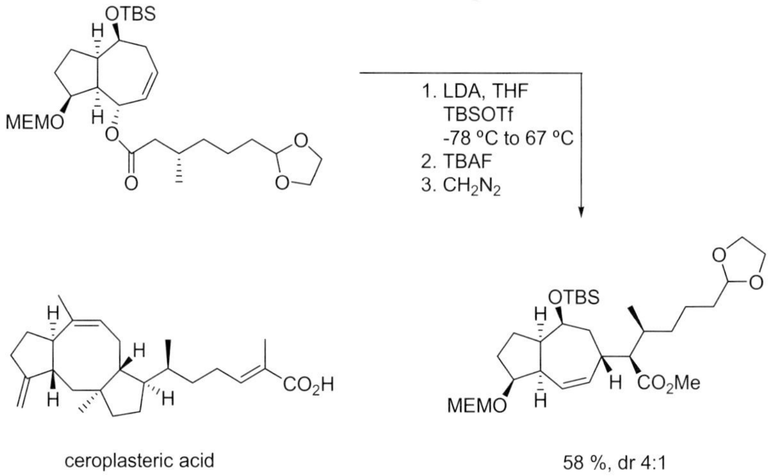

OTBS MEMO

1. LDA, THF TBSOTf -78 °C to 67 °C
2. TBAF
3. CH_2N_2

CO_2H

ceroplasteric acid

OTBS MEMO CO_2Me

58 %, dr 4:1

Scheme 4.118

4.9.9
Erythronolide A

Deslongchamps applied the equatorial selective Ireland–Claisen rearrangement described previously (cf. Scheme 4.41) in a formal synthesis of erthyronolide A (Scheme 4.119) [43]. The rearrangement of the *Z*-silyl ketene acetal proceeded via a chair transition state to establish the C4 and C5 stereocenters of erythronolide A.

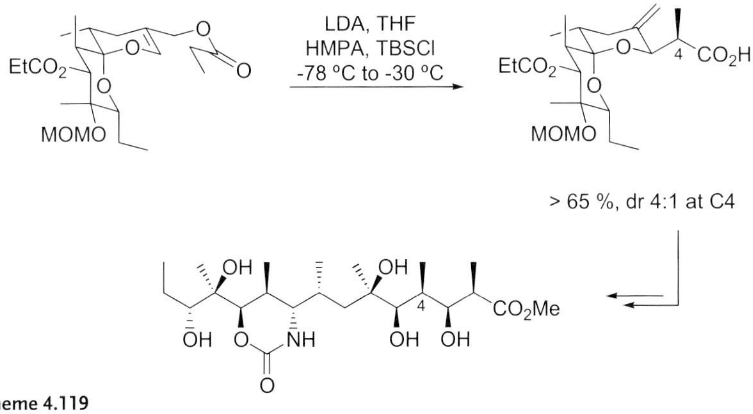

LDA, THF
HMPA, TBSCl
-78 °C to -30 °C

> 65 %, dr 4:1 at C4

Scheme 4.119

4.9.10
Ebelactone A and B

Paterson and Hulme used a chemoselective deprotonation of a ketoester to generate a silyl ketene acetal in a Claisen approach to (–)-ebelactones A and B (Scheme 4.120) [49]. The reaction of the *E*-silyl ketene acetal proceeded via the expected chair transition state to afford the pentenoate ester with very high diastereoselectivity.

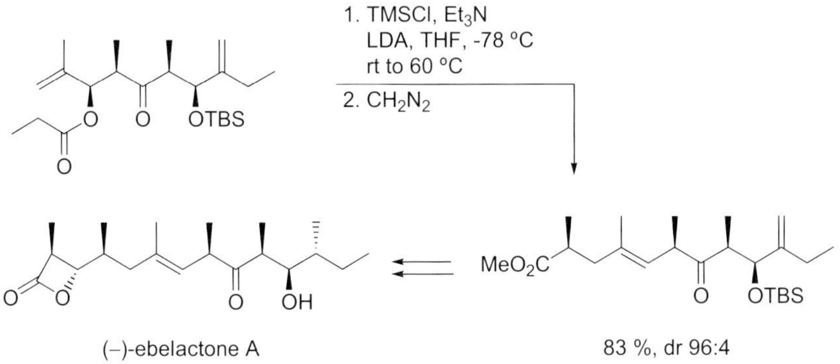

1. TMSCl, Et_3N
LDA, THF, -78 °C
rt to 60 °C
2. CH_2N_2

(–)-ebelactone A

83 %, dr 96:4

Scheme 4.120

4.9.11
25-OH Vitamin D2 Grundmann Ketone

Wilson and Jacob used a γ-deprotonation of an allyl acrylate to generate a *Z*-silyl ketene acetal in a Claisen approach to 25-hydroxy vitamin D2 Grundmann ketone (Scheme 4.121) [115]. The rearrangement proceeded via the expected chair transition state to afford the desired product as a single stereoisomer.

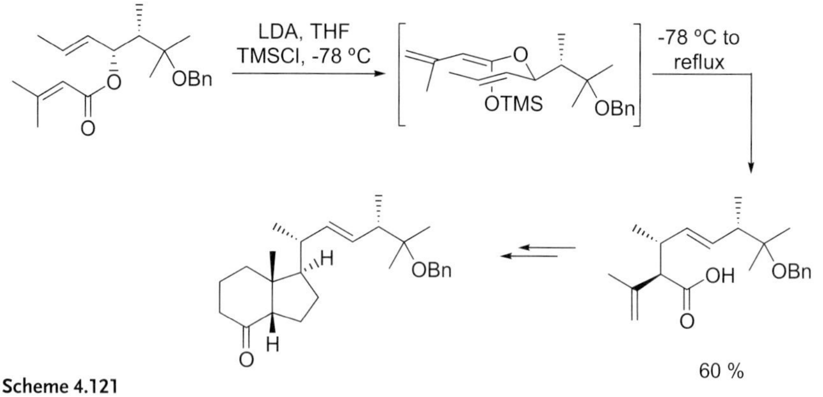

LDA, THF
TMSCl, -78 °C
OBn
OTMS
OBn
-78 °C to
reflux
OH
OBn
60 %
H
H
OBn

Scheme 4.121

4.9.12
Zincophorin

Kallmerten and Cywin used the glycolate Claisen rearrangement in an approach to zincophorin (Scheme 4.122) [116]. Rearrangement of the *Z*-silyl ketene acetal of the glycolate ester via a chair transition state afforded the *syn* stereochemistry in the product. The ester was further elaborated to the pyran which constitutes the C1–C11 subunit of zincophorin.

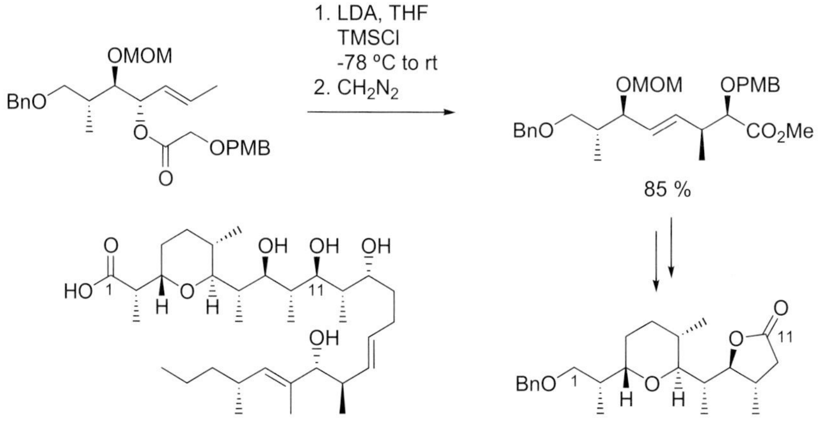

OMOM
BnO
OPMB
1. LDA, THF
TMSCl
-78 °C to rt
2. CH2N2
OMOM
OPMB
BnO
CO2Me
85 %
HO
1
OH
OH
OH
11
OH
BnO
1
11

Scheme 4.122

4.9.13
Steroid Side Chain Homologation

Nakai et al. used the glycolate Claisen to elaborate steroid side chains (Scheme 4.123) [117]. The rearrangement of the *Z*-silyl ketene acetal proceeded via a chair transition state to give the expected C20,C21 (steroid numbering) *syn* diastereomer in high yield.

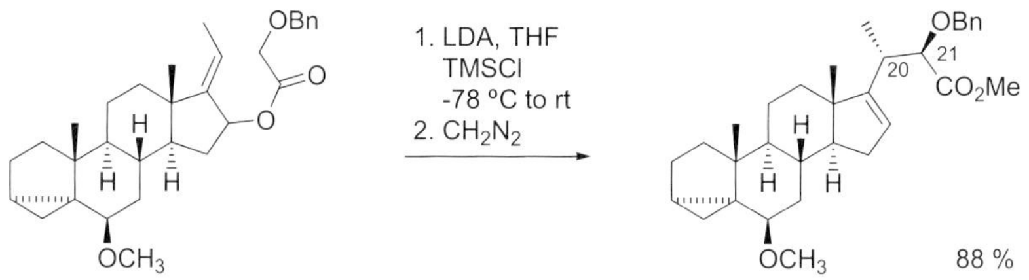

1. LDA, THF TMSCl -78 °C to rt
2. CH_2N_2

88 %

Scheme 4.123

4.9.14
Pseudomonic Acid C

Curran and Suh reported an approach to the pseudomonic acids that employed a novel mono-Claisen rearrangement of a *bis*-silyl ketene acetal (Scheme 4.124) [118]. Treatment of the bis-ester afforded the bis-silyl ketene acetal. Heating of the crude product to 60 °C resulted in mono-Ireland–Claisen rearrangement of to give the 2-substituted pyran. The authors attributed the selectivity to a vinylogous anomeric effect. The authors surveyed several related substrates and determined that the relative rate of the two Claisen rearrangements varied from 20–575:1. The rate difference is consistent with observed the accelerating effect of C6 oxygen substitution in the parent Claisen rearrangement. The carboxylic acid was elaborated to intercept an advanced intermediate in previous syntheses of pseudomonic acid C.

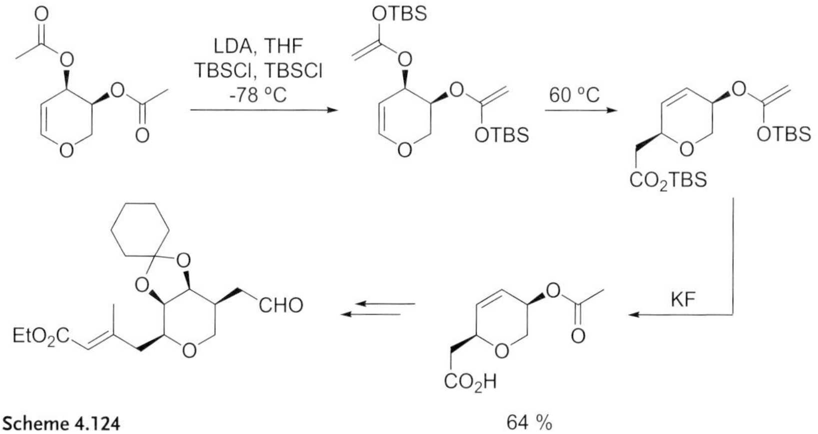

LDA, THF TBSCl, TBSCl -78 °C

60 °C

KF

64 %

Scheme 4.124

Barrish et al. used a glycolate Claisen rearrangement to establish the C12 and C13 stereocenters and the C14–C15 *E*-alkene in their total synthesis of pseudomonic acid C (Scheme 4.125) [119]. Because they used the *Z*-allylic alkene and *Z*-silyl ketene acetal, the rearranged product possessed the *anti* configuration.

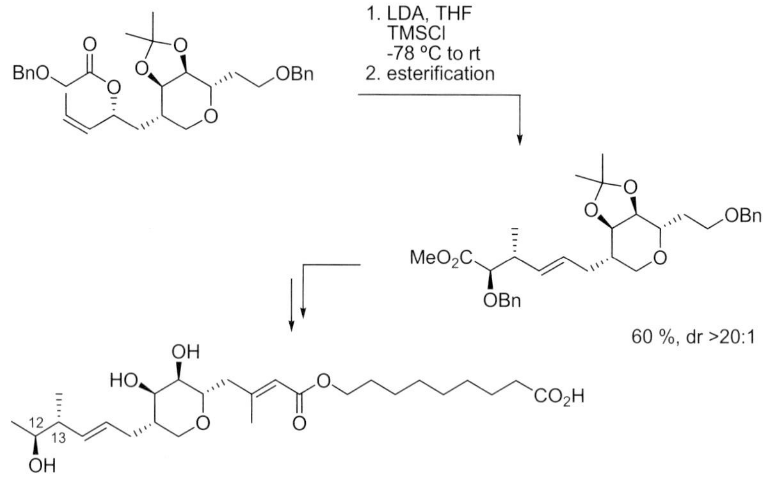

1. LDA, THF TMSCl -78 °C to rt
2. esterification

60 %, dr >20:1

Scheme 4.125

4.9.15
Pine Sawfly Pheromone

Kallmerten et al. used iterative glycolate Claisen rearrangements to establish the three stereocenters in pine sawfly pheromone (Scheme 4.126) [120]. An initial Ireland–Claisen rearrangement of the *Z*-allylic alkene afforded the *anti* pentenoic

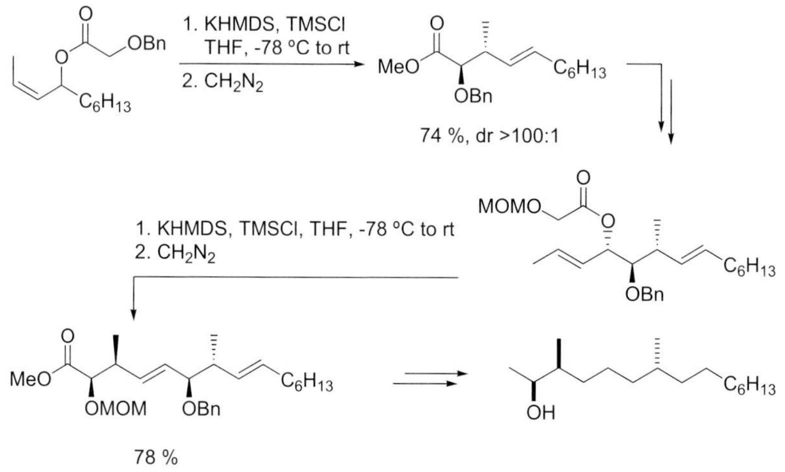

1. KHMDS, TMSCl THF, -78 °C to rt
2. CH_2N_2

74 %, dr >100:1

1. KHMDS, TMSCl, THF, -78 °C to rt
2. CH_2N_2

78 %

Scheme 4.126

ester in good yield and excellent *anti* diastereoselectivity. Reduction, olefination and acylation of the pentenoate ester a homologated *E*-allylic glycolate. Ireland–Claisen rearrangement afforded then afforded the *syn* diastereomer. A series of reductions then yielded the natural product.

4.9.16
Asteltoxin

Mulzer and Mohr used a glycolate Claisen to establish the C6, C7 stereocenters in an asymmetric synthesis of the asteltoxin *bis*-tetrahydrofuran fragment (Scheme 4.127) [121]. Rearrangement occurred via the expected *Z*-silyl ketene acetal and chair transition state to afford the adjacent carbinol and 4° carbon stereocenters. The high stereoselectivity of the rearrangement using a tetrasubstituted allylic alkene is noteworthy.

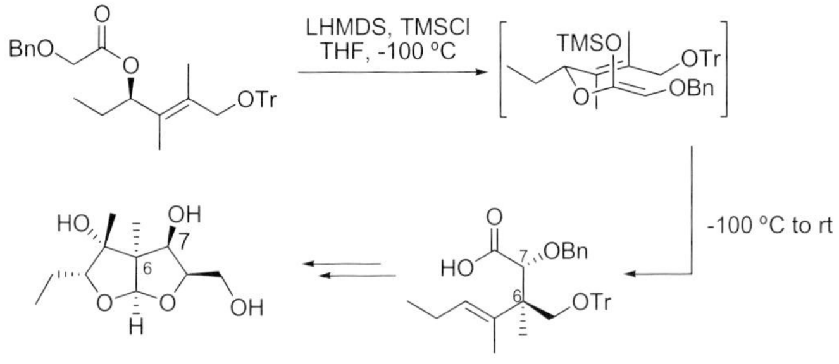

LHMDS, TMSCl
THF, -100 °C
TMSO
OTr
OBn
BnO
O
OTr
-100 °C to rt
HO
OH
7
6
H
OH
OBn
OTr
91 %, dr >98:2

Scheme 4.127

4.9.17
Breynolide

Burke et al. employed a cyclic variant of the glycolate Ireland–Claisen rearrangement in the asymmetric synthesis of (+)-breynolide (Scheme 4.128) [122]. The rearrangement of the *Z*-silyl ketene acetal via a boat transition state generated the C3,C4 stereochemistry of the natural product in high yield and stereoselectivity.

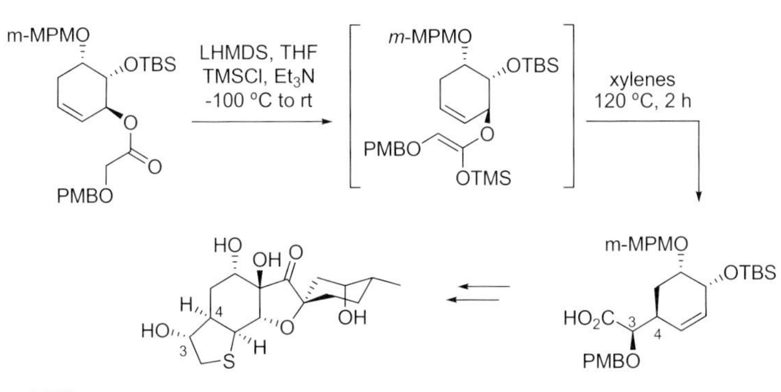

m-MPMO
OTBS
PMBO
LHMDS, THF
TMSCl, Et3N
-100 °C to rt
m-MPMO
OTBS
PMBO
OTMS
xylenes
120 °C, 2 h
HO
OH
H
HO
S
H
OH
m-MPMO
OTBS
HO2C
PMBO
> 85 %

Scheme 4.128

4.9.18
Methyl Ydiginate

Boeckman et al. made use of a glycinate variant of the Ireland–Claisen rearrangement to install the vicinal stereocenters in the imide ring of (–)-methyl ydiginate (Scheme 4.129) [123]. The allylic glycinate, prepared from the corresponding allylic alcohol in *ca.* 90% optical purity, was treated under the Bartlett conditions to afford the corresponding *syn* amino acid derivative. Oxidative cleavage of the alkene and further elaboration led to (–)-methyl ydiginate, also in 90% optical purity, indicating that chirality transfer via the Claisen rearrangement was essentially complete.

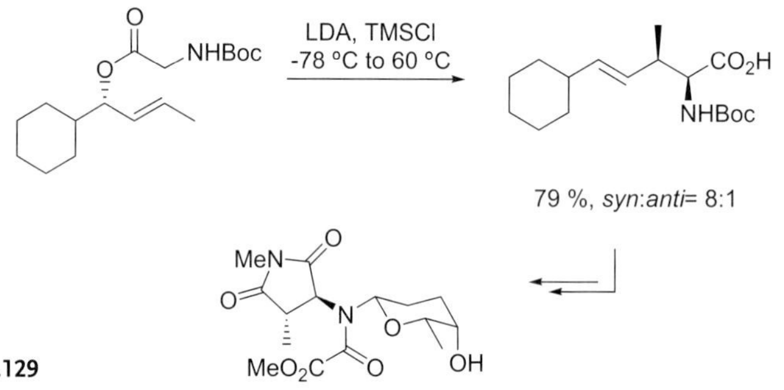
NHBoc
LDA, TMSCl
-78 °C to 60 °C
CO_2H
NHBoc
79 %, *syn*:*anti*= 8:1
MeN
MeO_2C
OH

Scheme 4.129

4.9.19
(–)-Petasinecine

Mulzer and Shanyoor prepared (–)-petasinecine using the Ireland–Claisen rearrangement to install the vicinal stereocenters of the B-ring of the alkaloid (Scheme 4.130) [36]. Rearrangement of the *Z*-silyl ketene acetal of the glycolate ester occurred with complete relative and 1,2-stereocontrol to afford the desired pentenoic acid. The transition state of the rearrangement presumably adopted the conformation shown with the nitrogen oriented antiperiplanar to the forming C–C bond (cf. Scheme 4.34).

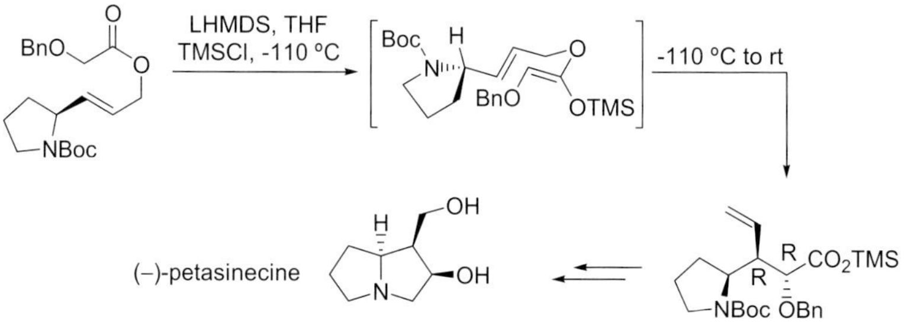
BnO
NBoc
LHMDS, THF
TMSCl, -110 °C
Boc
H
BnO
OTMS
-110 °C to rt
CO_2TMS
NBoc OBn
(–)-petasinecine
OH
OH

Scheme 4.130

4.9.20
β-Elemene

Corey et al. used the asymmetric Ireland–Claisen rearrangement in the synthesis of β-elemene (Scheme 4.131) [124]. Rearrangement of the *Z*-(*O*)-*B*-ketene acetal via the chair transition state afforded the trienic acid in good yield with complete diastereo- and enantioselectivity.

L_2BBr, Et_3N
toluene
-78 °C to 4 °C
O
O
O
L_2BO
O
OH
70 %, >99 % ee

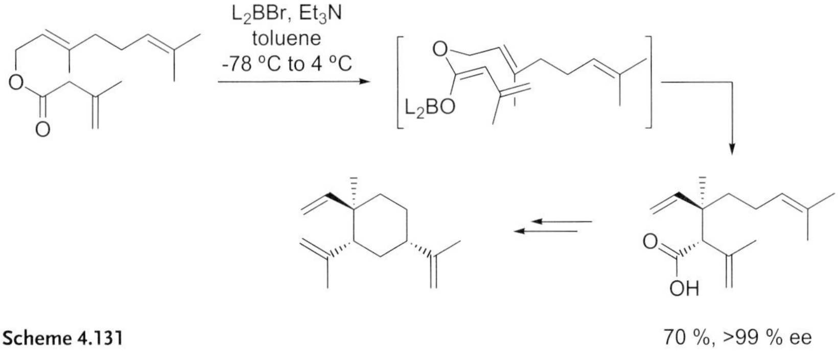

Scheme 4.131

4.9.21
(+)-Dolabellatrienone

Corey and Kania later prepared the bicyclic marine diterpenoid (+)-dolabellatrienone via the Ireland–Claisen rearrangement (Scheme 4.132) [125]. Ring contraction of the 15-membered ring lactone occurred with both high diastereo- and enantioselectivity via the *Z*-(*O*)-*B*-ketene acetal. The alkene and carboxylic acid group were manipulated in several steps to form the isopropylidene cyclopenta-

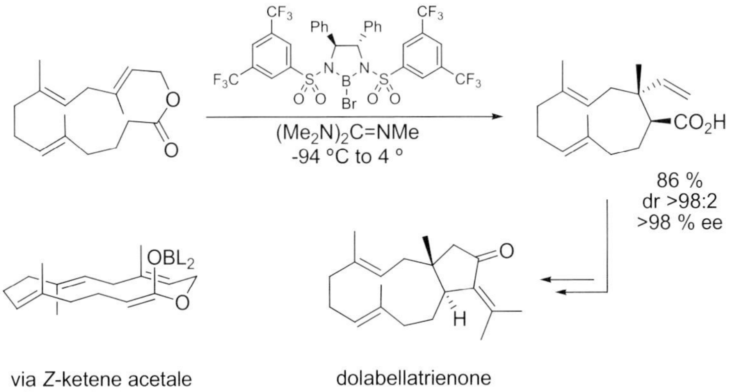

CF_3
Ph
Ph
CF_3
F_3C
CF_3
Br
$(Me_2N)_2C{=}NMe$
-94 °C to 4 °
CO_2H
86 %
dr >98:2
>98 % ee
OBL_2
O
H
via *Z*-ketene acetale
dolabellatrienone

Scheme 4.132

none ring of the natural product. As noted previously, boron ketene acetals have been used very infrequently in Ireland–Claisen rearrangements, although the work of Corey [47] and Oh [64] demonstrate their synthetic utility.

4.9.22 2-Keto-3-Deoxy-Octonic Acid (KDO)

Burke and Sametz have used an Ireland–Claisen rearrangement to desymmetrize a C_2 symmetric mannitol-derived divinyl diol to yield an enantiopure 2,6-disubstituted pyran (Scheme 4.133) [126]. The pyran was an intermediate in the total synthesis of KDO.

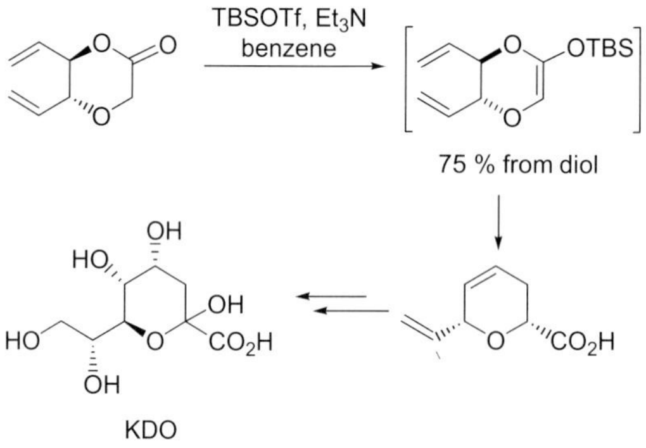

TBSOTf, Et_3N
benzene
OTBS
75 % from diol
HO
OH
OH
CO_2H
HO
OH
KDO
CO_2H

Scheme 4.133

4.9.23 Methylenolactocin

Garcia et al. used a related strategy in the total synthesis of methylenolactocin (Scheme 4.134) [127]. The C_2 symmetric *bis*-silyl ketene acetal underwent Ireland–Claisen rearrangement to the corresponding pentenoic acid. The superfluous pentyl side chain was oxidatively cleaved to install the *β*-carboxylic acid group.

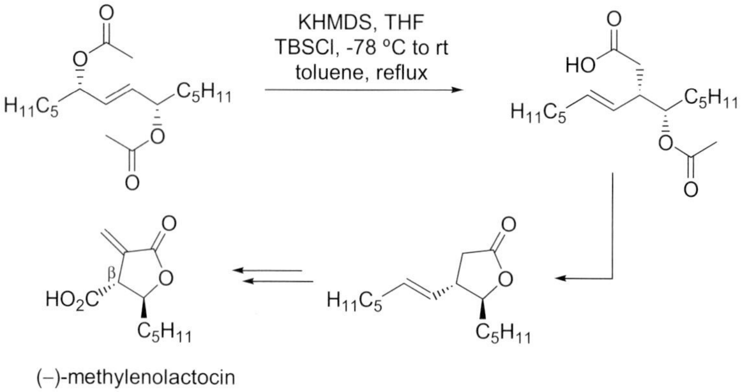

KHMDS, THF
TBSCl, -78 °C to rt
toluene, reflux
$H_{11}C_5$
C_5H_{11}
HO
HO_2C
β
(–)-methylenolactocin

Scheme 4.134

4.9.24
Eupomatilones

McIntosh et al. have applied the Ireland–Claisen rearrangement of *bis*-allyl silyl ketene acetals in studies directed toward the synthesis of the eupomatilones (Scheme 4.135) [128]. The 1,2-transposition of the alkene, which occurred in the rearrangement afforded a reactive vinyl epoxide (cf. Scheme 4.83). Stereoselective cyclization of the carboxylic acid onto the vinyl epoxide generated the 5-aryl lactone, which was further manipulated to the putative structure of 5-*epi*-eupomatilone-6.

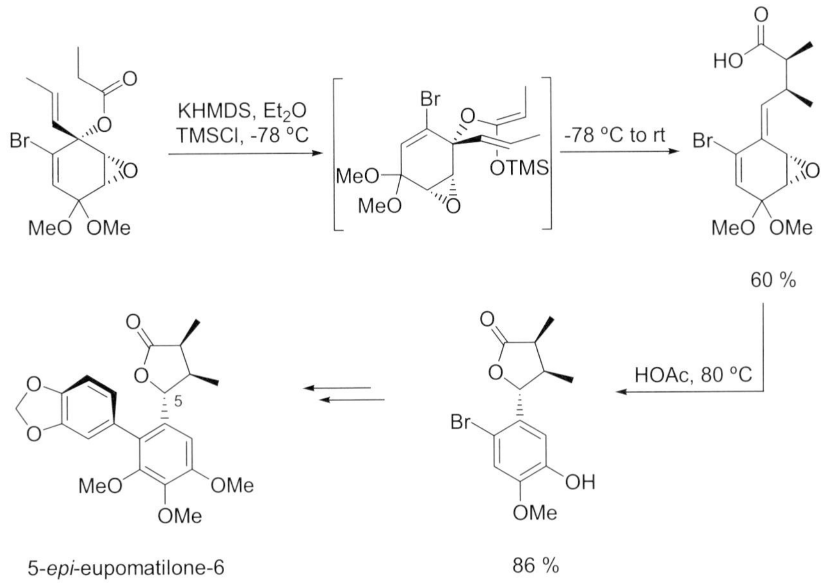

Scheme 4.135

4.9.25
Trichothecenes

As mentioned previously (Section 4.4.2), Ireland had noted early on that increasing the number of alkyl substituents at C1 and/or C6 of the allyl silyl ketene acetal resulted in an increase in the rate of rearrangement. Although not investigated by Ireland, the obvious extension of the trend would be that rearrangement of ketene acetals bearing disubstitution at both C1 and C6 would afford vicinal 4° centers.

In 1986, Kraus [129], Gilbert [130] and VanMiddlesworth [131] all reported application of the Ireland–Claisen rearrangement in syntheses of the trichothecenes trichodiene and bazzanene. In each case the diastereoselectivity of the rearrangement was minimal as a result of a lack of stereoselectivity in the enolization of the ester. The Kraus synthesis is illustrative (Scheme 4.136). Treatment of the cyclohexenyl ester with LDA and TBSCl presumably yielded a *ca.* 1:1 mixture of *E*- and *Z*-silyl ketene acetals. Heating of the reaction mixture to 65 °C afforded after desilylation a 1.5:1 mixture of pentenoic acids. The enoic acids were elaborated to trichoenone and the diastereomeric bazzanenone (not shown), thus completing formal syntheses of both compounds. Gilbert later resolved the issue of enolization stereochemistry in an enantioselective synthesis of trichodiene (cf. Scheme 4.26) [33].

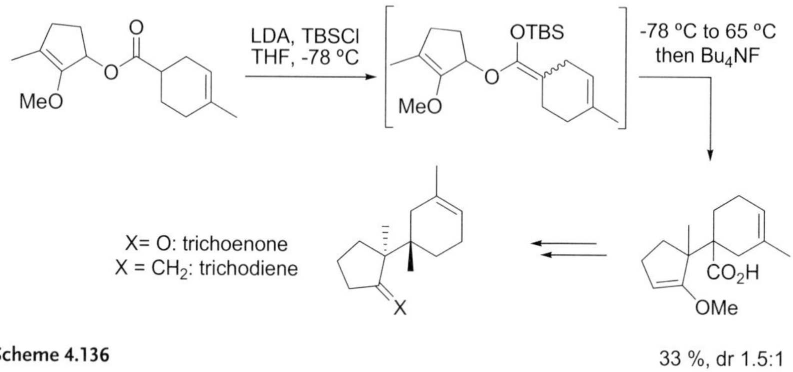

LDA, TBSCl THF, -78 °C; -78 °C to 65 °C then Bu_4NF

X= O: trichoenone
X = CH_2: trichodiene

33 %, dr 1.5:1

Scheme 4.136

4.9.26
(±)-Widdrol

The first example of an Ireland–Claisen rearrangement of an allyl lactone employed in natural product synthesis was that of Danishefsky and Tsuzuke in 1980 (Scheme 4.137) [132]. The allyl lactone possessed an exocyclic alkene, and hence resulted in a net 7-membered ring lactone to carbocycle conversion. A Bayer Villiger reaction converted the pentenoic acid to the 3° hydroxyl group of widdrol.

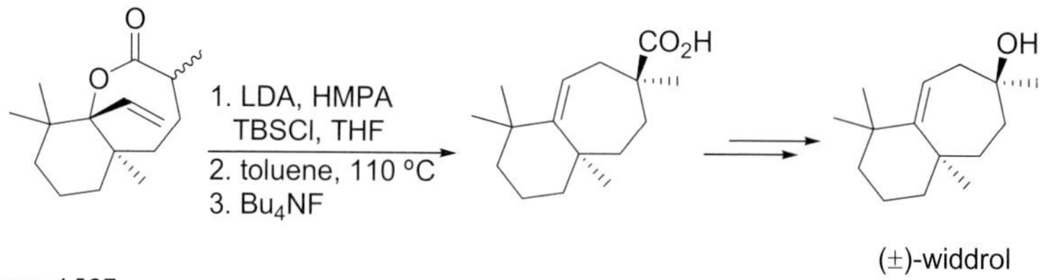

1. LDA, HMPA TBSCl, THF
2. toluene, 110 °C
3. Bu_4NF

(±)-widdrol

Scheme 4.137

4.9.27 Equisetin

Danishefsky et al. later used a related strategy in the synthesis of equisetin (Scheme 4.138) [133]. The presence of a second enolizable group was dealt with by its conversion to a silyl enol ether and subsequent hydrolysis. Danishefsky and Simoneau used a similar approach in syntheses of compactin and ML-236A [134].

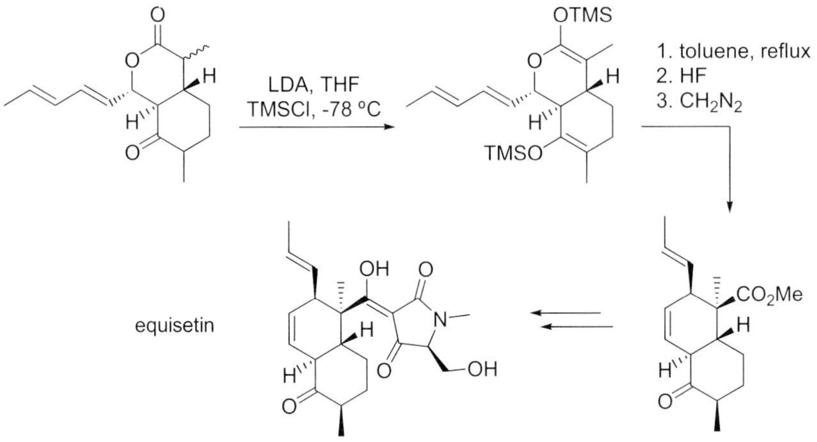

Scheme 4.138

4.9.28 Muscone

Brunner and Borschberg [135] employed an Ireland–Claisen rearrangement of a macrocyclic lactone bearing an exocyclic alkene in a synthesis of (±)-muscone (Scheme 4.139). Treatment of the lactone with LICA and TESOTf gave a 76:8 ratio of *E*- and *Z*-cycloalkenes, which were both elaborated to muscone.

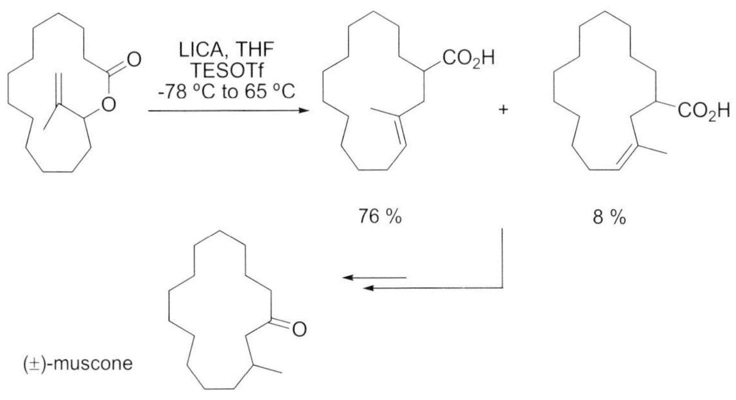

Scheme 4.139

4.9.29
Quadrone

Funk and Abelman used the Ireland–Claisen rearrangement in a formal synthesis of the tetracyclic terpenoid quadrone (Scheme 4.140) [136]. Treatment of the lactone under typical conditions resulted in facile rearrangement at or below room temperature to give the bicyclic acid after silyl ester hydrolysis. Oxidative cleavage of the alkenes then intercepted an intermediate previously employed by Schlessinger in a total synthesis of quadrone.

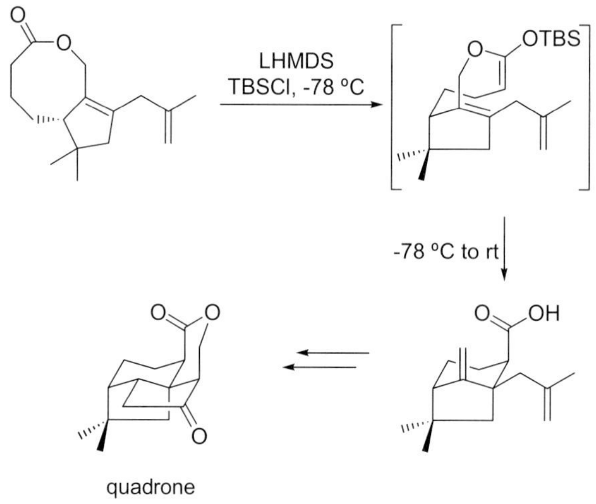

LHMDS
TBSCl, -78 °C
OTBS
-78 °C to rt
OH
quadrone

Scheme 4.140

4.9.30
Ingenanes

Funk and Olmstead employed a ring contraction variant of the Ireland–Claisen rearrangement to generate an in-out-bicyclo[4.4.1]undecan-7-one in an approach to ingenol (Scheme 4.141) [137a]. The rearrangement of the 11-membered ring lactone afforded the 7-membered ring carbocycle.

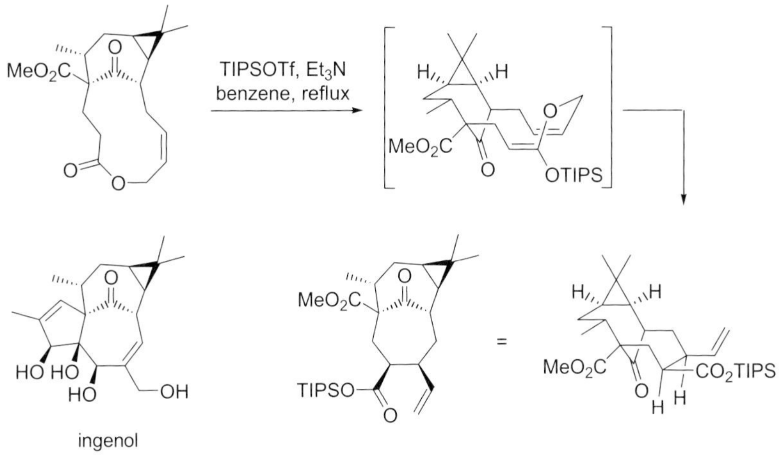

Scheme 4.141

Funk et al. subsequently used a modification of this strategy to prepare the complete ingenane skeleton (Scheme 4.142) [137b]. In this case a smaller 9-membered ring lactone was employed to afford a cycloheptane more readily adaptable for installation of the remaining cyclopentenone ring. The rearrangement proceeded in high yield to afford the β-C4 stereochemistry.

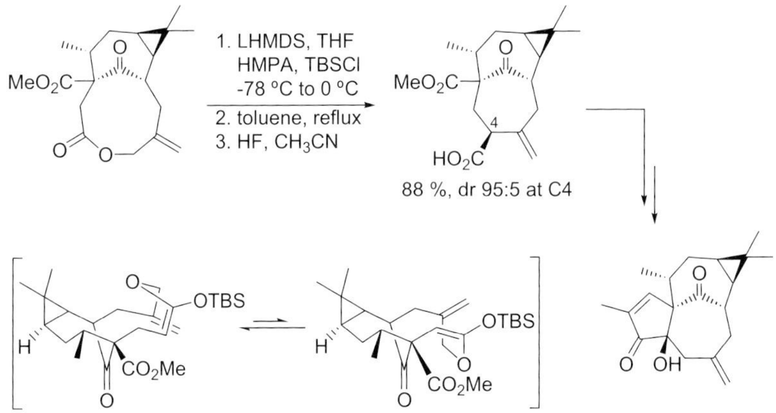

Scheme 4.142

4.9.31
(±)-Samin

Knight et al. have employed a ring contraction via the Ireland–Claisen rearrangement of an aryl lactone to generate a 2,3,4-trisubstituted tetrahydrofuran intermediate in the synthesis of (±)-samin (Scheme 4.143) [138]. The rearrangement proceeded via a boat transition state of the cyclic *E*-silyl ketene acetal.

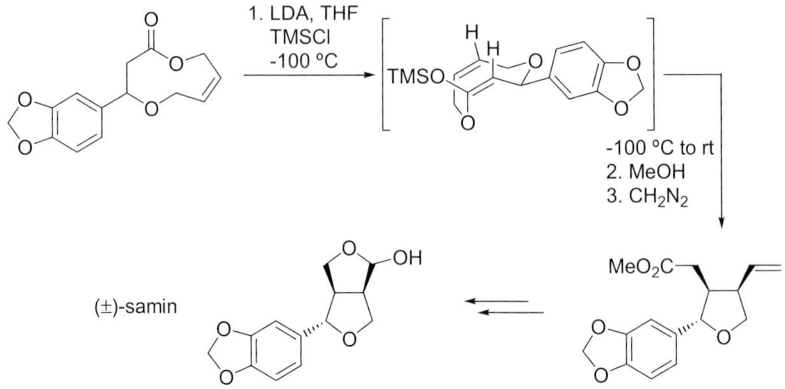
1. LDA, THF
TMSCl
-100 °C
TMSO
-100 °C to rt
2. MeOH
3. CH_2N_2
MeO_2C
OH
(±)-samin

Scheme 4.143

4.9.32
(+)-Monomorine

Angle and Breitenbucher reported the total synthesis of (+)-monomorine using the Ireland–Claisen rearrangement of a vinyl morpholinone to generate the requisite *cis*-2,6-disubstituted pipecolic ester intermediate (Scheme 4.144) [139]. The silyl ester was homologated and cyclized to generate the indolizidine bicycle.

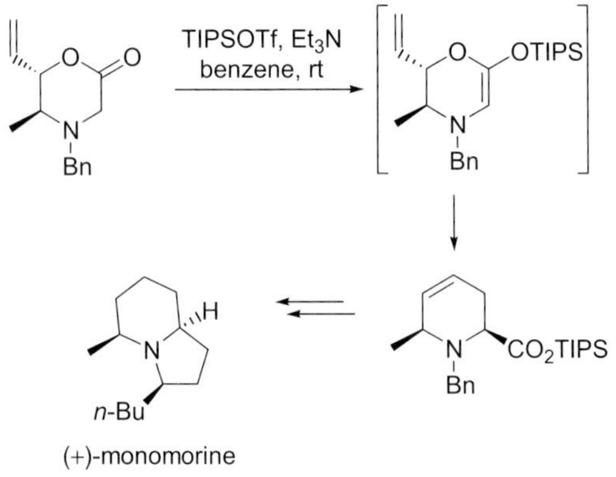
TIPSOTf, Et_3N
benzene, rt
OTIPS
Bn
CO_2TIPS
n-Bu
(+)-monomorine

Scheme 4.144

4.9.33
Dictyols

Knight and co-workers reported the use of a ring contraction variant of the Ireland–Claisen rearrangement to prepare the [5.3.0] ring system of the dictyols (Scheme 4.145) [140]. Rearrangement of the bicyclic lactone under typical Ireland conditions resulted in formation of the bicyclic skeleton of the dictyols.

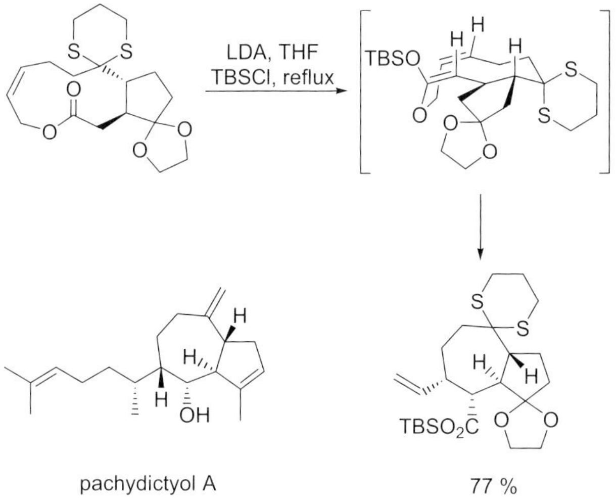

LDA, THF
TBSCl, reflux

pachydictyol A

77 %

Scheme 4.145

4.10
Propargyl Esters

The product of Ireland-Claisen rearrangement of a propargylic ester is a penta-3,4-dienic acid. In contrast to rearrangements of allyic esters, there has been considerably less work on propargylic esters [141].

Fujisawa et al. prepared allenic acids with high levels of diastereoselectivity in the rearrangements of silyl ketene acetals derived from propargyl glycolates (Scheme 146) [142]. The authors argue that the stereoselectivity is principally due to a diaxial-like interaction between the pseudo-axial H and OTMS substituents that is present in the chair-like transition state.

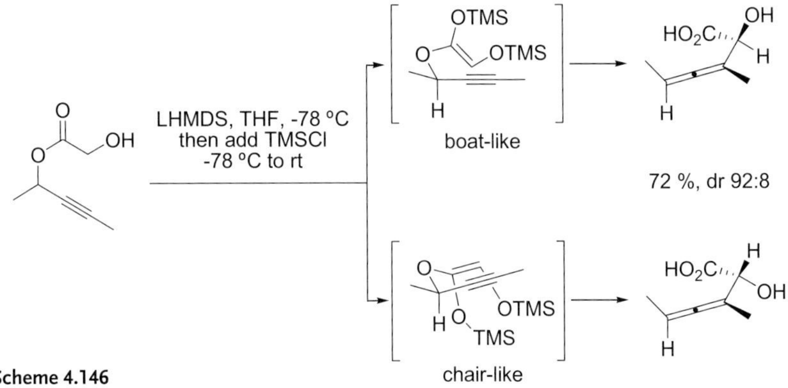

LHMDS, THF, -78 °C then add TMSCl -78 °C to rt

boat-like

72 %, dr 92:8

chair-like

Scheme 4.146

Kuwajima reported one example of an Ireland-Claisen rearrangement of a propargyl acrylate that was initiated by Cu(I)-catalyzed conjugate addition of MeMgBr (Scheme 147) [57].

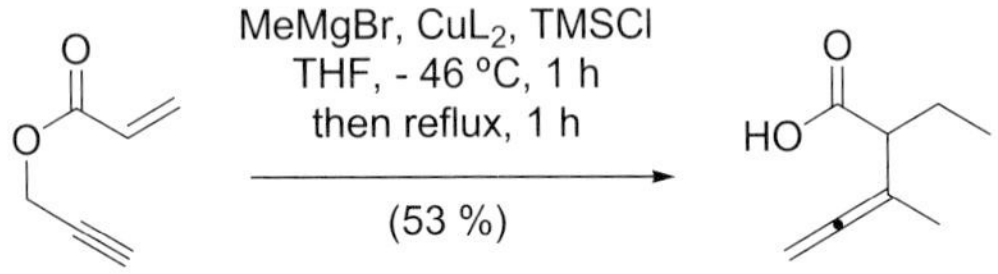

MeMgBr, CuL_2, TMSCl THF, - 46 °C, 1 h then reflux, 1 h (53 %)

Scheme 4.147

Brummond et al reported the rearrangement of a propargyl ester containing two terminal alkynes under Nakai conditions (Scheme 148) [143].

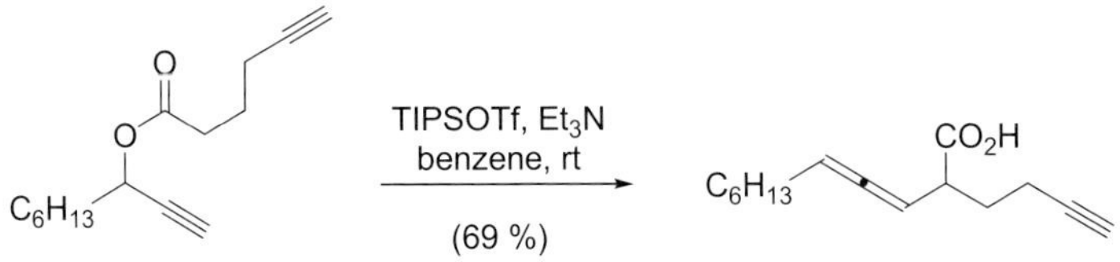

TIPSOTf, Et_3N benzene, rt (69 %)

Scheme 4.148

Wang et al employed the propargylic Ireland-Claisen rearrangement to initiate a cascade reaction that ultimately delivered a hexacyclic adduct (Scheme 149) [144]. The rearrangement was followed by a Schmittel cyclization, radical coupling and prototropic rearrangement to afford the hexacyclic acid after silyl ester hydrolysis.

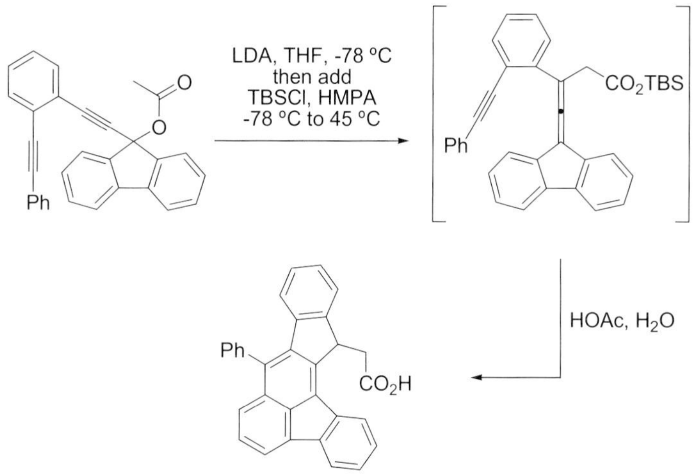

Scheme 4.149

4.11 Conclusion

As this chapter illustrates, the Ireland–Claisen rearrangement and variants thereof continue to offer an efficient and flexible means of preparing a wide variety of pentenoic acid derivatives with regio-, diastereo- and enantiocontrol and the imaginative uses to which the pentenoic acids have been put in the assembly of natural products and other targets.

In spite of the impressive variety of studies on the Ireland–Claisen rearrangement, several significant limitations to the reaction remain. A general solution to the problem of stereocontrolled formation of C1,C1-disubstituted silyl ketene acetals has yet to be reported. There is as yet no general catalytic enantioselective variant of the Claisen rearrangement. There are as yet no reports of stereoselective generation of acyclic tetrasubstituted alkenes.

The conversion of an allylic ester to the corresponding pentenoic acid is formally an isomerization. From an atom economy standpoint [145], the Ireland–Claisen rearrangement suffers from the need for stoichiometric (or superstoichiometric) base and silylating reagent. A truly catalytic variant of the rearrangement would constitute a significant advance.

Doubtless creative chemists will continue to develop novel variants and applications of the Claisen rearrangement.

List of Abbreviations

Ax	Axial
Boc	*t*-Butoxycarbonyl
BPS	t-Butyldiphenylsilyl
BSA	N,O-*Bis*(trimethylsilyl)acetamide
BuLi	*n*-Butyllithium
Cy	Cyclohexyl
DBU	Diazabicycloundecane
DMAP	4-(*N,N*-Dimethylamino)pyridine
DMF	*N,N*-Dimethylformamide
DMPU	Dimethylpropylene urea
dr	Diastereomeric ratio
ds	Diastereoselectivity
EDCI	1-Ethyl-3-(3-dimethylaminopropyl)-carbodiimide
eq	Equatorial
HMPA	Hexamethylphosphoramide
ICl	Iodine monochloride
IPC	Isopinylcampheyl
KHMDS	Potassium hexamethyldisilylamide
LDA	Lithium diisopropylamide
LHMDS	Lithium hexamethyldisilylamide
LICA	Lithium isopropylcyclohexylamide
MEM	Methoxyethoxymethyl
ms	Molecular sieves
MsCl	Methanesulfonyl chloride
NaH	Sodium hydride
NMI	*N*-Methylimidazole
NMO	*N*-Methylmorpholine-*N*-oxide
PCy_3	Tricyclohexylphosphine
PhH	Benzene
PhMe	Toluene
Piv	Pivaloyl
PMB	*p*-Methoxybenzyl
PNB	*p*-Nitrobenzyl
RCM	Ring closing metathesis
rs	Regioselectivity
rt	Room temperature
ska	Silyl ketene acetal
TBAF	Tetrabutylammonium fluoride
TBS	*t*-Butyldimethylsilyl
TBSOTf	*t*-Butyldimethylsilyl trifluoromethanesulfonate
TDS	Thexyldimethylsilyl
TESCl	Triethylsilyl chloride
TES	Triethylsilyl

THF	Tetrahydrofuran
TIPSOTf	Triisopropylsilyl trifluoromethanesulfonate
TIPS	Triisopropylsilyl
TMG	Tetramethylguanidine
TMS	Trimethylsilyl
TMSCl	Trimethylsilyl chloride
TMSI	Trimethylsilyl iodide
TMSOTf	Trimethylsilyl trifluoromethanesulfonate
TPP	Tetraphenylporphyrin
TPS	*t*-Butyldiphenylsilyl

Acknowledgments

We thank the NIH and the Arkansas Biosciences Institute for support of our program.

References

1 (a) Ireland, R. E.; Mueller, R. H. *J. Am. Chem. Soc.* **1972**, *94*, 5897–5898. (b) Ireland, R. E.; Mueller, R. H.; Willard, A. K. *J. Am. Chem. Soc.* **1976**, *98*, 2868–2877. (c) Ireland, R. E.; Wipf, P.; Armstrong, J. D. *J. Org. Chem.* **1991**, *56*, 650–657. (d) Ireland, R. E.; Wipf, P.; Xiang, J.-N. *J. Org. Chem.* **1991**, *56*, 3572–3582.

2 For reviews that include the Ireland–Claisen rearrangement, see: (a) Ziegler, F. E. *Acc. Chem. Res.* **1977**, *10*, 1423–1452. (b) Bennett, G. B. *Synthesis* **1977**, 589–606. (c) Hill, R. K. in *Asymmetric Synthesis*, Morrison, J. D. (ed.) Academic Press, **1984**; Vol. 3, pp 503–572. (d) Blechert, S. *Synthesis* **1989**, 71–82. (e) Wipf, P. in *Comprehensive Organic Synthesis*, Trost, B. M. (ed.) Pergamon Press, Oxford, 1991; Vol. 5, pp 827–873. (f) Tadano, K. in *Studies in Natural Products Chemistry*, Rahman, A.-U., (ed.) Elsevier, Amsterdam, Netherlands, **1992**, pp 405–455. (g) Pereira, S.; Srebnik, M. *Aldrichimica Acta* **1993**, *26*, 17–29. (h) Frauenrath, H. in *Stereoselective Synthesis*, 4th edn., Helchen, G.; Hoffmann, R. W.; Mulzer, J.; Schaumann, E. (eds.) Thieme Verlag: Stuttgart, **1995**; Vol. E21d, pp 3301–3756. (i) Chai, Y.; Hong, S.-P.; Lindsay, H. A.; McFarland, C.; McIntosh, M. C. *Tetrahedron* **2002**, *58*, 2905–2928. (k) Martin Castro, A. M. *Chem. Rev.* **2004**, *104*, 2939–3002.

3 Tseou, H.-F.; Wang, Y.-T. *J. Chin. Chem. Soc.* **1937**, *5*, 224–229.

4 Morimoto, Y.; Mikami, A.; Kuwabe, S.; Shirahama, H. *Tetrahedron Lett.* **1991**, *32*, 2909–2912.

5 Hodgson, D. M.; Gibbs, A. R. *Synlett* **1997**, 657–658.

6 (a) Vance, R. L.; Rondan, N. G.; Houk, K. N.; Jensen, F.; Borden, W. T.; Komornicki, A.; Wimmer, E. *J. Am. Chem. Soc.* **1988**, *110*, 2314–2315. (b) Aviyente, V.; Yoo, H. Y.; Houk, K. N. *J. Org. Chem.* **1997**, *62*, 6121–6128. (c) Yoo, H. Y.; Houk, K. N. *J. Am. Chem. Soc.* **1997**, *119*, 2877–2884. (d) Aviyente, V.; Houk, K. N. *J. Phys. Chem. A* **2001**, *105*, 383–391.

7 Wilcox, C. S.; Babston, R. E. *J. Am. Chem. Soc.* **1986**, *108*, 6636–6642.

8 Yamazaki, T.; Shinohara, N.; Kitazume, T.; Sato, S. *J. Org. Chem.* **1995**, *60*, 8140–8141.

9 For a recent review, see: Hiersemann, M.; Abraham, L. *Eur. J. Org. Chem.* **2002**, 1461–1471.

10 (a) Schenck, T. G.; Bosnich, B. *J. Am. Chem. Soc.* **1985**, *107*, 2058–2066. (b) Overman, L. E.; Hollis, T. K. *J. Organomet. Chem.* **1999**, *576*, 290–299.

11 Koch, G.; Janser, P.; Kottirsch, G.; Romero-Giron, E. *Tetrahedron Lett.* **2002**, *43*, 4837–4840.

12 Hanamoto, T.; Baba, Y.; Inanaga, J. *J. Org. Chem.* **1993**, *58*, 299–300.

13 Smith, P. M.; Thomas, E. J. *J. Chem. Soc., Perkin Trans. 1* **1998**, 3541–3556.

14 Gajewski, J. J.; Emrani, J. *J. Am. Chem. Soc.* **1984**, *106*, 5733–5734.

15 Kupczyk-Subotkowska, L.; Saunders, W. H., Jr.; Shine, H. J.; Subotkowski, W. *J. Am. Chem. Soc.* **1994**, *116*, 7088–7093.

16 (a) Ganem, B. *Angew. Chem. Int. Ed. Engl*.**1996**, *35*, 936–945. (b) Gajewski, J. J. *Acc. Chem. Res.* **1997**, *30*, 219–225.

17 Meyer, M. P.; DelMonte, A. J.; Singleton, D. A. *J. Am. Chem. Soc.* **1999**, *121*, 10865–10874.

18 Khaledy, M. M.; Kalani, M. Y. S.; Khuong, K. S.; Houk, K. N.; Aviyente, V.; Neier, R.; Soldermann, N.; Velker, J. *J. Org. Chem.* **2003**, *68*, 572–577.

19 Ireland, R. E.; Willard, A. K. *Tetrahedron Lett.* **1975**, 3975–3978.

20 (a) Bartlett, P. A.; Pizzo, C. F. *J. Org. Chem.* **1981**, *46*, 3896–3900. (b) Bartlett, P. A.; Barstow, J. F. *J. Org. Chem.* **1982**, *47*, 3933–3941.

21 (a) Ireland, R. E.; Wilcox, C. S.; Thaisrivongs, S.; Vanier, N. R. *Can. J. Chem.* **1979**, *57*, 1743–1745. (b) Ireland, R. E.; Thaisrivongs, S.; Vanier, N.; Wilcox, C. S. *J. Org. Chem.* **1980**, *45*, 48–61. (c) Ireland, R. E.; Daub, J. P. *J. Org. Chem.* **1981**, *46*, 479–485.

22 Katzenellenbogen, J. A.; Christy, K. J. *J. Org. Chem.* **1974**, *39*, 3315–3318.

23 Ireland, R. E.; Varney, M. D. *J. Am. Chem. Soc.* **1984**, *106*, 3668–3670.

24 (a) Panek, J. S.; Yang, M. *J. Org. Chem.* **1991**, *56*, 5755–5758. (b) Panek, J. S.; Clark, T. D. *J. Org. Chem.* **1992**, *57*, 4323–4326. (c) Panek, J. S.; Yang, M.; Solomon, J. S. *J. Org. Chem.* **1993**, *58*, 1003–1010.

25 Heathcock, C. H.; Finkelstein, B. L.; Jarvi, E. T.; Radel, P. A.; Hadley, C. R. *J. Org. Chem.* **1988**, *53*, 1922–1942.

26 (a) Zhang, X.; McIntosh, M. C. *Tetrahedron Lett.* **1998**, *39*, 7043–7046. (b) Hong, S.-P.; Lindsay, H. A.; Yaramasu, T.; Zhang, X.; McIntosh, M. C. *J. Org. Chem.* **2002**, *67*, 2042–2055.

27 For reviews, see: (a) Hoffman, R. W. *Chem. Rev.* **1989**, *89*, 1841–1860. (b) Hoveyda, A. H.; Evans, D. A.; Fu, G. C. *Chem. Rev.* **1993**, *93*, 1307–1370.

28 For a review, see: Cieplak, A. S. *Chem. Rev.* **1999**, *99*, 1265–1336.

29 Gilbert, J. C.; Yin, J.; Fakhreddine, F. H.; Karpinski, M. L. *Tetrahedron* **2004**, *60*, 51–60.

30 (a) Pratt, L. M.; Bowles, S. A.; Courtney, S. F.; Hidden, C.; Lewis, C. N.; Martin, F. M.; Todd, R. S. *Synlett* **1998**, 531–533. (b) Pratt, L. M.; Beckett, R. P.; Bellamy, C. L.; Corkill, D. J.; Cossins, J.; Courtney, P. F.; Davies, S. J.; Davidson, A. H.; Drummond, A. H.; Helfrich, K.; Lewis, C. N.; Mangan, M.; Martin, F. M.; Miller, K.; Nayee, P.; Ricketts, M. L.; Thomas, W.; Todd, R. S.; Whittaker, M. *Bioorg. Med. Chem. Lett.* **1998**, *8*, 1359–1364.

31 (a) Knight, D. W.; Share, A. C.; Gallagher, P. T. *J. Chem. Soc. Perkin Trans. 1* **1991**, 1615–1616. (b) Knight, D. W.; Share, A. C.; Gallagher, P. T. *J. Chem. Soc. Perkin Trans. 1* **1997**, 2089–2097.

32 Fujisawa, T.; Tajima, K.; Ito, M.; Sato, T. *Chem. Lett.* **1984**, 1169–1172.

33 (a) Gilbert, J. C.; Kelly, T. A. *J. Org. Chem.* **1986**, *51*, 4485–4488. (b) Gilbert, J. C.; Selliah, R. D. *Tetrahedron Lett.* **1992**, *33*, 6259–6262. (c) Gilbert, J. C.; Selliah, R. D. *Tetrahedron* **1994**, *50*, 1651–1664.

34 Betson, M. S.; Fleming, I. *Org. Biomol. Chem.* **2003**, *1*, 4005–4016.

35 Ishizaki, M.; Niimi, Y.; Hoshino, O. *Chem. Lett.* **2001**, 546–547.

36 Mulzer, J.; Shanyoor, M. *Tetrahedron Lett.* **1993**, *34*, 6545–6548.

37 Kahn, S. D.; Hehre, W. J. *J. Org. Chem.* **1988**, *53*, 301–305.

38 For a review, see: Anh, N. T. *Top. Curr. Chem.* **1980**, *88*, 145–161.

39 Hauske, J. R.; Julin, S. M. *Tetrahedron Lett.* **1993**, *34*, 4909–4912.

40 Cha, J. K.; Lewis, S. C. *Tetrahedron Lett.* **1984**, *25*, 5263–5266.

41 Shu, A. Y. L.; Djerassi, C. *J. Chem. Soc. Perkin Trans. 1* **1987**, 1291–1305.

42 Ireland, R. E.; Varney, M. D. *J. Org. Chem.* **1983**, *48*, 1829–1833.

43 (a) Sauve, G.; Schwartz, D. A.; Ruest, L.; Deslongchamps, P. *Can. J. Chem.* **1984**, *62*, 2929–2935. (b) Bernet, B.; Bishop, P. M.; Caron, M.; Kawamata, T.; Roy, B. L.; Ruest, L.; Sauve, G.; Soucy, P.; Deslongchamps, P. *Can. J. Chem.* **1985**, *63*, 2810–14.

44 Kallmerten, J.; Gould, T. J. *J. Org. Chem.* **1986**, *51*, 1152–1155.

45 Matsui, S.; Oka, N.; Hashimoto, Y.; Saigo, K. *Enantiomer* **2000**, *5*, 105–108.

46 Bartlett, P. A.; Tanzella, D. J.; Barstow, J. F. *Tetrahedron Lett.* **1982**, *23*, 619–622.

47 Corey, E. J.; Lee, D. H. *J. Am. Chem. Soc.* **1991**, *113*, 4026–4028.

48 Burke, S. D.; Fobare, W. F.; Pacofsky, G. J. *J. Org. Chem.* **1983**, *48*, 5221–5228.

49 (a) Paterson, I.; Hulme, A. N. *Tetrahedron Lett.* **1990**, *31*, 7513–7516. (b) Paterson, I.; Hulme, A. N.; Wallace, D. J. *Tetrahedron Lett.* **1991**, *32*, 7601–7604. (c) Paterson, I.; Hulme, A. N. *J. Org. Chem.* **1995**, *60*, 3288–3300.

50 (a) Burke, S. D.; Pacofsky, G. J.; Piscopio, A. D. *Tetrahedron Lett.* **1986**, *27*, 3345–3348. (b) Burke, S. D.; Pacofsky, G. J.; Piscopio, A. D. *J. Org. Chem.* **1992**, *57*, 2228–2235.

51 Gil, S.; Lazaro, M. A.; Parra, M.; Breitmaier, E.; Mestres, R. *Synlett* **1998**, 70–72.

52 Yokozawa, T.; Nakai, T.; Ishikawa, N. *Tetrahedron Lett.* **1984**, *25*, 3991–3994.

53 (a) Welch, J. T.; Samartino, J. S. *J. Org. Chem.* **1985**, *50*, 3663–3665. (b) Araki, K.; Welch, J. T. *Tetrahedron Lett.* **1993**, *34*, 2251–2254.

54 Kobayashi, M.; Masumoto, K.; Nakai, E.-i.; Nakai, T. *Tetrahedron Lett.* **1996**, *37*, 3005–3008.

55 Hu, Y.; Porco, J. A., Jr. *Tetrahedron Lett.* **1999**, *40*, 3289–3292.

56 Ito, Y.; Higuchi, N.; Murakami, M. *Tetrahedron Lett.* **1988**, *29*, 5151–5154.

57 Aoki, Y.; Kuwajima, I. *Tetrahedron Lett.* **1990**, *31*, 7457–7460.

58 (a) Eriksson, M.; Nilsson, M.; Olsson, T. *Synlett* **1994**, 271–227. (b) Eriksson, M.; Hjelmencrantz, A.; Nilsson, M.; Olsson, T. *Tetrahedron* **1995**, *51*, 12631–12644.

59 Takai, K.; Ueda, T.; Kaihara, H.; Sunami, Y.; Moriwake, T. *J. Org. Chem.* **1996**, *61*, 8728–8729.

60 Miller, S. P.; Morken, J. P. *Org. Lett.* **2002**, *4*, 2743–2745.

61 Troll, T.; Wiedemann, J. *Tetrahedron Lett.* **1992**, *33*, 3847–3850.

62 (a) Velker, J.; Roblin, J.-P.; Neels, A.; Tesouro, A.; Stoeckli-Evans, H.; Klaerner, F.-G.; Gehrke, J.-S.; Neier, R. *Synlett* **1999**, 925–929. (b) Soldermann, N.; Velker, J.; Vallat, O.; Stoeckli-Evans, H.; Neier, R. *Helv. Chim. Acta* **2000**, *83*, 2266–2276.

63 Rowley, M.; Tsukamoto, M.; Kishi, Y. *J. Am. Chem. Soc.* **1989**, *111*, 2735–2737.

64 Oh, T.; Wrobel, Z.; Devine, P. N. *Synlett* **1992**, 81–83.

65 Khripach, V. A.; Zhabinskii, V. N.; Konstantinova, O. V.; Khripach, N. B. *Tetrahedron Lett.* **2000**, *41*, 5765–5767.

66 Bartlett, P. A.; Tanzella, D. J.; Barstow, J. F. *J. Org. Chem.* **1982**, *47*, 3941–3945.

67 Kallmerten, J.; Gould, T. J., *Tetrahedron Lett.* **1983**, *24*, 5177–5180.

68 Sato, T.; Tajima, K.; Fujisawa, T. *Tetrahedron Lett.* **1983**, *24*, 729–730.

69 (a) Hattori, K.; Yamamoto, H. *J. Org. Chem.* **1993**, *58*, 5301–5303. (b) Hattori, K.; Yamamoto, H. *Tetrahedron* **1994**, *50*, 3099–3112.

70 Burke, S. D.; Hong, J.; Lennox, J. R.; Mongin, A. P. *J. Org. Chem.* **1998**, *63*, 6952–6967.

71 Wallace, G. A.; Scott, R. W.; Heathcock, C. H. *J. Org. Chem.* **2000**, *65*, 4145–4152.

72 Miller, J. F.; Termin, A.; Koch, K.; Piscopio, A. D. *J. Org. Chem.* **1998**, *63*, 3158–3159.

73 Burke, S. D.; Ng, R. A.; Morrison, J. A.; Alberti, M. J. *J. Org. Chem.* **1998**, *63*, 3160–3161.

74 Hatakeyama, S.; Sugawara, M.; Kawamura, M.; Takano, S. *J. Chem. Soc. Chem. Comm.* **1992**, 1229–1231.

75 Picoul, W.; Urchegui, R.; Haudrechy, A.; Langlois, Y. *Tetrahedron Lett.* **1999**, *40*, 4797–4800.

76 Barrett, A. G. M.; Ahmed, M.; Baker, S. P.; Baugh, S. P. D.; Braddock, D. C.; Procopiou, P. A.; White, A. J. P.; Williams, D. J. *J. Org. Chem.* **2000**, *65*, 3716–3721.

77 (a) Norley, M. C.; Kocienski, P. J.; Faller, A. *Synlett* **1994**, 77–78. (b) Norley, M.; Kocienski, P.; Faller, A. *Synlett* **1996**, 900–901.

78 Eshelby, J. J.; Parsons, P. J.; Sillars, N. C.; Crowley, P. J. *J. Chem. Soc. Chem. Comm* .**1995**, 1497–1498.

79 Sato, T.; Tsunekawa, H.; Kohama, H.; Fujisawa, T. *Chem. Lett.* **1986**, 1553–1556.

80 Ritter, K. *Tetrahedron Lett.* **1990**, *31*, 869–872.

81 Vidal, T.; Haudrechy, A.; Langlois, Y. *Tetrahedron Lett.* **1999**, *40*, 5677–5680.

82 Danishefsky, S.; Funk, R. L.; Kerwin, J. F., Jr. *J. Am. Chem. Soc.* **1980**, *102*, 6889–6891.

83 (a) Nemoto, H.; Shitara, E.; Fukumoto, K.; Kametani, T. *Heterocycles* **1985**, *23*, 1911–1913. (b) Nemoto, H.; Shitara, E.; Fukumoto, K.; Kametani, T. *Heterocycles* **1987**, *25*, 51–53.

84 Ireland, R. E.; Varney, M. D. *J. Org. Chem.* **1986**, *51*, 635–648.

85 (a) Burke, S. D.; Armistead, D. M.; Fevig, J. M. *Tetrahedron Lett.* **1985**, *26*, 1163–1166. (b) Burke, S. D.; Armistead, D. M.; Schoenen, F. J.; Fevig J. M., *Tetrahedron* **1986**, *42*, 2787–2801. (c) Burke, S. D.; Piscopio, A. D.; Kort, M. E.; Matulenko, M. A.; Parker, M. H.; Armistead, D. M.; Shankaran, K. *J. Org. Chem.* **1994**, *59*, 332–347.

86 Angle, S. R.; Breitenbucher, J. G.; Arnaiz, D. O. *J. Org. Chem.* **1992**, *57*, 5947–5955.

87 Abelman, M. M.; Funk, R. L.; Munger, J. D., Jr., *J. Am. Chem. Soc.* **1982**, *104*, 4030–4032.

88 Cameron, A. G.; Knight, D. W. *Tetrahedron Lett.* **1982**, *23*, 5455–5458.

89 Magriotis, P. A.; Kim, K. D. *J. Am. Chem. Soc.* **1993**, *115*, 2972–2973.

90 Vourloumis, D.; Kim, K. D.; Petersen, J. L.; Magriotis, P. A. *J. Org. Chem.* **1996**, *61*,4848–4852.

91 Roush, W. R.; Works, A. B. *Tetrahedron Lett.* **1996**, *37*, 8065–8068.

92 (a) Cooper, J.; Knight, D. W.; Gallagher, P. T. *J. Chem. Soc. Chem. Comm.* **1987**, 1220–1222. (b) Cooper, J.; Knight, D. W.; Gallagher, P. T. *J. Chem. Soc. Perkin Trans. 1* **1992**, 553–559.

93 Still, W. C.; Schneider, M. J. *J. Am. Chem. Soc.* **1977**, *99*, 948–950.

94 Johnson, W. S.; Buchanan, R. A.; Bartlett, W. R.; Thaqt, F. S.; Kullnigt, R. K. *J. Am. Chem. Soc.* **1993**, *115*, 504–515.

95 Overman, L. E.; Lin, N. H. *J. Org. Chem.* **1985**, *50*, 3669–3670.

96 Davidson, A. H.; Eggleton, N.; Wallace, I. H. *J. Chem. Soc. Chem. Comm.* **1991**, 378–380.

97 Enev, V.; Stojanova, D.; Bienz, S. *Helv. Chim. Acta* **1996**, *79*, 391–404.

98 (a) Reed, S. F., Jr. *J. Org. Chem.* **1965**, *30*, 1663–1665. (b) Cresson, P.; Lacour, L. *Compt. Rend. Acad. Sci. Série C* **1966**, *262*, 1157–1160. (c) Cresson, P.; Bancel, S. *Compt. Rend. Acad. Sci. Série C* **1968**, *266*, 409–412. (d) Bancel, S.; Cresson, P. *Compt. Rend. Acad. Sci. Série C* **1969**, *268*, 1535–1537. (e) Parker, K. A.; Farmar, J. G. *Tetrahedron Lett.* **1985**, *26*, 3655–3658. (f) Maruoka, K.; Banno, H.; Nonoshita, K.; Yamamoto, H. *Tetrahedron Lett.* **1989**, *30*, 1265–1266. (g) Hudlicky, T.; Kwart, L. D.; Tiedje, M. H.; Ranu, B. C.; Short, R. P.; Frazier, J. O.; Rigby, H. L. *Synthesis* **1986**, 716–727.

99 Parker, K. A.; Farmar, J. G. *J. Org. Chem.* **1986**, *51*, 4023–4028.

100 Tanis, S. P.; McMills, M. C.; Scahill, T. A.; Kloosterman, D. A. *Tetrahedron Lett.* **1990**, *31*, 1977–1980.

101 Roush, W. R.; Works, A. B. *Tetrahedron Lett.* **1997**, *38*, 351–354.

102 Magnus, P.; Westwood, N. *Tetrahedron Lett.* **1999**, *40*, 4659–4662.

103 Ireland, R. E.; Mueller, R. H.; Willard, A. K. *J. Org. Chem.* **1976**, *41*, 986–996.

104 (a) Ireland, R. E.; Vevert, J. P. *J. Org. Chem.* **1980**, *45*, 4259–4260. (b) Ireland, R. E.; Vevert, J. P. *Can. J. Chem.* **1981**, *59*, 572–583.

105 Ireland, R. E.; Anderson, R. C.; Badoud, R.; Fitzsimmons, B. J.; McGarvey, G. J.; Thaisrivongs, S.; Wilcox, C. S. *J. Am. Chem. Soc.* **1983**, *105*, 1988–2006.

106 Ireland, R. E.; Wuts, P. G. M.; Ernst, B. *J. Am. Chem. Soc.* **1981**, *103*, 3205–3207.

107 Ireland, R. E.; Smith, M. G. *J. Am. Chem. Soc.* **1988**, *110*, 854–860.

108 (a) Aicher, T. D.; Buszek, K. R.; Fang, F. G.; Forsyth, C. J.; Jung, S. H.; Kishi, Y.; Scola, P. M. *Tetrahedron Lett.* **1992**, *33*, 1549–1552. (b) Fang, F. G.; Kishi, Y.; Matelich, M. C.; Scola, P. M. *Tetrahedron Lett.* **1992**, *33*, 1557–1560.

109 (a) Ireland, R. E.; Norbeck, D. W. *J. Am. Chem. Soc.* **1985**, *107*, 3279–3285. (b) Ireland, R. E.; Haebich, D.; Norbeck, D. W. *J. Am. Chem. Soc.* **1985**, *107*, 3271–3278. (c) Ireland, R. E.; Norbeck, D. W.; Mandel, G. S.; Mandel, N. S. *J. Am. Chem. Soc.* **1985**, *107*, 3285–3294.

110 (a) Ireland, R. E.; Armstrong, J. D., III; Lebreton, J.; Meissner, R. S.; Rizzacasa, M. A. *J. Am. Chem. Soc.* **1993**, *115*, 7152–7165. (b) Ireland, R. E.; Meissner, R. S.; Rizzacasa, M. A. *J. Am. Chem. Soc.* **1993**, *115*, 7166–7172.

111 Di Florio, R.; Rizzacasa, M. A. *J. Org. Chem.* **1998**, *63*, 8595–8598.

112 (a) Gable, R. W.; McVinish, L. M.; Rizzacasa, M. A. *Aust. J. Chem.* **1994**, *47*, 1537–1544. (b) McVinish, L. M.; Rizzacasa, M. A. *Tetrahedron Lett.* **1994**, *35*, 923–926.

113 Martinez, G. R.; Grieco, P. A.; Williams, E.; Kanai, K.-i.; Srinivasan, C. V. *J. Am. Chem. Soc.* **1982**, *104*, 1436–1438.

114 Rigby, J. H.; McGuire, T.; Senanayake, C.; Khemani, K. *J. Chem. Soc. Perkin Trans. 1* **1994**, 3449–3457.

115 Wilson, S. R.; Jacob, L. *J. Org. Chem.* **1992**, *57*, 4380–4385.

116 Cywin, C. L.; Kallmerten, J. *Tetrahedron Lett.* **1993**, *34*, 1103–1106.

117 Mikami, K.; Kawamoto, K.; Nakai, T. *Tetrahedron Lett.* **1986**, *27*, 4899–4902.

118 (a) Curran, D. P.; Suh, Y. G. *Tetrahedron Lett.* **1984**, *25*, 4179–4182. (b) Curran, D. P.; Suh, Y. G. *Carbohydr. Res.* **1987**, *171*, 161–192.

119 Barrish, J. C.; Lee, H. L.; Baggiolini, E. G.; Uskokovic, M. R. *J. Org. Chem.* **1987**, *52*, 1372–1375.

120 Kallmerten, J.; Balestra, M. *J. Org. Chem.* **1986**, *51*, 2855–2857.

121 Mulzer, J.; Mohr, J.-T. *J. Org. Chem.* **1994**, *59*, 1160–1165.

122 Burke, S. D.; Letourneau, J. J.; Matulenko, M. A. *Tetrahedron Lett.* **1999**, *40*, 9–12.

123 Boeckman, R. K., Jr.; Potenza, J. C.; Enholm, E. J. *J. Org. Chem.* **1987**, *52*, 469–472.

124 Corey, E. J.; Roberts, B. E.; Dixon, B. R. *J. Am. Chem. Soc.* **1995**, *117*, 193–196.

125 Corey, E. J.; Kania, R. S. *J. Am. Chem. Soc.* **1996**, *118*, 1229–1230.

126 Burke, S. D.; Sametz, G. M. *Org. Lett.* **1999**, *1*, 71–74.

127 (a) Ariza, X.; Garcia, J.; Lopez, M.; Montserrat, L. *Synlett* **2001**, 120–122. (b) Ariza, X.; Fernandez, N.; Garcia, J.; Lopez, M.; Montserrat, L.; Ortiz, J. *Synthesis* **2004**, 128–134.

128 (a) Hong, S.; McIntosh, M. C., *Org. Lett.* **2002**, *4*, 19–21. (b) Hutchison, J.; Hong, S.-p.; McIntosh, M. C. *J. Org. Chem.* **2004**, *69*, 4185–4191.

129 Kraus, G. A.; Thomas, P. J. *J. Org. Chem.* **1986**, *51*, 503–505.

130 Gilbert, J. C.; Kelly, T. A. *J. Org. Chem.* **1986**, *51*, 4485–4488.

131 VanMiddlesworth, F. L. *J. Org. Chem.* **1986**, *51*, 5019–5021.

132 Danishefsky, S.; Tsuzuki, K. *J. Am. Chem. Soc.* **1980**, *102*, 6891–6893.

133 Turos, E.; Audia, J. E.; Danishefsky, S. J. *J. Am. Chem. Soc.* **1989**, *111*, 8231–8236.

134 Danishefsky, S. J.; Simoneau, B. *J. Am. Chem. Soc.* **1989**, *111*, 2599–2604.

135 Brunner, R. K.; Borschberg, H. J. *Helv. Chim. Acta* **1983**, *66*, 2608–2614.

136 Funk, R. L.; Abelman, M. M. *J. Org. Chem.* **1986**, *51*, 3247–3248.

137 (a) Funk, R. L.; Olmstead, T. A.; Parvez, M. *J. Am. Chem. Soc.* **1988**, *110*, 3298–3300. (b) Funk, R. L.; Olmstead, T. A.; Parvez, M.; Stallman, J. B. *J. Org. Chem.* **1993**, *58*, 5873–5875.

138 (a) Hull, H. M.; Knight, D. W. *J. Chem. Soc. Perkin Trans. 1* **1997**, 857–863. (b) Hull, H. M.; Jones, E. G.; Knight, D. W. *J. Chem. Soc. Perkin Trans. 1* **1998**, 1779–1788.

139 Angle, S. R.; Breitenbucher, J. G. *Tetrahedron Lett.* **1993**, *34*, 3985–3988.

140 (a) Begley, M. J.; Cameron, A. G.; Knight, D. W. *J. Chem. Soc. Perkin Trans. 1* **1986**, 1933–1938. (b) Begley, M. J.; Cameron, A. G.; Knight, D. W. *J. Chem. Soc. Chem. Comm.* **1984**, 827–829.

141 (a) See also, Baldwin, J. E.; Bennett, P. A. R.; Forrest, A. K. *J. Chem. Soc. Chem. Commun.* **1987**, 250–251; (b) Castelhano, A. L.; Horne, S.; Taylor, G. J.; Billedeau, R.; Krantz, A. *Tetrahedron* **1988**, *44*, 5451–5466 and ref 56.

142 Fujisawa, T.; Maehata, E.; Kohama, H.; Sato, T. *Chem. Lett.* **1985**, 1457–1458.

143 Brummond, K. M.; Chen, H.; Sill, P.; You, L. *J. Am. Chem. Soc.* **2002**, *124*, 15186–15187.

144 Yang, Y.; Petersen, J. L.; Wang, K. K. *J. Org. Chem.* **2003**, *68*, 3545–3549.

145 (a) Trost, B. M. *Science* **1991**, *254*, 1471–1477. (b) Trost, B. M. *Angew. Chem. Int. Ed. Engl.* **1995**, *34*, 259–281. (c) Trost B.M. *Acc. Chem. Res.* **2002**, *35*, 695–705.

5
Simple and Chelate Enolate Claisen Rearrangement

5.1
Simple Enolate Claisen Rearrangement

Mukund G. Kulkarni

5.1.1
Introduction

Since the first report of the rearrangement of allyl ethers of phenols and enols to the corresponding C-allyl derivatives by Claisen [1] in 1912, the mechanistic, stereochemical and synthetic aspects of the reaction have been extensively explored [2]. In modern terminology, the reaction, normally referred to as the Claisen rearrangement, is referred to as a [3,3]-sigmatropic rearrangement. It has been demonstrated [2] that this reaction, especially the aliphatic version, tolerates various substituents. This fact, coupled with accurate predictability of the stereochemical outcome of the reaction, has resulted in this reaction becoming one of the very important and quite frequently employed methods in the synthesis of complex molecules.

The Claisen rearrangement is almost a certainty [3] once the allyl vinyl ether is generated. Therefore, the application of the Claisen rearrangement is restricted only by the availability of the appropriate allyl vinyl ether. In the aromatic version of the Claisen rearrangement, the O-allylated phenols offer this structural feature easily. Comparatively, the generation of the allyl vinyl ether in the aliphatic version of this rearrangement is not as straightforward as in the case of its aromatic version. Therefore, development of methods to generate allyl vinyl ethers, especially in stereochemically pure form, is considered to be at the heart of a successful application of the aliphatic Claisen rearrangement. As a result several methods for preparing allyl vinyl ethers have been documented in the literature. This has expanded the scope of application of the Claisen rearrangement to a great extent. The important and more popular methods for the generation of allyl vinyl ethers are: the Marbet–Saucy method [4] and its variants based on addition and/or elimination sequences, the Carroll rearrangement [5], the Johnson orthoester method [6], the Eschenmoser method [7], the enolate method [8], the Wittig olefination method [9] and the Zwitterionic method [10].

The Claisen Rearrangement. Edited by M. Hiersemann and U. Nubbemeyer

ISBN: 978-3-527-30825-5

Recently, the enolate method has emerged as a widely and frequently employed method in the synthesis for reasons like the easy access to the starting materials, simplicity and mildness of the experimental procedures, accurate predictability of the stereochemistry of the products and good to excellent realization of the stereoselection. This method can be further sub-divided into sub-topics like the simple enolate method [11], the Ireland silyl ketene acetal method, the chelated ester enolate method, imidate method, and the N,O- and N,S-acetal method. The present discussion will be restricted to simple enolate method as the remaining sub-topics are covered separately in this book.

5.1.2 History

Historically, as early as 1937, the first enolate Claisen rearrangement, though it was not formally recognized as such, was documented. Tseou and Wang [12] reported the formation of 4-pentenoic acid in low yields on heating neat allyl acetate with sodium metal for 3 h to ca. 100°C while attempting the acetoacetic ester synthesis (Scheme 5.1.1).

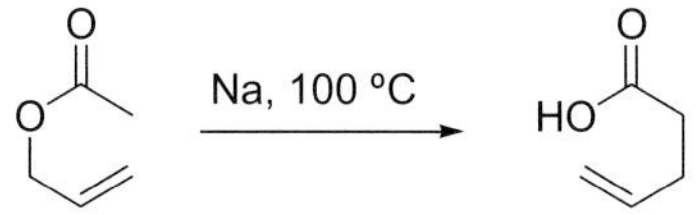

Na, 100 °C

Scheme 5.1.1

In 1949 Arnold and Searles [13] reported that allyl, crotyl and 1-methylallyl diphenylacetates, when treated with mesitylmagnesium bromide or sodium hydride, rearrange to the respective α-substituted diphenylacetic acids. Rearrangement of allyl α-ethyl phenyl acetate could also be effected with stronger base like $MgBrNEt_2$, albeit in very low yields (Scheme 5.1.2). The method was further extended to the synthesis of 9-fluorenecarboxylic acid [14] (Scheme 5.1.3).

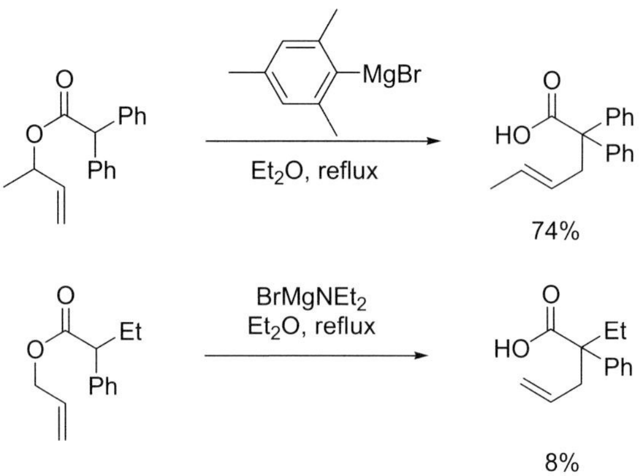

MgBr, Et_2O, reflux — 74%

$BrMgNEt_2$, Et_2O, reflux — 8%

Scheme 5.1.2

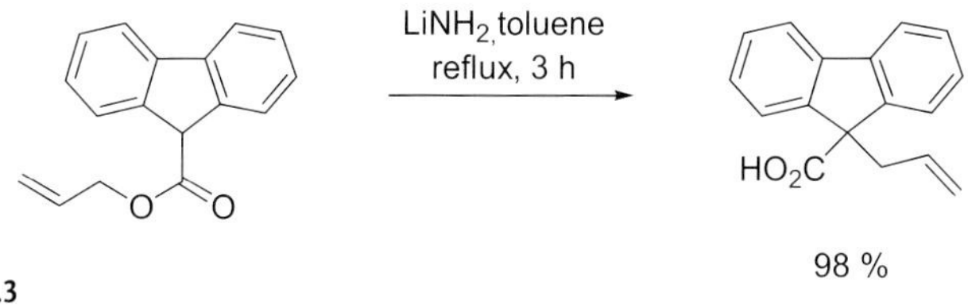

$LiNH_2$, toluene
reflux, 3 h

HO_2C

98 %

Scheme 5.1.3

By effecting the rearrangement of allyl and methallyl isobutyrates with sodium hydride in refluxing toluene, in 1960 Brannock et al. [15] showed that the rearrangement can be effected in good yield even in absence of an aryl group in the α-position of the acid part of the ester (Scheme 5.1.4).

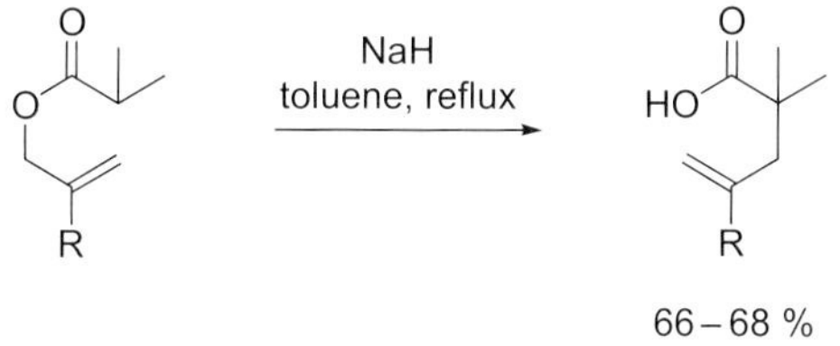

NaH
toluene, reflux

66 – 68 %

Scheme 5.1.4

Julia's synthesis of (+)-chrysanthemic acid [16] was the first application of the enolate Claisen rearrangement in the natural product synthesis. This was also the first example of an enolate Claisen rearrangement of a tertiary allylic ester (Scheme 5.1.5).

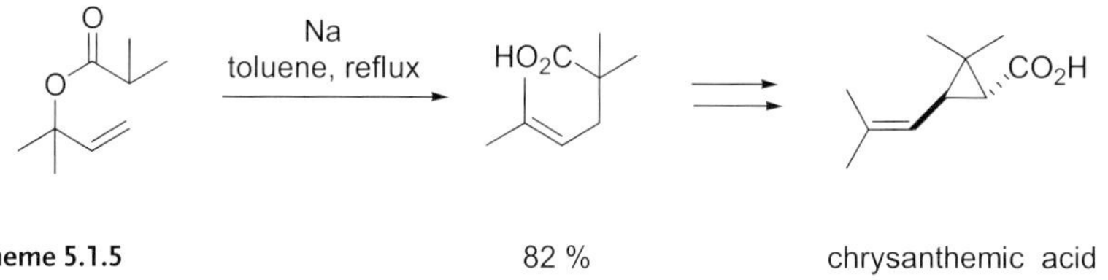

Na
toluene, reflux

HO_2C

CO_2H

82 %

chrysanthemic acid

Scheme 5.1.5

In 1972, Arnold and Hoffman [17] described the enolate Claisen rearrangement of many allylic esters of alicyclic carboxylic acids (Scheme 5.1.6). Interestingly, a Claisen rearrangement of allylic esters, through their enolates, was also observed in a mass spectrometer [18].

NaH
toluene, reflux

43 %

Scheme 5.1.6

Although it was possible to effect the enolate Claisen rearrangement of allylic esters of simple carboxylic acids, the reaction conditions were rather harsh and so the methods might not be extendable to more complex and sensitive substrates. Further, low yields of these rearrangements severely limited their usefulness.

5.1.3 Simple Enolates of Allylic Esters

In 1973, Baldwin and Walker [19] described the Claisen rearrangement of the allylic and acetylenic esters of α-bromo carboxylic acids with zinc at higher temperatures through the Reformatsky procedure. Though the method was better, it was severely restricted as it becomes mandatory to prepare the corresponding bromoester from the system of interest before the rearrangement could be effected (Scheme 5.1.7).

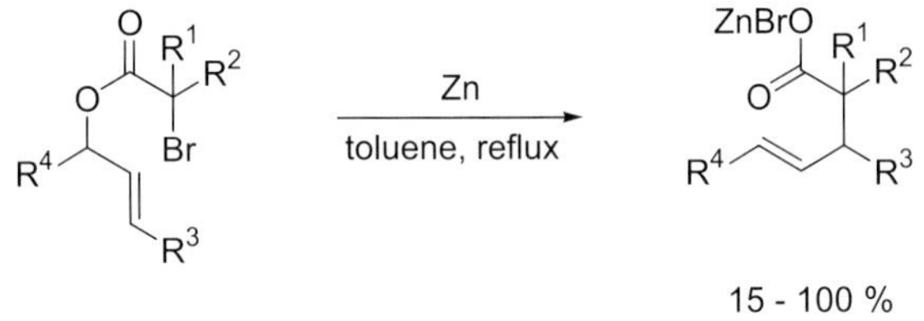

Zn
toluene, reflux
15 - 100 %

Scheme 5.1.7

Rathke and Lindert [20] showed that enolates of esters having α-hydrogen can be generated under very mild conditions using bases like lithium cyclohexyl isopropylamide (LICA) and lithium di-isopropylamide (LDA) under kinetic conditions. These enolates are generated in a quantitative manner and free from the competing aldol type condensation reaction. Through deuterium incorporation experiments they showed that the enolates are generated kinetically as, after their generation using excess base, they do not equilibrate and scramble the geometry of the enolate. Based on these results, Ireland et al. [8] in 1972, demonstrated for the first time that employing such strong bases the enolates of allylic esters of simple aliphatic carboxylic acids could be generated even at very low temperature. It was further shown that these enolates undergo the Claisen rearrangement efficiently and with great ease even at low to ambient temperatures to furnish the corresponding γ,δ-unsaturated carboxylic acids in good yield (Scheme 5.1.8). This was quite striking, particularly, in view of the alternative procedures. This efficacy of the process is evident from the half-life times of some of the silyl ethers of the corresponding enolates (Table 5.1.1).

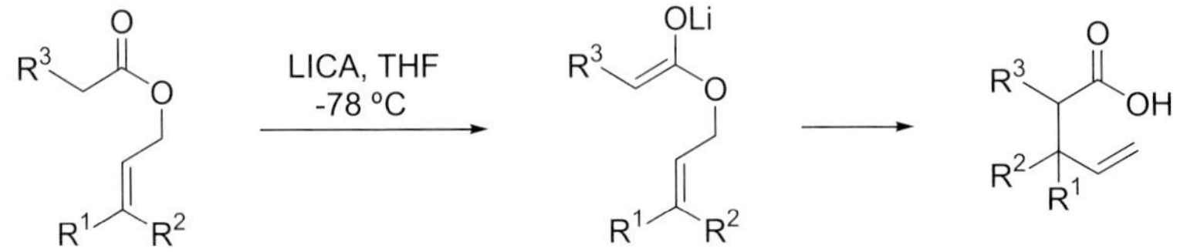

LICA, THF
-78 °C

Scheme 5.1.8

Tab. 5.1.1 Half-life times for rearrangement of silyl ketene acetals at 32°C.

Ketene acetal	$t_{1/2}$ [min]	Ketene acetal	$t_{1/2}$ [min]
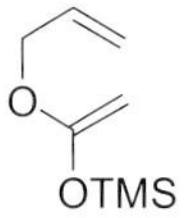 O, OTMS	210±30	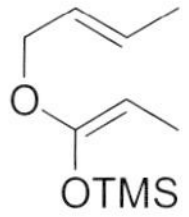 O, OTMS	5±1
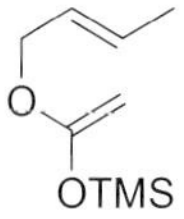 O, OTMS	150±30	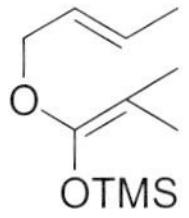 O, OTMS	≪1
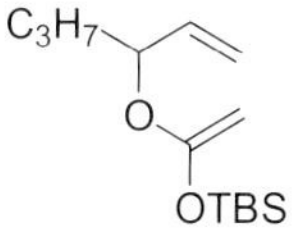 C_3H_7, O, OTBS	6±1	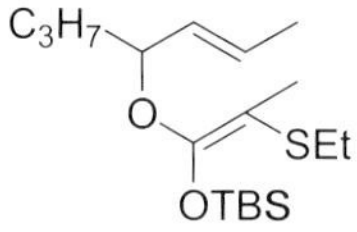 C_3H_7, O, SEt, OTBS	<1

To demonstrate the viability of the method even in the presence of acid labile groups, Ireland applied this method to the synthesis of dihydrojasmone from the ester **1** (Scheme 5.1.9).

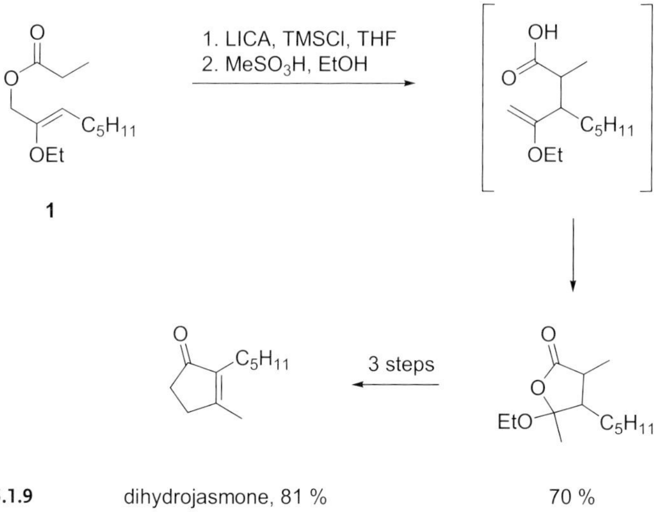

1. LICA, TMSCl, THF
2. $MeSO_3H$, EtOH

O, O, C_5H_{11}, OEt, **1**

OH, O, C_5H_{11}, OEt

O, C_5H_{11} — 3 steps — O, O, EtO, C_5H_{11}

Scheme 5.1.9 dihydrojasmone, 81 % 70 %

To establish the "top half" and the "bottom half" of a prostanoid molecule [21], the Ireland–Claisen rearrangement of the ester **2** was carried out at sub-zero temperature. The reaction conditions were so mild that the resulting Claisen product did not undergo a Cope rearrangement even though structurally it was feasible, as the latter rearrangement requires temperatures in excess of 100 °C (Scheme 5.1.10).

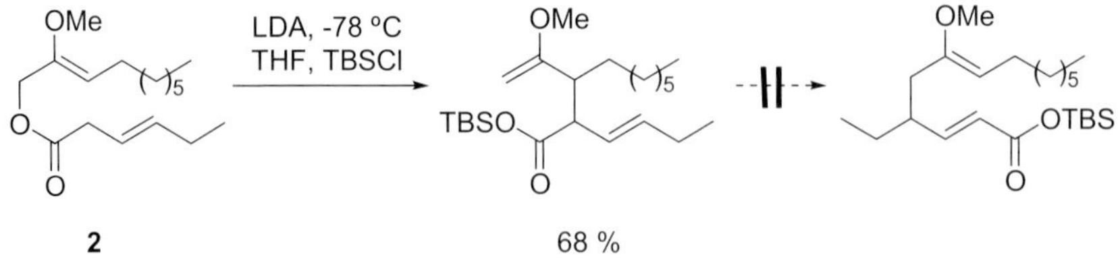

Scheme 5.1.10

Ireland and Mueller [8] found that in the case of the enolates of simple allyl acetates, the side reactions like decomposition of the enolate via ketene pathway and aldol like condensation compete with the Claisen rearrangement and this could best be avoided if the enolate is trapped via silylation to get the corresponding silyl ketene acetals, prior to effecting the Claisen rearrangement. Due to the instability of simple allylic esters, Ireland made it a practice of trapping the enolates as silyl ethers before effecting the rearrangement, but others have shown that this is not a mandatory requirement and in many instances, the enolates as such could be subjected to the rearrangement. This is particularly true in the case of esters with substituents in the acid part and/or the alcohol part of the allylic esters and in such cases the corresponding enolates could be fairly stable which, as such, may undergo the Claisen rearrangement in good yields. This avoided the extra step of protecting the enolate as the silyl ether prior to the rearrangement.

Wuts and Sutherland [22] treated the allylic esters derived from pyran-2-carboxylic acids with lithium hexamethyldisilazide (LHMDS) and allowed the enolate solution to warm up to room temperature when the enolates smoothly rearranged to give the products in good yield (Scheme 5.1.11).

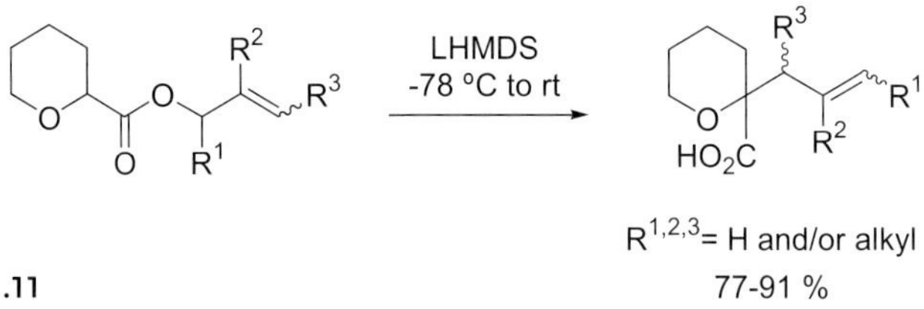

Scheme 5.1.11

While studying the regioselectivity in the Claisen rearrangement of 1,4-diene-3-ols, Parker and Farmar [23] found that the enolates derived from the esters of these allylic alcohols, using LDA, rearranged to afford the corresponding carboxylic acids in good yields (Scheme 5.1.12). The reaction showed good regioselectivity favoring the participation of the β-substituted olefin over the α one, apparently for steric reasons.

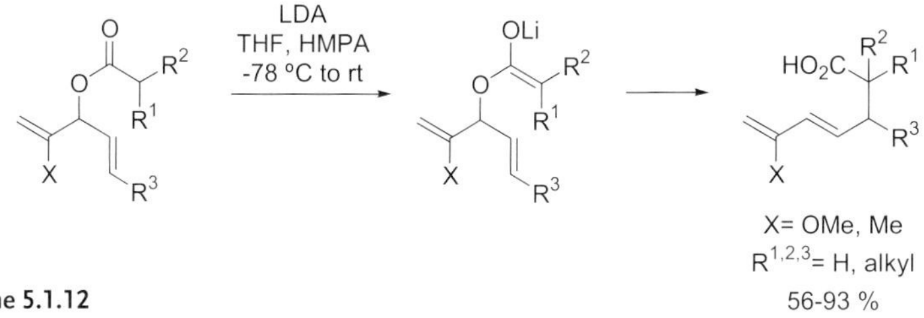

LDA, THF, HMPA, -78 °C to rt

X= OMe, Me
$R^{1,2,3}$= H, alkyl
56-93 %

Scheme 5.1.12

In order to effectively obtain products of homo-Claisen rearrangement [24], Wilson et al. prepared the propionate esters of the allylic alcohol **3** and treated them with LDA. The enolate so generated, on warming to room temperature, gave the δ, ε-unsaturated acid in good yield after heating the rearrangement product with methanolic hydrochloric acid (Scheme 5.1.13).

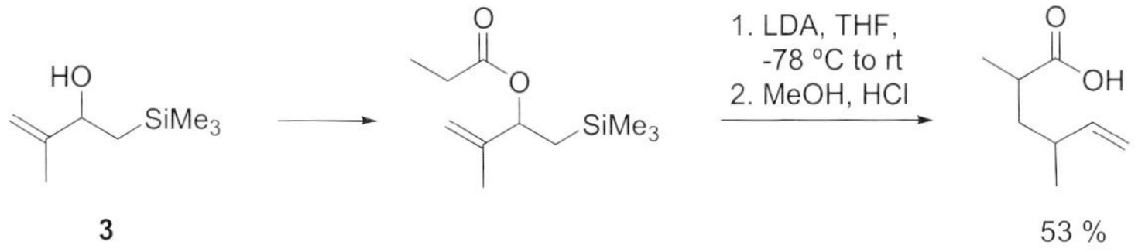

1. LDA, THF, -78 °C to rt
2. MeOH, HCl

3

53 %

Scheme 5.1.13

In a non-resolutive approach to chiral compounds, Hart et al. [25] treated the ester **5**, derived from the enantiomerically pure allylic alcohol **4**, with LICA and effected the rearrangement of the enolate so generated at room temperature to get the rearrangement products in good yields. In this way the chirality in the substrate was effectively transferred to the newly formed chiral center. Ozonolysis of the product of rearrangement gave optically pure lactol **6** with the recovery of the chiral auxiliary (Scheme 5.1.14).

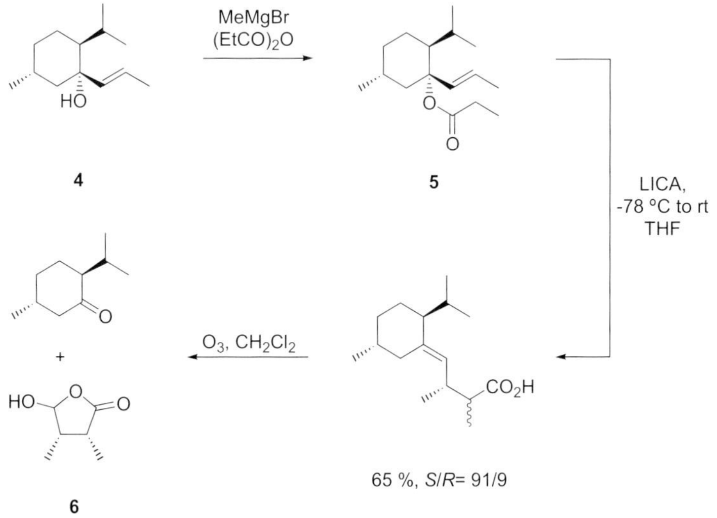

MeMgBr, $(EtCO)_2O$

4 **5**

LICA, -78 °C to rt, THF

O_3, CH_2Cl_2

65 %, *S*/*R*= 91/9

6

Scheme 5.1.14

Esters derived from thiophene-2-carbinols [26] were rearranged through the corresponding enolates generated using LICA. Resulting 2-alkyl thiophene-3-acetic acids were obtained in fairly good yield. Enolate Claisen rearrangement of the lactone **7** under the same conditions gave the bicyclic thiophene derivative **8** (Scheme 5.1.15).

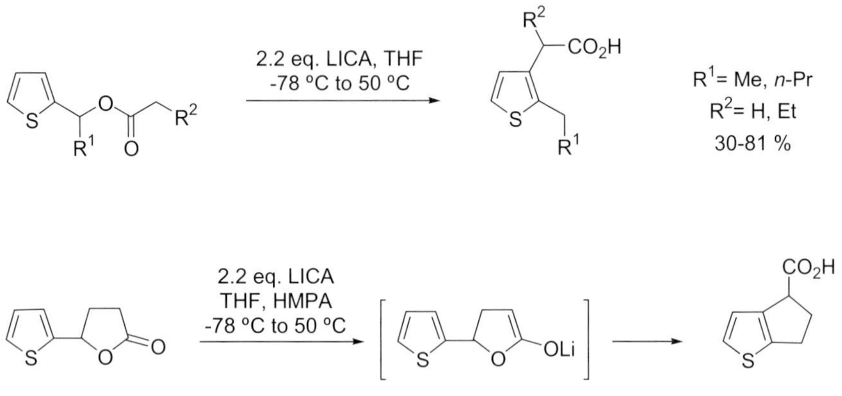

2.2 eq. LICA, THF
-78 °C to 50 °C

R^1= Me, *n*-Pr
R^2= H, Et
30-81 %

2.2 eq. LICA
THF, HMPA
-78 °C to 50 °C

7 **8**

Scheme 5.1.15

In the asymmetric synthesis of asteriscanolide, Wender et al. treated iso-butyrate **9**, obtained from the 1,4-diene-3-ol, with LDA to get the enolate, which rearranged at 0 °C [27]. The resulting dienoic acid was converted to tetraene **10**. Ni(0) mediated-[4+4]-cycloaddition of this tetraene gave the tricyclic lactone **11** which was converted to the sesquiterpene (+)-asteriscanolide (Scheme 5.1.16).

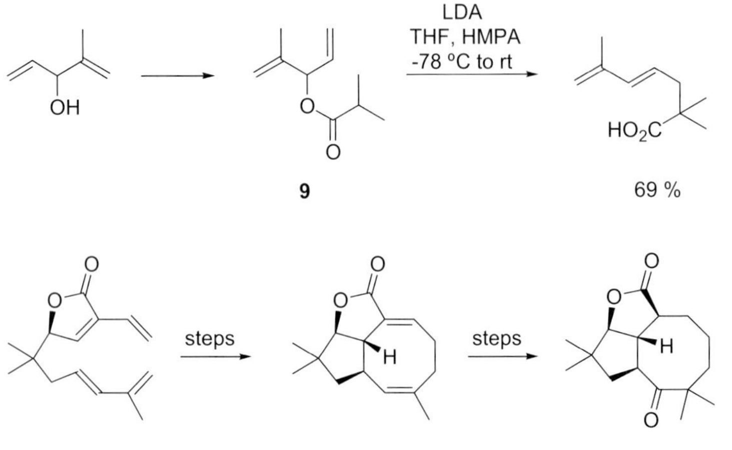

LDA
THF, HMPA
-78 °C to rt

9 69 %

steps steps

10 **11** (+)-asteriscanolide

Scheme 5.1.16

The Claisen rearrangement of the enolate of the ester **12** generated with LDA, afforded the γ,δ-unsaturated acid **13** which served as an intermediate in the total synthesis of (+)-artemisinin (Scheme 5.1.17) [28].

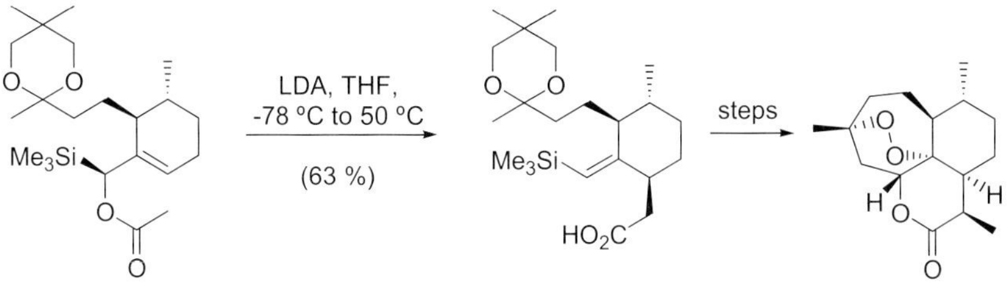

LDA, THF, -78 °C to 50 °C (63 %)

steps

12 (from (+)-pulegone) **13** (+)-artemisinin

Scheme 5.1.17

Substituted glutaric acids [29] were obtained from the ethyl malonic acid ester of the Baylis–Hillman product **14**. The enolate of this ester, generated with LDA at –78 °C rearranged to give the substituted glutaric acids in good yield (Scheme 5.1.18).

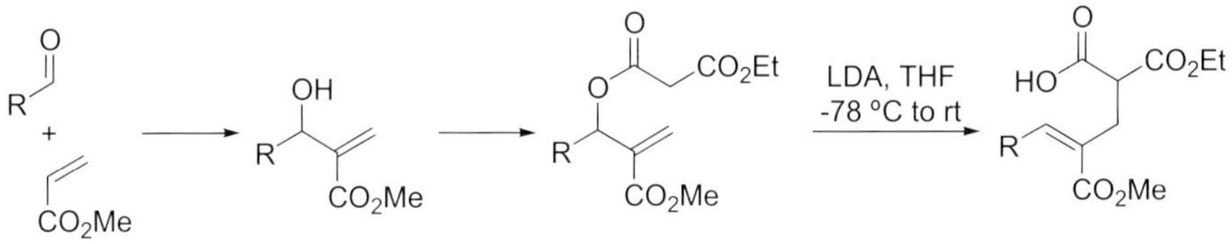

LDA, THF -78 °C to rt

14 78-92 %

Scheme 5.1.18

In the enantioselective synthesis of natural dolabellane [30], the chiral boron Lewis acid **15** was used to generate the enolate of the macrocyclic lactone **16**. The boron enolate rearranged smoothly to give the cycloundecidienoic acid **17**. This acid was further converted to natural dolabellane (Scheme 5.1.19).

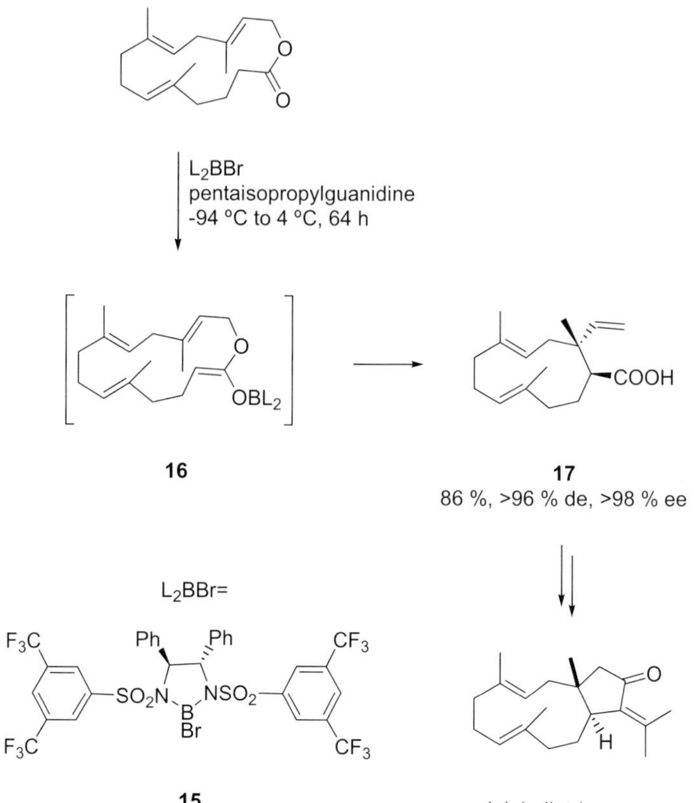

L$_2$BBr
pentaisopropylguanidine
-94 °C to 4 °C, 64 h

16

17
86 %, >96 % de, >98 % ee

L$_2$BBr=

15

dolabellatrienone

Scheme 5.1.19

5.1.4
Stereoselectivity in Enolate Formation

Long before the development of the enolate Claisen rearrangement, it was established theoretically [31] and with sufficient experimental precedence [6, 32] that, lacking the structural constraints in the allyl vinyl ether system, the chair-like transition state is preferred for the rearrangement. This is understandable since between the two possible transition states A and B, the transition state A is of lower energy due to the equatorial disposition of the R group.

Ireland found this predominance of the chair like transition state in the case of the rearrangement of the enolates of the esters **18** and **19** as well, which gave respectively the *E*-configured acids **20** and **21** with >99% diastereoselectivity.

Another consequence of the chair-like transition state is that in the rearrangement of the enolates, the relative configuration of the newly formed stereogenic centers as well as the configuration of the newly developed double bond would depend upon the double bond configuration of the starting allyl vinyl ether that is the 1,5-diene system. Therefore the knowledge of the configuration of the 1,5-diene system may allow one to predict the stereochemical outcome of the reaction with great accuracy. As the double bond configuration in the allyl part of the starting ester can be fixed in advance, method for stereoselective formation of either the *E*- or *Z*-enolate would allow one to correctly predict the stereochemistry of the rearrangement products.

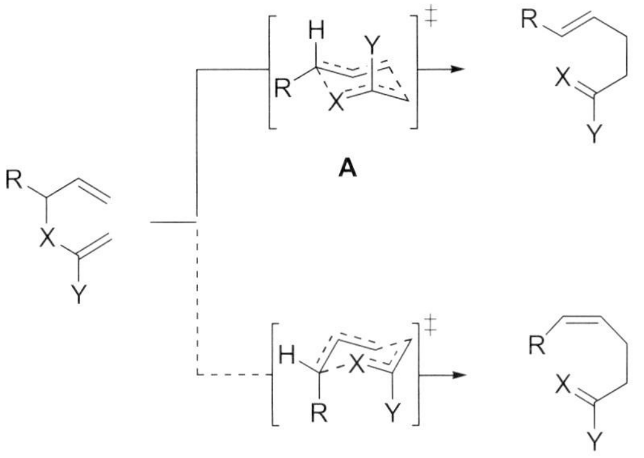

Scheme 5.1.20

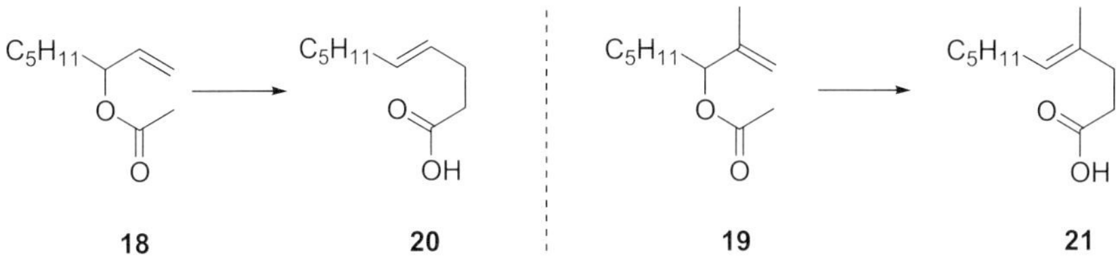

Scheme 5.1.21

The *E*-enolate of an ester with an *E*-configured allylic ether double bond would yield *anti*-configured acid, while the *Z*-enolate would give *syn*-configured acid. On the other hand in case of an ester with a *Z*-configured allylic ether double bond, the *E*-enolate would give the *syn*-configured acid, while the *Z*-enolate would give the *anti*-configured acid (Scheme 5.1.22) [33]. While addressing the problem of determining the geometry of the enolate it was reasoned that assuming a chair like transition state, the *syn/anti* ratio of the product would reflect upon the geometry of the enolate double bond. Rearrangement of *E*-and *Z*-crotyl propionates was used to probe and verify the above assumption [21]. Ireland et al. [21, 33] discovered that, and it was a very crucial finding, a singular factor of solvent polarity affected the *syn/anti* ratio of the products obtained from the rearrangement. The *E*-crotyl ester, when deprotonated in THF and allowed to rearrange, selectively gave the *syn* acid. On the other hand, when more coordinating solvent system, 23% HMPA-THF was employed, enolate formation took an alternative route and the *anti* acid predominated. The rearrangement of the *Z*-crotyl propionate gave the expected complete reversal of the product ratios (Table 5.1.2). This verified that

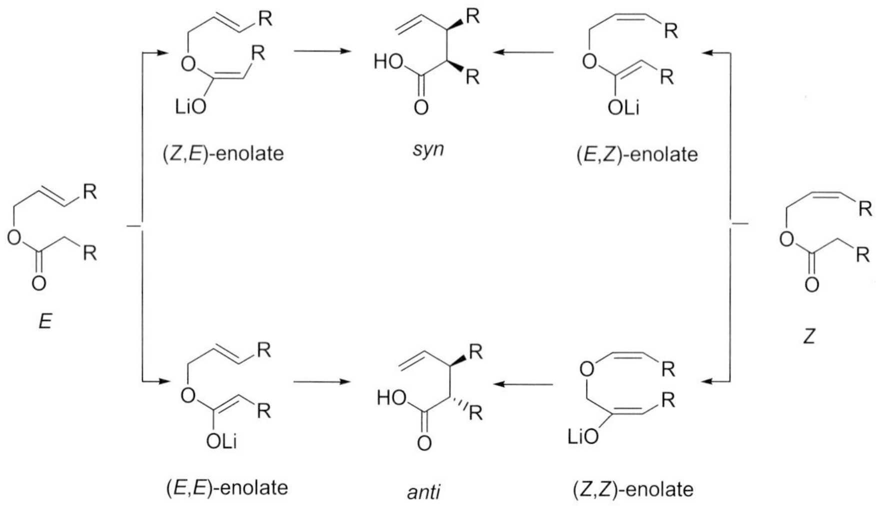

Scheme 5.1.22

the product determining step is indeed enolate formation and the course of the deprotonation is influenced by the solvent polarity. It was further shown that the stereochemical integrity of the enolate after its generation is not scrambled during the subsequent events.

Tab. 5.1.2 Effect of solvent on rearrangement of *E*- and *Z*-crotyl propionates.

Ester	Conditions	Enolate geometry	*syn/anti*
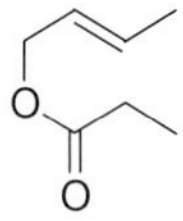	LDA, –78°C, THF	*Z*	92:8
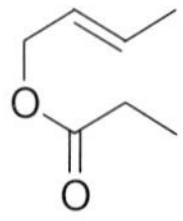	LDA, –78°C, 23% HMPA-THF	*E*	13:87
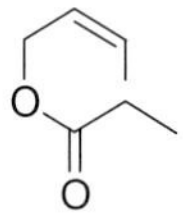	LDA, –78°C, THF	*Z*	11:89
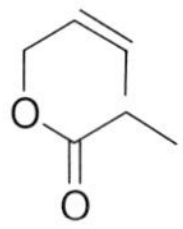	LDA, –78°C, 23% HMPA-THF	*E*	86:14

The stereochemical course of the rearrangement may be explained based on the assumption of two competing pericyclic, six-membered chair-like transition states (**I** and **II**) for the deprotonation (Scheme 5.1.23). Transition state **I**, leading to the *Z*-enolate, would be destabilized by an unfavorable 1,2-interaction between R and OR′. On the other hand, Transition state **II**, leading to the *E*-enolate, would be destabilized by an unfavorable 1,3-diaxial interaction between R and the axial isopropyl group of the base. In the absence of HMPA, the 1,3-diaxial interactions predominate. In the presence of HMPA, the 1,2-interactions predominate, which may be rationalized by the assumption that the lithium-enolate oxygen distance is increased due to the decreased Lewis acidity of the lithium cation in the presence of external Lewis basic ligands like HMPA. A "looser" cyclic transition state would weaken the transannular interactions and the 1,2-interactions are more important.

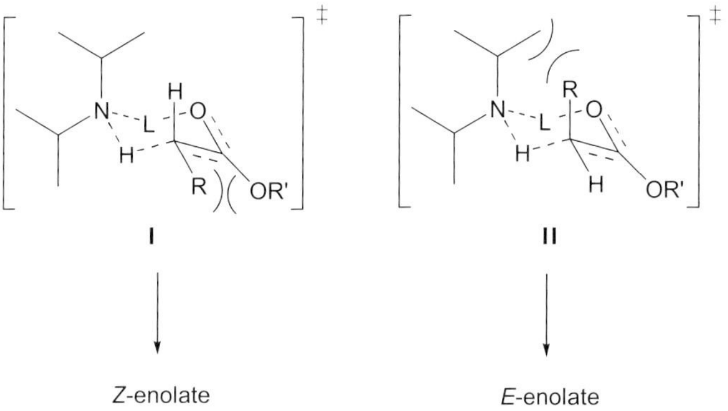

I → *Z*-enolate

II → *E*-enolate

Scheme 5.1.23

The same concept was found to be applicable to the deprotonation of ketones as well. This expanded the scope of this deprotonation method still further.

5.1.5
Simple Enolates of Allylic Esters of α-Hetero Acids

In 1977, Whitesell and Helbling [34] described the Claisen rearrangement of an allylic ester of α-methoxy acetic acid, through its enolate, generated using LDA, to the corresponding α-hetero-β,γ-unsaturated acid in good yield. Internal solvation of the counter ion through α-heteroatom to favor Z-enolate was found to be unimportant in this case (Scheme 5.1.24).

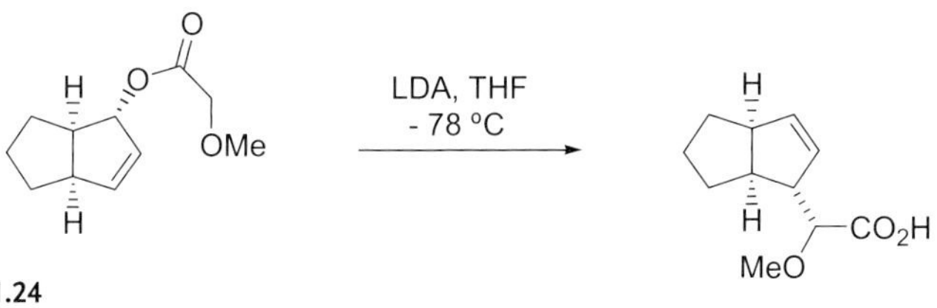

LDA, THF
- 78 °C

Scheme 5.1.24

Ager and Cookson [35] showed that the lithium enolates of the allylic ester of α-phenylthioacetic acid, generated using LDA, rearrange as such at room temperature to yield the products of the rearrangement in good yield. It was further shown that even the glycolic acid esters of allylic alcohols rearrange on treatment with two equivalents of LDA (Scheme 5.1.25). However, in these cases the configuration of the products has not been investigated.

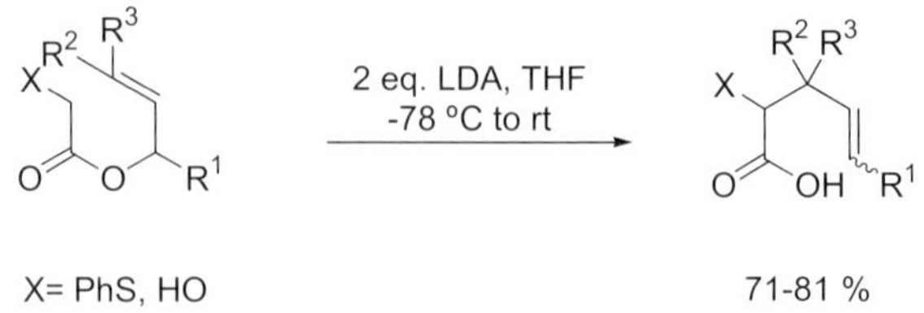

2 eq. LDA, THF
-78 °C to rt

X= PhS, HO

71-81 %

Scheme 5.1.25

Fujisawa et al. [36] investigated the rearrangement of *E* and *Z*-crotyl esters of glycolic acid and showed that the *E*-isomer gives the *syn/anti* products in a ratio of 92:8 while the *Z*-isomer gave the *anti* product in a ratio of 97:3. Protection of the hydroxyl group resulted in decreased diastereoselectivity. These results were explained on the basis of preferential formation of the enolate with a stable chelate structure. Use of polar solvent like HMPA-THF decreased the chemical yield of the reaction, however its effect on the enolate double bond configuration has not been commented upon. Bartlett [37] observed that the enediolate derived from crotyl mandelate rearranged in better yield and good stereoselection than the corresponding silylketene acetals.

In the rearrangement of the *E*-crotyl esters of 3-hydroxy butanoate, Fujisawa et al. [38] obtained, of the possible four, three diastereomers and discovered that the enediolate rearranges to give better yield as well as good diastereoselectivity as compared to the rearrangement of the corresponding silyl ketene acetal. The results were explained on the assumption that the enediolate has *E*-configuration due to the chelation effect. On the basis of NMR experiments the major diastereomer was assigned the *syn-anti* configuration. In case of the *Z*-enediolate the major diastereomer was assigned *anti-anti* configuration (Scheme 5.1.26).

In an independent study of the Claisen rearrangement of the *E*- and *Z*-crotyl esters of the 3-hydroxy butanoate, Kurth and Yu [39] isolated a mixture of two diastereomers each. NMR analysis of the 1,3-dioxanes, derived from the rearrangement products, was used to assign the stereochemistry at the 3,4 positions on the basis of coupling constants in the NMR spectra. To establish the stereochemistry at the 4,5 positions, (±)-botryodiplodin, a mycotoxin with both antibiotic and antileukemic properties, and its isomer 3-epibotryodiplodin were synthesized from the rearrangement products (Scheme 5.1.27) [39].

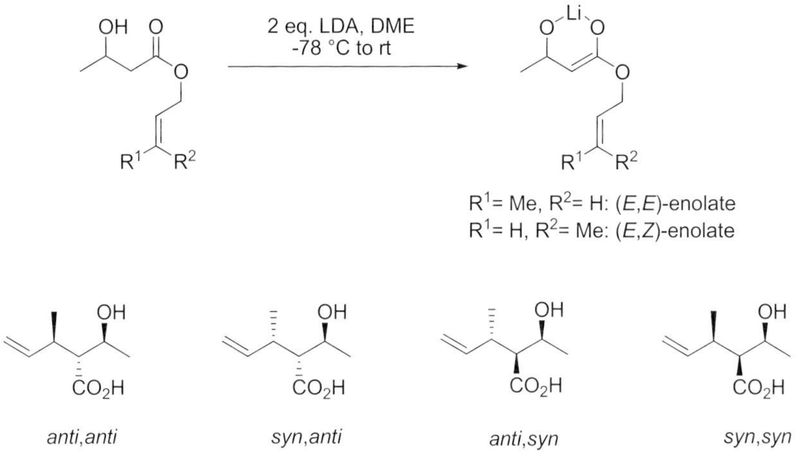

OH O
2 eq. LDA, DME
-78 °C to rt
Li
R^1= Me, R^2= H: (*E*,*E*)-enolate
R^1= H, R^2= Me: (*E*,*Z*)-enolate
OH
CO_2H
anti,anti
syn,anti
anti,syn
syn,syn

Fujisawa
(*E*,*E*)-enolate: *syn,anti* (major) + *anti,anti* (minor) 58 % (85:11)
(*E*,*Z*)-enolate: *anti,anti* (major) + *syn,anti* (minor) 31 % (75:19)

Kurth
(*E*,*E*)-enolate: *anti,syn* (major) + *anti,anti* (minor) 44 % (81:19)
(*E*,*Z*)-enolate: *anti,anti* (major) + *anti,syn* (minor) 36 % (85:15)

Scheme 5.1.26

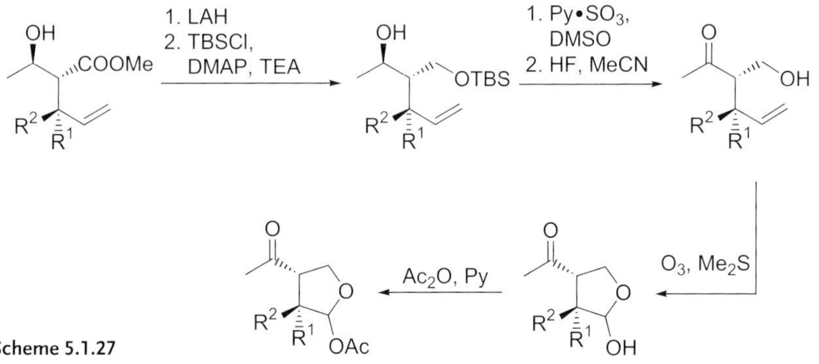

OH
COOMe
1. LAH
2. TBSCl,
DMAP, TEA
OTBS
1. Py•SO_3,
DMSO
2. HF, MeCN
OH
O_3, Me_2S
Ac_2O, Py
OAc

Scheme 5.1.27

Based on these studies Kurth showed that in the rearrangement of the *E*-crotyl ester *anti,syn* and *anti,anti* isomers were obtained and that the *anti,syn* isomer is the major isomer. In case of the *Z*-crotyl ester the same two isomers were obtained but the *anti,anti* isomer was the major product. He further showed that the enolization proceeds with excellent stereoselection in favor of the E-configured enolate (Scheme 5.1.24). Kurth's conclusions about the relative configuration of the product appear to be based on a more sound footing.

5.1.6
Simple Enolates of *N*-Allyl Amides

In 1990, Tsunoda et al. [40] described the enolate Claisen rearrangement of *N*-alkyl, *N*-allyl amides. Unlike the ester enolates, the decomposition of amide enolate via the ketene pathway was not found to be a competitive side reaction. Therefore there is no need to prepare the corresponding silyl acetals prior to the rearrangement. However the amide enolates require higher temperature (> 100°C) for the rearrangement (Scheme 5.1.28). An important aspect of this rearrangement is that the stereoselectivity of the enolate formation is quite high (often > 99:1) in favor of the *Z*-configured enolate double bond. This is a consequence of the steric bulk of the dialkyl amino group, which destabilizes the transition state that leads to the corresponding *E*-enolate due to 1,3-allylic strain.

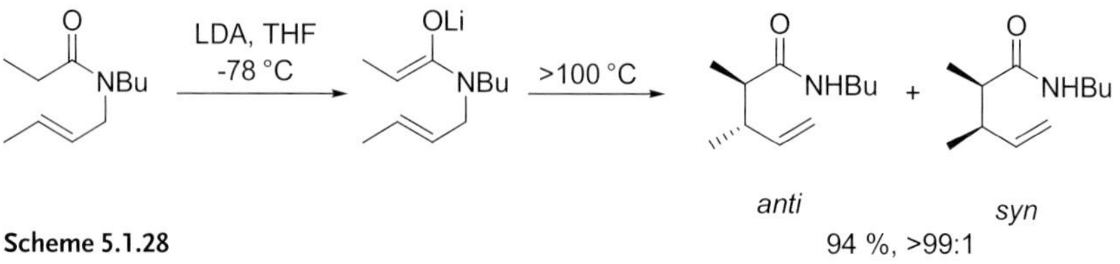

LDA, THF, -78 °C; >100 °C

NBu, OLi, NHBu, *anti*, *syn*

94 %, >99:1

Scheme 5.1.28

Use of chiral amines like α-phenethyl amine by Tsunoda et al. [41] in the amide enolate Claisen rearrangement, an advantage absent with the esters, results in asymmetric induction, though to a moderate level, in favor of the *R,S* enantiomer of the *syn* diastereomer (Scheme 5.1.29). Application of this method to α-hetero substituted (OH, NH_2) acetamides also gave the analogous products in good yield and with good stereoselection [42]. However the diastereoselectivity depended upon the nature of the substituent X. Tsunoda et al. used the asymmetric Claisen rearrangement of allyl amides in the syntheses of (–)-verrucarinolactone and *D*-allo-isoleucine [42].

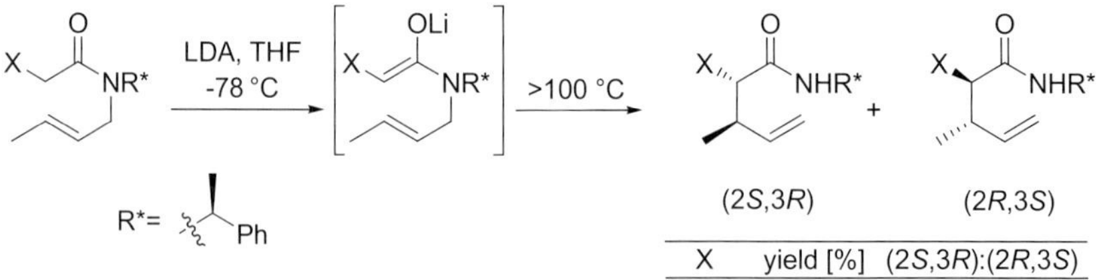

LDA, THF, -78 °C; >100 °C

R* = Ph

(2*S*,3*R*) (2*R*,3*S*)

X	yield [%]	(2*S*,3*R*):(2*R*,3*S*)
Me	85	11:89
OH	95	87:13
OTBS	62	67:33
NH_2	69	11:89

Scheme 5.1.29

Tsunoda et al. [43] have used this asymmetric amide enolate Claisen rearrangement in the total synthesis of (–)-iridomyrmecin **22** (Scheme 5.1.30). Rearrangement of the *Z*-enolate of the amide **23** via a chair like transition state generated, with a very high diastereoselectivity, the asymmetric centers α and β to the carbonyl group of the natural product.

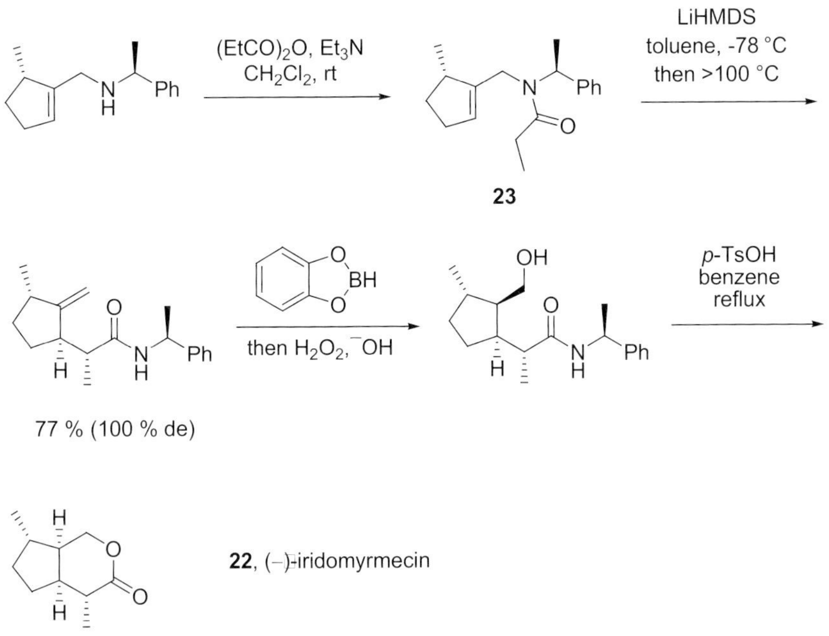

Scheme 5.1.30

Recently, Tsunoda et al. [44] reported a chiral synthesis of (–)-antimycin A_{3b} using an amide enolate Claisen rearrangement. The rearrangement of the allylic amide **24**, prepared by addition of *N*-trimethylsilyl (*R*)-α-methylbenzylamine to acrolein followed by acylation, gave the (*7S,8R*)-configured pentenoic amide **25** as an inseparable mixture with its (*7R,8S*) diastereomer (78% combined yield). This amide, which constitutes the C-6 to C-9 fragment of antimycin A_{3b}, was used to complete the synthesis (Scheme 5.1.31).

Suh et al. [45] have reported an intermolecular version of the amide enolate Claisen rearrangement to prepare macrocyclic lactams. Acylation of 2-vinyl *N*-heterocycles followed by deprotonation with a strong base provided the corresponding enolate, which on heating afforded the macrocyclic lactams in good yield. Suh et al. [46] applied this method to the synthesis of macrolactam, fluvirucinin A_1 (Scheme 5.1.32). The rearrangement of 2-vinyl piperidine, occurring through a chair-like transition state of the *Z*-amide enolate, gave the *E*-configured alkene with requisite *cis*-configuration at C-6 and C-10 positions.

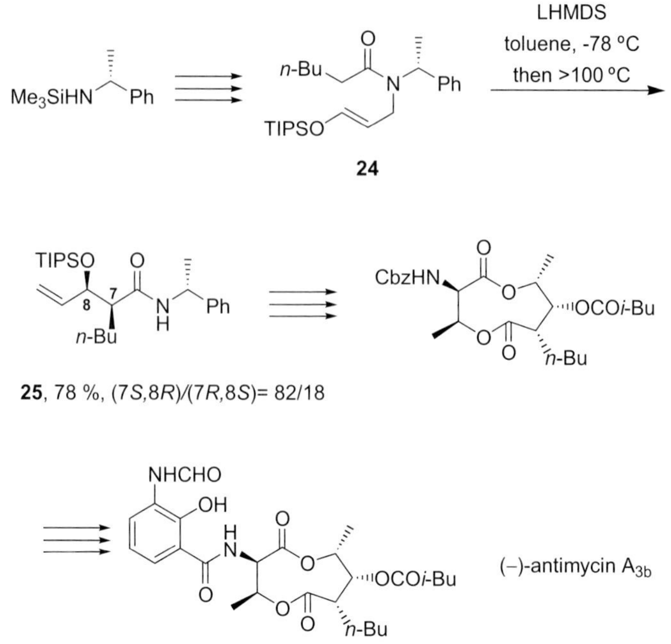

Scheme 5.1.31

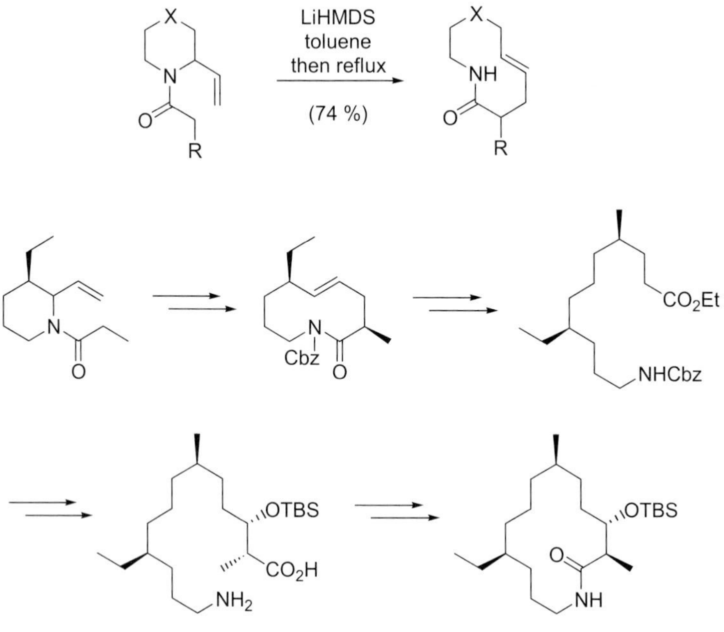

Scheme 5.1.32

Amides derived from vinyl aziridines have been used by Somfai and Lindström [47, 48] in amide enolate Claisen rearrangement to get substituted caprolactams. Due to the ring strain the reaction appears to be adopting a boat conformation in the transition state and secondly, the rearrangement, otherwise requiring high temperature, takes place at a significantly lower temperature. However *cis*-disubstituted aziridines, for the steric reasons failed to undergo the rearrangement (Scheme 5.1.33).

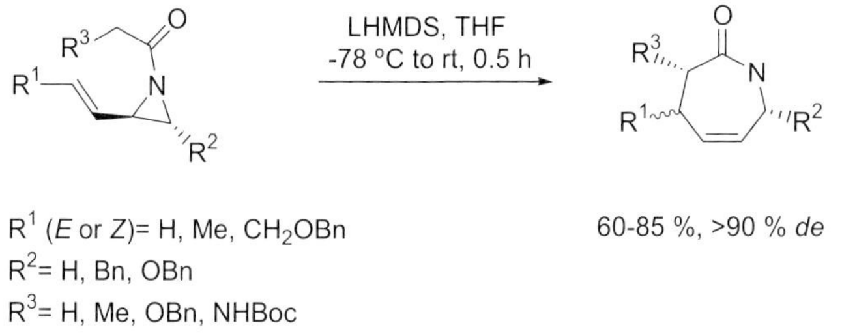

LHMDS, THF
-78 °C to rt, 0.5 h

R^1 (*E* or *Z*)= H, Me, CH_2OBn
R^2= H, Bn, OBn
R^3= H, Me, OBn, NHBoc

60-85 %, >90 % *de*

Scheme 5.1.33

5.1.7 Miscellaneous Enolates

Apart from the Claisen rearrangement of simple enolates of carboxylic acid derivatives there are a few isolated reports of the Claisen rearrangement of enolates derived from ketones. Koreeda and Luengo [49] showed, in an isolated report, that the allyl vinyl ethers could be generated through the deprotonation of α-allyloxy ketones bearing α-acidic hydrogen and these, like ester enolates, rearrange at sub-zero temperature to afford the α-hydroxy-γ,δ-unsaturated ketones. It was further shown that the required regioselective enolates of α-allyloxy ketones could as well be obtained via a Michael addition to an α-allyloxy-α,β-unsaturated ketone (Scheme 5.1.34).

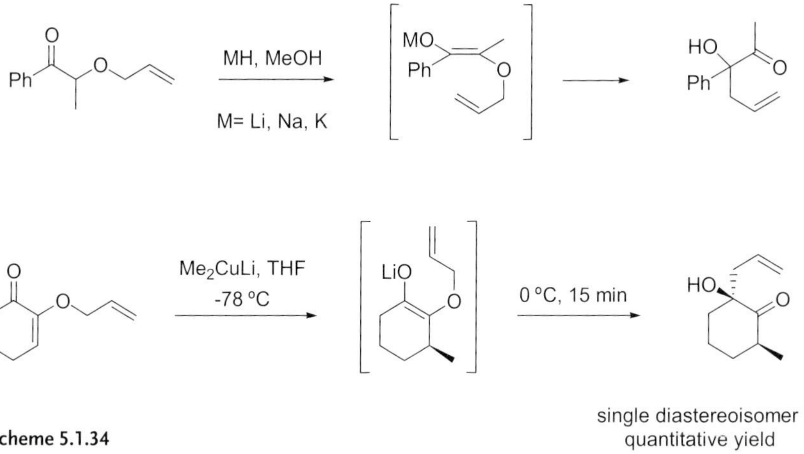

MH, MeOH
M= Li, Na, K

Me_2CuLi, THF
-78 °C

0 °C, 15 min

single diastereoisomer
quantitative yield

Scheme 5.1.34

In another isolated report [50], α-aryloxy-cyclohexenone was shown to undergo an enolate Claisen rearrangement when treated with potassium hydride to afford 6-aryl-cyclohex-2,3-enone (Scheme 5.1.35).

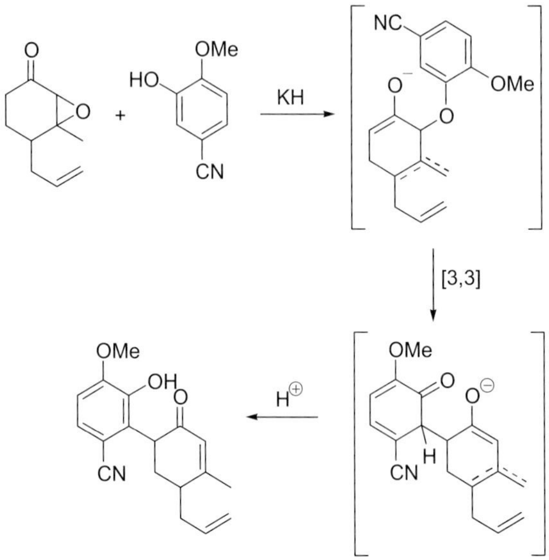

Scheme 5.1.35

5.1.8
Conclusion

Ireland's groundbreaking work on optimizing the reaction conditions for the stereoselective generation of the enolates from allylic esters has vastly expanded the applicability of the Claisen rearrangement in organic synthesis, especially in the area of stereoselective synthesis of complex natural products. Though Ireland made it a practice to trap the enolate as silyl ethers, as is seen from the above discussion, this need not be a mandatory step before effecting the Claisen rearrangement. This became more apparent with the development of the amide enolate Claisen rearrangement by Tsunoda, where the prior conversion of the amide enolate to the silyl ether is not necessary. Coupled with this the development of a mild procedure for the hydrolysis of amides by Tsunoda could increase the utility of the amide enolate method further.

Acknowledgment

The invaluable help rendered by Dr. (Miss) Saryu I. Davawala in the preparation of this manuscript is duly acknowledged.

References

1 Claisen, L. *Ber. Dtsch. Chem. Ges.* **1912**, *45*, 3157–3166.

2 For reviews that include enolate Claisen rearrangement see: a) Ziegler, F. E. *Acc. Chem. Res.* **1977**, *10*, 227–232. b) Bennett, G. B. *Synthesis* **1977**, 589–606. c) Hill, R. K. in *Asymmetric Synthesis*, Morrison, J. D., (ed.), Academic: N.Y. **1984**, Vol. 3, pp 503–572. d) Blechert, S. *Synthesis* **1989**, 71–82. e) Wipf, P. in *Comprehensive Organic Synthesis*, Trost, B. M. (ed.), Pergamon: Oxford, **1991**, Vol. 5, pp 827–873. f) Tadano, K. in *Studies in Natural Products Chemistry*, Rahman, A.-U., (ed.); Elsevier, Amsterdam, **1992**, pp 405–455. g) Pereira, S.; Srebnik, M. *Aldrichimica Acta*, **1993**, *26*, 17–29. h) Frauenrath, H. in *Stereoselective Synthesis*; Helchen, G., Hoffmann, R. W., Mulzer, J., Schaumann, E., (eds.), Georg Thieme: Stuttgart, **1995**, Vol. E21d, pp 3301–3756. i) Enders, D.; Knopp, M.; Schiffers, R. *Tetrahedron: Asymmetry* **1996**, *7*, 1847–1882. j) Ito, H.; Taguchi, T. *Chem. Soc. Rev.* **1999**, *28*, 43–50.

3 The Claisen rearrangement of allyl vinyl ethers is a typical [3,3]-sigmatropic process. However, whenever this process is energetically and/or sterically unfavorable (or for other reasons), the allyl vinyl ether undergoes [1,3]-sigmatropic rearrangement. Grieco, P. A.; Clark, J. D.; Jagoe, C. T. *J. Am. Chem. Soc.* **1991**, *113*, 5488–5489.

4 a) Marbet, R.; Saucy, G. *Helv. Chim. Acta* **1967**, *50*, 2091–2100. b) Dauben, W. G.; Dietsche, T. J. *J. Org. Chem.* **1972**, *37*, 1212–1216. c) Petrzilka, M. *Helv. Chim. Acta* **1978**, *61*, 2286–2289.

5 a) Carroll, M. F. *J. Chem. Soc.* **1940**, 704–706. b) Carroll, M. F. *J. Chem. Soc.* **1941**, 507–510.

6 Johnson, W. S.; Werthemann, L.; Bartlett, W. R.; Brocksom, T. J.; Li, T. T.; Faulkner, D. J.; Peterson, M. R. *J. Am. Chem. Soc.* **1970**, *92*, 741–743.

7 a) Wick, A. E.; Felix, D.; Steen, K.; Eschenmoser, A. *Helv. Chim. Acta* **1964**, *47*, 2425–2429. b) Wick, A. E.; Felix, D.; Gschwend-Steen, K.; Eschenmoser, A. *Helv. Chim. Acta* **1969**, *52*, 1030–1042.

8 Ireland, R. E.; Mueller, R. H. *J. Am. Chem. Soc.* **1972**, *94*, 5897–5898.

9 Kulkarni M. G.; Pendharkar, D. S.; Rasne, R. M. *Tetrahedron Lett.* **1997**, *38*, 1459–1462.

10 a) Diederich, M.; Nubbemeyer, U. *Angew. Chem. Int. Ed. Engl.* **1995**, *34*, 1026–1028. b) Nubbemeyer, U. *J. Org. Chem.* **1995**, *60*, 3773–3780.

11 For the purpose of this article the simple enolate method is defined as the enolate having an alkali or an alkaline earth metal as a counterion.

12 Tseou, H.-F.; Wang, Y. -T. *J. Chin. Chem. Soc.* **1937**, *5*, 224–229.

13 Arnold, R. T.; Searles, S. Jr. *J. Am. Chem. Soc.* **1949**, *71*, 1150–1151.

14 Arnold, R. T.; Parham, W. E.; Dodson, R. M. *J. Am. Chem. Soc.* **1949**, *71*, 2439–2440.

15 Brannock, K. C.; Pridgen, H. S.; Thompson, B. *J. Org. Chem.* **1960**, *25*, 1815–1816.

16 Julia, S.; Julia, M.; Linstrumelle, G. *Bull. Soc. Chim. Fr.* **1964**, *25*, 2693–2694.

17 Arnold, R. T.; Hoffman, C. *Synth. Commun.* **1972**, *2*, 27–33.

18 Eichinger, P. C. H.; Bowie, J. H. *J. Org. Chem.* **1987**, *52*, 5224–5228.

19 Baldwin, J. E.; Walker, J. A. *J. Chem. Soc. Chem. Commun.* **1973**, 117–118.

20 Rathke, M. W.; Lindert, A. *J. Am. Chem. Soc.* **1971**, *93*, 2318–2320.

21 Ireland, R. E.; Mueller, R. H.; Willard, A. K. *J. Am. Chem. Soc.* **1976**, *98*, 2868–2877.

22 Wuts, P. G. M.; Sutherland, C. *Tetrahedron Lett.* **1982**, *23*, 3987–3990.

23 Parker, K. A.; Farmar, J. G. *Tetrahedron Lett.* **1985**, *26*, 3655–3658.

24 Wilson, S. R.; Price, M. F. *J. Am. Chem. Soc.* **1982**, *104*, 1124–1126.

25 Hart, D. J.; Chillous, S. E.; Hutchinson, D. K. *J. Org. Chem.* **1982**, *47*, 5418–5420.

26 Kumamoto, T.; Sado, M.; Abe, H.; Inuzuka, N.; Shirai, K. *Bull. Chem. Soc. Jpn.* **1983**, *56*, 1665–1668.

27 Wender, P. A.; Ihle, N. C.; Correia, C. R. D. *J. Am. Chem. Soc.* **1988**, *110*, 5904–5906.

28 Avery, M. A.; Chong, W. K. M.; Jennings-White, C. *J. Am. Chem. Soc.* **1992**, *114*, 974–979.

29 Emslie, N. E.; Mason, P. H.; Drewes, S. E. *Synth. Commun.* **1995**, *25*, 183–190.

30 Corey, E. J.; Kania, R. S. *J. Am. Chem. Soc.* **1996**, *118*, 1229–1230.

31 a) Gill, G. B. *Quart. Rev. (London)*, **1968**, 22, 338–389. b) Hansen, H. J.; Schmid, H. *Tetrahedron*, **1974**, *30*, 1959–1969.

32 a) Takahashi, H.; Oshima, K.; Yamamoto, H.; Nozaki, H. *J. Am. Chem. Soc.* **1973**, *95*, 5803–5804. b) Doering, W. v. E.; Roth, W. R. *Tetrahedron* **1962**, *18*, 67–74. c) Vittoreli, P.; Winkler, T.; Hansen, H. J.; Schmid, H. *Helv. Chim. Acta* **1968**, *51*, 1457–1461. d) Sucrow, W.; Richter, W. *Chem. Ber.* **1971**, *104*, 3679–3688. e) Faulkner, D. J.; Petersen, M. R. *Tetrahedron Lett.* **1969**, *38*, 3243–3246.

33 Ireland, R. E.; Willard, A. K. *Tetrahedron Lett.* **1975**, *46*, 3975–3978.

34 Whitesell, J. K.; Helbling, A. M. *J. Chem. Soc., Chem. Commun.* **1977**, 594–595.

35 Ager, D. J.; Cookson, R. G. *Tetrahedron Lett.* **1982**, *23*, 3419–3420.

36 Fujisawa, T.; Sato, T.; Tajima, K. *Tetrahedron Lett.* **1983**, *24*, 729–730.

37 Bartlett, P. A.; Tanzella, D. J.; Barstow, J. F. *J. Org. Chem.* **1982**, *47*, 3941–3945.

38 Fujisawa, T.; Tajima, K.; Ito, M.; Sato, T. *Chem. Lett.* **1984**, 1169–1172.

39 a) Kurth, M. J.; Yu, C. -M. *Tetrahedron Lett.* **1984**, *25*, 5003–5006. b) Kurth, M. J.; Yu, C.-M. *J. Org. Chem.* **1985**, *50*, 1840–1845. c) Kurth, M. J.; Beard, R. L. *J. Org. Chem.* **1988**, *53*, 4085–4088.

40 Tsunoda, T.; Sasaki, O.; Ito, S. *Tetrahedron Lett.* **1990**, *31*, 727–730.

41 Tsunoda, T.; Sasaki, O.; Sakai, M.; Sako, Y.; Hondo, Y.; Ito, S. *Tetrahedron Lett.* **1992**, *33*, 1651–1654.

42 Tsunoda, T.; Tatsuki, S.; Shiraishi, Y.; Akasaka, M.; Ito, S. *Tetrahedron Lett.* **1993**, *34*, 3297–3300.

43 Tsunoda, T.; Tatsuki, S.; Kataoka, K.; Ito, S. *Chem. Lett.* **1994**, 543–546.

44 Tsunoda, T.; Nishii, T.; Yoshizuka, M.; Yamasaki, C.; Suzuki, T.; Ito, S. *Tetrahedron Lett.* **2000**, *41*, 7667–7671.

45 Suh, Y.-G.; Lee, J.-Y.; Kim, S.-A.; Jung, J.-K. *Synth. Commun.* **1996**, *26*, 1675–1680.

46 Suh, Y.-G.; Kim, S.-A.; Jung, J.-K.; Shin, D.-Y.; Min, K.-H.; Koo, B.-A.; Kim, H.-S. *Angew. Chem. Int. Ed.*, **1999**, *38*, 3545–3547.

47 Lindström, U. M.; Somfai, P. *J. Am. Chem. Soc.* **1997**, *119*, 8385–8386.

48 Lindström, U. M.; Somfai, P. *Chem. Eur. J.* **2001**, *7*, 94–98.

49 Koreeda, M.; Luengo, J. I. *J. Am. Chem. Soc.* **1985**, *107*, 5572–5573.

50 Schultz, A. G.; Napier, J. *J. Chem. Soc., Chem. Commun.* **1981**, 224–225.

5.2 Chelate Enolate Claisen Rearrangement

Uli Kazmaier

5.2.1 Introduction

The high popularity of [3,3]-sigmatropic rearrangement processes results from the fact that these processes proceed via highly ordered transition states, what results in high levels of stereocontrol [1]. Especially with (*E*)-configured substituted allylic esters **I**, the chair-like transition state is highly favored over the boat-like transition state, and the simple diastereoselectivity of the rearrangement depends nearly exclusively on the enolate geometry [2]. Ireland developed the well known Ireland–Claisen rearrangement during his studies on enolate formation [3], and important information about the enolate geometry could be obtained from the *syn/anti* ratio of the rearrangement products [4]. This is possible because the conversion of the enolate into the corresponding silylketene acetal occurs with retention of the enolate geometry, and therefore the enolate formation is the stereocontrolling step. The formation of (*E*)-enolates **II** is favored in THF, whereas the addition of HMPA or other metal chelating solvents reverses the enolate configuration to the (*Z*)-product **III**. In principle this enolate can directly undergo Claisen rearrangement, but side reactions such as enolate decomposition might cause serious problems. On the other hand, silylation provides stable compounds **IV** and **V**, which can be heated to reflux for completion of the rearrangement.

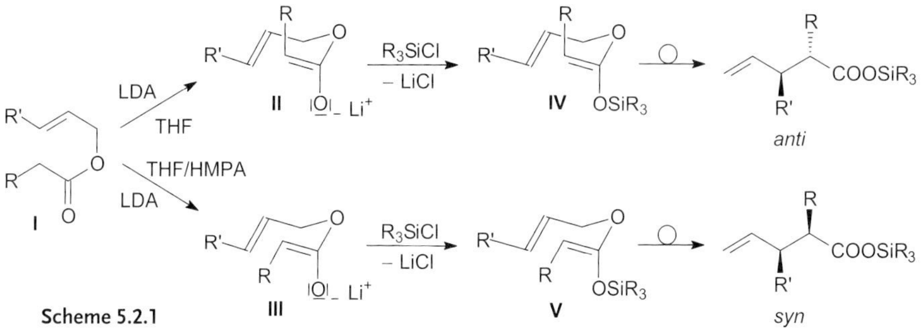

Scheme 5.2.1

High selectivities with respect to (*Z*)-enolate formation (even without HMPA) can be obtained if R contains substituents with free electron pairs (O, N, S), which are able to coordinate to the enolate metal [1k]. α-Substituted esters give rise to five-membered (**A**), and β-substituted esters to six-membered chelate rings (**B**) (Figure 5.2.1). The same is true for the enolates obtained from secondary allyl amides (aza-Claisen rearrangement) [5].

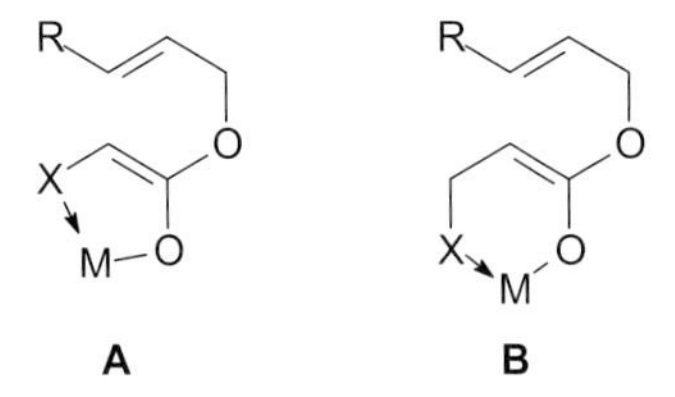

Fig. 5.2.1 Chelate enolates.

Chelate formation in general results in a stabilization of the enolate, which allows a direct rearrangement of the metal enolate [6]. On the other hand, silylation and rearrangement of the silylketene acetal obtained, according to the Ireland protocol, is possible as well.

This review describes rearrangements of allylic esters bearing oxygen, amino and sulfur substituents in the α- or β-position of the acid moiety. In principle, chelate formation can also be proposed in the rearrangement of β-ketoesters, but this so-called Carrol rearrangement is described in Chapter 8.

5.2.2 Claisen Rearrangements of Substrates with Chelating Substituents in the α-Position

5.2.2.1 Rearrangement of α-Hydroxy Substituted Allylic Esters

Chelation effects in general override the usual preferences in the formation of lithium ester enolates, and the (*Z*)-configured enolates are obtained nearly exclusively. Therefore the stereochemical outcome of the rearrangement should only be controlled by the olefin geometry in the allyl moiety and by the transition state (chair vs. boat). If substituted allylic esters of glycolic acid or related α-hydroxyacids are subjected to rearrangement, synthetically valuable unsaturated α-hydroxyacids are obtained, albeit the yield and stereoselectivity strongly depends on the substrate and the reaction conditions used.

Rearrangement via Dianionic Intermediates

If allyl glycolate **1a** is treated with at least two equivalents of LDA in THF at –78 °C and the mixture is warmed to room temperature, the dianionic intermediate **1a′** undergoes rearrangement (Scheme 5.2.2). Similar results were reported by Ager et al. for substituted allylic esters such as **1b** and **1c** (Table 5.2.1, entries 2 and 3) and the corresponding α-mercaptoester (see Section 5.2.1.4) [7]. Bartlett et al. intensively investigated the rearrangement of α-hydroxy acids as well as the *O*-protected derivatives [8]. They found that for α-alkoxy esters the enolate generation time is much longer than for propionate or α-acylamido esters, and that for α-hydroxy esters the situation is even worse. Good yields are obtained in the rearrangement of crotyl mandelate **1d** (entry 4), which gives a relatively well-stabilized enediolate, while the yields with other crotylesters were very low. Surprisingly, the *syn*-diastereoselectivity in the rearrangement of **1d** was very high in contrast to the corresponding Ireland–Claisen rearrangement, which provided a nearly 1:1 diastereomeric mixture. Similar observations were made by Fujisawa et al. in the rearrangement of crotyl glycolate **1e** (entry 5) [9]. In this case the yield could be increased by using LHMDS (entry 6), which is the base of choice for this system, especially if the Ireland version is used.

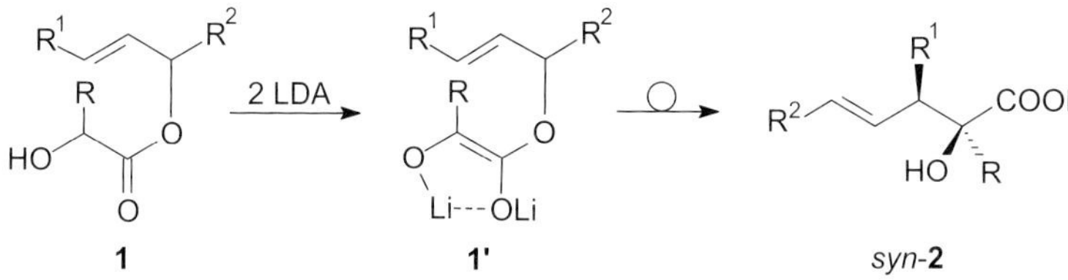

Scheme 5.2.2

Tab. 5.2.1 Chelate Claisen rearrangement of α-hydroxy allylic esters **1**.

Entry	Substrate	R	R^1	R^2	Yield [%]	ds [%]	Ref.
1	**1a**	H	H	H	79	–	7
2	**1b**	H	H	Me	75	–	7
3	**1c**	H	H	Ph	79	–	7
4	**1d**	Ph	Me	H	59	92	8
5	**1e**	H	Me	H	13	92	9
6	**1e**	H	Me	H	28[a)]	90	9

a) LHMDS was used as base.
ds = diastereoselectivity.

The major side reaction in these processes appeared to be ester cleavage under liberation of the corresponding alcohol [8]. This mode of decomposition presumably involves formation of a ketene intermediate, which then undergoes further side reactions [10]. The increased prevalence of this side reaction with the enediolate is not surprising in view of the enolate-destabilizing effect the ionized α-hydroxyl is expected to have.

Ireland–Claisen Rearrangements of α-Hydroxy Esters

Trapping of the enediolate intermediates **1′** with silylchlorides should give rise to α-silyloxy-substituted silylketene acetals **1″**, which then undergo an Ireland–Claisen rearrangement (Scheme 5.2.3) [11]. This should help to reduce the side reactions discussed above, and one might expect significantly better results in comparison to the enediolate rearrangement. In general this is true with respect to the yield, but often not for the stereoselectivity. For example, the yield obtained in the

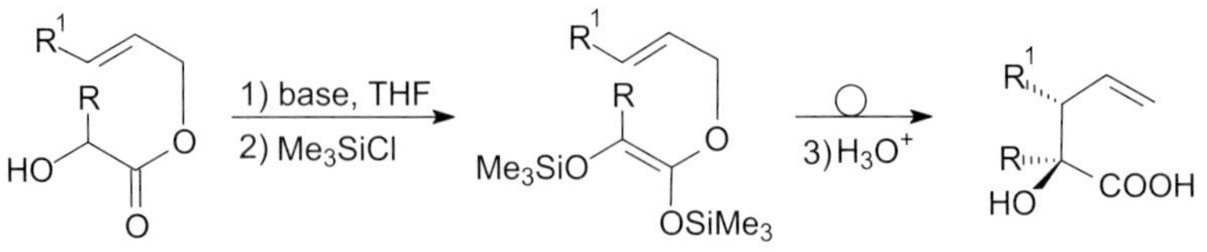

Scheme 5.2.3

rearrangement of **1d** under Ireland conditions (Table 5.2.2, entry 1) was comparable to the results of the enediolate rearrangement, but the diastereoselectivity was not only significantly worse, the *anti* product was formed preferentially [8]. A similar selectivity for the *syn* stereoisomer was obtained in the rearrangement of **1e** if lithium isopropylcyclohexylamide (LICA) was used as a base [8]. Fujisawa et al. investigated carefully the influence of the base and the solvents on the outcome of the reaction [9]. They found that lithium hexamethyldisilazide (LHMDS) is superior to LDA both with respect to the yield and the diastereoselectivity. Considering an *E*/*Z* ratio of the starting material **1e** (93:7) the stereoselectivity was 98%. THF was found to be the solvent of choice, while with other solvents such as ether, dimethoxyethane or THF-HMPA much lower yields were obtained [9].

The rearrangement of crotyl lactates **1f** was investigated by Bartlett et al. No significant effect was observed if HMPA was used as cosolvent, but if 1 equiv $MgCl_2$ was added to the ester, the stereoselectivity could be increased significantly, although unfortunately not the yield [8].

Tab. 5.2.2 Ireland–Claisen rearrangement of α-hydroxy allylic esters **1**.

Entry	Substrate	R	R^1	Conditions	Yield [%]	ds [%]	Ref.
1	**1d**	Ph	Me	2.25equiv LICA/Me_3SiCl	50	40	8
2	**1e**	H	Me	2.25equiv LICA/Me_3SiCl	38	58	8
3	**1e**	H	Me	3 equiv LDA/Me_3SiCl	40	70	9
4	**1e**	H	Me	3 equiv LHMDS/Me_3SiCl	84	92	9
5	**1f**	Me	Me	2.25equiv LICA/Me_3SiCl	30	80	8
6	**1f**	Me	Me	2.25equiv LICA/Me_3SiCl 20% HMPA	28	77	8
7	**1f**	Me	Me	2.25equiv LICA/Me_3SiCl 1 equiv $MgCl_2$	20	> 90	8

The optimized reaction conditions were also used for the rearrangement of (*Z*)-configured ester **3** (Scheme 5.2.4). The expected *anti*-product **4** was obtained after esterification in good yields and high stereoselectivity [9].

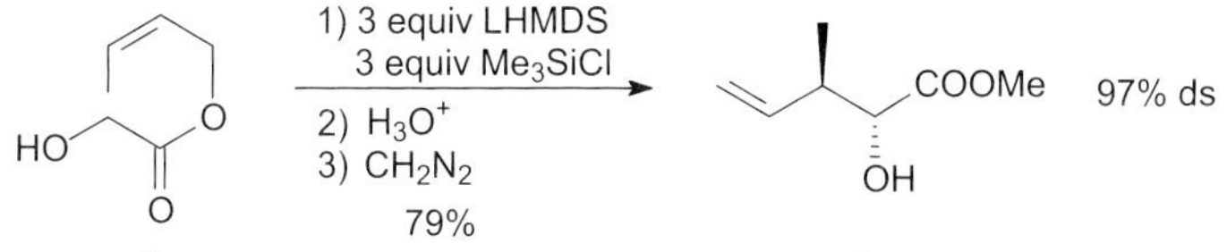

HO
O
O
1) 3 equiv LHMDS
3 equiv Me3SiCl
2) H3O+
3) CH2N2
79%
COOMe
OH
97% ds

3 4

Scheme 5.2.4

If chiral allylic alcohols are used, the rearrangement occurs with excellent chirality transfer (Scheme 5.2.5) [11]. The *α*-hydroxy acid **6**, obtained from **5**, was further converted into several pheromones such as **7** and **8** [11]. The butyl-substituted derivative **9** rearranges to **10**, which is an intermediate in the synthesis of blastmycinone **11**, a key building block of the antibiotic blastmycin [12]. The rearrangement of silylsubstituted *α*-hydroxy and *α*-alkoxy esters was investigated by Panek et al. [13]. In the rearrangement of hydroxyester **12** the expected *syn*-product **13** was obtained exclusively if Me_3SiCl was used for silylation in the presence of pyridine. In contrast, with sterically more demanding silyl groups such as TBDMS, obviously the (*E*)-configured silylketene intermediate is formed, and the *anti*-product is obtained preferentially.

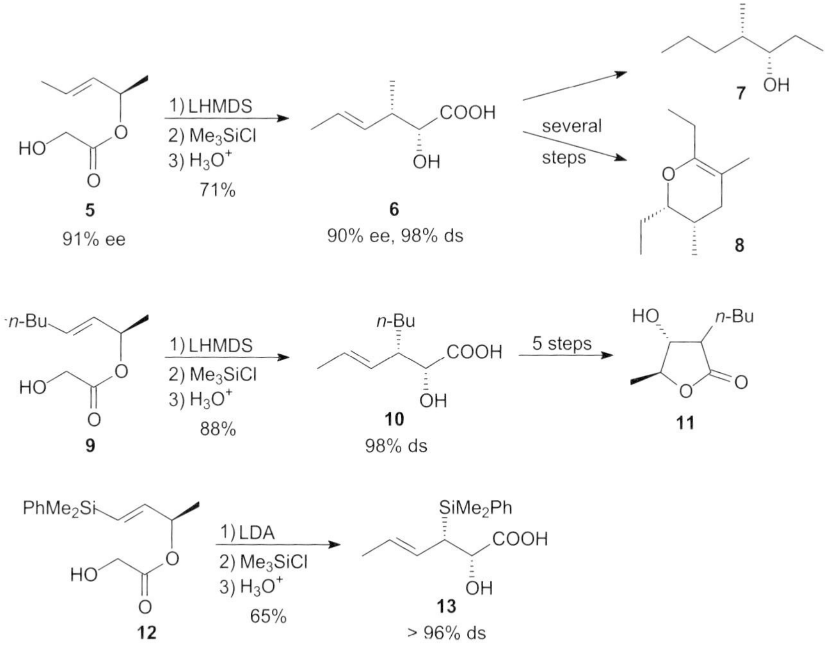

HO
O
O
1) LHMDS
2) Me3SiCl
3) H3O+
71%
91% ee
COOH
OH
90% ee, 98% ds
several
steps
OH
O
n-Bu
1) LHMDS
2) Me3SiCl
3) H3O+
88%
n-Bu
COOH
OH
98% ds
5 steps
HO
n-Bu
O
O
PhMe2Si
1) LDA
2) Me3SiCl
3) H3O+
65%
SiMe2Ph
COOH
OH
> 96% ds

5 6 7 8 9 10 11 12 13

Scheme 5.2.5

Representative Procedures for the Claisen Rearrangements of Vinylsilanes (12)
To a solution of diisopropylamine (12.22 mmol, 1.71 mL) in 23 mL of freshly distilled THF at 0 °C was added *n*-BuLi (11.51 mmol, 5.75 mL, 2.0 M in hexanes). The solution was stirred at 0 °C for 30 min before cooling to –78 °C and adding a solution of Me_3SiCl (10.8 equiv, 77.65 mmol, 9.85 mL) and pyridine (11.9 equiv, 85.56 mmol, 6.92 mL) in dry THF (17 mL). After 5 min a solution of (*E*)-1-(dimethylphenylsilyl)-1-buten-3-ol 2-hydroxyacetate (**12**) (2.0 g, 7.19 mmol, in 47.9 mL of dry THF) was added. The solution was stirred at –78 °C for 5 min before warming to 0 °C for 1 h. The solution was then diluted with 10% aqueous HCl (50 mL). The solution was extracted with EtOAc (2 × 50 mL) and dried ($MgSO_4$), and the solvent was removed *in vacuo* to afford a crude yellow oil. Purification on SiO_2 (100% petroleum ether-80% ethyl acetate in petroleum ether gradient elution) afforded **13** as a clear oil (1.30 g, 65%) in a *syn*:*anti* ratio of > 25:1 as determined by ^1H NMR analysis

The Ireland–Claisen rearrangement is not limited to allylic esters, but can also be applied to propargylic glycolates [14]. In the rearrangement of several propargylic esters **14** the diastereoselectivity (ds), the allenic α-hydroxyacids **15** are formed with, was always very high (Scheme 5.2.6, Table 5.2.3). In this reaction, LHMDS in THF is the best base/solvent combination, LDA gave no desired product, and the use of ether and dimethoxyethane resulted in a decrease of both yield and diastereoselectivity.

1) LHMDS
2) Me_3SiCl
3) H_3O^+

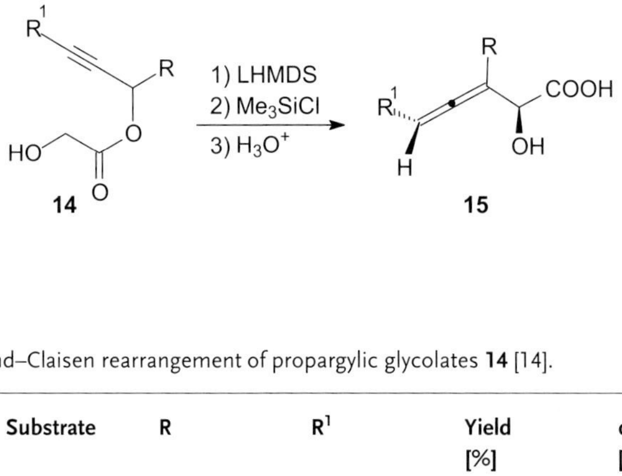

Scheme 5.2.6

Tab. 5.2.3 Ireland–Claisen rearrangement of propargylic glycolates **14** [14].

Entry	Substrate	R	R^1	Yield [%]	ds [%]
1	**14a**	H	Me	71	
2	**14b**	H	Me_3Si	47	
3	**14c**	Me	Me	72	92
4	**14d**	Me	Me_3Si	88	95
5	**14e**	Me	*n*-Bu	59	94
6	**14f**	Et	Me_3Si	69	94

5.2.2.2 Rearrangement of α-Alkoxy-Substituted Allylic Esters

Like α-hydroxy substituted allylic esters the corresponding α-alkoxy derivatives can also be rearranged either via the chelated metal enolate or by the Ireland version. The later protocol is by far the most applied rearrangement.

Ireland–Claisen Rearrangement of α-Alkoxy Substituted Allylic Esters

Better results with respect to the diastereoselectivity and yield in comparison to the glycolates were obtained in the Ireland–Claisen rearrangement of *O*-protected derivatives. This reaction was carefully investigated by Kallmerten [15] and Burke et al. [16]. Because of the fixed enolate geometry (by chelation) both stereoisomeric rearrangement products can be obtained, depending on the olefin geometry of the allylic ester used. The diastereomeric ratio depends on the *O*-protecting group and the reaction conditions used (Scheme 5.2.7, Table 5.2.4). For *O*-protected crotyl glycolates **16a-f** the product ratios range from 10.2:1 to 7.2:1 for the (*E*)- and from 23:1 to 11.4:1 for the (*Z*)-configured crotyl esters in the order $R = CH_3 > Bn > MEM$ [16]. The unprotected glycolates show no significant selectivity (see Table 5.2.2). In some cases higher selectivities are obtained if a solution of the base is added to the ester at –78 °C (inverse mode method B), but the influence of these "thermodynamic conditions" is not significant [15]. The counterion of the base also does not play an important role and KHMDS (method C) can be used instead of LDA or LHMDS (entries 12, 15). A notable increase in diastereoselectivity results from introduction of a α-substituent into the allyl system (entries 14–19) and in general only one stereoisomer is obtained.

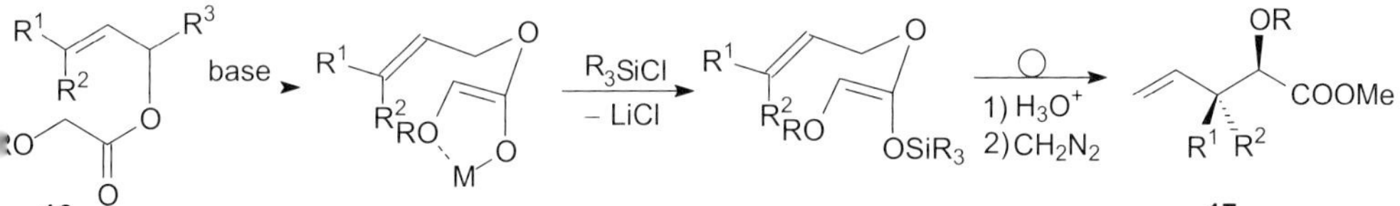

R^1 R^2 R^3 RO O O **16** — base → R^1 R^2 RO O M O — R_3SiCl / – LiCl → R^1 R^2 RO O $OSiR_3$ — 1) H_3O^+ 2) CH_2N_2 → OR COOMe R^1 R^2 **17**

Scheme 5.2.7

Method A: Synthesis of Methyl (*R,S*)-2-Methoxy-3-Methyl-4-Pentenoate (*syn*-17a)

To a solution of LDA (3.9 mmol) in THF (25 mL) at –100 °C the ester **16a** (0.50 g, 3.5 mmol) was added in THF (6 mL). After the reaction mixture had stirred for 1 h at –100 °C, there was added 1.0 mL of the supernatant from the centrifugation of a 1:1 mixture of Me_3SiCl and Et_3N. The mixture was allowed to stir an additional 1 h at –100 °C, was then allowed to warm to 25 °C, and was stirred for 12 h. The solution was then poured into 5% aqueous NaOH (75 mL) and stirred for 10 min, and the aqueous layer was washed with ether. The aqueous phase was acidified with concentrated HCl at 0 °C and was extracted repeatedly with CH_2Cl_2. The combined extracts were dried ($MgSO_4$) and concentrated. The crude carboxylic acid was dis-

Tab. 5.2.4 Ireland–Claisen rearrangements of α-alkoxy ester **16**.

Entry	Subst.	R	R[1]	R[2]	R[3]	Method	Yield [%]	Ratio *syn:anti*	Ref.
1	**16a**	Me	Me	H	H	A	65	10.2:1	16
2	**16b**	Bn	Me	H	H	A	77	9.6:1	16
3	**16c**	MEM	Me	H	H	A	70	7.2:1	16
4	**16d**	Me	H	Me	H	A	64	1:23	16
5	**16e**	Bn	H	Me	H	A	79	1:18.6	16
6	**16f**	MEM	H	Me	H	A	70	1:11.4	16
7	**16a**	Me	Me	H	H	B	95	8:1	15b
8	**16b**	Bn	Me	H	H	B	90	>40:1	15b
9	**16d**	Me	H	Me	H	B	60	1:30	15b
10	**16e**	Bn	H	Me	H	B	91	1:13	15b
11	**16g**	Me	Ph	H	H	B	36	>40:1	15b
12	**16g**	Me	Ph	H	H	C	64	>40:1	15b
13	**16h**	Bn	Ph	H	H	B	67	>100:1	15b
14	**16i**	Me	Me	H	Me	B	91	13:1	15b
15	**16i**	Me	Me	H	Me	C	72	>40:1	15b
16	**16k**	Bn	Me	H	Me	B	90	>40:1	15b
17	**16l**	Bn	H	Me	Me	B	44	1:100	15b
18	**16m**	Bn	Me	H	iBu	B	71	100:1	15b
19	**16n**	Bn	H	Me	iBu	B	81	1:100	15b

Method A: 1.1 equiv LDA, THF, –100 °C; **16**, 1 h; Me_3SiCl/NEt_3, –100 °C, 1 h; –100 °C → rt.
Method B: **16**, THF, –78 °C; LDA, 2–5 min; Me_3SiCl, –78 °C → rt, 1 h.
Method C: KHMDS instead of LDA.

solved in Et_2O (50 mL) at 0 °C and was esterified with an ethereal solution of diazomethane. Chromatography on silica gel (elution with 1:9 ether-hexanes) gave 330 mg (65%) of the rearrangement product *syn*-**17**, together with the diastereomeric ester *anti*-**17** as a minor product. The diastereomer ratio was found to be 10.2:1 by glass capillary GLC.

Method B: Methyl (2*R*,3*S*)-3-Methyl-2-(Phenylmethoxy)hex-4-Enoate (*syn*-17k)

To a solution of ester **16k** (1.10 g, 6.96 mmol) in THF (40 mL) at –78 °C was added LDA (17.6 mL, 0.5 M in THF, 8.80 mmol) over 30 s. The reaction was stirred 1.5 min and trimethylsilyl chloride (2.60 mL, 20.5 mmol) was added. The reaction was held at –78 °C for 10 min and then was allowed to warm to 25 °C for 2 h. The reaction was quenched by addition of saturated aqueous NH_4Cl (8 mL). The mixture was acidified with 5% HCl (8 mL) and extracted with ether (3 × 25 mL). The organic fractions were treated with diazomethane, dried over $MgSO_4$, and chromatographed on silica gel (30:1 hexane/ether) to yield a 40:1 mixture of *syn*-**17k** and *anti*-**17k** as a colorless oil (0.98 g, 89%).

Bartlett et al. investigated the rearrangement of *O*-protected crotyl lactates (Scheme 5.2.8) and found that these require much longer enolate generation times and higher temperature for rearrangement [8]. The best conditions found for the rearrangement of the benzyl ether **18a**, for example, involve the addition of the ester to a solution of LICA in THF at –78 °C, followed 30 min later by Me_3SiCl and slow warming to reflux. Subsequent workup provided the *syn*-configured product **19a** as the major diastereomer. Under the same reaction conditions the secondary allylic ester **18b** gave the rearrangement product **19b** as a 2:1 diastereomeric mixture (configuration not determined) in comparable yield [8].

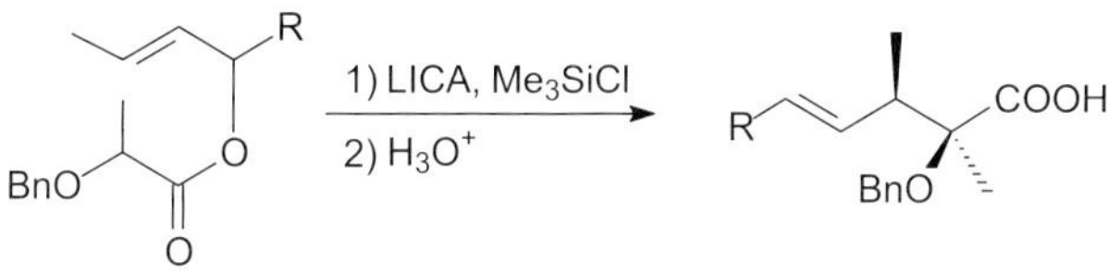

1) LICA, Me_3SiCl
2) H_3O^+

18 a R = H
b R = Et

19 a 67%, 75% ds
b 69%, 67% ds

Scheme 5.2.8

If chiral allylic alcohols are used, the chirality can easily be transferred to the α- and β-position of the α-alkoxy acid [17]. Kallmerten et al. also described another possibility for the asymmetric Ireland–Claisen rearrangement. They used substrates with chiral α-alkoxy substituents such as **20** and investigated the influence of this chiral center on the stereochemical outcome of the rearrangement (Scheme 5.2.9) [18]. In contrast to the well-established chirality transfer within the cyclic framework of the Claisen rearrangement (chiral allylic esters) this approach with the chirality "outside of the pericyclic area" is largely unexplored. But this approach offers unique synthetic advantages, since the stereodirecting center is retained in the rearrangement product. In contrast, the directing stereogenic center within the pericyclic framework is destroyed during the rearrangement step. A critical requirement for good asymmetric induction is a favored stereodifferentiating transition state conformation, in which the external auxiliary directs the facial selectivity of addition to the enolate π-system. Out of four possible stereoisomeric rearrangement products two major products are formed, both with *syn* configuration, and therefore the product ratio **21**/**22** reflects the degree of asymmetric

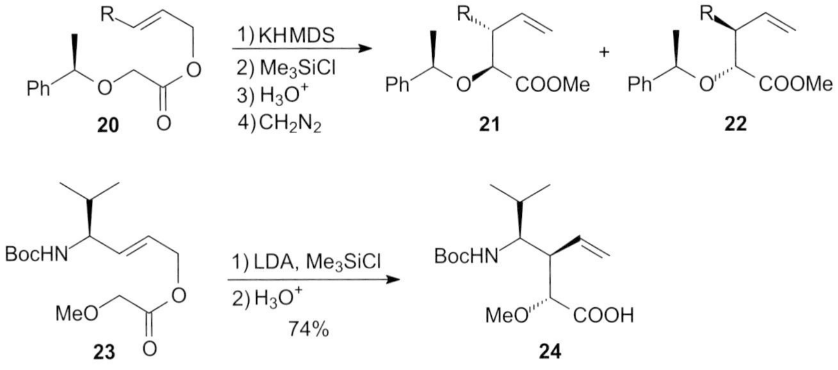

Scheme 5.2.9

induction attributable to the external auxiliary. The influence of the substitution pattern on the stereochemical outcome of the reaction is shown in Table 5.2.5. In general, the diastereoselectivities observed were moderate (entries 1 and 3), but in the case of the cinnamyl ester **20b** (entry 2) the selectivity was synthetically useful. In contrast, a (*Z*)-configured crotyl ester gave a 1:1 diastereomeric mixture.

Hauske et al. subjected ester **23** with the chirality center also outside of the pericyclic framework to the Ireland–Claisen rearrangement and observed a very strong influence of the stereogenic center. **24** was obtained as a single stereoisomer in good yield [19].

Tab. 5.2.5 Auxiliary-controlled Ireland–Claisen rearrangements of esters **20**.

Entry	Substrate	R	Yield [%]	Ratio 21:22
1	20a	Me	82	3.0:1
2	20b	Ph	77	6.1:1
3	20c	*i*Pr	60	2.4:1

The Ireland–Claisen rearrangement of silylated substrates was described by Panek et al. (Scheme 5.2.10) [20]. With (*E*)-vinylsilanes such as **25** the expected *syn*-configured silylated product **26** was obtained, while the corresponding (*Z*)-silane **27** gave rise to the opposite diastereomer **28**.

Similar results are obtained with fluorine-substituted allylic esters as reported by Kitazume et al. [21]. Ireland–Claisen rearrangement of the difluoromethylated esters **29** and **31** provided the desired *syn*- or *anti*-configured unsaturated acids, which were directly reduced to the corresponding alcohols in overall high yield.

Ritter investigated the Claisen rearrangement of stannylated allylic esters [22]. Both the (*Z*)- as well as the (*E*)-vinylstannanes (**33**) gave the expected *anti*- or *syn*-rearrangement product **34** with excellent diastereoselectivity.

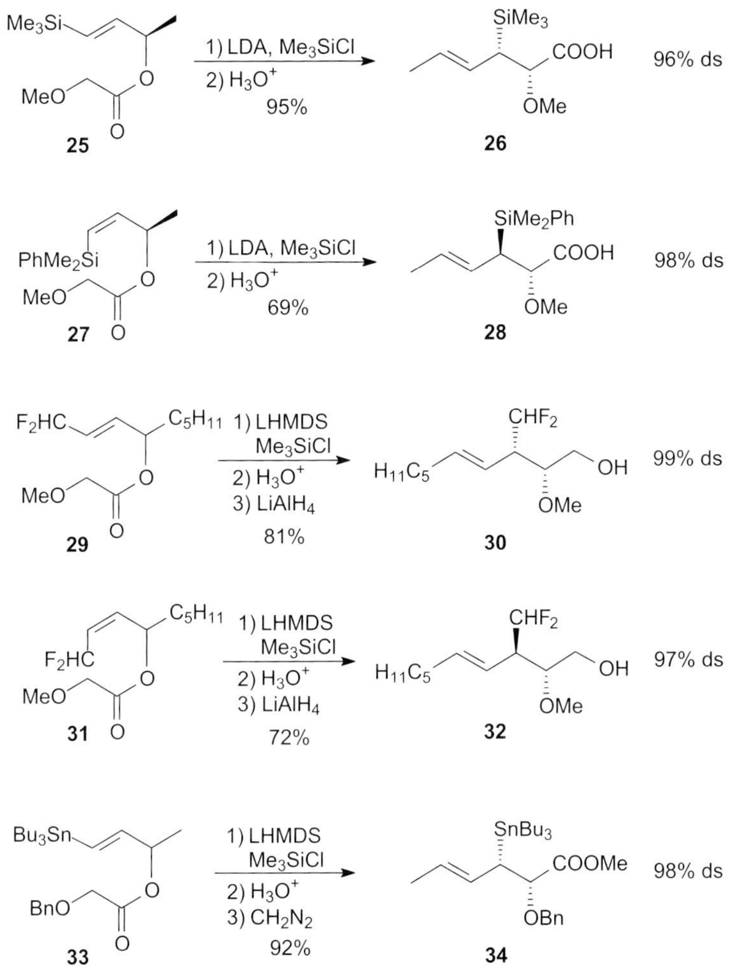

Scheme 5.2.10

Applications in Natural Product Synthesis

Meanwhile, the Ireland–Claisen rearrangement of protected α-hydroxy allylic esters found various applications in natural product syntheses, what clearly underlines the potential of this synthetic protocol. Several examples are described by Kallmerten et al. For example, rearrangement of (*Z*)-configured ester **35** gave rise to the required *anti* rearrangement product **36**, which was converted in five steps into the dilactone **37**, a precursor of the antifungal metabolite avenaciolide (Scheme 5.2.11) [23].

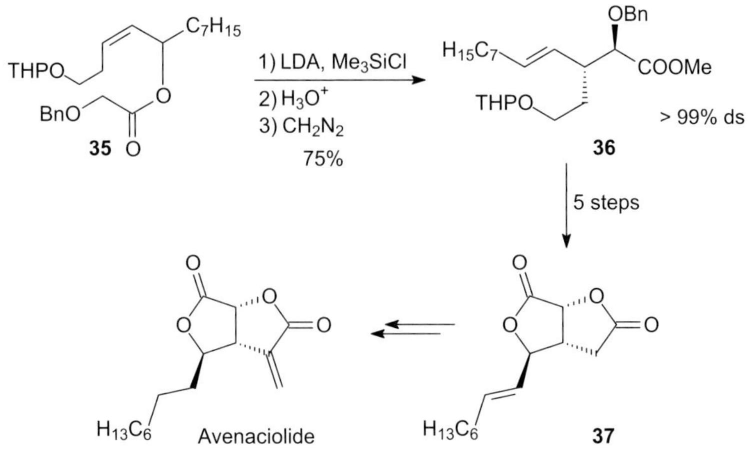

Scheme 5.2.11

An iterative Claisen rearrangement starting from allylic ester **38** was used in the synthesis of the tocopherol side chain **42** and pine sawfly pheromones (Scheme 5.2.12) [24]. Ireland–Claisen rearrangement of (*Z*)-allylic ester **38** and *in situ* reduction of the silyl ester afforded alcohol **39** as a single product. Completion

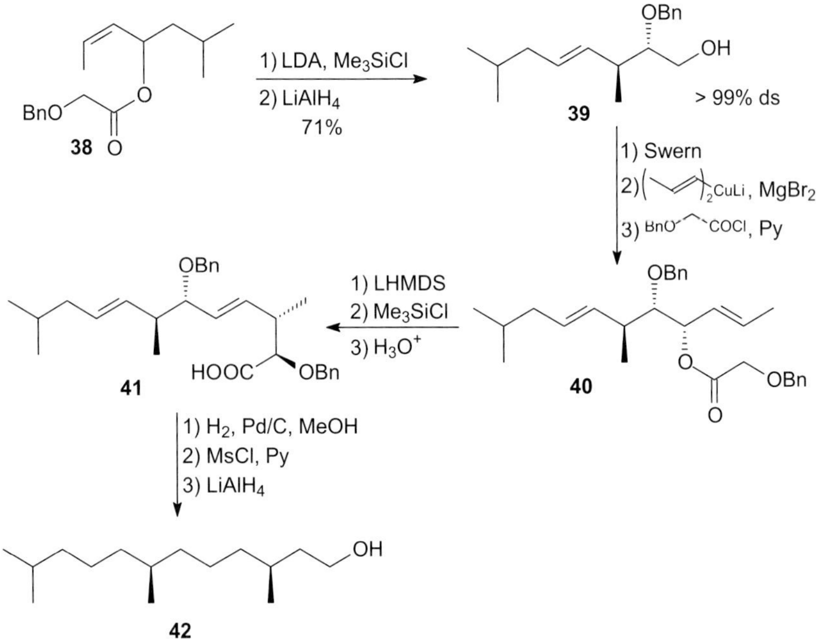

Scheme 5.2.12

of the first iterative cycle was accomplished by Swern oxidation of **39** to the aldehyde and chelation-directed addition of (*E*)-propenyl cuprate. Subsequent acylation gave rise to allylic ester **40**, the required substrate for the second sigmatropic event. Again no second diastereomer could be determined, and the tocopherol side chain **42** was obtained after reductive removal of the functionalities.

Nakai et al. reported on an Ireland–Claisen rearrangement in steroid syntheses [25]. The stereogenic information of the C-16–OH group in **43** was nicely transferred into the steroidal side chain giving the *syn* rearrangement product **44** as a single stereoisomer. The "opposite" *anti* product could be obtained from the same steroidal alcohol via a [2,3]-Wittig rearrangement.

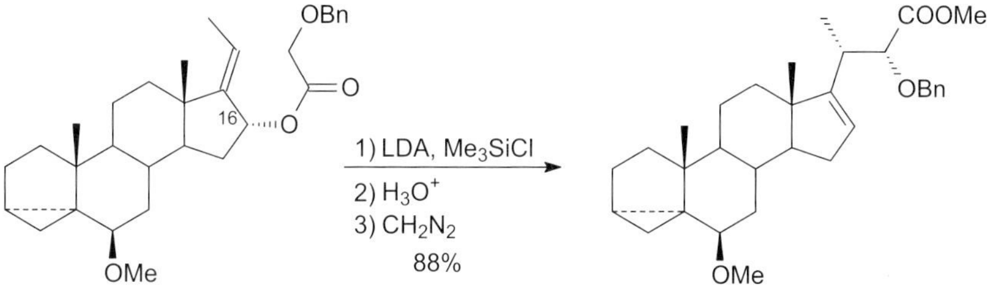

Scheme 5.2.13

In the total synthesis of pseudomonic acid C, Barrish et al. used the rearrangement of ester **45** as a key step [26]. With the (*Z*)-configured ester **45** the *anti*-product was obtained nearly exclusively (> 95% ds), while the corresponding (*E*)-ester showed a much lower diastereoselectivity.

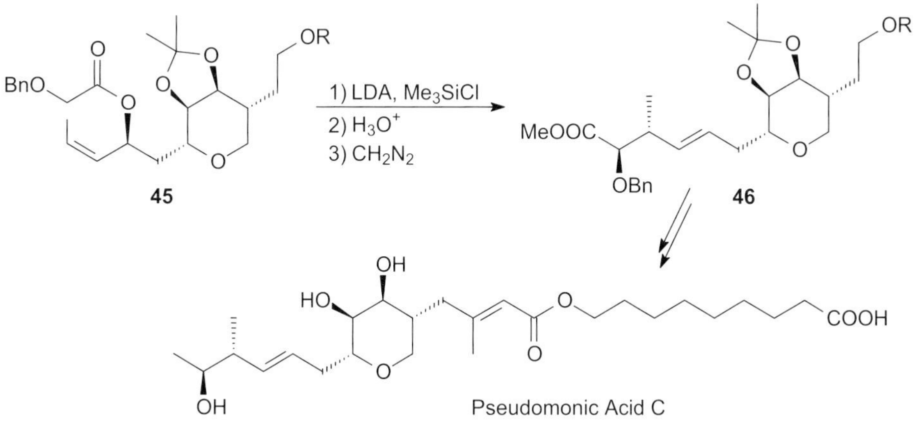

Scheme 5.2.14

Mulzer et al. used the rearrangement of **47** as a central step in their synthesis of an asteltoxin key intermediate (Scheme 5.2.15) [27]. The rearrangement product **48** was obtained as a single stereoisomer. The stereoselectivity remarkably hinged

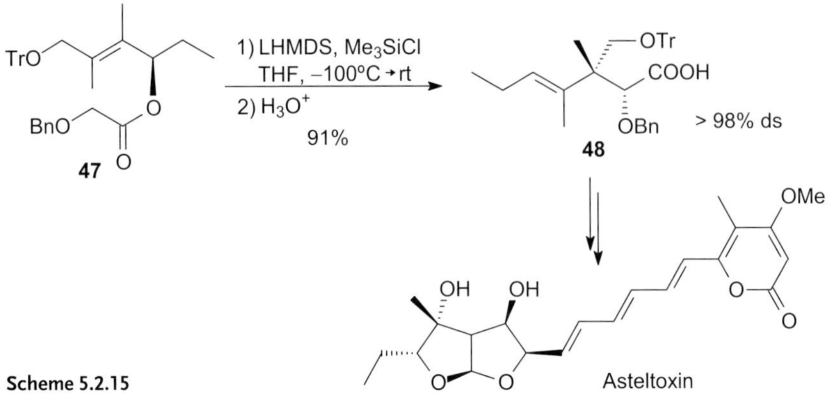

TrO, BnO, O, O, 47; 1) LHMDS, Me_3SiCl, THF, −100°C → rt; 2) H_3O^+; 91%; OTr, COOH, OBn, 48, > 98% ds; OH, OH, O, O, OMe, O, O, Asteltoxin

Scheme 5.2.15

on the trityl protecting group. With TBDPS instead of trityl both yield (83%) and stereocontrol (86:14 mixture of epimers) were much lower.

For their first step in the synthesis of the taxane skeleton **51**, Shea et al. used the glycolate Ireland–Claisen rearrangement of ester **49** (Scheme 5.2.16) [28]. The silylketene acetal, when warmed up to room temperature, gave, after hydrolysis and esterification, the desired α-benzyloxyester **50** as a 3:1 stereoisomeric mixture. The lower stereoselectivity in the case of this cyclic allylic ester is not surprising, because cyclic esters show a higher tendency to rearrange via a boat-like transition state, which gives rise to the "opposite" diastereomer [29].

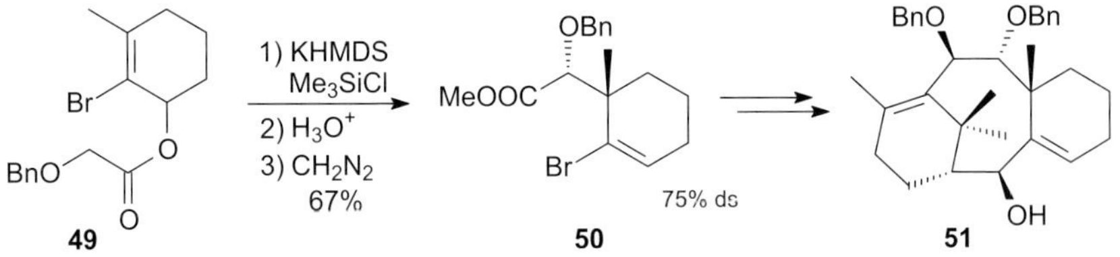

Br, BnO, O, O, 49; 1) KHMDS, Me_3SiCl; 2) H_3O^+; 3) CH_2N_2; 67%; OBn, MeOOC, Br, 50, 75% ds; BnO, OBn, OH, 51

Scheme 5.2.16

Kim et al. used the chelation-controlled Ireland–Claisen rearrangement twice in their synthesis of brefeldin A (Scheme 5.2.17) [30]. This synthesis nicely illustrates the power of [3,3]-sigmatropic rearrangements in natural product synthesis. In one of the first steps, glycolate ester **52** was subjected to the Ireland–Claisen rearrangement to furnish the γ,δ-unsaturated glycolate **53** in a highly stereoselective manner. After removal of the superfluous double bond by a diimide reduction, the resulting ester was converted to the more highly elaborated unsaturated glycolate **54** by a reiterative three-step sequence in overall good yield and stereoselectivity. The second Ireland–Claisen rearrangement provided **55**. Deprotection of the PMB group with DDQ generated the secondary allylic alcohol, which underwent a smooth Johnson orthoester Claisen rearrangement to give the corresponding diester **53**, which was converted into brefeldin A.

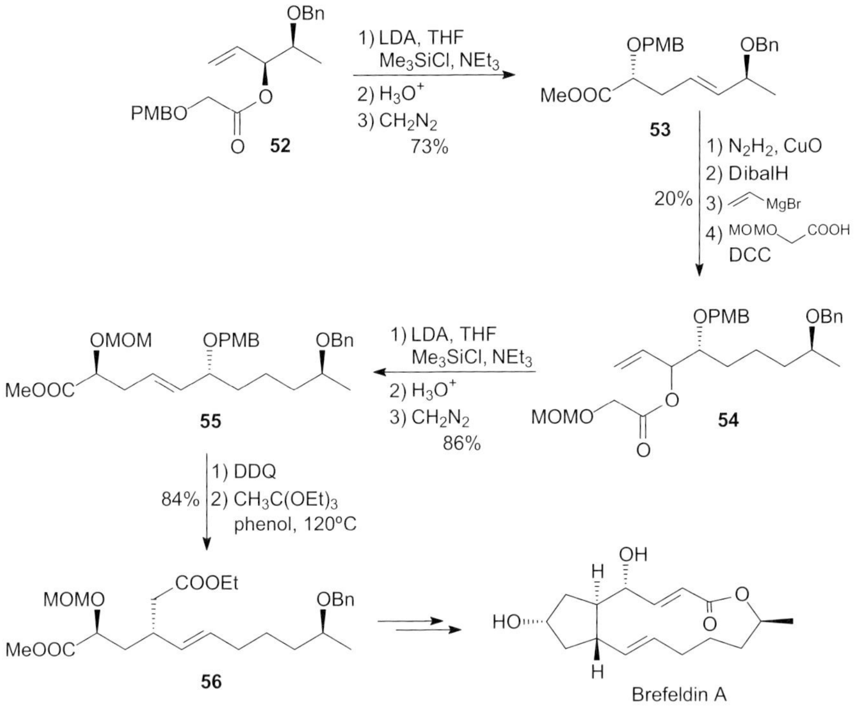

Scheme 5.2.17

Several applications of the chelation-controlled Ireland–Claisen rearrangement in natural product syntheses were reported by Burke et al. The syntheses of a series of bicyclic bilactone natural products such as ethisolide and related structures started with an epimeric mixture of β-hydroxy-α-methylene lactone **57** (Scheme 5.2.18) [31].

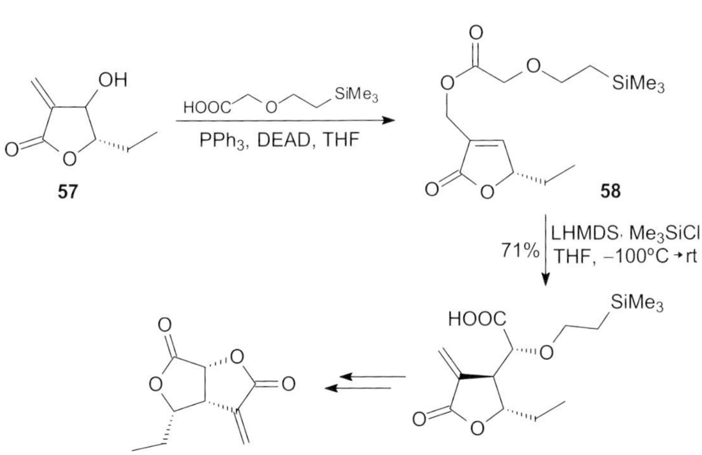

Scheme 5.2.18

Esterification under Mitsunobu conditions [32] gave rise to the required ester **58** as a sole product. Clean S_N2'-displacement was observed, facilitated by the unhindered nature of the olefin-terminus and the Michael acceptor properties of the *α*-methylene lactone **57**. Deprotonation of the *α*-alkoxy ester **58** at –100 °C in the presence of Me_3SiCl and warming to room temperature afforded only one diastereomer **59**, which was further converted into ethisolide.

Towards the syntheses of hydropyranes, an important structural unit found in many natural ionophores and polyether macrolides, a domino glycolate Claisen rearrangement/ring-closing metathesis was used as a key sequence (Scheme 5.2.19) [33]. As a simple example in a racemic series, the diallylated substrate **60** was subjected to standard Ireland–Claisen conditions giving rise to racemic **61** after esterification. Exposure of **61** to Grubbs catalyst [34] gave the required dihydropyrane derivative **62** in 73% yield. Further examples using this protocol are summarized in Table 5.2.6.

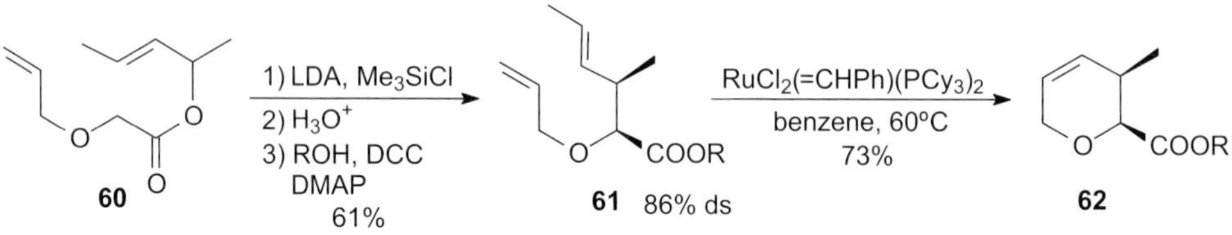

1) LDA, Me_3SiCl
2) H_3O^+
3) ROH, DCC DMAP
61%

60 → **61** 86% ds (COOR)

$RuCl_2(=CHPh)(PCy_3)_2$
benzene, 60°C
73%

→ **62** (COOR)

Scheme 5.2.19

Control of the relative configuration at the stereogenic centers is exerted by the choice of (*Z*)- or (*E*)-allylic alcohols (entries 2 and 3), and the absolute configuration at those centers is transferred from the stereogenic center in the allylic substrate as discussed before. The yields and stereoselectivities obtained with the higher oxygenated products (entries 2–4) in general were worth in comparison to the simplier systems (entry 1). A substantial amount of C-silylated byproducts were obtained under standard conditions. Fortunately, adding 2 equiv of LDA to a solution of the allylic ester and Me_3SiCl in 20% HMPA/THF at –100 °C for 1 h and then warming up to room temperature [35] afforded moderate to high stereoselectivities and better rearrangement yields. The simpler substrates (**60** and **63**) generally underwent a ring-closing metathesis under milder conditions than the substrates of entries 2–4. The latter have allylic branding at both alkene termini, which is known to retard the metathesis reaction [36]. Reaction times of several days were required for these substrates to provide moderate to good yields. The dihydropyrane derivatives **68** and **71** possess the correct stereochemistry of the ionophore antibiotics zincophorin [37] and indanomycin [38].

The same authors used the chelation-controlled glycolate Claisen rearrangement also for the synthesis of the C(12)–C(20) subunit of rhizoxin (Scheme 5.2.20) [39].

Tab. 5.2.6 Domino glycolate Claisen rearrangement/ring-closing metathesis.

Entry	Substrate	Rearrangement product	Yield [%]	ds [%]	Metathesis	Yield [%]
1	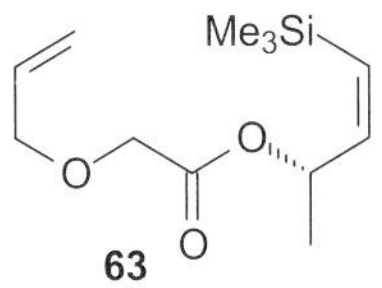 63	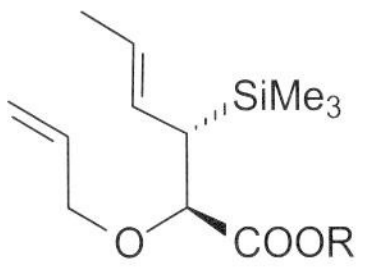 64	82	93	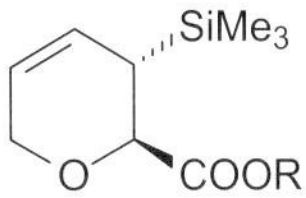 65	68
2	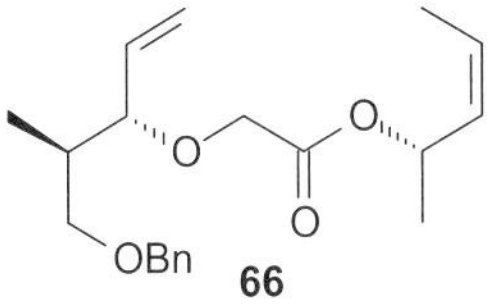 66	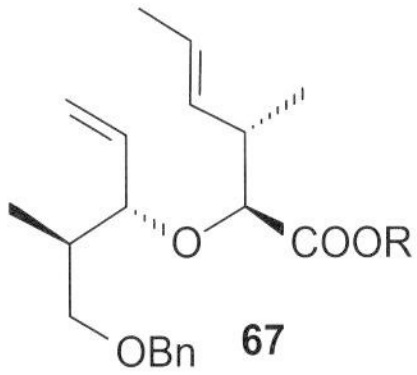 67	56	83	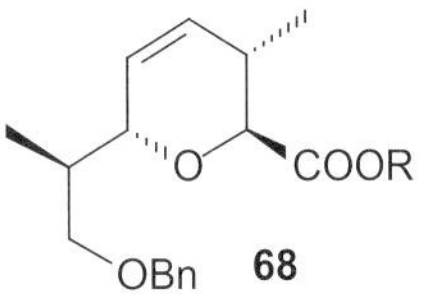 68	48
3	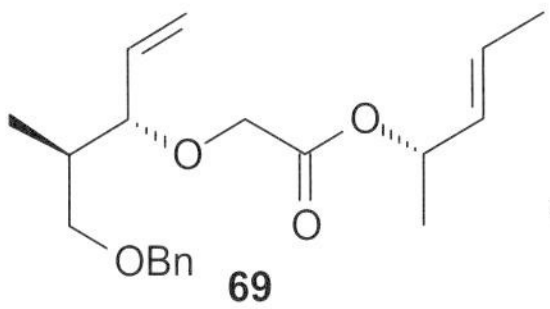 69	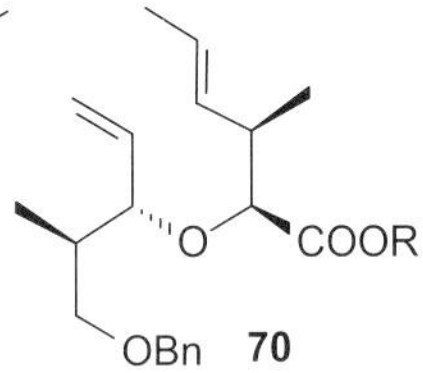 70	52	86	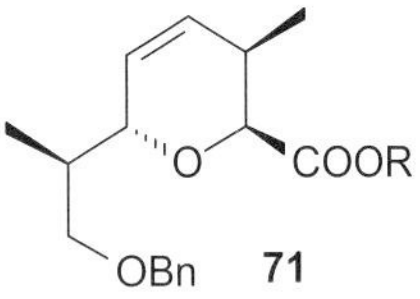 71	51
4	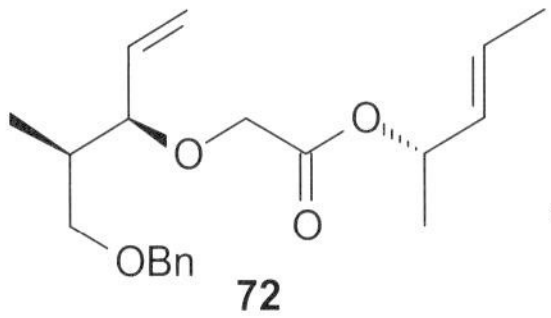 72	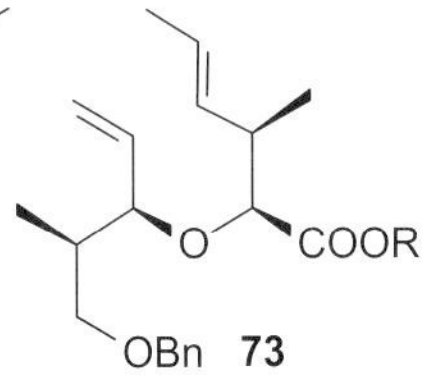 73	47	> 95	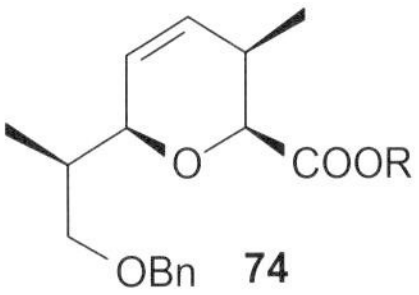 74	85

Starting from diastereomerically pure glycolate **75** the required rearrangement product **76** was obtained with excellent diastereoselectivity. Subsequent bromolactonization under kinetic conditions gave a nearly 1:1 mixture of bromolactones **77** and **78**, which could be separated by flash chromatography. The "wrong bromolactone" **78** was converted into **76** with zinc dust, and **77** was subjected to radical dehalogenation giving rise to lactone **79**, a key building block of rhizoxin. The configuration at C(15) was interconverted under Mitsunobu conditions with *p*-nitrobenzoic acid [40] during the further synthesis.

Langlais et al. also used the glycolate Claisen rearrangement in combination with a ring closing metathesis in their studies towards the synthesis of fumagillin [41], an important inhibitor of angiogenesis (Scheme 5.2.21) [42]. The key step of this synthesis was the rearrangement of ester **80**, which was intensively studied and optimized. The reaction conditions and results are summarized in Table 5.2.7. Interestingly, the unexpected product **82**, obtained via a nonchelated (*E*)-enolate intermediate, was obtained under standard conditions. Therefore, four parame-

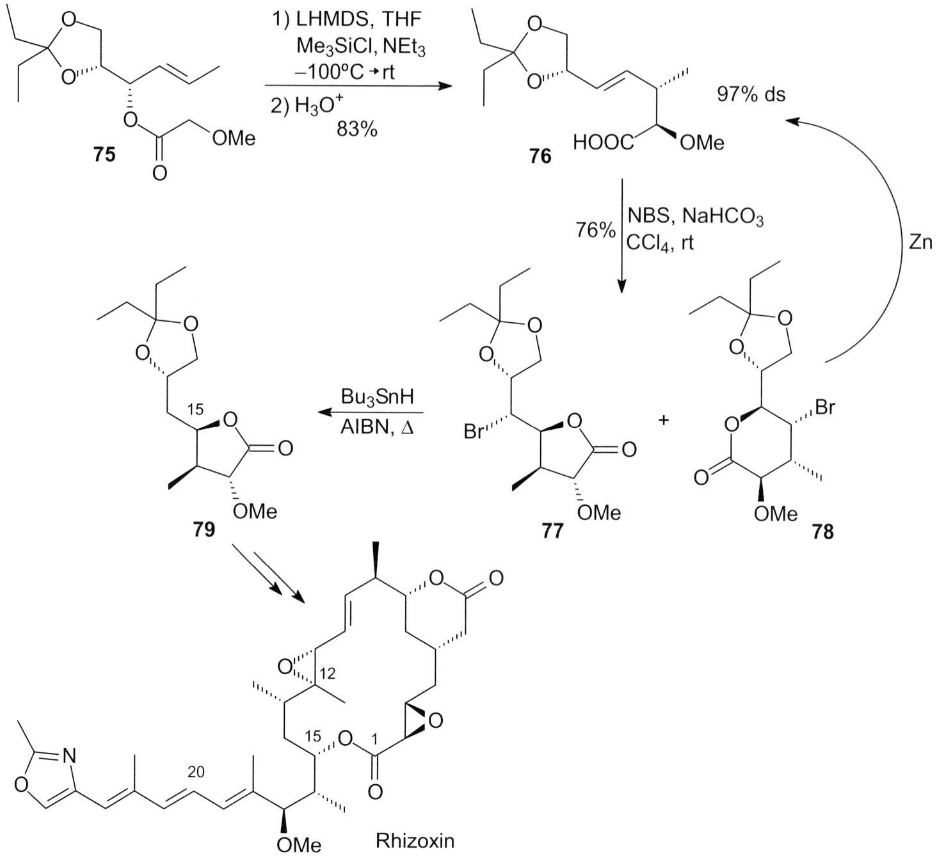
1) LHMDS, THF
Me3SiCl, NEt3
–100°C → rt
2) H3O+
83%
75
OMe
76
HOOC
OMe
97% ds
NBS, NaHCO3
CCl4, rt
76%
Zn
Bu3SnH
AIBN, Δ
15
OMe
79
Br
77
78
12
15
1
20
Rhizoxin

Scheme 5.2.20

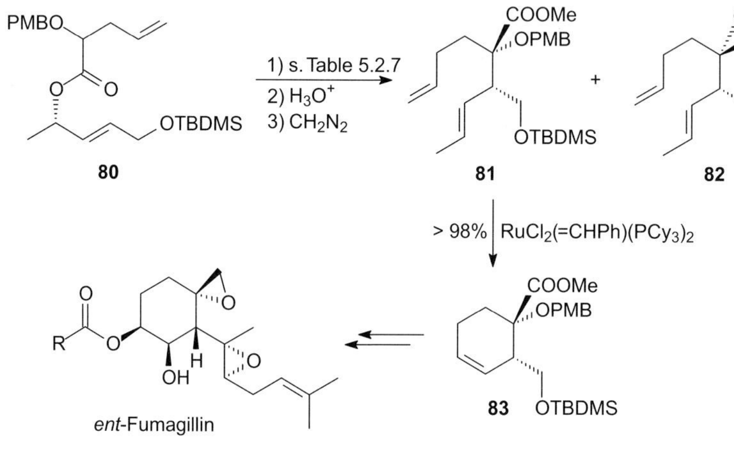
PMBO
OTBDMS
80
1) s. Table 5.2.7
2) H3O+
3) CH2N2
COOMe
OPMB
OTBDMS
81
82
> 98%
RuCl2(=CHPh)(PCy3)2
83
ent-Fumagillin

Scheme 5.2.21

Tab. 5.2.7 Claisen rearrangement of ester **80**.

Entry	Solvent	Base	*T* [°C]	*t*[a)] [min]	Ratio 81/82	Yield[b)] [%]
1	THF	LDA	–78	30	25/75	50
2	THF	LDA	–78	10	27/73	52
3	THF	LDA	–78	90	27/73	46
4	THF	LDA	–90	30	50/50	53
5	Ether	LDA	–78	30	61/39	53
6	Toluene	LDA	–78	30	83/17	42
7	THF/HMPA	LDA	–78	30	70/30	60
8	Toluene	KHMDS	–78	30	>95/5	90

a) Time of deprotonation before introducing Me_3SiCl.
b) Yields not optimized except for entry 8.

ters were successively submitted to variation. The time between deprotonation and Me_3SiCl quenching did not show any influence on the diastereoselectivity of the reaction (entries 1–3). However, when the deprotonation was performed at lower temperature, –90 °C vs. –78 °C, the stereoselectivity was significantly lower (entries 1 and 4). For the same base, LDA, the diastereoselectivity depends on the nature of the solvent. Better results and reversal of the diastereoselectivity were obtained in less polar solvents such as ether or toluene (entries 5 and 6). Curiously, the same reversal of selectivity was observed using the polar solvent mixture THF/HMPA (entry 7). Finally, a dramatic improvement in diastereoselectivity was observed with potassium bis(trimethylsilyl)amide in toluene (entry 8).

Under these conditions the ester **81** was obtained in 90% yield with recovery of the starting material **80** in 10% yield. Obviously the less polar solvent toluene favors the formation of a chelated enolate, and the higher selectivity obtained with the bulkier potassium counter ion is explained by an additional chelation with the oxygen of the silyl ether, which can reinforce the stability of the chelated transition state **80′** (Figure 5.2.2). Metathesis with the Grubbs catalyst [34] afforded in nearly quantitative yield the anticipated cyclohexene derivate **83**, which was reduced and dihydroxylated to provide the key structures of fumagillin.

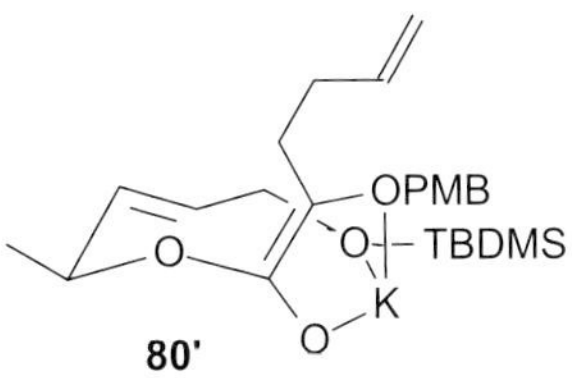

OPMB
O–TBDMS
O
K
O
80'

Fig. 5.2.2 Rearrangement via chelated enolate.

Rizzacasa et al. utilized the Ireland–Claisen rearrangement in the presence of a γ-leaving group for the synthesis of 2,2-disubstituted furanosid natural products such as sphydrofurane [43], cinatrin B [44], and the core of zaragozic acid [45]. This approach involves enolization and silylation of an allylic ester **I**, derived from a furanesiduronic acid, and subsequent Claisen rearrangement to give tetrahydrofurane derivative **II** (Scheme 5.2.22). Critical to the success of this approach is the mode of addition in which a solution of the ester is added to a solution of the base/silylating mixture (Me_3SiCl/NEt_3) in THF at –110 °C in the presence of HMPA as cosolvent. Under these conditions, β-elimination does not occur and the rearrangement can proceed via the less hindered face opposite to the β-OR-group in high yield.

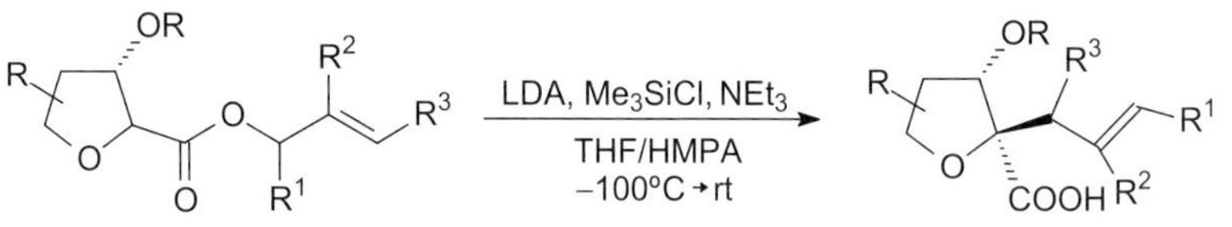

LDA, Me_3SiCl, NEt_3; THF/HMPA; –100°C → rt

I → **II**

Scheme 5.2.22

The route to cinatrin B began with the synthesis of the enantiopure acid **84**, which was obtained from arabinose (Scheme 5.2.23) [44]. Esterification with racemic alcohol **85** afforded the rearrangement substrate **86**. Addition of ester **86** to a solution of LDA and the centrifugate of a 1:1 mixture of Me_3SiCl/NEt_3 [33, 46] in THF/HMPA at –100 °C followed by warming to room temperature afforded, after hydrolysis and esterification, esters **87** and **88** in excellent yield but as a 43:57 mixture. The low stereoselectivity was attributed to the fact that facial selectivity of the rearrangement is mainly governed by the allyl ester stereocenter, which overrides the influence of the β-stereocenter of the tetrahydrofuran ring observed previously. With the assumption that a (*Z*)-O-silyl ketene acetal is formed preferentially under chelation control, the (*R*)-epimer should provide ester **87**, while the ester **88** should be obtained in the rearrangement of the (*S*)-configured allylic ester. Indeed, the selectivity **88** was formed (73% ds) was better with the optically pure (*S*)-ester, but the selectivity for this matched case was still moderate. It appears that the ratio of 27:73 (**87**/**88**) corresponds to the original *E*/*Z*-enolate ratio obtained since the (*E*)-silylketene acetal would give **87** via a chair-like transition state. This is in close agreement with the observation that enolization of ethyl 2-tetrahydrofuroate with LDA in THF/HMPA followed by silylation gives a 33:67 ratio of (*Z*)- and (*E*)-silylketene acetals, respectively [47].

As expected, exposure of the corresponding diastereomer (*R*)-**86** to the rearrangement conditions gave ester **87** as the major product but with lower selectivity (mismatched case). This suggests that the β-stereocenter in the tetrahydrofurane ring may have some effect on the stereochemical outcome of the reaction. Rearrangement product **88** was further converted into the spirolactone cinatrin B.

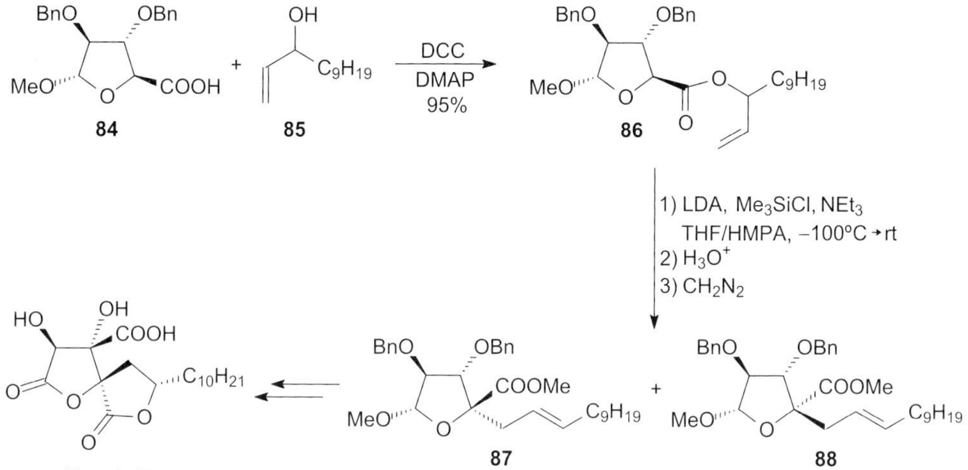

Scheme 5.2.23

Very recently, Hong et al. [48] used the chelation-controlled Ireland–Claisen rearrangement of **89** in their synthesis of carbovir analogues **91**. Herein, again the ring was closed via ring-closing metathesis.

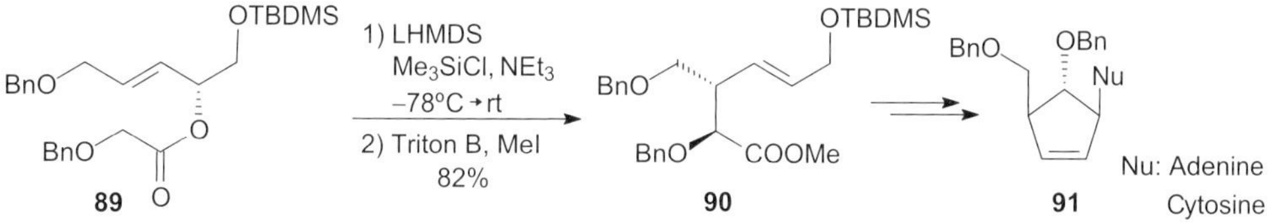

Scheme 5.2.24

As illustrated by these examples, the chelation-controlled Claisen rearrangement in combination with subsequent modifications of the resulting double bond, e.g., by metathesis or by halolactonization is a powerful synthetic tool. On the other hand, the resulting carboxylic acid can be also subjected to further transformations.

Whitesell and Helbling [49] reported on a conversion of allylic alcohols into β,γ-unsaturated esters based on a Claisen rearrangement/oxidative degradation (Scheme 5.2.25). Thus, Claisen rearrangement of ester **92** as the lithium enolate or as the silylenolether afforded α-methoxycarboxylate **93** as a diastereomeric mixture. Degradation by one carbon was effected by the sequence of Wasserman [50] wherein the dianion **94** formed, by subsequent deprotonation with additional LDA, was oxidized with molecular oxygen.

The resulting α-methoxy-α-hydroperoxy carboxylic acid **95** formed upon acidification spontaneously lost CO_2 and water to form the desired ester **96**. This sequence was applied to a wide range of substrates.

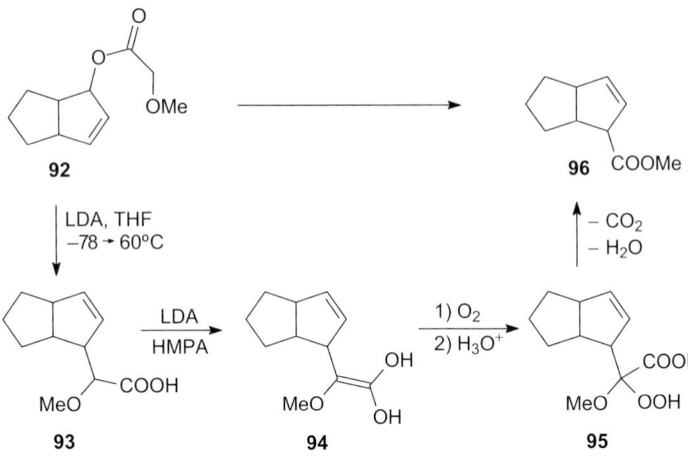

Scheme 5.2.25

An electrochemical oxidative decarboxylation in combination with an ester enolate Claisen rearrangement was reported by Wuts et al. (Scheme 5.2.26) [51]. A variety of allylic esters such as **97** was subjected to an Ireland–Claisen rearrangement, and the resulting acids (**98**) obtained were submitted to electrolytic decarboxylation in a divided cell to afford ketals **99**. The use of the divided cell was necessary to suppress side reactions such as alkene reduction.

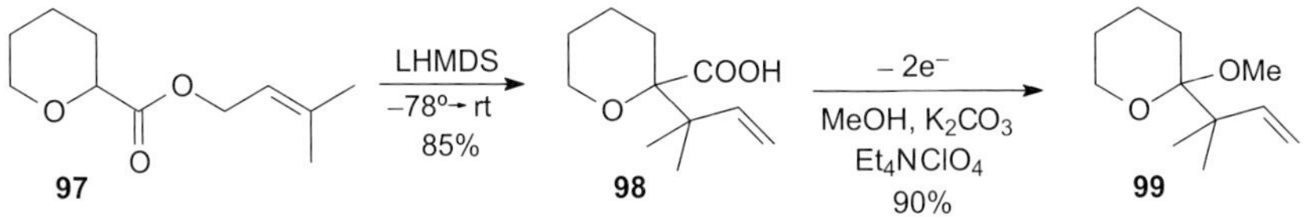

Scheme 5.2.26

Chelate Claisen Rearrangements of Metal Enolates of α-Alkoxy Allyl Esters

Although the Ireland version of the Claisen rearrangement is by far the most used application for the rearrangement of α-alkoxy-substituted esters, several other metals can be used for chelation. Oh et al. investigated the influence of the chelating metal on the rearrangement with respect to an asymmetric Claisen rearrangement [52]. The chelated enolates were obtained from the esters **100** and

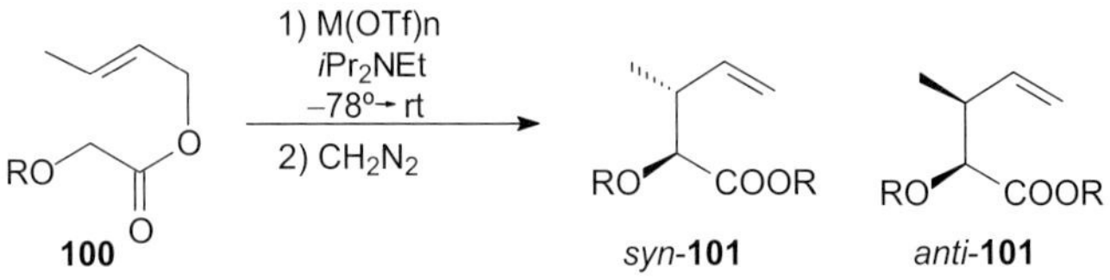

Scheme 5.2.27

Tab. 5.2.8 Chelate Claisen rearrangements of metal enolates.

Entry	Substrate	Metal salt	Ratio 101a:101b	Yield [%]
1	**100a**	*n*-Bu_2BOTf	86:14	32
2		*n*-Bu_2BOTf	77:23	48[a)]
3		$(IPC)_2BOTf_2$	92:8	60
4		$Sn(OTf)_2$	82:18	49
5		$Zn(OTf)_2$	–	traces
6	**100b**	*n*-Bu_2BOTf	63:37	34[a)]
7		$(IPC)_2BOTf$	91:9	38
8	**100c**	*n*-Bu_2BOTf	3:97	32
9		$(IPC)_2BOTf$	0.5:99.5	51
10	**100d**	$(IPC)_2BOTf$	< 0.5:> 99.5	64
11	**100e**	$(IPC)_2BOTf$	94: 6	59
12	**100f**	$(IPC)_2BOTf$	9:91	55
13	**100g**	$(IPC)_2BOTf$	47:53	29

a) Reaction was carried out in hexane.

the corresponding metal triflates (MOTf) in the presence of Hünig's base (Scheme 5.2.27), and the results are summarized in Table 5.2.8.

The rearrangement of boron [53] and tin enolates [54] gave acceptable yields and generally good vicinal diastereoselectivities (entries 1, 3 and 4). The enolate generated with diisopinocamphenylboron triflate ($(IPC)_2BOTf$) gave higher ratios than with *n*-Bu_2BOTf (entries 3, 7, 9 vs. 2, 6, 8). Zinc triflate, on the other hand, yielded only traces of product (entry 5).

In most cases, the *cis*-olefins gave better selectivities than the corresponding *trans*-olefines (**100c,d** vs. **100a,b**). Normal carboxylic acid esters such as **100g** without the possibility for chelation gave significantly worse results (entry 13). The yield and diastereoselectivity of the rearrangement was generally solvent-insensitive (CH_2Cl_2, Et_2O, $C_2H_4Cl_2$, toluene) except when hexanes were used. In hexanes, the rearrangement gave a slightly higher yield but with lower diastereoselectivity (entries 1 and 2). The product ratios of rearrangements involving $Sn(OTf)_2$ and n-Bu_2BOTf are comparable to rearrangements via silyl ketene acetals, in which the enolate geometry is controlled in a similar manner. However, the rearrangements involving $(IPC)_2BOTf$ gave higher product ratios. With these high diastereoselectivities obtained it is surprising that the enantioselectivities of these reactions were only in the range of 0–10% ee.

5.2.2.3 α-Amido Substituents

If allylic esters of *N*-protected amino acids are subjected to Claisen rearrangements, γ,δ-unsaturated amino acids are obtained. The first synthesis of allylic amino acids by Claisen rearrangement was described in 1975 by Steglich et al. [55]. Treatment of *N*-benzoyl amino acid allylic esters **102** with dehydrating agents such as phosgene results in the formation of allyloxazolinone **104** via Claisen rearrangement of the primarily formed oxazole intermediate **103** (Scheme 5.2.28). The reaction is especially suitable for the synthesis of α-alkylated allylic amino acids ($R^1 \neq H$) because, in this case, epimerization of the α-chiral center via enolization of the oxazolone **104** is not possible [56].

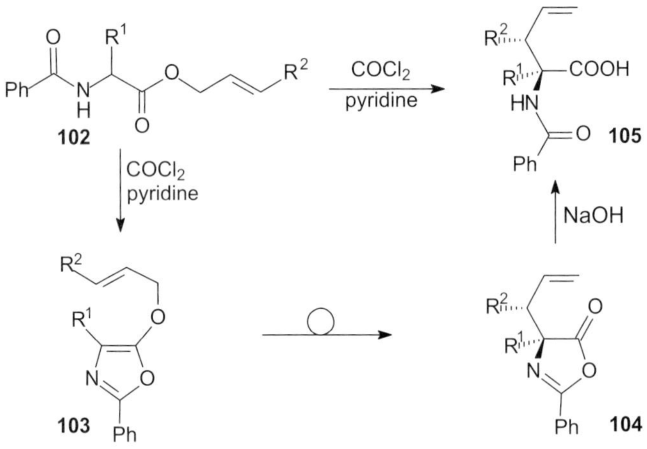

Scheme 5.2.28

As a result of the fixed olefin geometry in the oxazole ring, and a strong preference for the chair-like transition state, these α-alkylated amino acids **105** are formed in a highly diastereoselective fashion. Unfortunately, this elegant methodology is limited to *N*-benzoyl amino acid esters (or related aromatic or heteroaromatic *N*-acyl derivatives), while other common protecting groups like carbamates (Z, BOC, FMOC, etc.) do not allow the cyclization to the corresponding oxazole. For these derivatives another rearrangement process has to be applied.

Ireland–Claisen Rearrangement of α-Amido Substituted Allylic Esters

The Ireland–Claisen rearrangement [3a] of different *N*-protected glycine allylic esters was intensively studied by Bartlett et al. in 1982 [57]. This method allows great flexibility in view of protecting group variation and can be applied to various amino acids. However, the diastereoselectivity (ds) of the rearrangement strongly depends on various reaction parameters, such as solvent or the base used, the *N*-protecting group and the substitution pattern of the allylic ester moiety.

The influence of the reaction conditions were investigated using crotyl esters **106** of protected amino acids (R = CH_3) (Scheme 5.2.29) [57]. The results are summarized in Table 5.2.9. In general, the *syn* products **107** were obtained preferentially in all cases, but the relative ratio varied dramatically.

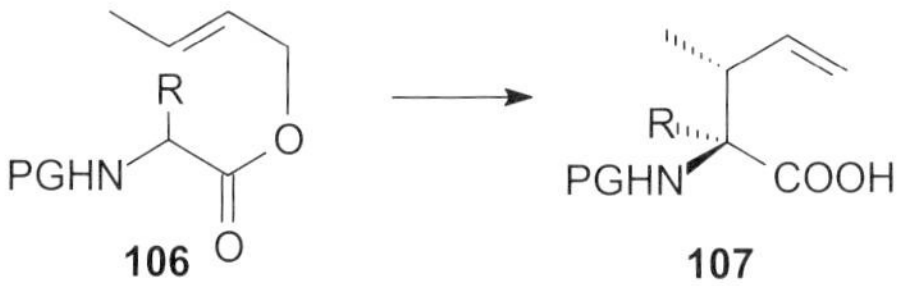

R
O
PGHN
O
106
R
PGHN
COOH
107

Scheme 5.2.29

Tab. 5.2.9 Influence of reaction conditions and protecting groups on the Ireland–Claisen rearrangement of crotyl esters **106**.

Entry	Substrate	Reaction conditions	PG	R	Yield [%]	ds [%]
1	**106a**	Standard cond.[a)]	Boc	H	60–65	90
2	**106a**	Ether as solvent	Boc	H	45	91
3	**106a**	20% HMPT/THF	Boc	H	51	80
4	**106a**	KDA as base	Boc	H	0	
5	**106a**	+1.1 equiv $MgCl_2$	Boc	H	42	91
6	**106b**	Standard cond.	Z	H	65	80
7	**106c**	Standard cond.	Bz	H	65	84
8	**106d**	Standard cond.	TFA	H	58	60
9	**106e**	Standard cond.	Boc	Me	59	75
10	**106f**	Standard cond.	Bz	Me	71	80

a) 2.1 equiv LDA, THF, –75 °C, 10 min; Me_3SiCl, –75 °C → 65 °C, 1 h.

The reaction conditions were optimized carefully with respect to the yield and diastereoselectivity. The “standard conditions” proved to be the best by both criteria. Deprotonation of the ester **106a** with LDA (2.1 equiv) in THF at –75 °C, silylation with Me_3SiCl after 10 min, and warming to reflux for 1 h provided the rearrangement product **107a** as a 9:1 diastereomeric mixture in 60–65% yield. Shorter (2.5 min) or longer (40 min) enolate generation times had no significant influence on yield or selectivity. Nor does the use of *tert*-butyldimethylsilyl chloride as silylating agent offer any advantages. Reaction in less polar solvents such as ether (entry 2) resulted in lower yields, but the selectivity was not affected. In contrast, the more highly dissociating system with hexamethylphosphoric triamide (HMPT) as a cosolvent (entry 3) showed both lower yield and selectivity. Attempts with counterions other than lithium met with mixed success: While with potassium diisopropylamide (KDA, entry 4) only cleavage of the allylic ester was observed, the corresponding magnesium salt (entry 5) gave the same selectivity as the lithium salt, albeit with reduced yield. With respect to the protecting groups used (entries 1, 6–8), all groups investigated gave the expected rearrangement product in comparable yield, but the best diastereoselectivity was obtained with the Boc-protected derivative **106a** (entry 1). Besides glycine esters, also the corresponding alanine derivatives **106e** and **106f** could be used in this rearrangement with comparable success, but with sterically more demanding amino acids, such as valine, no rearrangement product could be obtained. In addition to variations of the *N*-protecting groups the influence of the substitution pattern in the allyl moiety was investigated (Scheme 5.2.30, Table 5.2.10). For the Boc-protected crotyl **106a** and cinnamyl esters of glycine (**106h**) the expected dependence of the stereochemical outcome of the rearrangement on the olefin geometry was observed, although a significant reduction in selectivity was seen for the *cis*-cinnamyl substrate **106i** (entry 4) [3c]. This can be explained by steric interactions in the chair-like transition state, which results in its destabilization and favoring of a boat-like transition state. This effect is becoming stronger with increasing sterical demand of the *cis*-oriented substituent as illustrated with the *cis*-crotyl and -cinnamyl esters (entries 2 and 4).

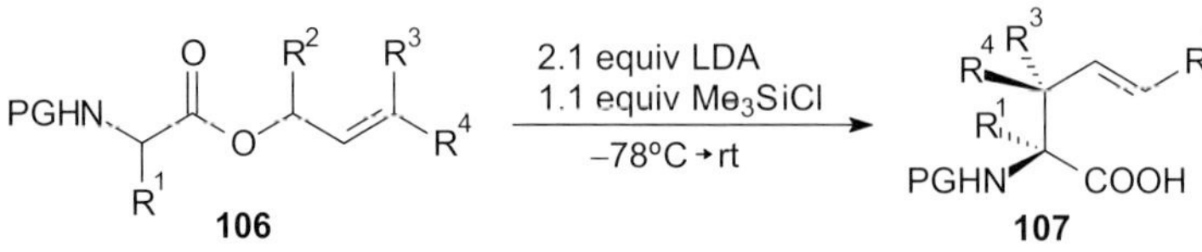

Scheme 5.2.30

The Ireland–Claisen rearrangement could also be applied to the synthesis of amino acids containing quaternary centers. Rearrangement of the dimethylated ester **106k** gave rise to amino acid **107k** in high yields. That the quaternary center can also be obtained in a highly stereoselective fashion was illustrated with the rearrangement of geranyl esters **106l** and **106m** of alanine (entries 6 and 7). Herein two quaternary stereogenic centers are generated in one step while only one stereoisomer could be detected. Similar results are obtained in the oxazole rearrangement as reported by Steglich [55].

Tab. 5.2.10 Influence of the substitution patterns.

Entry	Subst.	PG	R^1	R^2	R^3	R^4	Yield [%]	ds [%]
1	**106a**	Boc	H	H	H	Me	60–65	90
2	**106g**	Boc	H	H	Me	H	60	85
3	**106h**	Bz	H	H	H	Ph	68	>95
4	**106i**	Bz	H	H	Ph	H	50	70
5	**106k**	Boc	H	H	Me	Me	77	–
6	**106l**	Bz	Me	H	Me	C_6H_{11}	80	>95
7	**106m**	Boc	Me	H	Me	C_6H_{11}	56	>95

General Procedure for Ester–Enolate Claisen Rearrangements: 3-Methyl-2-[[(1,1-Dimethylethoxy)carbonyl]amino]-4-Pentenoic Acid (107a)
To a solution of diisopropylamine (0.38 mL, 1.98 mmol) in dry THF (6.5 mL) at 0 °C was added *n*-butyllithium (0.80 mL, 2.4 M in hexane, 1.93 mmol). After 10 min, the solution was cooled to –75 °C, and ester **106a** (211 mg, 0.92 mmol) in THF (1 mL) was added over ca. 40 s. After stirring for 10 min, trimethylsilyl chloride (0.25 mL, 1.93 mmol) was added, and the solution was stirred for 5 min before being allowed to warm to 21 °C over a 15-min period. The mixture was then kept at 55–60 °C for 1 h, cooled, and diluted with methanol (5 mL) to hydrolyze the silyl esters (5 min). The solution was diluted with ether and extracted with 2 N NaOH (four 3-mL portions), the combined aqueous layer was then acidified and extracted with $CHCl_3$ (three 3-mL portions). The organic layer was dried ($MgSO_4$), filtered, and evaporated at reduced pressure to provide rearranged acid **107a** (137 mg, 65% yield) as a 9:1 mixture of diastereomers.

During their synthesis of streptolydigin Boeckman et al. used the Ireland–Claisen rearrangement of chiral ester **108** as a key step (Scheme 5.2.31) [58]. Rearrangement under "standard conditions" proceeded smoothly to provide the protected *syn*-amino acid **109** in 79% yield and 89% dr. with complete chirality transfer. **109** was further converted into aminosuccinimide **110**, a synthetic intermediate towards streptolydigin.

Baldwin et al. applied the Ireland–Claisen rearrangement for the synthesis of deuterated allylglycines **111** and **113** (Scheme 5.2.32) [59]. Although the yields obtained were moderate, high diastereoselectivities were observed.

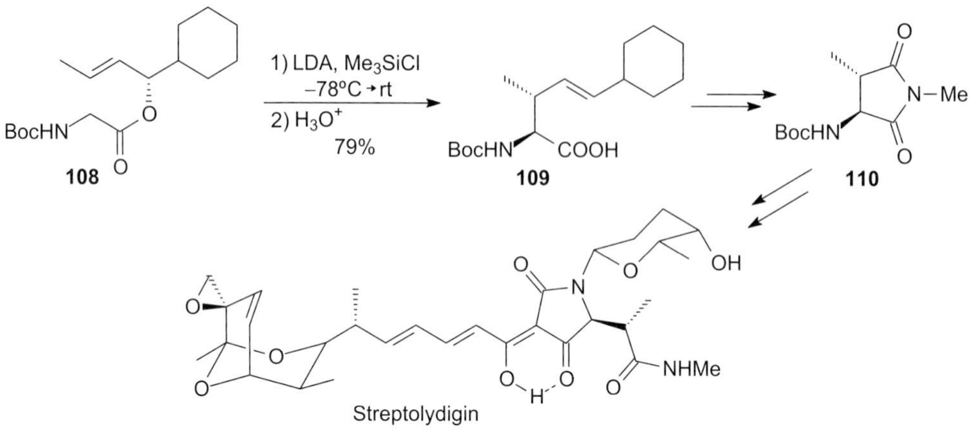

1) LDA, Me_3SiCl
−78°C → rt
2) H_3O^+
79%

BocHN 108 → BocHN COOH 109 → BocHN N−Me 110 → Streptolydigin

Scheme 5.2.31

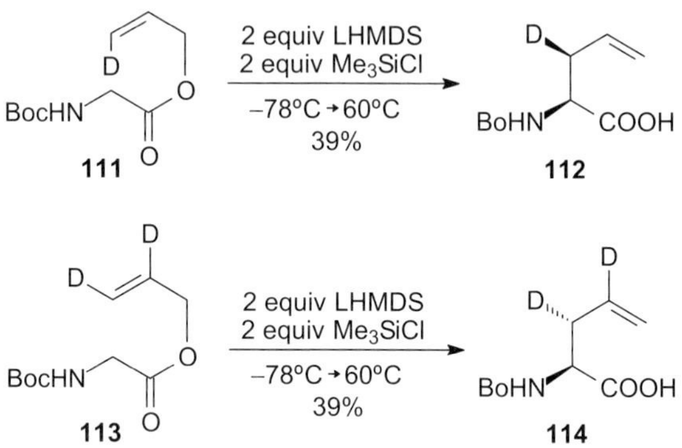

2 equiv LHMDS
2 equiv Me_3SiCl
−78°C → 60°C
39%

BocHN 111 → BoHN COOH 112

2 equiv LHMDS
2 equiv Me_3SiCl
−78°C → 60°C
39%

BocHN 113 → BoHN COOH 114

Scheme 5.2.32

Towards the synthesis of fluorinated amino acids, Konno et al. investigated allylic esters such as **115** (Scheme 5.2.33) [60]. In view of the findings of Bartlett et al. [57] that the Ireland–Claisen rearrangement of non-fluorinated allyl esters proceeds stereoselectively, it was very surprising, that the products **116a** and **116b** were obtained as a diasteromeric mixture in a ratio of 62:38 (Table 5.2.11, entry 1).

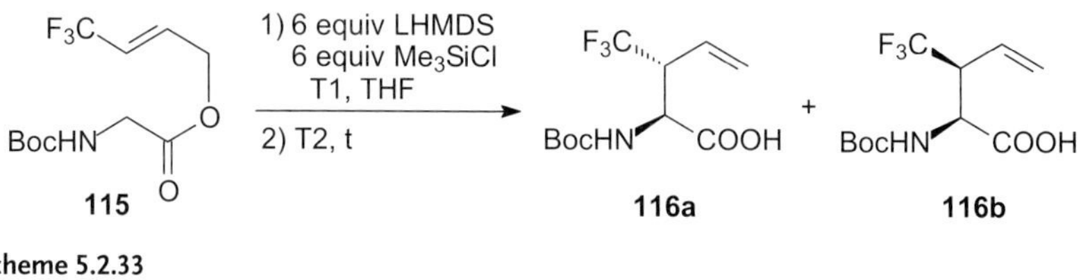

1) 6 equiv LHMDS
6 equiv Me_3SiCl
T1, THF
2) T2, t

F_3C BocHN 115 → F_3C BocHN COOH 116a + F_3C BocHN COOH 116b

Scheme 5.2.33

Tab. 5.2.11 Rearrangement of fluorinated allylic ester **115**.

Entry	MX_n	*T1* [°C]	*T2* [°C]	*t* [h]	Yield [%]	Ratio 116a:116b
1	Me_3SiCl	–78	rt	6	48	62:38
2	Me_3SiCl	–78	rt	20	45	54:46
3	Me_3SiCl	–78	reflux	6	94	73:27
4	–	–78	rt	20	49	100: 0
5	Me_3SiCl	0	reflux	6	84	91:9
6	$ZnCl_2$	0	reflux	6	82	100:0

Thus, the reaction conditions were examined in detail. These results are summarized in Table 5.2.11. As shown in entry 2, prolonging the reaction time (*T2*) for the [3,3]-sigmatropic rearrangement did not lead to a dramatic change in either the yield or stereoselectivity. When the mixture was stirred for 6 h under reflux, the yield could be increased to 94% (entry 3), but the stereoselectivity was still moderate. Excellent diastereoselectivity (100% ds) was observed in the reaction without Me_3SiCl (entry 4), albeit in moderate yield. It was observed that generation of the enolate and its trapping with Me_3SiCl at 0 °C improved the yield and diastereoselectivity of the reaction (entry 5). Even better results are obtained after addition of $ZnCl_2$ (entry 6) and the chelate Claisen rearrangement, according to Kazmaier et al. [6].

Brook et al. investigated the Ireland–Claisen rearrangement of silyl substituted allylic esters **117** [61] with the intention to use the silylated amino acids obtained for further modifications via well-documented allylsilane chemistry (Scheme 5.2.34) [62]. In general, high diastereoselectivities and yields were obtained under standard conditions (Table 5.2.12, entry 1), but a strong influence of the reaction conditions was observed. Yield, as well as selectivity, varied depending on the silylchloride used (entries 2–5). The rearrangement product was directly converted into the corresponding methyl ester. Attempts to improve the reaction by changing the ratio base/ester in the enolization step proved to be difficult. Increasing the quantity of base led to severely diminished yields. In the case of **117a** (R = H),

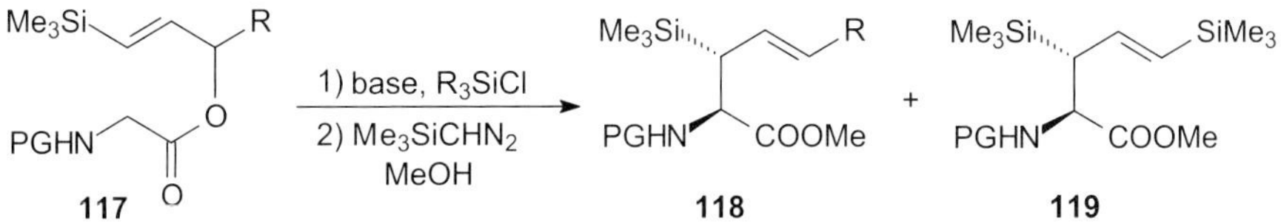

Me_3Si R O PGHN O **117**
1) base, R_3SiCl
2) Me_3SiCHN_2, MeOH
Me_3Si R PGHN COOMe **118** + Me_3Si $SiMe_3$ PGHN COOMe **119**

Scheme 5.2.34

Tab. 5.2.12 Ireland–Claisen rearrangement of silylated allyl esters **117**.

Entry	Substrate	R	PG	Reaction conditions	Yield [%]	ds [%]
1	**117a**	H	Boc	2,5 equiv LDA, NEt_3 3 equiv Me_3SiCl	85	97
2	**117a**	H	Boc	LDA, tBuMe$_2$SiCl	55	83
3	**117a**	H	Boc	LDA, Ph_2SiCl_2	42	75
4	**117a**	H	Boc	LDA, Cl_3SiH	40	83
5	**117a**	H	Boc	LDA, $PhSiCl_3$	34	75
6	**117a**	H	Boc	LHMDS, Me_3SiCl	79	95
7	**117a**	H	Boc	NEt_3, tBuMe$_2$SiCl	28	72
8	**117b**	CH_3	Boc	LDA, Me_3SiCl	92	78
9	**117b**	CH_3	Boc	KHMDS, Me_3SiCl	65	98
10	**117c**	CH_3	Z	LDA, Me_3SiCl	80	91
11	**117d**	CH_3	Bz	LDA, Me_3SiCl	71	89

excess base led to the formation of compound **119a** in 20% yield, which results from double deprotonation and silylation of **117** followed by rearrangement. In general, this unusual side reaction is not observed in Ireland–Claisen rearrangements. Additional base (3.5 equiv) led not only to **119**, but also to decomposition of the enolate intermediate via β-elimination. Elimination of the allyl ether moiety has been reported to compete in certain cases with the rearrangement [63]. But with 2.5 equiv LDA or LHMDS these side reactions could be suppressed. Under these conditions the rearrangement took place at –20 °C, and the yields and selectivities were moderate to excellent. The effect of different silyl groups was studied intensively. It was expected that greater steric bulk on the silane should lead to an enhanced stereoselectivity and to fewer degrees of freedom in the transition state of the Claisen rearrangement. But in all cases examined, the reaction of Me_3SiCl-substituted **117** gave rise to highest selectivities and yields. The *N*-protecting group also had a significant impact on the stereoselectivity. Boc-protected esters performed much better than Bz- or Z-protected esters (entries 10 and 11). When more than 3 equiv of base were employed in the enolization step, deprotection of the Boc group was observed at room temperature.

Regarding the yield, the Ireland–Claisen rearrangement of the silylated allylic ester **117a** under standard conditions was superior to the chelate Claisen rearrangement discussed in the next chapter.

Ireland–Claisen Rearrangement of Cyclic Allylic Esters

α-Cycloalkenyl amino acids are interesting bacterial growth inhibitors, and the Ireland–Claisen rearrangement has obvious potential for their stereoselective synthesis (Scheme 5.2.35) [64]. As indicated in Table 5.2.13 for the glycine derivatives **120a–d**, the *N*-Boc protecting group (entry 2) offers obvious advantages over the *N*-benzoyl group (entry 1) in terms of yield and stereoselectivity. The stereochemical outcome of the reaction can be explained by a preference for the boat-like transition state. In contrast to the glycine series, the *N*-benzoyl protecting group is necessary for the rearrangement of the corresponding alaninates (entry 5). No rearrangement was observed in this case with the corresponding Boc-derivatives (entry 6).

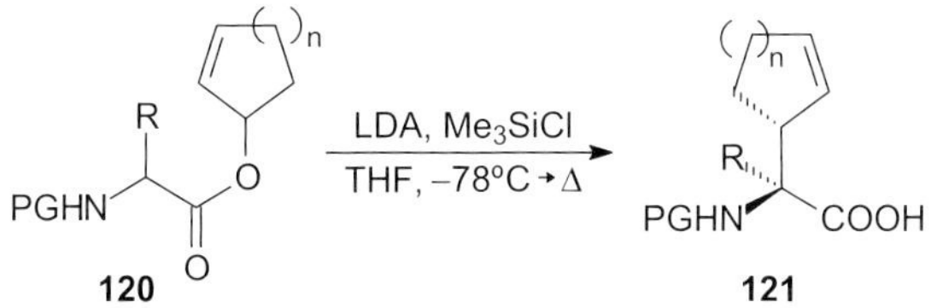

Scheme 5.2.35

Tab. 5.2.13 Ireland–Claisen rearrangement of cyclic allylic esters **120**.

Entry	Substrate	PG	R	n	Yield [%]	ds [%]
1	120a	Bz	H	1	45	50
2	120b	Boc	H	1	73	75
3	120c	Bz	H	2	22	60
4	120d	Boc	H	2	40	95
5	120e	Bz	Me	2	64	66
6	120f	Boc	Me	2	0	–

Chelate Claisen Rearrangements of *α*-Amido Substituted Allylic Esters

Like *α*-alkoxy substituted allylic esters, the *α*-amido substituted analogues are also able to form chelated ester enolates such as **122** (Fig. 5.2.3).

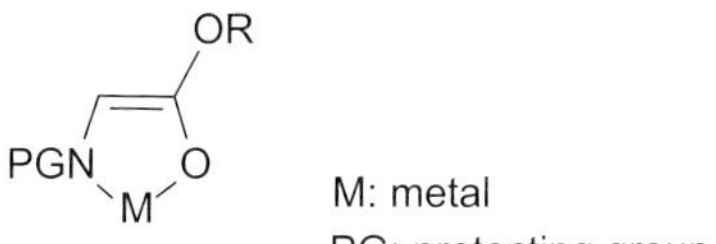

Fig. 5.2.3 Chelated amino acid ester enolate.

These chelated enolates have several advantages in comparison to their non-chelated analogues [65]:

The chelated enolates are significantly more stable than the corresponding, probably non-chelated, lithium enolates. In contrast to these, the chelate enolates can be warmed to room temperature without decomposition. Side reactions such as ketene formation via elimination, etc., can be suppressed in most cases.

Because of the fixation of the enolate geometry by chelation, many reactions of these enolates proceed with a high degree of diastereoselectivity.

Many manipulations on these enolates are possible. Besides variations of the protecting group PG, an excessive metal tuning should allow the modification of the reactivity and selectivity of these enolates. Because, in most cases, the coordination sphere of the metal ion M is not saturated in the bidentate enolate complex, the additional coordination of external ligands on the chelated metal is possible. Ligand tuning with chiral ligands should probably allow enolate reactions to be carried out not only in a diastereoselective, but also in an enantioselective way. Claisen rearrangements of such chelated ester enolates were intensively studied by Kazmaier et al. [6].

Rearrangement of Acyclic Allylic Esters

Deprotonation of *N*-protected amino acid allylic esters such as **123** with LDA at –78 °C, and subsequent addition of a metal salt (MX_n), presumably results in the formation of a chelated metal enolate **123′**, which undergoes Claisen rearrangement upon warming to room temperature, giving rise to unsaturated amino acid **124** (Scheme 5.2.36) [66].

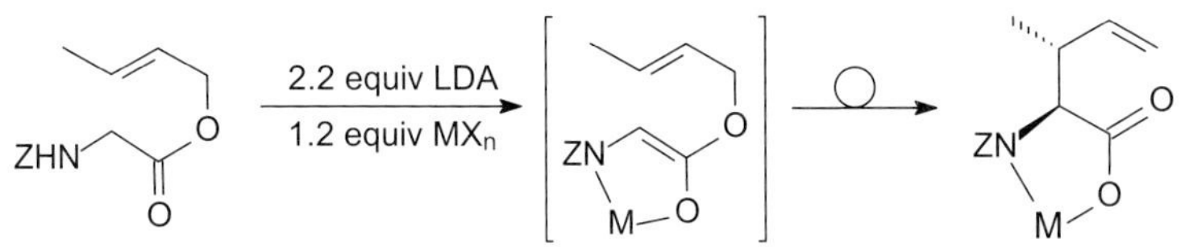

2.2 equiv LDA
1.2 equiv MX_n

ZHN, O, O, **123**; ZN, M–O, O, **123'**; ZN, M, O, O, **124**

Scheme 5.2.36

In contrast to the corresponding lithium enolates, which do not show this rearrangement because they decompose during the warming, the chelate-enolates are much more stable. Many different metal salts can be used for chelation, and in general, the yields (75–90%) and diastereoselectivities (90–95% ds) are high (Table 5.2.14). Normally the best results are obtained with zinc chloride (entry 1), but magnesium chloride (entry 3) gave comparable results in many cases. Otherwise, the metal enolates are clearly superior to silylketene acetals, both in terms of their reactivity and selectivity (entries 6 and 7). The driving force behind the accelerated rearrangement of the chelate enolates is probably the transformation of the high-energy ester enolate **123′** into a chelate bridged, stabilized carboxylate **124**.

Tab. 5.2.14 Rearrangement of *N*-benzyloxycarbonylglycine crotyl ester **123**.

Entry	MX_n	Yield [%]	ds [%]
1	$ZnCl_2$	90	95
2	$CoCl_2$	78	93
3	$MgCl_2$	85	91
4	$Al(OiPr)_3$	75	90
5	$Ti(OiPr)_4$	50	90
6	Me_2SiCl_2	50	85
7	Me_3SiCl	60	83

The formation of the *syn* product **124** can be explained by a preferential rearrangement via the chair-like transition state **A**, avoiding the sterical interactions between the pseudoaxial hydrogen and the chelate complex in the boat-like transition state **B** (Fig. 5.2.4).

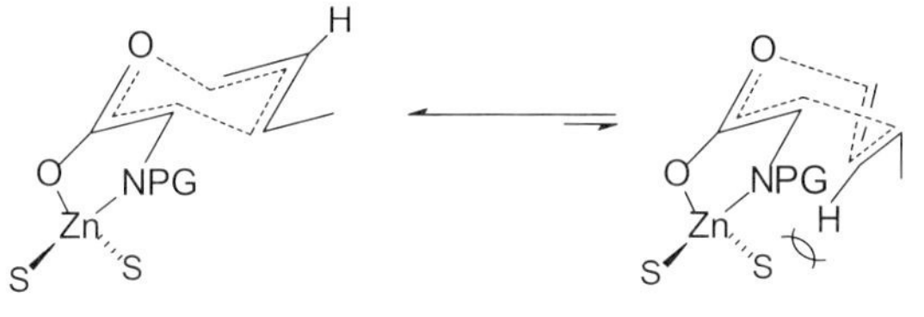

Fig. 5.2.4 Transition states for chelate Claisen rearrangement.

The yields and selectivites are generally high independent on the protecting groups (PG) used and the substitution pattern of the allylic double bond (Scheme 5.2.37, Table 5.2.15). While (*E*)-configured allylic esters give rise to the *syn* product, the corresponding *anti*-derivatives are obtained from (*Z*)-allyic esters. If secondary allylic esters ($R^1 \neq H$) are subjected to rearrangement, the (*E*)-configured unsaturated amino acid is obtained nearly exclusively (entries 4 and 5).

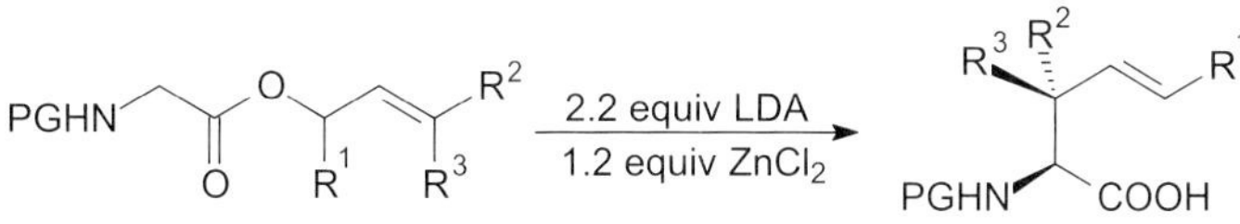

Scheme 5.2.37

Tab. 5.2.15 Influence of the substitution pattern, the olefin geometry and the protecting group in rearrangements of **125**.

Entry	Substrate	PG	R^1	R^2	R^3	Yield [%]	ds [%]	Ref.
1	**125a**	Z	H	H	H	88	–	[66]
2	**125a**	Z	H	H	H	78	–	[66]
3	**125b**	Z	H	Pr	H	76	95	[66]
4	**125c**	Z	Et	Me	H	98	95	[66]
5	**125d**	Z	Et	H	Bu	73	95	[66]
6	**125e**	Boc	H	Pr	H	78	96	[66]
7	**125f**	TFA	H	Pr	H	79	95	[66]
8	**125g**	Z	H	Ph	H	78	n.r.	[67]
9	**125h**	Boc	H	$SiMe_3$	H	50	96	[61]
10	**125i**	Boc	Me	$SiMe_3$	H	57	96	[61]
11	**125k**	Z	H	Me	Me	81	–	[66]
12	**125l**	Z	H	$-(CH_2)_5-$		64	–	[66]
13	**125m**	Z	H	$-(CH_2)_6-$		45	–	[66]

n.r.: not reported

Hruby et al. used the rearrangement of cinnamyl ester **125g** (entry 8) as a key step in their synthesis of conformationally constrained reverse turn dipeptide mimetics [67]. Piscopio et al. applied the rearrangement of crotyl esters in combination with an *N*-allylation and subsequent ring closing metathesis towards the synthesis of pipecolinic acid derivatives **A** [68], while Morimoto et al. used the rearrangement products for the synthesis of cyclopropyl amino acids **B** (Scheme 5.2.38) [69].

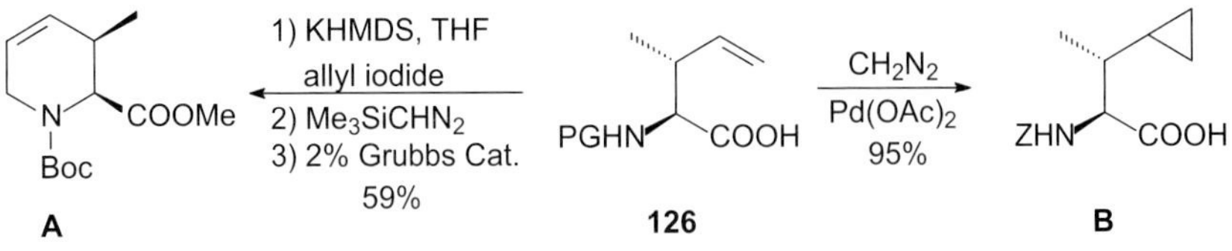

Scheme 5.2.38

The rearrangement of silyl substituted allylic esters was investigated by Brook et al. [61] (entries 9 and 10). The formation of product was observed during warming to temperatures of about –20 °C, but workup at room temperature gave only a 30% yield, and a substantial amount of decomposition of the ester was observed. The yield could be improved to 50% when the reaction time was reduced and when the workup was performed at 5 °C (entry 9). Similar reaction conditions were used for the rearrangement of **125i** (entry 10), and in both cases the diastereoselectivities obtained were excellent. With respect to the yield, the Ireland variation of the rearrangement was superior with these substrates, while the chelate Claisen rearrangement was more reliable in terms of selectivity.

Application of the chelate–enolate rearrangement to the rearrangement of highly substituted allylic esters (R^2, $R^3 \neq H$), gave rise to amino acids containing quaternary *β*-carbon centers (entries 11 to 13) [70]. In addition to the symmetrically substituted allylic esters ($R^1 = R^2$), the unsymmetrically substituted esters ($R^1 \neq R^2$) are especially interesting substrates (Scheme 5.2.39). The rearrangement of the *E*-configured ester **127a** resulted in the formation of the *syn*-configurated amino acid **128a** in a highly diastereoselective fashion. The observed diastereoselectivity is in good agreement with the selectivities observed for other *E*-configurated allylic esters, and results from a preferential rearrangement via a chair-like transition state. On the other hand, the rearrangement of the *Z*-configurated ester **127b** gave rise to the opposite diastereomer, although with a significantly lower degree of selectivity. The lower diastereoselectivity may result from interactions of the *cis*-oriented side chain and the chelated enolate. These interactions should destabilize the chair-like transition state with the consequence that the boat-like transition state should become more favored.

127 → **128**

a $R^1 = (CH_2)_2C{=}CMe_2$, $R^2 = Me$ — 68% 95% ds

b $R^1 = Me$, $R^2 = (CH_2)_2C{=}CMe_2$ — 64% 64% ds

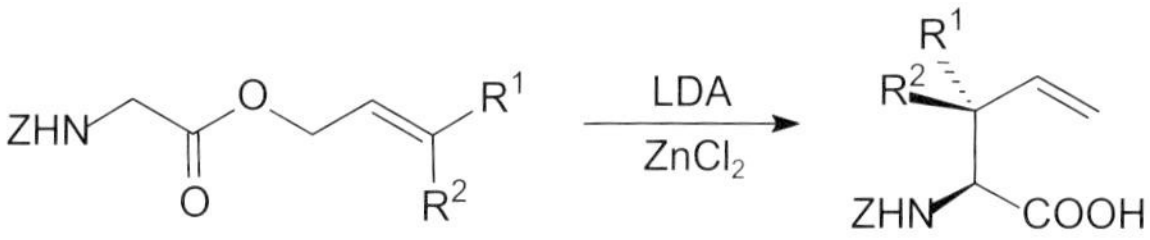

Scheme 5.2.39

That this is indeed the case was shown by rearrangement of the acetylenic substrates **129a** and **129b** (Scheme 5.2.40). While the *E*-configurated ester **129a** gave the expected product **130** in a highly diastereoselective fashion, the oppositely configurated *Z*-ester **129b** unexpectedly yielded the same product **130** – although less selectively. Introduction of bulky substituents R (aryl or trialkylsilyl groups) on the acetylenic moiety had no further influence on the reaction. Yield and diastereoselectivity were comparable to the results obtained with the esters **129a** and **129b**.

129a **130** **129b**

R = H, $SiMe_3$, Ar

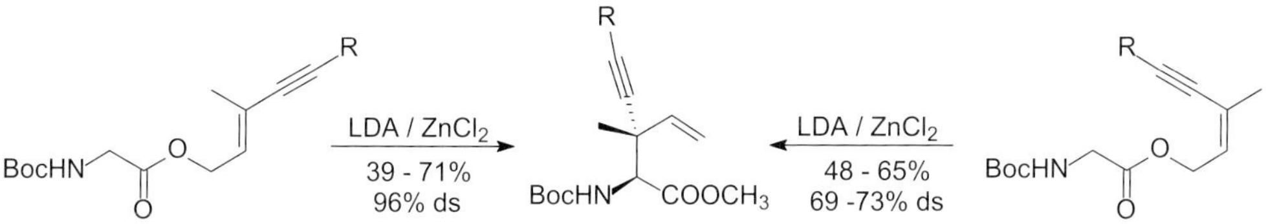

Scheme 5.2.40

The formation of the same product from these complementary precursors can be explained by a change of the transition state geometry (Scheme 5.2.41). The *E*-configurated ester **129a** rearranges preferentially via the expected chair-like transition state **A**. In the case of the *Z*-configurated ester **129b**, strong steric interactions can arise in the corresponding transition state **C** between the triple bond and the presumably solvated chelated metal. The system can switch to the boat-like transition state **D** to avoid these possible interactions, accepting the less dramatic interaction of the axial hydrogen and the chelate. This has the consequence that the *syn* product is also produced preferentially.

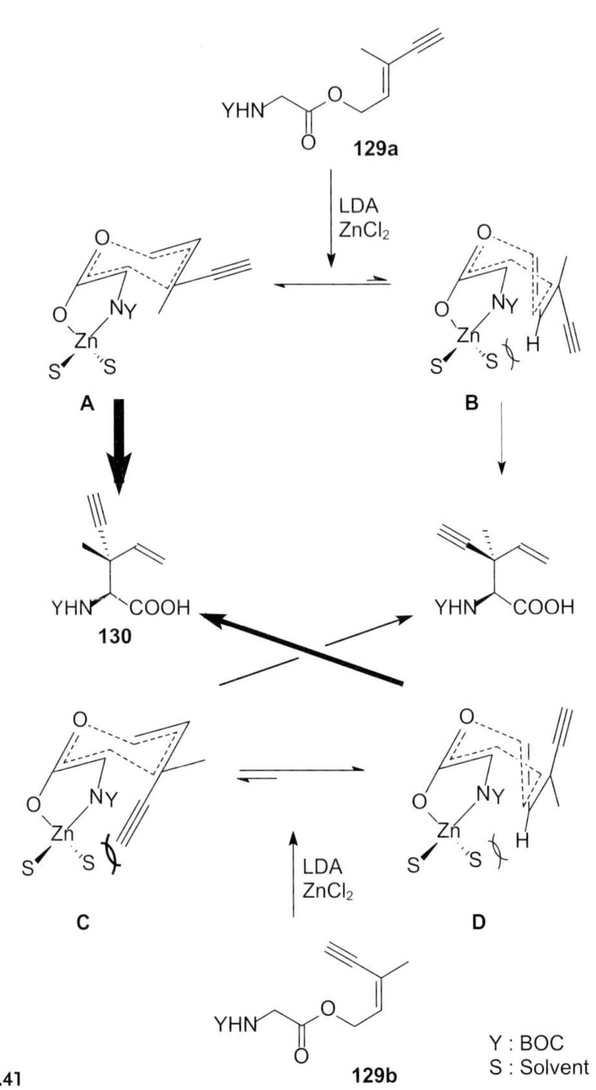

YHN 129a
LDA
$ZnCl_2$
A B
YHN COOH 130
YHN COOH
C D
LDA
$ZnCl_2$
YHN 129b
Y : BOC
S : Solvent

Scheme 5.2.41

Switching from glycine esters to allylic esters **131** of other amino acids allows the synthesis of α-alkylated amino acids [71] such as **132–135** (Scheme 5.2.42).

If highly substituted allylic esters are used, even two adjacent quaternary centers can be generated in one step (**134**). These examples also show that this method is not restricted to amino acids with aliphatic or aromatic side chains, but can also be applied to derivatives of functionalized amino acids such as tryptophan (**132**) or lysine (**133**). These examples had been chosen, because especially the tryptophan and lysine derivatives are critical substrates for α-alkylation reactions [72].

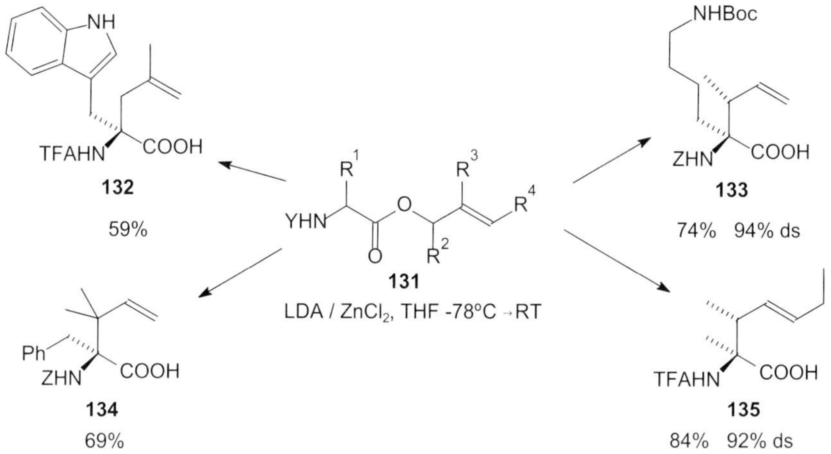

Scheme 5.2.42

α-Vinyl-substituted amino acids are important enzyme inhibitors, especially of amino acid decarboxylases [73]. If the chelate–enolate Claisen rearrangement is applied to allylic esters of α,β-didehydro amino acids such as **136**, which are easily obtained by a method developed by Schmidt [74], amino acids containing an allylic as well as a vinylic side chain (Scheme 5.2.43) are generated (**137**) [75].

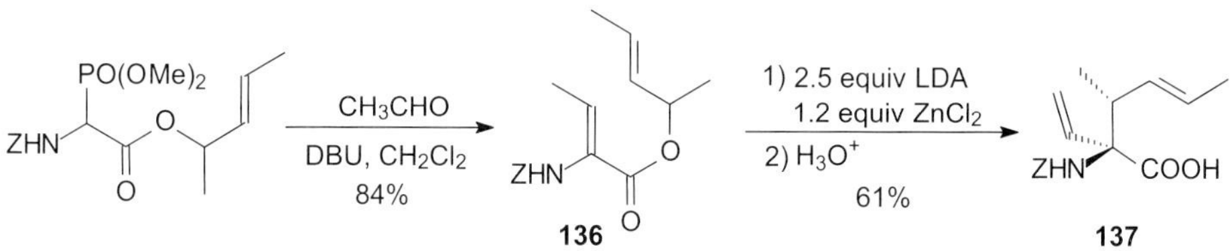

Scheme 5.2.43

Deprotonation with an excess of LDA probably results in the formation of the chelate bridged dienolate which undergoes the Claisen rearrangement during warming. Again, the *syn*-product is formed with a high degree of diastereoselectivity (94–96% ds) and with a clean *trans* olefin geometry in the allylic side chain.

The products obtained by these chelate enolate Claisen rearrangements are not only interesting because of their potential biological activity, but also as intermediates for the synthesis of even more complex amino acids via regioselective modifications of the different unsaturated side chains. Iodolactonization, for example, occurs in a highly diastereoselective fashion and exclusively in a 5-*exo-trig*-mode [76] at the allylic double bond. On the other hand, the Heck reaction [77] of vinyl derivatives proceeds regioselectively at the least sterically hindered position of the unsubstituted vinylic double bond [78], giving rise to the *all-trans*-configured amino acid.

Recently, applications of the chelate Claisen rearrangement towards the synthesis of fluorinated amino acids were reported (Scheme 5.2.44). Konno et al. investigated the rearrangement of esters such as **139** under certain reaction conditions and found that the chelate Claisen rearrangement was superior to the Ireland version with respect to both yield and selectivity [60]. Treatment of ester **139** with $ZnCl_2$ and LHMDS at 0 °C for 0.5 h, followed by heating the mixture under reflux for 6 h, yielded the expected amino acid **140** in 82% yield as a single isomer. Ester **139** was prepared *in situ* as a single regioisomer via palladium-catalyzed allylic alkylation using Boc-glycinate as a nucleophil. If chiral mesylates such as **138** are used as substrates a high degree of chirality transfer is observed.

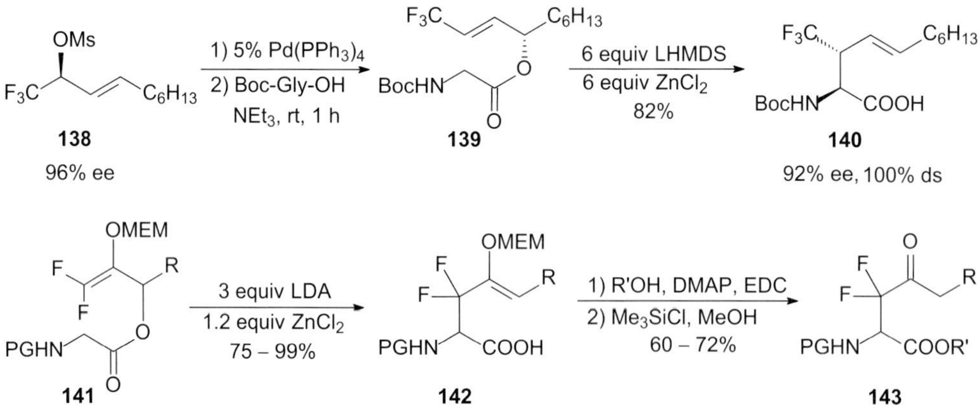

Scheme 5.2.44

Percy et al. reported on a synthetic protocol towards the synthesis of fluorinated γ-oxoamino acids **142** via rearrangement of allylic esters **141** [79]. Subsequent esterification of **142** and hydrolysis of the silylenolether subunit gave rise to the functionalized amino acid derivatives **143**.

Because of the good results obtained with acyclic substrates, the chelate ester enolate rearrangement was also applied by Kazmaier et al. [80] to the rearrangement of cycloalkenyl glycinates **144** (Scheme 5.2.45). The influence of the ring size as well as the metal salt used for chelation of the ester enolate was investigated and the results are listed in Table 5.2.16. The crude amino acids, obtained by

1) 2.5 equiv LDA
1.2 equiv MX_n
2) CH_2N_2

BocHN ... 144 → BocHN, COOMe 145a + BocHN, COOMe 145b

Scheme 5.2.45

Tab. 5.2.16 Rearrangement of *N*-Boc protected glycine cycloalkenyl esters **144**.

Ester	n	$ZnCl_2$ Yield [%]	$ZnCl_2$ Ratio 145a:145b	$MgCl_2$ Yield [%]	$MgCl_2$ Ratio 145a:145b	$Al(O\textit{i}Pr)_3$ Yield [%]	$Al(O\textit{i}Pr)_3$ Ratio 145a:145b
144a	1	79	80:20	57	79:21	47	75:25
144b	2	83	90:10	94	92:8	91	90:10
144c	3	73	92:8	79	92:8	69	89:11
144d	4	57	86:14	78	91:9	42	79:21

the rearrangement process, were directly converted into the corresponding methyl esters **145a** and **b**.

The best results concerning yield as well as stereoselectivity are obtained with cyclohexenyl ester **144b** (n= 2). In this case all metal salts gave the product **145** in excellent yield. The same high degree of diastereoselectivity was obtained in the rearrangement of the homologous cycloheptenyl ester **144c** (n= 3), while with the smaller ester **144a** (n= 1) and the larger and probably more flexible cyclooctenyl ester **144d** (n= 4) the selectivity decreases.

The product formation as well as the high diastereoselectivity observed in the rearrangement of the six- and seven-membered allylic esters can be explained by rearrangement via a boat-like transition state **A**, which is discussed frequently for cyclic allylic substrates [1g]. Steric interactions between the axial proton of the cycloalkenyl ring and the probably solvated chelating metal obviously disfavor the chair-like transition state **B** (Fig. 5.2.5).

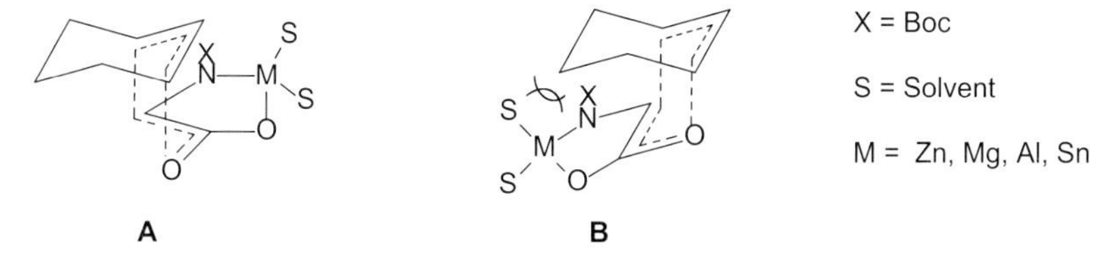

Fig. 5.2.5 Transition states for the rearrangement of cyclic esters **144**.

Chelate Claisen Rearrangement of Amino Acid Propargylic Esters

α-Allenic α-amino acids are attractive candidates for the specific inhibition of vitamin B_6 linked decarboxylases [81] and the [3,3]-sigmatropic rearrangement processes are particularly suitable for translocation of a three-carbon unit with concomitant interconversion of propargylic and allenic moieties [82].

A first application of this methodology was developed by Steglich et al. using the Oxazole–Claisen approach [83, 56a]. The Ireland version of the Claisen rearrangement was investigated by Castelhano and Krantz for amino acid propargylic esters [56c], but unfortunately the yields are not satisfying, even under the relatively mild conditions of the Ireland–Claisen rearrangement. Therefore the chelate–enolate Claisen rearrangement of various propargylic esters **146** (Scheme 5.2.46) for the syntheses of these very sensitive amino esters **147** was investigated by Kazmaier et al. [84]. The propargylester rearrangement can also be applied to various types of amino acids giving rise to α-alkylated amino acids such as **148**.

As in the allylic ester rearrangement, no dependence on the *N*-protecting group was observed and the *syn*-product was formed preferentially [85]. According to a suggestion of Hoppe [86], the *syn*/*anti* terminology [87] is used for the easy description of the allenic rearrangement products.

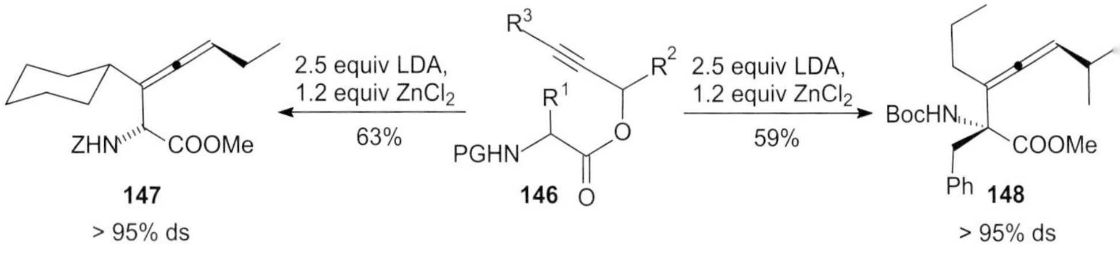

Scheme 5.2.46

The stereochemical outcome of the reaction can be explained by comparison of the relevant transition states of the rearrangement (Scheme 5.2.47). If secondary propargylic alcohols are used, the rearrangement preferentially proceeds via transition state **A**, where the substituent R^2 is oriented in an equatorial position, in comparison to **B** with R^2 in an axial position (Scheme 5.2.47). As expected, the selectivity increases as the size of R^2 increases, because of stronger sterical interactions of the axial R^2 and the chelated enolate in transition state **B**.

In comparison to allylic esters, the propargylic esters are less reactive as illustrated in comparative rearrangements of enyne esters **149** (Scheme 5.2.48) [88]. Rearrangement under standard conditions gave rise to the unsaturated acids, which were directly converted into the corresponding methyl esters. In general, the rearrangements proceeded with good yields and very good chemoselectivities (Table 5.2.17). Because of the slightly higher reactivity of allylic esters in comparison to propargylic esters, mainly the rearrangement products **150** were obtained. The diastereoselectivity of the reaction was excellent as well. No second set of signals could be found in the NMR spectra.

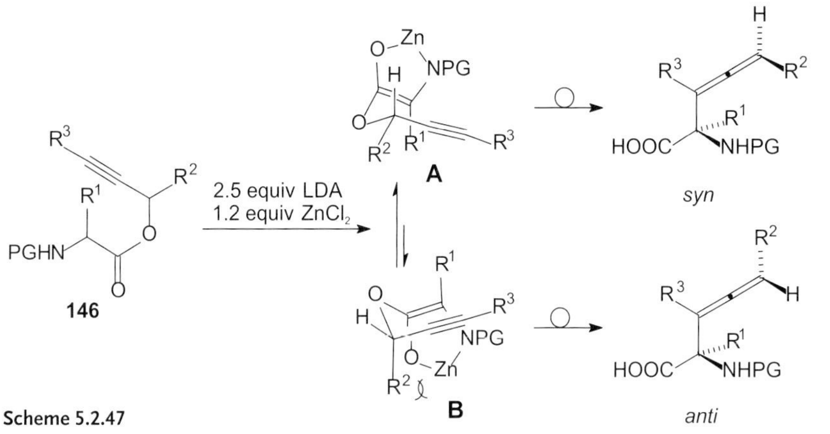

Scheme 5.2.47

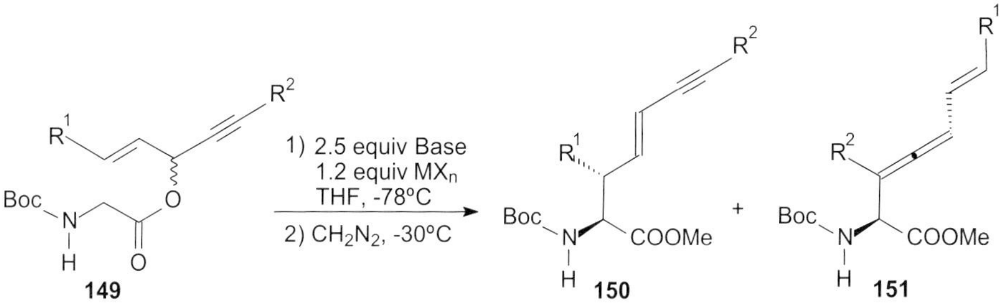

Scheme 5.2.48

Tab. 5.2.17 Chelate Claisen rearrangement of enyne ester **149**.

Entry	Ester	R^1	R^2	Base	MX_n	Yield [%]	Ratio 150:151
1	**149a**	Me	Me	LDA	$ZnCl_2$	65	90:10
2	**149a**	Me	Me	LHMDS	$ZnCl_2$	68	95:5
3	**149a**	Me	Me	LHMDS	–	75	> 95:5
4	**149b**	Me	*n*-Pr	LHMDS	$ZnCl_2$	71	> 95:5
5	**149b**	Me	*n*-Pr	LHMDS	–	76	> 95:5
6	**149d**	H	*n*-Pr	LHMDS	$ZnCl_2$	70	> 95:5
7	**149d**	H	*n*-Pr	LHMDS	–	70	> 95:5
8	**149e**	Me	$SiMe_3$	LHMDS	$ZnCl_2$	66	> 95:5
9	**149e**	Mc	$SiMe_3$	LHMDS	–	73	> 95:5

During the attempts to improve the chemoselectivity of the reaction it was found that the selectivity could be increased to >95:5 in favor of product **150** if LHMDS was used as a base. In addition, the reaction already proceeded at –78 °C and was completed within 2 h, in comparison to the rearrangement of normal allylic esters, where the rearrangement started around –20 °C during the warm up. As a consequence, the pure lithium enolates could not be used in those reactions, as decomposition of the lithium enolates takes place, whereas chelate enolates are stable up to room temperature. But with this low rearrangement temperature one might expect that the stabilizing effect of chelation might not be necessary and that the lithium enolate might react as well. Indeed, deprotonating the amino acid esters with LHMDS, stirring for one or two hours, and subsequent workup provided **150** in good yields.

Asymmetric Claisen Rearrangements

As illustrated, it is possible to carry out the ester enolate Claisen rearrangement in a highly diastereoselective fashion via control of the enolate geometry. However, from a synthetic point of view, the generation of racemic products is not fully satisfying. For an application to amino acid synthesis, it is important not only to control the relative, but also the absolute configuration of the stereogenic centers [1i].

Therefore with respect to asymmetric Claisen rearrangements three different approaches have been investigated (Fig. 5.2.6).

1) The classical approach for the introduction of chirality is the rearrangement of chiral allylic esters. If these esters are used, chirality transfer occurs not only from the ester moiety (C-4) to C-6 but also to the α-position (C-1) of the carboxylic acid subunit (1,4-chirality transfer).
2) Another possibility is the introduction of chirality on the amino terminus of the amino acid ester. The easiest way to do so is the incorporation of the amino acid allylic ester onto a peptide chain.
3) If chelate bridged species really exist, there should be at least some induction if the reaction is carried out in the presence of chiral ligands. In contrast to many other auxiliary-directed asymmetric syntheses, in this case the chiral information is bound temporarily by coordination, and therefore this approach is the most attractive one, especially if the ligand can easily be recovered after the reaction.

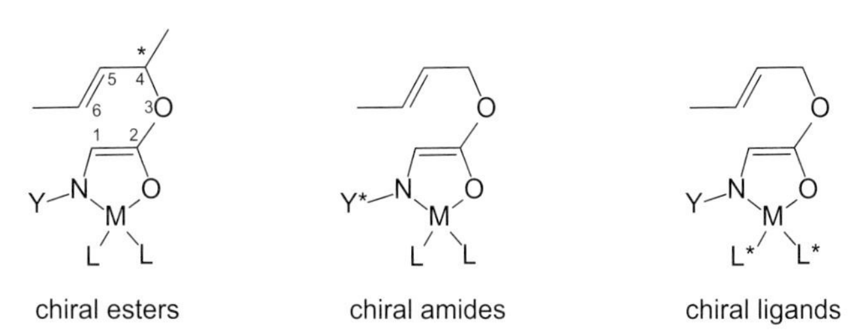

Fig. 5.2.6 Asymmetric Claisen rearrangements.

Rearrangement of Chiral Allylic Esters

The easiest way to transfer chirality within the Claisen system is the rearrangement of chiral secondary allylic alcohols. Like in most sigmatropic rerrangements, the chair-like transition state is highly preferred (Scheme 5.2.49), whereas two diastereomeric chairs are possible for esters of secondary allylic alcohols **152**. In the favored transition state **A**, the substituent R is orientated in an equatorial position, while in transition state **B** strong steric interactions between the axial R and the chelate-complex may occur. As a result of the highly ordered transition state, the [3.3]-sigmatropic rearrangement also determines the double bond geometry in the rearranged product. *E*-Olefins **153** are obtained if the reaction proceeds via transition state **A**, while the disfavored transition state **B** gives rise to *Z*-configurated olefins **154**. The exclusive formation of **153** clearly demonstrates the high preference for the favored transition state **A**. Because of the fixed enolate geometry, the reaction proceeds with a very high degree of chirality transfer (Table 5.2.18). As illustrated, the rearrangement can be applied to α-alkylated amino acids as well (**153e**), and also works with allylic esters containing sterically high demanding substituents (**152d**). Therefore, it is possible to introduce protected polyhydroxylated substituents into amino acids in one step and in a highly diastereoselective fashion [89].

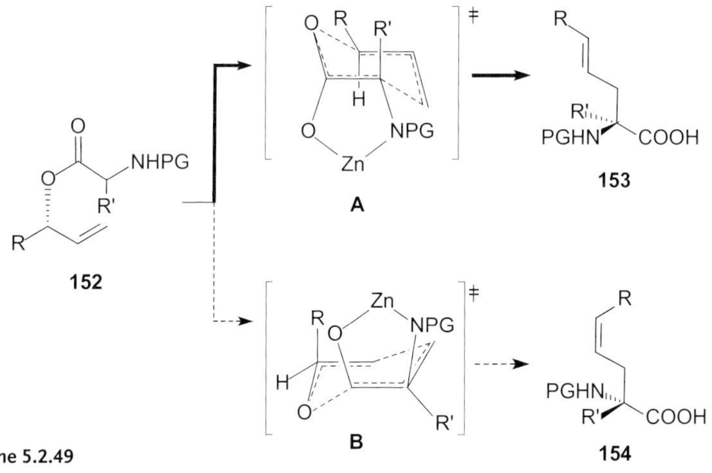

Scheme 5.2.49

The Rearrangement of chiral ester **155** to amino acid **156** was a central step in the synthesis of azasugars such as **157** (Scheme 5.2.50) [90]. Williams et al. started their synthesis of Cylindrospermopsin also with a chelate Claisen rearrangement of chiral ester **158** [91].

Toogood et al. used the rearrangement of the chiral (*Z*)-allylic ester **160** as a key step during their synthesis of microcystines and related toxins (Scheme 5.2.51). The rearrangement product **161** was used as starting material for the synthesis of two unusual amino acids obtained in these cyclopeptides [92]. All these peptides contain the amino acids α-methyl-D-aspartic acid (**162**) and (2*S*,3*S*,8*S*,9*S*,4*E*,6*E*)-3-amino-9-methoxy-2,6,8-trimethyl-10-phenyldeca-4,6 dienoic acid (ADDA, **163**), which can be obtained from a common precursor **161**.

Tab. 5.2.18 Amino acids obtained via asymmetric Claisen rearrangement.

Substrate	R	R'	PG	Yield [%]	Selectivity
152a	Ph	H	Z	86	98.7% ee
152b	$BnOCH_2$	H	Boc	95	>98% ee
152c		H	Z	85	94% de
152d		H	Z	80	94% de
152e		Me	Z	80	94% de

Scheme 5.2.50

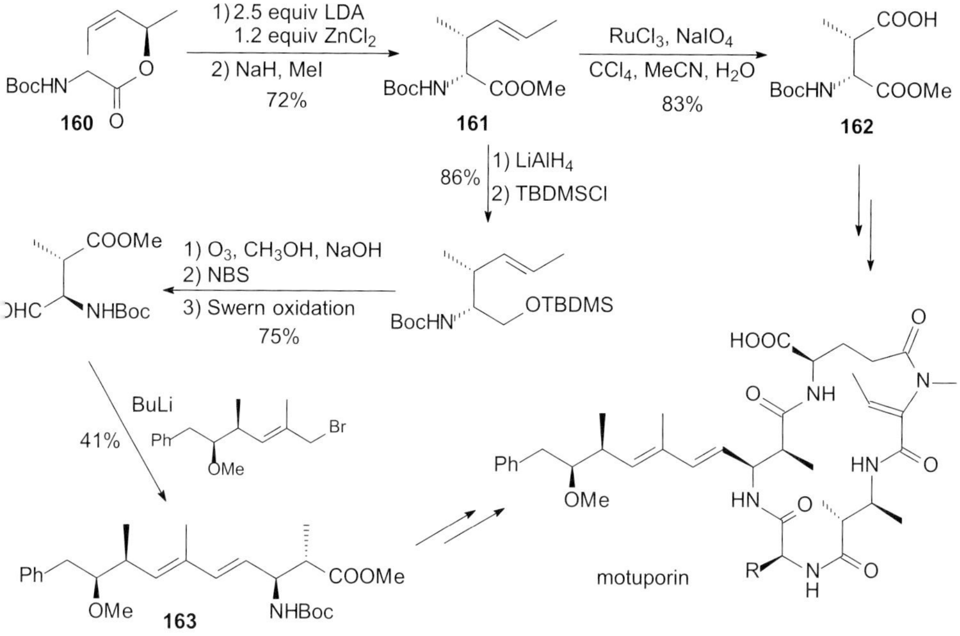

Scheme 5.2.51

Rearrangement of Peptide Allylic Esters

If the Claisen rearrangement is carried out with peptide allylic esters, the transfer of an allylic side chain to the α-position of the C terminal amino acid results in a modification of the peptide chain. This concept is comparable to the alkylations of peptide enolates described by Seebach et al. [93]. If it is possible to carry out the rearrangement not only with amino acids but also with peptide esters, the question arises if it is possible to transfer the chiral information from the peptide chain to the new chiral center formed during the rearrangement process, probably via some peptide metal enolate complexes.

As a first example the rearrangement of allylic ester **164** (Scheme 5.2.52) was investigated in the presence of zinc chloride [94]. Subsequent esterification of the rearrangement product with diazomethane resulted in the formation of unsaturated dipeptide **165** in only moderate yields and selectivity. Since the tremendous work done by Overman et al. on rearrangement catalysis [95, 96], many applications, especially of Pd(II)-catalyzed rearrangements, have been described in the literature. In this case, addition of 5 mol% Pd(II) resulted in a decrease in yield without affecting the stereoselectivity. Because Pd(II) is known to form stable complexes with peptides [97], the formation of a less reactive palladium enolate, which does not rearrange, is probable. In contrast, Pd(0) (10 mol%) catalyzed the reaction in the expected manner, and the desired product was obtained in excellent yield (80%) but without any selectivity.

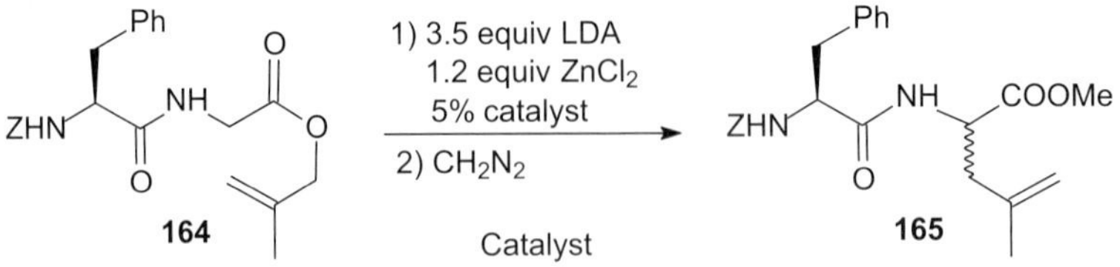

1) 3.5 equiv LDA
1.2 equiv $ZnCl_2$
5% catalyst
2) CH_2N_2

164 → **165**

Catalyst	
—	28%, 63% ds
$PdCl_2(COD)$	20%, 65% ds
$Pd(PPh_3)_4$	80%, 50% ds

Scheme 5.2.52

Unfortunately, under these conditions the allylic amino acid was obtained not via Claisen rearrangement, but an intermolecular palladium-catalyzed allylic alkylation. For example, in reactions of crotyl esters two regioisomers are obtained. This fact, as well as the nearly complete lack of diastereoselectivity in the newly generated amino acid of **25** (*syn*/*anti* 2:1) is a clear indication for an intermolecular process and the appearance of π-allylpalladium intermediates.

Although the application of palladium catalysts resulted in an increase of yield, it was also responsible for a decrease of regio- and diastereoselectivity as well. Therefore, an intensive metal tuning was undertaken to find suitable chelate complexes which undergo Claisen rearrangement without assistance of a palladium catalyst. By far the best results were obtained if manganese salts were used for chelation (Scheme 5.2.53) [98]. Independent of the protecting groups used, the yields obtained with these manganese enolates were always excellent (78–98%), with esters of terminal allylic alcohols and with *trans* configured substituted alcohols (**166**) as well. In all examples investigated so far, the simple diastereoselectivity of the rearrangement was very high (≥ 95% *syn*) and comparable to the results obtained with amino acids, but no significant induced diastereoselectivity was observed in all reactions carried out with these enolates. Obviously the *N*-terminal amino acid has no notable influence on the rearrangement. This is also reflected in the high yields obtained, which are also nearly independent of the peptide used. Because the influence of the adjacent amino acid on the rearrangement can be neglected, this allows for the stereoselective synthesis of peptides if esters of chiral allylic alcohols such as **168** are used. The corresponding dipeptides **169** were obtained not only in good to excellent yields, but also in a highly diastereoselective fashion. Depending on the chiral alcohol used, both configurations can be obtained on request. This protocol is also suitable for the direct introduction of α-alkylated amino acids into peptides (**171**). These derivatives show higher resistances towards proteases, and therefore they are interesting for the development of peptide-based pharmaceuticals.

Using the manganese enolates, the chirality is transfered from the chiral ester moiety towards the peptide backbone, but not inside the peptide backbone. Therefore, the influence of the chelating metal salt in the rearrangement of several dipeptide crotyl esters was investigated.

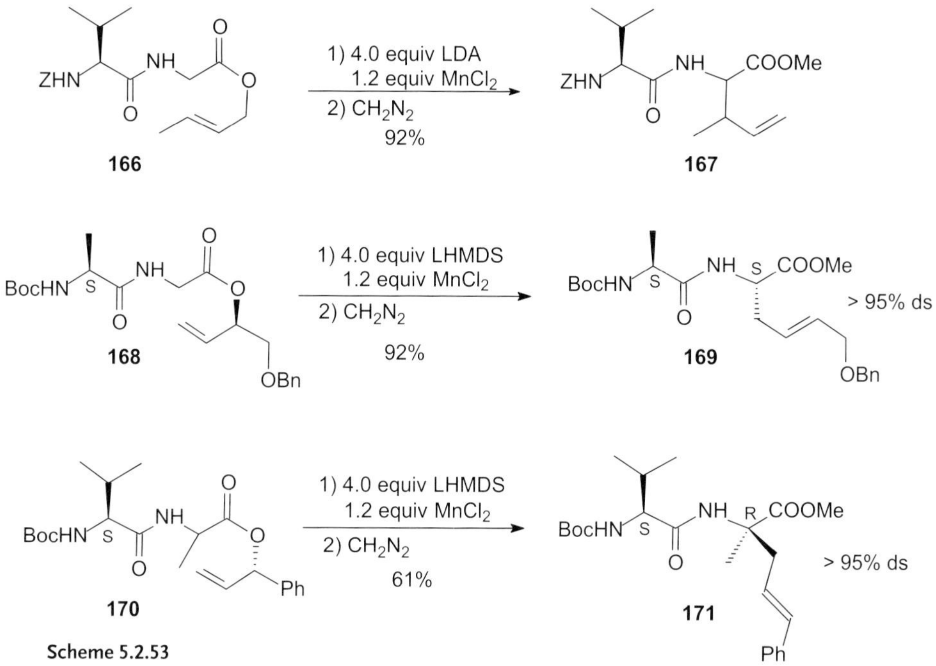

Scheme 5.2.53

Starting with the Boc-protected phenylalanine derivative **172**, good selectivities were found in the presence of $Ti(OiPr)_4$ and $NiCl_2$ as well (Scheme 5.2.54). Replacing phenylalanine by the sterically more bulky valine resulted in an increase of selectivity. Other metal salts such as CuBr gave comparable results. Switching to

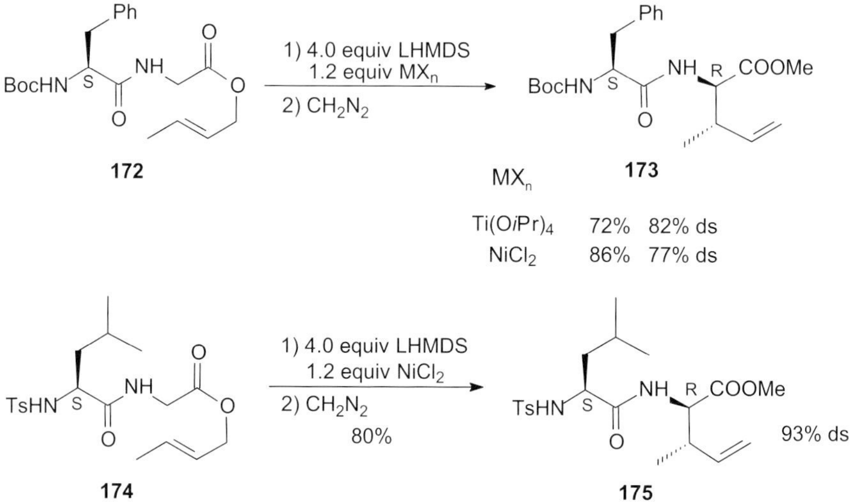

MX_n		
$Ti(OiPr)_4$	72%	82% ds
$NiCl_2$	86%	77% ds

Scheme 5.2.54

other carbamate protecting groups like Z had no significant effect on the diastereoselectivity. Introduction of the tosyl group (Ts) (**174**), however, increased the ds dramatically, up to 96% in some cases. Tosyl-protected peptides are therefore the substrates of choice; in general, the selectivites were in the range of 90 to 96 ds, with yields up to 90% [99].

In all examples investigated so far, an (*R*)-amino acid was formed during the Claisen rearrangement, if an (*S*)-amino acid was placed in the peptide chain. For the reactions with $NiCl_2$, this stereochemical outcome can be rationalized by the formation of a square planar chelate complex **176**, in which one face of the enolate is shielded by the sidechain R of the (*S*)-amino acid (Figure 5.2.7). The rearrangement occurs on the sterically less hindered "opposite" (*unlike*) face of the enolate, giving rise to the (*R*)-amino acid. Although this model is only a working hypothesis, it was very helpful for further developments and investigations in this area.

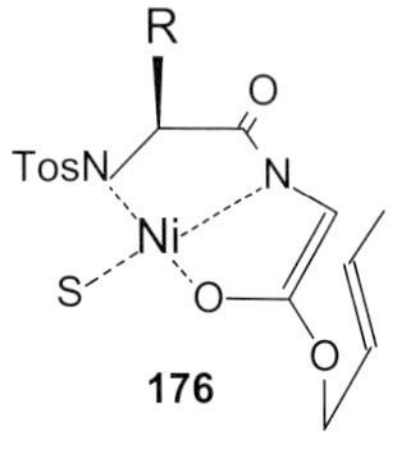

R
O
TosN
N
Ni
S
O
O
176

Fig. 5.2.7 Metal peptide complex.

Rearrangement in the Presence of Chiral Ligands

Besides the described substrate controlled asymmetric rearrangements, rearrangements in the presence of chiral ligands were also investigated. This approach is of special interest because not many examples of asymmetric rearrangements of this type are described in the literature so far. The first rearrangements in the presence of chiral Lewis acids were reported by Yamamoto et al. in 1990 [100], while the first ester enolate rearrangement, proceeding via chiral boron-enolates, was described by Corey et al. [101]. Both procedures need at least equimolar amounts of the chiral ligand. Overman et al. were the first reporting on a Pd(II)-catalyzed asymmetric allylimidate rearrangement [102], and the only catalytic asymmetric C–C-coupling was reported recently by Hiersemann et al. [103].

In view of the many different metal salts that can be used in the chelate enolate rearrangement, and the even higher number of chiral ligands commonly used in asymmetric synthesis, Kazmaier et al. undertook an excessive screening of the reaction conditions. As a model reaction the rearrangement of TFA-protected glycine crotyl ester **177** was investigated (Scheme 5.2.55) and the results are collected

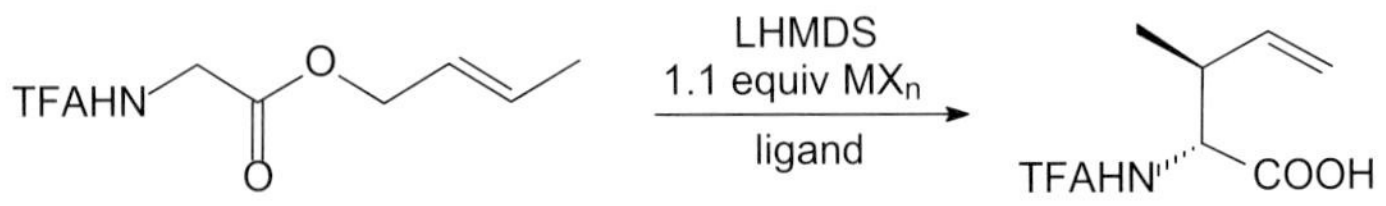

TFAHN
O
O
177
LHMDS
1.1 equiv MX_n
ligand
TFAHN
COOH
178

Scheme 5.2.55

Tab. 5.2.19 Rearrangement of crotyl esters **177** in the presence of chiral ligands.

Entry	MX_n	Ligand	Equiv	Yield [%]	ds [%]	ee [%]	Config.
1	$Al(iOPr)_3$	(*S*)-valinol	1.5	73	97	4	(2*R*,3*S*)
2	$Al(iOPr)_3$	(–)-ephedrine	1.2	72	96	27	(2*S*,3*R*)
3	$Al(iOPr)_3$	(+)-ephedrine	1.2	70	96	27	(2*R*,3*S*)
4	$ZnCl_2$	quinine	2.5	95	90	10	(2*R*,3*S*)
5	$CaCl_2$	quinine	2.5	73	96	65	(2*R*,3*S*)
6	$MgCl_2$	quinine	2.5	98	91	69	(2*R*,3*S*)
7	$Mg(OEt)_2$	quinine	2.5	98	96	83	(2*R*,3*S*)
8	$Al(iOPr)_3$	quinine	2.5	98	98	86	(2*R*,3*S*)
9	$Al(iOPr)_3$	quinidine	2.5	96	98	86	(2*S*,3*R*)

in Table 5.2.19 [104]. Lithiumhexamethyldisilazide (LHMDS) as a base gave better yields than LDA, which is usually applied. A cleavage of the TFA protecting group was sometimes observed with LDA, a side reaction that is completely eliminated using LHMDS.

Among the ligands investigated, the best results were obtained with amino alcohols. Diol- or diamino ligands, which are frequently used in asymmetric catalysis [105], showed no significant induction (less than 5% ee), independent of the metal salt used. Simple amino alcohols, easily obtained by reduction of amino acids, are comparable to these ligands (entry 1). Introduction of a second center of chirality (α to the OH-group) resulted in a dramatical increase in the selectivity. The switch from the simple alcohols like valinol to the ephedra alkaloids increased the enantioselectivity from 4% to 27% ee (entries 2 and 3). The big advantage of these alkaloids is the possibility to generate both enantiomers of the amino acids, because both enantiomeric ligands are commercially available. The same is true with the cinchona alkaloids quinine and quinidine, which are by far the ligands of choice. While (*R*)-amino acids are obtained in the presence of quinine (entries 4 to 9), the isomeric quinidine gives rise to the corresponding (*S*)-amino acids (entry 10).

Many variations were undertaken concerning the metal salts used for chelation. Zinc chloride, which normally gives the best results, provided the rearrangement product in excellent yields but with only moderate ee's (entry 4). Application of calcium chloride (entry 5) or magnesium chloride (entry 6) provided ee's between 65 and 70%, but the metal salts of choice were $Mg(OEt)_2$ (entry 7) and $Al(OiPr)_3$ (entry 8). With these salts excellent yields and diastereoselectivities were obtained (up to 98% each) as well as a very high induction (up to 86% ee), especially if 2 to 2.5 equiv of quinine were used. Unfortunately this excess of chiral ligand is necessary because the coordination of the ligand to the chelated metal decreases the reactivity of the enolate. Interestingly the "counterion" also had an influence (entry 6 vs. 7). Obviously alcoholates are the metal salts of choice giving by far the best results (entries 7 and 8). This was quite surprising, because alcoholates generally have no influence on reactions of these enolates. For example, in aldol reactions, Michael additions or Claisen rearrangements without chiral ligand the addition of these alcoholates had no influence on the outcome of the reaction, in comparison to the reaction of the lithium enolate. Therefore the rearrangement of the lithium enolate was investigated (entry 9). Without a chiral ligand, these lithium enolates decompose and do not undergo a Claisen rearrangement. But in the presence of quinine the rearrangement product was obtained in excellent yield and high stereoselectivity. Although the ee's were higher in the presence of the magnesium and aluminum alcoholates, the results were in the same range. Obviously the chiral ligand is able to stabilize the lithium enolate, and it is assumed, that it is this lithium enolate that undergoes the Claisen rearrangement.

General Procedure for Claisen Rearrangements in the Presence of Chiral Ligands
A LHMDS solution was prepared by adding 1.55 M *n*-buthyllithium in hexane (1.6 mL, 2.5 mmol) at −20 °C under argon to hexamethyldisilazane (HMDS) (470 mg, 2.9 mmol) in dry THF (1.5 mL) and stirring for 20 min. The *N*-protected glycine crotyl ester **177** (0.5 mmol), $Al(OiPr)_3$ (0.55 mmol) and the ligand (1–1.25 mmol) were dissolved under argon in 5 mL dry THF. The mixture was cooled to −78 °C, and the freshly prepared LHMDS solution was added slowly. The reaction mixture was allowed to warm to room temperature within 12 h. After diluting with 50 mL of ether, the reaction mixture was hydrolyzed by addition of 1 M aqueous $KHSO_4$ (25 mL). The organic layer was washed again with 1 M $KHSO_4$, before the reaction product was extracted with three portions of saturated aqueous $NaHCO_3$ solution (25 mL each). The basic solution was subsequently acidified by careful addition of solid $KHSO_4$ to pH 1 and extracted with three portions of ether (25 mL each). The combined etheral extracts were dried with Na_2SO_4 and the solvent was evaporated under reduced pressure. For the determination of the enantiomeric and diastereomeric ratios of the product, the residue was treated with diazomethane in etheral solution. Subsequent chromatography on silica gel gave the corresponding methyl esters.

In terms of an easy and reliable determination of the stereochemical outcome of the reaction by GC the trifluoroacetyl group was chosen as protecting group (PG) on the nitrogen [106]. But one might expect a strong influence of this protecting group on the reaction. Therefore, the influence of the steric and electronic effect of several protecting groups was investigated (Scheme 5.2.56, Table 5.2.20) [107]. By far the best results were obtained with the initially applied TFA-group (entry 1), while the other protecting groups gave significantly worse results, both in terms of yield and selectivity. Good yields were also obtained with Z- (entry 3) and tosyl-protected crotyl esters (entry 4), while the benzoyl protected derivatives gave good selectivities (entry 2). Quite surprising was the big difference between the trifluoroacetyl and the acetyl protected esters (entry 1 vs. 6). From a steric point of view there should be no big difference between these two acyl groups and obviously the strong electron withdrawing effect of the fluorine atoms is responsible for the high selectivities. Therefore, the rearrangement of fluorinated benzoyl protected esters was investigated. Introduction of one fluorine atom (entry 7) into the *p*-position of the benzoyl group resulted in an increased yield and diastereoselectivity, although the enantiomeric excess was nearly constant. But the pentafluorobenzoyl (Pfb) and the trifluoroacetyl protected derivate gave comparable results (entry 8).

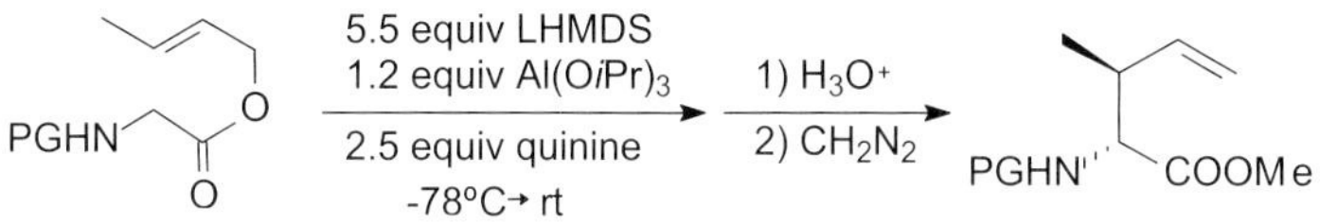

PGHN O O **179a-h** → 5.5 equiv LHMDS, 1.2 equiv $Al(OiPr)_3$, 2.5 equiv quinine, -78°C→ rt → 1) H_3O^+ 2) CH_2N_2 → PGHN COOMe **180a-h**

Scheme 5.2.56

Tab. 5.2.20 Influence of the protecting group (PG) on chelate Claisen rearrangements of ester **179**.

Entry	Substrate	PG	Yield [%]	ds [%]	ee [%]
1	**179a**	TFA	98	98	87
2	**179b**	Bz	50	95	60
3	**179c**	Z	88	90	49
4	**179d**	Ts	98	90	46
5	**179e**	Boc	78	90	16
6	**179f**	Ac	23	85	13
7	**179g**	*p*-F-Bz	77	99	58
8	**179h**	Pfb	93	99	87

Probably a bimetallic complex **181** is formed with the bidentate ligand quinine (or quinidine, respectively) coordinating to the lithium enolate (Figure 5.2.8). The incorporation of a second metal ion M (Li^+, Al^{3+}, Mg^{2+}) into the complex should stabilize this complex by forming a very rigid structure, in which one face of the enolate is shielded by the bicyclic substructure of the quinine. This would explain the high ee's obtained with this system. Obviously, the alcoholates are not involved in enolate formation directly, but they have an influence on the reaction via the bimetallic complex. The similar ee values obtained might be explained by similar ion radii of the metal ions (Li^+: 0.60 Å, Mg^{2+}: 0.65 Å, Al^{3+}: 0.51 Å).

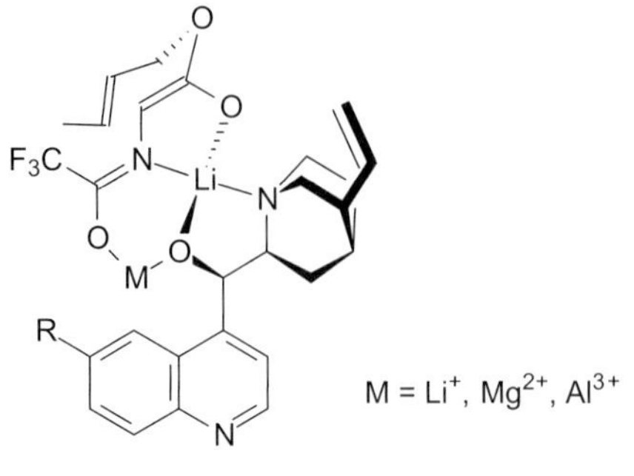

F$_3$C, N, Li, N, O, O, M, O, R, N
M = Li$^+$, Mg^{2+}, Al^{3+}
181

Fig. 5.2.8 Proposed transition state for ligand controlled rearrangement.

If this working model is correct, one should find strong effects if the chelating lithium ion is replaced by other metal ions. Indeed, if the reaction was carried out in the absence of lithium, the selectivity dropped dramatically. No rearrangement product at all was obtained when KHMDS was used as a base. With NaHMDS the yield (18%) and selectivities (85% ds, 14% ee) were low, but could be increased (yield: 78%, 96% ds, 59% ee) by adding LiCl to the reaction mixture. This clearly demonstrates the importance of the lithium ion for the complex formation.

To prove the position of the lithium ion, the reaction conditions were modified in order to generate the aluminum enolate (Scheme 5.2.57). For this purpose LHMDS was added to a suspension of $AlCl_3$ in THF at –20 °C. This mixture was stirred for 10 min at room temperature to form aluminum amide complexes before the quinine was added. After stirring the mixture for a further 2 h at room temperature, the clear pale yellow solution was added to the crotyl ester at –78 °C. Under these modified conditions highly surprising results were obtained: The rearrangement product was formed in good yield (74%) with only moderate diastereoselectivity (85% ds). The enantiomeric excess was rather high (71% ee), but what was most astonishing, the opposite enantiomer of the amino acid was obtained as in the reaction carried out under standard conditions. Exactly the same was true with the pentafluorobenzoyl protected derivative, which gave comparable results.

Obviously the chiral aluminum enolate complex prefers another complex geometry than the corresponding lithium complex. Therefore it is now possible to generate both enantiomers of the rearrangement product by using the same sub-

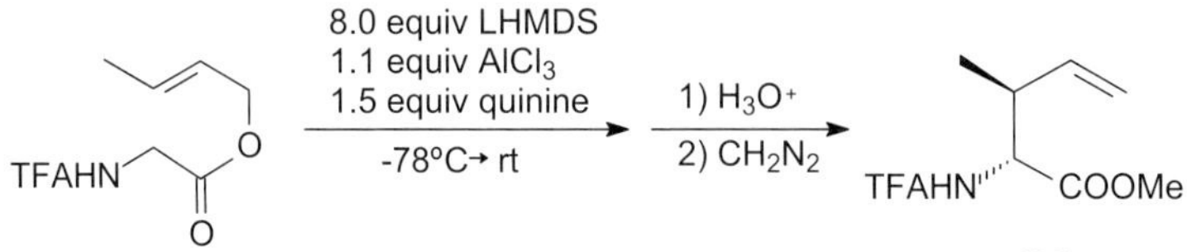

TFAHN, O, O
179a
8.0 equiv LHMDS
1.1 equiv AlCl$_3$
1.5 equiv quinine
-78°C→ rt
1) H$_3$O$^+$
2) CH$_2$N$_2$
TFAHN, COOMe
***ent*-180a**

Scheme 5.2.57

strate, the same chiral ligand, the same base and the same metal ions by simply changing the reaction conditions.

In general, the chelate Claisen rearrangement can be applied not only to esters of glycine, but also to those of most other amino acids, even those with functionalized side chains [71]. In the rearrangement without a chiral ligand, the yields and selectivities were comparable to glycine esters. But if the assumption of a bimetallic enolate complex such as **181** is correct, a substituent at the α-position of the enolate should interact with the *N*-protecting group in this planar complex, resulting in a destabilization of the complex and a breakdown of the enantiomeric excess. Exactly this effect was observed in the rearrangement of the corresponding alanine esters **182** (Scheme 5.2.58). Although the yields were good, the diastereoselectivity was moderate to good and the same as in the reactions without quinine (control experiment). The very low enantiomeric excesses ($\leq$ 10%) were a clear indicate for a collapse of a chiral complex proposed.

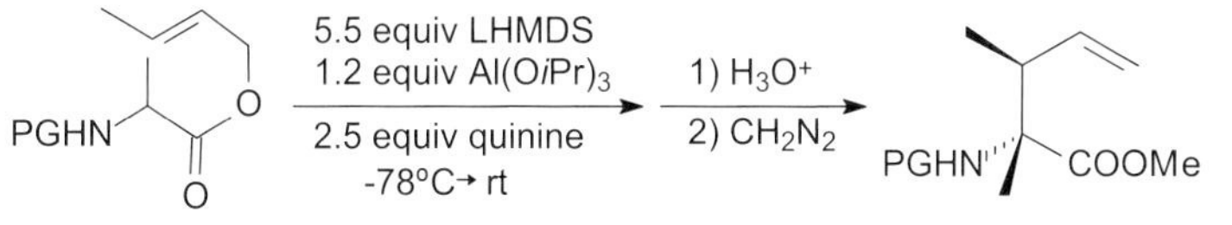

PGHN, O, O
182a,b

5.5 equiv LHMDS
1.2 equiv Al(O*i*Pr)$_3$
2.5 equiv quinine
-78°C→ rt

1) H_3O^+
2) CH_2N_2

PGHN, COOMe
183a,b

	PG			
a	TFA	86%	85% ds	9% ee
b	Pfb	95%	95% ds	10% ee

Scheme 5.2.58

On the other hand, if the steric interaction described in here destabilizes the complex, a connection between the *N*-protecting group and the side chain should favor the bimetallic complex by fixation of the spatial orientation of the amino acid ester. For this purpose crotyl esters were synthesized starting from pyroglutaminic acid **(184a)** and 6-oxopipecolinic acid **(184b)** and subjected to the rearrangement conditions (Scheme 5.2.59). Indeed, the enantioselectivity could be increased significantly in comparison to the open chain derivatives **182**, even those bearing fluorinated protecting groups. Although the ee's were "only" in the range of 30%, one should keep in mind that these derivatives do not contain electron withdrawing groups and that the results should be compared with those of the acetyl derivatives (Table 5.2.20, entry 6) rather than the trifluoroacetylated esters.

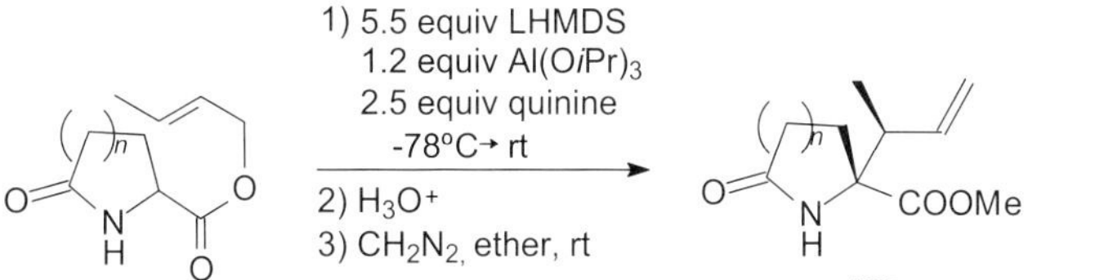

O, N, H, *n*, O, O
184

1) 5.5 equiv LHMDS
1.2 equiv Al(O*i*Pr)$_3$
2.5 equiv quinine
-78°C→ rt
2) H_3O^+
3) CH_2N_2, ether, rt

O, N, H, *n*, COOMe
185

a	*n*=1	58%	75% ds	28% ee
b	*n*=2	98%	95% ds	33% ee

Scheme 5.2.59

The asymmetric chelate enolate Claisen rearrangement could be applied to various types of substrates **186**, while with *E*-configurated allylic esters the enantioselectivities observed were always in the range between 80 and 90% ee (Scheme 5.2.60). If highly substituted allylic esters were used, amino acids with quaternary carbon centers were obtained (**189**) [108]. Even amino acids with two quaternary centers in a row (**190**) could be generated, while in the example shown the highest ee-value to date (93% ee) was obtained. Enantiomerically pure amino acids can be obtained by a single crystallization step using optically active *α*-phenethylamine for the diastereomeric salt formation.

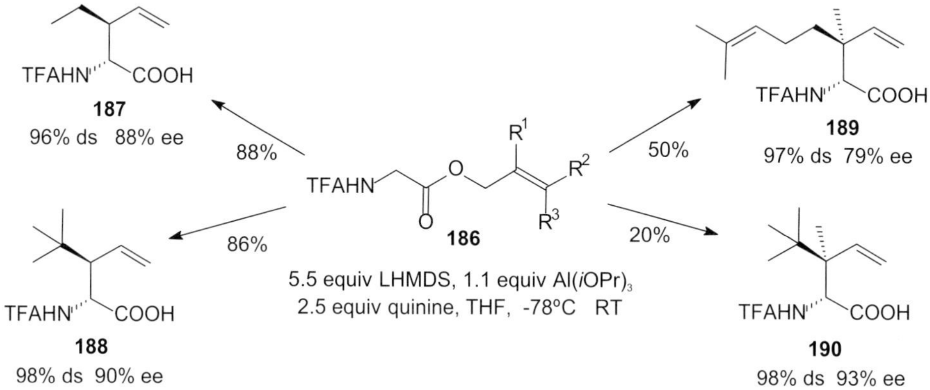

TFAHN COOH
187
96% ds 88% ee
88%
TFAHN COOH
188
98% ds 90% ee
86%
R1 O R2 TFAHN O R3
186
5.5 equiv LHMDS, 1.1 equiv Al(*i*OPr)$_3$
2.5 equiv quinine, THF, -78°C RT
50%
TFAHN COOH
189
97% ds 79% ee
20%
TFAHN COOH
190
98% ds 93% ee

Scheme 5.2.60

The quinine induced Claisen rearangement was also used for the synthesis of more complex amino acids (Scheme 5.2.61). Starting from TFA-protected glycine crotyl ester (**179a**) the suitably protected, enantiomerically pure *β*-hydroxy-*γ*-amino acid isostatine (**193**) was synthesized in only four steps, and the amino acid obtained was directly introduced into peptides (**194**) [109]. Isostatin is an essential amino acid of the didemnine [110], a group of cyclic peptides showing strong antitumor, antiviral as well as immunosuppressive activity [111].

On the other hand, rearrangement product **180a** was used as a substrate for the synthesis of hydroxylated proline and their direct incorporation into peptide [112]. Starting from crude rearrangement products **180a**, the purity was also increased by addition of (*S*)-phenethyl amine (PEA) to a solution of the amino acid in ether. Addition of an excess iodine to a solution of the salts in THF gave iodolactones **195** in excellent yields and acceptable diastereoselectivities (Scheme 5.2.62) [113]. Addition of DBU to **195** resulted in a deprotonation of the rather acidic TFA-amide, and subsequent attack of the amide on the primary iodide gives rise to lactone **196**. In this step only the *syn*-product could cyclize, and therefore enantiomerically and diastereomerically pure lactones **196** were obtained. Both **195** and **196** were important intermediates for the synthesis of unnatural and unusual amino acids. Especially interesting are reactions of bicyclic lactones **196** because they provide stereoisomerically pure products for the reasons discussed. These bicyclic

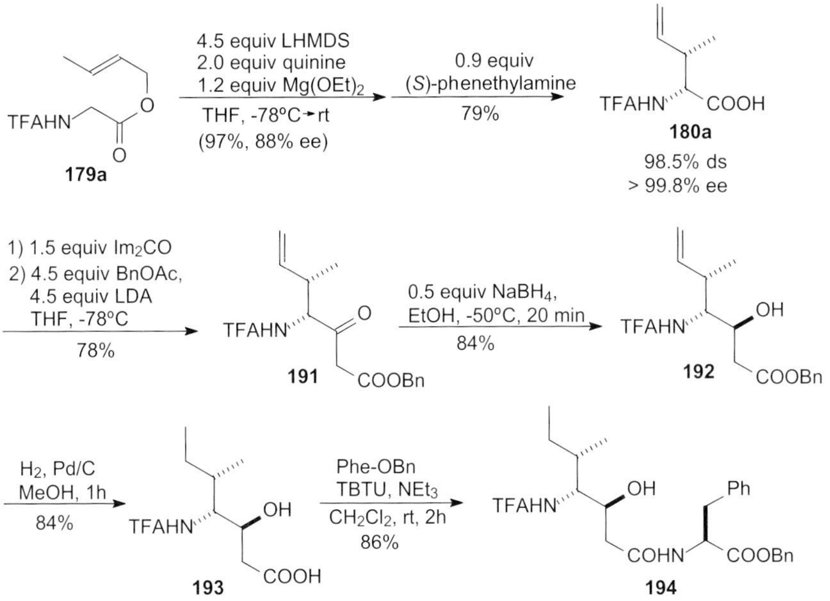

Scheme 5.2.61

lactones are rather sensitive towards nucleophilic lactone opening, probably because of ring strain in the bicyclic system. If an amino acid ester is used as a nucleophile, this protocol allows a direct incorporation of the modified proline into peptides. Steric hindrance obviously does not play a significant role. Even larger peptides with sterically demanding *N*-terminal amino acids, such as valine or isoleucine provide the corresponding enlarged peptide **197** in almost quantitative yield. The TFA-group can be removed easily by saponification [114] or reduction [115], allowing the prolongation of the peptide chain at the amino terminus.

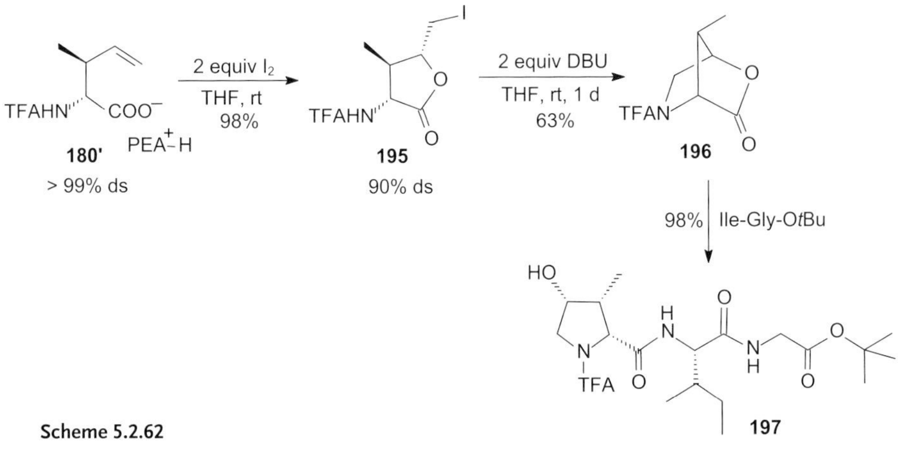

Scheme 5.2.62

5.2.2.4 Rearrangement of α-Thio Substituted Allylic Esters

In contrast to the α-oxy- or α-amino-substituted allylic esters, not many examples for α-thio substituted substrates and their application in Claisen rearrangements are described so far (Scheme 5.2.63). Lythegoe et al. applied the Ireland–Claisen rearrangement to the synthesis of α-mercapto acids such as **199** starting from allylic esters **198** [116]. **199** was oxidized and further converted into α,β-unsaturated esters or subjected to an oxidative degradation. Jones et al. used the Ireland–Claisen rearrangement of **200** in their syntheses of thietan oxides [117].

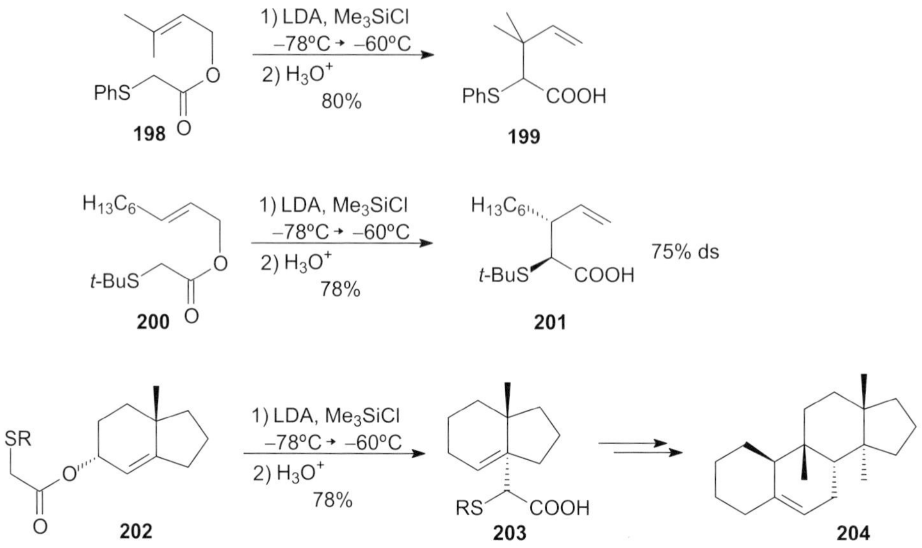

1) LDA, Me_3SiCl, −78°C → −60°C; 2) H_3O^+, 80%: PhS ... **198** → PhS ... COOH **199**

1) LDA, Me_3SiCl, −78°C → −60°C; 2) H_3O^+, 78%: $H_{13}C_6$ ... *t*-BuS ... **200** → $H_{13}C_6$ ... *t*-BuS ... COOH **201**, 75% ds

SR ... **202** → 1) LDA, Me_3SiCl, −78°C → −60°C; 2) H_3O^+, 78% → RS ... COOH **203** → **204**

Scheme 5.2.63

General Procedure. 2-*tert*-Butylthio-3-Vinylnonanoic Acid (201)

n-Butyllithium (222 mmol) in hexane (231 mL) was added to diisopropylamine (32.7 mL, 232 mmol) in THF (150 mL) with stirring at −10 °C under nitrogen. The solution was cooled to −78 °C, and a solution of (*E*)-non-2-enyl-*tert*-butylthioacetate (**200**) (57.3 g, 211 mmol) in THF (180 mL) was added over 20 min with stirring. After a further 30 min, Me_3SiCl (27.5 mL, 218 mmol) was added drop-wise. The mixture was allowed to warm to room temperature, and then heated at 60 °C for 2 h. After cooling, diethyl ether (1.2 L) was added and the solution was washed with 2N hydrochlorid acid (2 × 200 mL). The aqueous washings were extracted with diethyl ether (2 × 200 mL) and the combined extracts were dried ($MgSO_4$) and evaporated. Chromatography of the residue on silica (400 g), (diethyl ether/light petroleum, 1/9 and then diethyl ether as eluants) gave a mixture of *syn-* and *anti-***201** (44.7 g, 78%) in a 3:1 ratio.

In connection with a synthesis directed towards the lanostane skeleton **204**, Watt et al. used the Ireland–Claisen rearrangement of **202** as key step to introduce the quaternary center in **203** stereoselectively [118]. The diastereoselectivity of the rearrangement was uninteresting in this case, because the carboxylic group was removed during the further synthesis.

5.2.3
Claisen Rearrangements of Substrates Bearing Chelating Substituents in the β-Position

5.2.3.1 β-Hydroxy Substituents

To expand the scope of the diastereoselective C–C bond formation by the ester enolate Claisen rearrangement, the rearrangement of allylic esters of β-hydroxy acids is noted to introduce three continuous asymmetric centers. Fujisawa et al. investigated the rearrangement of *E*- and *Z*-crotylesters **205** under different reaction conditions [119].

Treatment of **205a** with 3 equiv LHMDS in THF at −30 °C and, after warming to room temperature, refluxing for 2 h gave a mixture of three diastereomers of which **206b** was formed preferentially (Table 5.2.21, entry 1). To obtain a higher diastereoselectivity and yield, several conditions for the base and solvent were examined. LDA and lithium 2,2,6,6-tetramethyl piperidide (LTMP) as a base gave the rearrangement product in only 33 or 24% yield with even lower diastereoselectivity (77 and 65%). A retro-aldol reaction was found to occur simultaneously as a side reaction. The use of a nonpolar solvent, such as hexane and toluene, resulted in a higher yield (71–76%) but lower diastereoselectivity (1:1 mixture). In contrast, more polar solvents such as dimethoxyethane (DME) gave the expected product

R^2 R^1 O OH O **205** 1) LHMDS 2) H_3O^+ 3) CH_2N_2 → OH COOMe **206a** + OH COOMe **206b** + OH COOMe **206c** + OH COOMe **206d**

Scheme 5.2.64

Tab. 5.2.21 Ester enolate Claisen rearrangement of esters **205**.

Entry	Ester	R^1	R^2	Solvent	Additive	Yield [%]	Ratio 206a : 206b : 206c : 206d
1	**205a**	H	Me	DME	–	58	11 : 85 : 4 : 0
2	**205b**	Me	H	DME	–	31	75 : 19 : 4 : 2
3	**205a**	H	Me	THF	Me_3SiCl	37	3 : 18 : 72 : 7
4	**205b**	Me	H	THF	Me_3SiCl	24	25 : 3 : 4 : 68

206b in 58% and 58% ds. Rearrangement of the corresponding (*Z*)-ester **205b** gave rise to stereoisomer **206a** preferentially in comparable selectivity (entry 2). The variation of the diastereoselectivity was further examined using the Ireland version of the Claisen rearrangement. Interestingly, the stereoisomers **206c** and **206d** were formed preferentially but with lower yield and selectivity (entries 3 and 4). This can be explained by different transition states for the dianion or silylketene acetal intermediates.

Similar results were obtained by Kurth et al., who also investigated the influence of the stereogenic center at the β-position of the hydroxy acid [120]. Replacing the methyl group at the double bond in **205** by an isopropyl group had no significant influence on the stereochemical outcome of the reaction. But excellent selectivities were obtained in the rearrangement of cyclic allylic esters such as **207** [121]. Upon treatment of **207a** with 2 equiv of LDA in THF at –78 °C and subsequent warming to 50 °C for 12 h gave rearrangement products **208a** and **208b** (after esterification) in a 98:2 ratio. It is unlikely that the improved diastereoselectivity with **207** in comparison to **205** is due to increased selectivity for the formation of the (*E*)-configured enolate in the rearrangement of **207**, in that both dianionic Claisen rearrangements were performed under the same conditions. Rather, these results are consistent with improved chair selectivity for **207** and, assuming **208** is exclusively the result of a chair-like transition state, indicate that an *E*/*Z*= 98/2 ratio for the enolate formation under these conditions. This can be explained by chelate complex formation as illustrated in Scheme 5.2.65.

As anticipated on the basis of steric bulk in the allyl moiety, the dianionic Claisen rearrangement of **207b** was slightly less selective than that of **207a**, but much

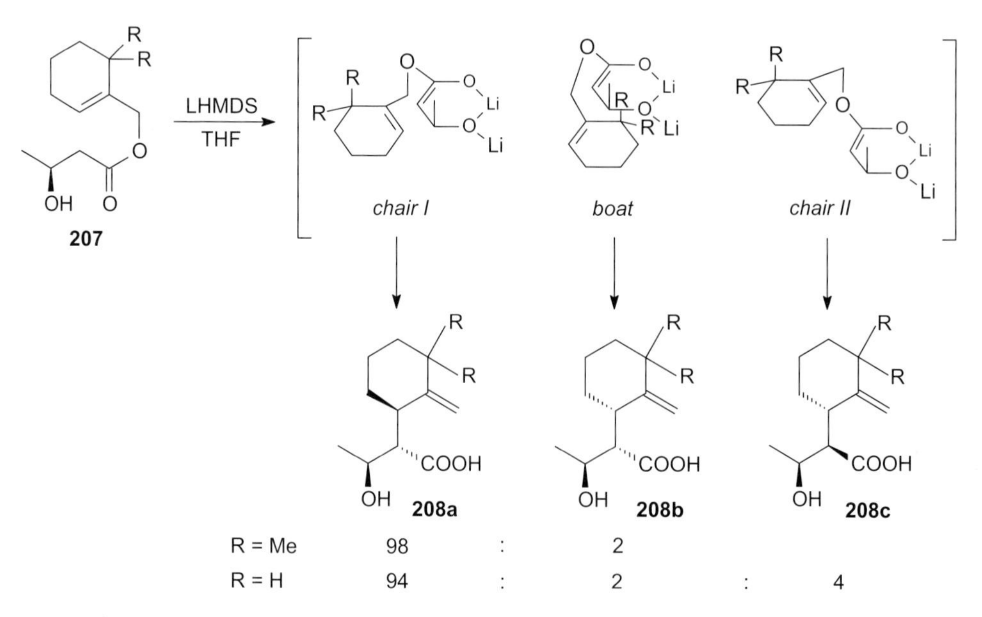

Scheme 5.2.65

more selective than that of **205b**. A comparison of the diastereoselectivities obtained with **205** and **207** corroborate the conclusion, that (*E*)-enolate selectivity is nearly complete because of chelation, while the chair/boat selectivity is variable. Moreover, it can be concluded that the (*E*)-enolate of **207a** is ~100% face selective at Cα, while the substrate-induced diastereoselectivities for the (*E*)-enolates of **205** and **207b** are lower. Hence, the formation of **206c** from **205** and **208c** from **207** is a consequence of an incomplete substrate-induced diastereoselectivity and is not due to a limited *E*/*Z* selectivity of the enolate formation.

5.2.3.2 β-Alkoxy Substitutents

Not many examples of rearrangements of β-alkoxy substituted esters are described, probably because of major side reactions, such as elimination of the alkoxy group from the enolate formed. During their synthesis of the sesquiterpene trichodiene Gilbert et al. used the rearrangement of ester **209** as a key step (Scheme 5.2.66) [122]. If the formation of the silylketene intermediate was carried out at –110 °C, the undesired β-elimination could be suppressed completely, and the rearrangement products **210a** and **210b** are formed in high yield and diastereoselectivity. The formation of **210a** can be explained by a chair-like transition state and **210b** by a boat-like transition state.

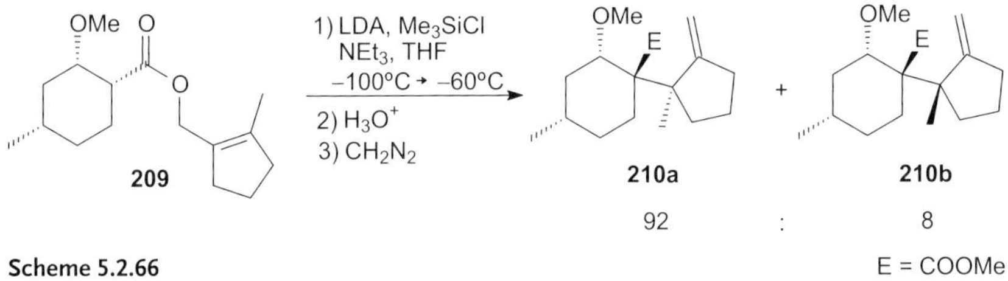

1) LDA, Me_3SiCl, NEt_3, THF, –100°C → –60°C; 2) H_3O^+; 3) CH_2N_2

209 → **210a** + **210b**

92 : 8

E = COOMe

Scheme 5.2.66

5.2.2.3 β-Amino Substituted Substrates

The rearrangement of allylic esters obtained from β-amino acids was investigated by Knight et al. during their syntheses of kainic acid derivatives [123]. They subjected different protected β-alanine esters **211** to the typical Ireland–Claisen conditions, and by far the best results are obtained with the *N*-Boc-protected substrates (Scheme 5.2.67, Table 5.2.22). In the case of the *Z*-protected derivatives only low yields are isolated along with other materials which appeared to arise from metal-

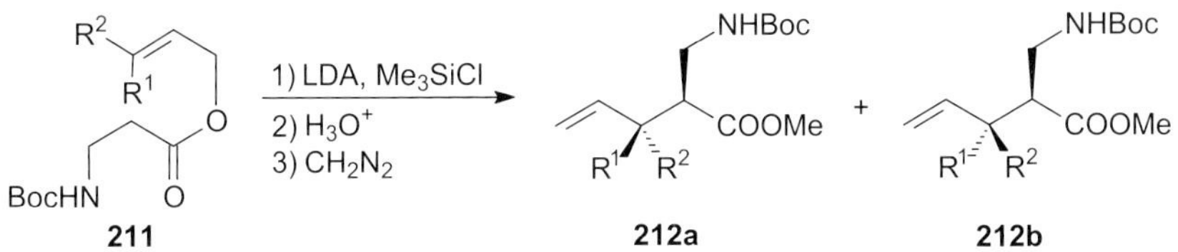

1) LDA, Me_3SiCl; 2) H_3O^+; 3) CH_2N_2

211 → **212a** + **212b**

Scheme 5.2.67

Tab. 5.2.22 Ireland–Claisen rearrangement of β-amino acid esters **208**.

Entry	Substrate	R^1	R^2	Yield [%]	Ratio **212a : 212b**
1	**211a**	Me	H	88	86 : 14
2	**211b**	H	Me	73	12 : 88
3	**211c**	$CH_2OTBDMS$	H	77	92 : 8
4	**211d**	H	$CH_2OTBDMS$	68	6 : 94
5	**211e**	Ph	H	31	55 : 45
6	**211f**	H	Ph	40	77 : 23
7	**211g**	*i*Pr	H	85	80 : 20
8	**211h**	H	*i*Pr	81	81 : 19

lation at the benzylic positions, whereas the N-phthaloyl derivatives gave virtually no rearrangement product, probably because of decomposition via β-elimination.

A number of alternative reaction conditions were also examined using ester **211**, but in all cases these resulted in either lower yield and/or reduced selectivity. Trapping the enolate with TBDMSCl gave only a 2:1 ratio of diastereomers as did treating the ester with a pre-mixed solution of LDA/Me_3SiCl. Slightly higher diastereoselectivities (91% ds) were obtained under the usual conditions at higher temperature, but the yields are considerably reduced (25%). The use of KHMDS as base resulted in high yields (80%), but with virtually no diastereoselectivity. The substitution pattern on the allylic moiety was also investigated and the results are shown in Table 5.2.22. The likely involvement of a single predominant transition state in most cases is indicated by the direct relationship between the olefin geometry and the major diastereoisomer (entries 1–4). An exception to this is, when the allylic alcohol residue is branched α to the six-centered transition state (entries 5–8). This implicates that in these (*Z*)-isomers, the chair conformation is rather crowded and the rearrangement proceeds instead via a boat-like transition state.

Finally it was observed that the *N,N*-dimethylated amino acid ester could also be rearranged in this way to give a 70:30 stereoisomeric mixture in 70% yield. However, as the dimethylamino function is much less useful in terms of subsequent manipulations, the use of this group was not further evaluated.

But Still et al. used such a dialkylamino acid ester rearrangement during their elegant synthesis of frullanolide [124]. They investigated several cyclic allylic esters such as **213** (Scheme 5.2.68).

The crude rearrangement product was directly subjected to elimination giving rise to the α-methylene ester **211**. Although a number of steps are involved in the transformation **213**→**214**, the entire operation may be conducted in a single flask.

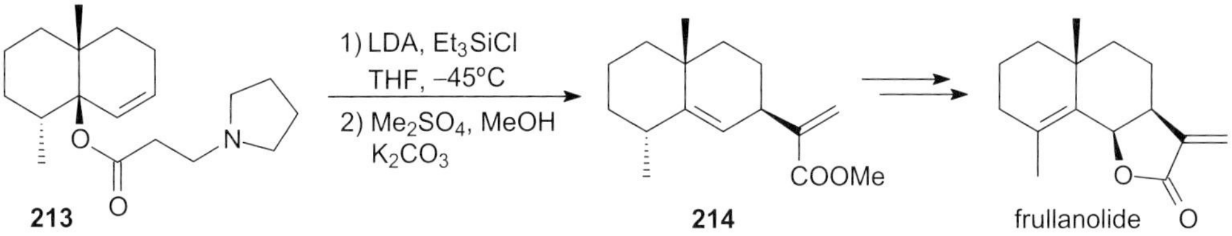

Scheme 5.2.68

5.2.4
Chelation Controlled Aza-Claisen Rearrangements

In contrast to the well-investigated chelation controlled ester enolate Claisen rearrangments, the corresponding rearrangement of *N*-allylated amides is much less applied. Tsunoda et al. evaluated the synthetical potential of *N*-crotyl glycolamides and glycinamides **215** [125] and the results obtained are listed in Table 5.2.23.

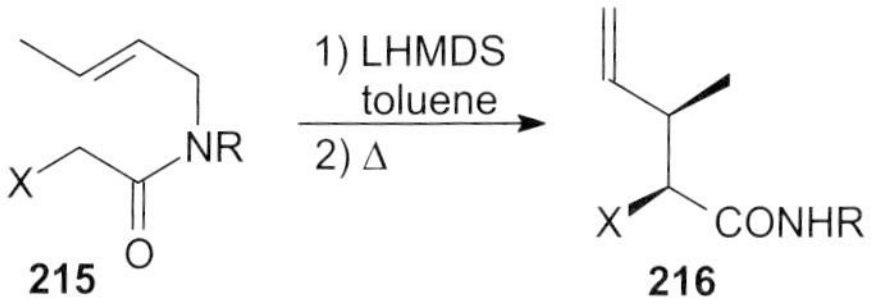

Scheme 5.2.69

Tab. 5.2.23 Aza–Claisen rearrangement of amides **215**.

Entry	Substrate	X	R	*T* [°C]	*t* [h]	Yield [%]	ds [%]
1	**215a**	Me	*n*-Bu	135	4	92	99.5
2	**215b**	OH	*n*-Bu	80	15	74	98
3	**215c**	NH_2	*n*-Bu	rt	15	81	98
4	**215d**	OTBDMS	*n*-Bu	100	15	59	98
5	**215e**	NHBoc	*n*-Bu	140	2	–	–
6	**215f**	OH	(*R*)-phenethyl	80	15	95	86 (2*S*,3*R*)
7	**215g**	NH_2	(*S*)-phenethyl	rt	15	89	89 (2*R*,3*S*)
8	**215h**	OTBDMS	(*R*)-phenethyl	120	6	62	67 (2*S*,3*R*)

Excellent *syn* diastereoselectivities were observed for nearly all cases. The presence of an OH or NH_2 group at the β-position (entries 2 and 3) facilitated the rearrangement in comparison to unfunctionalized carboxamides (entry 1), possibly due to the intermediary of dianioic species. The protection of the free OH or NH_2 group decelerates the reaction (entries 4 and 5) contrary to the rearrangement of ester enolates [8, 57]. This is especially dramatic for amino protected substrates (entry 5). Trapping of the enolate as TMS ether and subsequent thermal rearrangement also decelerates the reaction and lowers the yield and selectivity as in α-alkyl derivatives. Thus, the best method is simply to prepare amide enolates without any protection of the OH and NH_2 group, and, if necessary, heat them under basic conditions.

The substrate-induced diastereoselectivity was investigated using an (*R*)- or (*S*)-1-phenylethyl group as a chiral auxiliary on the amide nitrogen. The reactions with the unprotected compounds went smoothly and diastereomeric mixtures were obtained in excellent yield and *syn* diastereoselectivity (> 99% *syn*) (entries 6–8). The results were better than those obtained with the butylamides. The facial selectivity was in the range of about 8:1. The protection of the hydroxyl group lowered both the facial selectivity and the yield (entry 8).

List of Abbreviations

Bn	Benzyl
Boc	*t*-Butyloxycarbonyl
Bz	Benzoyl
COD	1,5-Cyclooctadiene
DDQ	2,3-Dichloro-5,6-dicyano-*p*-benzoquinone
DEAD	Diethyl azodicarboxylate
Dr	Diastereomeric ratio
Ds	Diastereoselectivity
E	Methoxycarbonyl
EDC	*N'*-(3-Dimethylaminopropyl)-*N*-ethylcarbodiimide
Ee	Enantiomeric excess
Fmoc	Fluorenylmethoxycarbonyl
HMPA	Hexamethyl phosphoric acid triamide
Im_2CO	Carbonylimidazol
IPC	Isopinocamphenyl
KHMDS	Potassium hexamethyldisilazide
LDA	Lithiumdiisopropylamide
LHMDS	Lithium hexamethyldisilazide
LICA	Lithium isopropylcyclohexylamide
MEM	Methoxyethoxymethyl
MOM	Methoxymethyl
Ms	Mesyl (methansulfonyl)
NBS	*N*-Bromo succinimide
Pfb	Pentafluorobenzoyl
PMB	*p*-Methoxybenzyl
Rt	Room temperature

TBDMS *tert*-Butyldimethylsilyl
TBTU *O*-1-H-(Benzotriazo-1-yl)-*N*,*N*,*N'*,*N'*-tetramethyluronium tetrafluoroborate
TFA Trifluoroacetyl
Tms Trimethylsilyl
Triton B Benzyltrimethylammonium hydroxide
Ts Tosyl
Z Benzyloxycarbonyl

References

1 General reviews on [3,3]-sigmatropic rearrangements, including stereochemical aspects: a) S. J. Rhoads, N. R. Raulins, *Org. React.* **1975**, *22*, 1–252; b) F. E. Ziegler, *Acc. Chem. Res.* **1977**, *10*, 227–232; c) G. B. Bennett, *Synthesis* **1977**, 589–606; d) P. A. Bartlett, *Tetrahedron* **1980**, *36*, 1–72; e) R. K. Hill in *Asymmetric Synthesis* J. D. Morrison (ed.) Vol. 3, p 503ff., Academic, **1984**; f) S. Blechert, *Synthesis* **1989**, 71–82; g) P. Wipf in *Comprehensive Organic Synthesis* Vol. 5, B. M. Trost, I. Fleming (eds.) Pergamon Press, New York, **1991**, p 827ff; h) H.-J. Altenbach in *Organic Synthesis Highlights* J. Mulzer, H.-J. Altenbach, M. Braun, K. Krohn, H.-U. Reissig (eds.) Wiley-VCH, Weinheim, **1991**, p 111f, p 116f; i) S. Pereira, M. Srebnik, *Aldrichimica Acta* **1993**, *26*, 17; k) H. Frauenrath in *Houben-Weyl E 21d* G. Helmchen, R. W. Hoffmann, J. Mulzer, E. Schaumann; (eds.) Thieme, Stuttgart, **1995**, p 3301ff; l) P. Metzner, *Pure Appl. Chem.* **1996**, *68*, 863–868 m) D. Enders, M. Knopp, R. Schiffers, *Tetrahedron Asym.* **1996**, *7*, 1847–1882; n) L. A. Paquette, *Tetrahedron* **1997**, *53*, 13971–14020; o) H. Ito, T. Taguchi, *Chem. Soc. Rev.* **1999**, *28*, 43–50; p) S. M. Allin, R. D. Baird, *Curr. Org. Chem.* **2001**, 395–415; q) M. Hiersemann, L. Abraham, *Eur. J. Org. Chem.* **2002**, 1461–1471; r) Y. Chai, S. P. Hong, H. A. Lindsay, C. McFarland, M. C. McIntosh, *Tetrahedron* **2002**, *58*, 2905–2928; s) U. Nubbemeyer, *Synthesis* **2003** 961–1008.

2 See Chapters 1 and 4.

3 a) R. E. Ireland, R. H. Mueller, *J. Am. Chem. Soc.* **1972**, *94*, 5897–5898; b) R. E. Ireland, R. H. Mueller, A. K. Willard, *J. Am. Chem. Soc.* **1976**, *98*, 2868–2877; c) R. E. Ireland, C. S. Wilcox, Jr., *Tetrahedron Lett.* **1977**, 2839–2842.

4 C. H. Heathcock in *Asymmetric Synthesis* J. J. Morrison (ed.) Academic Press, New York, **1984**, Vol. 3, Chapter 2.

5 a) S. Ito, T. Tsunoda, *Pure Appl. Chem.* **1990**, *62*, 1405–1408; b) T. Tsunoda, S. Tatsuki, Y. Shiraishi, M. Alasaka, S. Ito, *Tetrahedron Lett.* **1993**, *34*, 3297–3300.

6 U. Kazmaier, *Lieb. Ann./Recl.* **1997**, 285–295.

7 D. J. Ager, R. C. Cookson, *Tetrahedron Lett.* **1982**, *33*, 3419–3420.

8 a) P. A. Bartlett, D. J. Tanzella, J. F. Barstow, *Tetrahedron Lett.* **1982**, *23*, 619–622; b) P. A. Bartlett, D. J. Tanzella, J. F. Barstow, *J. Org. Chem.* **1982**, *47*, 3941–3945.

9 T. Sato, K. Tajima, T. Fujisawa, *Tetrahedron Lett.* **1983**, *24*, 729–730.

10 D. F. Sullivan, R. P. Woodbury, M. W. Rathke, *J. Org. Chem.* **1977**, *42*, 2038–2039.

11 T. Fujisawa, K. Tajima, T. Sato, *Chem. Lett.* **1984**, 1669–1672.

12 T. Fujisawa, H. Kohama, K. Tajima, T. Sato, *Tetrahedron Lett.* **1984**, *25*, 5155–5156.

13 M. A. Sparks, J. S. Panek, *J. Org. Chem.* **1991**, *56*, 3431–3438.

14 T. Fujisawa, E. Maehata, H. Kohama, T. Sato, *Chem. Lett.* **1985**, 1457–1458.

15 a) J. Kallmerten, T. J. Gould, *Tetrahedron Lett.* **1983**, *24*, 5177–5180; b) J. Kallmerten, M. D. Wittman, *Tetrahedron Lett.* **1986**, *27*, 2443–2446; c) T. J. Gould, M. Balestra, M. D. Wittman, J. A. Gary, L. T. Rossano, J. Kallmerten, *J. Org. Chem.* **1987**, *52*, 3889–3901.

16 S. D. Burke, W. F. Fobare, G. J. Pacofsky, *J. Org. Chem.* **1983**, 5221–5228.

17 M. Nagatsuma, F. Shirai, N. Sayo, T. Nakai, *Chem. Lett.* **1984**, 1393–1396.

18 J. Kallmerten, T. J. Gould, *J. Org. Chem.* **1986**, *51*, 1155–1157.

19 J. R. Hauske, S. M. Julin, *Tetrahedron Lett.* **1993**, *34*, 4909–4012.

20 J. S. Panek, T. D. Clark, *J. Org. Chem.* **1992**, *57*, 4323–4326.

21 T. Konno, T. Kitazume, *Chem. Commun.* **1996**, 2227–2228.

22 K. Ritter, *Tetrahedron Lett.* **1990**, *31*, 869–872.

23 J. Kallmerten, T. J. Gould, *J. Org. Chem.* **1985**, *50*, 1128–1131.

24 J. Kallmerten, M. Balestra, *J. Org. Chem.* **1986**, *51*, 2855–2857.

25 K. Mikami, K. Kawamoto, T. Nakai, *Tetrahedron Lett.* **1986**, *27*, 4899–4902.

26 J. C. Barrish, H. L. Lee, E. G. Baggiolini, M. R. Uskokovic, *J. Org. Chem.* **1987**, *52*, 1375–1378.

27 J. Mulzer, J.-T. Mohr, *J. Org. Chem.* **1994**, *59*, 1160–1165.

28 R. W. Jackson, K. J. Shea, *Tetrahedron Lett.* **1994**, *35*, 1317–1320.

29 R. E. Ireland, P. Wipf, J.-N. Xiang, *J. Org. Chem.* **1991**, *56*, 3572–3582.

30 D. Kim, J. I. Lim, *Tetrahedron Lett.* **1995**, *36*, 5035–5036.

31 a) S. D. Burke, G. J. Pacofsky, A. D. Piscopio, *Tetrahedron Lett.* **1986**, *27*, 3345–3348. b) S. D. Burke, G. J. Pacofsky, A. D. Piscopio, *J. Org. Chem.* **1992**, *57*, 2228–2235.

32 O. Mitsunobu, *Synthesis* **1981**, 1–28.

33 S. D. Burke, R. A. Ng, J. A. Morrison, M. J. Alberti, *J. Org. Chem.* **1998**, *63*, 3160–3161.

34 a) G. C. Fu, S. T. Nguyen, R. H. Grubbs, *J. Am Chem. Soc.* **1993**, *115*, 9856–9857; b) P. Schwab, M. B. France, J. W. Ziller, R. H. Grubbs, *Angew. Chem.* **1995**, *107*, 2179–2181; *Angew. Chem. Int. Ed.* **1995**, *34*, 2039–2041; c) E. L. Dias, S. T. Nguyen, R. H. Grubbs, *J. Am. Chem. Soc.* **1997**, *119*, 3887–3897.

35 R. E. Ireland, R. S. Meissner, M. A. Rizzacasa, *J. Am. Chem. Soc.* **1993**, *115*, 7166–7172.

36 a) A. Fürstner, K. Langemann, *J. Org. Chem.* **1996**, *61*, 3942–3943; b) A. Fürstner, K. Langemann, *Synthesis* **1997**, 792–803.

37 U. Gräfe, W. Schade, M. Roth, L. Radics, M. Incze, K. Kjszászy, *J. Antiobiot.* **1984**, *37*, 836–841.

38 J. W. Westley, R. H. Evans, C.-M. Liu, T. Hermann, J. F. Blount, *J. Am. Chem. Soc.* **1978**, *100*, 6784–6786.

39 S. D. Burke, J. Hong, J. R. Lennox, A. P. Mongin, *J. Org. Chem.* **1998**, *63*, 6952–6967.

40 S. F. Martin, J. A. Dodge, *Tetrahedron Lett.* **1991**, *32*, 3017–3020.

41 W. Picoul, R. Urchegui, A. Haudrechy, Y. Langlois, *Tetrahedron Lett.* **1999**, *40*, 4797–4800.

42 Review: A Giannis, F. Rübsam, *Angew. Chem. Int. Ed.* **1997**, *109*, 606–609; *Angew. Chem. Int. Ed.* **1997**, *36*, 588–590.

43 R. Di Florio, M. A. Rizzacasa, *J. Org. Chem.* **1998**, *63*, 8595–8598.

44 A. N. Cuzzupe, R. Di Florio, M. A. Rizzacasa, *J. Org. Chem.* **2002**, *67*, 4392–4398.

45 a) L. M. McVinish, M. A. Rizzacasa, *Tetrahedron Lett.* **1994**, *35*, 923–926; b) R. W. Gable, L. M. McVinish, M. A. Rizzacasa, *Aust. J. Chem.* **1994**, *47*, 1537–1544; c) R. K. Mann, J. G. Parsone, M. A. Rizzacasa, *J. Chem. Soc., Perkin Trans 1* **1998**, 1283–1293.

46 a) R. E. Ireland, D. W. Norbeck, *J. Am. Chem. Soc.* **1985**, *107*, 3279–3285; b) I. Paterson, A. W. Hulme, *J. Org. Chem.* **1995**, *60*, 3288–3300.

47 R. E. Ireland, P. Wipf, J. D. Armstrong III, *J. Org. Chem.* **1991**, *56*, 650–657.

48 J. H. Hong, C.-H. Oh, J.-H. Cho, *Tetrahedron* **2003**, *59*, 6103–6108.

49 J. K. Whitesell, A. M. Helbling, *J. Org. Chem.* **1980**, *45*, 4135–4139.

50 a) H. H. Wasserman, B. H. Lipshutz, *Tetrahedron Lett.* **1975**, 4611–4614; b) H. H. Wasserman, B. H. Lipshutz, *Tetrahedron Lett.* **1976**, 4613–4616.

51 P. G. M. Wuts, C. Sutherland, *Tetrahedron Lett.* **1982**, *23*, 3987–3990.

52 T. Oh, Z. Wrobel, P. N. Devine, *Synlett* **1992**, 81–83.

53 For the generation of boron enolates see: a) T. Inque, T. Mukaiyama, *Bull. Chem. Soc. Jpn.* **1980**, *53*, 174–183; b) D. A. Evans, J. V. Nelson, E. Vogel, T. R. Taber, *J. Am. Chem. Soc.* **1981**, *103*, 3099–3111.

54 For the generation of tin enolates see: a) N. Iwasawa, T. Mukaiyama, *Chem. Lett.* **1982**, 1441–1444; b) T. Mukaiyama, N. Iwasawa, R. W. Stevens, T. Haga, *Tetrahedron* **1984**, *40*, 1381–1390.

55 a) B. Kübel, G. Höfle, W. Steglich, *Angew. Chem.* **1975**, *87*, 64–67; *Angew. Chem. Int. Ed.* **1975**, *14*, 58–60; b) N. Engel, B. Kübel, W. Steglich, *Angew. Chem.* **1977**, *89*, 408–410; *Angew. Chem. Int. Ed.* **1977**, *16*, 349–351.

56 a) J. Fischer, C. Kilpert, U. Klein, W. Steglich, *Tetrahedron* **1986**, *42*, 2063–2074; b) K. Burger, K. Geith, K. Gaa, *Angew. Chem.* **1988**, *100*, 860–861; *Angew. Chem. Int. Ed.* **1988**, *27*, 848–851; c) A. L. Castelhano, S. Horne, G. J. Taylor, R. Billedeau, A. Krantz, *Tetrahedron*, **1988**, *44*, 5451–5466; d) L. Colombo, G. Casiraghi, A. Pittalis, *J. Org. Chem.* **1991**, *56*, 3897–3900; e) M. W. Holladay, A. M. Nadzan, *J. Org. Chem.* **1991**, *56*, 3900–3905.

57 P. A. Bartlett, J. F. Barstow, *J. Org. Chem.* **1982**, *47*, 3933–3941.

58 R. K. Boeckman, Jr., J. C. Potenza, E. J. Enholm, *J. Org. Chem.* **1987**, *52*, 469–472.

59 J. E. Baldwin, M. Bradley, N. J. Turner, R. M. Adlington, A. R. Pitt, H. Sheridan, *Tetrahedron* **1991**, *47*, 8203–8222.

60 T. Konno, T. Daitoh, T. Ishihara, H. Yamanaka, *Tetrahedron Asym.* **2001**, *12*, 2743–2748.

61 a) M. Mohamed, M. A. Brook, *Tetrahedron Lett.* **2001**, *42*, 191–193; b) M. Mohamed, M. A. Brook, *Helv. Chim. Acta* **2001**, *85*, 4165–4181.

62 I. Fleming, A. Barbero, D. Walter, *Chem. Rev.* **1997**, *97*, 2063–2192.

63 H. M. Hull, D. W. Knight, *J. Chem. Soc., Perkin Trans. I*, **1997**, 857–863.

64 P. A. Bartlett, J. F. Barstow, *Tetrahedron Lett.* **1982**, 623–626.

65 For reviews on reactions of chelated enolates see: a) U. Kazmaier, *Recent Res. Devel. in Organic Chem.* **1998**, *2*, 351–358; b) U. Kazmaier in *Bioorganic Chemistry*, Wiley-VCH, Weinheim, **1999**, 201–206; c) U. Kazmaier, *J. Indian. Chem. Soc.* **1999**, *76*, 631–639; d) U. Kazmaier, S. Maier, F. L. Zumpe, *Synlett* **2000**, 1523–1535.

66 U. Kazmaier, *Angew. Chem.* **1994**, *106*, 1046–1047.

67 W. Qui, X. Gu, V. A. Soloshonok, M. D. Carducci, V. J. Hruby, *Tetrahedron Lett.* **2001**, *42*, 145–148.

68 J. F. Miller, A. Termin, K. Koch, A. D. Piscopio, *J. Org. Chem.* **1998**, *63*, 3158–3159.

69 Y. Morimoto, M. Takaishi, T. Kinoshita, K. Sakaguchi, K. Shibata, *Chem. Commun.* **2002**, 42–43.

70 a) U. Kazmaier, *Synlett* **1995**, 1138–1140; b) U. Kazmaier, *J. Org. Chem.* **1996**, *61*, 3694–3699.

71 a) U. Kazmaier, S. Maier, *J. Chem. Soc., Chem. Comm.* **1995**, 1991–1992; b) U. Kazmaier, S. Maier, *Tetrahedron* **1996**, *52*, 941–954.

72 M. Gander-Coquoz, D. Seebach, *Helv. Chim. Acta*, **1988**, 71, 224–236.

73 a) A. L. Maycock, S. D. Aster, A. A. Patchett, *Developments in Biochemistry* **1979**, *6*, 115–117; b) C. Danzin, P. Casara, N. Claverie, B. W. Metcalf, *J. Med. Chem.* **1981**, *24*, 16.

74 a) U. Schmidt, A. Lieberknecht, J. Wild, *Synthesis* **1984**, 53–60.

75 U. Kazmaier, *Tetrahedron Lett.* **1996**, *37*, 5351–5354.

76 J. E. Baldwin, *J. Chem. Soc., Chem. Commun.* **1976**, 734–736.

77 R. F. Heck, Palladium Reagents in Organic Syntheses, Academic Press, London, 1985.

78 G. T. Crisp, P. T. Glink, *Tetrahedron*, **1992**, *48*, 3541–3556.

79 a) M. J. Broadhurst, J. M. Percy, M. E. Prime, Tetrahedron Lett. **1997**, *38*, 5903–5906; b) J. M. Percy, M. E. Prime, M. J. Broadhurst, *J. Org. Chem.* **1998**, *63*, 8049–8051.

80 U. Kazmaier, *Tetrahedron* **1994**, *50*, 12895–12902.

81 a) H. Gehring, R. R. Rando, P. Christen, *Biochemistry* **1977**, *16*, 4832–4836;

b) J. J. Likos, H. Ueno, R. W. Feldhaus, D. E. Metzler, *Biochemistry* **1982**, *21*, 4377–4386.

82 G. B. Bennett, *Synthesis* **1977**, 589–606.

83 B. Kübel, P. Gruber, R. Hurnaus, W. Steglich, *Chem. Ber.* **1979**, *112*, 128–137.

84 U. Kazmaier, C. H. Görbitz, *Synthesis* **1996**, 1489–1493.

85 Determined by X-Ray-Structure.

86 D. Hoppe, C. Gonschorrek, E. Egert, D. Schmidt, *Angew. Chem.* **1985**, *97*, 706–707; *Angew. Chem. Int. Ed. Engl.* **1985**, *24*, 700–701.

87 S. Masamune, S. A. Ali, D. L. Snitman, D. S. Garvey, *Angew. Chem.* **1980**, *92*, 573–575; *Angew. Chem. Int. Ed. Engl.* **1980**, *19*, 654–656.

88 F. L. Zumpe, U. Kazmaier, *Synlett* **1998**, 434–436.

89 a) U. Kazmaier, C. Schneider, *Synlett*, **1996**, 975–977; b) U. Kazmaier, C. Schneider, *Synthesis* **1998**, 1321–1326.

90 a) U. Kazmaier, C. Schneider, *Tetrahedron Lett.* **1998**, *39*, 817–818; b) C. Schneider, U. Kazmaier, *Eur. J. Org. Chem.* **1998**, 1155–1159.

91 R. E. Looper, R. M. Williams, *Tetrahedron Lett.* **2001**, *42*, 769–771.

92 a) H. Y. Kim, P. L. Toogood, *Tetrahedron Lett.* **1996**, *37*, 2349–2352; b) H. Y. Kim, K. Stein, P. L. Toogood, *Chem. Commun.* **1996**, 1683–1684; c) R. Samy, H. Y. Kim, M. Brady, P. L. Toogood, *J. Org. Chem.* **1999**, *64*, 2711–2728.

93 For reviews see: D. Seebach, *Angew. Chem.* **1988**, *100*, 1685–1715; *Angew. Chem. Int. Ed. Engl.* **1988**, *27*, 1624–1655. b) D. Seebach, *Aldrichim. Acta* **1992**, *25*, 59.

94 U. Kazmaier, *J. Org. Chem.* **1994**, *59*, 6667–6670.

95 a) L. E. Overman, C. B. Campbell, *J. Org. Chem.* **1976**, *41*, 3338–3340; b) L. E. Overman, C. B. Campbell, F. M. Knoll, *J. Am. Chem. Soc.* **1978**, *100*, 4822–4834; c) L. E. Overman, *Angew. Chem.* **1984**, *96*, 565–573; *Angew. Chem. Int. Ed.* **1984**, *23*, 579–591. See also: R. P. Lutz, *Chem. Rev.* **1984**, *84*, 205–247.

96 For mechanistic studies see: a) L. E. Overman, E. J. Jacobsen, *J. Am. Chem. Soc.* **1982**, *104*, 7225–7231; b) T. G. Schenck, B. Bosnich, *J. Am. Chem. Soc.* **1985**, *107*, 2058–2066; c) L. E. Overman, A. F. Renaldo, *J. Am. Chem. Soc.* **1990**, *112*, 3945–3949.

97 H. Sigel, B. Martin, *Chem. Rev.* **1982**, *82*, 385–426.

98 a) U. Kazmaier, S. Maier, *J. Chem. Soc., Chem. Commun.* **1998**, 2535–2536; b) U. Kazmaier, S. Maier, *Org. Lett.* **1999**, *1*, 1763–1766; c) S. Maier, U. Kazmaier, *Eur. J. Org. Chem.* **2000**, 1241–1251.

99 U. Kazmaier, S. Maier, *J. Org. Chem.* **1999**, *64*, 4574–4575.

100 a) K. Maruoka, H. Banno, H. Yamamoto, *J. Am. Chem. Soc.* **1990**, *112*, 7791–7793; b) K. Maruoka, S. Saito, H. Yamamoto, H. *J. Am. Chem. Soc.* **1995**, *117*, 1165–1166.

101 a) E. J. Corey, D.-H. Lee, *J. Am. Chem. Soc.* **1991**, *113*, 4026–4028; b) E. J. Corey, B. E. Roberts, B. R. Dixon, *J. Am. Chem. Soc.* **1995**, *117*, 193–196.

102 a) M. Calter, T. K. Hollis, L. E. Overman, J. Ziller, G. G. Zipp, *J. Org. Chem.* **1997**, *62*, 1449–1456 ; b) M. A. Calter, T. K. Hollis, L. E. Overman, *J. Org. Chem.* **1999**, *64*, 1428; c) L. E. Overman, C. E. Owen, M. M. Pavan, C. J. Richards, Org. Lett. 2003, 5, 1809–1812; d) C. E. Anderson, L. E. Overman, *J. Am. Chem. Soc.* **2003**, *125*, 12412–12413.

103 a) M. Hiersemann, L. Abraham, A. Pollex *Synlett* **2003**, 1088–1095; b) S. Kaden, M. Hiersemann, *Synlett* **2002**, 1999–2002; c) M. Hiersemann, L. Abraham, *Eur. J. Org. Chem.* **2002**, 1461–1471; d) L. Abraham, R. Czerwonka, M. Hiersemann, *Angew. Chem.* **2001**, *113*, 4835–4837; *Angew. Chem. Int. Ed.* **2001**, *40*, 4700–4703.

104 U. Kazmaier, A. Krebs, *Angew.Chem.* **1995**, *107*, 2213–2214; *Angew. Chem. Int. Ed. Engl.* **1995**, *34*, 2012–2013.

105 I. Ojima in *Catalytic Asymmetric Synthesis*, VCH, New York, 1993.

106 V. Schurig, *Angew. Chem.* **1984**, *96*, 733–752; *Angew. Chem. Int. Ed.* **1984**, *23*, 747–765.

107 U. Kazmaier, H. Mues, A. Krebs, *Chem. Eur. J.* **2002**, *8*, 1850–1855.

108 A. Krebs, U. Kazmaier, *Tetrahedron Lett.* **1996**, *37*, 7945–7946.

109 U. Kazmaier, A. Krebs, *Tetrahedron Lett.* **1999**, *40*, 479–482.

110 R. Sakai, J. G. Stroh,. D. W. Sullins, K. L. Rinehart, *J. Am. Chem. Soc.* **1995**, *117*, 3734–3748, and references cited.

111 R. Sakai, K. L. Rinehart, V. Kishore, B. Kundu, G. Faircloth, J. B. Gloer, J. R. Carney, M. Namikoshi, F. Sun, R. G. Hughes, Jr., D. G. Gravalos, T. G. de Quasada, G. R. Wilson, R. M. Heid, *J. Med. Chem.* **1996**, *39*, 2819–2834, and reference cited therein.

112 a) H. Mues, U. Kazmaier, *Synlett* **2000**, 1004–1006; b) H. Mues, U. Kazmaier *Synthesis* **2001**, 487–498.

113 For comparable iodolactonizations see: a) P. A. Bartlett, J. F. Barstow, *J. Org. Chem.* **1982**, *47*, 3933–3941; b) N. Kurokawa, Y. Ohfune, *J. Am. Chem. Soc.* **1986**, *108*, 6041–6043; c) O. Kitagawa, T. Sato, T. Taguchi, T. *Chem Lett.* **1991**, 177–180; d) U. Kazmaier, *Tetrahedron* **1994**, *50*, 12895–12902.

114 F. Weygand, W. Swodenk, W. *Chem Ber.* **1957**, *90*, 639–645.

115 F. Weygand, E. Frauendorfer, *Chem. Ber.* **1970**, *103*, 2437–2449.

116 a) B. Lythgoe, J. R. Milner, J. Tideswell, *Tetrahedron Lett.* **1975**, *30*, 2593–2596. b) B. Lythgoe, R. Manwaring, J. R. Milner, T. A. Moran, M. E. N. Nambudiry, J. Tideswell, *J. Chem. Soc. Perkin I* **1978**, 387–395.

117 D. N. Jones, T. P. Kogan, P. Murray-Rust, J. Murray-Rust, R. F. Newton, *J. Chem. Soc. Perkin Trans. I* **1982**, 1325–1332.

118 S. K. Richardson, M. R. Sabol, D. S. Watt, *Syn. Commun.* **1989**, *19*, 359–367.

119 T. Fujisawa, K. Tajima, M. Ito, T. Sato, *Chem. Lett.* **1984**, 1169–1172.

120 a) M. J. Kurth, C.-M. Yu, *Tetrahedron Lett.* **1984**, *25*, 5003–5006; b) M. J. Kurth, C. M. Yu, *J. Org. Chem.* **1985**, *50*, 1840–1845.

121 M. J. Kurth, R. L. Beard, *J. Org. Chem.* **1988**, *53*, 4085–4088.

122 J. C. Gilbert, R. D. Selliah, *J. Org. Chem.* **1993**, *58*, 6255–6265.

123 a) C. P. Dell, K. M. Khan, D. W. Knight, *J. Chem. Soc., Chem. Commun.* **1989**, 1812–1814; b) C. P. Dell, K. M. Khan, D. W. Knight, *J. Chem. Soc. Perkin Trans. I* **1994**, 341–349.

124 W. C. Still, M. J. Schneider, *J. Am. Chem. Soc.* **1977**, 948–950.

125 T. Tsunoda, S. Tatsuki, Y. Shiraishi, M. Akasaka, S. Ito, *Tetrahedron Lett.* **1993**, *34*, 3297–3300.

6
Claisen–Johnson Orthoester Rearrangement

Yves Langlois

6.1
Introduction

Among the various versions of the sigmatropic rearrangements, [3,3]-sigmatropic rearrangements became, in the past 20 years, a powerful tool for C–C bond formation. In synthesis, they are particularly the method of choice for 1,3-chirality transfer, for the stereoselective formation of di- and especially trisubstituted double bonds and for the construction of quaternary centers. The aim of this chapter is to emphasize the scope and limitation of the orthoester Claisen–Johnson rearrangement by comparison with the related rearrangements and to illustrate the utility of this reaction in synthesis [1].

6.2
Historical Overview

Since its discovery in 1912, the Claisen rearrangement [2] of aryl or allyl vinyl ethers became of increasing importance and various versions of this [3,3] sigmatropic rearrangement appeared in the second part of 20th century (Scheme 6.1).

From an historical point of view, in 1940, Carroll [3] reported the base-catalyzed rearrangement of acetoacetic esters of allylic alcohols afforded after decarboxylation of γ-unsaturated ketones (Scheme 6.1).

Twenty-seven years later, Saucy [4] described a new kind of Claisen rearrangement that occurred after condensation in acidic medium between a propargylic or allylic alcohol and vinyl ether affording propargylvinyl or allylvinyl ether, respectively. In the later case, γ-unsaturated ketones are isolated in good yields after rearrangement (Scheme 6.1).

In 1964, Eschenmoser [5] discovered that the exchange between amide acetals and allylic alcohols observed by Meerwein [6] afforded after rearrangement of γ,δ-unsaturated amides (Scheme 6.1).

The Claisen Rearrangement. Edited by M. Hiersemann and U. Nubbemeyer

ISBN: 978-3-527-30825-5

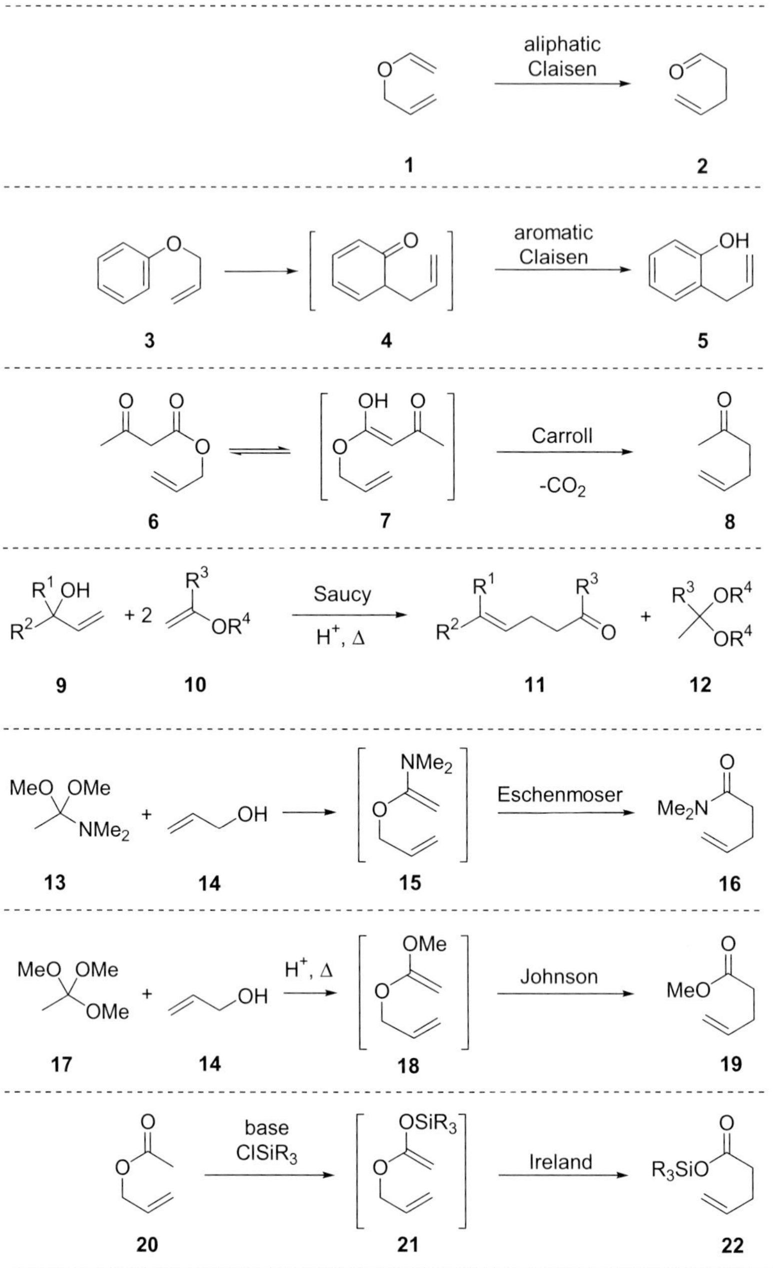
aliphatic
Claisen
1
2
3
4
aromatic
Claisen
5
Carroll
$-CO_2$
6
7
8
Saucy
H+, Δ
9
10
11
12
Eschenmoser
13
14
15
16
H+, Δ
Johnson
17
14
18
19
base
ClSiR3
Ireland
20
21
22

Scheme 6.1

The Claisen–Johnson rearrangement [7] is closely related to both Saucy vinyl allyl ether rearrangement and Eschenmoser rearrangement. The reaction proceeds via a ketene acetal, which results from the condensation between an orthoester and an allylic alcohol giving rise to a mixed orthoester followed by the elimination of the low-boiling-point alcohol. This ketene intermediate forms after rearrangement of a γ,δ-unsaturated ester (Scheme 6.1).

In this case, ketene acetal formation and rearrangement occurred in a single operation, whereas the allyl vinyl ether rearrangement is a two step process. However, the most powerful improvement in this type of [3,3] sigmatropic rearrangements was described by Ireland [8]. In the Claisen–Ireland reaction, deprotonation of allylic esters under kinetic conditions afforded the corresponding enolate, which are silylated *in situ* with trialkylchlorosilanes. Then the resulting silyl ketene acetals rearrange smoothly and afford after hydrolysis of γ,δ-unsaturated acids. The main advantage of this process is the possible selective preparation of *E*- or *Z*-configured ester enolates, which allowed a good control of the stereoselectivity of the rearrangement (Scheme 6.1).

6.3 Mechanistic Aspects

6.3.1 Reactivity

Claisen rearrangements are formally intramolecular $S_{N'}$ processes that involve concomitant σ bonds breaking and the formation as well as migration of π bonds. On the other hand, numerous investigations support a more or less concerted mechanism for this type of reaction depending on the nature of substituents on the 1,5 diene unit [1d].

From a reactivity point of view, the Claisen–Johnson rearrangement, like most of the other Claisen [3,3] sigmatropic rearrangements, is an irreversible exothermic reaction. This irreversibility is due to the large difference of stability between the reactants and products. However the exchange affording the mixed orthoester and the alcohol elimination is reversible and the equilibrium can be shifted towards the formation of ketene acetal intermediate by distillation of the low boiling point alcohol. The rearrangement itself could probably occur at room temperature as in the case of the Eschenmoser rearrangement illustrated in Scheme 6.2 [9].

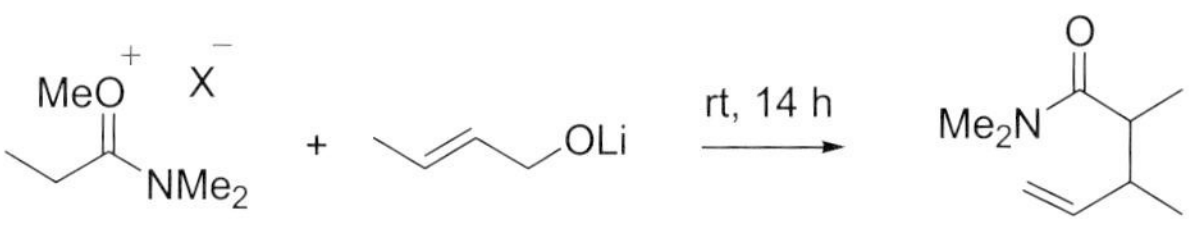

MeO⁺ X⁻ NMe$_2$ (23) + OLi (24) —rt, 14 h→ Me$_2$N, O (25)

Scheme 6.2

Many charge-accelerated reactions have been observed in [3,3] sigmatropic rearrangements [10]. In the Claisen–Johnson rearrangement, a combination of microwave irradiation and acidic catalysis (KSF clay) has been used successfully by Jones with several cyclic allylic alcohols [11]. Rate acceleration (9 min versus 12.5 h for conventional heating) and better yields were observed (Scheme 6.3).

OH

26 → 27, a: 73% b: 80% (CO_2Et)

OH

28 → 29, a: 53% b: 78% (CO_2Et)

OH

30 → 31, a: 47% b: 66% (CO_2Et)

OH

32 → 33, a: 83% b: 89% (CO_2Et)

OH

34 → 35 a: 68% b: 100% (CO_2Et)

Scheme 6.3 a $MeC(OEt)_3$, $EtCO_2H$, 140 °C, 12 h; ester was always contaminated with alcohol acetate. b $MeC(OEt)_3$, KSF, DMF, microwave irradiation, 9 min, 500 W.

Srikrihsna [12] also used the microwave heating technique and observed, with triethyl orthoacetate and propionic acid as catalyst in dimethyl formamide, a dramatic rate enhancement in the rearrangement of various allylic and propargylic alcohols (Scheme 6.4).

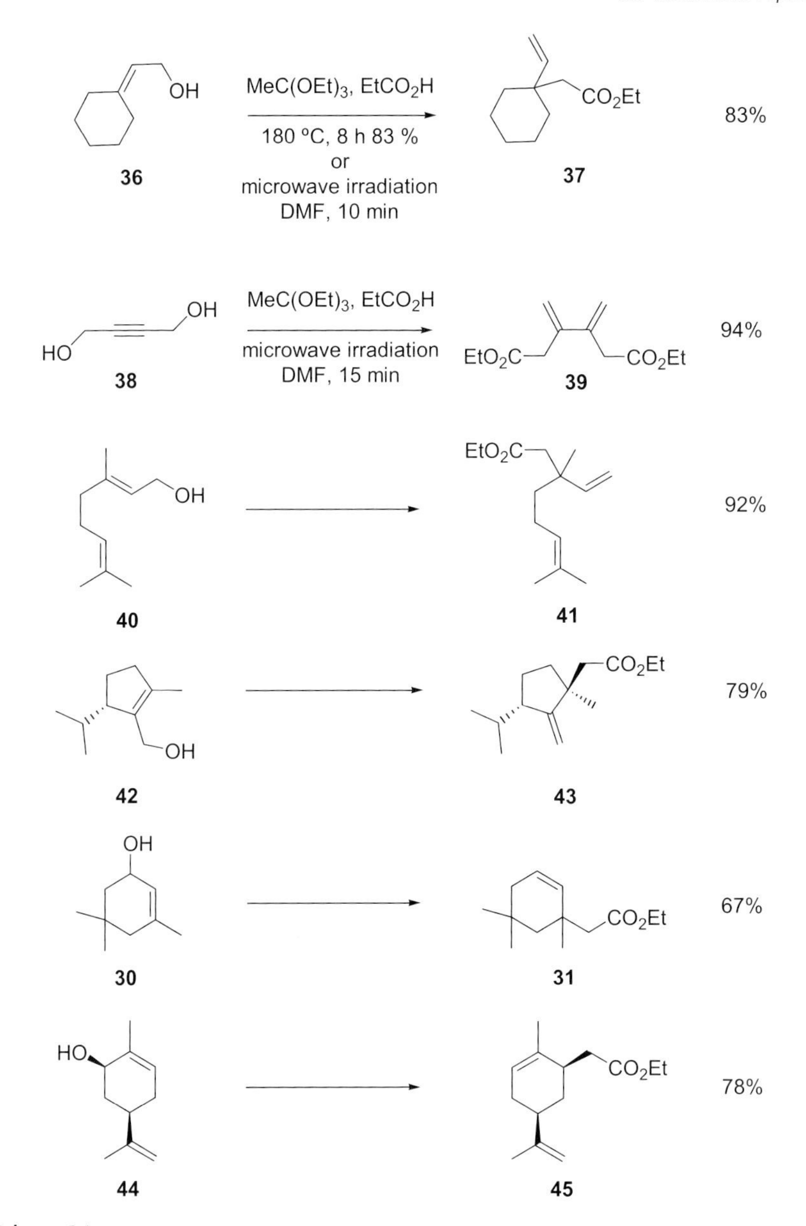
OH
36
MeC(OEt)3, EtCO2H
180 °C, 8 h 83 %
or
microwave irradiation
DMF, 10 min
CO2Et
37
83%
HO
OH
38
MeC(OEt)3, EtCO2H
microwave irradiation
DMF, 15 min
EtO2C
CO2Et
39
94%
OH
40
EtO2C
41
92%
OH
42
CO2Et
43
79%
OH
30
CO2Et
31
67%
HO
44
CO2Et
45
78%

Scheme 6.4

6.3.2
Stereoselectivity

As observed by Johnson in the original publication that described the first utilization of this reaction in a very elegant synthesis of squalene **50** with double chain extension [7], the orthoester rearrangement of 1,1-disubstituted allylic alcohols is highly stereoselective and gave rise to *E*-trisubstituted double bonds. Thus, successive rearrangements afforded efficiently a tetraene ester **49** as illustrated in Scheme 6.5.

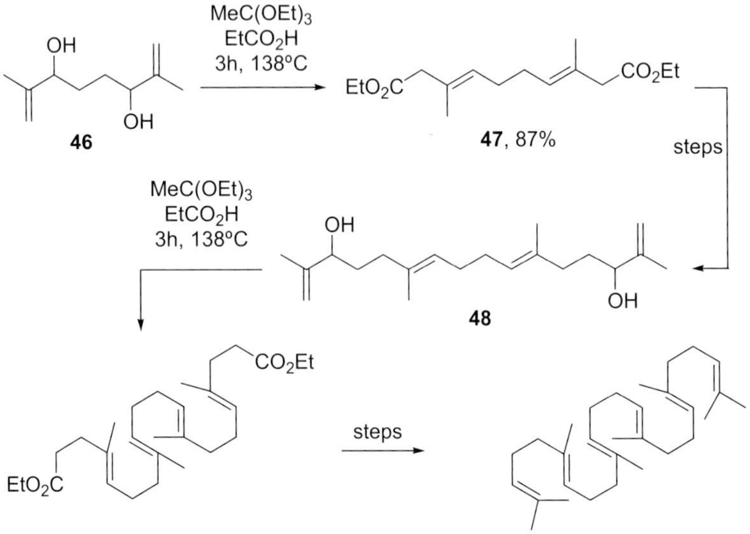

Scheme 6.5

However, contrary to the Eschenmoser rearrangement giving rise selectively to a *E*-aminoketene acetal [13], the Johnson rearrangement of substituted orthoacetates failed to give stereochemically defined ketene acetals. Accordingly, this reaction is more successfully applied to orthoacetate or orthoesters with α-stereogenic centers that are eliminated in following steps. However, the stereoselectivity of the sigmatropic rearrangement is generally excellent. This process, often called transfer of chirality, is illustrated for example in a synthesis of prostaglandin A_2 **56** reported by Stork [14]. This synthesis is also characterized by the use of two successive orthoester Claisen rearrangements as illustrated in Scheme 6.6.

A chair-like transition state in the Claisen–Johnson rearrangement as well as in other [3,3] sigmatropic rearrangements is the most common transition state. However, in some cases involving cyclic ketene acetal, for instance, a boat-like transition state is more likely. This particular stereochemical outcome has been discussed by Lythgoe [15] who studied in synthetic approaches oriented towards the total synthesis of vitamins D_2, the rearrangement of orthoester **58** with cyclohexen-5-ol **57** (Scheme 6.7). The rearrangement is highly stereoselective and afforded single diastereomer 59. Configurations were determined after chemical correlation and could result from a boat-shaped transition state **60**.

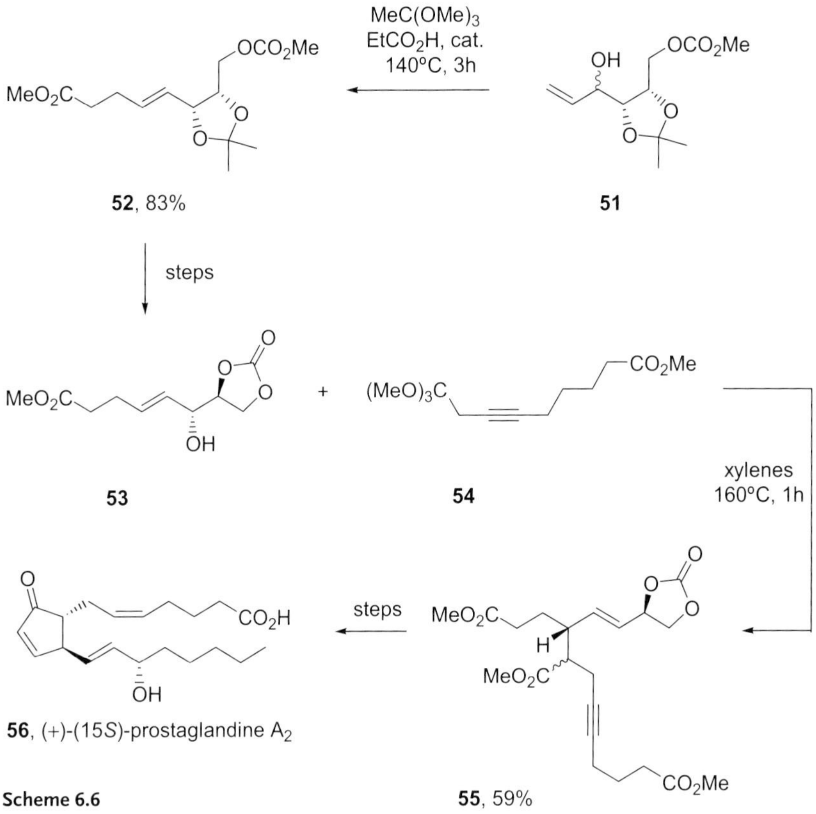

MeC(OMe)$_3$
EtCO$_2$H, cat.
140°C, 3h

52, 83%

51

steps

53

54

xylenes
160°C, 1h

steps

56, (+)-(15*S*)-prostaglandine A$_2$

55, 59%

Scheme 6.6

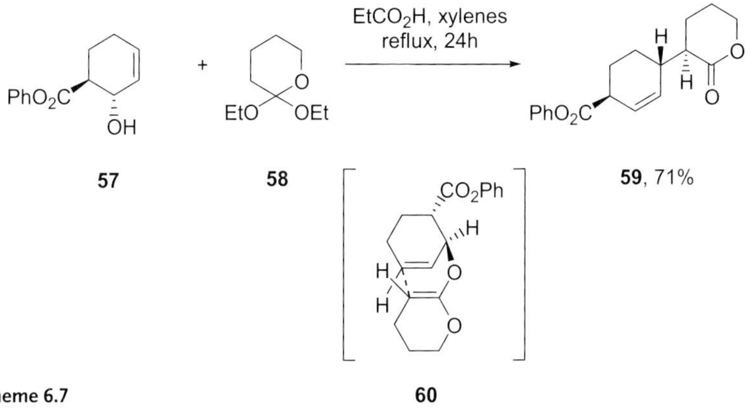

EtCO$_2$H, xylenes
reflux, 24h

57

58

59, 71%

60

Scheme 6.7

The same stereoselectivity due to a boat-like transition state has been observed by Lallemand [16] during an approach in the synthesis of antifeedent compound clerodine. Interestingly, a chemical correlation has been done with the products resulting from a Claisen–Ireland rearrangement in the open chain series. Accordingly, the *E*-ester enolate obtained after deprotonation and silylation of ester **64** [17] afforded, via a chair-like transition state, a compound which was correlated via **63** with the cyclic orthoester product **62** which resulted from a necessarily *Z*-ketene acetal. Consistently the *E*-ester enolate gave rise to a diastereomer **65** after the same sequence of reactions (Scheme 6.8).

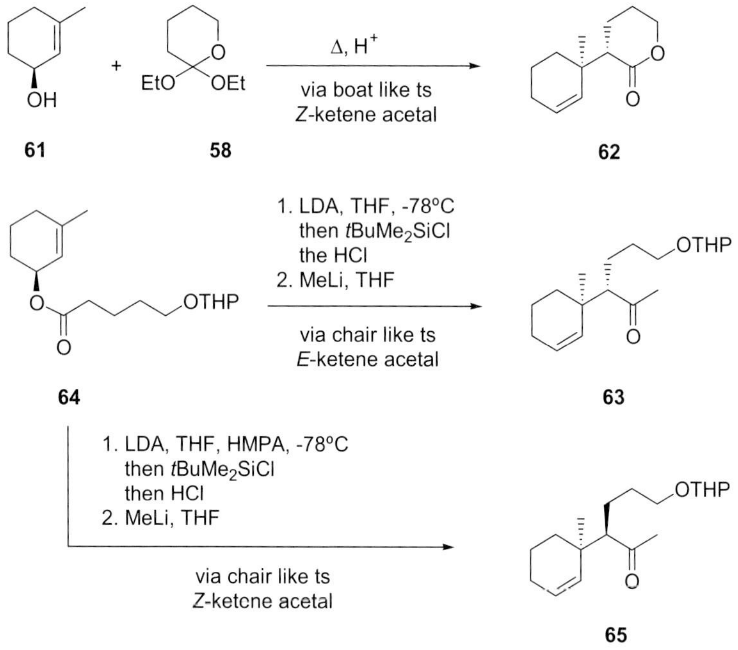

Scheme 6.8

An other example of particular stereoselectivity observed during a Claisen–Johnson rearrangement has been described by Yadav [18]. In this study, the observed stereoselection is both the result of stereoelectronic effects [19] and of *anti* selectivity by reference to the sulfur atom (Scheme 6.9). Axial attack is favored in all cases. Steric interactions in compound **69** induced a half-chair flipping (**70**) and consequently an attack *syn* to the sulfur atom. In compounds **66** and **72** both axial attack and *anti* sulfur selectivity afforded compounds **68** and **74**, respectively (Scheme 6.9).

In open chain series the Claisen–Johnson rearrangement is known to give rise to *E*-configuration for the resulting double bond [20]. An unprecedented stereochemical reversal from alkyl to aryl substituents has been described by Basavaiah

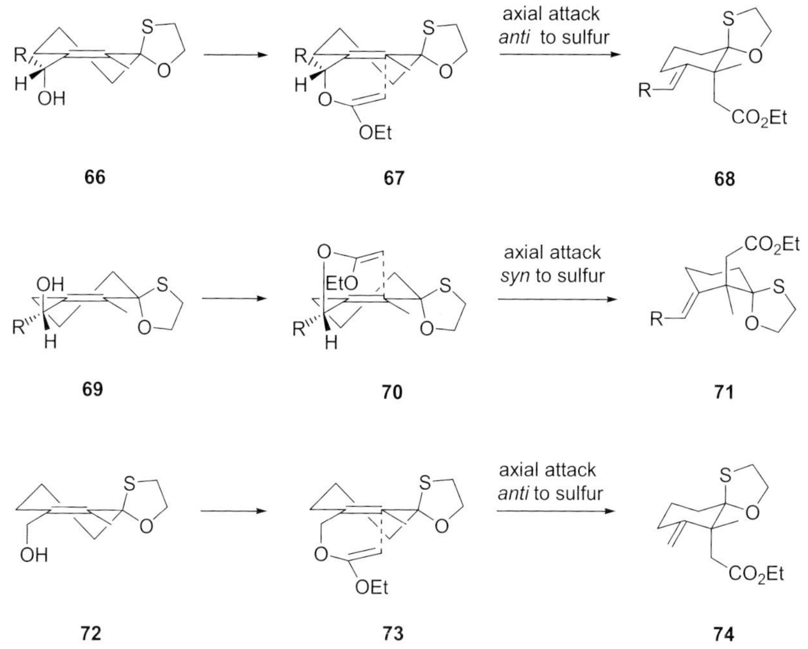

Scheme 6.9

[21]. This particular stereochemical outcome observed with Baylis–Hillman alcohols **75a** and **75b** can be explained by a predominant 1,3-diaxial interaction with alkyl groups that shift the conformational equilibrium to transition state **78b**, whereas an 1,2-allylic strain is more prominent with aryl groups and favor transition state **78c** (Scheme 6.10).

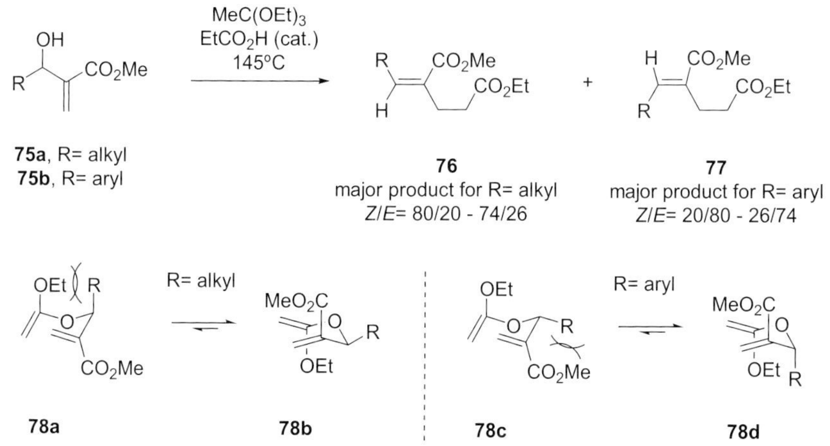

Scheme 6.10

6.3.3
Alternatives to the Orthoester Rearrangement

As pointed out earlier, ketene acetal involved in the Claisen–Johnson rearrangements is obtained after alcohol exchange in the starting orthoester and elimination of the low-boiling-point alcohol. The whole process requires generally acid catalysis, propionic acid being the most common catalyst, and heating or even distillation to shift the equilibrium to the formation of the ketene acetal. Obviously, these reaction conditions are not always compatible with sensitive compounds.

Alternative preparation of ketene acetal intermediates has been reported. Petrzilka [22] described a new high yield tandem reaction, generation of ketene acetal followed by Claisen rearrangement (Scheme 6.11). In this methodology, *in situ* elimination of benzeneseleninic acid from β-selenoxide acetal **82** afforded the anticipated ketene acetal **83**, which rearranged spontaneously under the reaction conditions affording γ-unsaturated ester **84**. This methodology has been also applied successfully for the preparation of large-membered ring lactones, such as **87** [22b].

Scheme 6.11

Later on, Holmes [23] prepared in good yields eight- and nine-membered ring lactones **90** and **93** in the same way and applied this methodology to the synthesis of natural products (Scheme 6.12).

On the other hand, Marples [24] reported a new base mediated generation of ketene acetal obtained by hydrobromic acid elimination in bromoacetals **96**. These compounds are prepared by condensation between allylic alcohols **95** and 1,2-dibromo-1-ethoxyethane **94**. However, the elimination of hydrobromic acid required rather harsh conditions (HMPA, Et_3N, 150–160 °C, overnight) (Scheme 6.13).

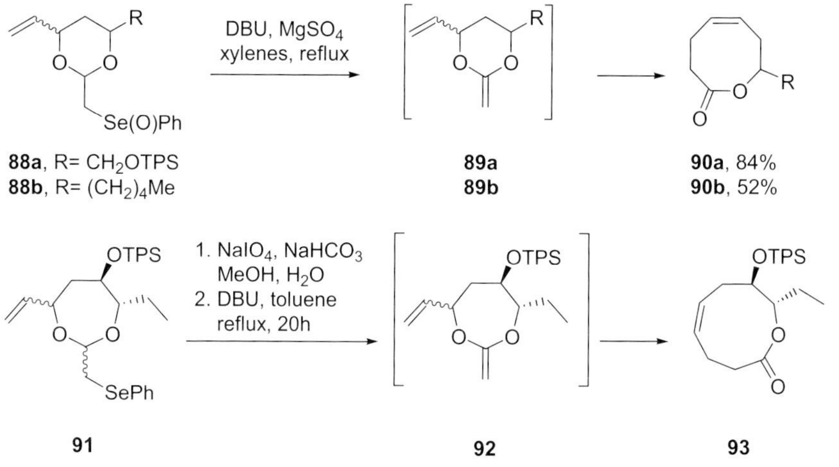

R
O O
Se(O)Ph

DBU, $MgSO_4$
xylenes, reflux

R
O O

O
O R

88a, R= CH_2OTPS
88b, R= $(CH_2)_4Me$

89a
89b

90a, 84%
90b, 52%

OTPS
O O
SePh

1. $NaIO_4$, $NaHCO_3$
MeOH, H_2O
2. DBU, toluene
reflux, 20h

OTPS
O O

OTPS
O
O

91

92

93

Scheme 6.12

Br
Br
OEt

+

R^2
R^1
ONa
R^3

R^2
R^1
O
Br
R^3
OEt

94

95

96

Et_3N, HMPA
150-160°C, 12 h

R^2
R^1
R^3
EtO_2C

R^2
R^1
O
R^3
OEt

98

97

OH

40

CO_2Et

41, 56%

OH

99

CO_2Et

100, 68%

HO

101

CO_2Et

102, 62%

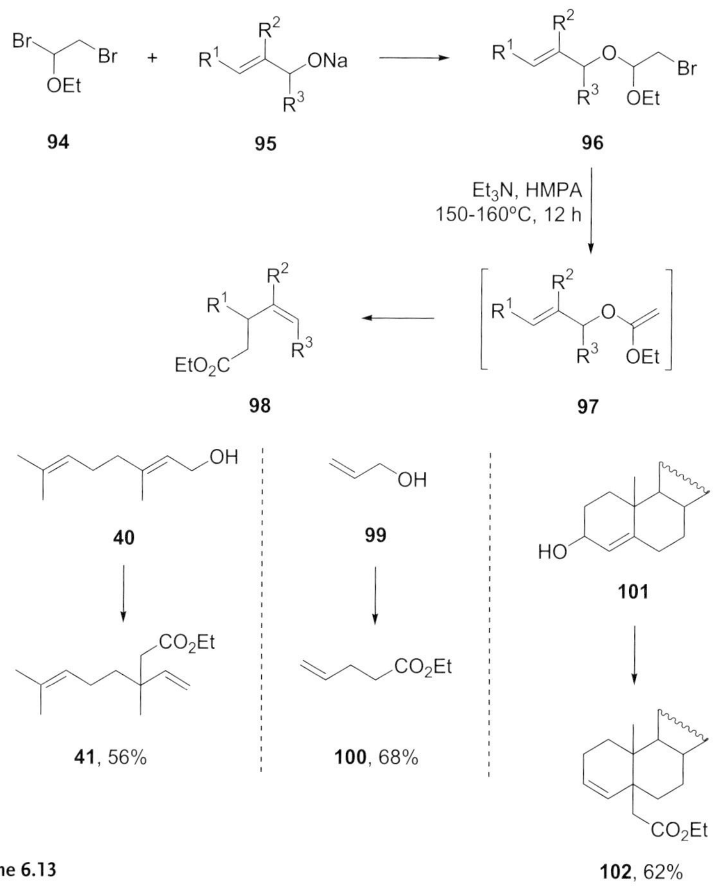

Scheme 6.13

6.4 Synthetic Applications

6.4.1 Terpenes, Fatty Acids, and Polyketide Derivatives

In synthetic studies concerning the side chain of tocopherol, Cohen and Saucy [25a] used a chemoenzymatic reaction for the control of the absolute configuration at C-2 and orthoester rearrangement for the control of the absolute configuration

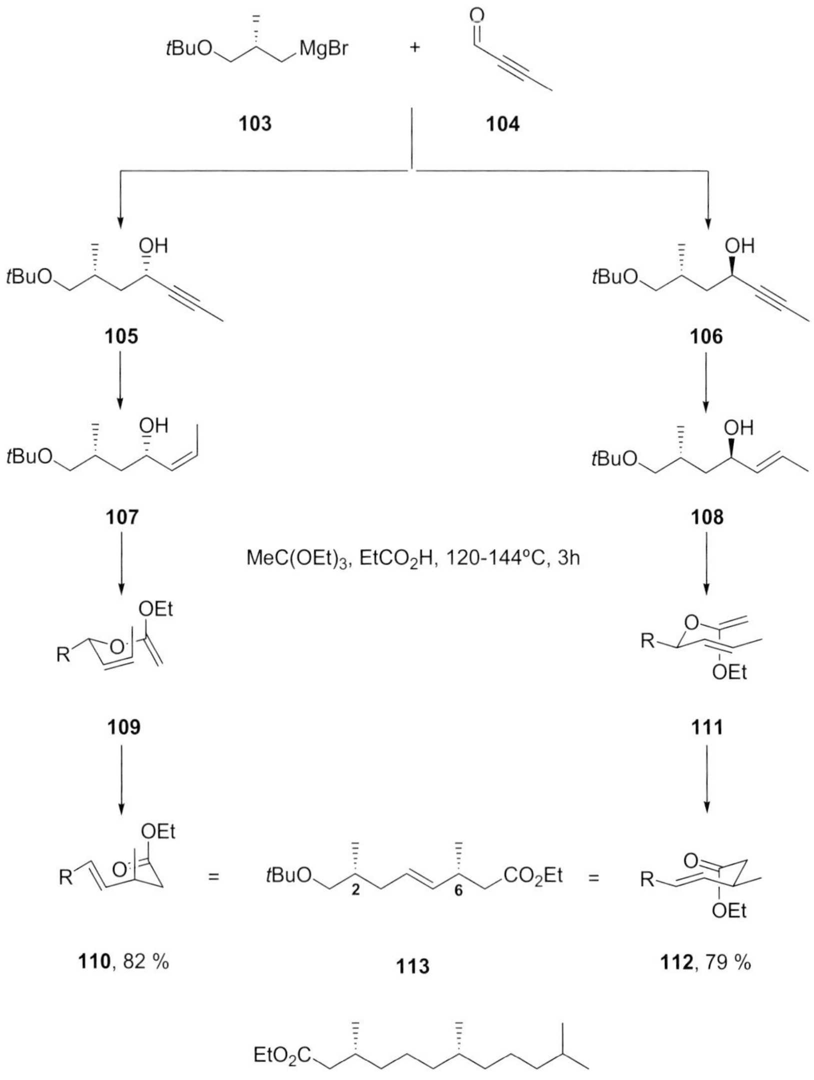

Scheme 6.14

at C-6 in compound **113** (Scheme 6.14). Thus, the diastereoisomeric propargylic alcohols **105** and **106** were reduced to the diastereomer *Z*- and *E*-configured allylic alcohols **107** and **108**, respectively. Each alcohol, after treatment with triethyl orthoacetate and propionic acid as catalyst, afforded in stereoconvergent rearrangements the single ester **113** (Scheme 6.14). The same strategy has been used in the preparation of ester **114** [25b].

The same group [26], in a new synthesis of tocopheryl acetate **115**, applied a stereoconvergent Claisen–Johnson rearrangement to compounds **116** and **117**, ester **118** was isolated in good yields in both cases (Scheme 6.15).

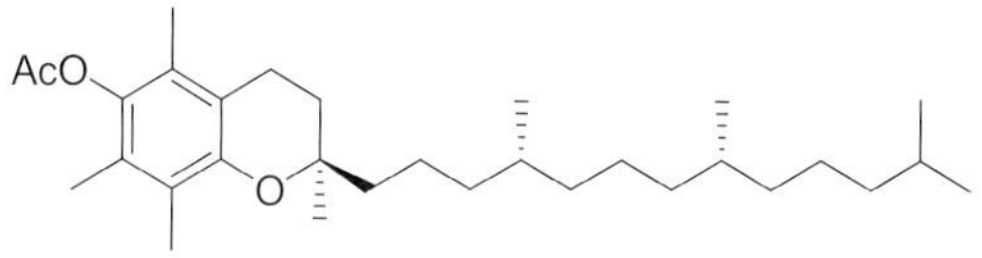

AcO

115, tocopheryl acetate (vitamin E acetate)

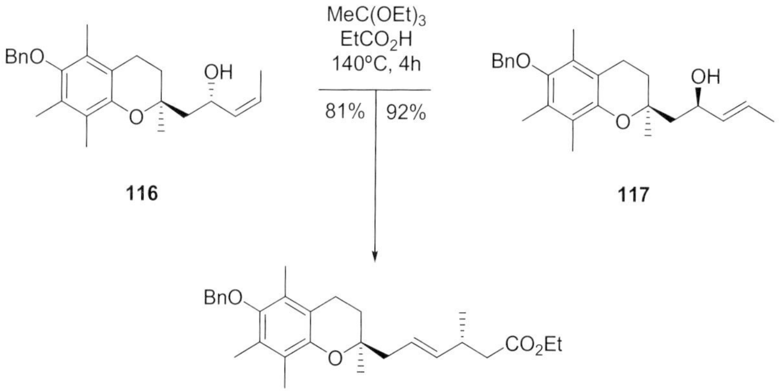

BnO OH **116**

$MeC(OEt)_3$
$EtCO_2H$
140°C, 4h
81% | 92%

BnO OH **117**

BnO CO_2Et **118**

Scheme 6.15

In a synthesis of (±)-geijerone **119** and (±)-γ-elemene **120**, orthoester rearrangement has been used by Kim [27]. Condensation of triethyl orthopropionate with primary allylic alcohol **121** afforded the rearranged ester **122** as a 1:1 mixture of diastereomers. Here the diastereoselectivity at C-3 and C-4 is controlled in a further stage during cyclization affording compound **123** (Scheme 6.16).

The Claisen–Johnson rearrangement is also a reaction of choice for the construction of quaternary centers in the synthesis of triquinanes and related terpenes. Accordingly, transposition of allylic alcohol **125** is a cornerstone reaction in the synthesis of tricyclic compound **124**, a key intermediate for the access to several triquinanes described by Iwata [28] (Scheme 6.17). Acid **126** was obtained in 57% yield after transposition and saponification. Claisen allyl vinyl ether rearrangement was also applied to allylic alcohol **125**. Oxidation of the aldehyde inter-

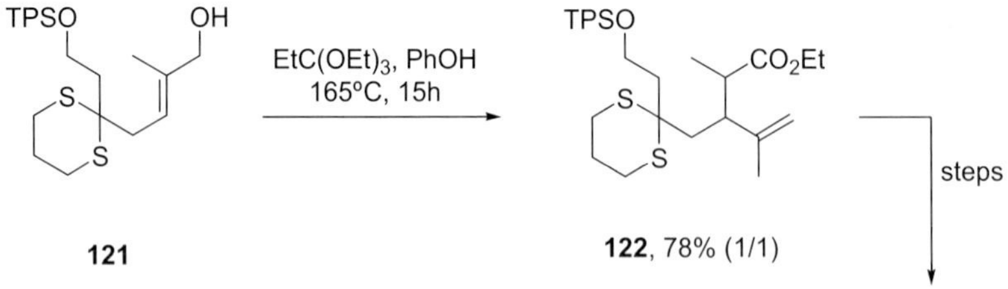

TPSO OH 121 $EtC(OEt)_3$, PhOH 165°C, 15h TPSO CO_2Et 122, 78% (1/1) steps

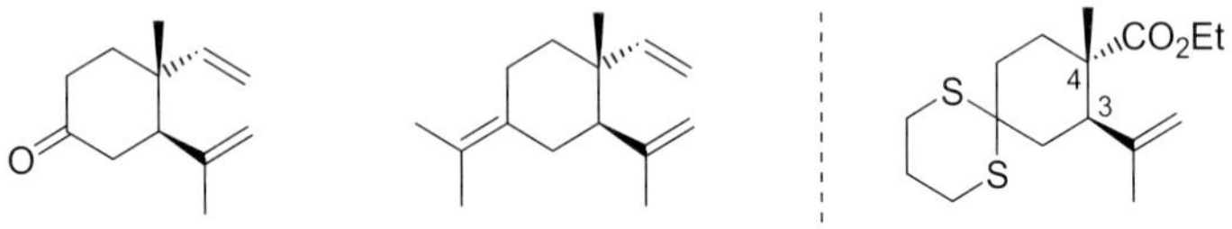

geijerone 119 γ-elemene 120 CO_2Et 4 3 123

Scheme 6.16

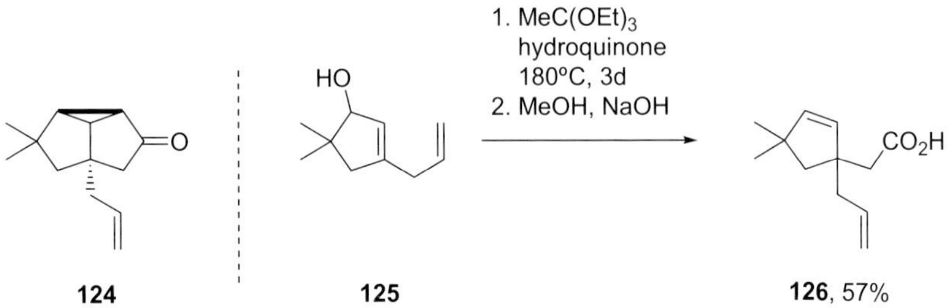

124 HO 125 1. $MeC(OEt)_3$ hydroquinone 180°C, 3d 2. MeOH, NaOH CO_2H 126, 57%

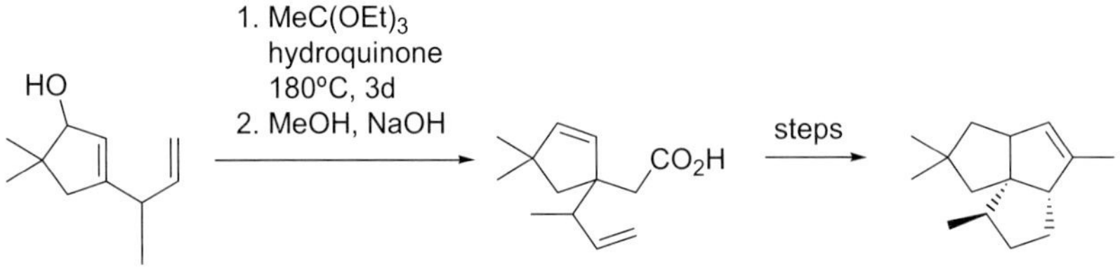

HO 127 1. $MeC(OEt)_3$ hydroquinone 180°C, 3d 2. MeOH, NaOH CO_2H 128, 57% steps pentalenene 129

Scheme 6.17

mediate afforded the same acid **126** in 43% yield. Orthoester rearrangement was still used by this group [29] in the synthesis of (±)-pentalenene **129** (Scheme 6.17).

In a stereoselective synthesis of (–)-chokol A **130**, Suzuki and Kametani [30] introduced the side chain by Claisen–Johnson rearrangement with a modest stereoselectivity. However the stereogenic center in β-position to the ester group in compound **132** is suppressed later in the synthesis (Scheme 6.18).

HO OH (–)-chokol A **130**

HO HO **131**

$MeC(OMe)_3$, $EtCO_2H$
xylenes
145-150°C, 1.5h

HO H CO_2Me
(*R*)-**132**, 51%
(*S*)-**132**, 19%

Scheme 6.18

During a synthesis of (±)-albene **133**, an orthoester rearrangement allowed the control in the formation of a stereogenic quaternary center in a bicyclo[2.2.1]heptane alcohol derivative. In this synthesis [31], Srikishna prepared the allylic alcohol **134** by a Diels–Alder cycloaddition followed by several transformations. Alcohol **134** afforded by rearrangement the ester **135** via a sterically preferred *exo* transition state (Scheme 6.19).

albene **133**

HO **134**

$MeC(OEt)_3$
$MeCO_2H$
sealed tube
180°C, 48h

MeO_2C **135**, 86%

Scheme 6.19

In the synthesis of (±)-cyclolaurene **136** and (±)-*β*-cuparenone **137**, the same author [32] used a Claisen–Johnson rearrangement affording ester **139** for the construction of one of the two quaternary centers as depicted in Scheme 6.20.

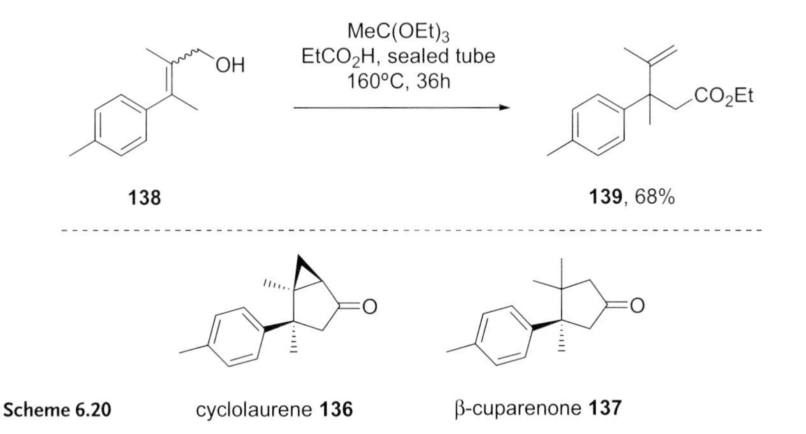

Scheme 6.20

The skeleton of the sesquiterpenes thapsene **140** and **141** contains three contiguous quaternary centers [33]. The synthesis of two biogenetical precursors of these terpenes started with cyclogeraniol **143**. Orthoester rearrangement afforded ester **144**. As in several other syntheses, the third quaternary center resulted from diazoketone cyclopropane insertion affording in this synthesis compound **146** (Scheme 6.21).

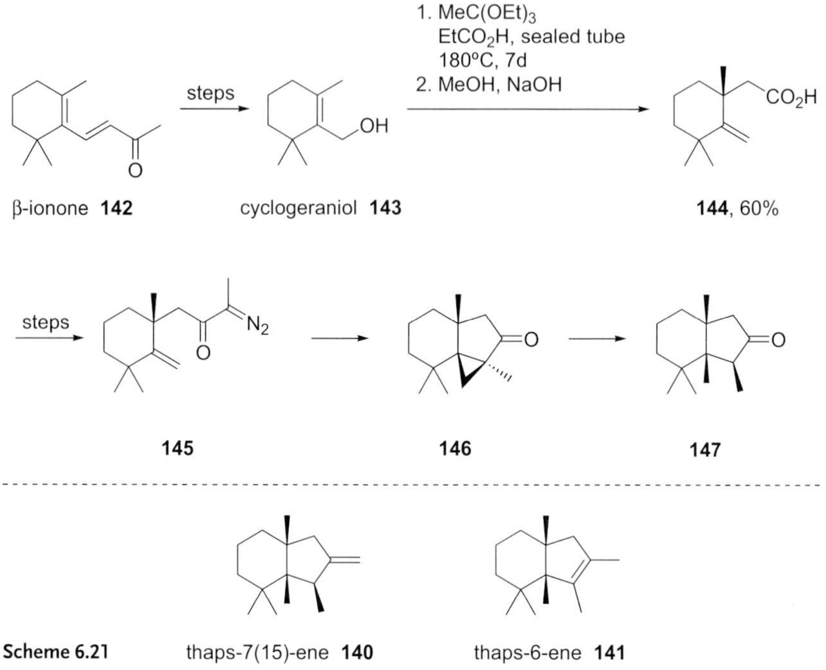

β-ionone **142** → (steps) → cyclogeraniol **143** → (1. $MeC(OEt)_3$, $EtCO_2H$, sealed tube, 180°C, 7d; 2. MeOH, NaOH) → **144**, 60%

→ (steps) → **145** → **146** → **147**

thaps-7(15)-ene **140** thaps-6-ene **141**

Scheme 6.21

Another example of orthoester rearrangement for the construction of a quaternary center is given in the enantioselective synthesis of (+)-valerane **148** [34]. In this synthesis, *R*-(–)-carvone **149** is used as starting material. Rearrangement of allylic alcohol **150** afforded ester **151**. Diazoketone cyclization is then followed by a ring enlargement affording the bicyclo[4.4.0]decane system (Scheme 6.22).

In three recent publications, Srikrishna described the syntheses of both racemic and enantiomerically pure pinguisenol **152** [35–36]. In the synthesis of the racemic terpene, alcohol **154** was prepared in three steps from the Hagemann ester **153**. Orthoester rearrangement afforded ester **155**. As above, the following steps included diazoketone cyclization and functional group manipulations. The synthesis of (+)-pinguisenol **152** began with (*R*)-carvone **149** as chiral pool starting material. A five-step sequence afforded alcohol **157**. Rearrangement under classical conditions, gave rise to ester **158**. This reaction was followed by diazoketone cyclization giving rise to compound **159**. (+)-Pinguisenol **152** is then synthesized in six steps (Scheme 6.23).

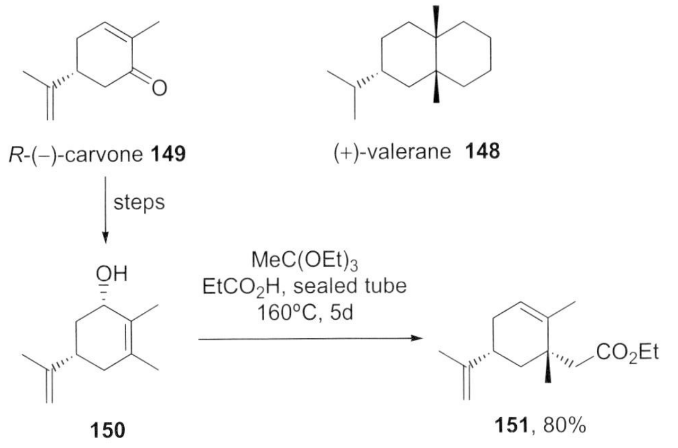

R-(−)-carvone **149**

(+)-valerane **148**

steps

OH

$MeC(OEt)_3$
$EtCO_2H$, sealed tube
160°C, 5d

CO_2Et

150

151, 80%

Scheme 6.22

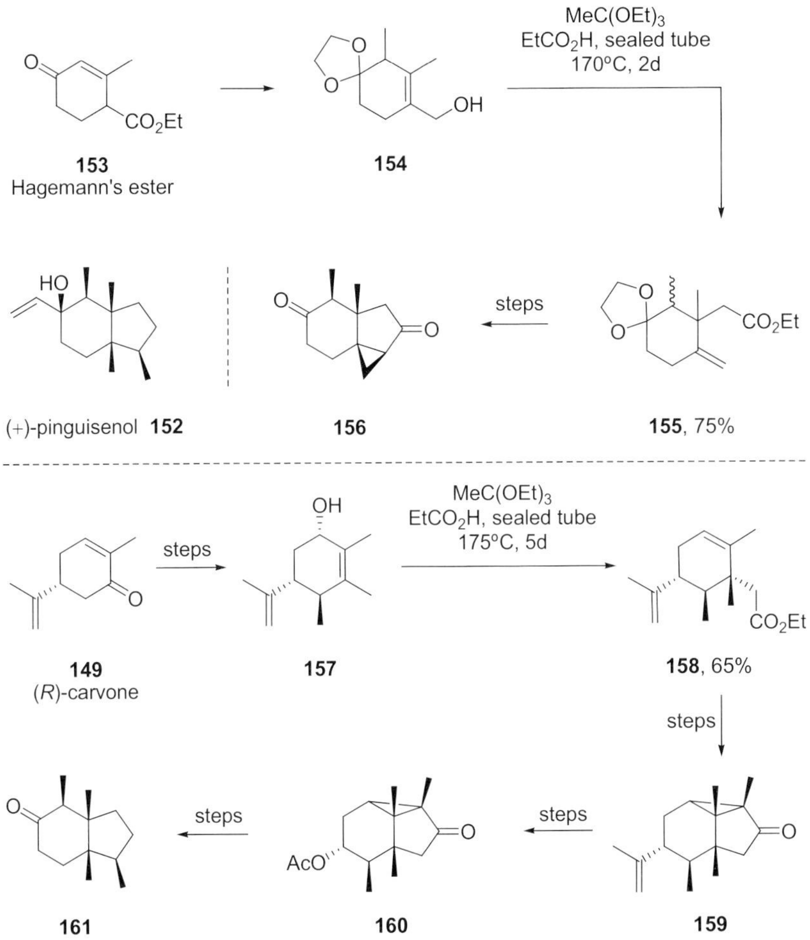

$MeC(OEt)_3$
$EtCO_2H$, sealed tube
170°C, 2d

CO_2Et

OH

153
Hagemann's ester

154

HO

steps

CO_2Et

(+)-pinguisenol **152**

156

155, 75%

OH

$MeC(OEt)_3$
$EtCO_2H$, sealed tube
175°C, 5d

steps

CO_2Et

149
(*R*)-carvone

157

158, 65%

steps

steps

steps

AcO

161

160

159

Scheme 6.23

In the synthesis of (±)-1,14-herbetenediol **162**, a Claisen–Johnson rearrangement ring-closing metathesis strategy was described by the same group [37]. Treatment of allylic alcohol **163** with triethyl orthoacetate afforded ester **164**. Allylation of compound **164** followed by ring-closing metathesis gave rise to compound **166**, a synthetic precursor of herbetenediol **162** (Scheme 6.24).

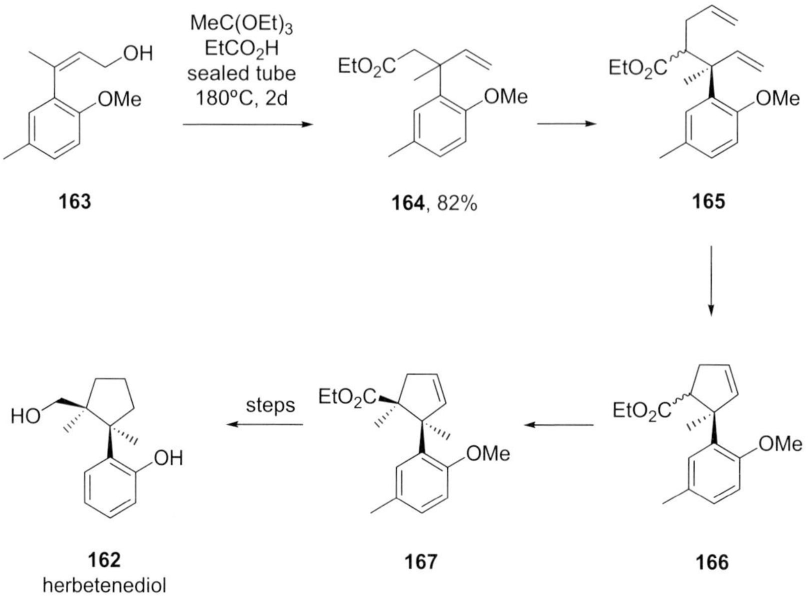

$MeC(OEt)_3$, $EtCO_2H$, sealed tube, 180°C, 2d

163 → **164**, 82% → **165** → **166** → **167** → (steps) **162** herbetenediol

Scheme 6.24

In the first enantioselective synthesis of triquinane (–)-cucumin H **168** [38], a quaternary center is introduced via an orthoester rearrangement. Allylic alcohol **170** is prepared from (*R*)-limonene **169** by a known sequence of reactions. Orthoester rearrangement occurred in a highly stereoselective manner *anti* to the isopropylidene side chain and afforded ester **171**. Then diazoketone cyclization followed by cationic cyclization gave rise to the desired tricyclic system (Scheme 6.25).

In syntheses of two marine sesquiterpenes, africanol **172** and dactylol **173**, Paquette [39] prepared the bicyclic allylic alcohol **175** in nine steps from 4,4-dimethyl cyclohexanone **174**. The Claisen–Johnson rearrangement was highly stereoselective and gave rise to ester **176** with the proper configuration of the side chain methyl in relation to dactylol **173** (Scheme 6.26).

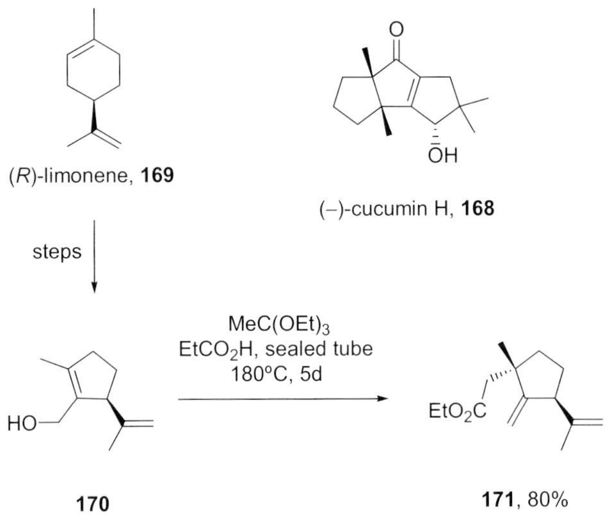

(R)-limonene, 169
(–)-cucumin H, 168
steps
MeC(OEt)3
EtCO2H, sealed tube
180°C, 5d
170
171, 80%

Scheme 6.25

africanol, 172
dactylol, 173
steps
MeC(OEt)3
cat. EtCO2H
xylenes, reflux, 3h
174
175
176, 97%
177
178

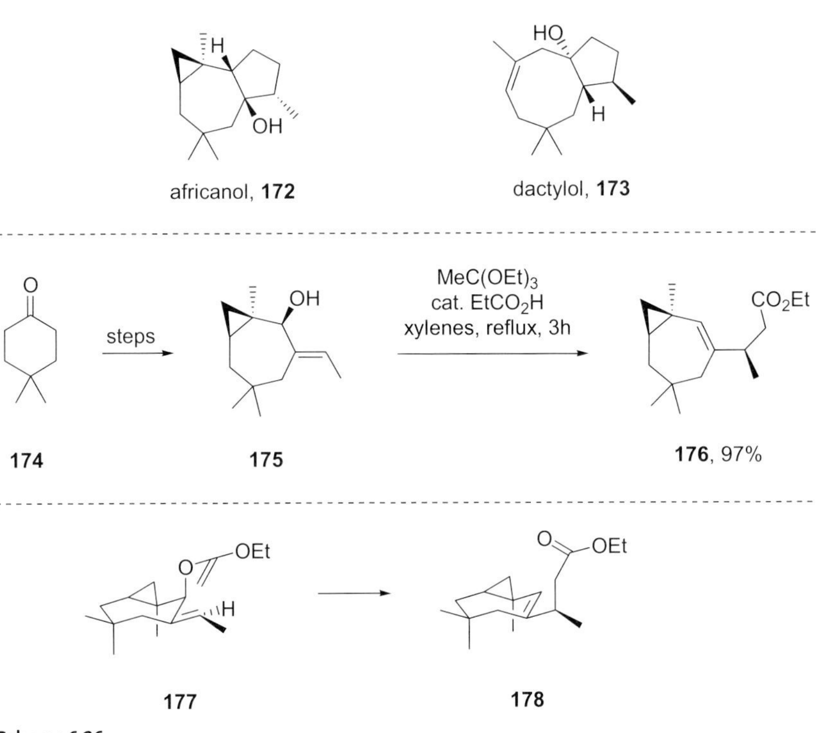

Scheme 6.26

In connection with structure activity relationship studies, Chapuis [40] prepared a series of sandalwood odorant alcohols. In this context, alcohol **181** has been prepared from (+)-*trans*-pinocarveol **179** in two steps. Accordingly, orthoester rearrangement of alcohol **179** afforded ester **180** which, after the Grignard reaction, gave tertiary alcohol **181** (Scheme 6.27).

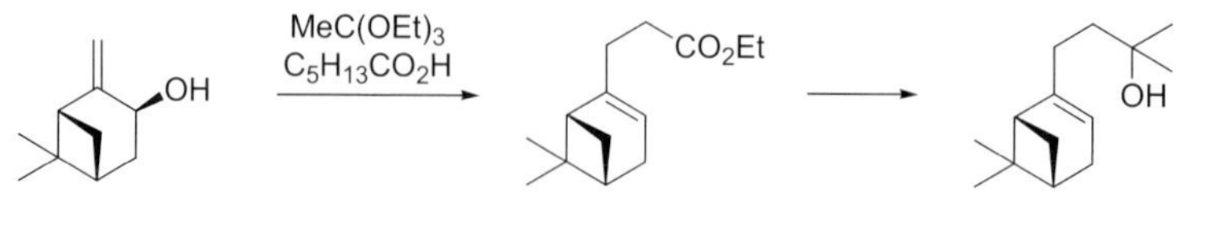

$MeC(OEt)_3$, $C_5H_{13}CO_2H$

179 (+)-*trans*-pinocarveol; **180**, 68%; **181**

Scheme 6.27

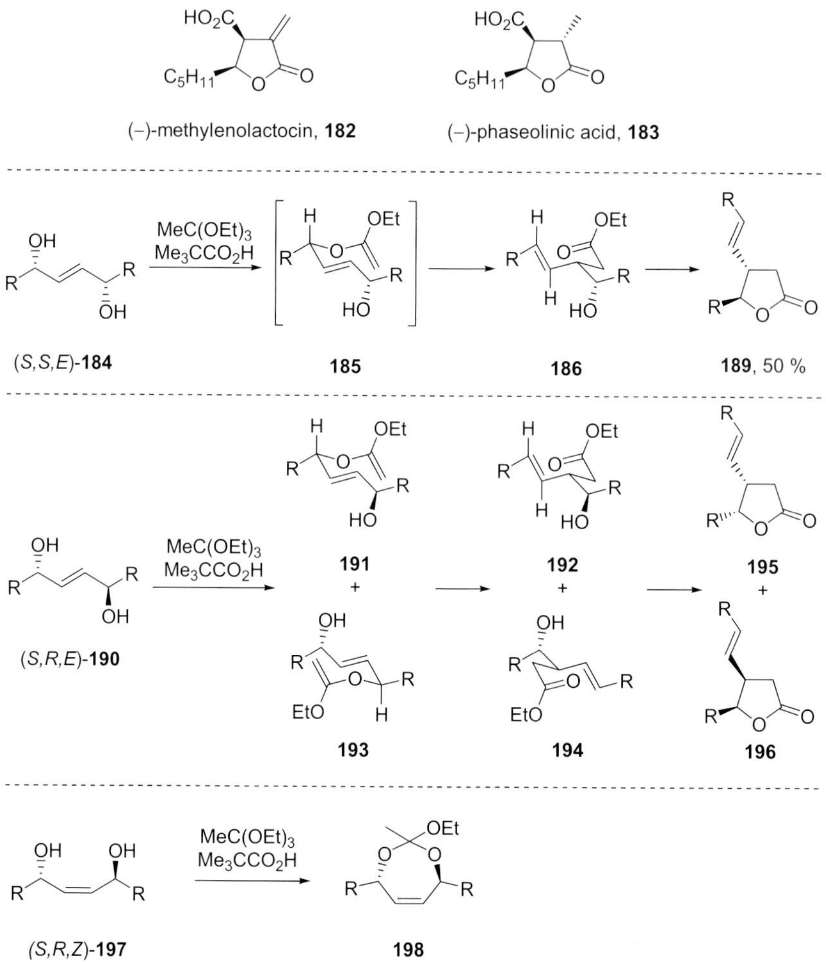

(–)-methylenolactocin, **182**; (–)-phaseolinic acid, **183**

$MeC(OEt)_3$, Me_3CCO_2H

(*S,S,E*)-**184**; **185**; **186**; **189**, 50 %

$MeC(OEt)_3$, Me_3CCO_2H

(*S,R,E*)-**190**; **191** + **193**; **192** + **194**; **195** + **196**

$MeC(OEt)_3$, Me_3CCO_2H

(*S,R,Z*)-**197**; **198**

Scheme 6.28

Concise formal syntheses of (–)-methylenolactocin **182** and (–)-phaseolinic acid **183** have been described by Garcia [41]. A stereoselective oxazaborolidine reduction of the corresponding acetylenic diketone, followed by lithium aluminum hydride or Lindlar reduction of the acetylenic diols afforded allylic diols **184**, **190** and **197**, respectively. As depicted in Scheme 6.28, the *S,S,E*-configured diol **184**, after rearrangement, gave rise to a single *trans*-lactone **189**. Whereas the *S,R,E*-configured diol **190** afforded racemic *cis*-lactone **195–196** via a competitive rearrangement involving both *R*- and *S*-configured alcohols in **190**. On the other hand, the *S,R,Z*-configured diol **197** gave the cyclic orthoester **198**. In addition, lactones **189** and **195** were prepared in three steps in slightly better yields with the Claisen–Ireland rearrangement of the corresponding bisacetates (Scheme 6.28).

A Claisen–Johnson rearrangement with transfer of chirality has been used by Pearson in a stereoselective synthesis of A-factor **199**, an autoregulator of the production of streptomycin in *Streptomyces griseus* [42]. Allylic alcohol **200** is prepared in 80% yield and 84% ee by oxazaborolidine reduction of the corresponding ketone. Orthoester rearrangement afforded ester **201** in 75% yield. Ozonolysis of unsaturated ester **201** is then followed by lactonization (Scheme 6.29).

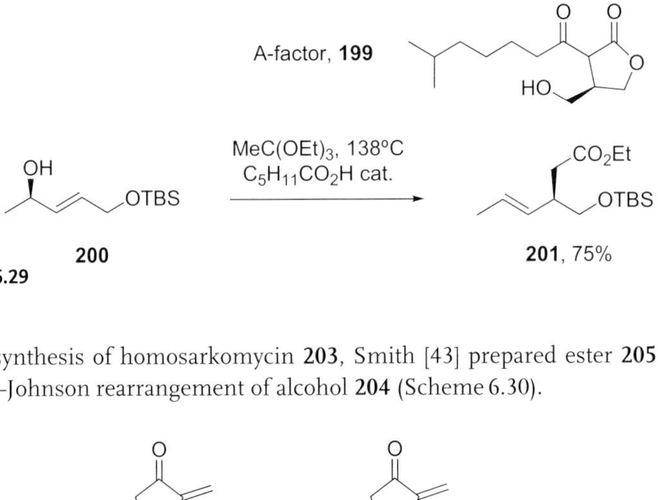

A-factor, **199**

OH OTBS **200** — $MeC(OEt)_3$, 138°C, $C_5H_{11}CO_2H$ cat. → CO_2Et OTBS **201**, 75%

Scheme 6.29

In a synthesis of homosarkomycin **203**, Smith [43] prepared ester **205** using a Claisen–Johnson rearrangement of alcohol **204** (Scheme 6.30).

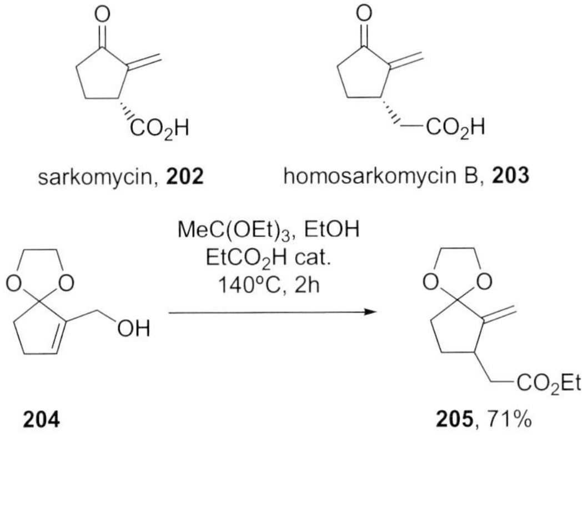

sarkomycin, **202**   homosarkomycin B, **203**

204 — $MeC(OEt)_3$, EtOH, $EtCO_2H$ cat., 140°C, 2h → **205**, 71%

Scheme 6.30

In a structure activity relationship study on polyether ionophore monensin **206**, Still [44] prepared side-chain-modified analogues like compound **207** of the antibiotic by the orthoester rearrangement (Scheme 6.31). The stability of this complex polyether under the reaction conditions is worthy of note.

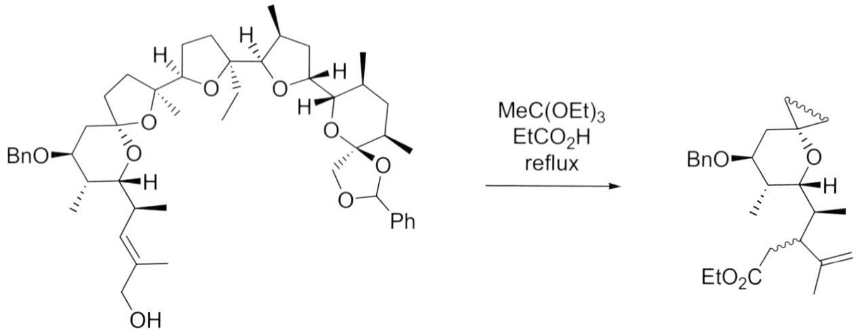

$MeC(OEt)_3$
$EtCO_2H$
reflux

monensine derivative, **206**

207, d.r.= 3-4:1

Scheme 6.31

One of the key steps in the stereoselective synthesis of the HMG-CoA reductase inhibitor mevilonin **208** described by Wovkulich and Uskokovic [45] involved a Claisen–Johnson rearrangement. Accordingly, allylic alcohol **210** prepared in few steps from (*S*)-pulegone **209** gave rise, after condensation with triethyl orthopropionate, to ester **211** as a 89/11 mixture of diastereomers (Scheme 6.32).

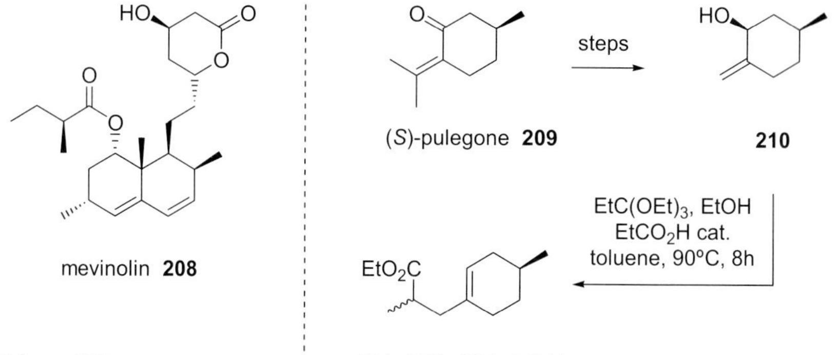

mevinolin **208**

(*S*)-pulegone **209**

steps

210

$EtC(OEt)_3$, EtOH
$EtCO_2H$ cat.
toluene, 90°C, 8h

211, 85%, *S*/*R*= 89/11

Scheme 6.32

The control of the configuration of the side chain at C-8 in a synthesis of *C*-glycoside antibiotic pseudonomic acid **212** described by DeShong [46] resulted from a Claisen–Johnson rearrangement. Dihydropyrane derivatives **213–214** after treatment with triethyl orthoacetate afforded an inseparable mixture of diastereomers. The anomeric center was then reduced and the side chain secondary alcohol oxidized. Diastereomeric ketones **215** and **216** were isolated at this stage in a 2:1 ratio (Scheme 6.33).

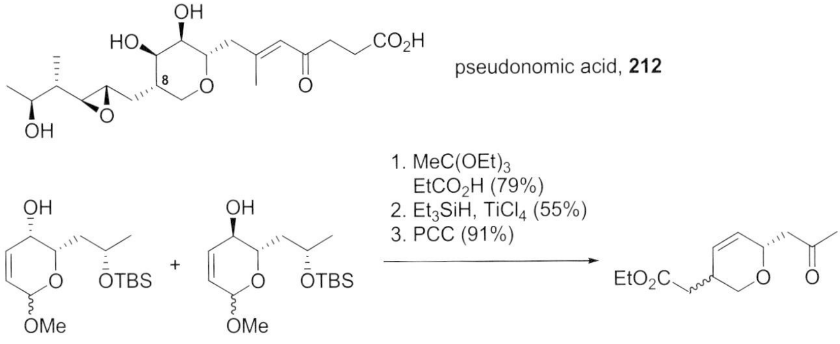

pseudonomic acid, **212**

1. $MeC(OEt)_3$, $EtCO_2H$ (79%)
2. Et_3SiH, $TiCl_4$ (55%)
3. PCC (91%)

213 + **214** → (*R*)-**215**/(*S*)-**216** = 2/1

Scheme 6.33

In the enantioselective synthesis of the C-1–C-28 portion of cytotoxic marine natural product amphidinolide B1 **217**, Lee [47] developed a strategy using the key steps of the Evans oxazolidinone alkylation, Sharpless epoxidation and orthoester Claisen rearrangement. Allylic alcohol **218** as a mixture of diastereomers gave rise after treatment with triethyl orthoacetate under acidic catalysis to ester **219** in 66% yield (Scheme 6.34). This result exemplifies the potency of the orthoester rearrangement even with such probably sensitive epoxyallylic alcohol.

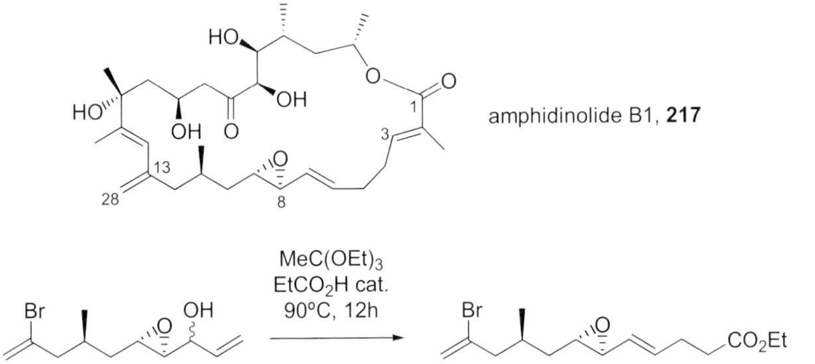

amphidinolide B1, **217**

$MeC(OEt)_3$, $EtCO_2H$ cat., 90°C, 12h

218 → **219**, 66%

Scheme 6.34

Rearrangement of propargylallyl alcohol **221** has been developed by Mulzer [48] in studies towards the synthesis of macrodiolide antibiotic tartrolone **220**. The regioselectivity of this reaction is noteworthy as eneyne ester **222**, a precursor of C-11–C-20 fragment of the molecule, was isolated in good yield (Scheme 6.35).

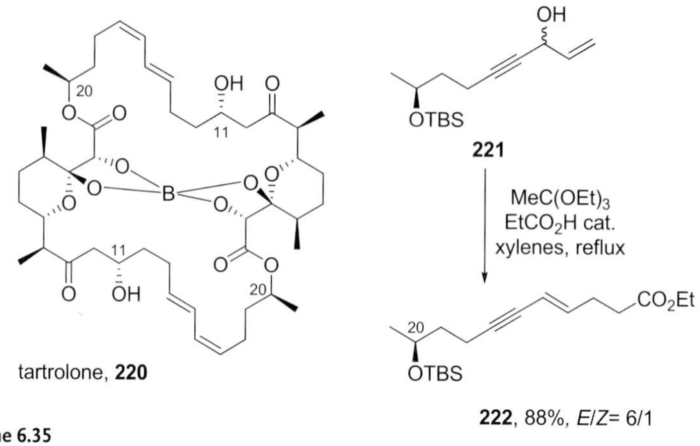

tartrolone, **220**

221

$MeC(OEt)_3$
$EtCO_2H$ cat.
xylenes, reflux

222, 88%, *E/Z*= 6/1

Scheme 6.35

Classical orthoester rearrangement is used in the initial steps of the preparation of a versatile intermediate **223** for the total synthesis of acetogenins [49] (Scheme 6.36). Esters **230–232**, which are intermediates in the synthesis of capsaicins **226–228** [50], important ingredients of spices, were obtained similarly in good yields (Scheme 6.36).

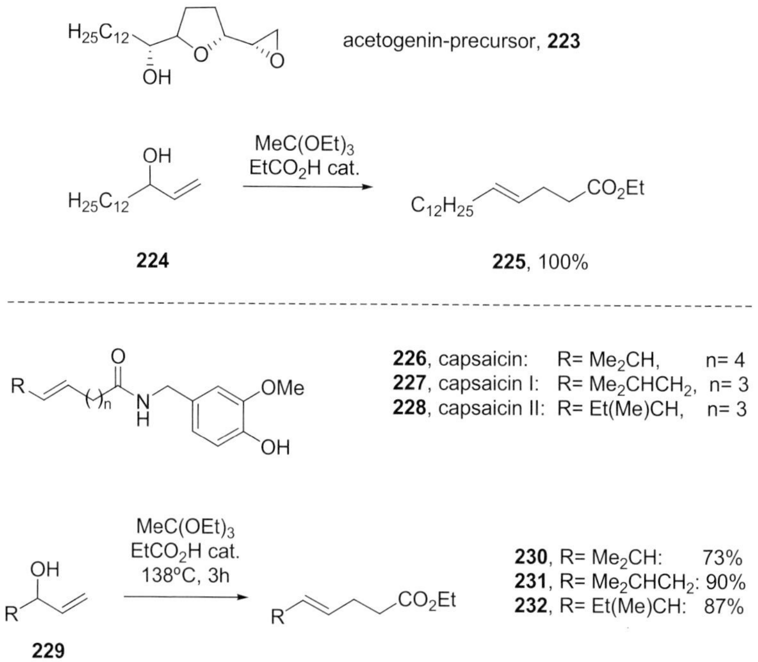

acetogenin-precursor, **223**

$MeC(OEt)_3$
$EtCO_2H$ cat.

224 → **225**, 100%

226, capsaicin: R= Me_2CH, n= 4
227, capsaicin I: R= Me_2CHCH_2, n= 3
228, capsaicin II: R= Et(Me)CH, n= 3

$MeC(OEt)_3$
$EtCO_2H$ cat.
138°C, 3h

229

230, R= Me_2CH: 73%
231, R= Me_2CHCH_2: 90%
232, R= Et(Me)CH: 87%

Scheme 6.36

Enprostil **233** is a potent inhibitor of gastric acid secretion. This compound is characterized by the presence of an allenic moiety on one of the side chains. Cooper [51] described the preparation of four stereoisomers of **223** using an ortho-ester Claisen rearrangement. It is known that enantiomerically pure propargylic alcohols allow the preparation of pure allene epimers of known configuration. In this study, for instance, treatment of propargylic alcohol **234** with triethyl orthoacetate afforded the anticipated allenic derivative **235**. Functional group manipulation afforded the target ester **238a**. Alternatively, ester **238b** can also be prepared by a S_N2' displacement on bromopropargylic derivative **236** in the presence of organo copper derivative **237** (Scheme 6.37).

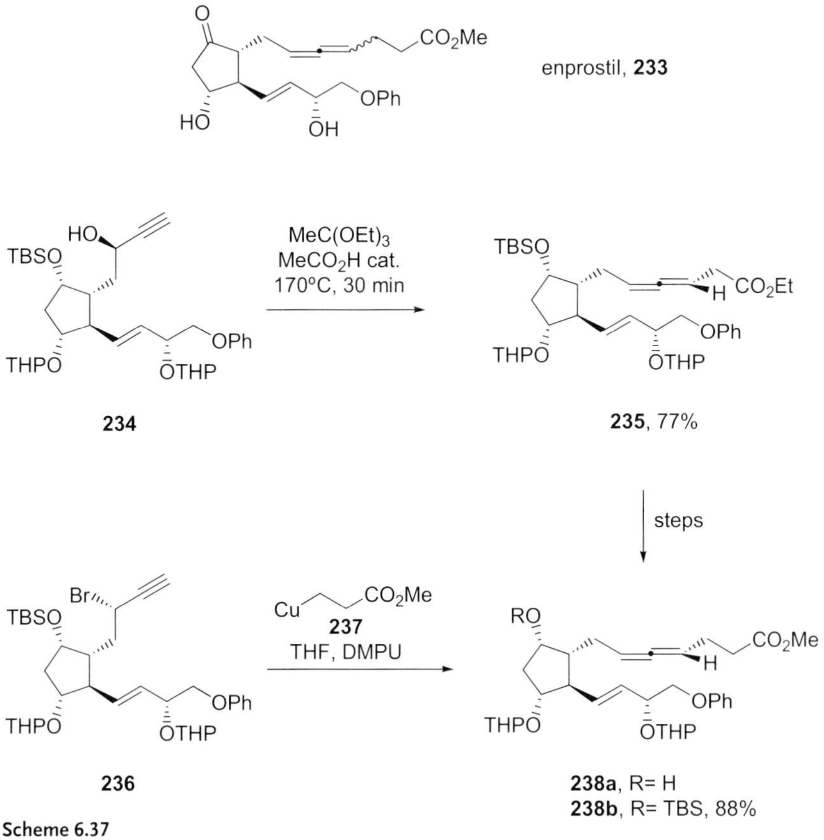

enprostil, **233**

$MeC(OEt)_3$, $MeCO_2H$ cat., 170°C, 30 min

234 → **235**, 77%

steps

$Cu\text{-}CH_2CH_2CO_2Me$ **237**, THF, DMPU

236 → **238a**, R= H; **238b**, R= TBS, 88%

Scheme 6.37

Still in the field of prostaglandins synthesis, Takahashi [52] took advantage of a 1,3 transfer of chirality to control the configuration at C-12 in a synthesis of the prostaglandin intermediate **240** (Scheme 6.38).

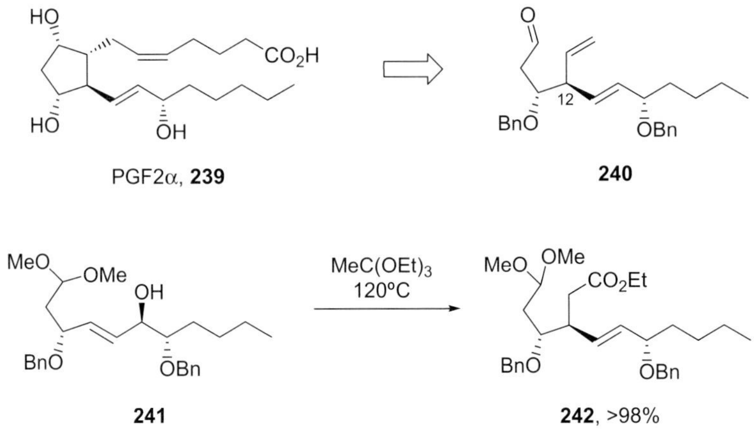

HO HO OH CO2H PGF2α, 239 O 12 BnO OBn 240

MeO OMe OH BnO OBn 241 MeC(OEt)3 120°C MeO OMe CO2Et BnO OBn 242, >98%

Scheme 6.38

In a more recent study [53], the same group, in an approach to carbacyclin **244**, described a high yielding Claisen–Johnson rearrangement using cyclic orthoester **246**, which allowed a nice control of the absolute configuration at the C-8 and C-12 centers in compound **248** (Scheme 6.39).

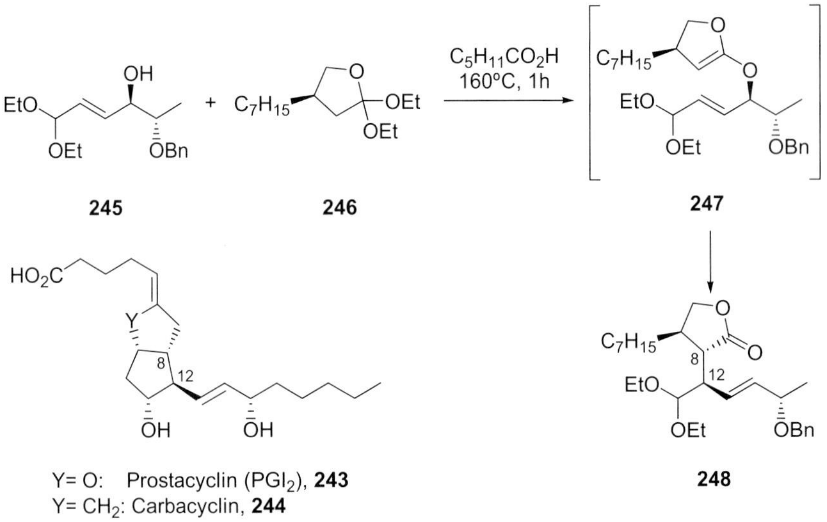

EtO OEt OH OBn 245 + C_7H_{15} O OEt OEt 246 $C_5H_{11}CO_2H$ 160°C, 1h [C_7H_{15} O O EtO OEt OBn] 247

HO_2C Y 8 12 OH OH

Y= O: Prostacyclin (PGI_2), **243**
Y= CH_2: Carbacyclin, **244**

C_7H_{15} O O 8 12 EtO OEt OBn 248

Scheme 6.39

In a synthesis of conformationally restricted LTB4 analogues **249** [54], the mono-protected diol diastereomers **250** and **251** were subjected to the conditions of a Claisen–Johnson rearrangement. Classical reaction conditions afforded the expected ester isomers **252** and **253**, respectively (Scheme 6.40).

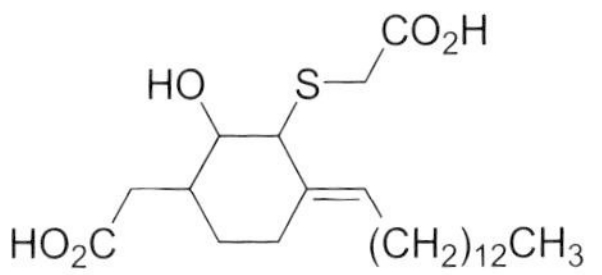

nor-LTB$_4$ analogue **249**

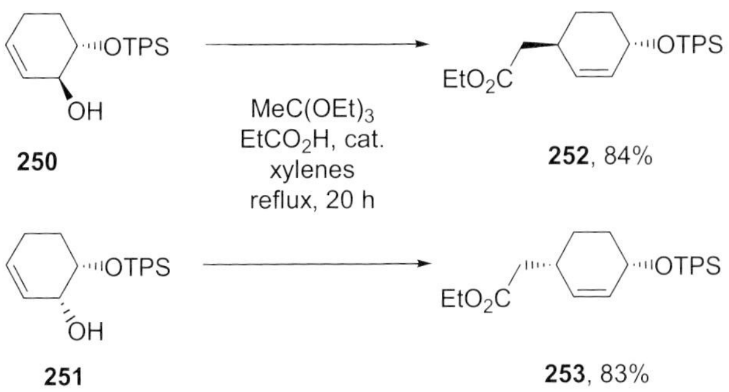

MeC(OEt)$_3$
EtCO$_2$H, cat.
xylenes
reflux, 20 h

250 → **252**, 84%

251 → **253**, 83%

Scheme 6.40

Mycophenolic acid **254** is an immunosuppressive metabolite of *Penicillium brevi-compactum*. The side chain of this compound, characterized by a trisubstituted double bond in the γ-position to an ester group, is an ideal target for ortho-ester rearrangement [55]. Accordingly, alcohol **255** after classical treatment afforded ester **256a** with small amount of the corresponding phenol **256b**. The *E*-alkene was obtained exclusively in this rearrangement (Scheme 6.41).

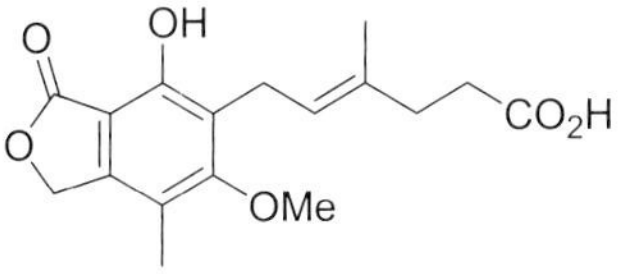

mycophenolic acid, **254**

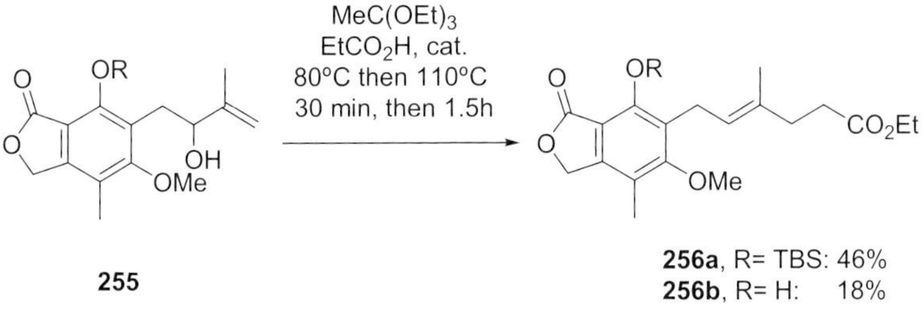

MeC(OEt)$_3$
EtCO$_2$H, cat.
80°C then 110°C
30 min, then 1.5h

255 →
256a, R= TBS: 46%
256b, R= H: 18%

Scheme 6.41

Hirota, in the total synthesis of quassinoid amarolide **257** [56], used a non-classical acidic catalysis with pentachlorophenol. Under these conditions and in the presence of 4 Å molecular sieves, the expected ester **259** was obtained in moderate yield. However, large amounts of starting material were recovered. The diastereoselectivity of this rearrangement is probably oriented by the presence of the angular methyl group at C8 (Scheme 6.42).

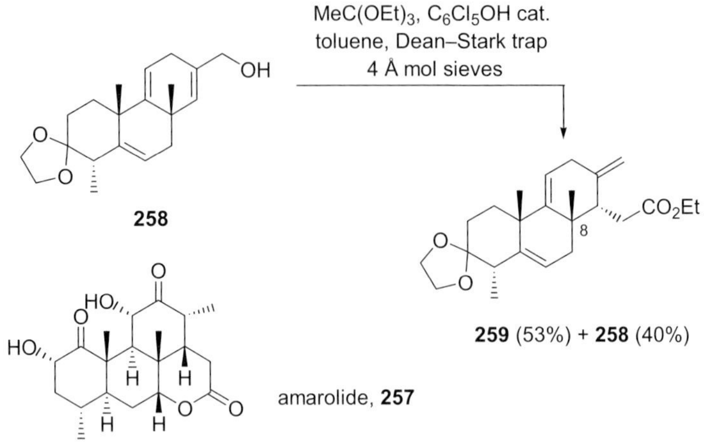

MeC(OEt)$_3$, C_6Cl_5OH cat.
toluene, Dean–Stark trap
4 Å mol sieves

258

259 (53%) + **258** (40%)

amarolide, **257**

Scheme 6.42

The Claisen–Johnson rearrangement has also been often used in pheromone synthesis, particularly with compounds characterized by the presence of one or several di- or trisubstituted double bonds. Descoins [57] in a selective synthesis of angoulure **260**, (7*Z*,11*E*)-1-acetoxyhexadecadiene, pheromone of a crop pest, used a Claisen–Johnson rearrangement for the introduction of the *E*-configured double bond (Scheme 6.43). The synthesis of the eneallenic pheromone of the male dried bean beetle **265** has been described by Mori [58]. In this synthesis, both enantiomers of propargylic alcohol **263** were submitted to the orthoester rearrangement and gave rise to the enantiomeric allenes. Completion of the synthesis from ester enantiomer **264** confirmed the *R* configuration for the natural product **265** (Scheme 6.43). An original synthesis of 1,4-dienes has been developed by Wilson [59] starting with 2-[(trimethylsilyl)methyl]cyclopropyl carbinol **269**. This method was then used in a synthesis of the melon fly pheromone **266**. The allylsilane ester **268**, which was subjected to the cyclopropanation reaction, is prepared by a Claisen–Johnson rearrangement of (*E*)-3-(trimethylsilyl)prop-2-en-1-ol **267** (Scheme 6.43).

In a short formal synthesis of the Californian red scale pheromone **271**, Claisen orthoester rearrangement has been employed as a key step [60] (Scheme 6.44). Stereoselective synthesis of the trisubstituted double bond in a component of Caribbean sponge **274** was accomplished with the Claisen–Johnson rearrangement as depicted in Scheme 6.44 [61]. A similar strategy was developed for the stereoselective synthesis of the pheromone **277** of square necked grain (Scheme 6.44) [62].

angoulure, **260**

$MeC(OEt)_3$, $EtCO_2H$, 145°C

261 → **262**, 90%

$MeC(OEt)_3$, $EtCO_2H$, 110°C, 7h

263 → (*aR*)-**264**, 88%

steps → **265**

pheromone of male dried bean beetle

melon fly pheromone, **266**

$MeC(OEt)_3$, $EtCO_2H$ then 140°C, overnight

267 → **268** 67%

steps → **269** → **270**

Scheme 6.43

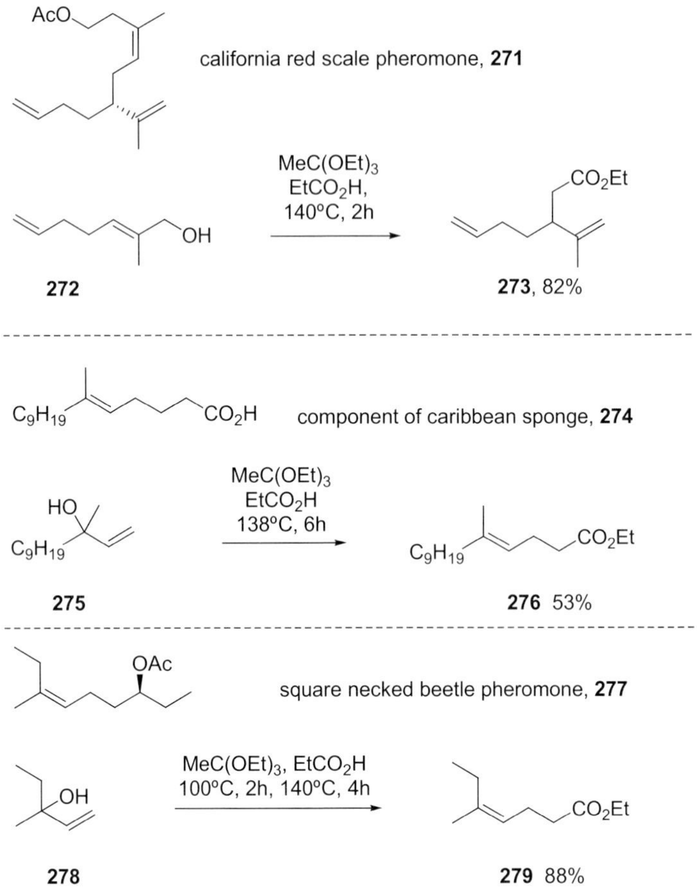

california red scale pheromone, **271**

$MeC(OEt)_3$, $EtCO_2H$, 140°C, 2h

272 → **273**, 82%

component of caribbean sponge, **274**

$MeC(OEt)_3$, $EtCO_2H$, 138°C, 6h

275 → **276** 53%

square necked beetle pheromone, **277**

$MeC(OEt)_3$, $EtCO_2H$, 100°C, 2h, 140°C, 4h

278 → **279** 88%

Scheme 6.44

Two pheromones **280** and **281** that contain a (*Z*,*Z*)-configured 1,4-diene moiety were prepared via a *cis*-hydrogenation of a 3,5-dien-1-yne in an ultimate step. Enyne intermediates **283a,b** in these syntheses resulted from Claisen–Johnson rearrangements of alcohols **282** (Scheme 6.45) [63]. A (*E*,*Z*)-1,3-dienic pheromone **286** was synthesized [64a] by a sequence of reactions involving a orthoester rearrangement of propargylic alcohol **287** followed by the isomerization of the resulting allene **288** following a process described by Takeda (Scheme 6.45) [64b].

The antitumor polyhalogenated compound halomon **290** is extracted from a red alga. Mioskowski [65] described an efficient synthesis of this intriguing compound. One of the key steps of the synthesis, a Claisen–Johnson rearrangement, afforded ester **293** allowing the introduction the stereogenic quaternary center (Scheme 6.46).

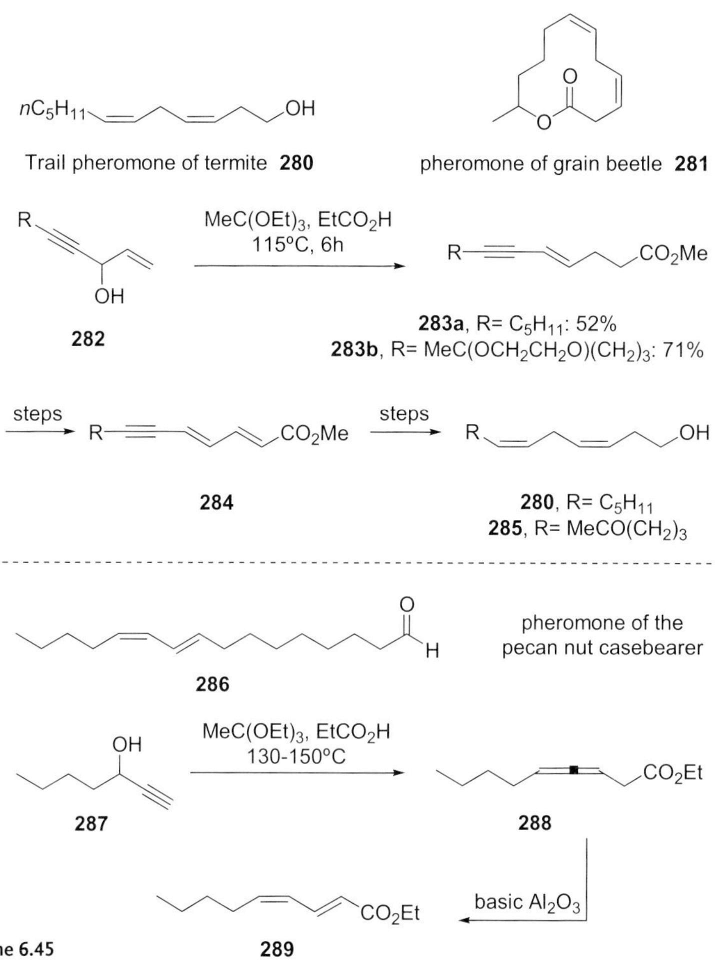

nC_5H_{11} ... OH

Trail pheromone of termite **280**

pheromone of grain beetle **281**

282 → $MeC(OEt)_3$, $EtCO_2H$, 115°C, 6h → R—≡—CH=CH—$CH_2CH_2CO_2Me$

283a, R= C_5H_{11}: 52%
283b, R= $MeC(OCH_2CH_2O)(CH_2)_3$: 71%

steps → **284** → steps → **280**, R= C_5H_{11}; **285**, R= $MeCO(CH_2)_3$

286 pheromone of the pecan nut casebearer

287 → $MeC(OEt)_3$, $EtCO_2H$, 130-150°C → **288** (CO_2Et) → basic Al_2O_3 → **289** (CO_2Et)

Scheme 6.45

291 TBSO—CH_2—≡—CH_2—OTBS → $Et_4N^+Cl_3^-$, CH_2Cl_2 → **292** (TBSO, Cl, Cl, OH) → $MeC(OEt)_3$, TsOH, 170°C → **293**, 59% (TBSO, Cl, CO_2Me, Cl)

halomon, **290** (Cl, Br, Cl, Br, Cl)

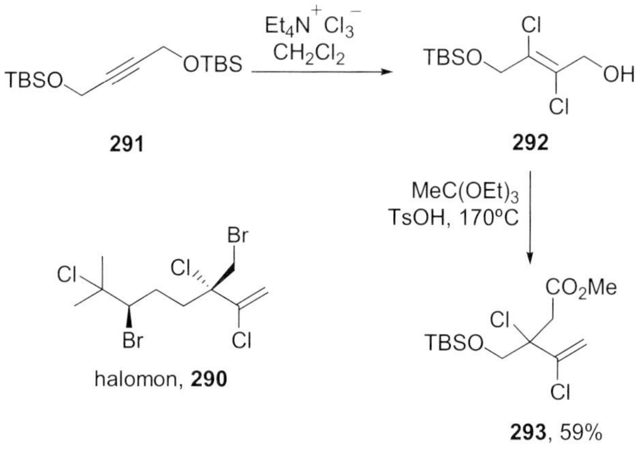

Scheme 6.46

6.4.2 Steroids

6.4.2.1 Syntheses of the Tetracyclic Core of Steroids

Following his impressive illustration of the usefulness of the orthoester Claisen rearrangement in the synthesis of squalene [7], Johnson [66] used this reaction thoroughly in several biomimetic syntheses of steroids. Some examples of these reactions are depicted in Scheme 6.47. Esters were generally obtained in high yield with complete stereoselectivity.

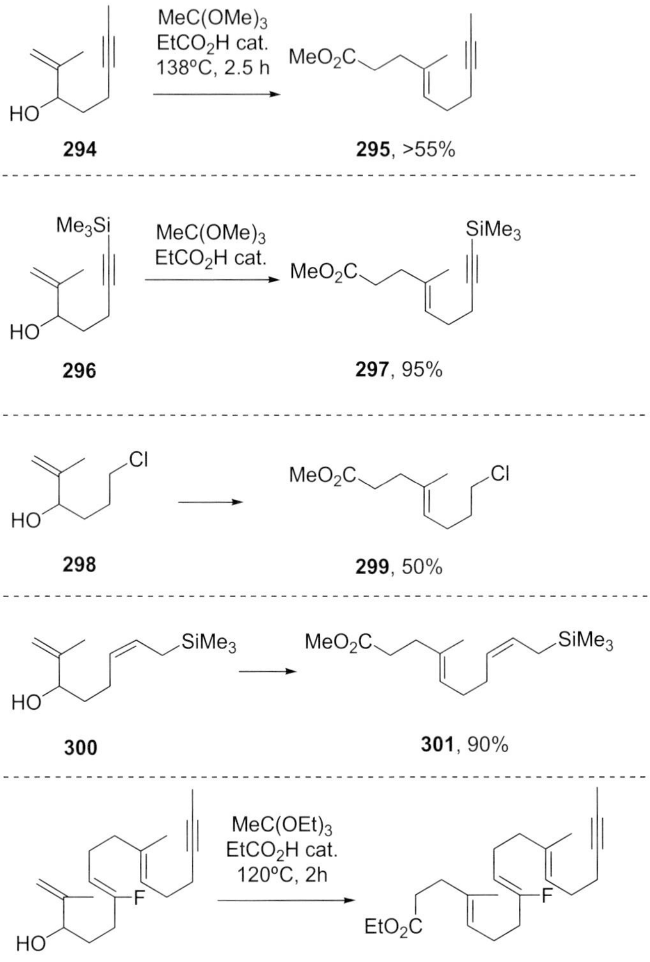

HO **294** → $MeC(OMe)_3$, $EtCO_2H$ cat., 138°C, 2.5 h → MeO_2C **295**, >55%

Me_3Si HO **296** → $MeC(OMe)_3$, $EtCO_2H$ cat. → MeO_2C $SiMe_3$ **297**, 95%

Cl HO **298** → MeO_2C Cl **299**, 50%

$SiMe_3$ HO **300** → MeO_2C $SiMe_3$ **301**, 90%

F HO **302** → $MeC(OEt)_3$, $EtCO_2H$ cat., 120°C, 2h → EtO_2C F **303**, 76%

Scheme 6.47

In a synthesis of 11-deoxy-19-norcorticosterone **304** using an orthoquinodimethane intramolecular Diels–Alder strategy, Fukumoto [67] developed an orthoester Claisen rearrangement starting from isopropylidene glyceraldehyde derivative **306**. Both yield and selectivity were modest. Claisen–Ireland rearrangement gave only slightly better results (Scheme 6.48).

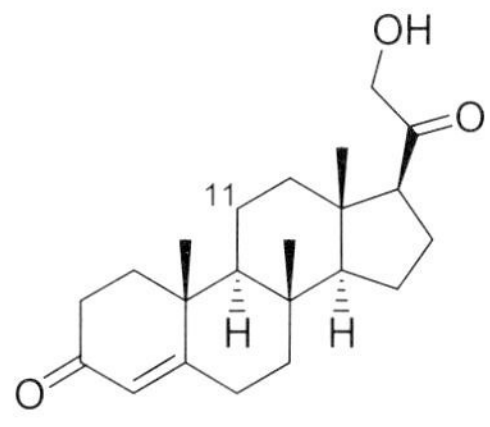

(+)-11-deoxy-19-norcorticosterone, **304**

305 → steps → **306** → $MeC(OEt)_3$, $EtCO_2H$, cat., 135°C, 10 min (Claisen–Johnson) → **307** 34 %, *R*/*S*= 53/47

308 → 1. LDA, THF, TMSCl, -78°C, then reflux 2. $LiAlH_4$, THF (Claisen–Ireland) → **309** 45 %, *R*/*S*= 66/33

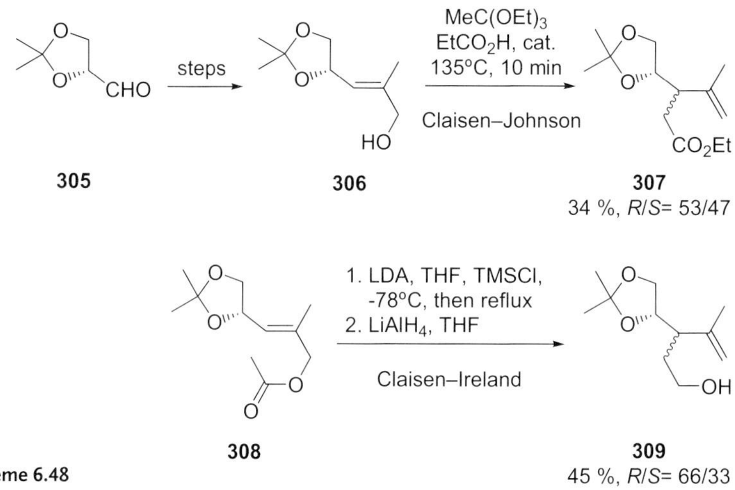

Scheme 6.48

Analogs of calcitriol **310** have been synthesized by Posner [68] who developed an elegant tandem reaction involving sequentially Claisen–Johnson rearrangement/sulfoxide elimination. Thus, the diene ester **314**, a synthetic precursor of **311**, was obtained directly (Scheme 6.49).

Tsuji [69], in a stereoselective synthesis of de-*AB*-cholestan-9-one **315**, studied an orthoester rearrangement followed by an allyl vinyl ether rearrangement affording the advanced intermediate **319** (Scheme 6.50).

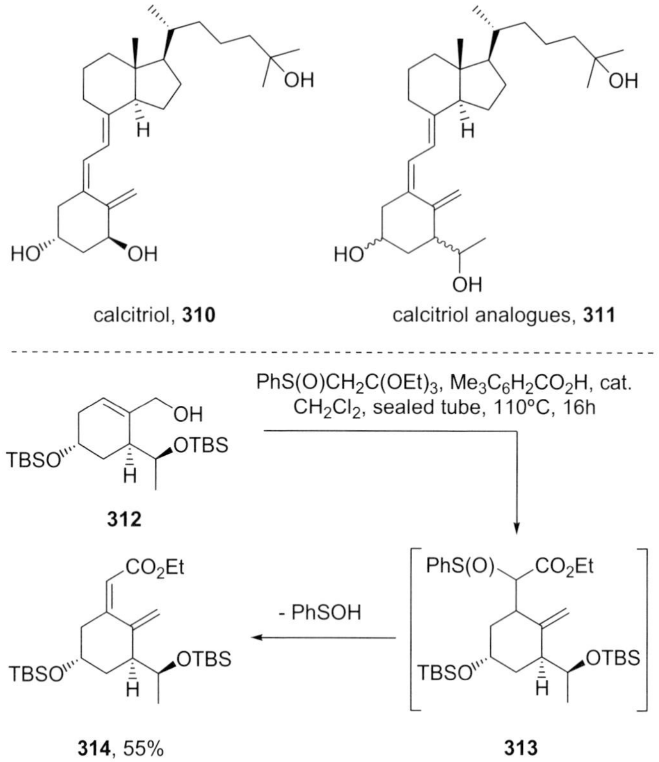

calcitriol, **310**

calcitriol analogues, **311**

$PhS(O)CH_2C(OEt)_3$, $Me_3C_6H_2CO_2H$, cat.
CH_2Cl_2, sealed tube, 110°C, 16h

312

- PhSOH

314, 55%

313

Scheme 6.49

316

$MeC(OEt)_3$
$EtCO_2H$
120°C, 3h

de-*AB*-cholestan-9-one, **315**

steps

317, 57%

318

319, >70%

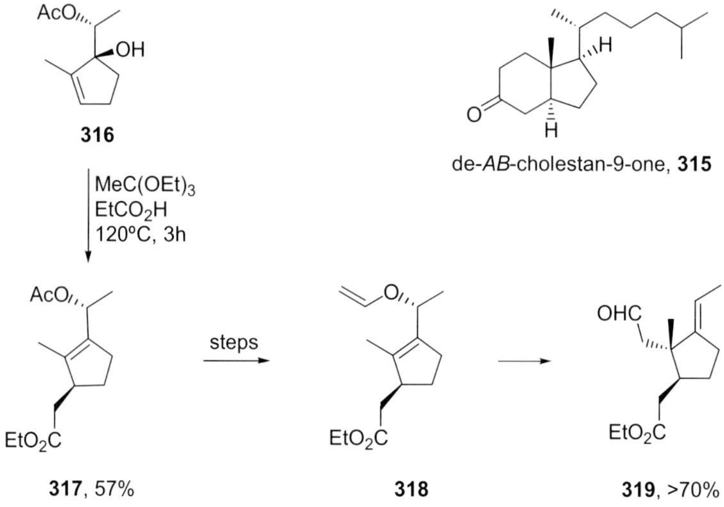

Scheme 6.50

6.4.2.2 Syntheses of Steroid Side Chains

The Claisen–Johnson rearrangement has often been used for the steroid side chain stereoselective functional group transformations in acyclic diasteroselective processes. Accordingly, McMorris and Djerassi independently determined the configuration at C-24 of the side chain of the steroidal hormone oogoniol **320** of the water mold [70, 71] (Scheme 6.51).

In a study concerning the insect steroid metabolism, Prestwich [72] developed the same strategy for the preparation of 29-fluorophytosterols **329** as depicted in Scheme 6.52.

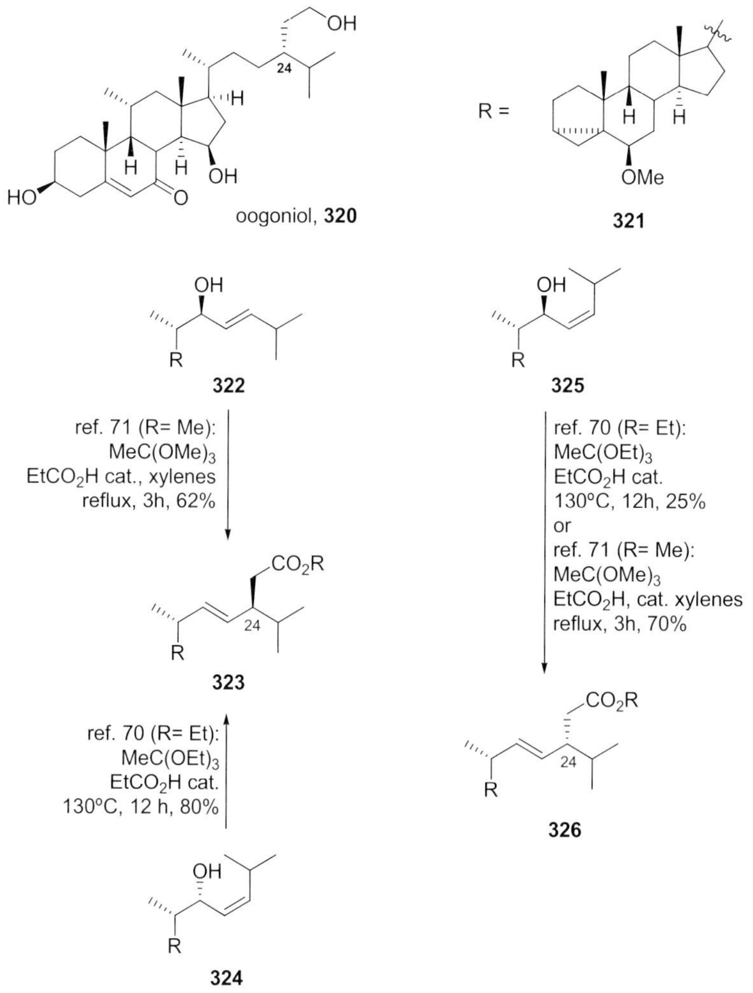

oogoniol, **320**

R = **321**

322 → **323**: ref. 71 (R= Me): $MeC(OMe)_3$, $EtCO_2H$ cat., xylenes, reflux, 3h, 62%

324 → **323**: ref. 70 (R= Et): $MeC(OEt)_3$, $EtCO_2H$ cat., 130°C, 12 h, 80%

325 → **326**: ref. 70 (R= Et): $MeC(OEt)_3$, $EtCO_2H$ cat., 130°C, 12h, 25% or ref. 71 (R= Me): $MeC(OMe)_3$, $EtCO_2H$, cat. xylenes, reflux, 3h, 70%

Scheme 6.51

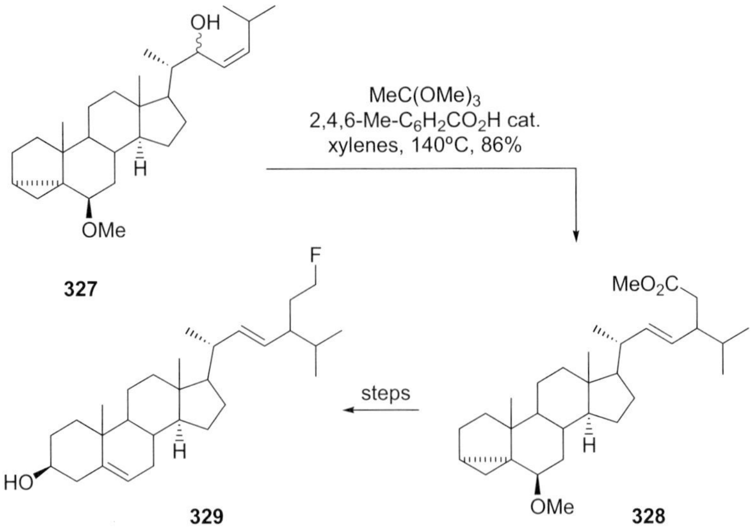
OH
MeC(OMe)3
2,4,6-Me-C6H2CO2H cat.
xylenes, 140°C, 86%
H
OMe
327
F
MeO2C
steps
HO
H
329
H
OMe
328

Scheme 6.52

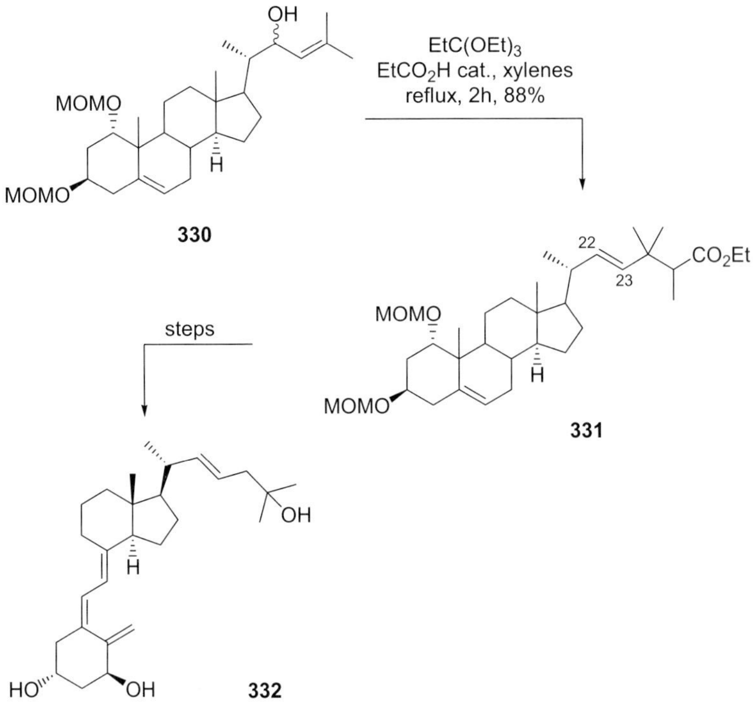
OH
EtC(OEt)3
EtCO2H cat., xylenes
reflux, 2h, 88%
MOMO
MOMO
H
330
22
CO2Et
23
MOMO
steps
MOMO
H
331
H
OH
HO
OH
332

Scheme 6.53

Ikekawa [73] described the synthesis of two new vitamin D_3 derivatives such as compound **332**. Introduction of a C^{22}/C^{23} double bond resulted from a Claisen–Johnson rearrangement (Scheme 6.53).

Isopropylidene glyceraldehyde **305** submitted to orthoester rearrangement was used by Suzuki in the initial steps of the synthesis of intermediate **333** leading to vitamin D_3 [74] (Scheme 6.54). Still in the field of vitamin D_3, a stereoselective synthesis of steroid side chain and CD rings has been described by Takahashi and Tsuji [75]. Triethylorthohexenoate **335** was used as an orthoester. The stereochemical control of carbon β to ester was good but as in other cases the α center was obtained as a mixture of diastereomers in ester **336** (Scheme 6.54).

The synthesis of steroid lactone **340** used, as the key step, the orthoester rearrangement [76] (Scheme 6.55).

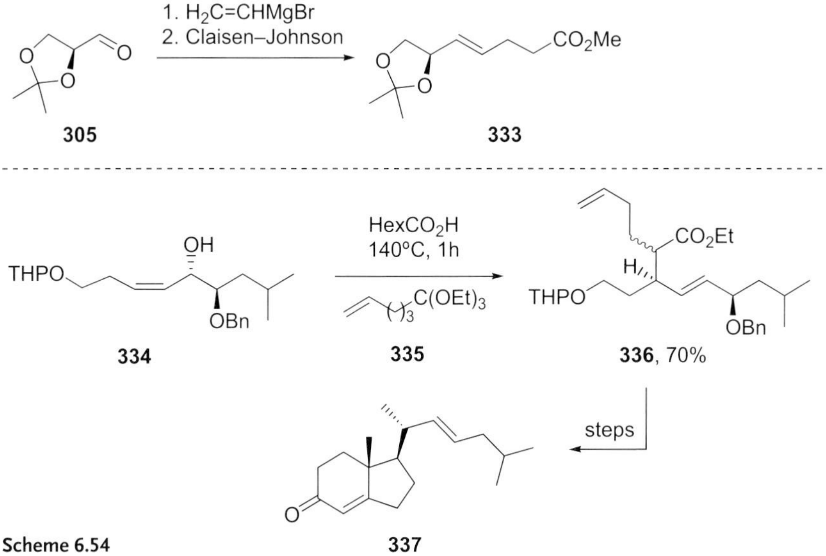

1. H2C=CHMgBr
2. Claisen–Johnson
CO2Me
305
333
HexCO2H
140°C, 1h
C(OEt)3
THPO
OH
OBn
CO2Et
334
335
336, 70%
steps
337

Scheme 6.54

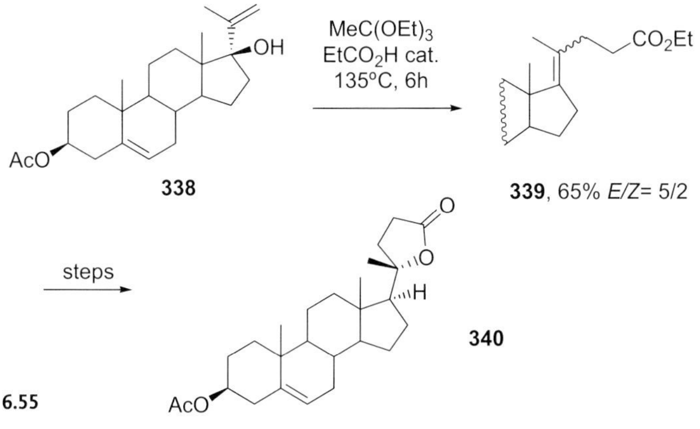

MeC(OEt)3
EtCO2H cat.
135°C, 6h
OH
AcO
CO2Et
338
339, 65% E/Z= 5/2
steps
O
H
340

Scheme 6.55

In several studies concerning the synthesis of brassinolides such as compound **343** described by Takatsuto [77, 78], the Claisen–Johnson rearrangement afforded stereoselectively functionalized side chains. As previously, formation of the center in β of ester group was stereoselective. A mixture of epimers at the α-stereocenter was formed during the process, but since this carbon atom was no longer stereogenic in compound **343**, this lack of stereoselectivity was of no consequence (Scheme 6.56).

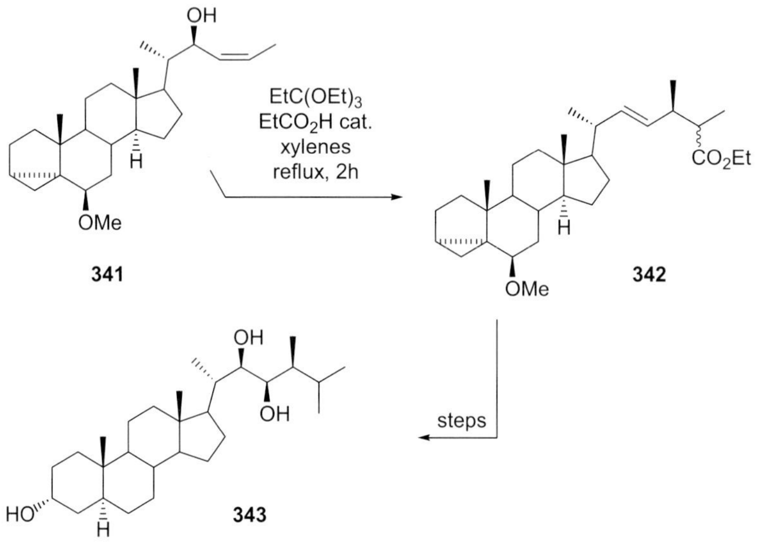

Scheme 6.56

The trifluoromethyl compound **345**, resulting from a [2+3] cycloaddition followed by functional group manipulations, was subjected to a highly diastereoselective orthoester rearrangement and afforded the ester **346**. The high selectivity observed in the formation of the two stereogenic centers in this example is noteworthy. Compounds **346** afforded the target molecule **344** after further functional group transformations (Scheme 6.57) [79].

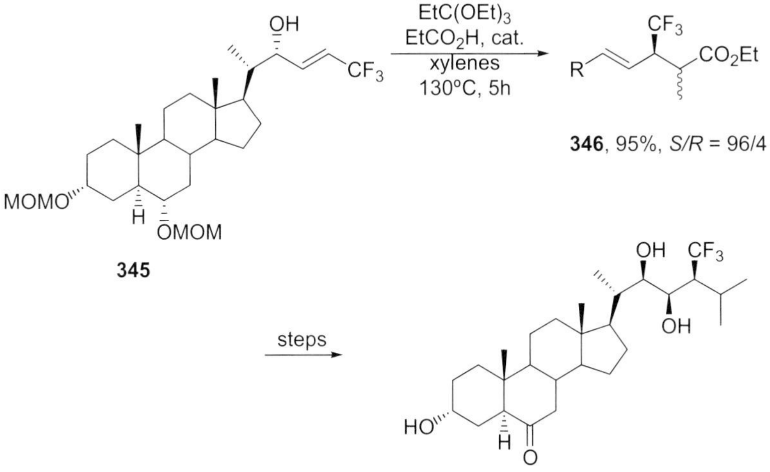

OH EtC(OEt)$_3$ EtCO$_2$H, cat. xylenes 130°C, 5h CF$_3$ R CF$_3$ CO$_2$Et

MOMO H OMOM

345

346, 95%, S/R = 96/4

steps

OH CF$_3$ OH HO H O

24-trifluoromethyl typhasterol, 344

Scheme 6.57

The configuration at C-24 in 24-propyl cholesterol **347** extracted from Texas brown tide, was established through a synthesis using in a main step a Claisen–Johnson rearrangement applied to allylic alcohol **348** [80] (Scheme 6.58).

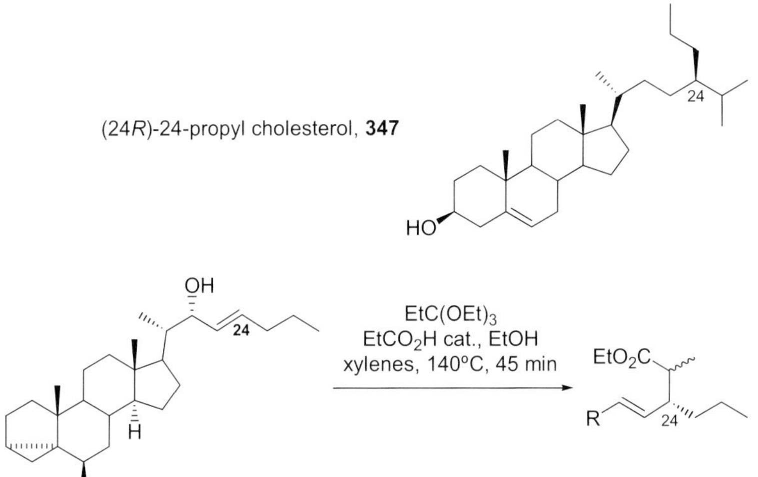

(24R)-24-propyl cholesterol, **347**

24 HO

OH 24 EtC(OEt)$_3$ EtCO$_2$H cat., EtOH xylenes, 140°C, 45 min EtO$_2$C R 24

H OMe **348**

349, 100% (1:1)

Scheme 6.58

6.4.3
Alkaloids

6.4.3.1 Indole Alkaloids

Few years after its disclosure, the potential of the Claisen–Johnson rearrangement in alkaloid synthesis was recognized. This reaction showed its utility especially in the field of indole alkaloids. Thus in 1973, Ziegler [81] applied this rearrangement as a corner stone step in the synthesis of tabersonine **350** for the construction of the C-20 quaternary center (Scheme 6.59). In this particular case, the Claisen–Johnson gave a better yield than the Eschenmoser rearrangement.

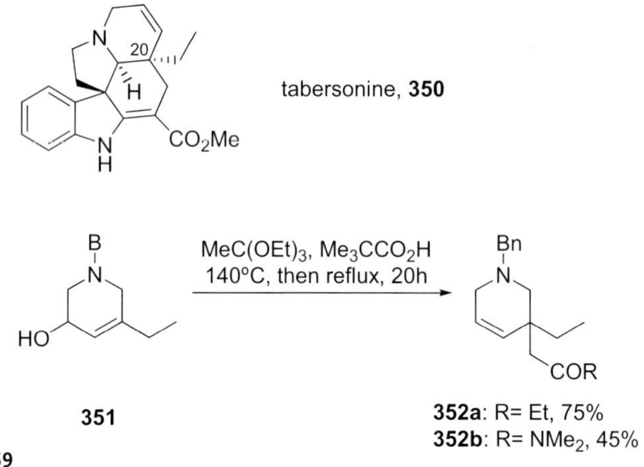

tabersonine, **350**

$MeC(OEt)_3$, Me_3CCO_2H
140°C, then reflux, 20h

351

352a: R= Et, 75%
352b: R= NMe_2, 45%

Scheme 6.59

Still with a tetrahydopyridine derivative alcohol **354**, a high-yielding orthoester rearrangement was used in a synthesis of antitumor alkaloid ellipticine **353** [82] (Scheme 6.60). The same kind of reaction was applied to a synthesis of descarbomethoxy vobasine **356b** (Scheme 6.60) [83]. This type of rearrangement was also used by Kametani [84] in an approach to ervitsine **359** (Scheme 6.60).

In a synthesis of heteroyohimbane alkaloids tetrahydroalstonine **362** and akuammigine **363**, Uskokovic prepared a chemoenzymatic process alcohol **364** [85, 86]. This compound afforded ester **366** with *R* configuration at C-15 after orthoester rearrangement through a chair-like transition state **365**. Functionalization at C-16 followed by a stereoselective hydroboration-oxidation afforded compound **367** with the DE rings framework of the alkaloids (Scheme 6.61). The same rearrangement has been used by Bosch in a synthesis of alkaloid tubifoline **368** [87] (Scheme 6.61).

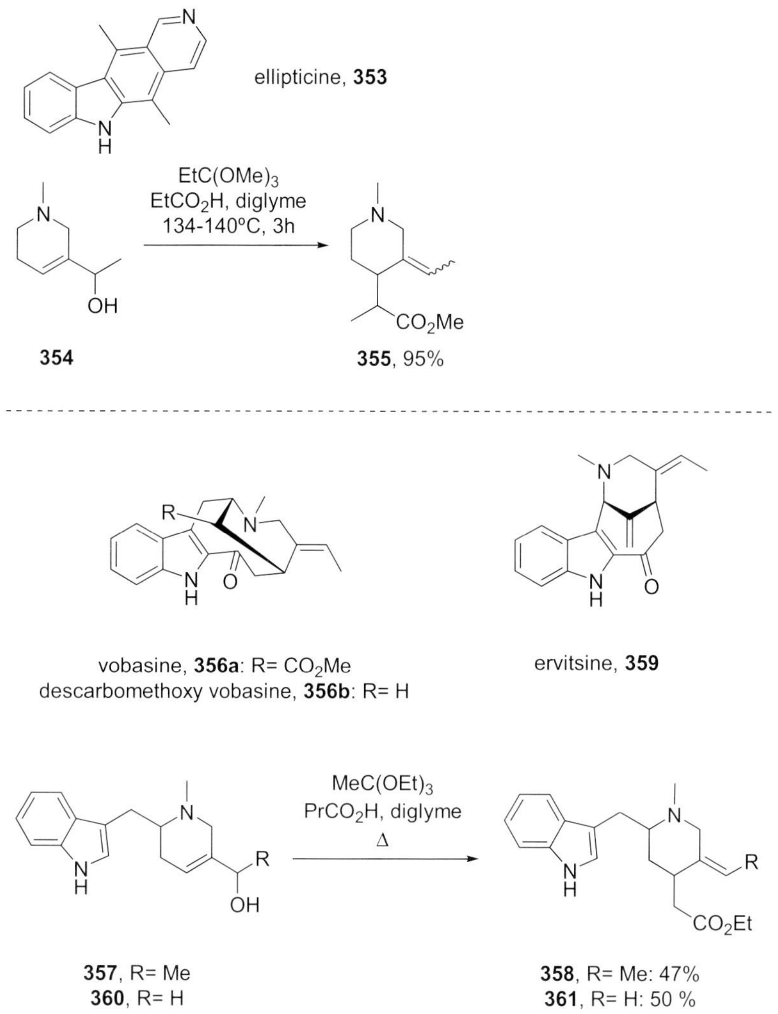

ellipticine, **353**

$EtC(OMe)_3$
$EtCO_2H$, diglyme
134-140°C, 3h

354

355, 95%

vobasine, **356a**: R= CO_2Me
descarbomethoxy vobasine, **356b**: R= H

ervitsine, **359**

$MeC(OEt)_3$
$PrCO_2H$, diglyme
Δ

357, R= Me
360, R= H

358, R= Me: 47%
361, R= H: 50 %

Scheme 6.60

Lounasmaa prepared various isomers in indoloquinolizidine series of alkaloids with orthoester rearrangement [88]. Isositsirikine **371** derivatives were synthesized by the same group and NMR methods were applied to determine the configuration at C-16 [89] (Scheme 6.62). Similarly, Lounasmaa extended these rearrangements to other functionalized orthoesters [90]. For instance, indolic ketene acetal **374** afforded after rearrangement a mixture of diastereoisomeric esters **376** and **377**. The major diastereomer resulted as expected from a chair-like transition state. However, with other orthoesters like triethyl 3,3-diethoxy orthopropionate **378**, a complex mixture of products was obtained.

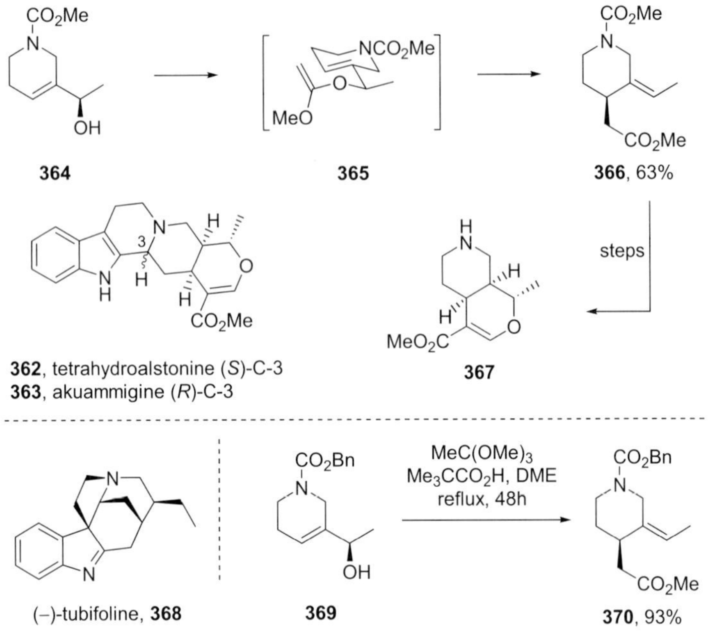

CO_2Me 364 → 365 → 366, 63%

steps

362, tetrahydroalstonine (*S*)-C-3
363, akuammigine (*R*)-C-3

367

(–)-tubifoline, 368

369 → $MeC(OMe)_3$, Me_3CCO_2H, DME, reflux, 48h → 370, 93%

Scheme 6.61

sitsirikine, 371

372 → $EtC(OMe)_3$, $MeCO_2H$, dioxane, 100°C, 72h → 373, 76% (1:1)

374 → dioxane, 100°C, 12h → 375a → 376; 375b → 377

60%, **376**/**377** = 75/25

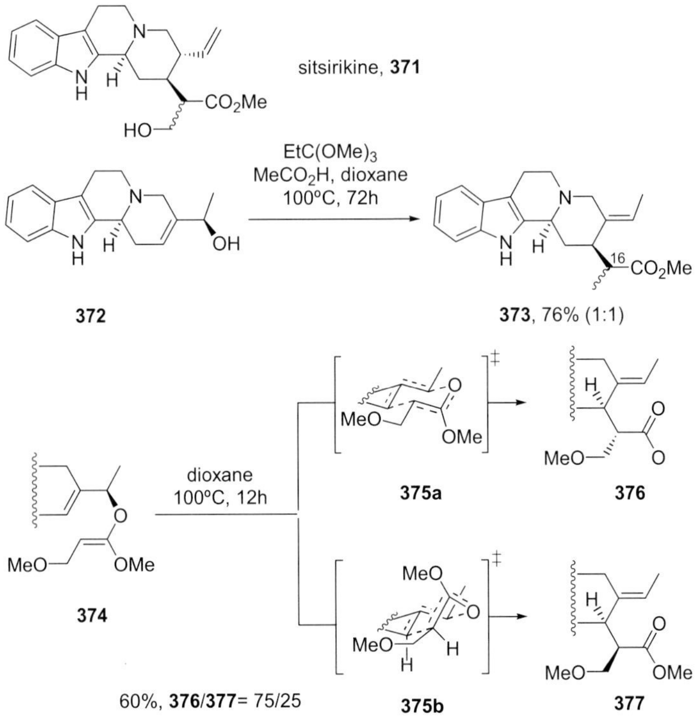

Scheme 6.62

With triethyl ortho-2-ethoxycarbonylacetate **379** (Scheme 6.63), diastereomeric alcohols (1*R*)-**372** and (1*S*)-**372** afforded compounds **380a,b**, **381** and **382** respectively [91].

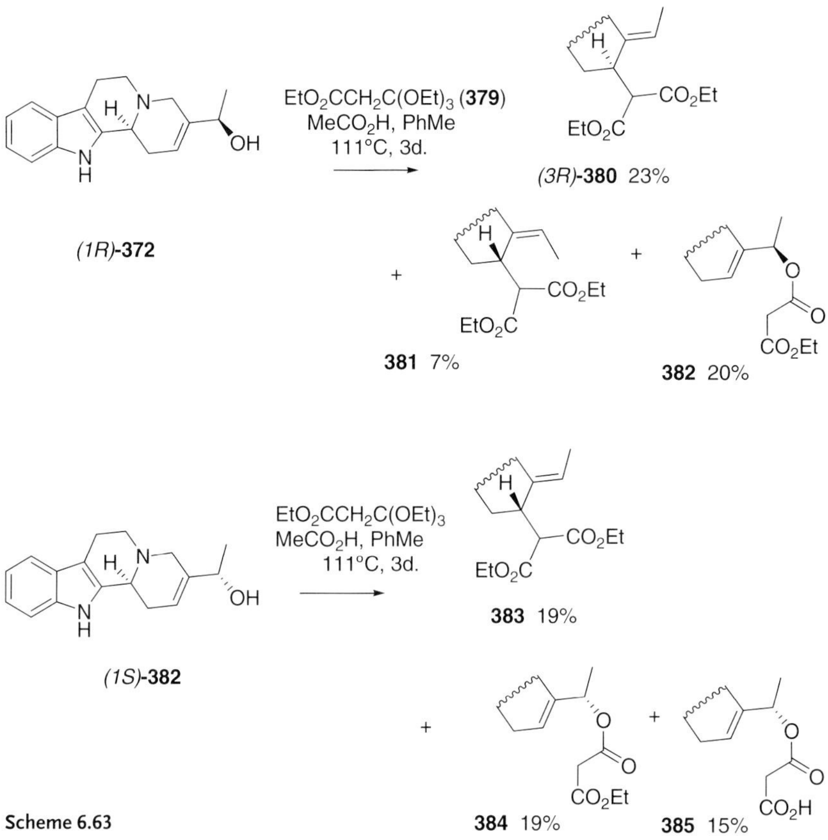

H N OH N H (1*R*)-**372**

$EtO_2CCH_2C(OEt)_3$ (**379**)
$MeCO_2H$, PhMe
111°C, 3d.

H CO_2Et EtO_2C (3*R*)-**380** 23%

+ H CO_2Et EtO_2C **381** 7%

+ O O CO_2Et **382** 20%

H N OH N H (1*S*)-**382**

$EtO_2CCH_2C(OEt)_3$
$MeCO_2H$, PhMe
111°C, 3d.

H CO_2Et EtO_2C **383** 19%

+ O O CO_2Et **384** 19%

+ O O CO_2H **385** 15%

Scheme 6.63

Cook, in an elegant synthesis of macroline alkaloid suaveoline **386**, developed a stereoselective Claisen–Johnson rearrangement [92a]. In this example, the major isomers *R*-**388** and *S*-**389** were formed via a boat-like transition state **391** (Scheme 6.64). Comparison between allyl vinyl ether rearrangement and Claisen–Johnson rearrangement showed that the former rearrangement occurred preferentially via a chair-like transition state (Scheme 6.64) [92b]. Other macroline alkaloids have been synthesized by the same group by the way of allyl vinyl ether rearrangement [93].

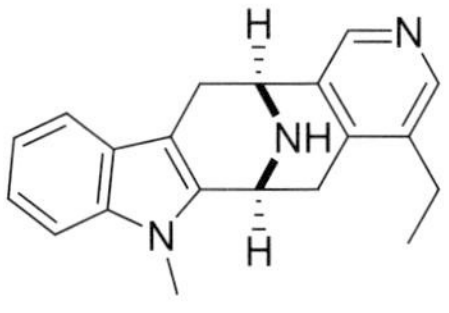

suaveoline, **386**

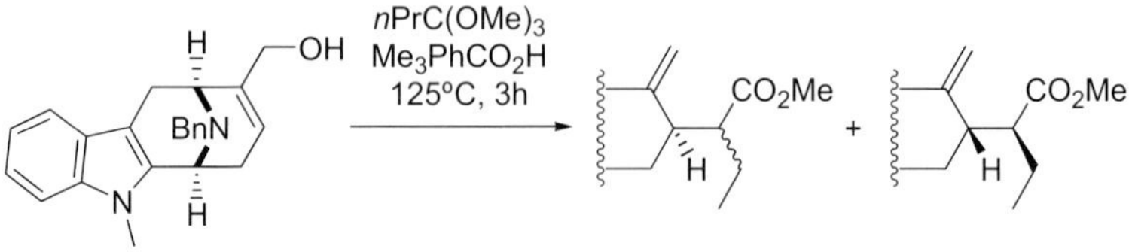

$nPrC(OMe)_3$
Me_3PhCO_2H
125°C, 3h

387

388, 93 %, *R/S*= 2/1
via boat-like ts

389, 7%

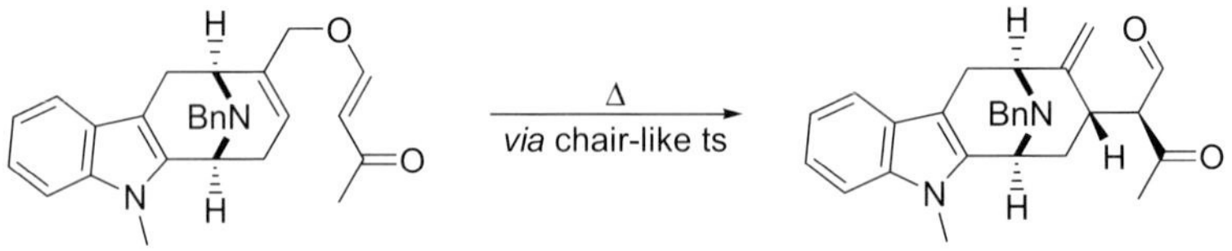

Δ
via chair-like ts

390

391

Scheme 6.64

In a recent synthesis of the aspidosperma alkaloid aspidophytine **396**, Fukuyama used a Claisen–Johnson rearrangement to introduce stereoselectively the quaternary center at C-20 (Scheme 6.65) [94]. In a previous synthesis of this alkaloid by Corey, the configuration of same stereogenic center was controlled via a Claisen–Ireland rearrangement [95].

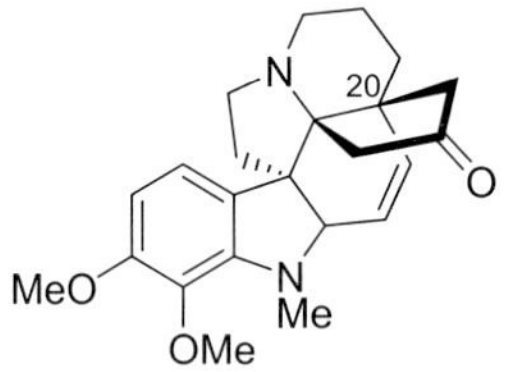

aspidophytine, **396**

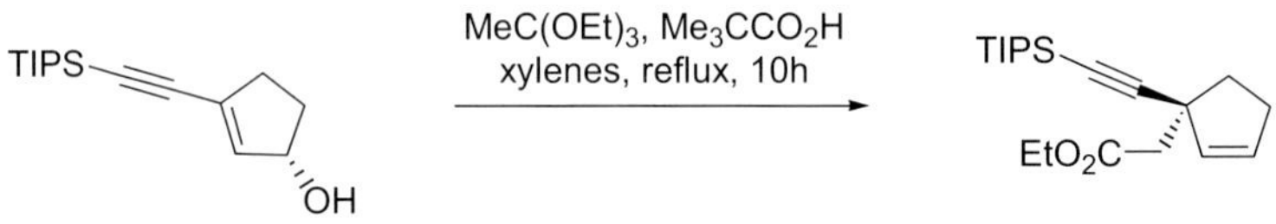

$MeC(OEt)_3$, Me_3CCO_2H
xylenes, reflux, 10h

397

398

Scheme 6.65

6.4.3.2 Other Alkaloids

Various codeine **399** derivatives were prepared for structure activity studies via a sequence of reactions that began with a Claisen–Eschenmoser rearrangement affording amide **401** from codeine **399** [96]. Unexpectedly, with the same alkaloid under the Claisen–Johnson conditions the orthoesters **402** and **403** were the only products isolated (Scheme 6.66).

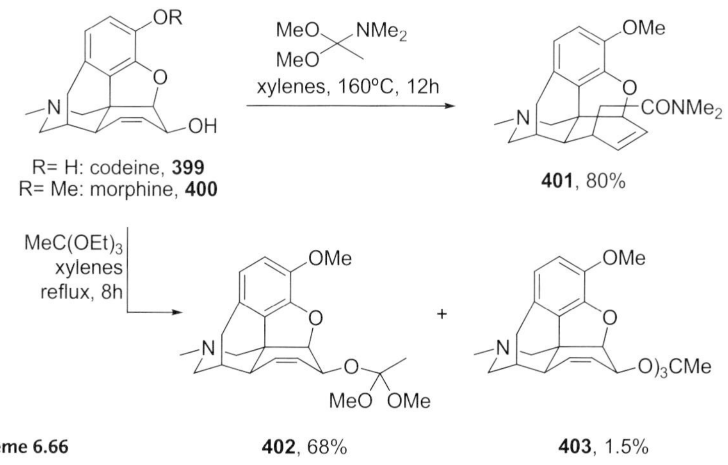

R= H: codeine, **399**
R= Me: morphine, **400**

$MeC(OMe)_2NMe_2$, xylenes, 160°C, 12h → **401**, 80%

$MeC(OEt)_3$, xylenes, reflux, 8h → **402**, 68% + **403**, 1.5%

Scheme 6.66

In the area of imminium salt photochemistry and the application of this methodology to erythrina alkaloids synthesis, Mariano [97a] prepared the functionalized allylic silane **405** via a Claisen–Johnson rearrangement (Scheme 6.67). Compound **405** has also been used by Ley in a synthesis of the immunosuppressant rapamycin [97b]. Moriwake and Saito [98] prepared two pyrrolizidine alkaloids such as (–)-isoretronecanol **406** through a Claisen–Johnson rearrangement on prolinol derived allylic alcohol **407** (Scheme 6.67). A mixture of diasteromeric esters **410** and **411** was obtained. Two chair-like transition states **408** and **409** could be considered and steric interaction explained the observed selectivity. The vinyl ether double bond approaches the allylic ether moiety from the less sterically hindererd face opposite to NBoc-group. Moreover, in the proposed models the C–N bond could be oriented orthogonal to the double bond to maximize the overlapping between σ^*(C–N) and π(C=C) orbitals. The depicted conformations are also favored based on the concept of minimized 1,3-allylic strain.

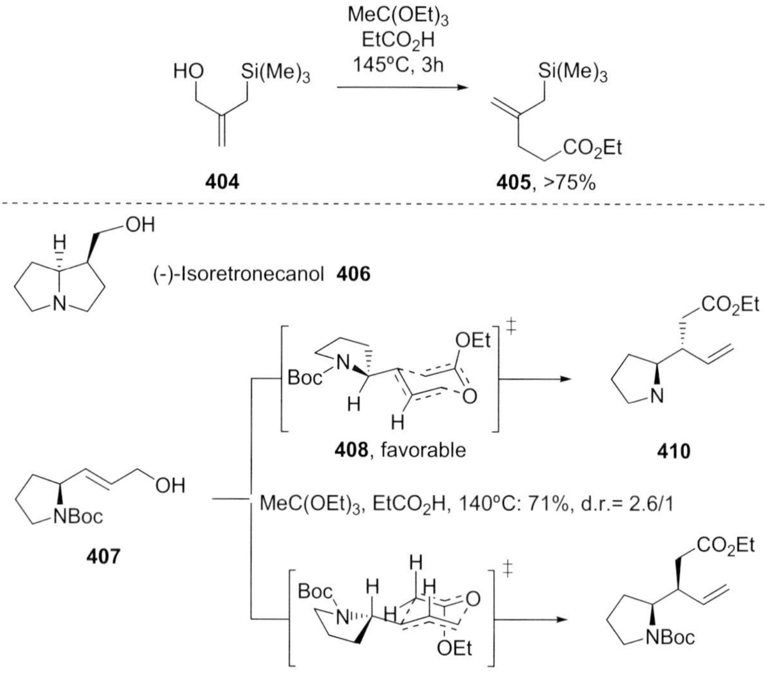

MeC(OEt)3
EtCO2H
145°C, 3h

HO Si(Me)3

Si(Me)3

CO2Et

404

405, >75%

(-)-Isoretronecanol 406

408, favorable

410

407

MeC(OEt)3, EtCO2H, 140°C: 71%, d.r.= 2.6/1

409, less favorable

411

Scheme 6.67

413

swainsonine, 412

steps

MeC(OMe)3
EtCO2H, toluene
reflux, 24h

414

415, 99%

MeC(OEt)3
EtCO2H cat.
130°C, 3h

417

418a, R= C3H7: 78%
418b, R= C6H13: 87%

416
ant venom alkaloid

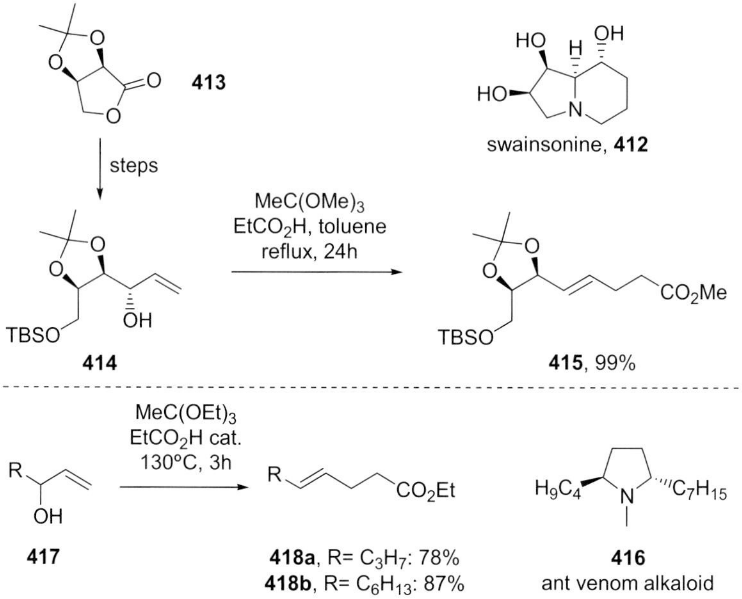

Scheme 6.68

The indolizidine alkaloid swainsonine **412**, a potential anticancer drug, has been the target of a number of syntheses. Pearson [99] started with 2,3-*O*-isopropylidene-D-erythronolactone **413**, which afforded in three steps the allylic alcohol **414**. This compound was in turn submitted to Claisen–Johnson rearrangement and afforded ester **415**. Selective dihydroxylation followed by functional group manipulations led to the target alkaloid (Scheme 6.68). An aminyl radical cyclization was the key step in an ant venom alkaloid **416** synthesis [100]. The Claisen–Johnson rearrangement was used in the preparation of both side chains of this pyrrolidine alkaloid (Scheme 6.68).

Porphobilinogen **419**, a biogenetic precursor of tetrapyrroles, has been prepared by Jacobi [101] through a sequence including Claisen–Johnson rearrangement, intramolecular Diels–Alder cycloaddition and oxidative cleavage. Interestingly, furano derivatives **420** under the rearrangement reaction conditions afforded directly the Diels–Alder adducts **421** and **422** with a yield that was dependent on the substituent R on the amide nitrogen atom. For R=dibenzosuberyl, the observed result was probably a consequence of a favorable conformation, the equilibrium retro-Diels–Alder/Diels–Alder reaction is shifted to the formation of the cycloaddition product (Scheme 6.69).

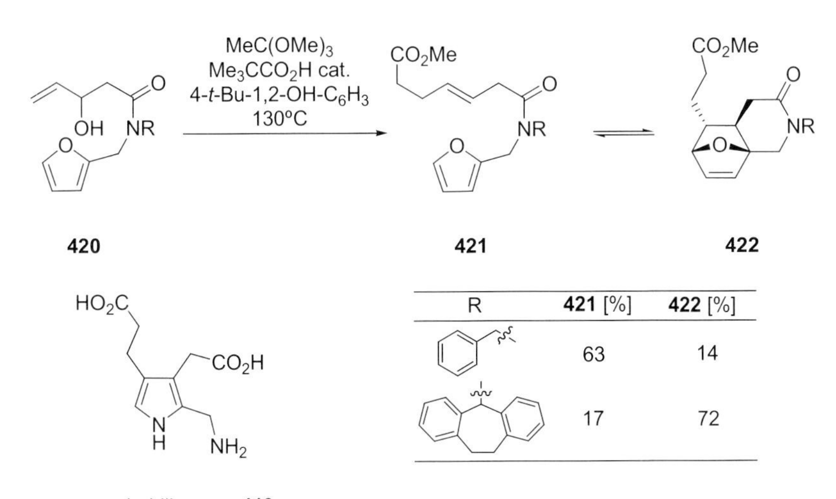

R	**421** [%]	**422** [%]
	63	14
	17	72

Scheme 6.69

6.4.4
Carbohydrates

In a pioneering work, Fraser-Reid developed a strategy using as the key step a [3,3]-sigmatropic rearrangement for the introduction of a quaternary center in modified sugar derivatives [102]. The allyl vinyl ether rearrangement was first used, but subsequently the same group also studied the Claisen–Eschenmoser and the Claisen–Johnson rearrangements routes to pyranoside diquinanes [103]. *Z* isomer (*Z*)-**425** led to a mixture of diastereomeric alcohols (**426**), whereas *E*-**425** afforded the *S*-configured diastereomer exclusively (Scheme 6.70).

$EtC(OEt)_3$
$EtCO_2H$

425 → **426**

(*Z*)-**425**, 85 %, *R*/*S*= 69/31
(*E*)-**425**, 85 %, *R*/*S*= 0/100

Scheme 6.70

Tadano [104] used the same reaction for the introduction of a quaternary stereogenic carbon atom in hexofuranoses **429** and **432**. As in the preceding study, configuration of the allylic alcohol double bond played a crucial role for both yield and diastereoselectivity. Thus, alcohol (*E*)-**429** afforded a mixture of esters **430** and **431**, whereas alcohol (*Z*)-**429** gave rise exclusively to ester **430**. The same behavior was observed with alcohols (*E*)- and (*Z*)-**432** (Scheme 6.71).

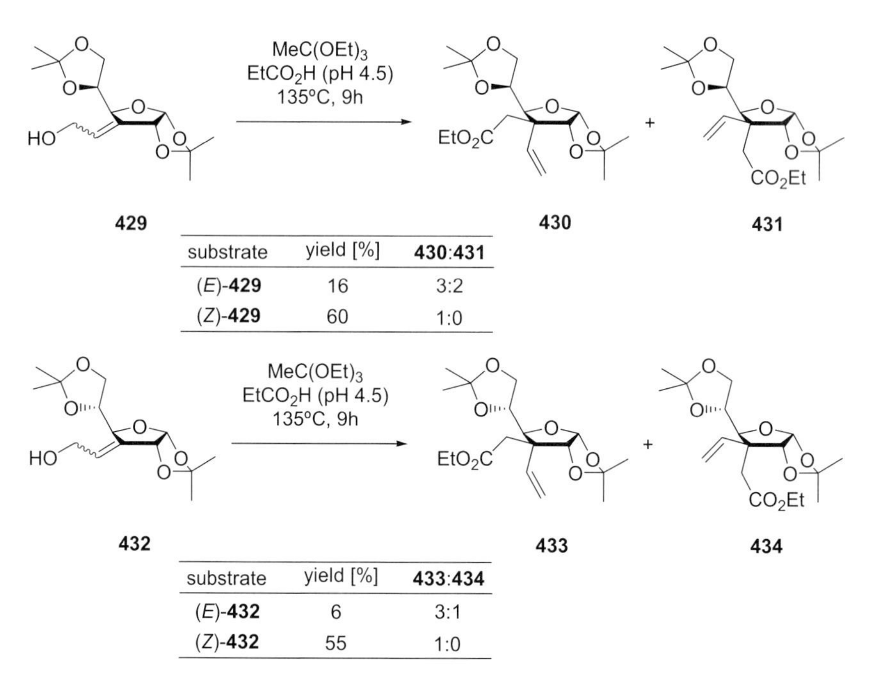

substrate	yield [%]	**430**:**431**
(*E*)-**429**	16	3:2
(*Z*)-**429**	60	1:0

substrate	yield [%]	**433**:**434**
(*E*)-**432**	6	3:1
(*Z*)-**432**	55	1:0

Scheme 6.71

In a recent study [105], the same group applied the Claisen–Johnson rearrangement to D-glucose derived spirocyclic substrate **437**. This reaction afforded a mixture of stereoisomers. It is worthy of note that the selectivity of the rearrangement is highly dependant upon the substitution of the orthoester (Scheme 6.72). The Claisen–Ireland rearrangement was inoperative with these glucose derivatives. This particular behavior could be due to chelation between lithium and oxygen atoms present in the molecule.

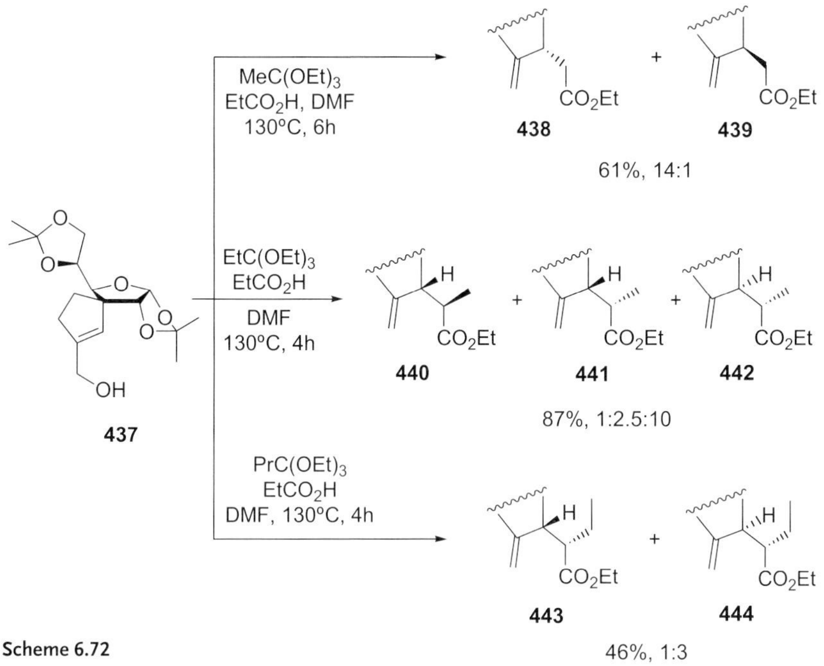

Scheme 6.72

6.4.5
Miscellaneous Compounds

(*E*)-Ethyl 3-methyl pentadienoate **450**, a diene reacting with maleic anhydride in a Diels–Alder cycloaddition, was prepared via an orthoester rearrangement followed by a bromation–dehydrobromation sequence [106] (Scheme 6.73). The Claisen–Johnson rearrangement of several allylic alcohols **451** followed by Wacker oxidation of the resulting unsaturated esters afforded δ- and γ-keto esters **453** or **454** [107] (Scheme 6.73). Rearrangement of 2-ethoxyallyl alcohols **457** provided a new entry to 1,4-dicarbonyl compounds **459** after acidic hydrolysis [108]. Baylis–Hillman derived 3-hydroxy-2-methylenealkanenitriles **460** gave rise with good yield and selectivity to ethyl (4*Z*)-4-cyano-4-enoates **461** via the Claisen–Johnson rearrangement [109] (Scheme 6.73). Similarly, Baylis–Hillman ester **462** subjected to the orthoester rearrangement gave rise to α-alkylidene ketoglutaric esters **463** [110] (Scheme 6.73).

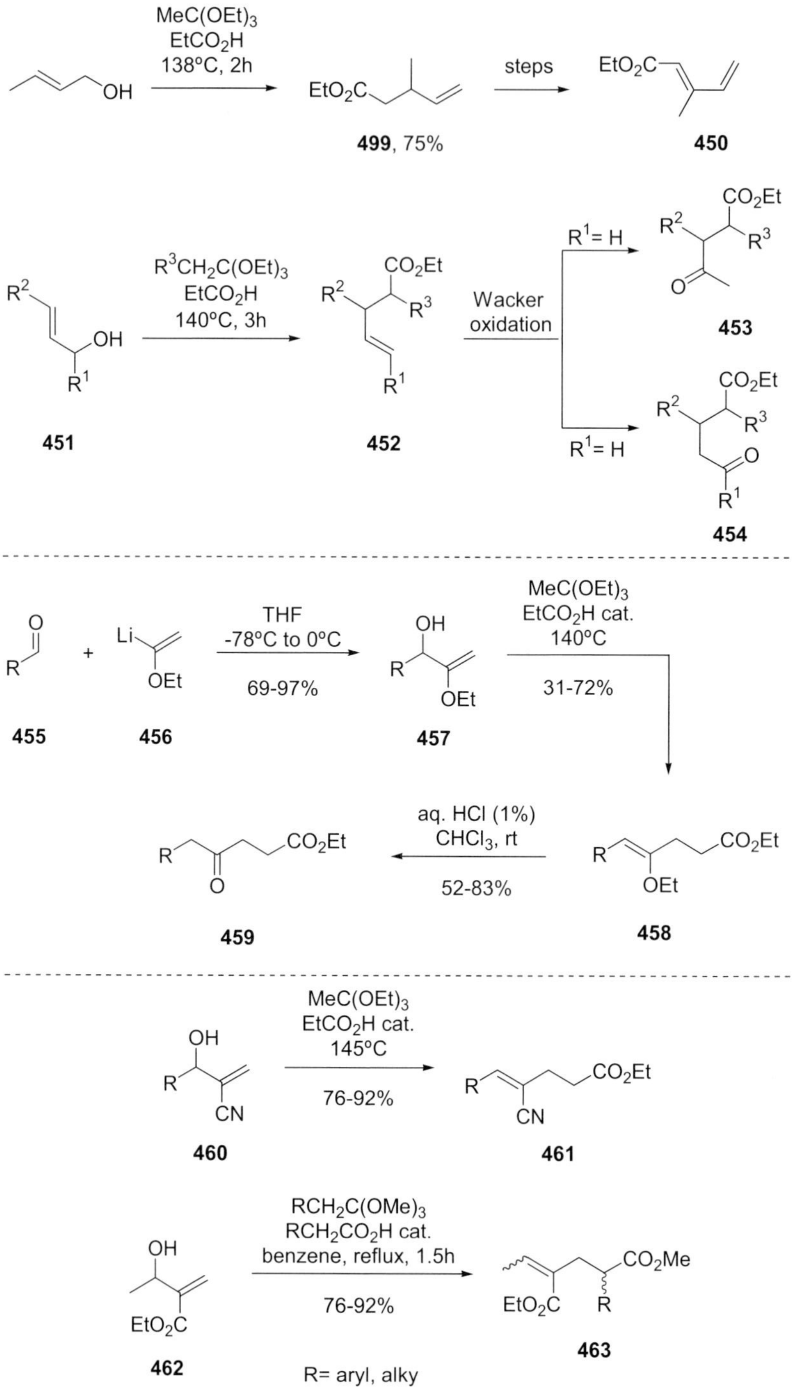

MeC(OEt)3
EtCO2H
138°C, 2h
EtO2C
499, 75%
steps
EtO2C
450

R3CH2C(OEt)3
EtCO2H
140°C, 3h
R2
OH
R1
451
CO2Et
R2
R3
R1
452
Wacker oxidation
R1= H
CO2Et
R2
R3
O
453
R1= H
CO2Et
R2
R3
O
R1
454

THF
-78°C to 0°C
69-97%
O
R
+
Li
OEt
455
456
OH
R
OEt
457
MeC(OEt)3
EtCO2H cat.
140°C
31-72%
R
CO2Et
OEt
458
aq. HCl (1%)
CHCl3, rt
52-83%
R
CO2Et
O
459

MeC(OEt)3
EtCO2H cat.
145°C
76-92%
OH
R
CN
460
R
CO2Et
CN
461

RCH2C(OMe)3
RCH2CO2H cat.
benzene, reflux, 1.5h
76-92%
OH
EtO2C
462
CO2Me
EtO2C
R
463
R= aryl, alky

Scheme 6.73

α-Hetero substituted orthoesters underwent Claisen–Johnson rearrangement in moderate to good yields [111] (Scheme 6.74). This reaction constitutes a new entry to α-heterosubstituted γ-unsaturated esters **465**. No stereoselectivity was observed when two stereogenic centers resulted from the rearrangement. γ-Hydroxyvinyl sulfones **466** afforded β-phenylsulfonyl-γ-unsaturated esters **467** after orthoester rearrangement [112]. These esters, after elimination of the sulfonyl group, gave rise to 2,4-dienoic esters **468** (Scheme 6.74). The same group [113] showed that unsaturated 3-hydroxy ketones **469** and unsaturated 3-hydroxy nitriles **471** are readily converted into 4-oxo esters **470** and 3-cyano esters **472** by means of orthoester rearrangement (Scheme 6.74).

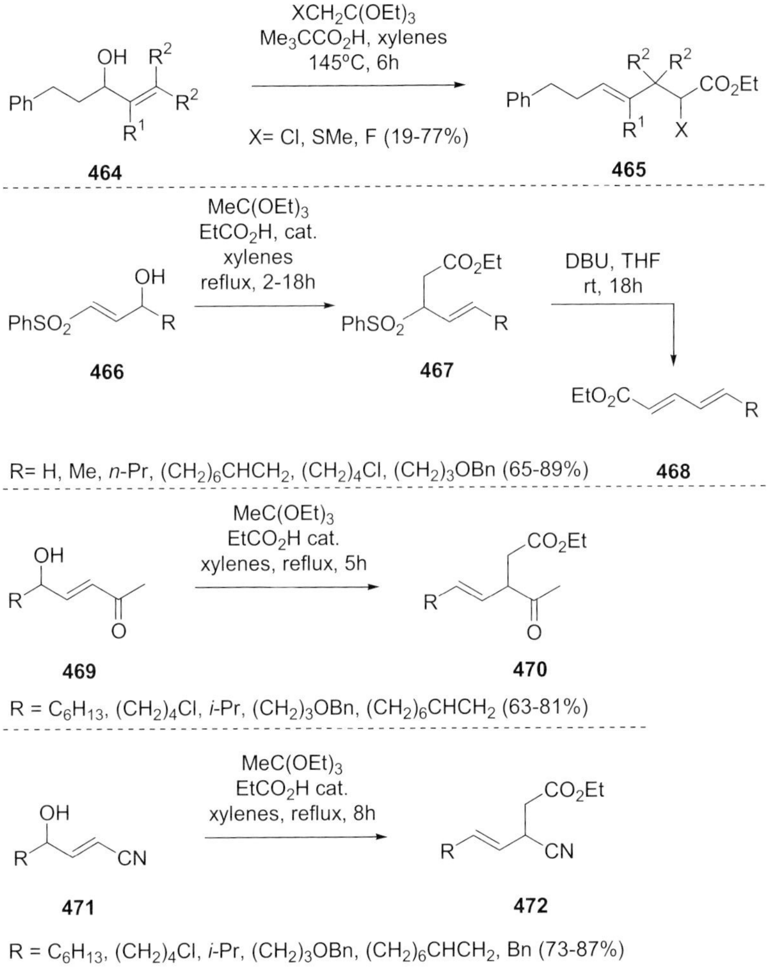

$XCH_2C(OEt)_3$, Me_3CCO_2H, xylenes, 145°C, 6h; X= Cl, SMe, F (19-77%)

464 → **465**

$MeC(OEt)_3$, $EtCO_2H$, cat., xylenes, reflux, 2-18h; DBU, THF, rt, 18h

466 → **467** → **468**

R= H, Me, *n*-Pr, $(CH_2)_6CHCH_2$, $(CH_2)_4Cl$, $(CH_2)_3OBn$ (65-89%)

$MeC(OEt)_3$, $EtCO_2H$ cat., xylenes, reflux, 5h

469 → **470**

R = C_6H_{13}, $(CH_2)_4Cl$, *i*-Pr, $(CH_2)_3OBn$, $(CH_2)_6CHCH_2$ (63-81%)

$MeC(OEt)_3$, $EtCO_2H$ cat., xylenes, reflux, 8h

471 → **472**

R = C_6H_{13}, $(CH_2)_4Cl$, *i*-Pr, $(CH_2)_3OBn$, $(CH_2)_6CHCH_2$, Bn (73-87%)

Scheme 6.74

Gore [114] prepared several hydoxyalkyl allenic esters **474** through a Claisen–Johnson rearrangement. After reduction of the ester group, the resulting diols were submitted to electrophilic cyclization and gave rise to dihydropyrane derivatives **475** (Scheme 6.75). Enantiomerically pure monoprotected diol (*R*)- and (*S*)-**476** was prepared from lactic acid. The Claisen–Johnson rearrangement afforded esters (*S*)- and (*R*)-**477**, respectively, with complete 1,3-chirality transfer [115] (Scheme 6.75). Glycidic esters such as **478** are readily isomerized to α-hydroxy β,γ-unsaturated esters **479** with ytterbium triflate or trimethylsilyl triflate. The resulting ester derivatives were in turn converted into rearranged products in presence of triethyl orthoacetate [116]. An example of this methodology applied to a triquinane precursor **480** is given in Scheme 6.75.

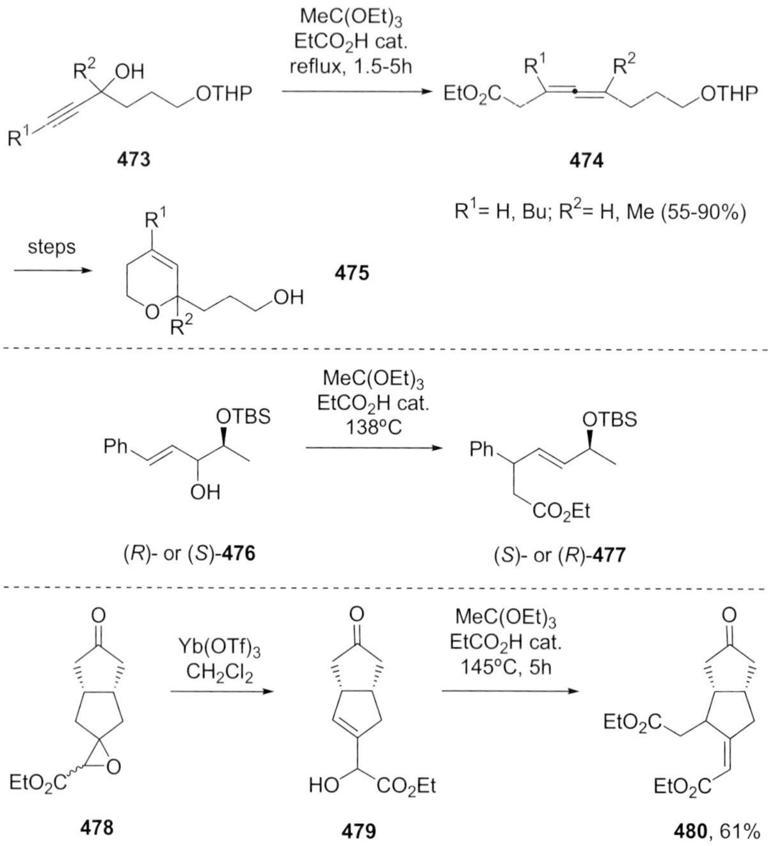

MeC(OEt)$_3$, EtCO$_2$H cat., reflux, 1.5-5h: **473** → **474**; R^1= H, Bu; R^2= H, Me (55-90%); steps → **475**

MeC(OEt)$_3$, EtCO$_2$H cat., 138°C: (*R*)- or (*S*)-**476** → (*S*)- or (*R*)-**477**

Yb(OTf)$_3$, CH$_2$Cl$_2$: **478** → **479**; MeC(OEt)$_3$, EtCO$_2$H cat., 145°C, 5h → **480**, 61%

Scheme 6.75

Marshall [117] prepared several enantiomerically pure allenylstannanes **485** either by cuprate S_N2' displacement of propargylic mesylate **486** or via a Claisen–Johnson rearrangement from alcohol **483**. After functional group transformations, correlation was achieved between the two series of compounds

(Scheme 6.76). Heathcock developed a synthetic strategy wherein the 1,2-stereoselection obtainable from aldol reaction is parlayed by a subsequent Claisen rearrangement into 1,4 and 1,5-stereoselection [118]. Claisen–Johnson, Claisen–Eschenmoser and Claisen–Ireland rearrangements were examined in this study. However the orthoester Claisen–Johnson rearrangement was not the most convenient in this case. The rather harsh reaction conditions often led to dehydration and other byproducts. An example of this rearrangement is given in Scheme 6.76. In this case allylic alcohol **487** afforded the rearranged ester **488** in moderate yield. In the development of a new strategy using Claisen–Ireland ring contraction of large ring lactone followed by transannular Diels–Alder cycloaddition, Roush [119] prepared the tetraene diester **490** through a Claisen–Johnson rearrangement (Scheme 6.76).

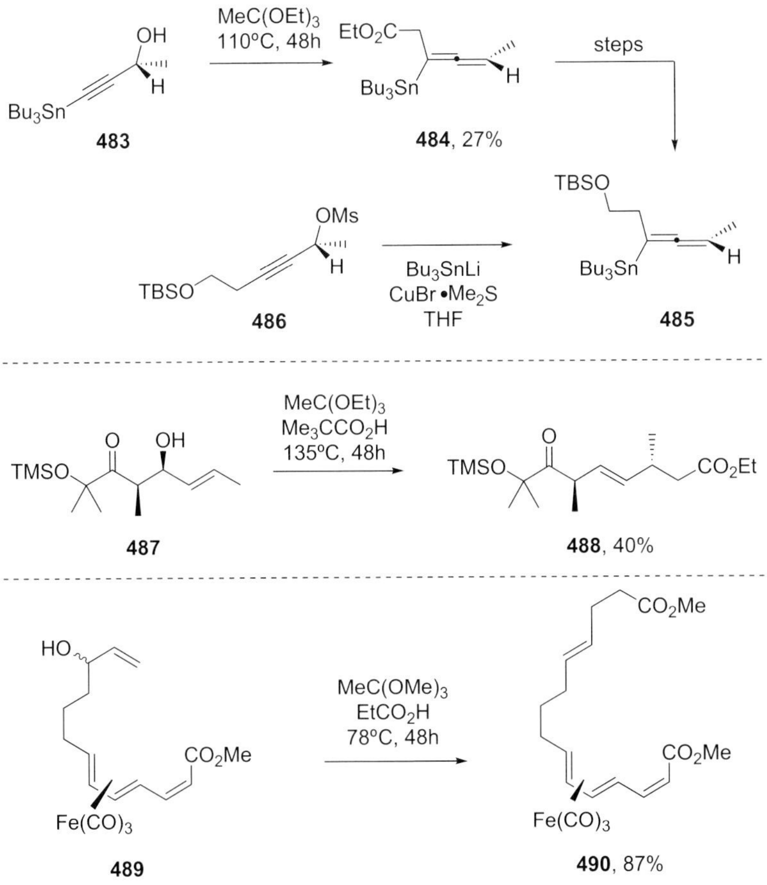

Scheme 6.76

In a synthesis of immunosuppressive and antiinflamatory agent RS-97613 **491**, related to mycophenolic acid **254**, a stereoselective Claisen–Johnson rearrangement afforded in very good yield ester **493** [120] (Scheme 6.77). It is noteworthy that in the following reaction, a methyl group is introduced stereoselectively by alkylation of the ester enolate intermediate affording compound **494**. The use of triethyl orthopropionate, which could afford the same product **494** in one step, should probably have given rise to a mixture of diastereoisomers.

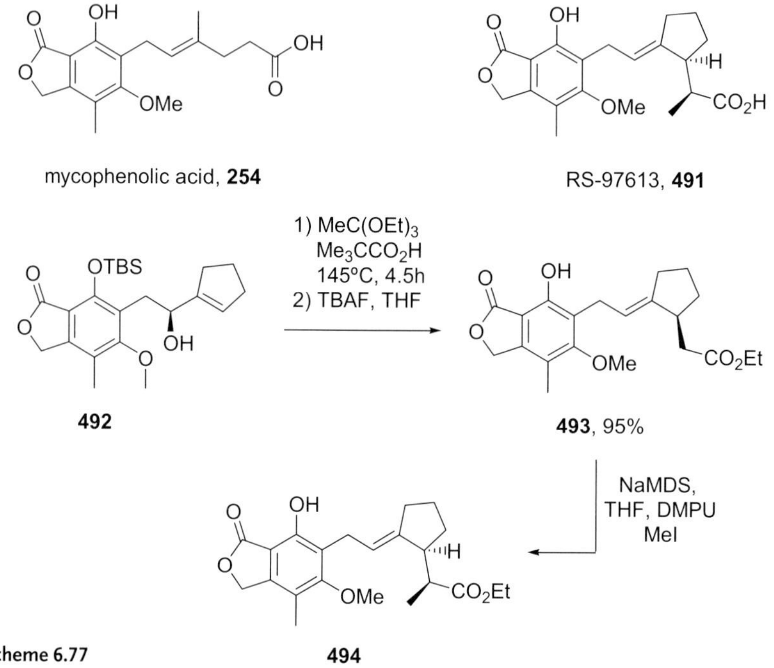

mycophenolic acid, **254**

RS-97613, **491**

492

1) $MeC(OEt)_3$, Me_3CCO_2H, 145°C, 4.5h
2) TBAF, THF

493, 95%

NaMDS, THF, DMPU, MeI

494

Scheme 6.77

A strategy based upon ring-closing metathesis was used for the preparation of novel carbocyclic nucleosides [121]. A Claisen–Johnson rearrangement with allylic alcohol **496** as substrate produced compound **497** with suitable functional groups (Scheme 6.78). A new class of aryl octahydrobenzazepine **498**, as a potential analgesic, was prepared using a strategy including orthoester rearrangement and intramolecular Diels–Alder cycloaddition [122] (Scheme 6.78).

496 **497**, 86% carbocyclic nucleoside **495**

499 **500** **498** **502** **501**

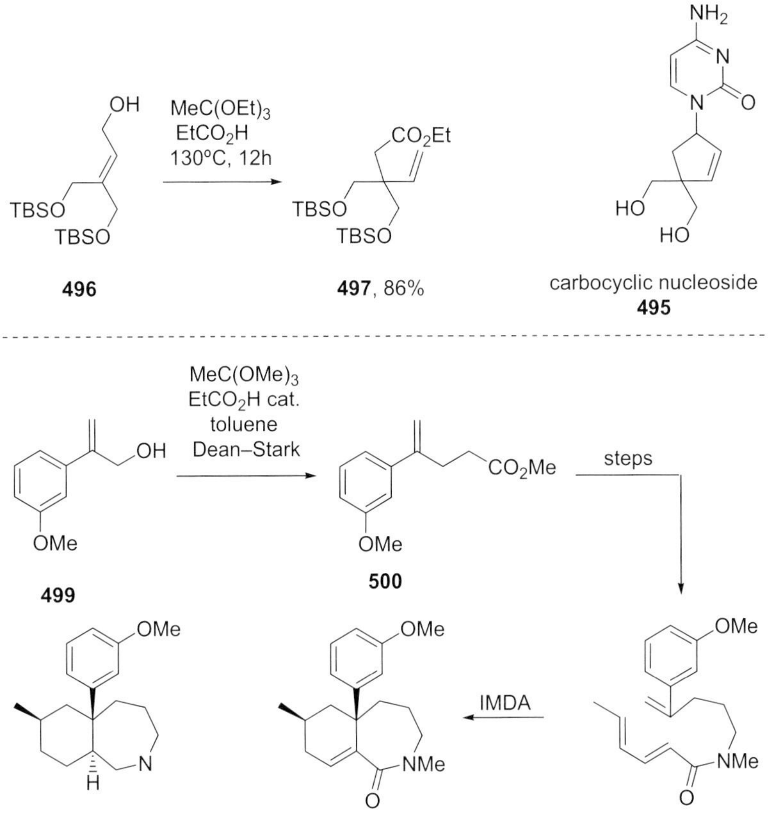

Scheme 6.78

A combination of enzymatic resolution and orthoester rearrangement has been used by Brenna in a short synthesis of (*R*)-(–)-baclofen **503** (Scheme 6.79) [123a]. (*R*) or (*S*)-3-Methyl-2-phenylbutylamine **506** were recently obtained using the same strategy (Scheme 6.79) [123b]. A Claisen–Johnson rearrangement of a benzyl vinyl ether studied by Raucher provided substituted arenes **510**. Better results were observed with electron donating substituted arenes. This reaction has been extended to indole derivatives **511** (Scheme 6.79) [124].

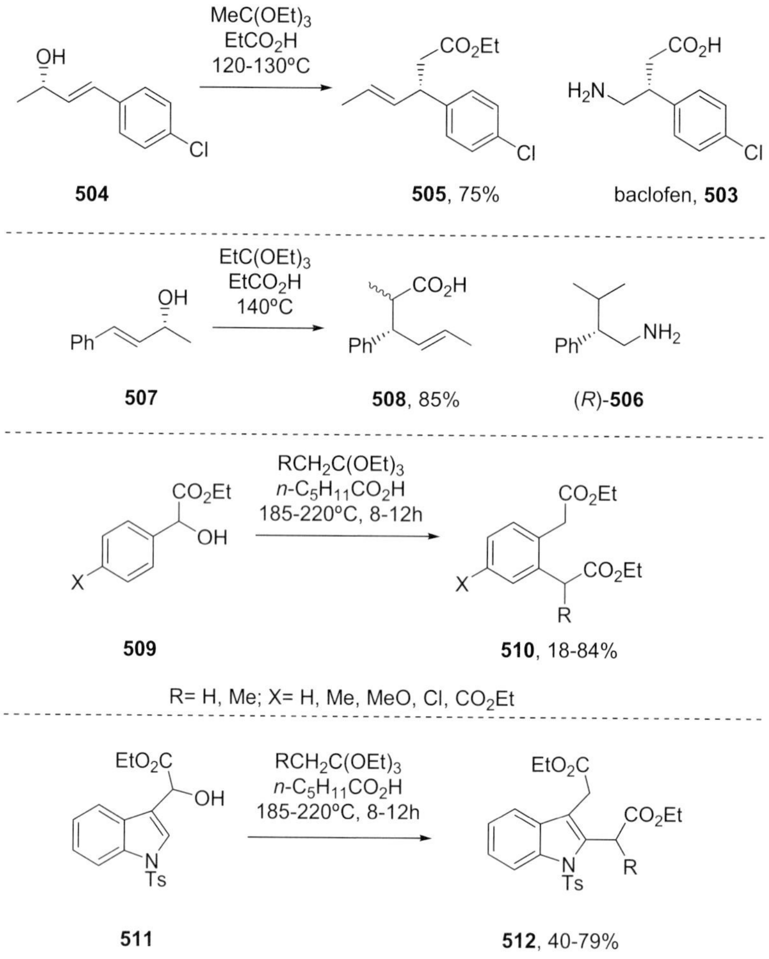

Scheme 6.79

The reactivity in orthoester rearrangement of allylic benzodioxinic alcohols **513** has been studied by Coudert [125] (Scheme 6.80). Better yields were obtained using the Claisen–Eschenmoser rearrangement. A variant of the Claisen rearrangement was developed by Rapoport for the synthesis of thiocoumarins [126] (Scheme 6.80). Triethyl orthopropenoate **516** and triethyl orthopropynoate **517** were used in these rearrangements. An efficient synthesis of β-alkylidene–α,α-disubstituted cyclopentanones **521** described by Burke [127] utilized as the main step the Claisen–Johnson rearrangement (Scheme 6.80).

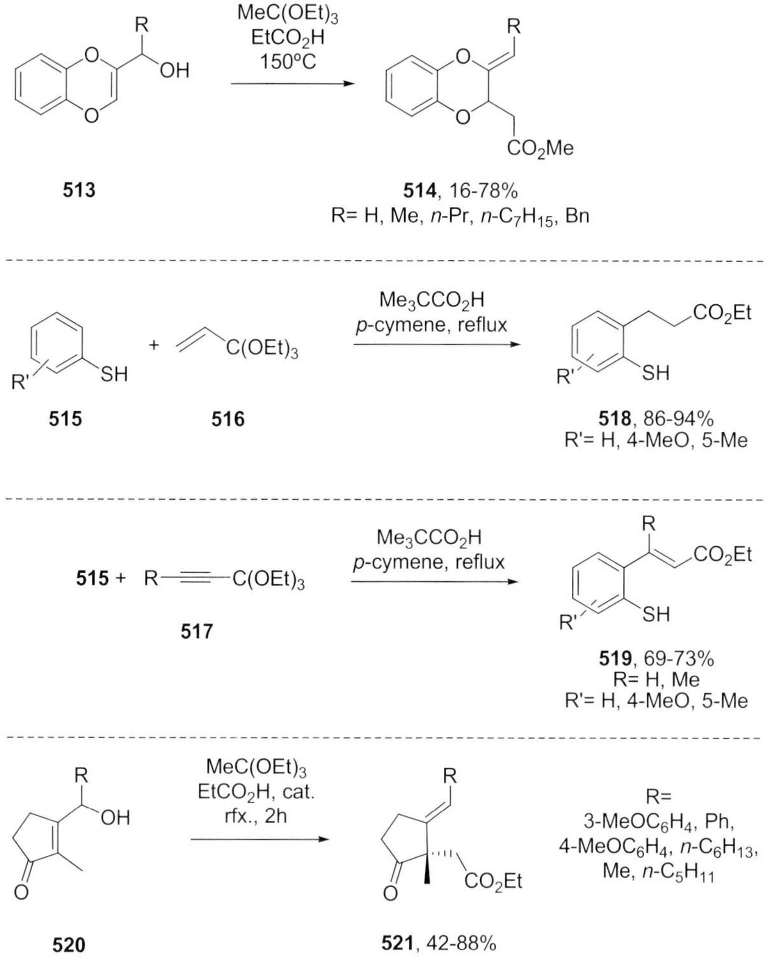

$MeC(OEt)_3$, $EtCO_2H$, 150°C

513 → **514**, 16-78%
R= H, Me, *n*-Pr, *n*-C_7H_{15}, Bn

515 + **516** (Me_3CCO_2H, *p*-cymene, reflux) → **518**, 86-94%
R'= H, 4-MeO, 5-Me

515 + R—≡—$C(OEt)_3$ **517** (Me_3CCO_2H, *p*-cymene, reflux) → **519**, 69-73%
R= H, Me
R'= H, 4-MeO, 5-Me

520 ($MeC(OEt)_3$, $EtCO_2H$, cat., rfx., 2h) → **521**, 42-88%
R= 3-$MeOC_6H_4$, Ph, 4-$MeOC_6H_4$, *n*-C_6H_{13}, Me, *n*-C_5H_{11}

Scheme 6.80

The quaternary center of a dendritic tetramines **522** and **523** described by Feldman [128] was established through a Claisen–Johnson rearrangement of allylic alcohol **524** (Scheme 6.81). One of the main steps in the synthesis of the protease inhibitor **526** [129] was achieved through a high-yield diastereoselective Claisen–Johnson rearrangement as described in Scheme 6.81. A stereoselective synthesis of the alkylidene dipeptide isosteres **530** and **532** proceeded via a Claisen–Johnson rearrangement. It is worthy of note that yields with *anti*- or *syn*-configured aminoalcohol derivatives **529** and **531** were quite different (Scheme 6.81) [130].

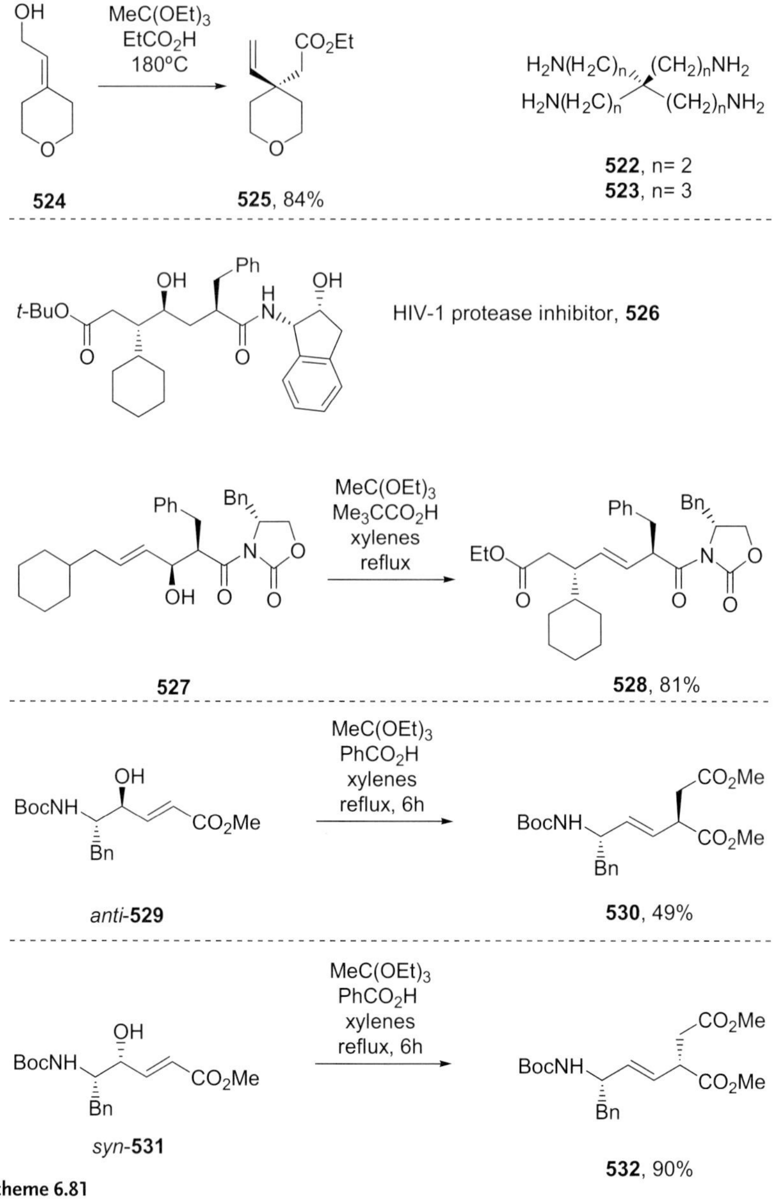

Scheme 6.81

Fluorinated analogues of biologically active molecules have received, in the past twenty years, an increasing interest. The Claisen–Johnson rearrangement applied to fluoro-substituted allylic alcohols appeared rapidly as a method of choice for the synthesis of various fluoro-polyfunctionalized compounds. Claisen rearrangement and Claisen–Johnson rearrangement of trifluoromethyl allylic alcohols **533** developed by Ishikawa [131] afforded selectively the anticipated esters **534** and **535**. *Z,syn*-configured diastereomers were generally obtained as major compounds (Scheme 6.82). Difluoromethyl *m*-tyrosine **536** was obtained by a sequence of reactions involving inter alia orthoester Claisen rearrangement, DAST fluorination and diastereoselective amination utilizing Evans' auxiliary [132] (Scheme 6.82). Kitazume prepared several 3-trifluoromethyl-4-alkenyl esters **541** and **544** from enantiomerically enriched 3-trifluoromethyl allylic alcohols **539** and **542** with Claisen–Johnson or Claisen–Eschenmoser rearrangements. The alcohols (*E,R*)-**539** and (*Z,R*)-**542** afforded respectively the (*E,S*)- and (*E,R*)-configured trifluoromethyl esters **541** and **544** through chair-like transition states **540** and **543** [133] (Scheme 6.82).

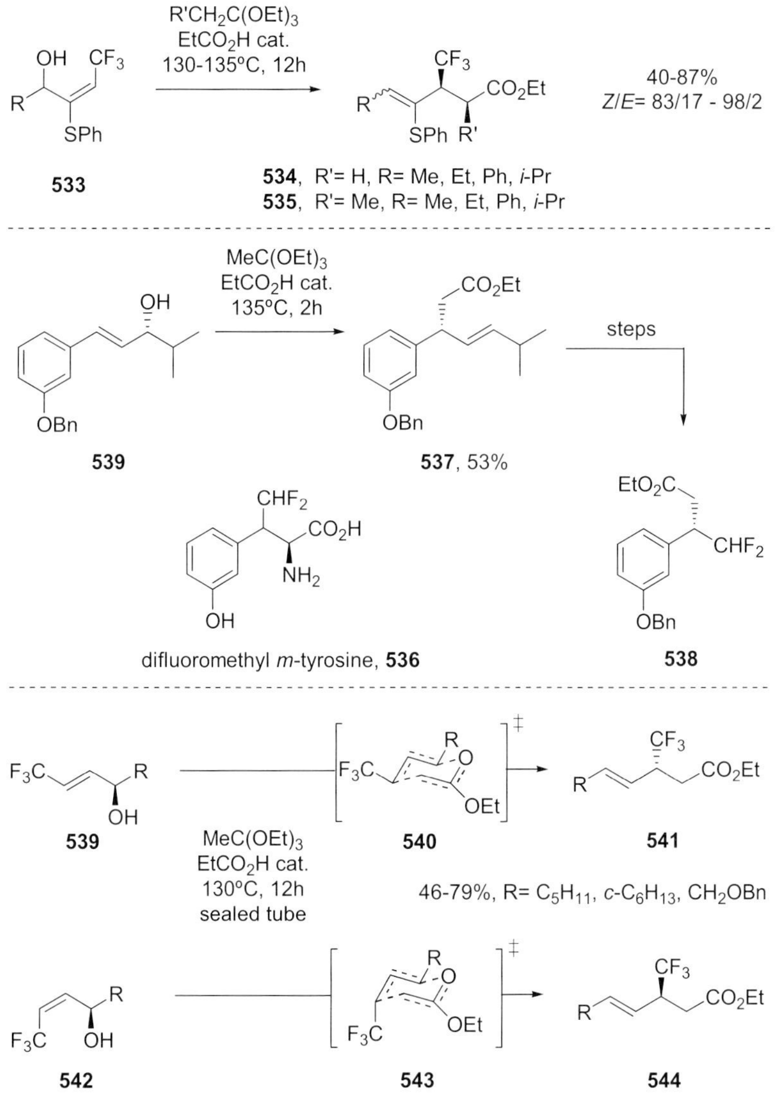

$R'CH_2C(OEt)_3$, $EtCO_2H$ cat., 130-135°C, 12h

533 → **534**, R'= H, R= Me, Et, Ph, *i*-Pr; **535**, R'= Me, R= Me, Et, Ph, *i*-Pr

40-87%, *Z*/*E*= 83/17 - 98/2

$MeC(OEt)_3$, $EtCO_2H$ cat., 135°C, 2h

539 → **537**, 53% → steps → **538**

difluoromethyl *m*-tyrosine, **536**

$MeC(OEt)_3$, $EtCO_2H$ cat., 130°C, 12h, sealed tube

539 → [**540**]‡ → **541**

46-79%, R= C_5H_{11}, *c*-C_6H_{13}, CH_2OBn

542 → [**543**]‡ → **544**

Scheme 6.82

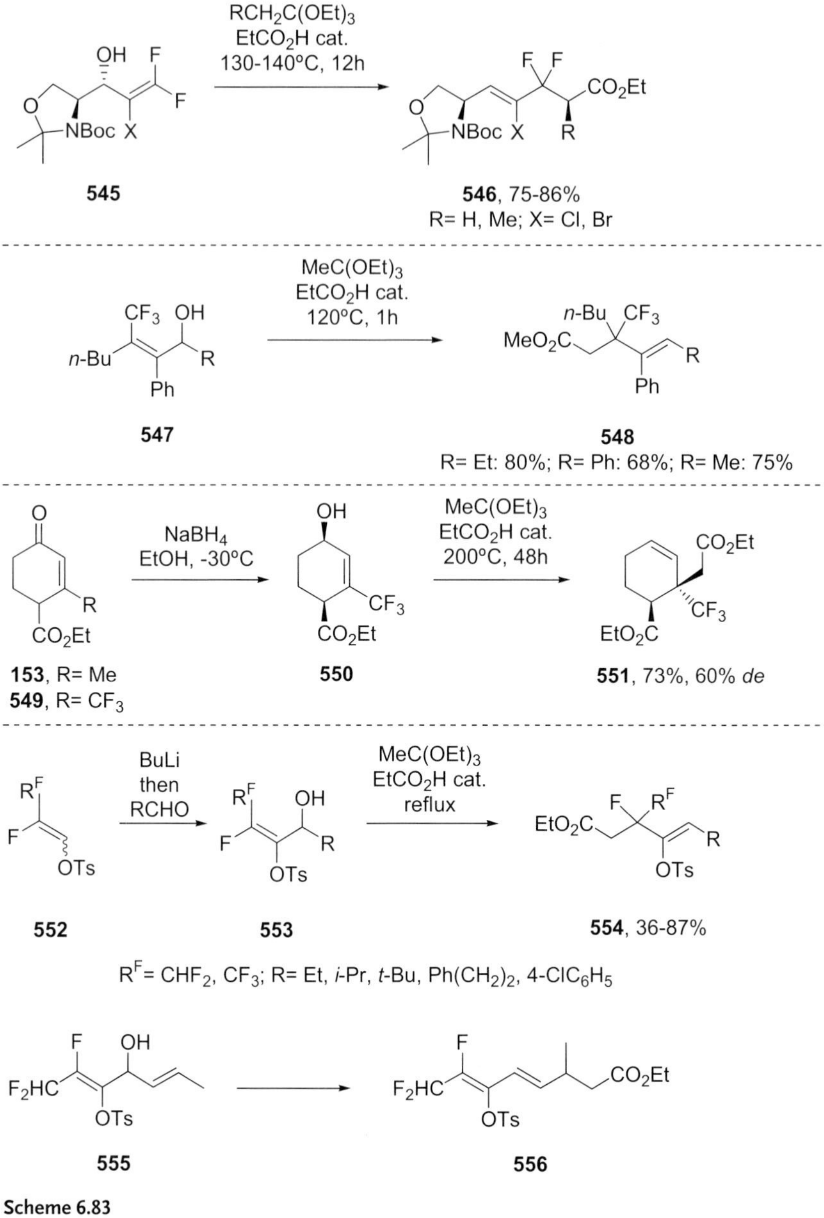

Scheme 6.83

CF_3CFBr_2 and CF_3CCl_3 reacted with Garner's aldehyde after metal–halogen exchange in the presence of Zn–Al. The resulting allylic alcohols **545** subjected to Claisen–Johnson reaction conditions gave rise stereoselectively to the anticipated rearranged esters **546** [132] (Scheme 6.83). The Claisen–Johnson rearrangement allowed the creation of trifluromethyl substituted quaternary carbons. Accordingly, allylic alcohols **547**, prepared by condensation of lithio trifluoromethyl vinyl derivatives with aldehyde, underwent a nice orthoester rearrangement as described by Bonnet–Delpon [135] (Scheme 6.83). The Claisen–Eschenmoser rearrangement has been applied to these alcohols as well. The trifluoro analogue **549** of the Hagemann's ester **153** prepared by the same group [136] afforded selectively the corresponding alcohol **550** after low-temperature reduction by sodium borohydride in ethanol. This allylic alcohol **550** in the presence of triethyl orthoacetate, under rather drastic conditions, gave rise to the rearranged ester **551** in good yield (Scheme 6.83). Funabiki [137] described recently a new example of Claisen–Johnson rearrangement applied to fluoro derivatives. Allylic alcohol **553** reacted with triethyl orthoacetate to afford the corresponding (Z)-β-fluoro-β-polyfluoromethyl-γ,δ-unsaturated esters **554** in fair to good yields. Interestingly, allylic alcohol **555** afforded, with an almost complete regioselectivity, the rearranged ester **556**.

6.5 Conclusion

The above examples illustrate the versatility of the Claisen–Johnson rearrangement. Despite acidic catalysis and a rather high temperature in most of the cases, the reaction is compatible with numerous functional groups including groups that are known to be acid-sensitive like silyl ethers. Another advantage of the Claisen–Johnson rearrangement is the simplicity of procedure. Orthoesters are either commercially available or easily prepared from imino ether hydrochlorides obtained with the Pinner reaction. In addition, the Claisen–Johnson rearrangement occurred in a one-step procedure in contrast with the allyl vinyl ether rearrangement, which requires the preparation of the ether before rearrangement. The use of strictly anhydrous conditions and strong base, as in the Claisen–Ireland rearrangement, is also avoided. In some cases the reaction gives better results than the Claisen–Eschenmoser or the Claisen–Ireland rearrangements. However, with alkyl orthoesters bearing other groups than methyl, the formation of the ketene acetal intermediate is generally not stereoselective and consequently the stereogenic center in the α-position to the rearranged ester is generated unselectively. This is the main disadvantage of this method when compared with the more commonly used Claisen–Ireland rearrangement.

References

1 For recent comprehensive reviews concerning the Claisen rearrangement and related reactions, see: (a) F. E. Ziegler, *Chem Rev.* 1988, *88*, 1423–1452. (b) S. Blechert, *Synthesis*, 1989, 71–82. (c) P. Wipf, in *Comprehensive Organic Syntheses*, vol. 5, 827–873, B. M. Trost, I. Fleming (eds.) Pergamon press, 1991. (d) H. Frauenrath, *Stereoselective Synthesis*, in *Formation of* C–C *bonds by sigmatropic rearrangements and electrocyclic reactions*, vol. E 21, 3301–3341, G. Helmchen, R. W. Hoffmann, J. Mulzer, E. Schaumann (eds) Thieme, 1996.

2 L. Claisen, *Chem. Ber.*, 1913, *45*, 3157–3166.

3 (a) M. F. Carroll, *J. Chem. Soc.*, 1940, 704–706. (b) M.F. Carroll, *J. Chem. Soc.*, 1940, 1266–1268. (c) M. F. Caroll, *J. Chem. Soc.*, 1941, 507–509.

4 (a) G. Saucy, R. Marbet, *Helv. Chim. Acta*, 1967, *50*, 1158–1167; (b) G. Saucy, R. Marbet, *Helv. Chim. Acta*, 1967, *50*, 2091–2100.

5 (a) A. E. Wick, D. Felix, K. Steen, A. Eschenmoser, *Helv. Chim. Acta*. 1964, *47*, 2425–2429. (b) D. Felix, K. Gschwend-Steen, A. E. Wick, A. Eschenmoser, *Helv. Chim. Acta*. 1969, *52*, 1030–1042.

6 H. Meerwein, W. Florian, N. Schon, G. Stopp, *Justus Liebig Ann. Chem.* 1961, *641*, 1–39.

7 W. S. Johnson, L. Werthemann, W. R. Bartlett, T. J. Brocksom, T.-T. Li, J. D. Faulkner, M. R. Petersen, *J. Am. Chem. Soc.* 1970, *92*, 741–743.

8 (a) R. E. Ireland, R. H. Mueller, *J. Am. Chem. Soc.* 1972, *94*, 5897–5898. (b) R. E. Ireland, R. H. Mueller, A. K. Willard, *J. Am. Chem. Soc.* 1976, *98*, 2868–2877.

9 J. T. Welch, S. Eswarakrishnan, *J. Org. Chem.* 1985, *50*, 5909–5910.

10 L. E. Overmann, *Angew. Chem. Int. Ed. Engl.* 1984, *23*, 579–585.

11 (a) R. S. Huber, G. B. Jones, *J. Org. Chem.* 1992, *57*, 5778–5780. (b) G. B. Jones, R. S. Hubert, S. Chau, *Tetrahedron*, 1993, *49*, 369–380.

12 (a) A. Srikrishna, S. Nagaraju, *J. Chem. Soc. Perkin Trans I*, 1992, 311–312. (b) A. Srikrishna, S. Nagaraju, P. Kondaiah, *Tetrahedron*, 1995, *51*, 1809–1816. (c) For an application of this technique to 2-butyne-1,4-diol for the synthesis of constrained α-amino acids, see: S. Kotha, N. Sreenivasachary, E. Brahmachary, *Tetrahedron*, 2001, *57*, 6261–6265. (d) For a mono Claisen–Johnson rearrangement with 2-butyne-1,4-diol affording an allenic ester, see: B.T. B. Hue, J. Dijkink, S. Kuiper, K. K. Larson, F. S. Guziec Jr., K. Goubitz, J. Fraanje, H. Schenk, J. H. van Maarseveen, H. Hiemstra, *Org. Biomol. Chem.* 2003, *1*, 4364–4366.

13 W. Sucrow, W. Richter, *Chem. Ber.* 1971, *104*, 3679–3688.

14 G. Stork, S. Raucher, *J. Am. Chem. Soc.* 1976, *98*, 1583–1584.

15 (a) C. B Chapleo, P. Hallett, B. Lytgoe, I. Waterhouse, P. W. Wright, *J. Chem. Soc. Perkin Trans I*, 1977, 1211–1218. (b) R. J. Cave, B. Lytgoe, D. A. Metclafe, I. Waterhouse, *J. Chem. Soc. Perkin Trans I*, 1977, 1218–1228.

16 H. Bouchard, P. Y. Renard, J.-Y. Lallemand, *Tetrahedron Lett.* 1991, *32*, 5953–5956.

17 *E* and *Z* isomers refer to the oxygen linked with the metal.

18 (a) V. K. Yadav, D. A. Jeyaraj, M. Parvez, R. Yamdagni, *J. Org. Chem.* 1999, *64*, 2928–2932. (b) For a recent related study, see: M. Parvez, G. Senthil, *Acta Crystallographica, Section C*, 2001, *57*, 79–81.

19 (a) P. Deslogschamps, in *Steroelectronic effects in Organic Chemistry*, Pregamon Press, New York, 1983. (b) For the *anti* sulfur atom selectivity, see: A.S. Cieplak, *J. Am. Chem. Soc.* 1981, *103*, 4540–4552.

20 *Z*-selectivity has been observed by Yamamoto during aluminum-derivative-catalyzed allylvinyl ether rearrangement, see: K. Maruoka, K. Nanoshita, H. Banno, H. Yamamoto, *J. Am. Chem. Soc.* 1988, 110, 7922–7924.

21 D. Basavaiah, S. Pandiaraju, M. Krishnamacharyulu, *Synlett*, 1996, 747–748.

22 (a) M. Petrzilka, *Helv. Chim. Acta*, 1978, *61*, 2286–2289. (b) M. Petrzilka, *Helv. Chim. Acta*, 1978, *61*, 3075–3078. (c) M. Pitteloud, M. Petrzilka, *Helv. Chim. Acta*, 1979, *62*, 1319–1325.

23 (a) R.B Carling, A. B Holmes, *J. Chem. Soc. Chem. Commun.* 1986, 325–326. (b) N. R. Curtis, A. B. Holmes, M. G. Looney, *Tetrahedron*, 1991, *47*, 7171–7178.

24 P. S. Lidbetter, B. A. Marples, *Synth. Commun.*, 1986, *16*, 1529–1534.

25 (a) N. Cohen, W. F. Eichel, R. J. Lopresti, C. Neukom, G. Saucy, *J. Org. Chem.* 1976, *41*, 3512–3515. (b) K.-K. Chan, N. Cohen, J. P. De Noble, A. C. Specian, G. Saucy, *J. Org. Chem.* 1976, *41*, 3497–3505.

26 K. K. Chan, A. C. Specian Jr., G. Saucy, *J. Org. Chem.* 1978, *43*, 3435–3440.

27 D. Kim, H. S. Kim, *J. Org. Chem.* 1987, *52*, 4634–4635.

28 T. Imanishi, M. Yamashita, M. Matsui, F. Ninbari, T. Tanaka, C. Iwata, *Chem. Pharm. Bull.* 1988, *36*, 1351–1357.

29 T. Imanishi, M. Yamashita, F. Ninbari, T. Tanaka, C. Iwata, *Chem. Pharm. Bull.* 1988, *36*, 1371–1378.

30 T. Suzuki, E. Sato. Y. Matsuda, H. Tada, S Koizumi, K. Unno, T. Kametani, *J. Chem. Soc. Chem. Commun.* 1988, 1531–1533.

31 A. Srikrishna, S. Nagaraju, *J. Chem. Soc. Perkin Trans I*, 1991, 657–658.

32 A. Srikrishna, K. Krishnan, *Tetrahedron*, 1992, *48*, 3429–3436.

33 A. Srikrishna, K. Krishnan, *J. Chem. Soc. Perkin Trans I*, 1993, 667–673.

34 A. Srikishna, R. Viswajanani, *Tetrahedron Lett.* 1996, *37*, 2863–2864.

35 (a) A. Srikrishna, D. Vijakumar, *J. Chem. Soc. Perkin Trans I*, 1997, 3295–3296. (b) A. Srikrishna, D. Vijakumar, *J. Chem. Soc. Perkin Trans I*, 1999, 1265–1271.

36 A. Srikrishna, D. Vijakumar, *J. Chem. Soc. Perkin Trans I*, 2000, 2583–2589.

37 A. Srikrishna, M. S. Rao, *Tetrahedron Lett.* 2002, *43*, 151–154.

38 A. Srikrishna, D. H. Dethe, *Org. Lett.* 2003, *5*, 2295–2298.

39 L. Paquette, W.H. Ham, *J. Am. Chem. Soc.* 1987, *109*, 3025–3036.

40 C. Chapuis, R. Brauchli, *Helv. Chim. Acta*, 1992, *75*, 1527–1546.

41 X. Ariza, J. Garcia, M. Lopez, L. Montserrat, *Synlett*, 2001, 120–122.

42 P. J. Pearson, P. Lacrouts, A. D. Buss, *J. Chem. Soc. Chem. Commun.* 1995, 437–438.

43 B. A. Wexler, B. H. Toder, A. B. Smith III, *J. Org. Chem.* 1982, *47*, 3333–3335.

44 P. W. Smith, W. C. Still, *J. Am. Chem. Soc.* 1988, *110*, 7917–7919.

45 P. M. Wovkulich, P. C. Tang, N. K. Chadha, A. D. Batcho, J. C. Barrish, M. R. Uskokovic, *J. Am. Chem. Soc.* 1989, *111*, 2596–2599.

46 Y. J. Class, P. DeShong, *Tetrahedron Lett.* 1995, *36*, 7631–7634.

47 D.-H. Lee, S.-W. Lee, *Tetrahedron Lett.* 1997, *38*, 7909–7910.

48 (a)J. Mulzer, M. Berger, *Tetrahedron Lett.* 1998, *39*, 803–806. (b) J. Mulzer, M. Berger, *J. Org. Chem.* 2004, *69*, 891–898.

49 S.-T Jan, K. Li, S. Vig, A. Rudolph, F. M. Uckun, *Tetrahedron Lett.* 1999, *40*, 193–196.

50 H. Kaga, K. Goto, T. Takahashi, M. Hino, T. Tokuhashi, K. Orito, *Tetrahedron*, 1996, *52*, 8451–8470.

51 G. F. Cooper, D. L. Wren, D. Y. Jackson, C. C. Beard, E. Galeazzi, A. R. Van Horn, T. T. Li, *J. Org. Chem.* 1993, *58*, 4280–4286.

52 T. Takahashi, T. Shimayama, M. Miyazawa, H. Yamada, K. Takatori, M. Kajiwara, *Tetrahedron Lett.* 1992, *33*, 5973–5976.

53 T. Takahashi, M. Miyazawa, Y. Sakamoto, H. Yamada, *Synlett*, 1994, 902–904.

54 F. Sabol, *Tetrahedron Lett.* 1989, *30*, 3377–3380.

55 J. W. Paterson, *Tetrahedron*, 1993, *49*, 4789–4798.

56 H. Hirota, A. Yokoyama, K. Miyashi, T. Nakamura, M. Igarashi, T. Takahashi, *J. Org. Chem.* 1991, *56*, 1119–1127.

57 A. Hamoud, C. Descoins, *Bull. Soc. Chim. France*,1978, 299–303.

58 K. Mori, T. Nukada, T. Ebata, *Tetrahedron*, 1981, *37*, 1343–1347.

59 S. R. Wilson, P. A Zucker, *J. Org. Chem.* 1988, *53*, 4682–4693.
60 S. M. Kher, G. H. Kulkarni, *Synth. Commun.* 1990, *20*, 495–501.
61 B. A. Kulkarni, A. Chattopadhyay, V. R. Mamdapur, *Synth. Commun.* 1992, *22*, 495–501.
62 A. S. Pawar, S. Chattopadhyay, *Tetrahedron: Asymmetry*, 1995, *6*, 463–468.
63 A.A. Vasil'ev, L. Engman, E. P. Serebryakov, *J. Chem. Res. (S)*, 1998, 706–707.
64 (a) L. C. Passaro, F. X. Webster, *Synthesis*, 2003, 1187–1190. (b) S. Tsuboi, T. Masuda, A. Takeda, *J. Org. Chem.* 1982, *47*, 4478–4482.
65 T. Schlama, R. Baati, V. Gouverneur, A. Valleix, J. R. Falck, C. Mioskowski, *Angew. Chem. Int. Ed.* 1998, *37*, 2085–2087.
66 (a) W. S. Johnson, M. B. Gravestock, B. E. Mc Carry, *J. Am. Chem. Soc.* 1971, *93*, 4332–4334. (b) W. S. Johnson, T. M. Yarnell, R.F. Myers, D. R Morton, S. G. Boots, *J. Org. Chem.* 1980, *45*, 1254–1259.(c) W. S. Johnson, Y.-Q. Chen, M. S. Kellog, *J. Am. Chem. Soc.* 1983, *105*, 6653–6656. (d) W. S. Johnson, V. R. Fletcher, B. Chenera, W. R. Bartlett, F. S. Tham, R. K. Kullnig, *J. Am. Chem. Soc.* 1993, *115*, 497–504.
67 H. Nemoto, A. Satoh, M. Ando, K. Fukumoto, *J. Chem. Soc. Perkin Trans I*, 1991, 1309–1314.
68 G. H. Posner, H. Dai, K. Afarinkia, N. N. Muthy, K. Z. Guyton, T. W. Kensler, *J. Org. Chem.* 1993, *58*, 7209–7215.
69 T. Tahahashi, H. Yamada, J. Tsuji, *J. Am. Chem. Soc.* 1981, *103*, 5259–5261.
70 M. W. Preus, T. C. McMorris, *J. Am. Chem. Soc.* 1979, *101*, 3067–3071.
71 J. R. Wiersing, N. Waespe-Sarcevic, C. Djerassi, *J. Org. Chem.* 1979, *44*, 3374–3382.
72 G. D. Prestwich, S. Phirwa, *Tetrahedron Lett.* 1983, *24*, 2461–2464.
73 H. Sai, S Takatsuto, N. Ikekawa, *Chem. Pharm. Bull.* 1985, *33*, 4815–4820.
74 T. Suzuki, E. Sato, K. Unno, *Chem. Pharm. Bull.* 1993, *41*, 244–247.
75 T. Takahashi, H. Ueno, M. Miyazawa, J. Tsuji, *Tetrahedron Lett.* 1985, *26*, 4463–4466.
76 M. Kocor, B. Berz, *Tetrahedron*, 1985, *41*, 197–203.
77 (a) S. Takatsuto, T. Watanabe, S. Fujioka, A. Sakurai, *J. Chem. Research* (*S*), 1997, 134–135.
78 For other related applications of the orthoester rearrangement, see: (a) S. Takatsuto, *J. Chem. Soc. Perkin Trans I*, 1986, 591–593. (b) Y. Fujimoto, M. Kimura, F. A. M. Khalifa, N. Ikekawa, *Chem. Pharm. Bull.* 1984, *32*, 4372–4381.
79 B. Jiang, Y. Liu, W.-S. Zhou, *J. Org. Chem.* 2000, *65*, 6231–6236.
80 J.-L. Giner, X. Li, *Tetrahedron*, 2000, *56*, 9575–9580.
81 F. E. Ziegler, G. B. Bennett, *J. Am. Chem. Soc.* 1973, *95*, 7458–7464.
82 Y. Langlois, N. Langlois, P. Potier, *Tetrahedron Lett.* 1975, *11*, 955–958.
83 Y. Langlois, P. Potier, *Tetrahedron*, 1975, *31*, 419–422.
84 T. Suzuki, E. Sato, K. Goto, K. Unno, T. Kametani, *Heterocycles*, 1980, *14*, 433–437.
85 M.R. Uskokovic, R. L. Lewis, J. J. Partridge, C. W. Despreaux, D. L. Pruess, *J. Am. Chem. Soc.* 1979, *101*, 6742–6744.
86 For related studies using the allyl vinyl ether rearrangement, see: G. Rackur, M. Stahl, M. Walkowiak, E. Winterfeldt, *Chem. Ber.* 1976, *109*, 3817–3824 and references therein.
87 M. Amat, M. A. Coll, D. Passarella, J. Bosch, *Tetrahedron: Asymmetry*, 1996, *7*, 2775–2778.
88 M. Lounasmaa, R. Jokela, B. Tirkonen, J. Miettinen, M. Halonen, *Heterocycles*, 1992, *34*, 321–339.
89 P. Hanhinen, T. Nurminen, R. Jokela, M. Lounasmaa, *Heterocycles*, 1994, *38*, 2027–2044.
90 M. Lounasmaa, P. Hanhinen, R. Jokela, *Tetrahedron*, 1995, *51*, 8623–8648.
91 M. Lounasmaa, P. Hanhinen, *Heterocycles*, 1996, *43*, 1981–1989.

92 (a) M. L. Trudell, J. M. Cook, *J. Am. Chem. Soc.* 1989, *111*, 7504–7507. L.-H. Zhang, M. L. Trudell, S. P. Hollinshead J. M. Cook, *J. Am. Chem. Soc.* 1989, *111*, 8263–8265. (c) M. Trudell, R. W. Weber, L. Hutchins, D. Grubisha, D. Bennett, J. M. Cook, *Tetrahedron*, 1992, *48*, 1805–1822.

93 L.–H. Zhang, J. M. Cook, *J. Am. Chem. Soc.* 1990, *112*, 4088–4090.

94 S. Sumi, K. Matsumoto, H Tokuyama, T. Fukuyama, *Org. Lett.* 2003, *11*, 1891–1893.

95 F. He, Y. Bo, J. D. Altom, E. J. Corey, *J. Am. Chem. Soc.* 1999, *121*, 6771–6772.

96 W. Fleischhacker, B. Richter, *Chem. Ber.* 1980, *113*, 3866–3880.

97 (a) R. Ahmed-Schofield, P. S. Mariano, *J. Org. Chem.* 1985, *50*, 5687–5677. (b) S. V. Ley, C. Kouklovsky, *Tetrahedron*, 1994, *50*, 835–848.

98 T. Moriwake, S.-i. Hamano, S. Saito, *Heterocycles*, 1988, *27*, 1135–1139.

99 W. H. Pearson, E. J. Hembre, *J. Org. Chem.* 1996, *61*, 7217–7221.

100 H. Senboku, H. Hasegawa, K. Orito, M. Tokuda, *Heterocycles*, 1999, *50*, 333–340.

101 P. A. Jacobi, Y. Li, *J. Am. Chem. Soc.* 2001, *123*, 9307–9312.

102 B. Fraser-Reid, R. Tsang, D. B. Tulshian, K. M. Mo, *J. Org. Chem.* 1981, *46*, 3764–3767.

103 H. Pak, J. K. Dickinson Jr., B. Fraser-Reid, *J. Org. Chem.* 1989, *54*, 5357–5364.

104 K.-i. Tadano, Y. Idogaki, H. Yamada, T. Suami, *J. Org. Chem.* 1987, *52*, 1201–1210.

105 (a) K.-i. Takao, H. Saegusa, G. Watanabe, K.-i. Tadano, *Tetrahedron: Asymmetry*, 2000, *11*, 453–464. (b) K.-i. Takao, H. Saegusa, K.-i. Tadano, *J. Carbohydr. Chem.* 2001, *20*, 57–69 and references therein.

106 H.-J. Liu, P. R. Pednekar, *Synth. Commun.* 1982, *12*, 395–400.

107 M. G. Kulkarni, M. T. Sebastian, *Synth. Commun.* 1991, *21*, 581–586.

108 R. Bao, S. Valverde, B. Herradon, *Synlett*, 1992, 217–219.

109 D. Basavaiah, S. Pandiaraju, *Tetrahedron Lett.* 1995, *36*, 757–758.

110 V. Helaine, J. Rossi, J. Bolte, *Tetrahedron Lett.* 1999, *40*, 6577–6580.

111 T. R. Elworthy, D. J. Morgans Jr., W. S. Palmer, D. B. Repke, D. B. Smith, A. M. Waltos, *Tetrahedron Lett.* 1994, *35*, 4951–4954.

112 R. Giovannini, E. Marcantoni, M. Petrini, *Tetrahedron Lett.* 1998, *39*, 5827–5830.

113 A. Giardina, E. Marcantoni, T. Mecozzi, M. Petrini, *Eur. J. Org. Chem.* 2001, 713–718.

114 J.-J. Chilot, A. Doutheau, J. Gore, *Bull. Soc. Chim. Fr.* 1984, II, 307–316.

115 T. Hiyama, K. Kobayashi, M. Fujita, *Tetrahedron Lett.* 1984, *25*, 4959–4962.

116 R. Kumaresawaran, S. P. Shahi, S. Rani, A. Gupta, K. P. Madhusudanan, Y. D. Vankar, *Arkivoc*, 2002, 126–138.

117 J. A. Marshall, X.-j. Wang, *J. Org. Chem.* 1992, *57*, 1242–1252.

118 C. H. Heathcock, B. L. Finkelstein, E. T. Jarvi, P. A. Radel, C. R. Hadley, *J. Org. Chem.* 1988, *53*, 1922–1942.

119 S. A. Frank, A. B. Works, W. R. Roush, *Can. J. Chem.* 2000, *78*, 757–771.

120 D. B. Smith, A. M. Waltos, D. G. Loughead, R. J. Weikert, D. J. Morgans Jr., J. C. Rohloff, J. O. Link, R.-r. Zhu, *J. Org. Chem.* 1996, *61*, 2236–2241.

121 O. H. Ko, J. H. Hong, *Tetrahedron Lett.* 2002, *43*, 6399–6402.

122 S. Handa, K. Jones, C. G. Newton, *Tetrahedron Lett.* 1988, *29*, 3841–3844.

123 (a) E. Brenna, N. Caraccia, C. Fuganti, D. Fuganti, P. Graselli, *Tetrahedron: Asymmetry*, 1997, *8*, 3801–3805. (b) E. Brenna, C. Fuganti, F. G. Gatti, M. Passoni, S. Serra, *Tetrahedron: Asymmetry*, 2003, *14*, 2401–2406.

124 S. Raucher, A. S.-T. Lui, *J. Am. Chem. Soc.* 1978, *100*, 4902–4903.

125 P. Moreau, M. Al Neirabeyeh, G. Guillaumet, G. Coudert, *Tetrahedron Lett.* 1991, *32*, 5525–5528.

126 J. A. Panetta, H. Rapoport, *J. Org. Chem.* 1982, *47*, 2626–2628.

127 S. D. Burke, S. A. Shearouse, D. J. Burch, B. W. Sutton, *Tetrahedron Lett.* 1980, *21*, 1285–1288.

128 K. S. Feldman, K. M. Master, *J. Org. Chem.* 1999, *64*, 8945–8947.

129 B. D. Dorsey, J. J. Plzak, R. G. Ball, *Tetrahedron Lett.* 1993, *34*, 1851–1854.

130 J. M. McKinney, D. F. Eppley, R. M. Keenan, *Tetrahedron Lett.* 1994, *35*, 5985–5988.
131 T. Yamazaki, N. Ishikawa, *Bull. Soc. Chim. Fr.* 1986, 937–943.
132 J. S. Sabol, N. W. Brake, I. A. McDonald, *Tetrahedron Lett.* 1994, *35*, 1821–1824.
133 T. Konno, H. Nakano, T. Kitazume, *J. Fluorine Chem.* 1997, *86*, 81–87.
134 S. Peng, F.-L. Qing, *J. Chem. Soc. Perkin I*, 1999, 3345–3348.
135 D. Bouvet, H. Sdassi, M. Ourévitch, D. Bonnet-Delpon, *J. Org. Chem.* 2000, *65*, 2104–2107.
136 T Barhoumi-Slimi, B. Crousse, M. Ourévitch, M. El Gaied, J.-P. Béguet, D. Bonnet-Delpon, *J. Fluorine Chem.* 2002, *117*, 137–141.
137 K Funabiki, N. Hara, M. Nagamori, K. Shibata, M. Matsui, *J. Fluorine Chem.* 2003, *122*, 237–242.

7
The Meerwein–Eschenmoser–Claisen Rearrangement

Stefan N. Gradl and Dirk Trauner

7.1
Definition, Discovery and Scope

The Meerwein–Eschenmoser–Claisen rearrangement is one of the most useful pericyclic reactions. In its basic form, it involves the conversion of an allylic alcohol **1** to a ketene *N,O*-acetal **2**, which undergoes rapid [3,3]-sigmatropic rearrangement to yield a γ,δ-unsaturated amide **3** (Scheme 7.1). In accordance with the general electronic effects observed in Claisen rearrangements, the presence of an electron-donating amino substituent on the ketene acetal intermediate substantially increases the rate of the pericyclic step.

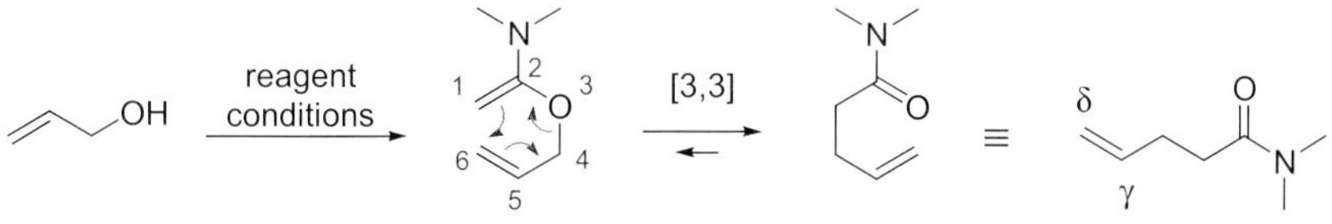

Scheme 7.1

Originally discovered by Meerwein in 1961 [1] during the investigation of amide acetals, it was not until Eschenmoser introduced a practical version and established its scope that the reaction became widely used [2, 3]. Eschenmoser reported that heating various allylic alcohols with dimethylacetamide dimethyl acetal **4** (DMADMA) afforded γ,δ-unsaturated amides in good to excellent yields (Scheme 7.2). Under these conditions, the ketene *N,O*-acetal intermediate **2** is formed *in situ* through alcohol exchange and elimination of methanol. Benzylic alcohols also underwent the Eschenmoser–Claisen rearrangement to yield aryl acetamides after rearomatization (Scheme 7.2, Eq. 3).

Since its original discovery, the Meerwein–Eschenmoser–Claisen rearrangement has proven to be a reliable reaction with considerable scope. Apart from the allylic and benzylic systems [4–6] shown below, propargylic alcohols [7, 8] and allenyl

The Claisen Rearrangement. Edited by M. Hiersemann and U. Nubbemeyer

ISBN: 978-3-527-30825-5

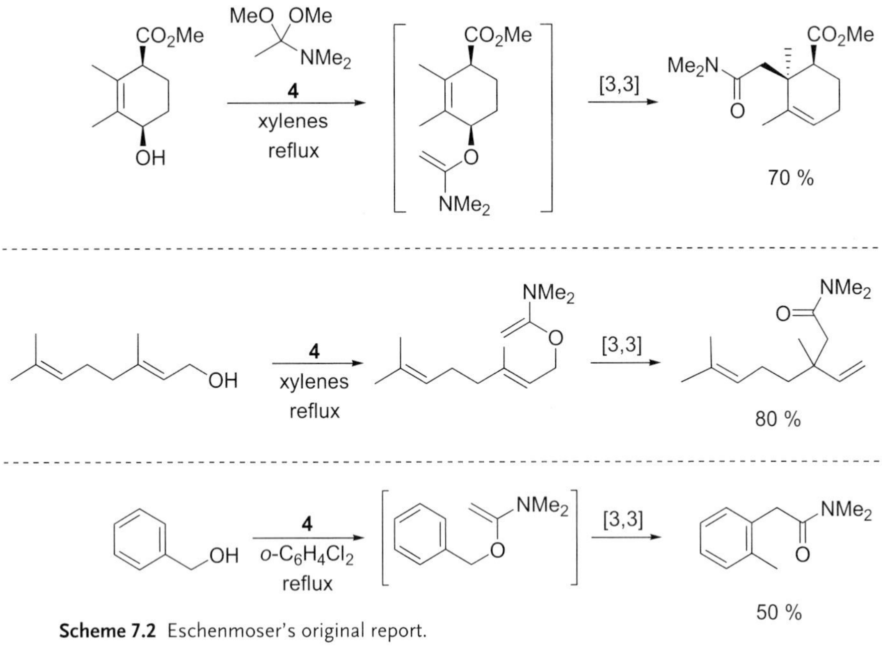

Scheme 7.2 Eschenmoser's original report.

carbinols [9] have also been employed as substrates (Scheme 7.3, Eqs. 1 and 2). The scope of the reaction has been further expanded to yield α-branched amides (Eq. 2 [9] and Eq. 4 [10]), and amides that are more complex and reactive than the *N,N*-dimethylamides usually obtained (Eq. 3) [11]. Rarely, however, has the reaction been extended to transfer more than two-carbon or three-carbon fragments, i.e., acetamide or propionamide moieties (Eqs. 1 to 3). One of the few exceptions is shown in Scheme 7.3, Eq. 4 [10].

When directly compared with the Johnson- or Ireland-variant of the Claisen rearrangement, the Eschenmoser version is often found to proceed with higher yields and better selectivities. The essentially neutral conditions used in the formation of the ketene *N,O*-acetal allow for the employment of sensitive substrates. Many functional groups, for instance tertiary amines, silyl ethers, acetals, esters and carbamates, are compatible with the reaction, provided they tolerate the relatively high temperatures necessary. Eliminations are rarely observed as side reactions. Unprotected non-allylic hydroxyl groups, however, undergo esterification under the Eschenmoser conditions. The rearrangement is performed as a one-pot procedure, i.e., the ketene *N,O*-acetal is formed *in situ*, which greatly benefits its operational simplicity.

Apart from its practicality, the synthetic value of the Meerwein–Eschenmoser–Claisen rearrangement lies in its ability to overcome considerable sterical hindrance, for instance the establishment of quaternary stereocenters (Fig. 7.1). The reaction allows for chirality transfer starting from readily accessible enantiomerically pure allylic and propargylic alcohols to afford amides with carbon-based

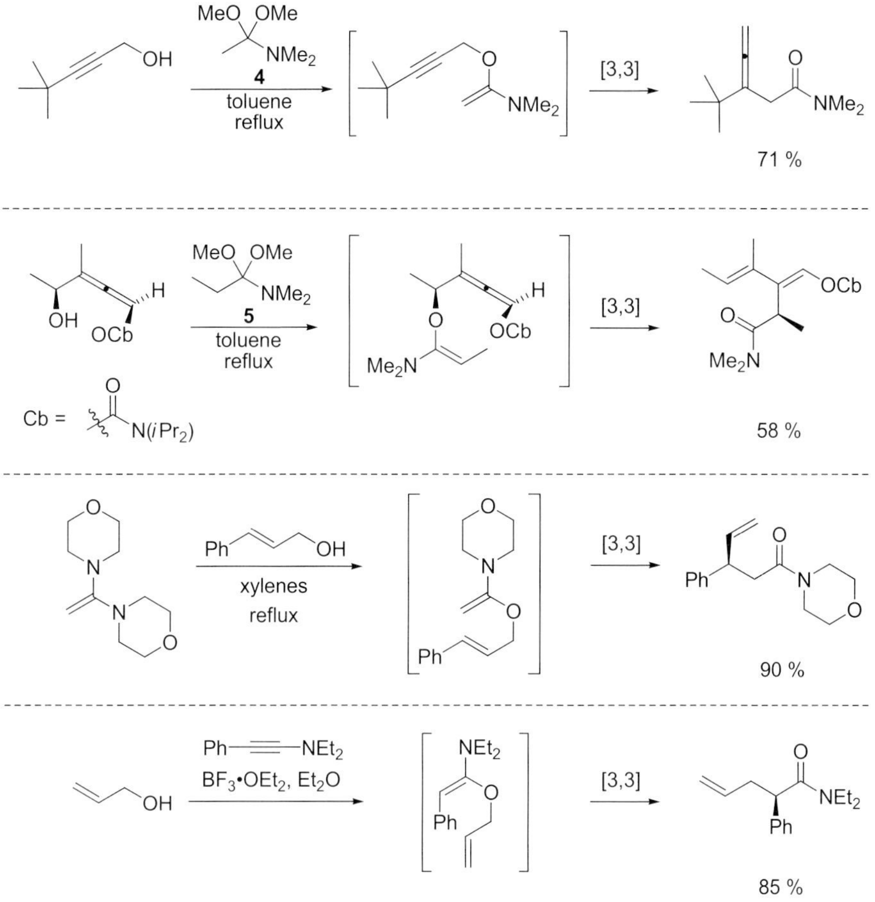

Scheme 7.3 Scope of the Meerwein–Eschenmoser–Claisen rearrangement.

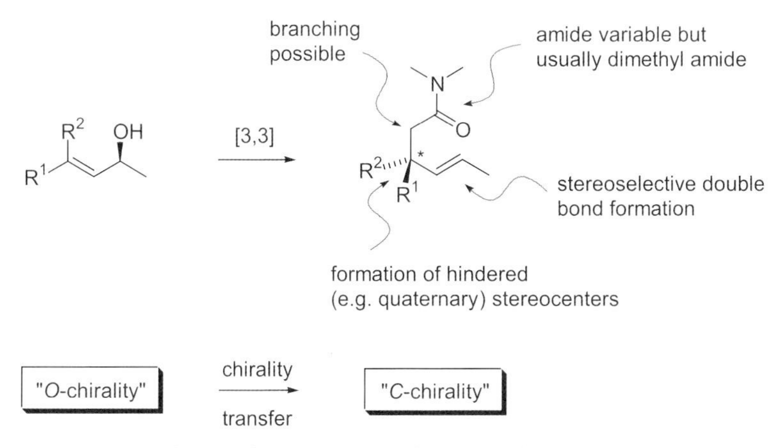

Fig. 7.1 Synthetic usefulness of the Meerwein–Eschenmoser–Claisen rearrangement.

chiral centers or axes (*O*-chirality to *C*-chirality) [12, 13]. Due to the highly ordered six-membered transition state of the sigmatropic rearrangement, the reaction typically proceeds with good stereocontrol both with respect to stereogenic centers (and axes) as well as olefin geometry. In fact, the Meerwein–Eschenmoser–Claisen rearrangement is a useful method for the stereoselective synthesis of di- and trisubstituted olefins.

From a retrosynthetic point of view, the Meerwein–Eschenmoser–Claisen rearrangement shares the basic Claisen retron, a γ,δ-unsaturated carbonyl compound, with other variants of the reaction. More specifically, its retron consists of a two-carbon chain branching off an allylic stereocenter and terminating in an amide or a functional group derived thereof. Such a motif can be readily identified in numerous natural products and other synthetic targets.

Several variants of the Meerwein–Eschenmoser–Claisen rearrangement have been reported, which mostly differ in the way the ketene *N,O*-acetal intermediate is formed. Following a review of this aspect, the regio- and stereoselectivity of the reaction is discussed. Finally, the usefulness of the reaction in the synthesis of complex target molecules is highlighted using selected examples, mostly from natural product syntheses.

7.2
Formation of Ketene *N,O*-Acetals

Meerwein–Eschenmoser–Claisen reactions proceed through the intermediacy of ketene *N,O*-acetals, which usually cannot be isolated. The following paragraphs give an overview of the various synthetic routes leading to these intermediates and include representative experimental procedures.

7.2.1
Condensation with Amide Acetals or Ketene Acetals (Eschenmoser–Claisen Rearrangement)

The most convenient and common way to carry out a Meerwein–Eschenmoser–Claisen rearrangement is by heating an allylic alcohol with a ketene acetal or amide acetal (Schemes 7.3 and 7.4). In this chapter, these conditions are referred to as the Eschenmoser–Claisen rearrangement *per se*.

With few exceptions, commercially available *N,N*-dimethylacetamide dimethyl acetal, (DMADMA, **4**) or dimethyl propionamide dimethyl acetal (**5**) are used. Upon heating these reagents with the allylic alcohol, a mixed *N,O*-acetal **6** is initially formed, which subsequently eliminates methanol to afford the reactive *N,O*-acetal intermediate **7** (Scheme 7.4). The high temperatures usually applied are required for the elimination steps. The sigmatropic rearrangement itself proceeds at comparatively low temperature once the allylic *N,O*-acetal intermediate is formed [14].

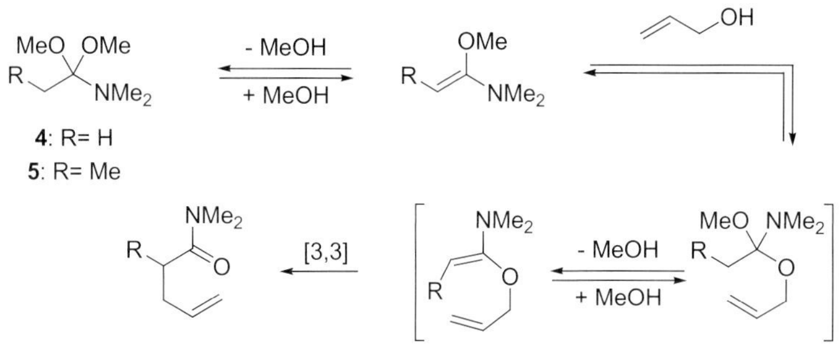

Scheme 7.4 Eschenmoser–Claisen rearrangement via condensation with amide acetals.

The Eschenmoser–Claisen reaction is typically carried out with excess dimethylacetamide dimethyl acetal (DMADMA, **4**) in toluene or xylenes at or near reflux. The reaction can be driven to completion by removal of methanol via distillation or by refluxing the solution over a Soxhlet extractor filled with 4 Å molecular sieves. On a small scale, this can be conveniently done by bubbling inert gas through the reaction mixture using a fine capillary or by gently blowing the gas over the solution just below its boiling point.

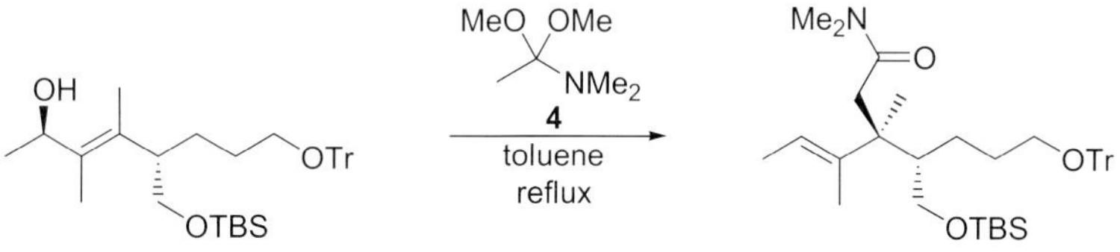

Scheme 7.5 Representative procedure for an Eschenmoser–Claisen reaction.

Representative procedure [15]: a solution of allyl alcohol **8** (4.38 g, 7.84 mmol) in 200 mL of toluene was treated at 110 °C with 7.67 mL (6.97 g, 47.04 mmol) of dimethylacetamide dimethyl acetal (90%) in 10 mL of toluene. During the following 4 h, a mild argon stream was introduced into the solution through a glass capillary to remove methanol. The solution was concentrated and purified by silica gel chromatography (3:1 hexane/EtOAc) to give 4.82 g (7.68 mmol, 98%) of amide **9** as a clear, viscous oil.

Gradl and Trauner recently reported the use of ketene *N,N*-acetals as precursors for the ketene *N,O*-acetal intermediate (Scheme 7.6) [11]. This version of the reaction allows for the incorporation of various amines, for instance morpholine or proline derivatives. The ketene *N,N*-acetals, e.g., **10**, can be conveniently prepared by heating amide acetal **4** with higher boiling amines such as morpholine.

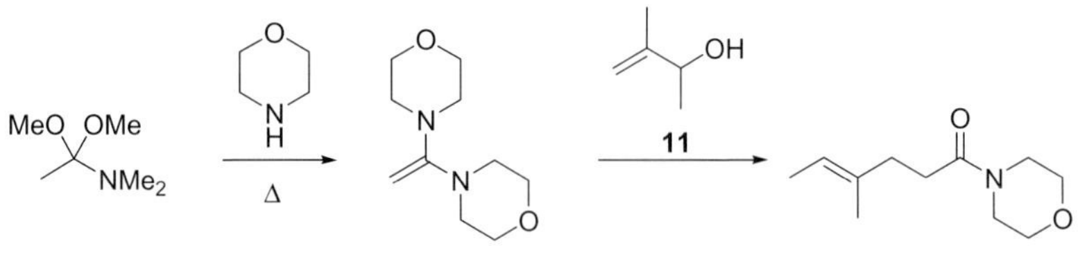

MeO OMe NMe2 4 → (morpholine, Δ) 10 → 11 (OH) → 12, 90 %

Scheme 7.6

Representative procedure [11]: a round-bottomed flask was charged with 8 mL morpholine (30 mmol) and dimethylacetamide dimethyl acetal (3.8 mL, 10 mmol) and the mixture was slowly heated to 190 °C over 5 h under a steady stream of N_2. After cooling to room temperature, dry benzene (10 mL) and 3-methyl-but-3-en-2-ol **11** (430 mg, 5 mmol) were added. The reaction mixture was transferred to a sealed tube and the solution was refluxed until the allylic alcohol was consumed (24 h) as determined by TLC (30% hexanes in ethyl acetate, v/v). The solvent was removed in vacuo and the residue was purified by column chromatography (hexanes:ethyl acetate 2:1 → 1:1) to afford the pure product **12** as a colorless oil (870 mg, 4.5 mmol, 90%).

7.2.2
Addition of Alkoxides to Amidinium Ions (Meerwein–Claisen Rearrangement)

As originally discovered by Meerwein in 1961 [1], ketene *N,O*-acetals can be prepared by treating amides with a strong alkylating agent such as *O*-ethyl-oxonium-tetrafluoroborate (Meerwein's salt, **13**) to afford amidinium ions **14**, followed by addition of sodium alkoxides (Scheme 7.7). If allylic alkoxides are used, the resulting mixed ketene acetals undergo elimination and rearrangement upon distillation.

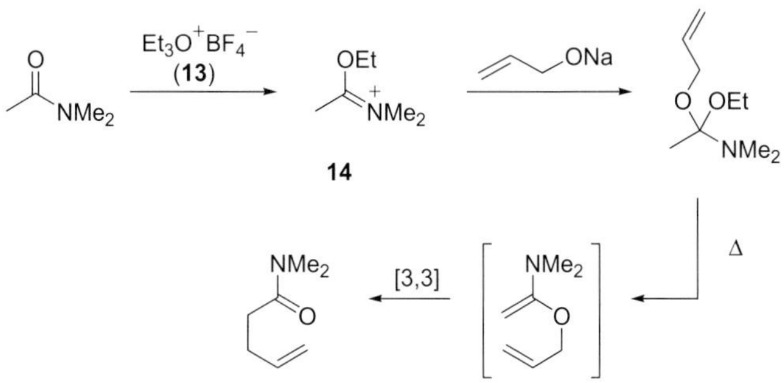

$Et_3O^+BF_4^-$ (**13**); OEt, $\overset{+}{N}Me_2$ **14**; ONa; O OEt NMe2; Δ; NMe2 O; [3,3]; NMe2 O

Scheme 7.7 The Meerwein–Claisen rearrangement.

This route to ketene *N,O*-acetals was further investigated by Welch who found that methylation of amides with methyl triflate, followed by addition of lithium allyl alkoxides afforded allylic ketene *N,O*-acetals that rearranged at room temperature [14]. Lactams also undergo this reaction (Scheme 7.8) [16].

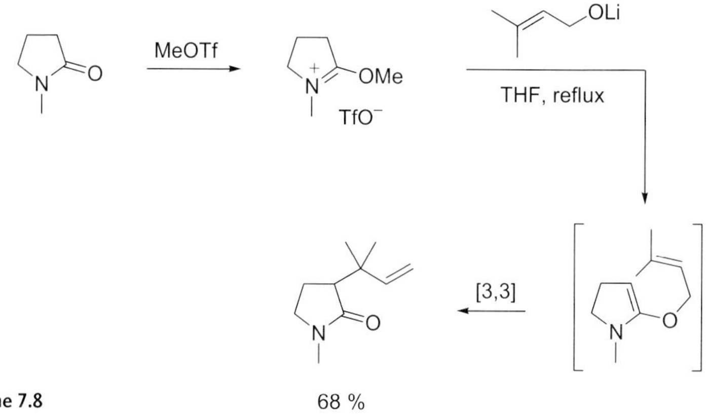

Scheme 7.8

68 %

7.2.3
Addition of Alcohols to Ynamines and Ynamides (Ficini–Claisen Rearrangement)

Shortly after Meerwein's and Eschenmoser's original reports, Ficini disclosed the use of ynamines as suitable precursors of allylic ketene *N,O*-acetals (Scheme 7.8) [10, 17]. Allylic alcohols add to ynamines (**15**) either in the presence of catalytic amounts of a Lewis acid, for instance $BF_3{\cdot}OEt_2$, or at elevated temperatures. This addition presumably proceeds through the intermediacy of a keteniminium ion **16**. The resulting ketene *N,O*-acetals then undergo the sigmatropic rearrangement to yield the corresponding amides.

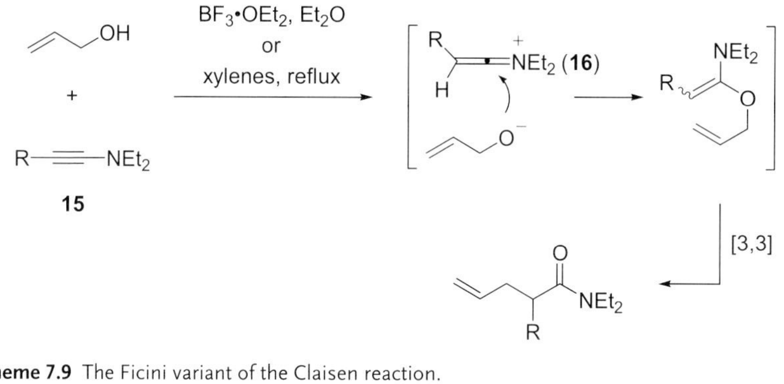

Scheme 7.9 The Ficini variant of the Claisen reaction.

Although the Ficini–Claisen reaction has shown a considerable scope of applications this method has been somewhat limited by the instability and synthetic inaccessibility of ynamines [18, 19]. Recently, ynam*ides* have emerged as alternatives to ynamines in the Ficini–Claisen reaction [20]. Ynamides are more stable and easier to handle than ynamines but also less reactive. Chiral ynamides, e.g., **17**, can

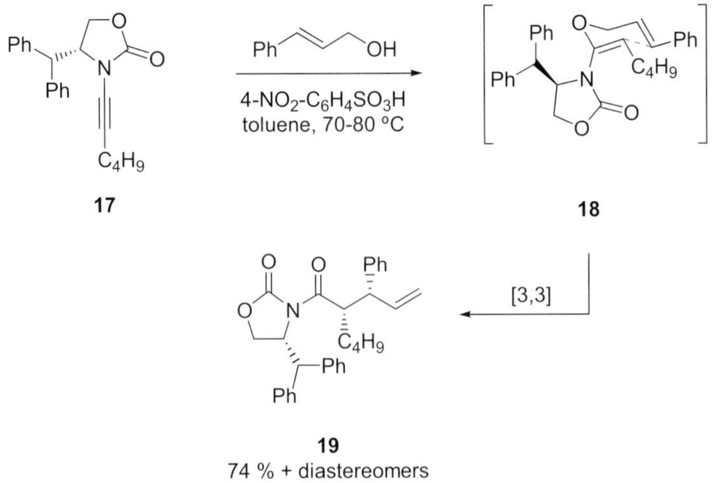

Scheme 7.10 Ynamide Ficini–Claisen rearrangement.

be conveniently prepared using copper-mediated cross-coupling reactions [21, 22] and have found uses in asymmetric Ficini–Claisen rearrangements (Scheme 7.10).

Representative procedure [20]: ynamide **17** (66 mg, 0.2 mmol), anhydrous *p*-nitrobenzenesulfonic acid (4 mg, 0.02 mmol), cinnamyl alcohol (40 mg, 0.3 mmol), and anhydrous toluene (4 mL) were combined in a flame-dried 25 mL sealed tube under nitrogen atmosphere. The tube was sealed and the reaction mixture was heated to 75°C for 15 h, cooled to room temperature, filtered through celite and concentrated *in vacuo*. The remaining residue was purified by flash silica gel chromatography (gradient solvent system: 0 to 25% EtOAc in hexanes) to provide the rearranged Claisen product **19** in 74% yield (69 mg, 14.8 mmol) and the three other possible diastereomers in a combined 3% yield.

7.2.4 Miscellaneous Methods

The isomerization of *N*-phenyl allylimidates (e.g., **20**) also provides access to allylic ketene *N,O*-acetals (Scheme 7.11) [23, 24]. Imidates of type **20** can be synthesized from phenyl amides by treatment with phosphorus pentachloride to yield the corresponding imidochloride followed by addition of the appropriate lithium allyl alkoxide. The thermal isomerization of **20** to the ketene *N,O*-acetal requires high temperature conditions (e.g., refluxing decalin).

Alternatively, the phenyl imidate **20** is deprotonated and trapped as the *N*-silylketene-*N,O*-acetal (**21**), which then rearranges smoothly at room temperature [25, 26]. This approach has also been further explored using chiral phenylimidates resulting in the formation of unsaturated amides with high diastereomeric excess (see Scheme 7.21) [27].

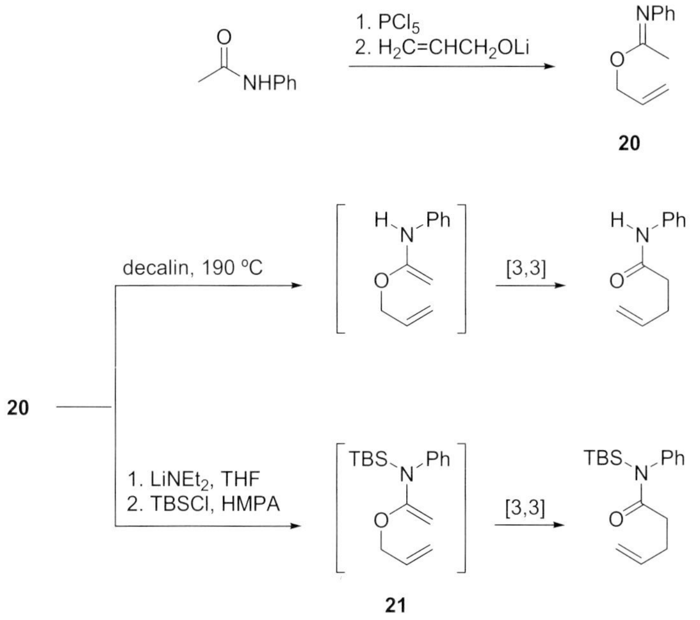

Scheme 7.11 Rearrangement via *N*-phenyl allylimidates.

An unusual approach to ketene *N,O*-acetals has been reported by Holmes (Scheme 7.12) [28]. Condensation of protected amino alcohol **22** with phenylselenyl acetal **23** afforded *N,O*-acetal **24**. Oxidation and elimination of phenyl selenic acid then lead to ketene *N,O*-acetal **25**, which underwent rearrangement at elevated temperatures to afford unsaturated caprolactam **26**.

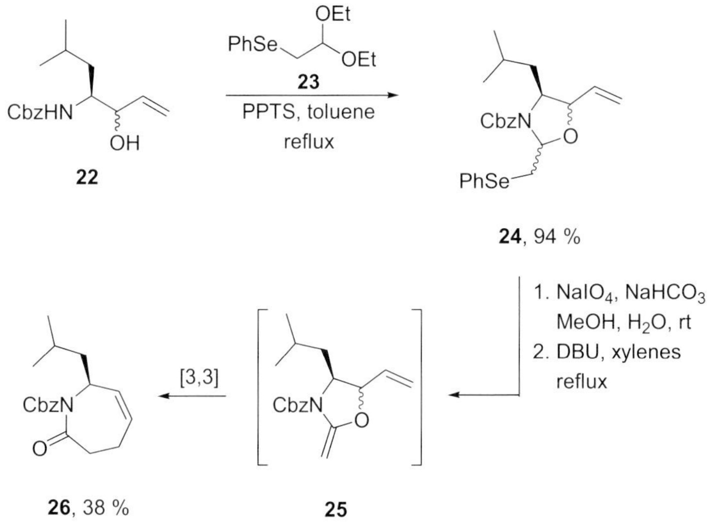

Scheme 7.12

7.3
Selectivity

In addition to its wide substrate scope and functional group compatibility, the Meerwein–Eschenmoser–Claisen rearrangement is marked by high regio- and stereoselectivity, notably in the case of acyclic substrates. In fact, the reaction provides some of the best examples for acyclic stereocontrol reported in the literature.

7.3.1
Regioselectivity

When divinyl carbinols are heated with dimethylacetamide dimethyl acetal (**4**), the less substituted and less electron-rich double bond engages in the rearrangement (Scheme 7.13, Eqs. 1 and 2) [29]. Vinyl-propargyl carbinols were found to rearrange with preferential participation of the double bond (Eq. 3) [30]. For another case of a highly regioselective Eschenmoser–Claisen rearrangement see Scheme 7.16, Eq. 4 [31].

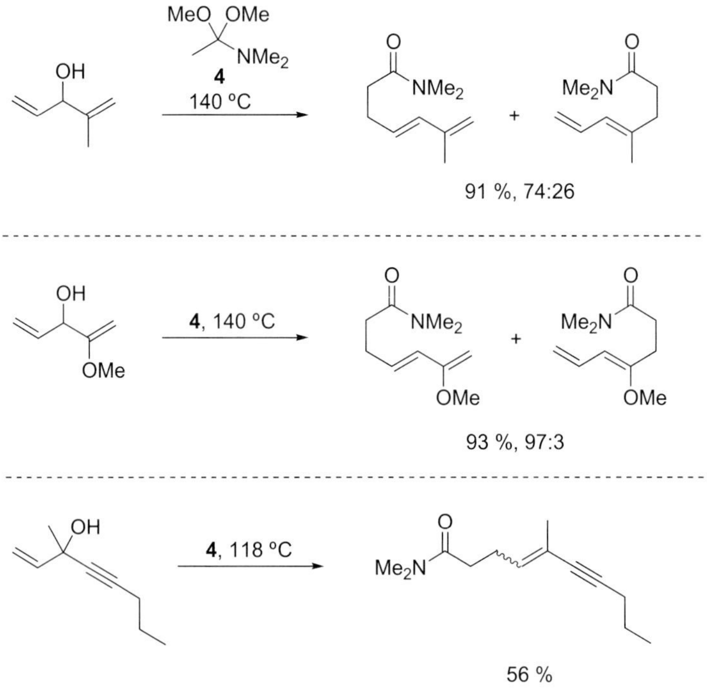

MeO OMe NMe$_2$ **4** 140 °C

OH → NMe$_2$ + Me$_2$N

91 %, 74:26

OH OMe **4**, 140 °C → NMe$_2$ OMe + Me$_2$N OMe

93 %, 97:3

OH **4**, 118 °C → Me$_2$N

56 %

Scheme 7.13 Regioselectivity in the Eschenmoser–Claisen rearrangement.

7.3.2 Stereoselectivity

Due to the highly ordered nature of its six-membered transition state, the Meerwein–Eschenmoser–Claisen rearrangement often proceeds with excellent stereocontrol. The stereochemical outcome of the rearrangement can be analyzed in terms of a chair-shaped six-membered transition state wherein steric interactions are minimized. As with other variants of the Claisen rearrangement, a boat-shaped transition state is sometimes invoked when cyclic allylic alcohols undergo the rearrangement. The geometry of the newly formed double bond as well as the absolute and relative configuration of emerging stereocenters can be simultaneously predicted using these models.

7.3.2.1 Cyclic Allylic Alcohols

The suprafacial nature of the sigmatropic rearrangement allows an efficient chirality transfer in cyclic allylic alcohols. For instance, the Eschenmoser–Claisen rearrangement of *trans* allyl alcohol **27** gave *trans* amide **28** (Scheme 7.14, Eq. 1),

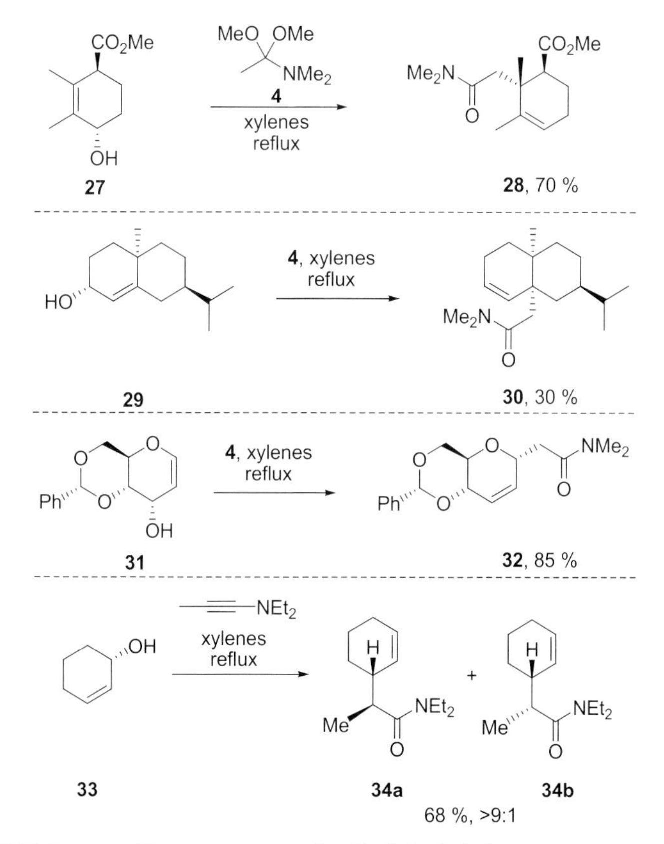

Scheme 7.14 Stereospecific rearrangements of cyclic allylic alcohols.

whereas the corresponding *cis*-cyclohexenol afforded the *cis* amide (cf. Scheme 7.2, Eq. 1) [2, 3]. Rearrangement of allylic alcohol **29** resulted in the formation of *cis*-decalin derivative **30** [32, 33]. Glucals, such as **31**, readily engage in the rearrangement to afford C-glycosides of type **32** in high yields (Eq. 3) [34]. Rearrangement of cyclohexenol **33** under Ficini–Claisen conditions gave a >9:1 mixture of amides **34a** and **34b** (Eq. 4). The high simple diastereoselectivity of this reaction was explained by invoking a boat-shaped transition state [19].

Numerous additional examples for Eschenmoser–Claisen rearrangements in cyclic systems can be found in Section 7.4.

7.3.2.2 Acyclic Allylic Alcohols

Whereas cyclic allylic alcohols of normal size can only yield (*Z*)-configured γ,δ-unsaturated amides, acyclic allylic alcohols preferentially give (*E*)-isomers. Indeed, the Meerwein–Eschenmoser–Claisen rearrangement is an excellent method for the stereoselective formation of di- and trisubstituted (*E*)-double bonds in acyclic systems. The high diastereoselectivities observed (dr > 95:5) can be explained by invoking a chair shaped transition state **35a** or **36a** (Scheme 7.15). This minimizes

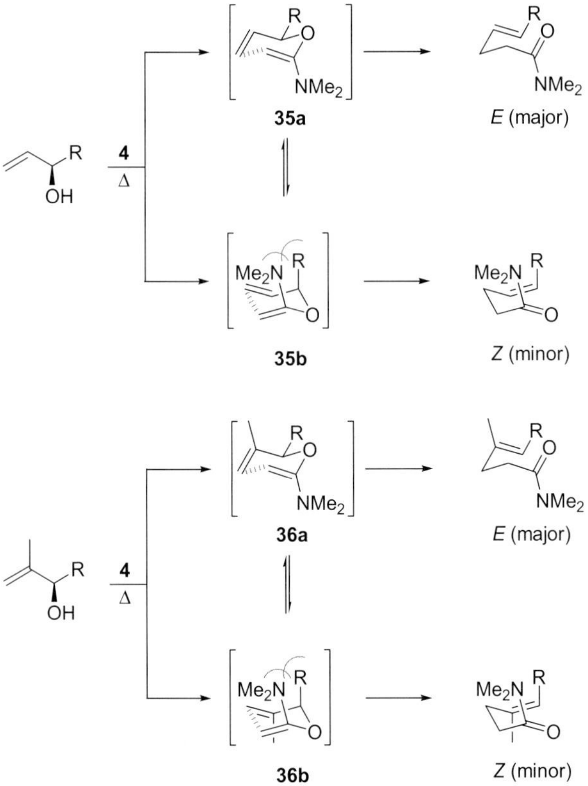

R OH 4 Δ NMe2 35a E (major) Me2N R 35b Z (minor)

R OH 4 Δ NMe2 36a E (major) Me2N R 36b Z (minor)

Scheme 7.15

pseudo 1,3-diaxial interactions between the bulky dimethyl amine moiety and residue R, which occur in the alternative transitions states **35b** and **36b**. Since the dimethyl amine moiety is sterically more demanding than corresponding substituents in other variants of the Claisen rearrangement, the Meerwein–Eschenmoser–Claisen reaction typically leads to higher selectivities than, for instance, the Johnson–Claisen rearrangement.

Numerous examples of the stereoselective formation of di- and trisubstituted double bonds through Meerwein–Eschenmoser–Claisen rearrangement have been reported (Scheme 7.16). In a pioneering study reported by Faulkner, rearrangement of allylic alcohol **37** afforded (*E*)-amide **38** contaminated with less than 0.6% of the corresponding (*Z*)-isomer (Scheme 7.16, Eq. 1) [35]. Trisubstituted olefins can also be prepared with high selectivity (**39** → **40**, Eq. 2) [36]. In the case of diol **41**, the intended product **42** was accompanied by substantial amounts of ester **43** (Eq. 3) [37]. The remarkable ability of the reaction to stereoselectively install double bonds is further demonstrated by the clean formation of (*Z*,*Z*)-configured iodo diene **45** starting from the sensitive divinyl carbinol **44** (Eq. 4). Also note the regioselectivity of the rearrangement [31].

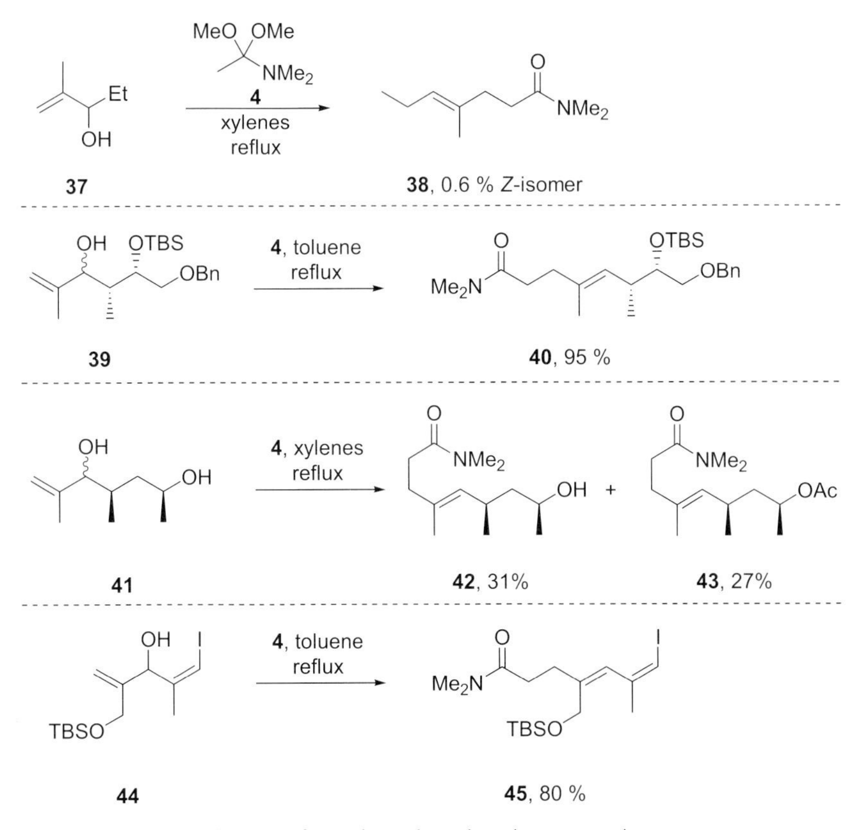

Scheme 7.16 Stereoselective olefin synthesis through Eschenmoser–Claisen rearrangement.

In accordance with the transition state model discussed above, tertiary allylic alcohols rearrange with little stereoselectivity unless the substituents at the quaternary stereocenter differ substantially in size. For instance, virtually no selectivity was observed by Johnson in a synthesis of fluorinated precursors for polyolefin cyclizations (Scheme 7.17) [38].

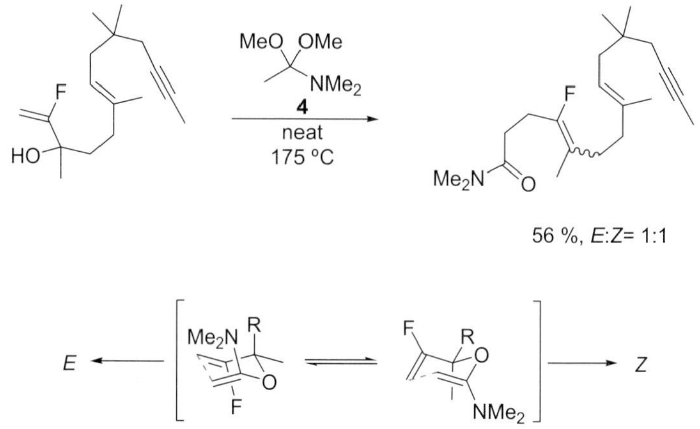

MeO OMe, NMe2, 4, neat, 175 °C

56 %, E:Z= 1:1

Scheme 7.17 Rearrangement of tertiary allylic alcohols.

Similar arguments can be made to explain the high degree of chirality transfer observed with acyclic allylic alcohols (Scheme 7.18) [12]. In the preferred transition state **46a**, unfavorable 1,3-diaxial interactions between the bulky dimethyl amine moiety and the R^1 group as well as allylic $A^{1,3}$-strain are avoided.

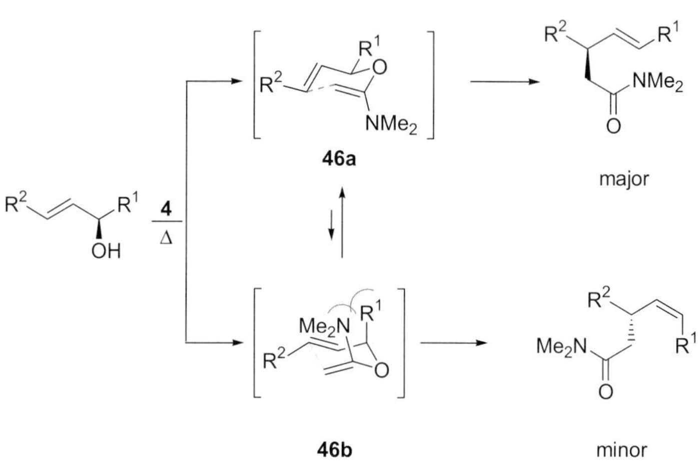

46a, major, 46b, minor

Scheme 7.18 Chirality transfer in acyclic allylic alcohols.

Selected examples of chirality transfer in acyclic allylic alcohols are shown in Scheme 7.19. Note that the formation of a new chiral center is often associated with stereoselective double bond formation. For instance, only one diastereomer (**48**) was observed when allylic alcohol **47** was subjected to Eschenmoser's conditions (Scheme 7.19, Eq. 1) [39]. The desymmetrization of C_2 symmetrical diol **49**

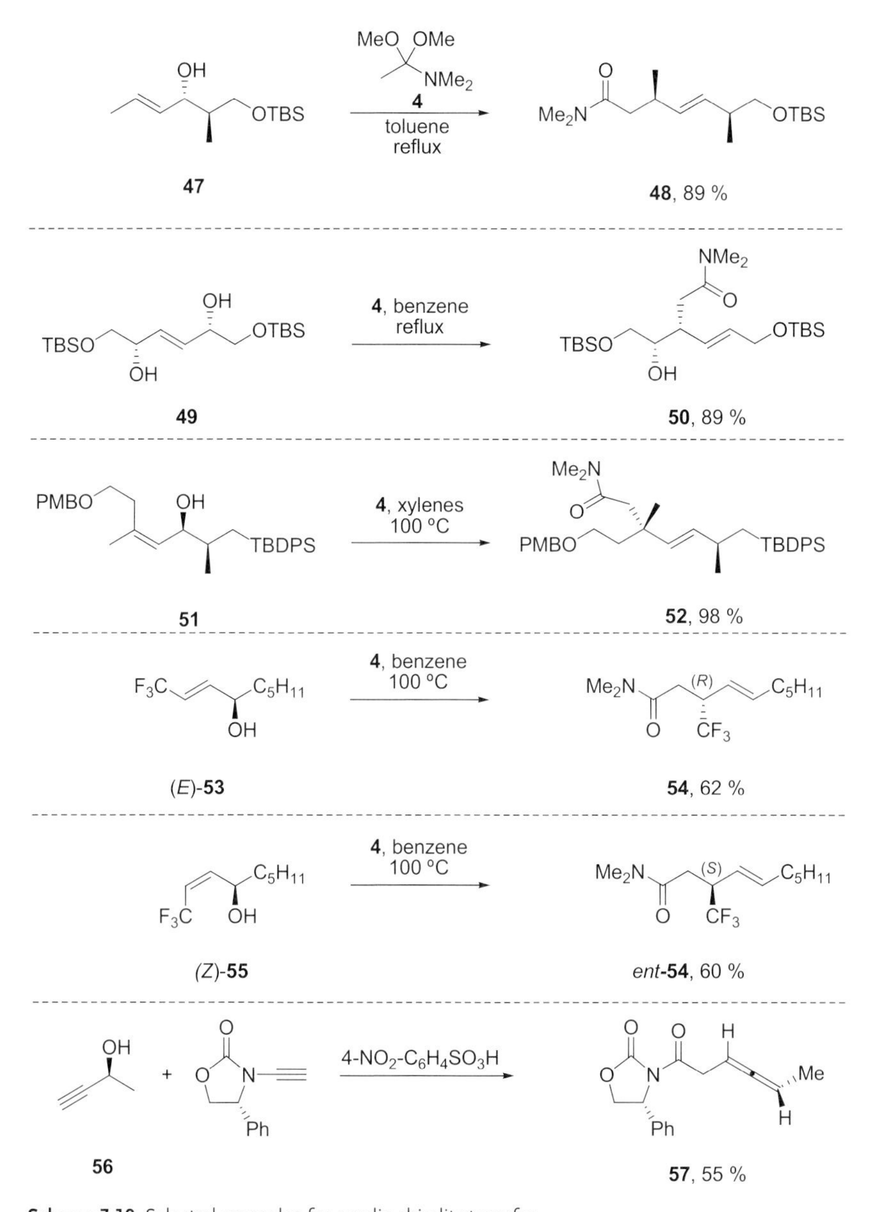

Scheme 7.19 Selected examples for acyclic chirality transfer.

resulted in the exclusive formation of **50** (Eq. 2) [40]. A quaternary stereocenter and a disubstituted double bond are stereoselectively formed in the rearrangement of **51** to afford **52** (Eq. 3) [41]. Enantiomeric products **54** and *ent*-**54** were formed from diastereomeric starting materials (*E*)-**53** and (*Z*)-**55**, respectively, in a synthesis of trifluoromethyl substituted amides (Eqs. 4 and 5) [42]. Again, the Eschenmoser protocol was found to provide slightly better chirality transfer than the Johnson protocol due to increased steric strain in the disfavored transition state. An example of chirality transfer from an enantiomerically pure propargylic alcohol (**56**) to an allene (**57**) is shown in Eq. 6 [43]. Note, however, that this is a case of double diastereoselection since the alkynylamide is chiral as well.

Relatively few examples of highly stereoselective Meerwein–Eschenmoser– Claisen rearrangements have been reported wherein the diastereoselectivity is controlled remotely, in other words wherein the carbon bearing the hydroxyl group is not a chiral center. An example involving a carbohydrate derived allylic alcohol is shown in Scheme 7.20 [44]. Rearrangement of **58** occurred on the less hindered diastereoface to afford **59** exclusively after hydrolysis of the acetate formed from the unprotected non-allylic primary alcohol. Other examples for this type of stereocontrol can be found in Danishefsky's synthesis of sesquicillin and Ogasawara's synthesis of arnicenone (Scheme 7.29, Eqs. 3 and 4).

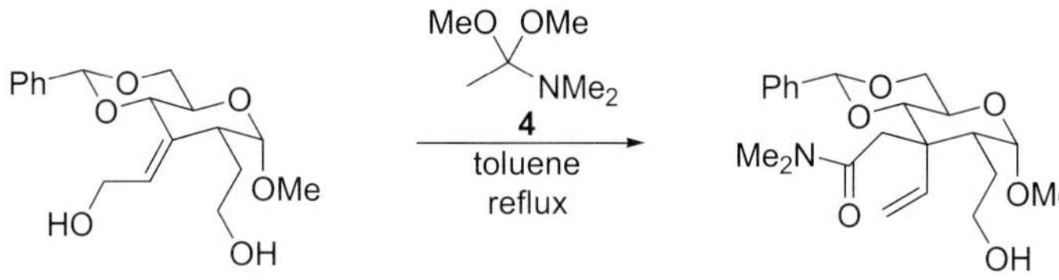

58 → (MeO OMe, NMe$_2$ **4**, toluene reflux) → **59**, 95 %

Scheme 7.20 Remote stereocontrol in the Eschenmoser–Claisen rearrangement.

The transfer of chirality from chiral amine moieties has been investigated (Scheme 7.21) [11, 45, 46]. Generally, the selectivities observed appear to be low, with the exception of a case involving binaphtylamine derived imidate **60** (Scheme 7.21, Eq. 4) [27]. Note that simple diastereoselectivity is also an issue in Eqs. 2 to 4. An analogous rearrangement involving a chiral ynamide has been presented in Scheme 7.10.

The issue of simple diastereoselectivity arises when both the allyl and vinyl moieties of the ketene *N,O*-acetal intermediate **2** are substituted at their terminus, leading to vicinal stereocenters in the products (Scheme 7.22). In analogy to the aldol reaction, the stereochemical outcome can be predicted in terms of a Zimmermann–Traxler type chair-shaped transition state. Accordingly, the *syn*/*anti* ratio of the products depends on double bond geometry. Whereas the geometry of the unsaturated alcohol is pre-determined and usually not subject to equilibration, the geometry of the ketene *N,O*-acetal moiety depends on the reaction conditions that lead to its *in situ* formation.

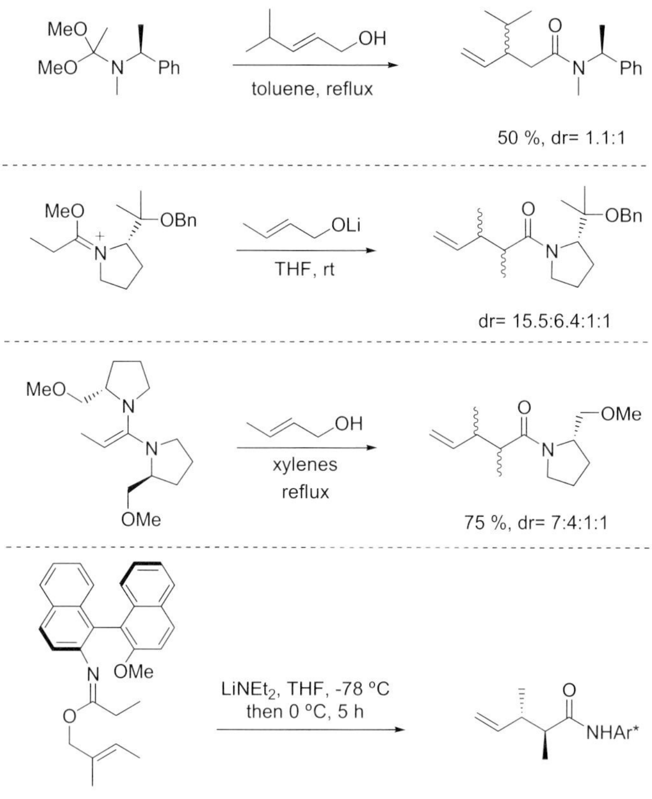

MeO MeO N Ph → OH, toluene, reflux → N Ph, O; 50 %, dr= 1.1:1

MeO OBn N+ → OLi, THF, rt → O OBn N; dr= 15.5:6.4:1:1

MeO N N OMe → OH, xylenes reflux → O OMe N; 75 %, dr= 7:4:1:1

N OMe O **60** → LiNEt$_2$, THF, -78 °C then 0 °C, 5 h → O NHAr*; 78 %, er= 97:3

Scheme 7.21 Chirality transfer from chiral amines.

In the case of substituted ketene *N,O*-acetals formed through equilibria, for instance under the classical Eschenmoser conditions, the thermodynamically more stable (*Z*)-ketene *N,O*-acetal **61a** is preferentially formed to avoid allylic strain between residue R^3 and the bulky dimethylamine moiety (Scheme 7.22). Invoking a chair-shaped transition state, the *anti*-isomer is then formed from an (*E*)-allylic alcohol in the course of the sigmatropic rearrangement. Note that transition states **61a** and **61b** are diastereomers and not conformational isomers.

Scheme 7.23 illustrates the diastereoselectivities observed under various conditions in the synthesis of 2,3-dimethyl pent-5-enamides from (*E*)-2-buten-1-ol [11, 14, 25, 26, 47]. The *anti* isomer usually predominates with the exception of the thermal Ficini–Claisen variant (Scheme 7.23, Eq. 2) [18]. In this case, slow addition of the allylic alcohol to the ynamine at elevated temperatures resulted in a 1:2 mixture of *anti*:*syn* products. This result can be explained by assuming that addition of the alcohol to the ketene iminium intermediate (cf. **16**, Scheme 7.9) occurs from the less hindered side and results in the preferential formation of the (*E*)-ketene *N,O*-acetal. This kinetic intermediate then undergoes rearrangement

R^2 R^1 OH + MeO OMe R^3 NMe_2

5 (R^3 = Me)

Δ

R^3 R^1 R^2 O NMe_2 (*Z*)-**61a** → R^1 R^3 NMe_2 R^2 O *anti* (major)

R^1 R^2 O R^3 NMe_2 (*E*)-**61b** → R^1 R^3 NMe_2 R^2 O *syn* (minor)

Scheme 7.22 Simple diastereoselectivity in Meerwein–Eschenmoser–Claisen rearrangements.

OH + OMe NMe_2 — xylenes reflux → O NMe_2

76 %, *anti:syn*= 20:1

OH + NEt_2 — xylenes reflux → O NEt_2

xylenes, reflux: 62 %, *anti:syn*= 1:2
$BF_3 \cdot OEt_2$, benzene, rt: 44 % *anti:syn*= 20:1

OLi + OMe + N — THF, rt → O N

82 %, *anti:syn*= 17:1

OH + O N N O — benzene 160 °C → O N O

84 %, *anti:syn*= 5.1:1

O NPh — $LiNEt_2$, THF then TBSCl, HMPA → O NHPh

67 %, *anti:syn*= 9.5:1

Scheme 7.23 Simple diastereoselectivity in the Eschenmoser–Claisen rearrangement.

through a chair-shaped transition state. By contrast, boron-trifluoride catalyzed addition yields the *anti* isomer with excellent diastereoselectivity (Eq. 2). Apparently, under these conditions equilibration to the thermodynamically favored (*Z*)-ketene *N,O*-acetal takes place faster than the rearrangement. In the case of ynamides, the predominance of the *syn* product points to the preferential formation of the kinetic (*E*)-ketene *N,O*-acetal intermediate **18** (see Scheme 7.10).

7.4 Applications in Synthesis

The Meerwein–Eschenmoser–Claisen rearrangement, in particular the Eschenmoser amide acetal version, has been extensively applied toward the synthesis of natural products and other complex target molecules. The literature is replete with cases where the reaction provided the only way to place a substituent in a sterically hindered environment. The following paragraphs provide selected examples of its use and also serve to highlight the further synthetic transformation of the unsaturated *N,N*-dimethylamides normally obtained. Perhaps the only drawback of the Eschenmoser–Claisen rearrangement is the stability of these amides, whose hydrolysis and reduction requires relatively harsh conditions. However, electrophilic activation *via* the γ,δ-double bond can be used to manipulate this functionality.

Muxfeldt reported one of the first applications of the reaction in the course of a total synthesis of the alkaloid crinine (Scheme 7.24) [48]. In its key step, cyclohexenol **62** was heated with dimethylacetamide dimethyl acetal (**4**) to yield amide **63** in 45% yield together with substantial amounts of an elimination product (not shown). Intramolecular transamidation under forcing conditions yielded lactam **64**, which was subsequently converted into the natural product.

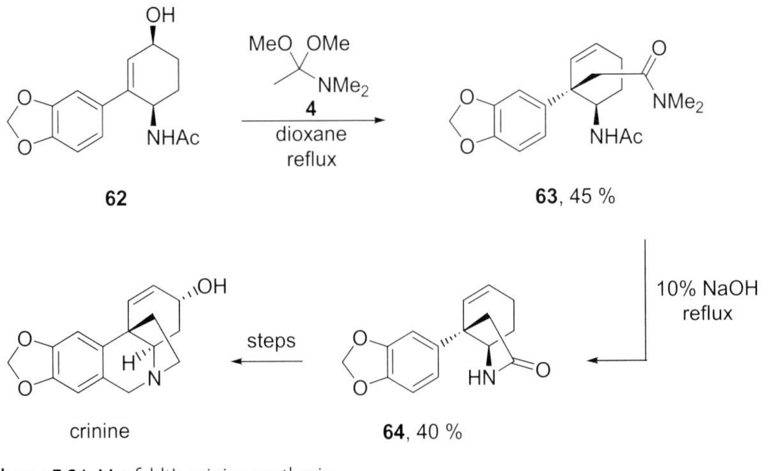

OH
NHAc
62
MeO OMe
NMe_2
4
dioxane
reflux
O
NMe_2
NHAc
63, 45 %
10% NaOH
reflux
HN
O
64, 40 %
steps
OH
H
N
crinine

Scheme 7.24 Muxfeldt's crinine synthesis.

Both Parsons [49] and Mulzer [50, 51] used related Eschenmoser–Claisen rearrangements to set a benzylic quaternary stereocenter in their approach to morphine alkaloids (Scheme 7.25) [5, 52, 53]. Reduction of cyclohexenone **65** followed by Eschenmoser–Claisen rearrangement gave unsaturated amide **66**, which was subsequently converted into a known precursor of morphine (Scheme 7.24, Eq. 1). Treatment of the acid sensitive phenanthrenol **67** with dimethylacetamide dimethyl acetal (**4**) afforded amide **68** comprising the entire carbon skeleton of the morphine (Eq. 2). The amide was subsequently reduced to a primary alcohol (**69**) using lithium triethylborohydride (Super-Hydride), the most suited reagent to perform this task. Previous total syntheses of the alkaloid were intercepted at the stage of dehydrocodeinone.

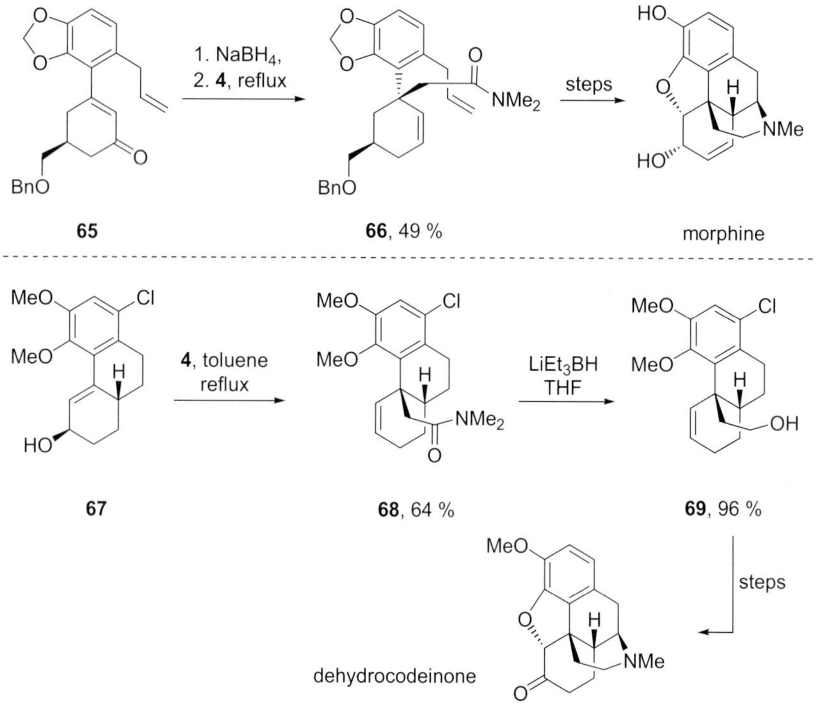

Scheme 7.25 The Eschenmoser–Claisen rearrangement in the synthesis of morphine alkaloids.

Danishefsky et al. took recourse to an Eschenmoser–Claisen rearrangement in a total synthesis of gelsemine (Scheme 7.26) [54]. Although the implementation of the reaction required the subsequent removal of one carbon, it proved to be the only viable way to install the spirocyclic stereocenter. A variety of alternative [2,3]- and [3,3]-sigmatropic rearrangements were found to be unsuccessful. Treatment of allylic alcohol **70** with DMADMA furnished amide **71**, which underwent condensation with the nearby benzyl carbamate to yield lactam **73** upon purification on silica gel in 45% overall yield. The unusual byproduct **72** could be recycled to **70** by treatment with aqueous acid.

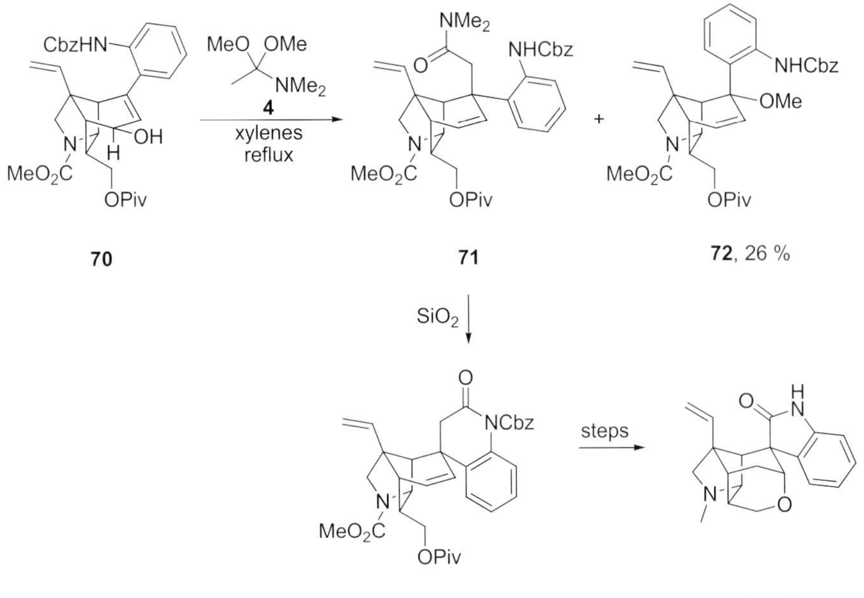

Scheme 7.26 Danishefsky's gelsemine synthesis.

Lounasmaa used the reaction in a synthesis of (*Z*)-geissoschizol [55] based on early precedence provided by Ziegler (Scheme 7.27) [56]. Stereoselective rearrangement of **74** gave unsaturated amide **75** featuring the trisubstituted (*Z*)-double bond of the natural product. Hydrolysis of the dimethyl amide required forcing conditions and was immediately followed by Fischer esterification.

Hart's approach toward the stemona alkaloids stenine serves to illustrate a common strategy for the further transformation of the unsaturated amide (Scheme 7.28) [57]. Rearrangement of **76** followed by treatment of product **77** with iodine in aqueous THF afforded iodolactone **78**, which was then converted into the natural product using the iodine substituent as a functional handle to introduce an ethyl side chain. Wipf used a similar strategy involving a selenolactonization to synthesize tuberostenine [58].

Other alkaloids that have been synthesized using a Eschenmoser–Claisen rearrangement include dehydroquebrachamine [56], tabersonine [59], nominine [60, 61], and aspidophytine [62].

The Eschenmoser–Claisen rearrangement has found numerous applications in the synthesis of terpenoids and steroids. A recent example is Loh's formal synthesis of deoxyanisatin (Scheme 7.29, Eq. 1) [63]. Treatment of **79**, the product of a Birch reduction, with DMADMA gave amide **80** after migration of a double bond. Other versions of the Claisen rearrangement lead only to decomposition or rearomatization. The allylic quaternary stereocenter of dysidiolide was installed through rearrangement of allylic alcohol **81** to afford octahydronaphtalene **82**

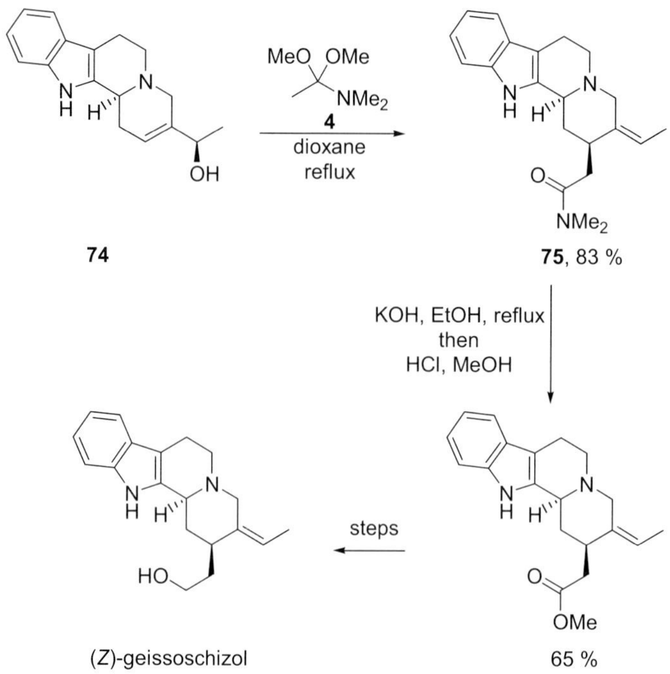
MeO OMe
NMe2
4
dioxane
reflux
OH
NMe2
74
75, 83 %
KOH, EtOH, reflux
then
HCl, MeOH
steps
HO
OMe
(Z)-geissoschizol
65 %

Scheme 7.27 Lounasmaa's synthesis of (*Z*)-geissoschizol.

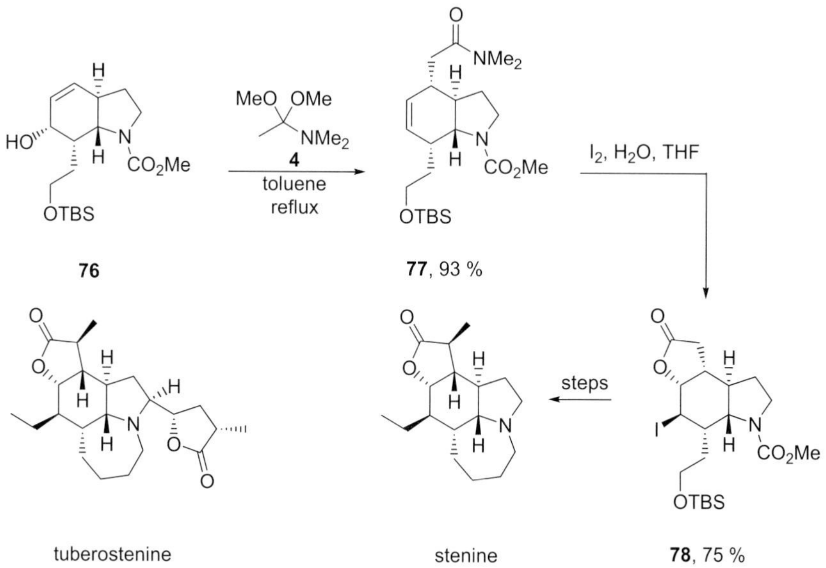
HO
OTBS
CO2Me
MeO OMe
NMe2
4
toluene
reflux
NMe2
CO2Me
OTBS
I2, H2O, THF
76
77, 93 %
steps
CO2Me
OTBS
tuberostenine
stenine
78, 75 %

Scheme 7.28 The Eschenmoser–Claisen rearrangement in the synthesis of stemona alkaloids.

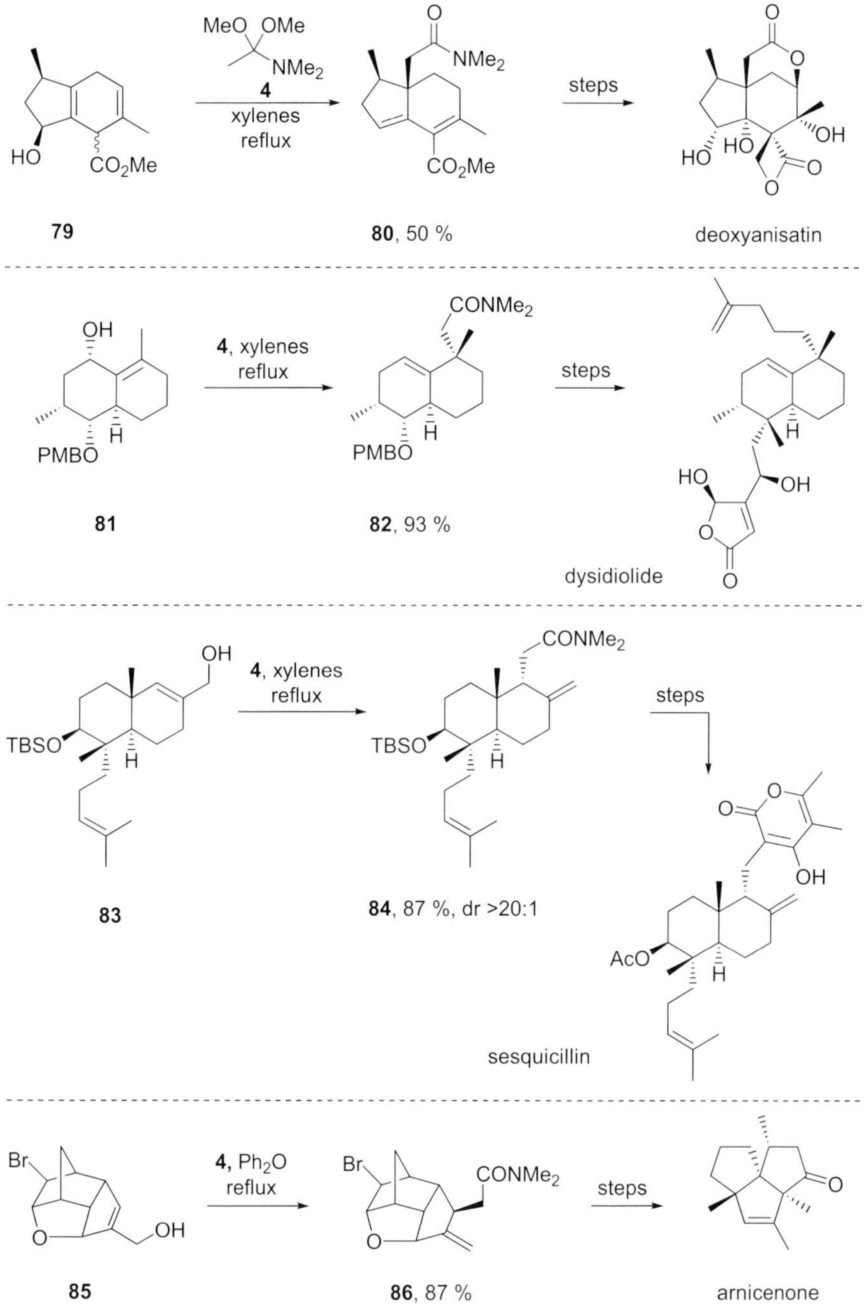
MeO OMe
NMe2
4
xylenes
reflux
steps
79
80, 50 %
deoxyanisatin
4, xylenes
reflux
steps
81
82, 93 %
dysidiolide
4, xylenes
reflux
steps
83
84, 87 %, dr >20:1
sesquicillin
4, Ph2O
reflux
steps
85
86, 87 %
arnicenone

Scheme 7.29 The Eschenmoser–Claisen rearrangement in the synthesis of terpenoids.

(Eq. 2) [64]. In Danishefsky's total synthesis of sesquicillin (Eq. 3), the side chain, which resides on the thermodynamically less favorable face of the decalin core, was introduced with high diastereoselectivity using a Eschenmoser–Claisen rearrangement (Scheme 7.19, Eq. 1) [65]. The mildness of the reaction conditions is reflected by the fact that the double bond did not migrate into the endocyclic position of **84**, which was otherwise found to pose a major problem. In a recent synthesis of the angular triquinane arnicenone, Eschenmoser–Claisen rearrangement proceeded on the convex face of the caged allylic alcohol **85** to afford amide **86** as a single diastereomer. Other terpenoids synthesized via Eschenmoser–Claisen rearrangement include calciferol [66], euryfuran [67], pravastatin [68], paniculide A [69], acetoxytubipofuran [70], and rhopaloic acid [71].

Sucrow provided one of the few examples for the use of an Eschenmoser–Claisen rearrangement involving simple diastereoselection in natural product synthesis (Scheme 7.30) [72, 73]. In a semi-synthesis of sigmastatrienol, the *syn* isomer **88** was preferentially formed as a consequence of the (*Z*)-configuration of allylic alcohol **87**. Subsequent reduction to the tertiary amine followed by Cope elimination completed the installation of the steroid's side chain [74].

The Eschenmoser–Claisen rearrangement has found applications in the synthesis of polyketides as well. For instance, in Nicolaou's synthesis of rapamycin, a building block **90** corresponding to the functionalized cyclohexane ring of the nat-

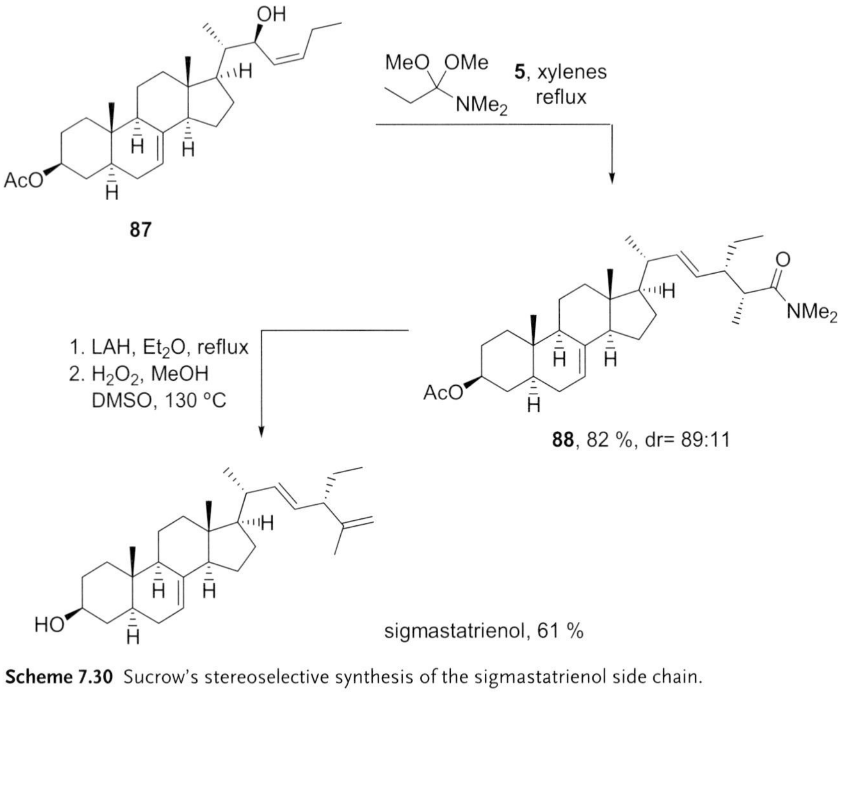

Scheme 7.30 Sucrow's stereoselective synthesis of the sigmastatrienol side chain.

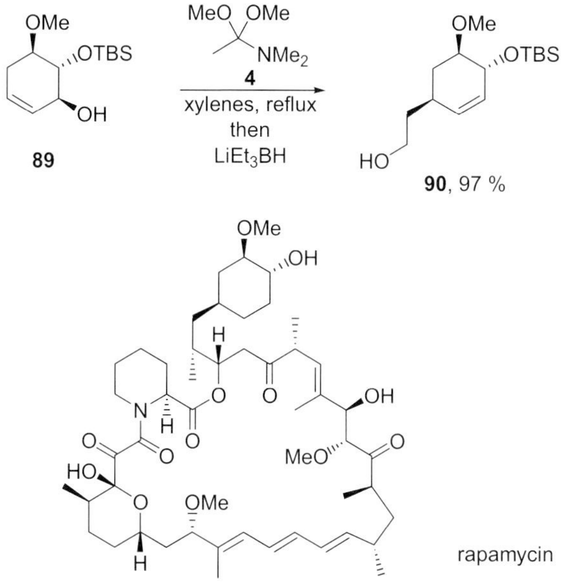

Scheme 7.31 The Eschenmoser–Claisen rearrangement in Nicolaou's synthesis of rapamycin.

ural product was synthesized via rearrangement of cyclic allylic alcohol **89** followed by reduction (Scheme 7.31) [75].

In Danishefsky's synthesis of phomoidride B (CP-225,917), rearrangement of bicyclo[4.3.1]decenol **91** gave amide **92**, allowing for the stereoselective installment of a side chain of the natural product (Scheme 7.32) [54].

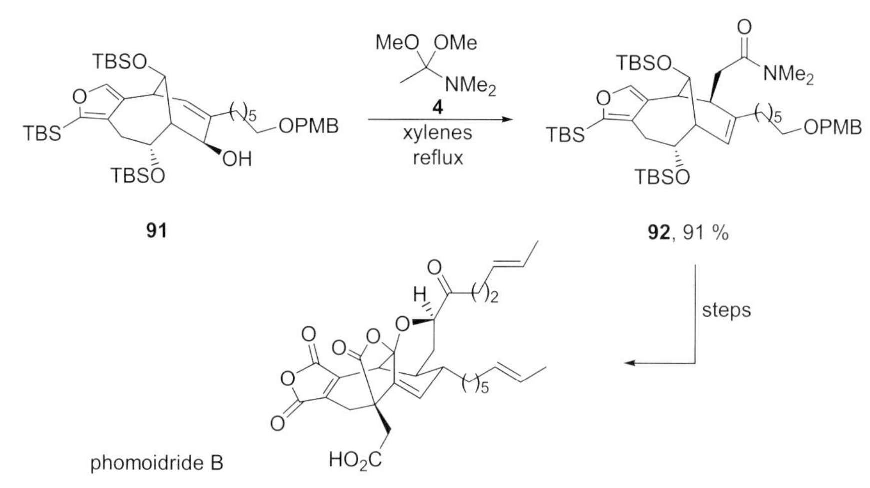

Scheme 7.32 The Eschenmoser–Claisen rearrangement in Danishefsky's synthesis of phomoidride B.

Corey's synthesis of thromboxan B_2 featured an Eschenmoser–Claisen rearrangement of glucose-derived allylic alcohol **93** to afford unsaturated amide **94** (Scheme 7.33). Iodolactonization followed by radical reduction then gave pyran building block **95** with correct configuration at the carbons bearing the side chains [76].

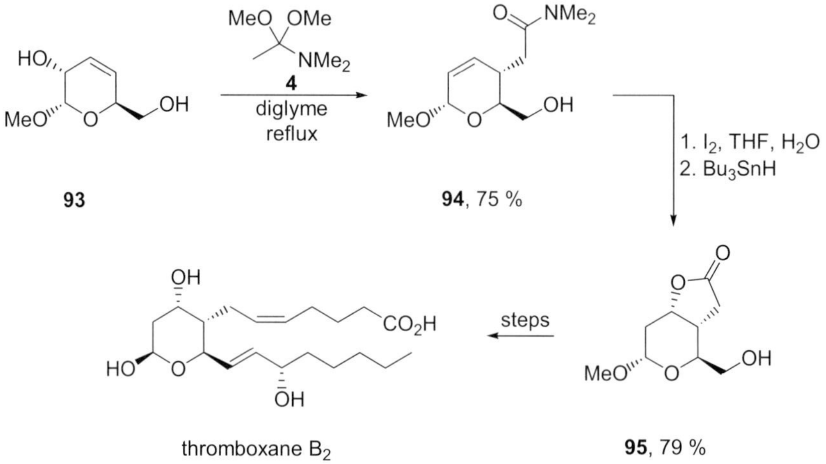

Scheme 7.33 The Eschenmoser–Claisen rearrangement in Corey's synthesis of thromboxan B_2.

Fleet reported an elegant twofold application of the Eschenmoser–Claisen rearrangement in a synthesis of pseudomonic acid A (Scheme 7.34) [77]. First, dihydropyranol **96** underwent Eschenmoser–Claisen rearrangement to afford unsaturated amide **97**. Subsequent iodolactonization gave **98**, which was converted into glycal **99** through elimination and a series of functional group manipulations. A second Eschenmoser–Claisen rearrangement then gave dihydropyran **100** featuring the desired *cis* configuration of the natural product. The unusual conversion of the dimethyl amide moiety into a methyl ketone **(100 → 101)** was presumably facilitated by its proximity to the pyran oxygen.

Further polyketides synthesized via Eschenmoser–Claisen rearrangement include milbemycin derivatives [78], jasplakinolide (Scheme 7.16, Eq. 2) [37], lonomycin A (Scheme 7.16, Eq. 3) [36], methylenolactocin (Scheme 7.19, Eq. 2) [40], and norzoanthamine (Scheme 7.19, Eq. 3) [41].

Somewhat ironically, the Eschenmoser–Claisen rearrangement was not employed in Woodward's and Eschenmoser's synthesis of Vitamin B_{12}, for which it was initially developed [79]. However, the reaction figured prominently in Mulzer's approach toward the molecule (cf. Scheme 7.5) [15] and has found extensive applications in Montfort's studies on bacterial chlorines [80–82]. For instance, in a synthesis of heme d1, a twofold Eschenmoser–Claisen rearrangement was used to convert porphyrin **102** into chlorin **103**, setting the quaternary stereocenters of the target (Scheme 7.35) [83].

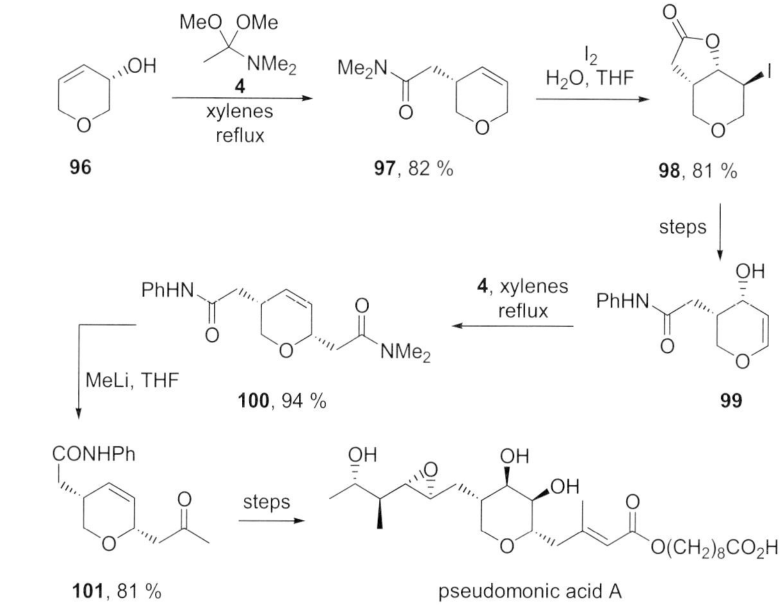

Scheme 7.34 Fleet's synthesis of pseudomonic acid A.

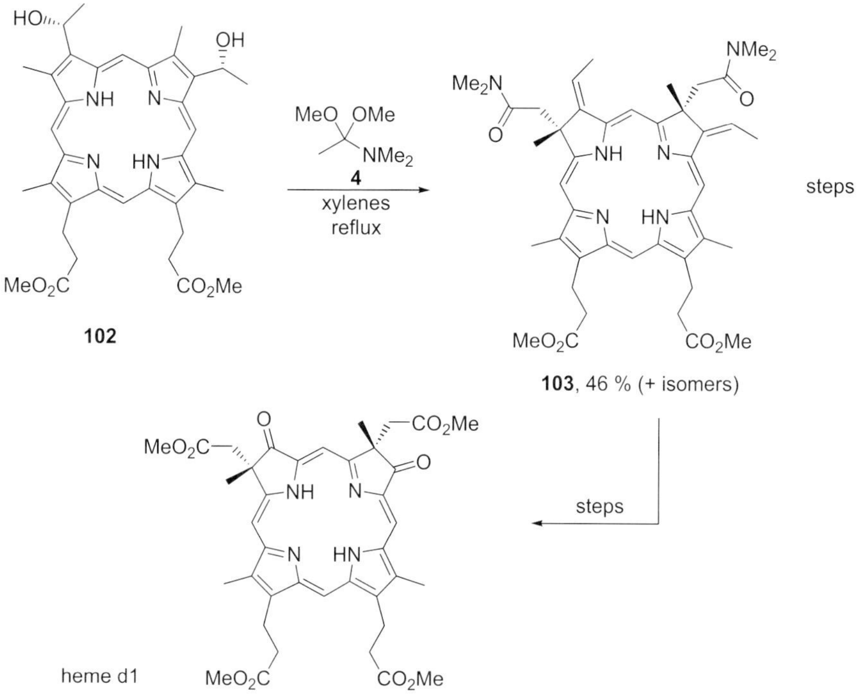

Scheme 7.35 The Eschenmoser–Claisen rearrangement in the synthesis of a chlorin.

Finally, the Eschenmoser–Claisen rearrangement has found applications in drug synthesis. Mulzer's synthesis of the antidepressant (*R*)-rolipram serves as an example (Scheme 7.36) [84]. Rearrangement of chiral cinnamyl alcohol **104** gave unsaturated amide **105** with little erosion of optical purity. Ozonolysis and reduction afforded alcohol **106** and set the stage for a subsequent Mitsunobu reaction with hydrazoic acid. Reduction of the azide function followed by intramolecular transamidation gave γ-lactam **107**, which was converted into (*R*)-rolipram in two straightforward steps.

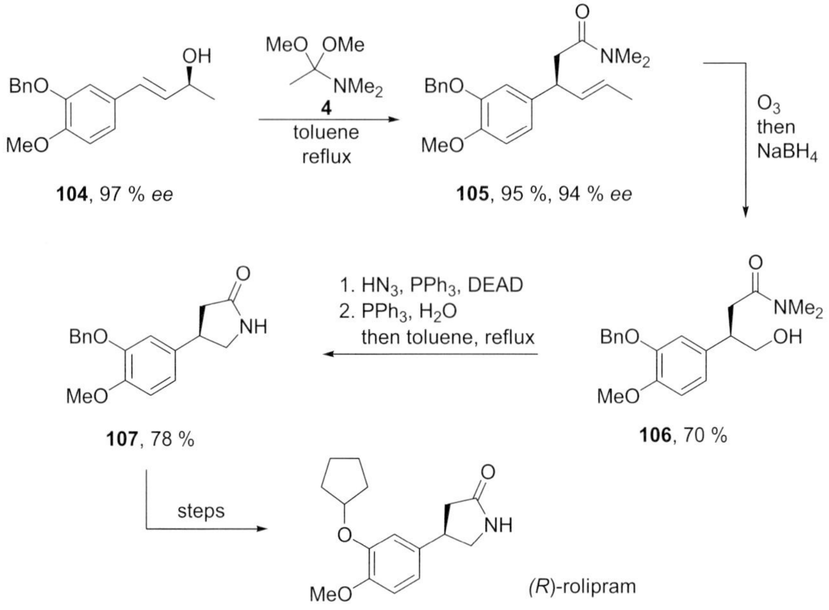

Scheme 7.36 Mulzer's rolipram synthesis.

References

1 Meerwein, H.; Stopp, G.; Florian, W.; Schon, N. *Liebigs Ann. Chem.* **1961**, *641*, 1–39.
2 Wick, A. E.; Steen, K.; Felix, D.; Eschenmoser, A. *Helv. Chim. Acta* **1964**, *47*, 2425–2429.
3 Felix, D.; Gschwend. K; Wick, A. E.; Eschenmoser, A. *Helv. Chim. Acta* **1969**, *52*, 1030–1042.
4 Kawasaki, T.; Ohtsuka, H.; Mihira, A.; Sakamoto, M. *Heterocycles* **1998**, *47*, 367–373.
5 Fleischhacker, W.; Richter, B. *Monatsh. Chem.* **1992**, *123*, 837–848.
6 Lee, T. J. *Tetrahedron Lett.* **1979**, 2297–2300.
7 Trost, B. M.; Pinkerton, A. B.; Seidel, M. *J. Am. Chem. Soc.* **2001**, *123*, 12466–12476.
8 Parker, K. A.; Petraitis, J. J.; Kosley, R. W.; Buchwald, S. L. *J. Org. Chem.* **1982**, *47*, 389–398.
9 Egert, E.; Beck, H.; Schmidt, D.; Gonschorrek, C.; Hoppe, D. *Tetrahedron Lett.* **1987**, *28*, 789–792.
10 Ficini, J.; Barbara, C. *Tetrahedron Lett.* **1966**, 6425–6429.

11 Gradl, S. N.; Kennedy-Smith, J. J.; Kim, J.; Trauner, D. *Synlett* **2002**, 411–414.
12 Hill, R. K.; Soman, R.; Sawada, S. *J. Org. Chem.* **1972**, *37*, 3737–3740.
13 Scott, J. W.; Valentin. D. *Science* **1974**, *184*, 943–952.
14 Welch, J. T.; Eswarakrishnan, S. *J. Org. Chem.* **1985**, *50*, 5909–5910.
15 Mulzer, J.; List, B.; Bats, J. W. *J. Am. Chem. Soc.* **1997**, *119*, 5512–5518.
16 Coates, B.; Montgomery, D.; Stevenson, P. J. *Tetrahedron Lett.* **1991**, *32*, 4199–4202.
17 Ficini, J. *Tetrahedron* **1976**, *32*, 1449–1486.
18 Bartlett, P. A.; Hahne, W. F. *J. Org. Chem.* **1979**, *44*, 882–883.
19 Bartlett, P. A.; Pizzo, C. F. *J. Org. Chem.* **1981**, *46*, 3896–3900.
20 Mulder, J. A.; Hsung, R. P.; Frederick, M. O.; Tracey, M. R.; Zificsak, C. A. *Org. Lett.* **2002**, *4*, 1383–1386.
21 Mulder, J. A.; Kurtz, K. C. M.; Hsung, R. P. *Synlett* **2003**, 1379–1390.
22 Dunetz, J. R.; Danheiser, R. L. *Org. Lett.* **2003**, *5*, 4011–4014.
23 Black, D. S.; Eastwood, F. W.; Poynton, A. J.; Welker, C. H.; Wade, A. M.; Okraglik, R. *Aust. J. Chem.* **1972**, *25*, 1483–1494.
24 Metz, P.; Mues, C. *Tetrahedron* **1988**, *44*, 6841–6853.
25 Metz, P.; Linz, C. *Tetrahedron* **1994**, *50*, 3951–3966.
26 Metz, P.; Mues, C. *Synlett* **1990**, 97–98.
27 Metz, P.; Hungerhoff, B. *J. Org. Chem.* **1997**, *62*, 4442–4448.
28 Evans, P. A.; Holmes, A. B.; Russell, K. *Tetrahedron-Asym.* **1990**, *1*, 593–596.
29 Parker, K. A.; Farmar, J. G. *Tetrahedron Lett.* **1985**, *26*, 3655–3658.
30 Parker, K. A.; Kosley, R. W. *Tetrahedron Lett.* **1975**, 3039–3040.
31 Beaudry, C. M. *PhD Thesis*, University of California at Berkeley, USA **2005**.
32 Dawson, D. J.; Ireland, R. E. *Tetrahedron Lett.* **1968**, 1899–1901.
33 Dauben, W. G.; Dietsche, T. J. *J. Org. Chem.* **1972**, *37*, 1212–1216.
34 Fraserreid, B.; Dawe, R. D.; Tulshian, D. B. *Can. J. Chem.* **1979**, *57*, 1746–1749.
35 Faulkner, D. J.; Petersen, M. R. *Tetrahedron Lett.* **1969**, 3243–3246.
36 Evans, D. A.; Ratz, A. M.; Huff, B. E.; Sheppard, G. S. *J. Am. Chem. Soc.* **1995**, *117*, 3448–3467.
37 Chu, K. S.; Negrete, G. R.; Konopelski, J. P. *J. Org. Chem.* **1991**, *56*, 5196–5202.
38 Johnson, W. S.; Buchanan, R. A.; Bartlett, W. R.; Tham, F. S.; Kullnig, R. K. *J. Am. Chem. Soc.* **1993**, *115*, 504–515.
39 Heathcock, C. H.; Finkelstein, B. L. *J. Chem. Soc., Chem. Commun.* **1983**, 919–920.
40 Masaki, Y.; Arasaki, H.; Itoh, A. *Tetrahedron Lett.* **1999**, *40*, 4829–4832.
41 Williams, D. R.; Brugel, T. A. *Org. Lett.* **2000**, *2*, 1023–1026.
42 Konno, T.; Nakano, H.; Kitazume, T. *J. Fluorine Chem.* **1997**, *86*, 81–87.
43 Frederick, M. O.; Hsung, R. P.; Lambeth, R. H.; Mulder, J. A.; Tracey, M. R. *Org. Lett.* **2003**, *5*, 2663–2666.
44 Dickson, J. K.; Tsang, R.; Llera, J. M.; Fraserreid, B. *J. Org. Chem.* **1989**, *54*, 5350–5356.
45 Bauermeister, S.; Gouws, I. D.; Strauss, H. F.; Venter, E. M. M. *J. Chem. Soc., Perkin Trans. 1* **1991**, 561–565.
46 Welch, J. T.; Eswarakrishnan, S. *J. Am. Chem. Soc.* **1987**, *109*, 6716–6719.
47 Sucrow, W.; Richter, W. *Chem. Ber.-Recl.* **1971**, *104*, 3679–3688.
48 Muxfeldt, H.; Schneider, R. S.; Mooberry J. B. *J. Am. Chem. Soc.* **1966**, *88*, 3670–3671.
49 Parsons, P. J.; Penkett, C. S.; Shell, A. J. *Chem. Rev.* **1996**, *96*, 195–206.
50 Mulzer, J.; Bats, J. W.; List, B.; Opatz, T.; Trauner, D. *Synlett* **1997**, 441–444.
51 Trauner, D.; Bats, J. W.; Werner, A.; Mulzer, J. *J. Org. Chem.* **1998**, *63*, 5908–5918.
52 Labidalle, S.; Min, Z. Y.; Reynet, A.; Moskowitz, H.; Vierfond, J. M.; Miocque, M.; Bucourt, R.; Thal, C. *Tetrahedron* **1988**, *44*, 1159–1169.
53 Fleischhacker, W.; Richter, B. *Monatsh. Chem.* **2000**, *131*, 997–1009.
54 Kwon, O. Y.; Su, D. S.; Meng, D. F.; Deng, W.; D'Amico, D. C.; Danishefsky, S. J. *Angew. Chem., Int. Ed. Engl.* **1998**, *37*, 1877–1880.

55 Lounasmaa, M.; Jokela, R.; Tirkkonen, B.; Miettinen, J.; Halonen, M. *Heterocycles* **1992**, *34*, 321–339.

56 Ziegler, F. E.; Bennett, G. B. *Tetrahedron Lett.* **1970**, 2545–2547.

57 Chen, C. Y.; Hart, D. J. *J. Org. Chem.* **1993**, *58*, 3840–3849.

58 Wipf, P.; Rector, S. R.; Takahashi, H. *J. Am. Chem. Soc.* **2002**, *124*, 14848–14849.

59 Ziegler, F. E.; Bennett, G. B. *J. Am. Chem. Soc.* **1973**, *95*, 7458–7464.

60 Muratake, H.; Natsume, M. *Tetrahedron Lett.* **2002**, *43*, 2913–2917.

61 Muratake, H.; Natsume, M. *Angew. Chem., Int. Ed. Engl.* **2004**, *43*, 4646–4649.

62 Sumi, S.; Matsumoto, K.; Tokuyama, H.; Fukuyama, T. *Tetrahedron* **2003**, *59*, 8571–8587.

63 Loh, T. P.; Hu, Q. Y. *Org. Lett.* **2001**, *3*, 279–281.

64 Piers, E.; Caille, S.; Chen, G. *Org. Lett.* **2000**, *2*, 2483–2486.

65 Zhang, F.; Danishefsky, S. J. *Angew. Chem., Int. Ed. Engl.* **2002**, *41*, 1434–1437.

66 Chapleo, C. B.; Hallett, P.; Lythgoe, B.; Waterhouse, I.; Wright, P. W. *J. Chem. Soc., Perkin Trans. 1* **1977**, 1211–1218.

67 Kanematsu, K.; Soejima, S. *Heterocycles* **1991**, *32*, 1483–1486.

68 Daniewski, A. R.; Wovkulich, P. M.; Uskokovic, M. R. *J. Org. Chem.* **1992**, *57*, 7133–7139.

69 Amano, S.; Takemura, N.; Ohtsuka, M.; Ogawa, S.; Chida, N. *Tetrahedron* **1999**, *55*, 3855–3870.

70 Kundig, E. P.; Cannas, R.; Laxmisha, M.; Liu, R. G.; Tchertchian, S. *J. Am. Chem. Soc.* **2003**, *125*, 5642–5643.

71 Kadota, K.; Ogasawara, K. *Heterocycles* **2003**, *59*, 485–490.

72 Sucrow, W.; Girgensohn, B. *Chem. Ber.-Recl.* **1970**, *103*, 750–756.

73 Sucrow, W.; Schubert, B.; Richter, W.; Slopiank, M. *Chem. Ber.-Recl.* **1971**, *104*, 3689–3703.

74 Sucrow, W. *Angew. Chem., Int. Ed. Engl.* **1968**, *7*, 629.

75 Nicolaou, K. C.; Piscopio, A. D.; Bertinato, P.; Chakraborty, T. K.; Minowa, N.; Koide, K. *Chem. Eur. J.* **1995**, *1*, 318–333.

76 Corey, E. J.; Shibasaki, M.; Knolle, J. *Tetrahedron Lett.* **1977**, 1625–1626.

77 Fleet, G. W. J.; Gough, M. J.; Shing, T. K. M. *Tetrahedron Lett.* **1983**, *24*, 3661–3664.

78 Naito, S.; Kobayashi, M.; Saito, A. *Heterocycles* **1995**, *41*, 2027–2032.

79 Furst, A.; Brubacher, G.; Meier, W.; Ruttimann, A. *Helv. Chim. Acta* **1993**, *76*, 1–59.

80 Gerlach, B.; Montforts, F. P. *Liebigs Ann.* **1995**, 1509–1514.

81 Hoper, F.; Montforts, F. P. *Liebigs Ann.* **1995**, 1033–1038.

82 Kusch, D.; Tollner, E.; Lincke, A.; Montforts, F. P. *Angew. Chem., Int. Ed. Engl.* **1995**, *34*, 784–787.

83 Romanowski, F.; Mai, G.; Kusch, D.; Montforts, F. P.; Bats, J. W. *Helv. Chim. Acta* **1996**, *79*, 1572–1586.

84 Mulzer, J. *J. Prakt. Chem.* **1994**, *336*, 287–291.

8
The Carroll Rearrangement

Mark A. Hatcher and Gary H. Posner

8.1
Introduction

The use of Claisen-type rearrangements has been a reliable and efficient means of constructing stereodefined intermediates useful in the rational design of complex organic molecules. The Carroll rearrangement is one Claisen-variant that has been shown to be a versatile complement to these important reactions. The Carroll reaction involves the thermal or anionic [3,3]-sigmatropic rearrangement of a β-keto-allylic ester to a β-keto acid, which upon decarboxylation, gives the corresponding γ,δ-unsaturated ketone (Scheme 8.1). The overall transformation in this reaction is the equivalent of a Claisen rearrangement, but at the ketone oxidation state.

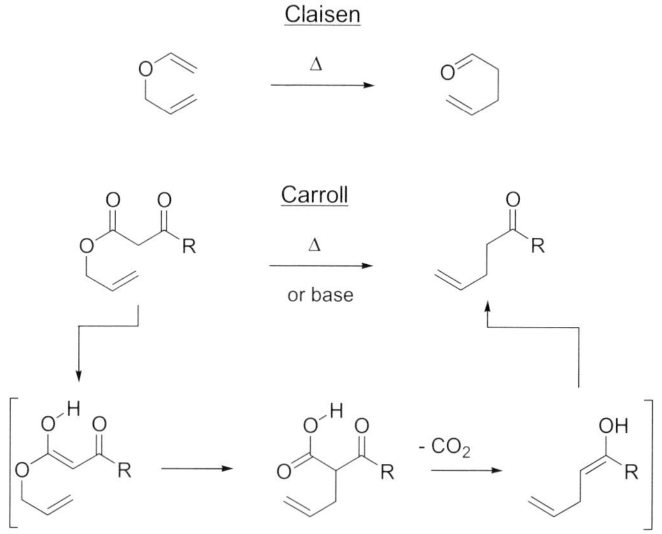

Claisen

Δ

Carroll

Δ

or base

$- CO_2$

Scheme 8.1

The Claisen Rearrangement. Edited by M. Hiersemann and U. Nubbemeyer

ISBN: 978-3-527-30825-5

Since the discovery of this reaction in 1940 by Carroll [1–3], there have been numerous examples of its synthetic application as well as many new insights into the mechanism, scope, limitations and efficient preparation of starting materials. Recently, a number of different Carroll-type reactions have emerged, which contain the basic allylic ester moiety, but differ by the nature of the adjacent β-ketone group. These Carroll-variants have interesting applications into more diverse molecules and have in some instances led to asymmetric Carroll-type reactions. Accordingly, the following discussion will focus on the mechanism, synthetic applications and advances that have been made since the inception of the Carroll rearrangement.

8.2 Mechanism

In an attempt to acetylate a variety of alcohols, Carroll discovered an unexpected side-reaction, which later became known as the Carroll rearrangement [1–3]. It was thought that by heating ethyl acetoacetate and certain alcohols in the presence of an alkaline catalyst, such as NaOAc, a reverse acetoacetic ester condensation would occur, giving the acylated alcohol and ethyl acetate as products (Scheme 8.2).

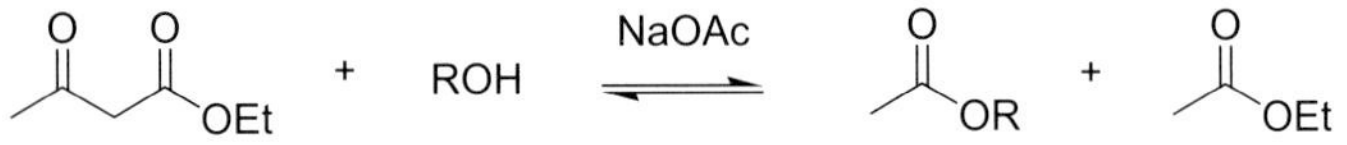

O O OEt + ROH NaOAc O OR + O OEt

Scheme 8.2

As it is now known, this reaction gives only the transesterified product from simple alcohols. In the course of this work, Carroll observed that when the allylic alcohol linalool was heated with acetoacetate at elevated temperatures, the desired geranyl acetate was not observed, but rather geranylacetone was formed instead. Carroll showed this to be the case for a number of different allylic alcohols [3]. He suggested the mechanism of this reaction to involve a Michael condensation addition reaction (Scheme 8.3). This mechanism proposed by Carroll had some serious flaws even though it accounted for the introduced acyl group at the γ-position.

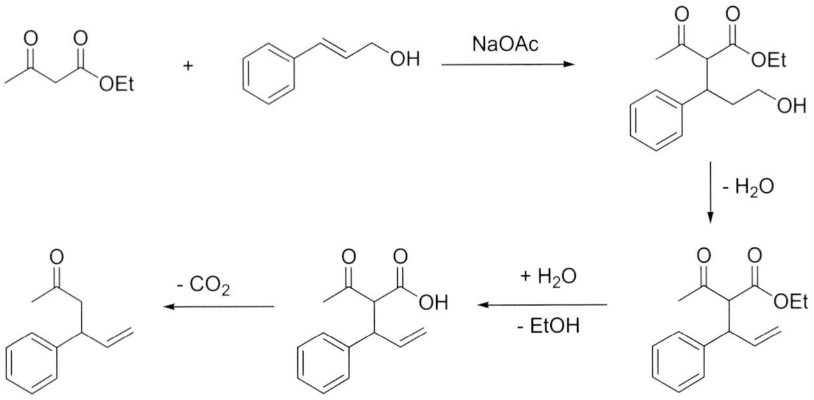

O O OEt + OH NaOAc O O OEt OH

$- H_2O$

O $- CO_2$ O O OH $+ H_2O$ $- EtOH$ O O OEt

Scheme 8.3

Table 8.1 Thermally-induced Carroll rearrangements using simple allylic alcohols and diketene.

Acetoacetate	T [°C]	t [h]	% CO_2	Name	Formula	Yield [%]
allyl	185–200	92	55	allylacetone	$CH_3COCH_2CH_2CH{=}CH_2$	31
methallyl	200–215	25	40	meythylallylacetone	$CH_3COCH_2CH_2(CH_3){=}CH_2$	26
crotyl	190–220	18	51	3-methyl-1-hexene-5-one	$CH_3COCH_2CH(CH_3)CH{=}CH_2$	37
methylvinyl-carbinyl	185–200	12	84	5-heptene-2-one	$CH_3COCH_2CH_2CH{=}CHCH_3$	80
cinnamyl	250	1.2	95	3-phenyl-hexene-5-one	$CH_3CCOCH_2CH(C_6H_5)CH{=}CH_2$	74
phenylvinyl-carbinyl	200–240	1	97	1-phenyl-1hexene-5-one	$CH_3COCH_2CH_2CH{=}CHC_6H_5$	88
linalyl	170–235	1.2	97	geranylacetone	$CH_3COCH_2CH_2CH{=}C(R)CH_3$	78
geranyl	220–230	8	66	geranylacetone	$CH_3COCH_2CH_2CH{=}C(R)CH_3$	23

R= $(CH_3)_2C{=}CHCH_2CH_2$-

These flaws were pointed out by Kimel and Cope in 1943 [4]. They observed that acetoacetates, derived from the reaction of allylic alcohols and diketene, and the benzoylacetates of several β,γ-unsaturated alcohols reacted at elevated temperatures to give γ,δ-unsaturated ketones and CO_2 (Table 8.1).

This observation led to the proposal of an alternative mechanism similar to the Claisen mechanism involving allyl enol ethers. They proposed that, in the case of Carroll's observations, the first step was a trans-esterification of the allylic alcohol, catalyzed by NaOAc, to the allyl acetoacetate, followed by a Claisen-type rearrangement to the β-keto acid, which subsequently decarboxylated under the elevated reaction temperature (Scheme 8.4).

This mechanism seemed to make more sense compared to the mechanism proposed by Carroll for a number of reasons. Kimel and Cope suspected that the

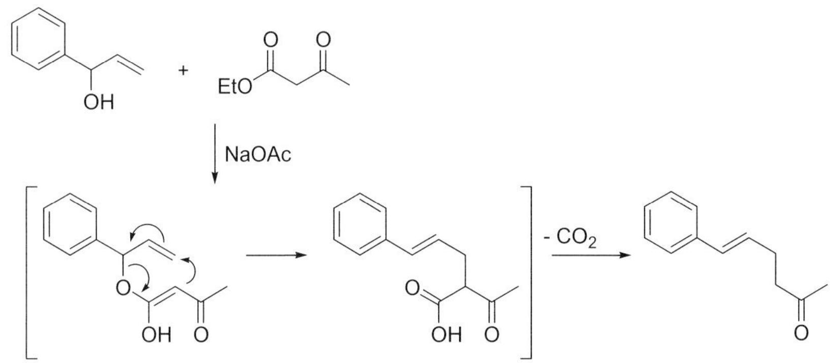

OH + EtO ... NaOAc ... OH O ... O OH O ... - CO_2 ... O

Scheme 8.4

initial Michael condensation was unlikely, due to the insufficient activation of the double bond on the allylic alcohol, as well as the unlikely dehydration of the primary alcohol (Scheme 8.3).

Additional evidence supporting the mechanism proposed by Kimel and Cope was reported many years later. Hill and Synerholm observed that the acetoacetate of optically active 2-cyclohexen-1-ol (**1**) underwent a stereospecific rearrangement to the optically active γ,δ-unsaturated ketone **2** (Scheme 8.5) [5]. They also observed that the acetoacetate **3** underwent the Carroll reaction to give predominantly the *trans* γ,δ-unsaturated ketone **4** (Scheme 8.5). These observations led Hill and Synerholm to conclude that the Carroll reaction, as proposed by Kimel and Cope, undergoes a concerted [3,3]-sigmatropic rearrangement through a chair-like transition state favoring the larger substituent in the equatorial position (Scheme 8.5).

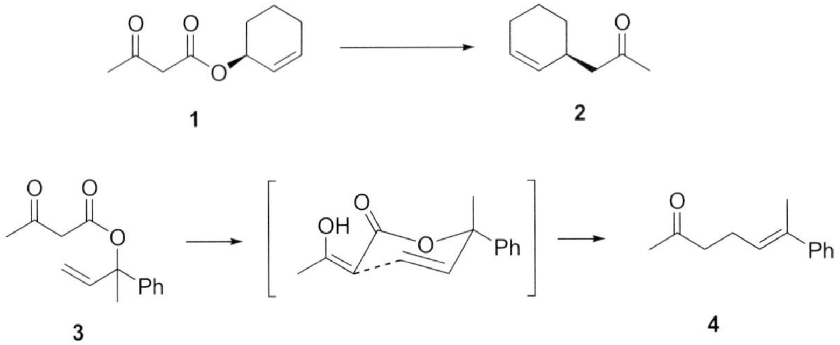

Scheme 8.5

More recent research has shown that the rate of the Carroll rearrangement is accelerated in the presence of a base. Wilson and Price reported that allylic acetoacetates treated with 2 eq. of lithium diisopropylamide (LDA) at –78 °C readily underwent the Carroll rearrangement to the desired β-keto acids when warmed to 65 °C (Scheme 8.6) [6].

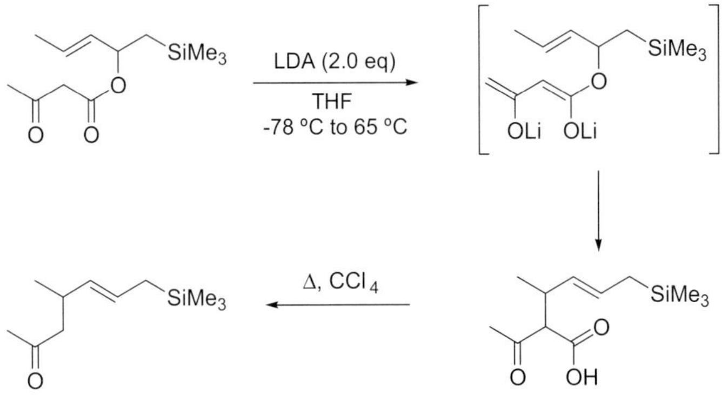

Scheme 8.6

The resulting β-keto acids were subsequently decarboxylated by heating in CCl_4 for 1 h. Some interesting observations resulted from this work. First, it was shown that 2 eq. equivalents of base were needed to promote the rearrangement, whereas 1 eq. of base gave only recovered starting material. From this observation, it was proposed that the dianion, resulting from 2 eq. of base, was able to undergo the reaction at lower temperatures due to the more nucleophilic nature of the reacting carbon. Second, these reaction conditions gave exclusively the products derived from a [3,3]-sigmatropic rearrangement, but under typical pyrolysis conditions the same molecules occasionally gave [1,3] rearrangement products (Scheme 8.7). Finally, Wilson and Price observed a reaction rate dependence based on the allylic substitution pattern of the acetoacetates, showing that primary allylic alcohols rearranged much more slowly than secondary and tertiary allylic alcohols. This type of reaction rate dependence has also been observed in other Claisen-type rearrangements, which further supports the proposed mechanism [7, 8].

8.3 Synthetic Applications

8.3.1 Tertiary and Quaternary Carbon Bond Formation

It was not long after the discovery of the Carroll rearrangement that its synthetic value in constructing complex organic molecules was realized. With the ready availability of allylic alcohols and their subsequent acetoacetates, the Carroll rearrangement has been shown to be a reliable protocol for the generation of carbon-carbon bonds in a stereospecific manner. Also interesting is its ability to generate adjacent tertiary and quaternary carbon centers, leading to congested units not easily prepared by other means.

As a test of the limitations of the Carroll reaction, Bryson and coworkers were able to demonstrate the versatility of the reaction when applied to hindered cyclic systems in the pursuit of newly formed tertiary and quaternary carbon–carbon bonds (Scheme 8.8) [9].

It was observed that when these allylic acetoacetates, prepared by condensation with derivatives of Meldrum's acid (Scheme 8.9) [10–13], were subjected to typical Carroll reaction conditions (110–250 °C, base), only starting material was recovered. However, by using the Wilson procedure (2 eq. of LDA) and forming the dianion, the hindered allylic acetoactetates readily underwent rearrangement and decarboxylation to give the desired γ,δ-unsaturated ketones in good to excellent yield. This work showed the versatility of the Carroll reaction in using hindered trisubstituted and tetrasubstituted allyl acetoacetates.

A general method for the diastereoselective formation of adjacent quaternary carbon atoms is a challenging process that has many inherent problems. Due to steric issues, alkylation and substitution reactions often fall short as sufficient means to construct these complex molecules [14]. Gilbert and Kelly were extremely

Entry	Substrate	Conditions	Ratio of products		Yield (%)
1		A	67	33	50
		B	56	44	50
		C	100	0	84
2		A	83	17	40
		C	100	0	40
		D	no reaction		-
		E	no reaction		-
3		F	100	0	37
		D	no reaction		
		C	100	0	95
4		F	54	46	67
		C	58	42	83
5			-	-	
		F	complex mixture		67
		C	no reaction		-
6		B	36	64	70
		D	complex mixture		-
		C	no reaction		-
7		B	100	0	58
		C	100	0	80

Scheme 8.7 Conditions: (A) 170 °C, 2 h, neat. (B) 200 °C, 2 h, neat. (C) LDA (2 eq.), THF, –70 to 65 °C (affording the β-keto acid); CCl_4, 77 °C, 1 h (decarboxylation). (D) LDA (1 eq.), THF, –70 to 65 °C. (E) KH, THF, 65 °C. (F) 190–220 °C, 18 h, Ph_2O.

Entry	Substrate	Product	Yield (%)
1	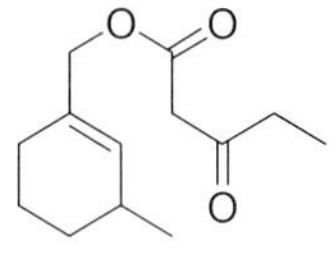	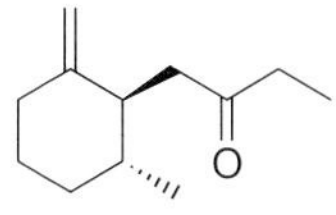	90
2	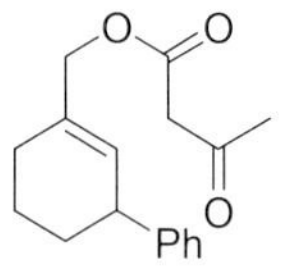 Ph	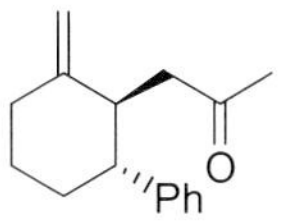 Ph	90
3	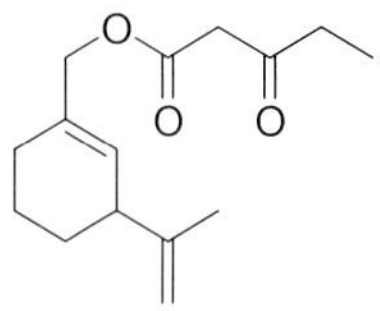	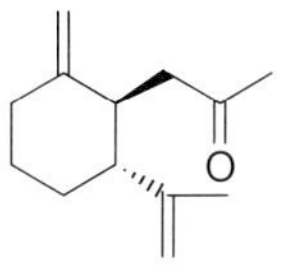	94
4	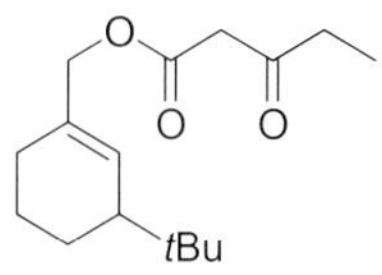 tBu	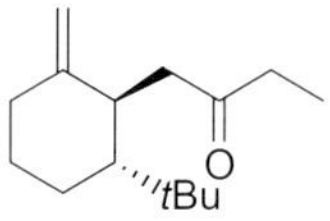 tBu	95
5	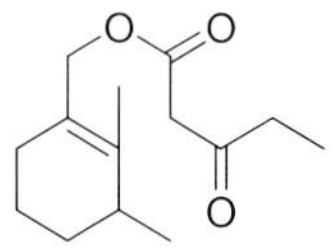	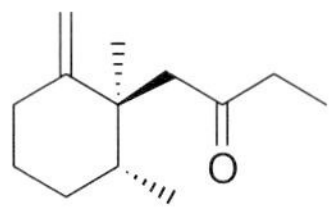	65–75
6	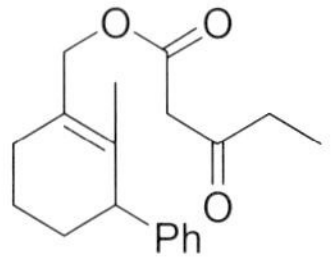 Ph	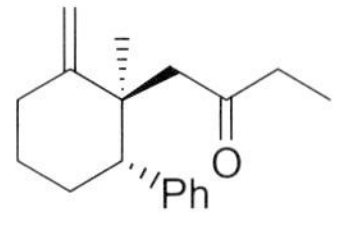 Ph	65–75

Scheme 8.8

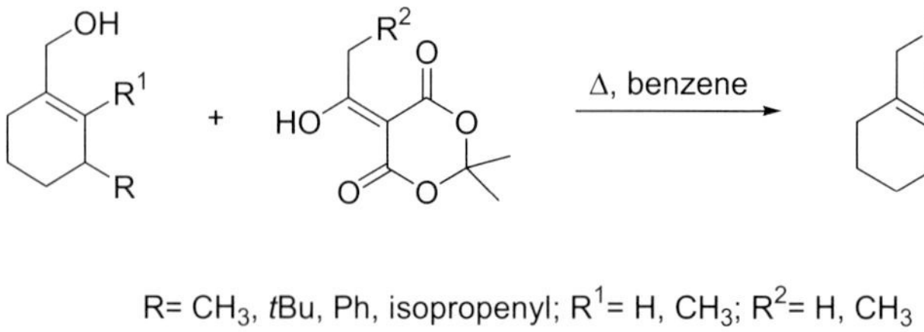

OH
R1
R
+
HO
R2
O
O
O
O
Δ, benzene
O
O
R1
R2
R
O
R= CH3, tBu, Ph, isopropenyl; R1= H, CH3; R2= H, CH3

Scheme 8.9

successful in the diasteroselective formation of adjacent quaternary carbon centers using the Carroll rearrangement [14]. Initially using the Wilson procedure [6], they formed the dianion of the highly substituted allyl acetoacetate **5**, which upon warming to 25 °C decomposed to a tar-like material (Scheme 8.10). Borrowing an idea from the Ireland–Claisen reaction, they decided to trap the dianion as the bis-silyl ether **7**. This proved to be successful in obtaining compound **8**, albeit in very low yield due to products resulting from silylation of the terminal carbon. Alternatively, Gilbert and Kelly obtained the desired bis-silyl enol ether **7** in a stepwise fashion rather than in one pot. To form the mono-silyl enol ether they adopted an efficient route, developed by Ainsworth, that involved refluxing hexamethyldisilazane (HMDS) in the presence of a catalytic amount of imidazole [15]. This procedure gave the desired mono-silyl enol ether **6** in 84% yield solely as the *E*-isomer. The silyl ketene acetal **7** was then prepared by treatment of **6** with 1.3 eq. of a 1:1 mixture of lithium tetramethylpiperidine (LTMP) and tetramethylethylenediamine (TMEDA) at −50 °C and then quenching of the anion with 2.0 eq. of a

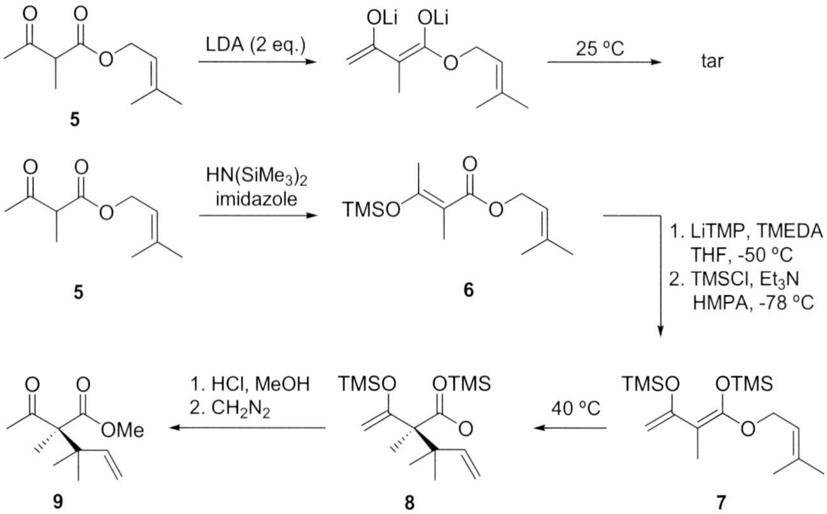

LDA (2 eq.)
OLi OLi
25 °C
tar
5
HN(SiMe3)2
imidazole
TMSO
5
6
1. LiTMP, TMEDA
THF, -50 °C
2. TMSCl, Et3N
HMPA, -78 °C
OMe
1. HCl, MeOH
2. CH2N2
TMSO OTMS
40 °C
TMSO OTMS
9
8
7

Scheme 8.10

1:1 mixture of trimethylsilylchloride (TMSCl) and triethylamine. It was observed that by adding hexamethylphosphoramide (HMPA) after the addition of TMSCl the yield increased from 55 to 76%. The Carroll rearrangement of the newly formed silyl ketene acetal proceeded smoothly after the reaction mixture was warmed to 40 °C and the resulting acid **8** was converted directly into the methyl ester **9** in good overall yield (76%).

The diastereoselectivity of this modified Carroll reaction was further examined through allylic esters **10** and **11**. When subjected to the above conditions, the esters cleanly rearranged into the new methyl esters **14** and **15**, respectively (Scheme 8.11). These rearranged products were determined to be a single diastereomer by ^{1}H NMR (>98:2). To distinguish between diastereomers **14** and **15**, a series of transformations were accomplished to form the racemic compound **16** and the *meso* compound **17**, which could be distinguished by NMR techniques (Scheme 8.11). It was shown that in the presence of a chiral shift reagent the resonances of the racemic compound **16** were split into two sets of peaks, whereas the *meso* compound **17** showed only one set of peaks for its resonances. These data showed that compounds **16** and **17** were assigned correctly and were formed as a result of the *E*-geometry of the enol ethers **12** and **13**, respectively.

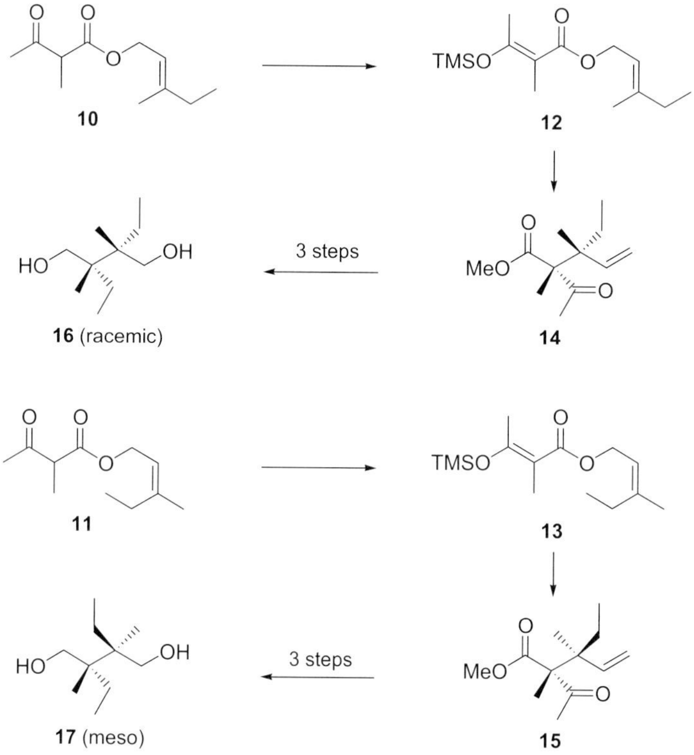

TMSO
10
12
3 steps
HO
OH
MeO
16 (racemic)
14
TMSO
11
13
3 steps
HO
OH
MeO
17 (meso)
15

Scheme 8.11

8.3.2
Natural Products

Natural product synthesis is an attractive and challenging area of research in synthetic organic chemistry. Natural products and pharmacologically active drugs typically contain a variety of functionalization with complex stereochemistry and therefore are quite challenging to synthesize. The regiocontrolled and stereocontrolled nature of the Carroll rearrangement provides an excellent means of preparing complex compounds from simple starting materials.

Using the Carroll rearrangement as the key step, Rodriguez and coworkers synthesized derivatives of the Prelog–Djerassi lactone (**18**) (Scheme 8.12) [16].

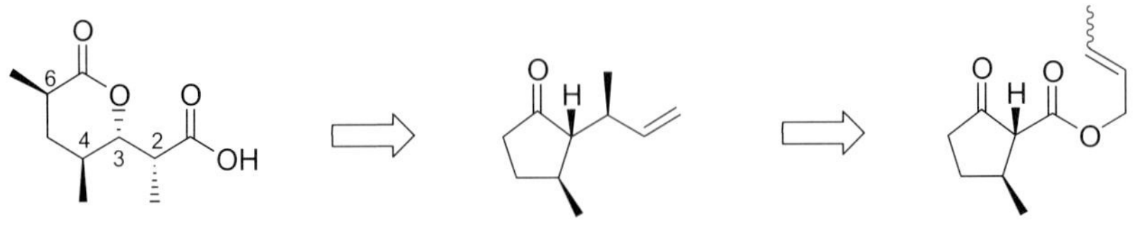

(2*R*,3*S*,4*S*,6*R*)-**18** **19** (*E*)- or (*Z*)-**20**

Scheme 8.12

This lactone is an important intermediate in the synthesis of several macrolide antibiotics [17] and was thought to be obtainable through a stereoselective [3,3]-sigmatropic rearrangement of allylic *β*-ketoester **20**. Upon reaction of (*E*)-**20** with 2 eq. of LDA, intermediate **21** was obtained that subsequently decarboxylated to yield ketone **22** as a 80/20 mixture of the *trans* and *cis* products in good yield (Scheme 8.13).

This ketone mixture was enolized in the presence of base (DMAP) to give exclusively the more stable *trans*-**22** product. The relative stereochemistry of *trans*-**22** was confirmed by X-ray crystallography of the carboxylic acid **23**, prepared by the oxidation of **22** with RuO_4. The synthesis was completed in three more steps to give the (2*R*,3*R*,4*R*,6*S*)-**18** derivative of the Prelog–Djerassi lactone (Scheme 8.13). This was accomplished by the regioselective Baeyer–Villiger reaction of **22** to give lactone **24**. Lactone **24** was treated with LDA at −78 °C and then alkylated with methyl iodide under thermodynamically equilibrated conditions to give **25** as a 6:1 mixture of the *α*:*β* methyl lactones, which were separated by medium pressure chromatography. Finally, the terminal alkene of lactone **25** was oxidized using RuO_4. The synthesis of the derivative (2*S*,3*R*,4*R*,6*S*)-**18** was also achieved using (*Z*)-**20**. Interestingly, this isomer gave only moderate diastereoselectivity (~40%). With improved conditions (Scheme 8.14), Rodriguez and coworkers were able to obtain *trans*-**26** as a 9/1 ratio with the undesired *trans*-**22** in 30% overall yield.

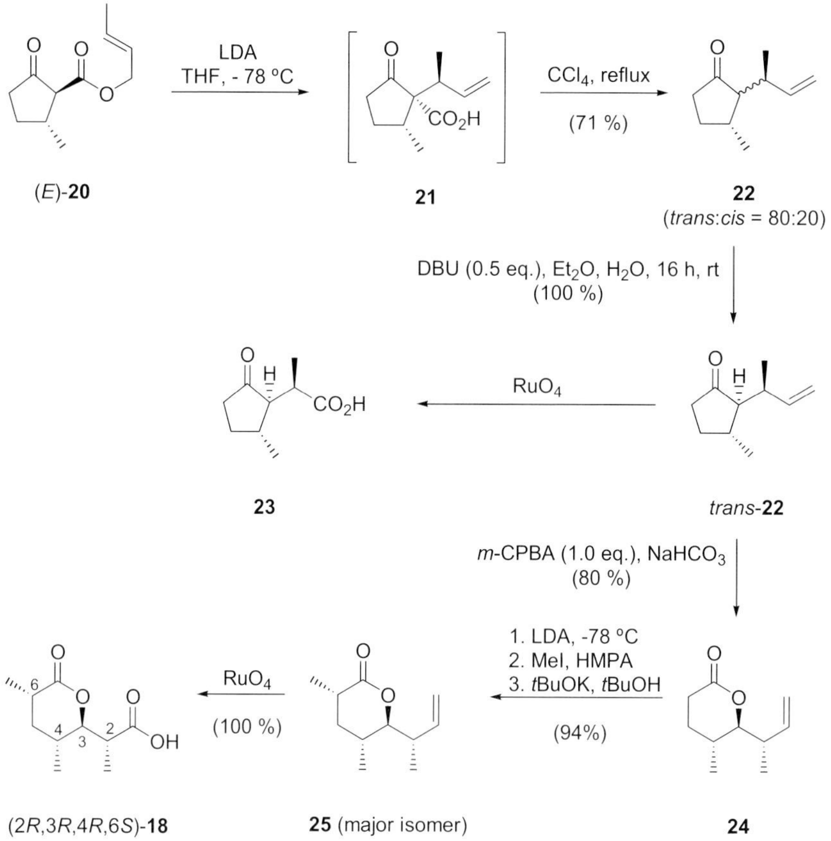

LDA, THF, - 78 °C; CCl$_4$, reflux (71 %)

(*E*)-**20** → **21** → **22** (*trans*:*cis* = 80:20)

DBU (0.5 eq.), Et$_2$O, H$_2$O, 16 h, rt (100 %)

RuO$_4$

23 ← *trans*-**22**

m-CPBA (1.0 eq.), NaHCO$_3$ (80 %)

1. LDA, -78 °C
2. MeI, HMPA
3. *t*BuOK, *t*BuOH

(94%)

RuO$_4$ (100 %)

(2*R*,3*R*,4*R*,6*S*)-**18** ← **25** (major isomer) ← **24**

Scheme 8.13

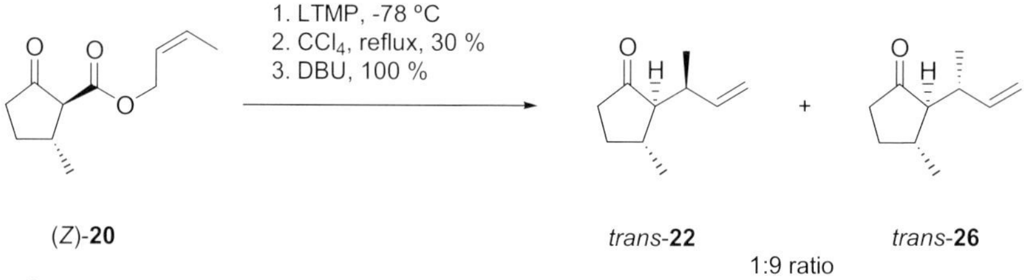

1. LTMP, -78 °C
2. CCl$_4$, reflux, 30 %
3. DBU, 100 %

(*Z*)-**20** → *trans*-**22** + *trans*-**26** (1:9 ratio)

Scheme 8.14

An attractive synthesis of natural products containing 1,5-dimethylated acyclic side chains (vitamin E, vitamin K and phytol) using the Carroll rearrangement as the key step was accomplished by Koreeda and Brown [18]. Starting with enantiomerically pure (+)-pulegone, chirality was transferred through the highly efficient Carroll rearrangement to provide the vitamin E side chain alcohol **27a** (3*R*, 7*R*) and its unknown C-7 diastereomer **27b** (3*R*, 7*S*) (Scheme 8.15).

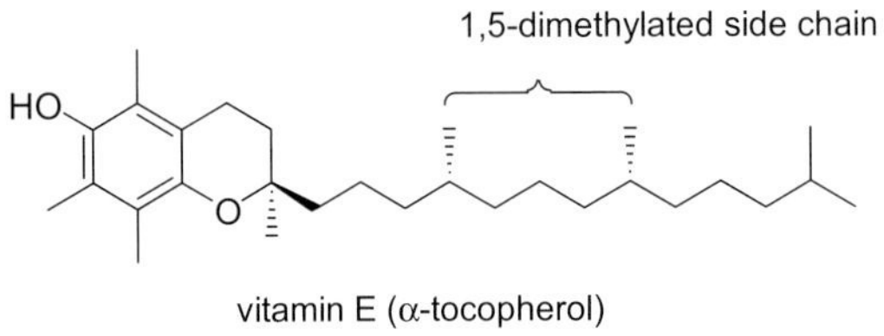

1,5-dimethylated side chain

vitamin E (α-tocopherol)

27a: R^1=CH_3, R^2=H
27b: R^1=H, R^2=CH_3

Scheme 8.15

The synthesis began with the formation of allylic alcohols **28** and **29** from the chiral starting material (+)-pulegone in 58 and 42% yields, respectively. These intermediate alcohols were then treated with 5-isovaleryl-Meldrum's acid to give the desired β-keto esters **30** and **31** (Scheme 8.16). After heating at 220 °C for 2 h,

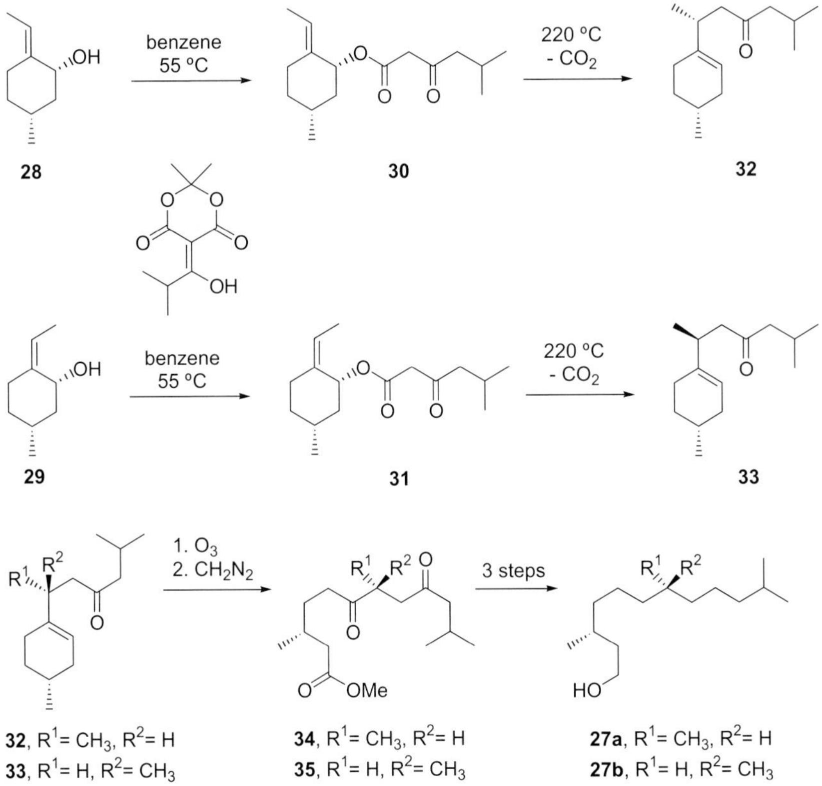

benzene 55 °C

220 °C −CO_2

28 **30** **32**

29 **31** **33**

1. O_3
2. CH_2N_2

3 steps

32, R^1= CH_3, R^2= H
33, R^1= H, R^2= CH_3

34, R^1= CH_3, R^2= H
35, R^1= H, R^2= CH_3

27a, R^1= CH_3, R^2= H
27b, R^1= H, R^2= CH_3

Scheme 8.16

30 and **31** smoothly rearranged into ketones **32** and **33** in 65 and 70% yields, respectively. Ozonolysis and subsequent methylation of the resulting acids gave the desired diketoesters **34** and **35**, which were converted into the enantiomerically pure side chain alcohols **27a** and **27b** in three additional steps.

The Carroll rearrangement is often used as one of many key steps in the preparation of natural products due to its reliability as a stereodifferentiating reaction. Snider and Beal recently showed this in their formal synthesis of isocomene (**41**) (Scheme 8.17) [19]. Ketone **36** was reduced to alcohol **37** and treated with diketene to form the β-keto ester **38**. The [3,3]-sigmatropic Carroll rearrangement was performed using Wilson's procedure (2 eq. of LDA, –78 to 65 °C) [6] to give the desired ketone **39** in 72% yield. This step was key in setting up the desired hindered carbon–carbon bond needed in the final product. An intramolecular cycloaddition and a few additional steps provided ketone **40**, which is a late intermediate in Wenkert's isocomene (**41**) synthesis [20].

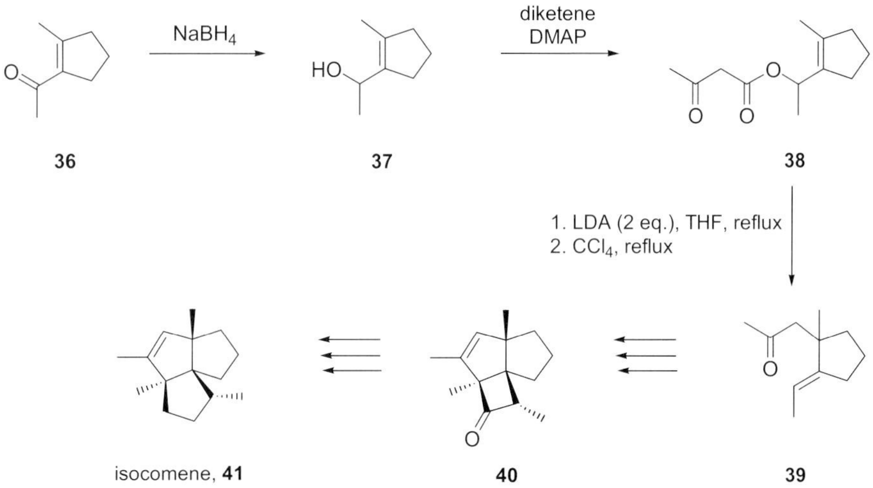

$NaBH_4$

diketene
DMAP

36 **37** **38**

1. LDA (2 eq.), THF, reflux
2. CCl_4, reflux

isocomene, **41** **40** **39**

Scheme 8.17

Commercial production of vitamin A, as well as other carotenoids, typically proceeds via the key intermediate pseudoinone (**48**) [21], which is prepared by condensation of acetone with the natural product citral. This commercial procedure has the drawback of relying on the purification of citral from lemon grass oil. In an attempt to relieve this dependence on citral, Kimel and coworkers at Hoffmann-La Roche developed an alternative fully synthetic route to pseudoinone (**48**) involving multiple Carroll rearrangements (Scheme 8.18) [22].

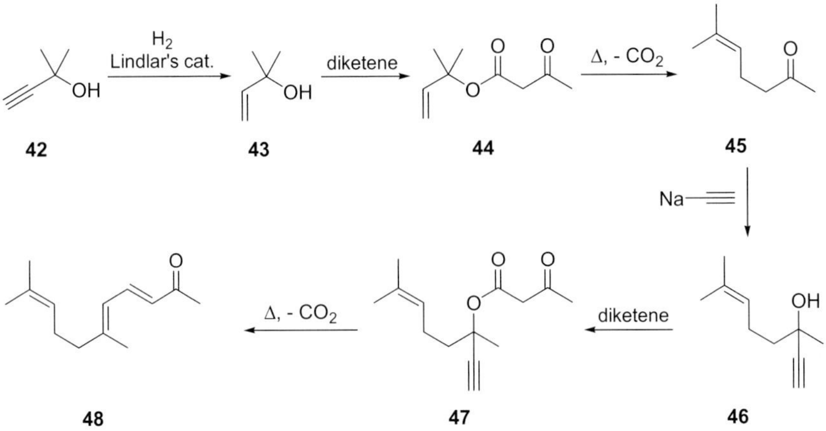

H_2 Lindlar's cat. | diketene | Δ, - CO_2 | Na—≡ | diketene | Δ, - CO_2

42 43 44 45 46 47 48

Scheme 8.18

Allylic alcohol **43** was obtained by the condensation of acetone with sodium acetylide to form 2-methyl-3-butyn-2-ol (**42**), which was then hydrogenated in the presence of Lindlar's catalyst to give allylic alcohol **43**. This allylic alcohol was then treated with diketene to give the *β*-keto ester **44** in quantitative yield. Upon pyrolysis of **44**, in the presence of base at 140–160 °C, the *β*-keto ester cleanly rearranged into 6-methyl-5-heptene-2-one (**45**). Condensation with sodium acetylide afforded the carbinol dehydrolinalool (**46**), which was converted to dehydrolinalyl acetoacetate (**47**) by reaction with diketene. When this *β*-keto ester was subjected to pyrolysis conditions, pseudoinone (**48**) was produced as the major product along with another ketone impurity. The overall yield of pseudoinone from acetone was 35%. The ketone impurity was determined to be compound **50**, which is thought to be derived from intermediate **49** via the proposed rearrangement reaction mechanism (Scheme 8.19).

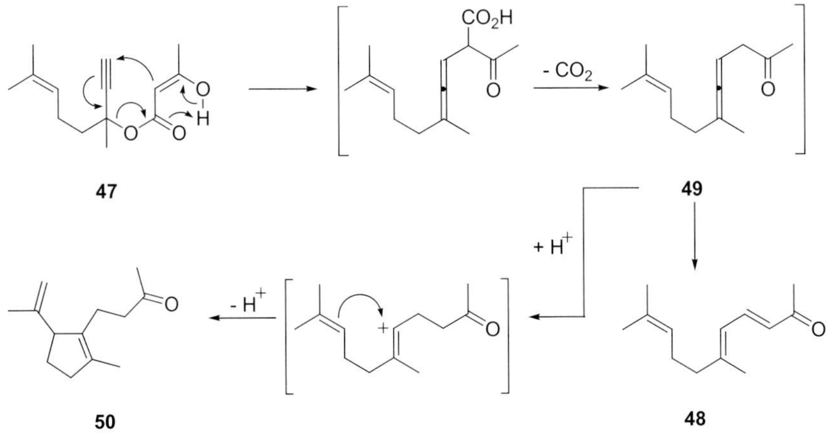

- CO_2 | + H^+ | - H^+

47 49 48 50

Scheme 8.19

The synthesis of the antibiotic acetomycin (**51**) and its diastereomers (Scheme 8.20) was achieved by Echavarren and coworkers using the Carroll rearrangement as the key step [23].

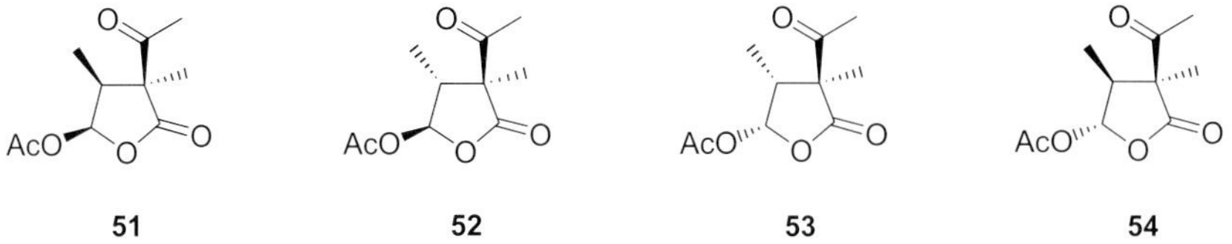

AcO 51 52 53 54

Scheme 8.20

The stereoselective synthesis of the 4-epimeric diastereomer **52** was accomplished via Carroll rearrangement of *β*-keto ester **55**, using Wilson's procedure [6], to give intermediate carboxylic acids **56** and **57** in a 20:1 ratio (Scheme 8.21). These crude acids were ozonylized at –78 °C and then acetylated *in situ* to give **52** and **53** as a 12:1 mixture. The formation of acid **56** as the major product was expected from the rearrangement of the *E*-enolate reacting through a chair-like conformation. The synthesis of acetomycin (**51**) was similarly attempted using the

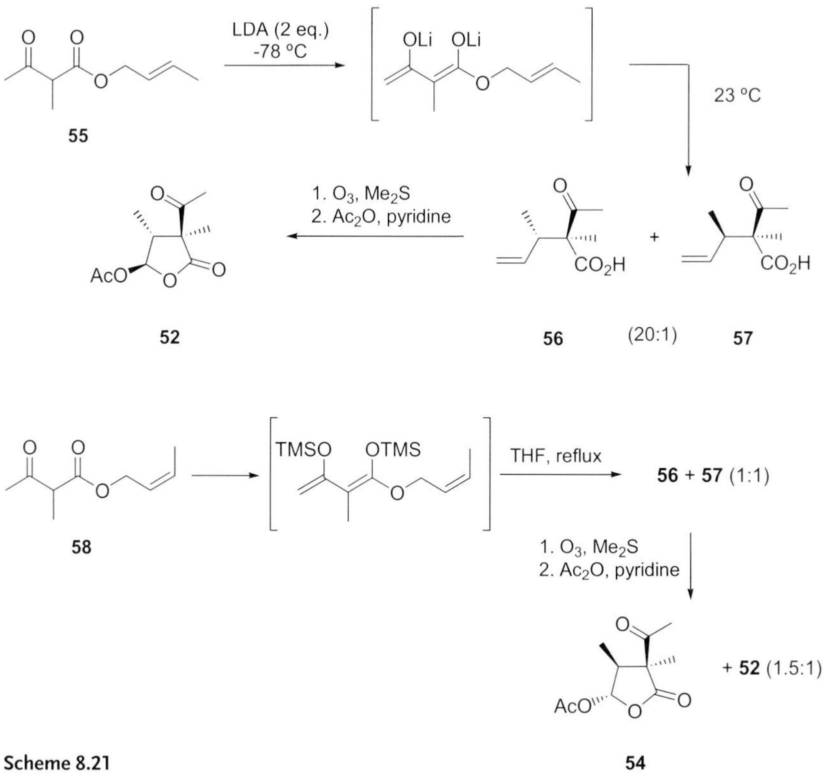

LDA (2 eq.) -78 °C; OLi OLi; 23 °C; **55**

1. O_3, Me_2S 2. Ac_2O, pyridine; **52**; CO_2H; **56** (20:1) **57**

TMSO OTMS; THF, reflux; **56** + **57** (1:1); **58**

1. O_3, Me_2S 2. Ac_2O, pyridine; AcO; **54** + **52** (1.5:1)

Scheme 8.21

Z-*β*-keto ester **58**, but proved to be unsuccessful. When **58** was treated with 2 eq. of LDA no desired product was obtained. When the silyl ketene acetal of **58** was formed and heated under reflux, only a very small amount of **56** and **57** was obtained. This mixture, when subjected to ozonolysis and acetylation, gave a 1.5:1 mixture of **52** and **54**.

8.3.3
Steroidal Side-Chain Formation

With the discovery of steroidal containing compounds possessing a wide variety of biological activities, there has been an increased push for novel synthetic preparations of these compounds. One area of interest lies in the preparation of synthetic analogs, which can produce desired biological activity while minimizing other unwanted side effects. The synthesis of these analogs has for the most part been directed toward the modification of the natural steroidal side chain. The Carroll rearrangement has been shown to be an excellent means of preparing these side-chain modifications, while maintaining the intact steroid moiety with its defined stereochemistry.

In 1980, Tanabe and Hayashi successfully used the Carroll rearrangement in the synthesis of cholesterol and its C-20 epimeric diastereomer, 20-isocholesterol, from the readily available 16*α*,17*α*-epoxy-20-keto-pregnane (**61**) (Scheme 8.22) [24].

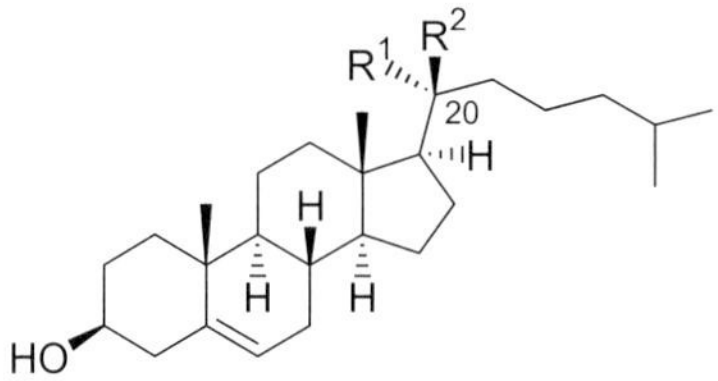

59, $R^1 = CH_3$, $R^2 = H$
60, $R^1 = H$, $R^2 = CH_3$

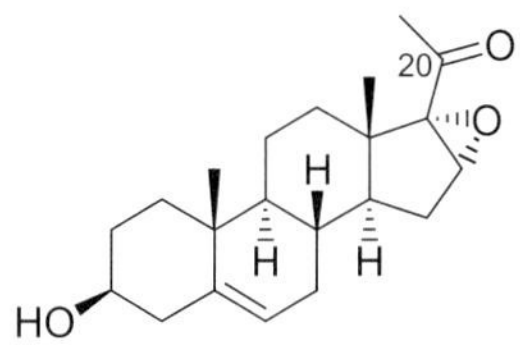

61

Scheme 8.22

The alcohol of **61** was protected as the 3*α*,5*α*-cycloether and subjected to Wharton [25] reaction conditions to give the allylic alcohol **62** as the major product with a minor amount of the geometric isomer **63** (Scheme 8.23).

The *β*-ketoacetates **64** and **65** were prepared by reaction of **62** and **63** with 5-isovaleryl Meldrum's acid in refluxing xylene. When the *β*-keto ester **64** was heated in boiling xylene, it cleanly rearranged to the ketone **66** in 90% yield as the single isomer at C-20. This observation reinforced the belief that the reaction proceeds through a highly ordered six-membered ring transition state (Scheme 8.24). Catalytic hydrogenation of the 16-ene moiety from the alpha face, removal of the cyclo protecting group with dilute sulfuric acid and finally a Wolff–Kishner reduction of the C-23 ketone gave the desired cholesterol (**59**). The 20-isocholesterol (**60**) was prepared in a similar manner using the unnatural ester **65**, with slightly lower yields (Scheme 8.23).

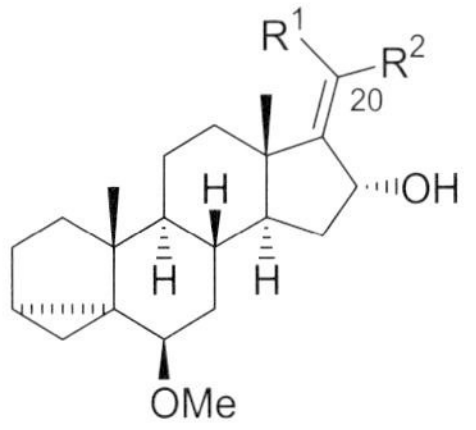

62, $R^1 = CH_3$, $R^2 = H$
63, $R^1 = H$, $R^2 = CH_3$

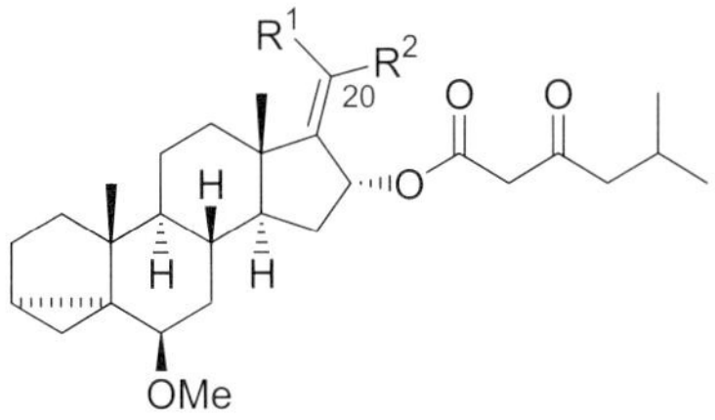

64, $R^1 = CH_3$, $R^2 = H$
65, $R^1 = H$, $R^2 = CH_3$

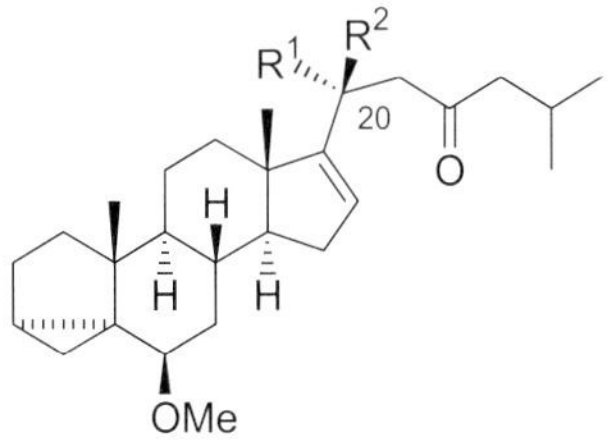

66, $R^1 = CH_3$, $R^2 = H$
67, $R^1 = H$, $R^2 = CH_3$

Scheme 8.23

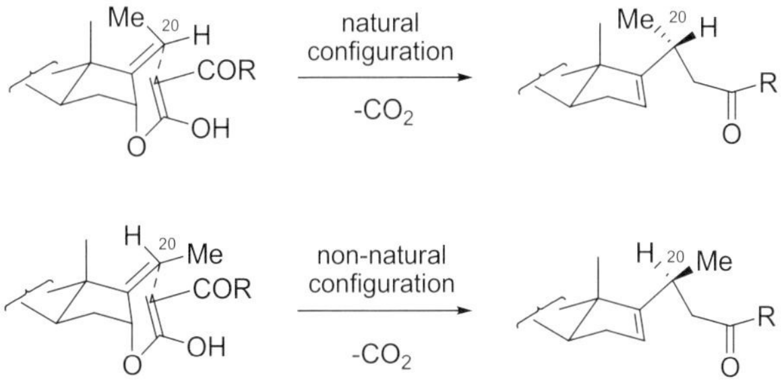

natural configuration
$-CO_2$

non-natural configuration
$-CO_2$

Scheme 8.24

Hatcher and Posner used a similar approach when constructing 16-ene C,D-ring side chains toward the synthesis of novel analogs of the biologically active hormone 1α,25-dihydroxyvitamin D_3 (calcitriol, **68**) (Schemes 8.25 and 26) [26].

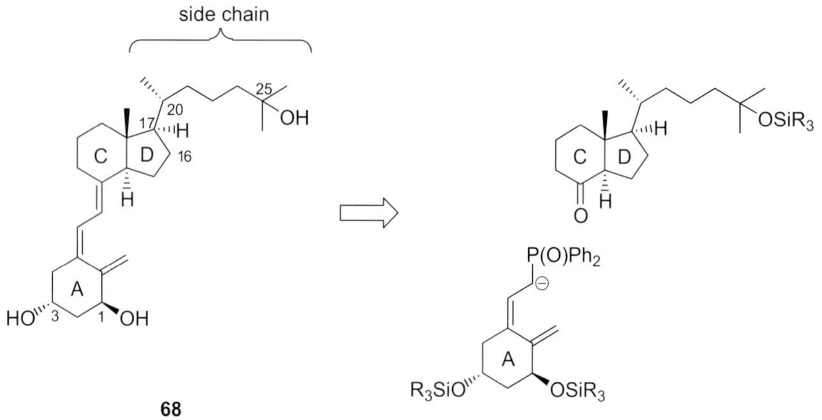

side chain

68
(1α,25-dihydroxyvitamin D_3, calcitriol)

Scheme 8.25

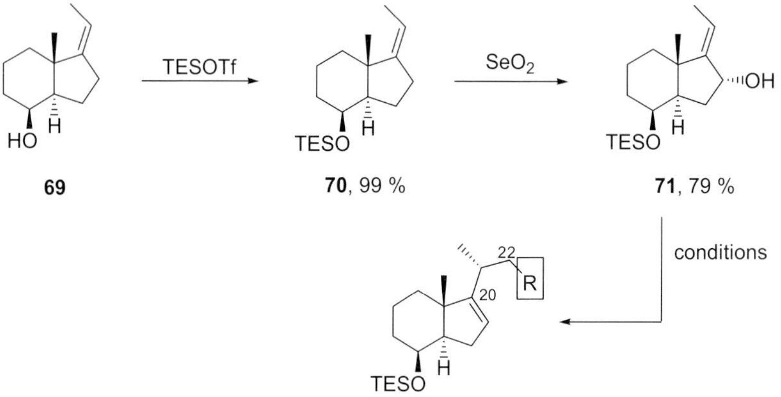

TESOTf

SeO_2

69

70, 99 %

71, 79 %

conditions

Conditions:
Claisen rearrangement: H_2C=CHOEt, $Hg(OAc)_2$, 120 °C: R= CHO (**72**, 97 %)
Johnson–Claisen rearrangement: $H_3CC(OCH_3)_3$, TMBA, 140 °C: R= CO_2CH_3 (**73**, 83 %)

Scheme 8.26

Known C,D-ring alcohol **69** was converted into the triethylsilyl ether **70** in near quantitative yield. This protected alcohol was then oxidized with SeO_2 to form the allylic alcohol **71**. The *ene*-like SeO_2 oxidation occurs exclusively on the α face of the ring due to steric hindrance from both the β-C-18 methyl group and the β-C-8 triethylsiloxy group. This allylic alcohol **71** proved to be a versatile intermediate for several [3,3]-sigmatropic rearrangements on the α face of the D-ring giving the natural C-20 stereochemistry and elaborated side chains. In particular, the Carroll rearrangement was performed by synthesis of the β-ketoacetate of **71** using diketene and pyridine, which was subsequently refluxed in xylene in the presence of base to give the 16-ene-C-23-ketone **74** in 90% overall yield (Scheme 8.27).

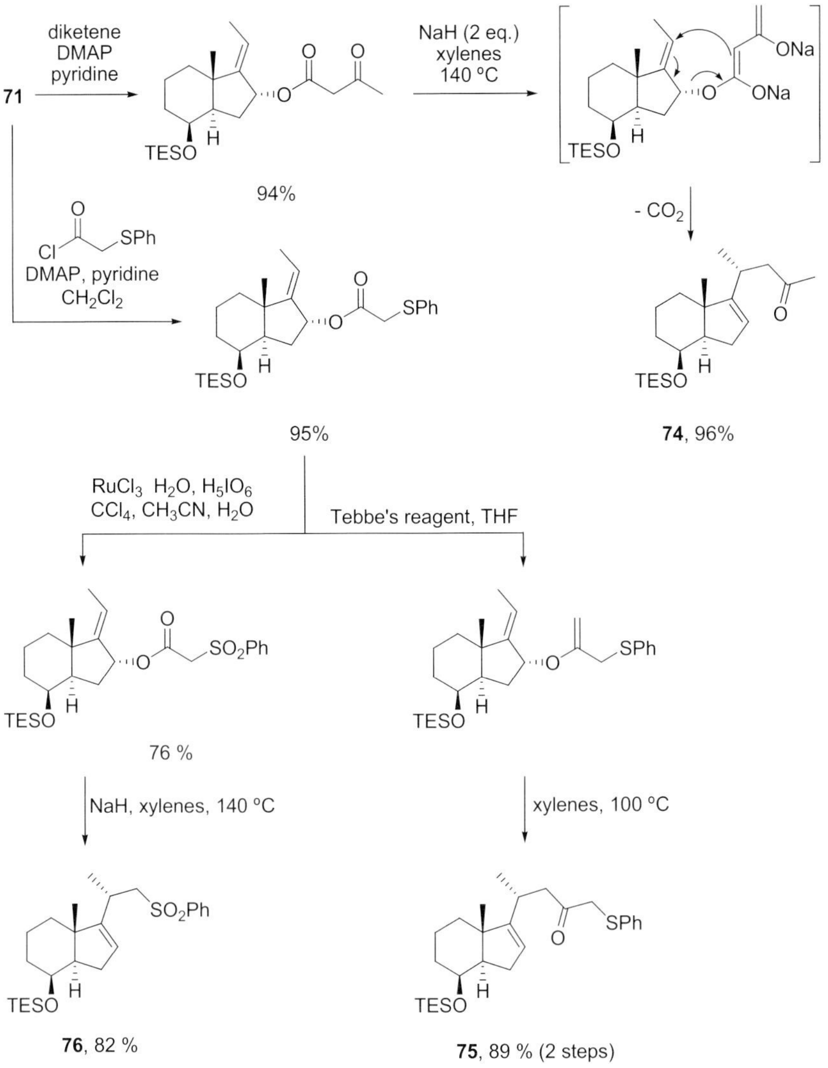

Scheme 8.27

8.3.4
Aromatic Carroll Rearrangements

Recently, the Carroll rearrangement has also been shown to be an effective means of installing carbon–carbon bonds on a variety of aromatic compounds. Sorgi and coworkers were successful in applying the Carroll rearrangement to a number of *p*-quinols to afford substituted arylacetone derivatives (Scheme 8.28) [27].

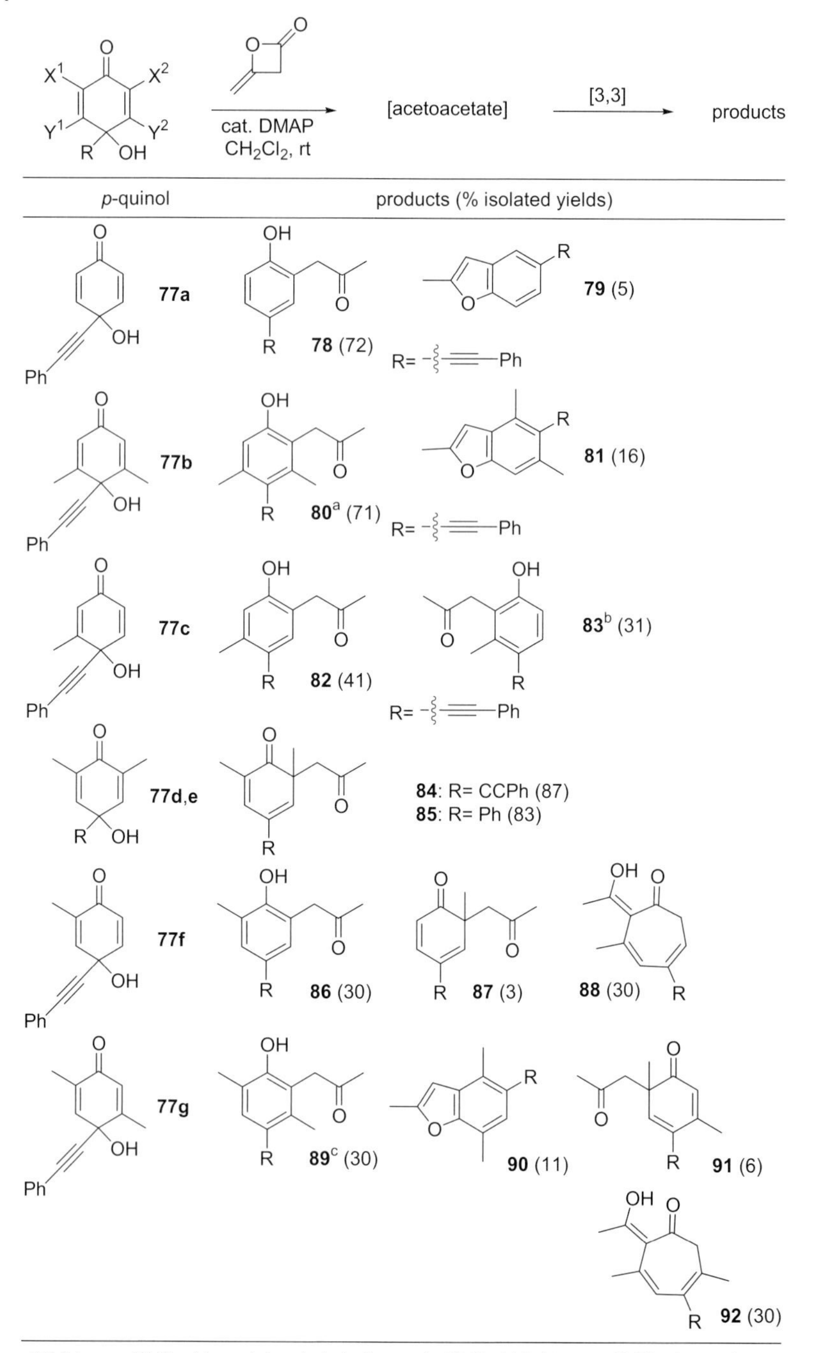

a) Exists as a 55:45 mixture of ring-chain tautomers in $CDCl_3$. b) Exists as a 63:33 mixture of ring-chain tautomers in $CDCl_3$. c) Exists as a 77:23 mixture of ring-chain tautomers in $CDCl_3$.

Scheme 8.28

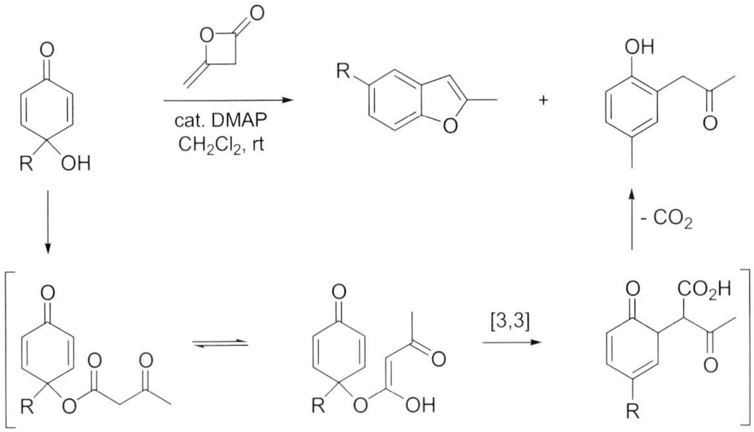

Scheme 8.29

A variety of *p*-quinols, when reacted with diketene and catalytic amounts of DMAP at room temperature, underwent facile [3,3]-sigmatropic rearrangements followed by rapid decarboxylation and rearomatization to give the desired arylacetone compounds (Scheme 8.29). Formation of the respective benzofurans was also observed. Regiocontrol of the reaction was studied by the reactions of *p*-quinol **77c**, **77f** and **77g**. As seen by the 1.3:1 ratio of **82** and **83**, the 10:1 ratio of **86** and **87** and the 5:2:1 ratio of **89**, **90** and **91**, the rearrangement favors the reaction at the less substituted carbon. In the reactions involving **77f** and **77g**, the unexpected cycloheptadienones **88** and **92** were also formed (Scheme 8.28). These products were thought to have been derived from the ring expansion of the respective norcaradiene-like intermediate shown in Scheme 8.30. The reaction of **77d** and **77e** showed the effectiveness of this reaction to form very hindered carbon–carbon bonds in excellent yield. The lack of products formed in the reaction with **77h** was explained by the steric crowding of the adjacent methyl groups with the phenyl group in the developing transition state.

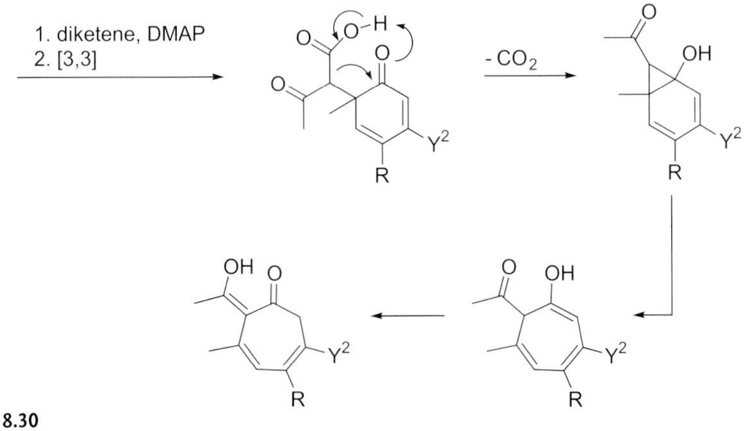

Scheme 8.30

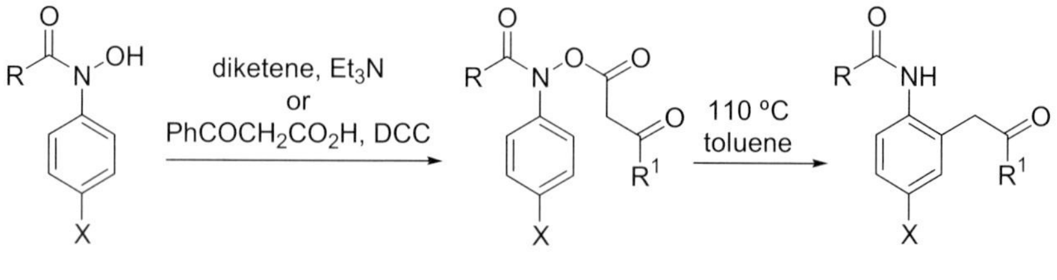

	X	R	R^1	**94** Yield (%)	**95** Yield (%)
a	H	Me	Me	71	55
b	H	Ph	Me	77	46
c	H	Me	Ph	a)	51 b)
d	Me	Me	Me	81	82
e	Me	Me	Ph	a)	50 b)
f	Me	Ph	Me	88	61
g	Cl	Me	Me	85	44
h	CO_2Me	Me	Me	70	40

a) Yield not determined at this stage. b) Overal yield from **93**.

Scheme 8.31

Synthesis of anilides containing *ortho*-substituted carbonyl-functionalized groups using a regiospecific Carroll-type reaction was accomplished by Coates and Said through a 1-aza-1′-oxa [3,3]-sigmatropic rearrangement using *N*-aryl-*N*,*O*-diacylhydroxylamines (Scheme 8.31) [28]. The *N*-aryl-*N*-hydroxyamides (**93a–h**) were prepared by partial reduction of the respective nitroarenes to form *N*-arylhydroxylamines, which were then selectively *N*-acylated. Preparation of the Carroll reaction precursor, *N*-aryl-*N*,*O*-diacylhroxylamines (**94a–h**), was accomplished by *O*-acylation using diketene and catalytic amounts of triethylamine (Scheme 8.31). Compounds **94c** and **94e** were prepared by the dicylcohexylcarbodimide (DCC) coupling of benzoylacetic acid and the respective hydroxyamides. Compounds **94a–94h** were heated in refluxing toluene for 30–90 min to give the desired *o*-(*N*-acylamino) aryl ketones (**95a–3h**) as the major products in 40–82% yield. The mechanism of the reaction was proposed as shown in Scheme 8.32.

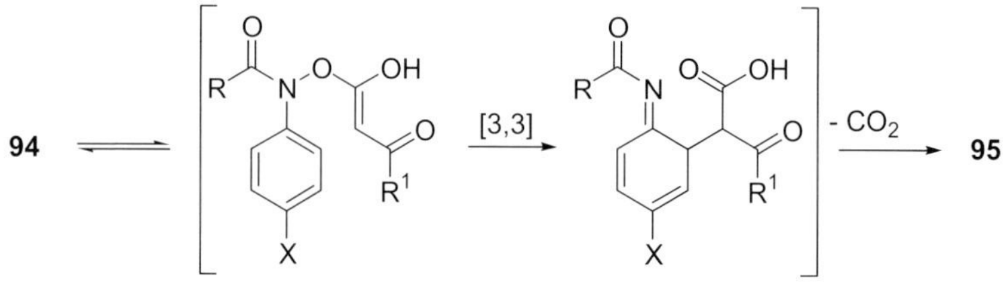

Scheme 8.32

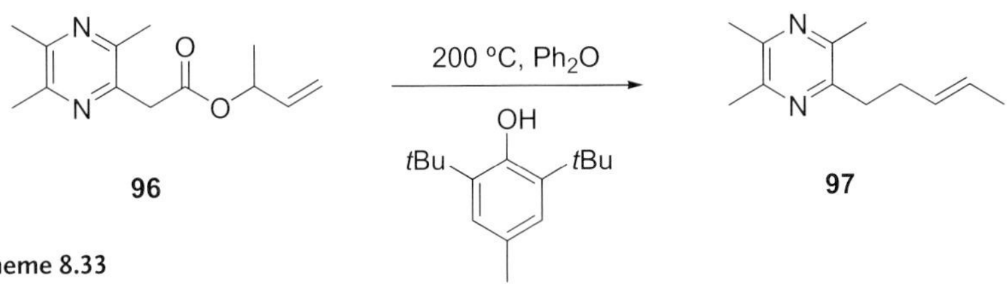

96

200 °C, Ph_2O

OH tBu tBu

97

Scheme 8.33

In an effort to obtain a practical synthesis of γ,δ-unsaturated pyrazines, Podraza was able to use the Carroll reaction to thermally rearrange a number of 3,5,6-trimethyl-2-pyrazinylacetates with modest yields (10–46%) [29]. For example (Scheme 8.33), the pyrazinylacetate **96**, using optimized conditions, underwent a [3,3]-sigmatropic rearrangement to give the γ,δ-unsaturated pyrazine **97** in 46% yield. No reaction was observed in the rearrangement of substrates using the enolate version of the Carroll reaction (LDA, –78 to 65 °C). This lack of reaction was proposed to be due to a larger contribution from resonance form **98a** than from resonance form **98b**; thereby impeded the desired [3,3]-sigmatropic rearrangement from occurring (Scheme 8.34).

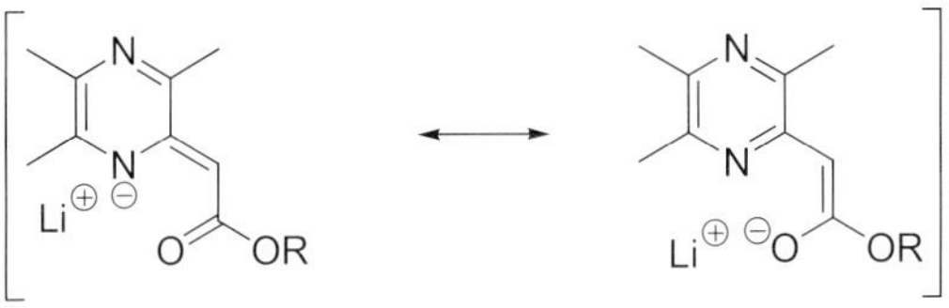

98a

98b

Scheme 8.34

8.4 Carroll Variants

With the synthetic potential of the Carroll rearrangement revealed, a number of new Carroll-variants have begun to emerge. These variants resemble the original reaction involving a [3,3]-sigmatropic rearrangement of an allylic ester group, but differ in the nature of the β-keto group. Building on the stereocontrolled nature of the Carroll rearrangement, these variants have been used to form complex molecules with new and interesting functionality.

8.4.1 α-Sulfonyl Carroll Rearrangement

Davidson and coworkers introduced a new Carroll-variant in which tertiary allylic β-sulfonyl esters underwent [3,3]-sigmatropic rearrangements to γ,δ-unsaturated sulfones [30]. As seen in Scheme 8.35, ambruticin, an antifungal agent, was envi-

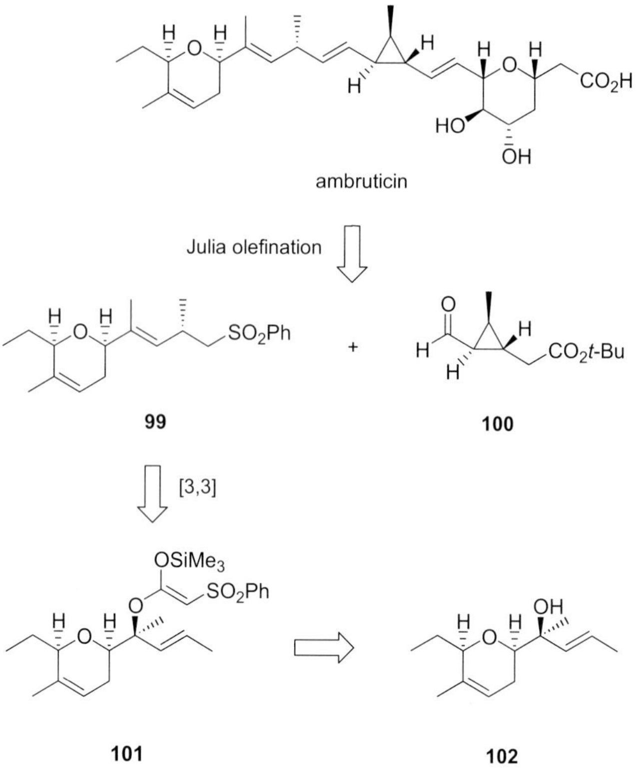
ambruticin
Julia olefination
SO2Ph
CO2H
CO2t-Bu
99
100
[3,3]
OSiMe3
101
102

Scheme 8.35

sioned to be formed via the Julia coupling between the γ,δ-unsaturated sulfone **99** and aldehyde **100**. This sulfone **99** could in turn be prepared from a Carroll-type rearrangement of the α-sulfonyl-silyl ketene acetal **101**, prepared from the tertiary alcohol **102**. To investigate this approach, Davidson and coworkers prepared a series of tertiary alcohols (**103–107**) and their respective sulfonyl esters (Scheme 8.36). The sulfonyl esters were treated with LDA and trimethylsilyl chloride (TMSCl)

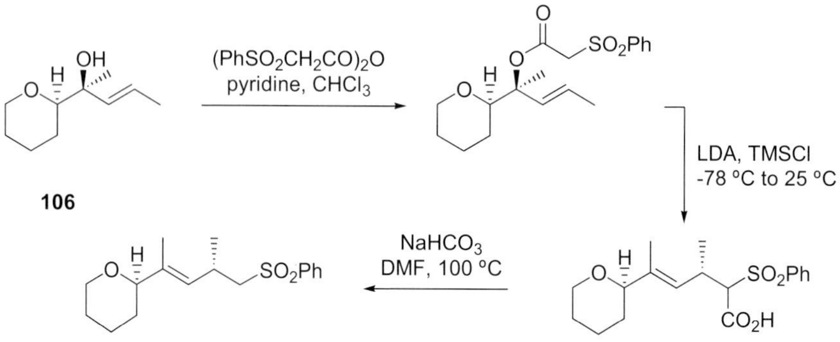
(PhSO2CH2CO)2O
pyridine, CHCl3
LDA, TMSCl
-78 °C to 25 °C
NaHCO3
DMF, 100 °C
106
SO2Ph
CO2H

Scheme 8.36

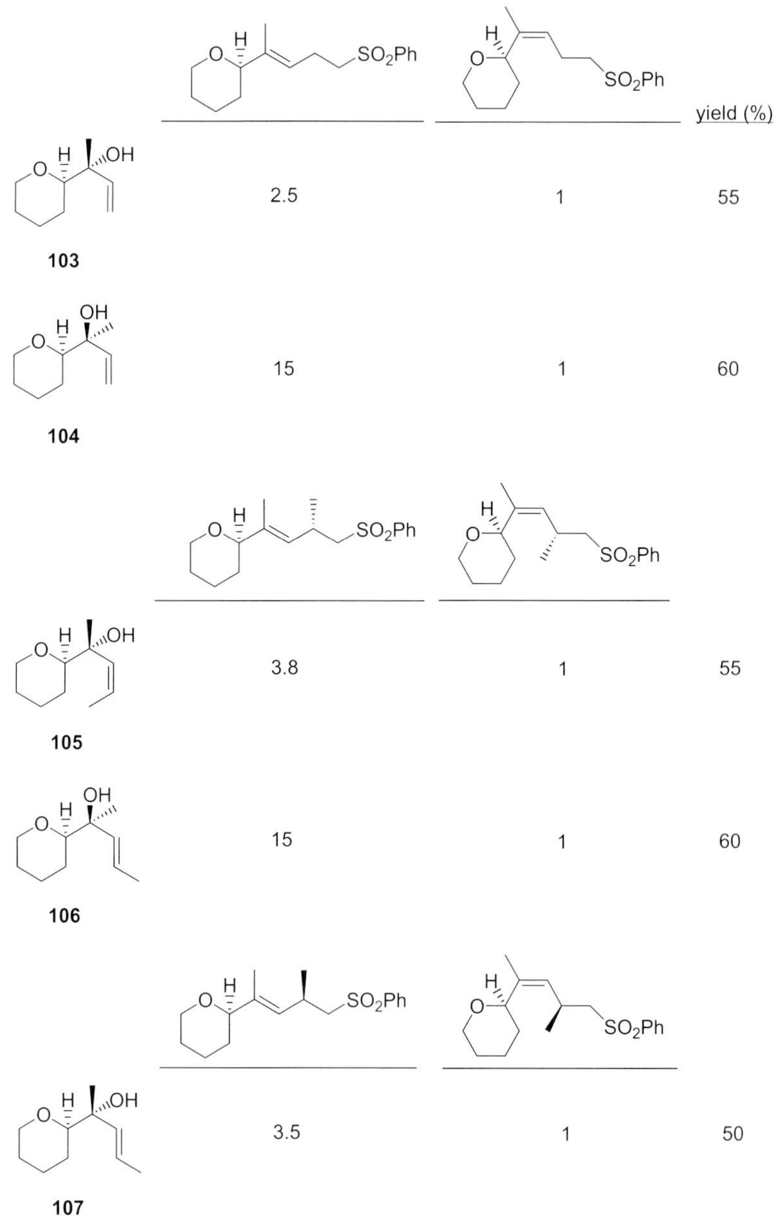

			yield (%)
103	2.5	1	55
104	15	1	60
105	3.8	1	55
106	15	1	60
107	3.5	1	50

SO$_2$Ph

Scheme 8.37

to form the silyl ketene acetals, which upon warming to room temperature underwent the Carroll rearrangement to give the α-sulfonyl carboxylic acids in high yields as mixtures of diastereomers. The carboxylic acids were then decarboxylated by reaction with sodium bicarbonate in dimethylformamide (DMF) at 100 °C. As seen in Schemes 8.36 and 8.37, the diastereomers **103**, **105**, and **107** gave poor *E*/*Z* ratios (2.5–3.8/1) whereas the diastereomers **104** and **106** gave excellent *E*/*Z* ratios (15/1). These results were in accordance with previous work [31], which showed that the stereochemical outcome of similar rearrangements was determined by the relative configuration of the adjacent chiral centers and not by the substitution or geometry of the double bonds.

Hatcher and Posner have also used an α-sulfonyl Carroll-variant in their synthesis of the vitamin D_3 side-chain unit **110** (Scheme 8.38) [26]. Allylic alcohol **71** was treated with (phenylthio)acetyl chloride to give the α-sulfide ester **108** in 95% yield that was then oxidized into the α-sulfonyl ester **109** in 76%. The α-sulfonyl ester **109** was treated with sodium hydride in refluxing xylenes promoting the rearrangement to give the desired 16-ene γ,δ-unsaturated phenylsulfone **110** in 82% yield. This side-chain unit was used to prepare the 16-ene-23-phenylsulfone vitamin D_3 analog **111** having significant *in vitro* antiproliferative activity [32].

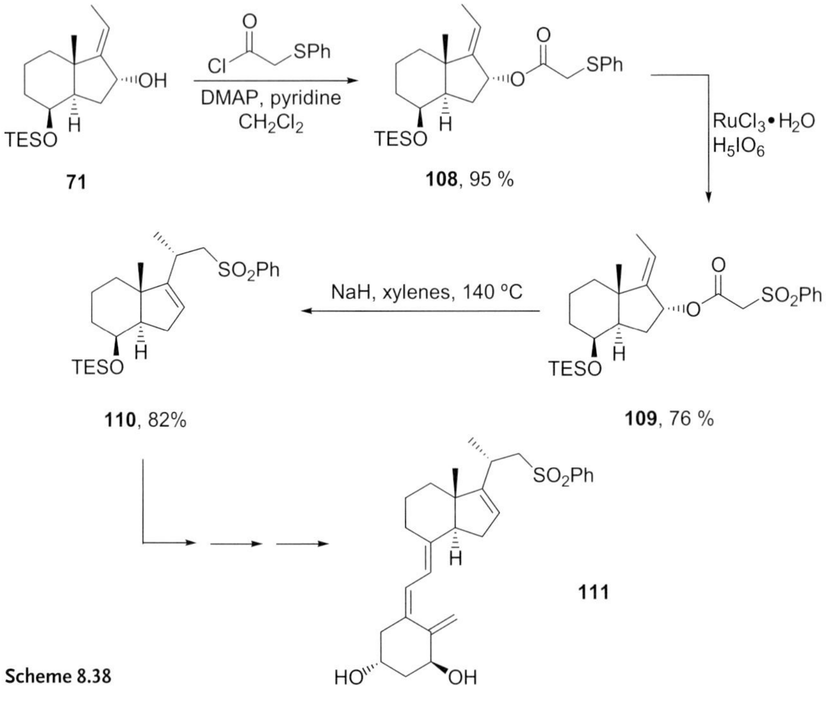

Scheme 8.38

8.4.2
Asymmetric Carroll Rearrangement

The Carroll rearrangement proceeds via a highly ordered chair-like transition state and therefore allows for a high degree of diastereoselectivity and asymmetric induction [33]. Recently, Enders and coworkers have developed an asymmetric dianionic version of the Carroll rearrangement by using the already successful SAMP/RAMP hydrazone chiral auxiliary methodology [33–36].

The replacement of a chiral hydrazone for the *β*-ketone moiety, forming *β*-hydrazono allylic esters, allowed for the efficient diastereo and enantioselective synthesis of functionalized ketones bearing vicinal quaternary and tertiary stereogenic centers (Scheme 8.39). The *β*-keto allylic esters **112a–n**, prepared by known procedures, were treated with (*S*)-1-amino-2-(methoxymethyl)pyrrolidine (SAMP) to form the respective hydrazones **113a–n** in good yields (83–99%) (Table 8.2). When hydrazones **113a–n** were treated with 2.6 eq. of LDA in THF/TMEDA (8:1) at −100 °C, the dianionic intermediate **114** was formed (Scheme 8.39), which upon warming to room temperature, smoothly underwent the Carroll rearrangement. The newly formed carboxylic acids were immediately reduced into the corresponding primary alcohols **115a–n** with $LiAlH_4$ in good yields over two steps (Table 8.2). As seen in Table 8.2, the dianionic rearrangement gave excellent diastereo- and enantioselectivities. The chiral auxiliary hydrazone was removed by chemoselective oxidative cleavage with ozone to give the *β*-hydroxy ketones **116a–n**. The relative and absolute configuration of the products was determined by X-ray crystallography. The stereochemical outcome of the rearrangement was in accordance with previous results and presumed to be due to the Li-chelation effect seen in intermediate **114**, which results in a sterically biased attack on the less hindered *Re*-face of the enolate double bond and a chair-like transition state during the Carroll rearrangement (Scheme 8.39).

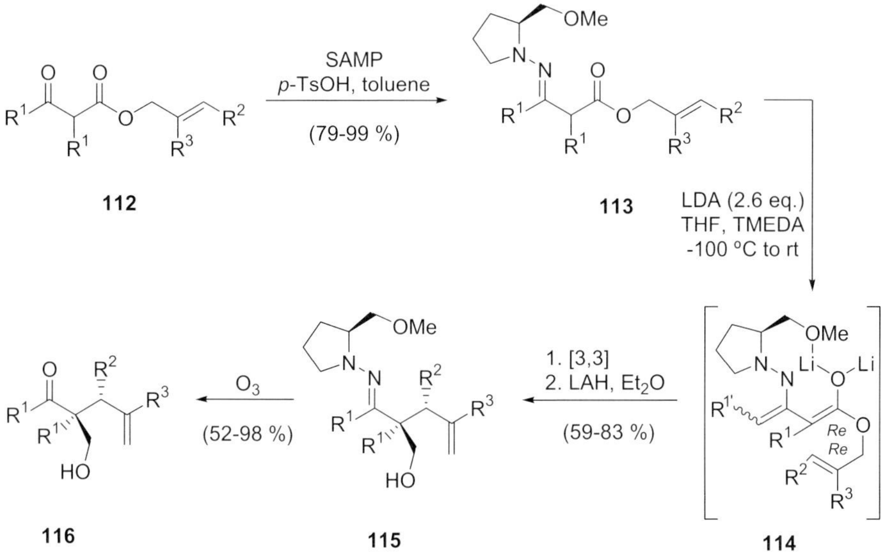

SAMP, p-TsOH, toluene (79-99 %); LDA (2.6 eq.), THF, TMEDA, -100 °C to rt; 1. [3,3] 2. LAH, Et2O (59-83 %); O3 (52-98 %)

112 113 114 115 116

Scheme 8.39

Table 8.2 Asymmetric dianionic (SAMP/RAMP) β-hydrazono allylic ester Carroll rearrangements.

113	R^1	R^2	R^3	116	Yield (%)	*de* (%)	*ee* (%)
a	$(CH_2)_3$	*n*-Pr	H	a	84	89	86
b	$(CH_2)_3$	*i*-Pr	H	b	89	96	94
c	$(CH_2)_3$	*n*-Bu	H	c	91	91	96
d	$(CH_2)_3$	*i*-Bu	H	d	96	94	>98
e	$(CH_2)_3$	*i*-Pe	H	e	93	93	>96
f	$(CH_2)_3$	*c*-Hex	H	f	86	91	82
g	$(CH_2)_3$	$(CH_2)_4$		g	95	89	>98
h	$(CH_2)_4$	*n*-Pr	H	h	92	88	94
I	$(CH_2)_4$	*i*-Pr	H	I	98	88	98
j	$(CH_2)_4$	*n*-Bu	H	j	>99	88	98
k	$(CH_2)_4$	*i*-Bu	H	k	52	90	96
l	$(CH_2)_4$	*i*-Pe	H	l	57	>98	>96
m	$(CH_2)_4$	*c*-Hex	H	m	89	93	60
n	Me	$(CH_2)_4$		n	90	90	74

Using a Lewis acid, Carroll rearrangement of these β-hydrazo allylic esters was also accomplished (Scheme 8.40). When hydrazones **113a–f** were treated with 1.3 eq. of *t*-butyldimethylsilyl triflate (TBSOTf) in the presence of base, the silyl ketene acetals **117a–f** were obtained, which underwent a Carroll rearrangement when warmed to room temperature (Table 8.3, Scheme 8.40). The silyl esters were subsequently reduced to the β-hydroxy hydrazones **118a–f** as a mixture of the four possible diastereomers. The desired β-hydroxy ketones **119a–f** were obtained by reaction with ozone. As seen in Table 8.3, the diastereo and enantiomeric ratios were much lower in comparison to the dianionic Carroll rearrangements. The diminished diastereomeric ratios were rationalized by the possible formation of both the *E* or *Z* silyl ketene acetal or possibly by a rearrangement through a boat-like transition state. It is noteworthy that the major diastereomer obtained from the Lewis acid mediated rearrangement was the minor diasteromer obtained in the anionic rearrangement.

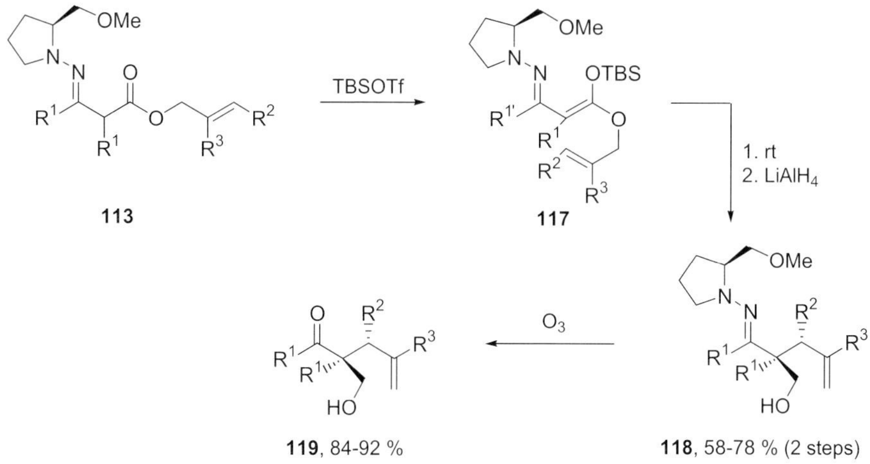

Scheme 8.40

Table 8.3 Lewis acid-assisted asymmetric (SAMP/RAMP) β-hydrazono allylic ester Carroll rearrangements.

4	R^1	R^2	R^3	Overall Yield (%)	*de* (%)	*ee* (%)
a	$(CH_2)_3$	*n*-Pr	H	84	19	57
c	$(CH_2)_3$	*n*-Bu	H	89	20	57
d	$(CH_2)_3$	*i*-Bu	H	90	16	60
I	$(CH_2)_4$	*i*-Pr	H	90	25	71
j	$(CH_2)_4$	*n*-Bu	H	86	5	67
k	$(CH_2)_4$	*i*-Bu	H	87	8	77
l	$(CH_2)_4$	*i*-Pe	H	87	41	76

Enders' Carroll rearrangement methodology was applied to the asymmetric synthesis of (–)-malyngolide **120** (Scheme 8.41) [36].

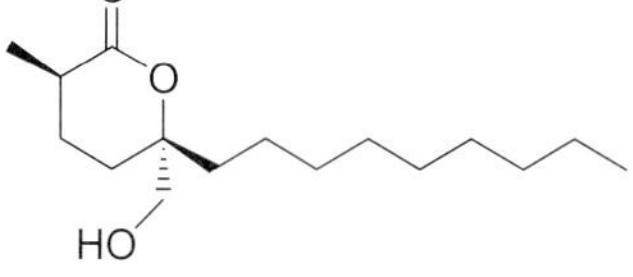

Scheme 8.41

120, (–)-(2*R*,5*S*)-malyngolide

Allyloxycyclopentanone **121** was treated with RAMP to afford the *β*-hydrazano ester **122**, which was then treated under the standard rearrangement conditions to give the *β*-hydroxy hydrazone **123** in 96% yield and greater than 96% *de* (Scheme 8.42). The hydrazone chiral auxiliary was removed by treatment with ozone and a number of synthetic transformations gave *β*-hydroxy ketone **124**. The final step was a Baeyer–Villiger reaction that proceeded with complete retention of absolute configuration to give the desired (–)-malyngolide-**120** in greater than 96% *ee*.

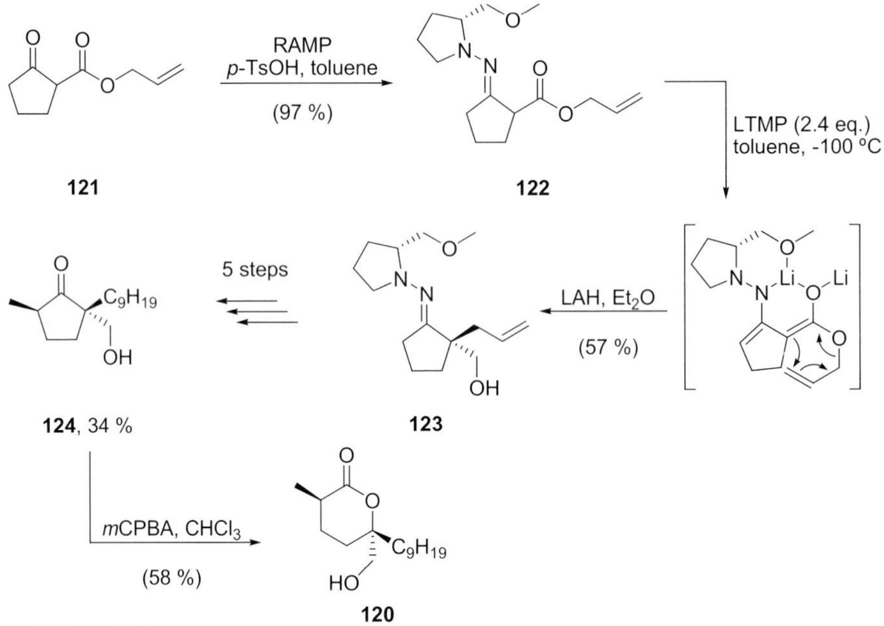

RAMP, *p*-TsOH, toluene (97 %); LTMP (2.4 eq.), toluene, -100 °C; LAH, Et_2O (57 %); 5 steps; *m*CPBA, $CHCl_3$ (58 %)

121, **122**, **123**, **124**, 34 %, **120**

Scheme 8.42

8.4.3
Metal-Catalyzed Carroll Rearrangement

A synthetic equivalent of the Carroll rearrangement, accomplished by Tsuda and coworkers, is represented by the Pd-catalyzed reaction shown in Scheme 8.43 [37]. The allyl cyclohexanone-2-carboxylate (**125**) was reacted in the presence of catalytic amounts of $Pd(PPh_3)_4$ at room temperature, which rapidly evolved CO_2 forming the allylpalladium(II)-enolate complex **126**. This Pd-enolate complex then recombined to produce the 2-allylcyclohexanone (**127**) in 88% yield.

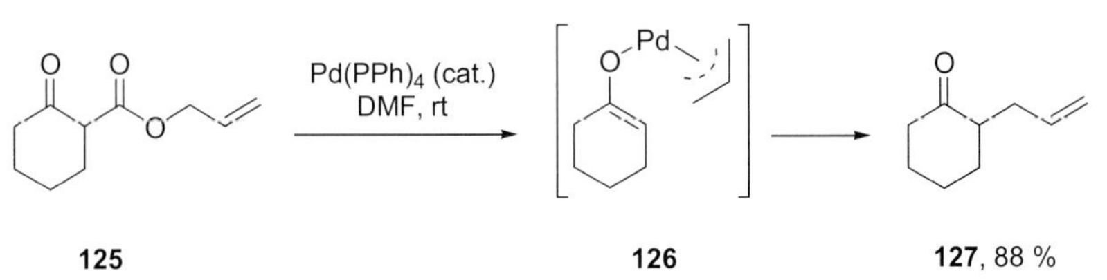

$Pd(PPh)_4$ (cat.), DMF, rt

125, **126**, **127**, 88 %

Scheme 8.43

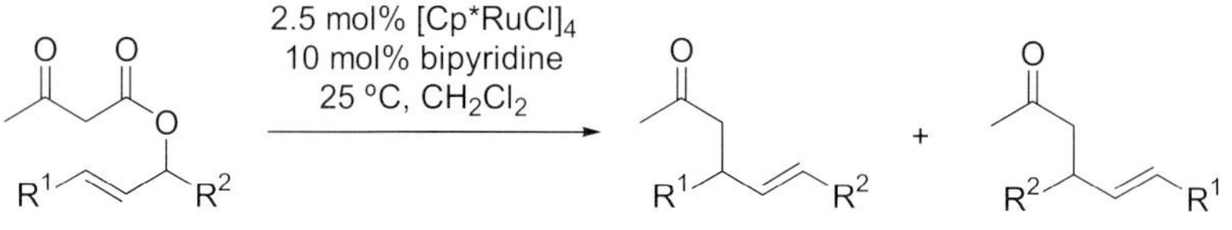

2.5 mol% $[Cp^*RuCl]_4$
10 mol% bipyridine
25 °C, CH_2Cl_2

128 → **129**, [3,3] + **130**, [1,3]

Substrate	R^1	R^2	t	Yield of **129** (%)
128a	Ph	H	1.5 h	94
128b	*p*-tolyl	H	2 h	96
128c	*o*-tolyl	H	5 d	81
128d	*p*-C_6H_4OMe	H	15 min	91
128e	*o*-C_6H_4OMe	H	15 min	93
128f	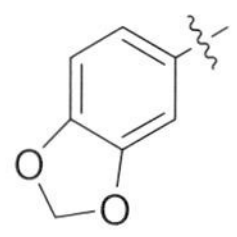 (3,4-methylenedioxyphenyl)	H	30 min	91
128g	*p*-C_6H_4Cl	H	4 h	96
128h	*p*-$C_6H_4CF_3$	H	40 h	90
128i	H	Ph	1.5 h	93 (**130**)
128j	Ph	Ph	1 h	67

Scheme 8.44

Using a similar approach, Burger and Tunge accomplished a number of formal [3,3]-sigmatropic Carroll rearrangements with catalytic amounts of ruthenium-bypridine complex [38]. As seen in Scheme 8.44, a number of substituted cinnamyl β-ketoesters (**128a–h** and **128j**) were reacted with 2.5 mol% of $[Cp^*RuCl]_4$ and 10 mol% of bypyridine at room temperature in dichloromethane to give the desired [3,3]-rearranged products in excellent yields (67–96%). Noteworthy is the isomeric molecule **128i**, which gave the formal [1,3]-rearranged product in 93%. This experiment suggested that these reactions likely involve C–O metal insertion to give a ruthenium-π-allyl intermediate and, with loss of CO_2, a ketone enolate that selectively reacts at the more substituted allyl terminus (Scheme 8.45). This proposed mechanism was further supported by the crossover experiment shown in Scheme 8.46. When equal quantities of β-keto esters **131** and **132** were reacted under the standard conditions, approximately equal amounts of desired products **133** and **134** and crossover products **135** and **136** were observed, giving evidence for a freely formed ketone enolate.

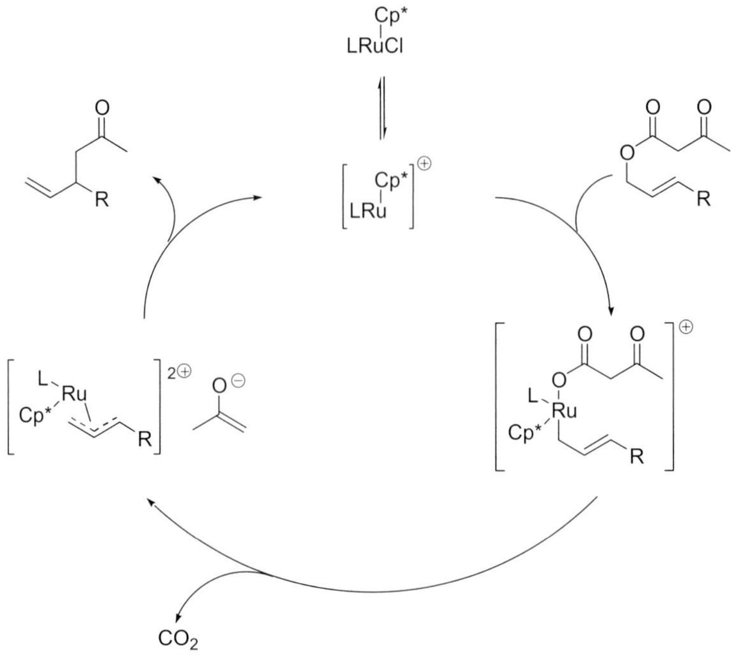

Scheme 8.45

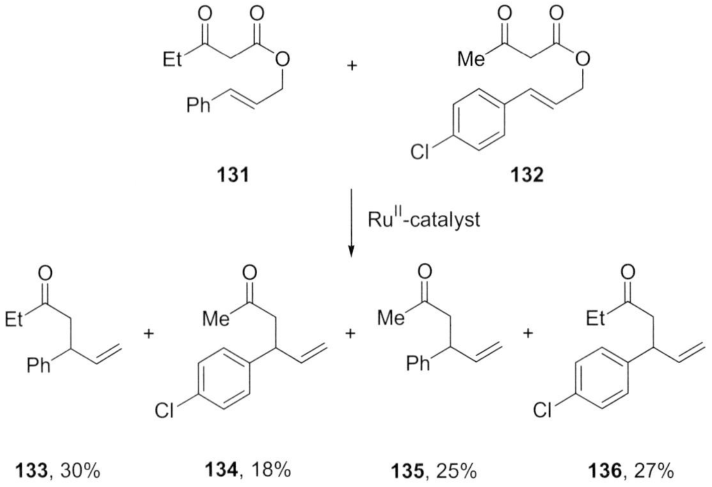

Scheme 8.46

8.5 Conclusion

The Carroll rearrangement has been a reliable and efficient reaction over the past 60 years. The chair-like transition state of this [3,3]-sigmatropic rearrangement has proved to be an excellent means of constructing carbon–carbon bonds in a stereocontrolled manner while giving access to molecules difficult to synthesize by other means. The ability to transfer chirality, be it from chiral-pool material or from the installation of a chiral auxiliary, through the concerted nature of the rearrangement has shown that this reaction can be an important asset in the asymmetric synthesis of chiral molecules.

References

1 Carroll, M.F. *J. Chem. Soc.* **1940**, 704–706.
2 Carroll, M.F. *J. Chem. Soc.* **1940**, 1266–1268.
3 Carroll, M.F. *J. Chem. Soc.* **1941**, 507–511.
4 Kimel, W.; Cope, A.C. *J. Am. Chem. Soc.* **1943**, *65*, 1992–1998.
5 Hill, R.K.; Synerholm, M.E. *J. Org. Chem.* **1968**, *33*, 925–927.
6 Wilson, S.R.; Price, M.F. *J. Org. Chem.* **1984**, *49*, 722–725.
7 Burrows, C.J.; Carpenter, B.K. *J. Am. Chem. Soc.* **1981**, *103*, 6983–6984.
8 Ireland, R.E.; Mueller, R.H.; Willard, A.K. *J. Am. Chem. Soc.* **1976**, *98*, 2868–2877.
9 Genus, J.F.; Peters, D.D.; Ding, J-f.; Bryson, T.A. *Synlett* **1994**, 209–210.
10 Meldrum, A.N. *J. Chem. Soc.* **1908**, *93*, 598.
11 Davidson, D.; Bernhard, S.A. *J. Am. Chem. Soc.* **1948**, *70*, 3426–3428.
12 Oikawa, Y.; Sugano, K.; Yonemitsu, O. *J. Org. Chem.* **1978**, *43*, 2087–2088.
13 Genus, J.F.; Peters, D.D.; Bryson, T.A. *Synlett* **1993**, 759–760.
14 Gilbert, J.C.; Kelly, T.A. *Tetrahedron* **1988**, *44*, 7587–7600.
15 Torkelson, S.; Ainsworth, C. *Synthesis* **1976**, 722–724
16 Ouvrard, N.; Rodriguez, J.; Santelli, M. *Tetrahedron Lett.* **1993**, *34*, 1149–1150.
17 Martin, S.F.; Guinn, D.E. *Synthesis* **1991**, 245–262.
18 Koreeda, M.; Brown, L. *J. Org. Chem.* **1983**, *48*, 2122–2124.
19 Snider, B.B.; Beal, R.B. *J. Org. Chem.* **1988**, *53*, 4508–4515.
20 Wenkert, E.; Arrhenius, T.S. *J. Am. Chem. Soc.* **1983**, *105*, 2030–2033.
21 Isler, O.; Huber, W.; Ronco, A.; Kofler, M. *Helv. Chem. Acta.* **1947**, *30*, 1911–1947
22 Kimel, W.; Sax, N.W.; Kaiser, S.; Eichmann, G.G.; Chase, G.O.; Ofner, A. *J. Org. Chem.* **1958**, *23*, 153–157.
23 Echavarren, A.M.; Mendosa, J.; Prados, P.; Zapata, A. *Tetrahedron Lett.* **1991**, *32*, 6421–6424.
24 Tanabe, M.; Hayashi, K. *J. Am. Chem. Soc.* **1980**, *102*, 862–863.
25 Wharton, P.S.; Bohlen, D.H. *J. Org. Chem.* **1961**, *26*, 3615–3616.
26 Hatcher, M.A.; Posner, G.H. *Tetrahedron Lett.* **2002**, *43*, 5009–5012.
27 Sorgi, K.L.; Scott, L.; Maryanoff, C.A. *Tetrahedron Lett.* **1995**, *36*, 3597–3600.
28 Coates, R.M.; Said, I.Md. *J. Am. Chem. Soc.* **1977**, *99*, 2355–2357.
29 Podraza, K.F. *J. Heterocyclic Chem.* **1986**, *23*, 581–583.
30 Davidson, A.H.; Eggleton, N.; Wallace, I.H. *J. Chem. Soc. Chem. Commun.* **1991**, 378–380.
31 Davidson, A.H.; Wallace, I.H. *J. Chem. Soc. Chem. Commun.* **1986**, 1759–1760.
32 Posner, G.H.; Crawford, K.R.; Yang, H.W.; Kahraman, M.; Jeon, H.B.; Li, H.; Lee, J.K.; Suh, B.C.; Hatcher, M.A.; Labonte, T.; Usera, A.; Dolan, P.M.; Kensler, T.W.; Peleg, S.; Jones, G.;

Zhang, A.; Korczak, B.; Saha, U.; Chuang, S.S. *J. Steroid. Biochem. Mol. Biol.* **2004**, *89–90*, 5–12.

33 Enders, D.; Knopp, M.; Runsink, J.; Raabe, G. *Liebigs Ann.* **1996**, 1095–1116.

34 Enders, D.; Knopp, M.; Runsink, J.; Raabe, G. *Angew. Chem., Int. Ed. Engl.* **1995**, *34*, 2278–2280.

35 Enders, D.; Knopp, M.; Runsink, J.; Raabe, G. *Angew. Chem.* **1995**, *107*, 2442.

36 Enders, D.; Knopp, M. *Tetrahedron* **1996**, *52*, 5805–5818.

37 Tsuda, T.; Yoshiki, C.; Nishi, S.; Tawara, K.; Saegusa, T. *J. Am. Chem. Soc.* **1980**, *102*, 6381–6384.

38 Burger, E.C.; Tunge, J.A. *Org. Letters* **2004**, *6*, 2603–2605.

9
Thio-Claisen Rearrangement

Stéphane Perrio, Vincent Reboul, Carole Alayrac, and Patrick Metzner

9.1
Introduction

The replacement of the oxygen atom by sulfur in a Claisen rearrangement substrate brings significant differences to the process. Besides giving access to thiocarbonyl compounds, it affords a kinetically facile transposition, with a changed thermodynamic pattern (Scheme 9.1).

Scheme 9.1

The development in the sulfur series has gone through roughly four stages from the 1960s until present:

1. Exploration of the aromatic version, with thermolysis of allyl aryl sulfides and isolation of sulfur heterocycles.
2. Introduction of the aliphatic version and easy synthesis of unsaturated thiocarbonyl compounds.
3. Investigation of the scope of the reaction with a variety of sulfides as substrates.
4. Relative and absolute stereocontrol.

9.1.1
Early Developments

9.1.1.1 Aromatic and Heteroaromatic Series

The first report found in this context is a contribution by Hurd and Greengard [1]. In 1930, they attempted the pyrolysis of allyl aryl sulfides (at 207–240 °C), in analogy to reported examples with ethers. They claimed the formation of 2-allyl thio-

The Claisen Rearrangement. Edited by M. Hiersemann and U. Nubbemeyer

ISBN: 978-3-527-30825-5

phenol, but it was disputed by Karaulova and coworkers [2, 3] in 1957, who proposed that under such conditions, a cyclization reaction took place to produce a five-membered ring sulfide. An isomeric thioether, a six-membered ring, was subsequently isolated by Meyers et al. [4].

In the 1960s, Kwart and his group studied this reaction extensively [5]. They demonstrated that the reaction course is largely dependent on the experimental conditions. The thermolysis of allyl phenyl sulfide, in the presence of a high boiling amine [6], gave optimum yields of five- and six-membered heterocycles (Scheme 9.2).

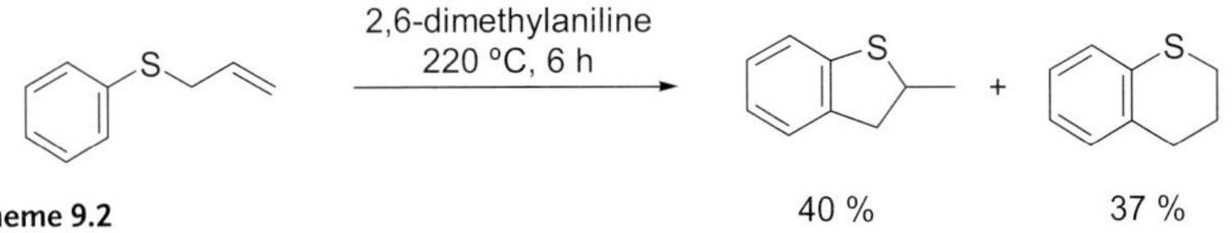

2,6-dimethylaniline
220 °C, 6 h

40 %   37 %

Scheme 9.2

They established that 2-allylthiophenol was a common intermediate [7]. The mechanism involves an initial [3,3] sigmatropic shift, prototropy of the thiocarbonyl compound with aromatization, and cyclization in 5-exo-trig or 6-endo-trig fashion (Scheme 9.3).

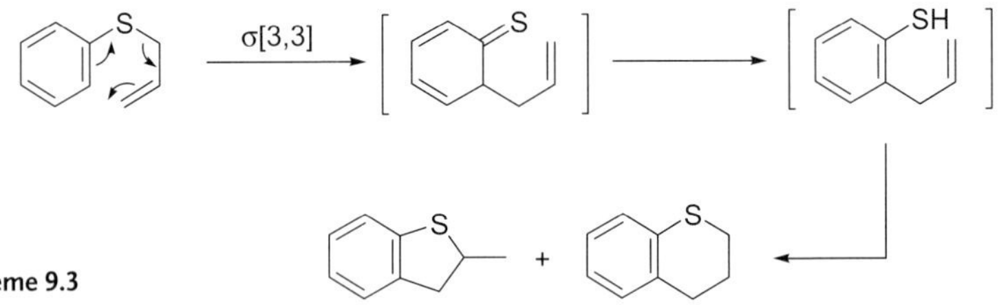

σ[3,3]

Scheme 9.3

The numerous papers by Kwart report on various aspects, including nucleophilic catalysis and mechanism [6, 8–10], evidence for the intermediate [11], and scope [7, 12–14].

Makisumi and coworkers [15] investigated the thermal behavior of allyl 4-quinolyl sulfides (Scheme 9.4). After heating at 200 °C, 2-methyl-2,3-dihydrothieno [3,2-c]quinoline was obtained in 73% yield (Scheme 9.4). This efficient heterocycle synthesis was extended to precursors bearing a substituted allyl or propargyl chain [16]. If an amine was used as solvent, a mixture of five- and six-membered ring products was obtained [17, 18].

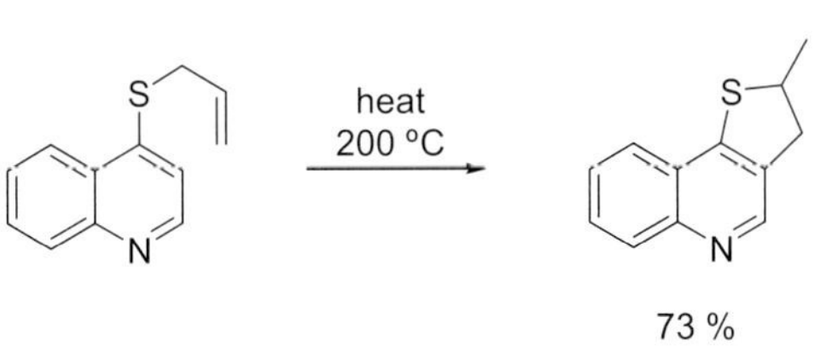

heat
200 °C

73 %

Scheme 9.4

At this stage, no specificity of the sulfur rearrangement had yet been disclosed, except the easy cyclization of allyl thiophenols, the postulated intermediates.

9.1.1.2 Aliphatic Series

The first isolation of the primary product from the transposition was effected in the aliphatic series and launched many developments. A pioneer contribution came from the group of Brandsma in the Netherlands, which started with a paper in 1968 [19]. The substrates were prepared by deprotonation of thionesters, dithioesters and thioamides with potassium amide and reaction with a variety of allyl bromides (Scheme 9.5). In some cases the initial allyl ketenethioacetal was observed and it was rearranged upon purification by distillation (b.p. range 66–152 °C). In most cases, the authors noted that the rearrangement had already taken place during work up at ambient temperature.

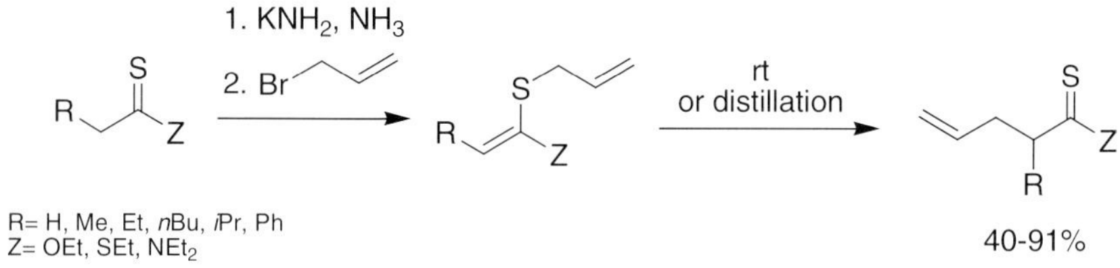

1. KNH_2, NH_3
2. Br
rt or distillation
R= H, Me, Et, *n*Bu, *i*Pr, Ph
Z= OEt, SEt, NEt_2
40-91%

Scheme 9.5

It appears that the first evidence for thio-Claisen transposition takes place at much lower temperature than with the oxygen analogue. The work also revealed that this route was an efficient path to aliphatic thiocarbonyl compounds, which then had a reputation of instability under a variety of conditions. One has to note that these results were met with substrates bearing a heteroatomic substituent: Z = alkoxy, alkylthio or dialkylamino. Replacement by other groups was attempted. A hydrogen substituent could not be used for some time. Lawesson extended the results of Brandsma [20–22] and used a vinyl sulfide derived from dimedone. Rearrangement was performed at 190–210 °C leading to a tautomeric enethiol, which could be trapped as a thiolester by reaction with acetic anhydride. Plieninger [23] observed a comparable example in the indole series by achieving allylation α to the initial thiol group. In another series, an α-thiouracil was obtained [24] as a result of the preferred stability of the thiocarbonyl form over the enethiol tautomer.

9.1.2 Specificities of the Sulfur Version – Kinetics Versus Thermodynamics

The remarkable feature that emerged from the 1970s reports was that the thio-Claisen rearrangement is more facile in kinetic terms than the oxygen analogue. Two studies gave access to activation parameters. The group of Brandsma [25] and our group [26, 27] have shown that the activation enthalpies of two basic examples, the dithioester and thioketone series, are close to 20 kcal/mol (Scheme 9.6). These values are 5–10 kcal/mol lower than observed in the oxygen series. This may be largely related to the facile carbon–sulfur breaking step. Indeed the energy of the C–S bond, approximately 65 kcal/mol, is lower than the C–O bond, 85 kcal/mol. *Ab initio* calculations have been performed [28] to compare the various versions and have shown that an element of the second-row of the periodical table is more favorable kinetically for this reaction. DFT calculations [29] were also carried out.

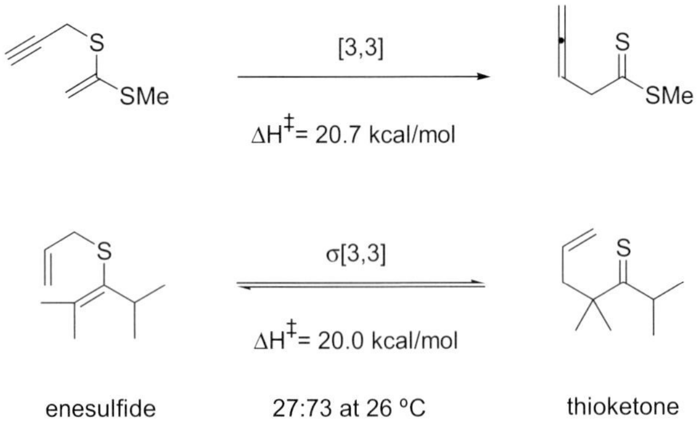

S SMe [3,3] $\Delta H^{\ddagger}$= 20.7 kcal/mol S SMe

S σ[3,3] $\Delta H^{\ddagger}$= 20.0 kcal/mol S

enesulfide 27:73 at 26 °C thioketone

Scheme 9.6

In thermodynamic terms, the situation is opposite. In principle, any sigmatropic shift should be a reversible process. However the Claisen rearrangement "equilibrium" is largely shifted towards the formation of the carbonyl compound. This is related to the strong stabilization of the carbonyl moiety versus the enol ether of the starting material. In the sulfur series, the situation largely depends on the substrates with a finely balanced situation. Our group [26] has shown that simple aliphatic allyl vinyl sulfides are in equilibrium with γ-unsaturated thioketones. In the example depicted in Scheme 9.6, a 27:73 mixture was observed at 26 °C. The enthalpy variation $\Delta H = -4.3$ kcal/mol, is much smaller than the one in the oxygen series, calculated as −19 kcal/mol. *Ab initio* calculations [28] confirmed that the sulfur version is less favorable than the oxygen series. The 2p–3p orbital overlap of carbon and sulfur atoms (different rows) is rather poor and does not provide a strong stabilization as for the carbonyl moiety (same row elements).

Other factors may tune the position of the equilibrium, which is nicely illustrated in the thioketone series by two extreme examples. The ring strain of a [2.2.1] bicyclic system (Scheme 9.7) drives the equilibrium completely towards the formation of 2-allyl thiocamphor (unpublished). In contrast, introduction of a phenyl group provides a better conjugation with the C=C bond of the starting vinyl sulfide than with the thiocarbonyl of the Claisen product, and shifts the equilibrium to the left.

Other examples of retro-Claisen rearrangement were observed in the aromatic quinoline series [30] and with bicyclic systems [31, 32]. Some specific cases have been predicted by calculations to be endothermic [33, 34].

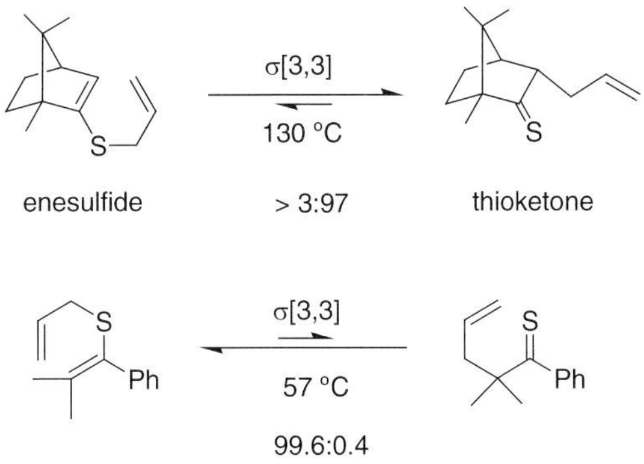

σ[3,3]
130 °C
enesulfide > 3:97 thioketone

σ[3,3]
57 °C
99.6:0.4

Scheme 9.7

On the other hand, the rearrangement leading to thiocarbonyl compounds, such as dithioesters, thionesters or thioamides, instead of thioketones, has been shown to be driven to the right side. The only exception was observed when the initial unsaturated sulfide exhibits an aromatic nature [35].

Together with the kinetic facility of the transposition, this process is productive in most cases and thus exploitable in synthesis.

9.1.3 Reviews

A limited number of accounts of the literature have appeared: Brandsma [36], Kwart [5], Paquer [37], Anisimov [38], our contribution in the context of thiocarbonyl compounds in synthesis [39] and a recent review by Majumdar [40] with an emphasis on the synthesis of sulfur heterocycles.

9.2 Basic Versions

9.2.1 Flexible Synthesis of the Substrates

The substrates for the thio-Claisen transposition may be prepared by a variety of routes. Interestingly, some of them are not readily available in the oxygen series.

The common method involves deprotonation of a thiocarbonyl compound and reaction of the intermediate enethiolate with an allyl halide (Scheme 9.8). This actually relies on two noticeable features of the sulfur series. (1) The proton located α to a thiocarbonyl group is much more acidic, by 7–10 pKa units, than the one of a carbonyl moiety [39, 41]. This may be related to the strong ability of the sulfur atom (polarizability) to stabilize the negative charge of the enethiolate. Presently, the preferred conditions involve LDA as a base for optimum deprotonation [42–45]. (2) The resulting anionic species are soft ambident nucleophiles. The regioselectivity of their reactions depends upon the nature of the opposed electrophile. Allyl halides, which exhibit a relatively soft acid character, react by the sulfur terminus of the nucleophile, to afford allyl vinyl sulfides. Similar results were obtained with allyl alcohols [46] and tosylates [47].

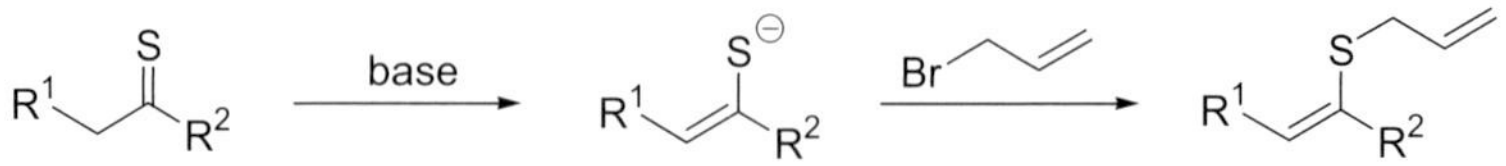

Scheme 9.8

It must be stressed that in the sulfur series, the reaction of crotyl halides takes place regioselectively, leading to *S*-crotyl derivatives (examples [42, 47]). The oxygen series stand in contrast, with reactions of enolates proceeding on the carbon atom, and not on oxygen. Furthermore, it is not regioselective with crotyl-type halides.

A second significant difference deals with the stereochemistry of the deprotonation [41]. The reaction of thiocarbonyl compounds leads to *cis* enethiolates (SLi and R^1 groups are located on the same side of the C=C bond) predominantly with dithioesters [48] or exclusively with thioamides [47, 49] and thioketones [50]. Under similar (kinetic) conditions, analogous esters or ketones lead to the *trans* enolates.

Thioamides exhibit a nucleophilic character that allows a reversed order of the two deprotonation and alkylation steps (Scheme 9.9). Addition of allyl halides to thioamides provided *S*-allyl thioimidate salts. Subsequent deprotonation by a variety of bases (notably *t*-BuOK or simple tertiary amines) led to transient keteneaminothiocetals, which underwent *in situ* rearrangement to the awaited γ-unsaturated thioamides [51–54].

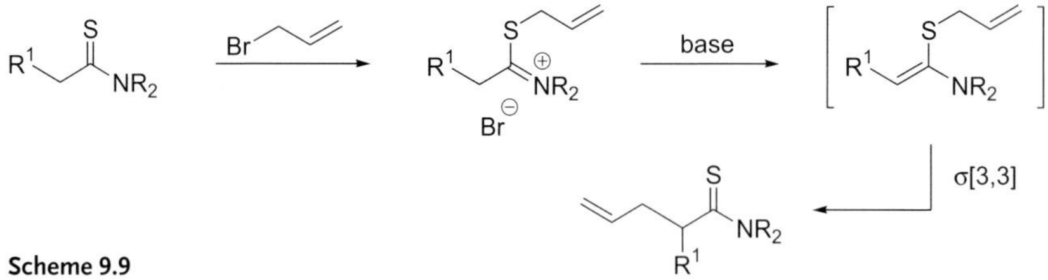

Scheme 9.9

Some other interesting routes to unsaturated sulfides have also been achieved:

- Intramolecular addition of a carbene to an allyl sulfide [55].
- Deprotonation of allyl vinyl sulfide in the position α to the sulfur atom and alkylation of the intermediate anion to provide a branched allyl moiety [56, 57], normally not accessible by direct allylation of an enethiolate species.
- In analogy to the Ireland version of the oxygen Claisen reaction, direct transposition of the enethiolate generated by deprotonation of allyl dithioesters [58].

9.2.2
Scope and Limitations, Reaction Conditions

9.2.2.1 Synthesis of Unsaturated Aldehydes (via Transient Thioaldehydes)

Thioaldehydes are highly reactive compounds, which cannot be isolated under classical conditions. Therefore, any Claisen transposition, that would lead to a thioaldehyde, can be efficient only if the expected product is converted *in situ* in an isolable derivative. As an example, thermolysis of vinyl sulfides in the presence either of mercuric oxide [59] or dimethyl sulfoxide and water [60] led to γ-unsaturated aldehydes (Scheme 9.10).

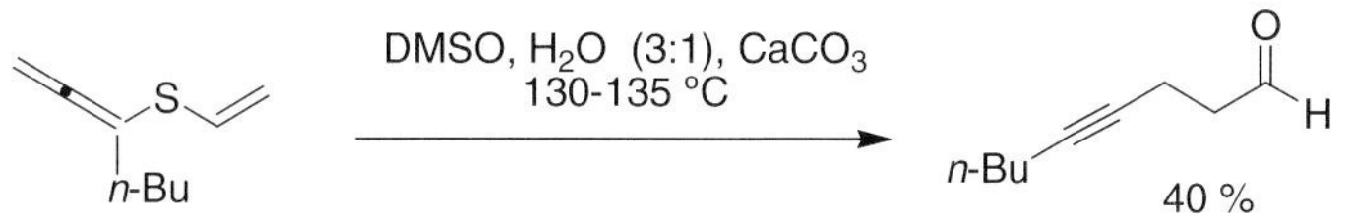

DMSO, H_2O (3:1), $CaCO_3$, 130-135 °C; *n*-Bu; 40 %

Scheme 9.10

9.2.2.2 Thioketones

As compared to thioaldehydes, these compounds are somewhat easier to isolate, their reactivity in solution or towards the oxygen in air being more moderate. The first evidence is due to Ojima and Kondo [55], who studied a simple allyl vinyl sulfide rearrangement (Scheme 9.11) at ambient temperature. By NMR, they detected an unsaturated thioketone, which was stable under a nitrogen atmosphere, but further converted to the corresponding ketone just by standing in air.

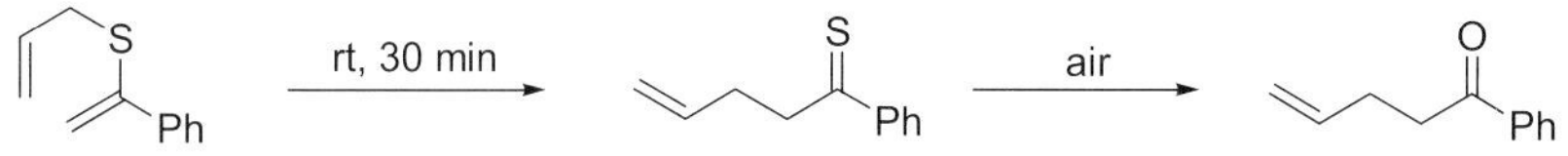

rt, 30 min; air; S; Ph; O

Scheme 9.11

A number of unsaturated thioketones, bearing only aliphatic substituents, could be isolated [61, 62]. When the products possess protons on the α carbons, they were accompanied by the tautomeric enethiols [62]. When this form was favored, some authors carried out the [3,3] shift in acetic anhydride as a solvent and isolated the corresponding thiolester [63, 64]. Recently, the rearrangement of tungsten(pentacarbonyl) complexes of sulfides was reported to lead to the corresponding thiocarbonyl derivatives [65].

9.2.2.3 Dithioesters

As the introduction of a heteroelement substituent on the thiocarbonyl provides enhanced stability, compounds as dithioesters have become popular in organic synthesis and a number of reports have dealt with their interest in the field of this review.

After their first report (see Scheme 9.5), Brandsma's group demonstrated that the reaction could be extended to *S*-allenyl [66] or *S*-propargyl [25] ketenedithioacetals (Scheme 9.12).

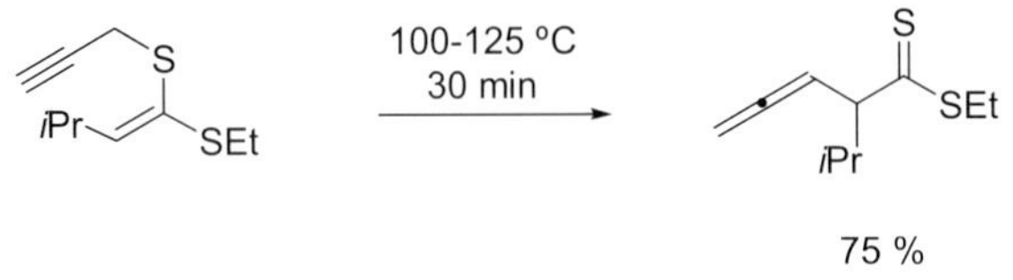

100-125 °C
30 min

*i*Pr SEt S → S SEt *i*Pr

75 %

Scheme 9.12

A variety of unsaturated dithioesters have been prepared in an analogous fashion (Fig. 9.1), incorporating *E*-trisubstituted double bonds [67], an additional double bond [68], carbonyl [68–70], hydroxyl [44], and sulfinyl groups [71, 72]. Unsaturated thionoselenoesters are also accessible [73].

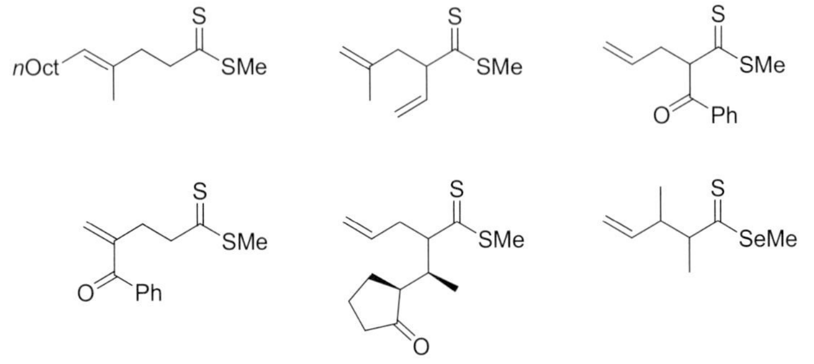

*n*Oct S SMe; S SMe; S SMe O Ph; S SMe O Ph; S SMe O; S SeMe

Fig. 9.1 Unsaturated dithioesters.

9.2.2.4 **Thionesters**

Though they are less popular, thionesters can serve as starting materials for ketenethioacetals, which, despite their sensitivity towards water, could be rearranged [74, 75]. A synthetically useful report [76] includes access to 2-allyl thionolactones (Scheme 9.13).

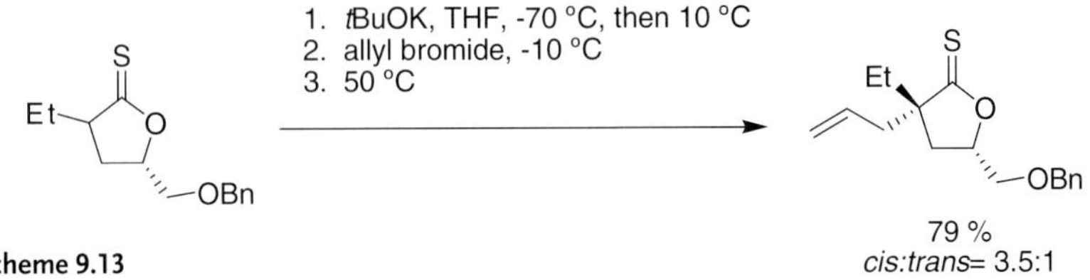

1. *t*BuOK, THF, -70 °C, then 10 °C
2. allyl bromide, -10 °C
3. 50 °C

Et S O OBn → Et S O OBn

79 %
cis:trans= 3.5:1

Scheme 9.13

9.2.2.5 **Thioamides**

This class is the most common of thiocarbonyl compounds, in respect to a nice stability. Many groups have utilized them for the Claisen transposition from the 1960s initial report [19]. A major contribution by Tamaru and Yoshida involves tertiary thioamides [47, 77–79], especially for efficient acyclic stereocontrol (see Section 9.3.1).

Secondary thioamides [52] can be used by a sequence of double deprotonation, treatment with allyl bromides and rearrangement simply at room temperature (Scheme 9.14).

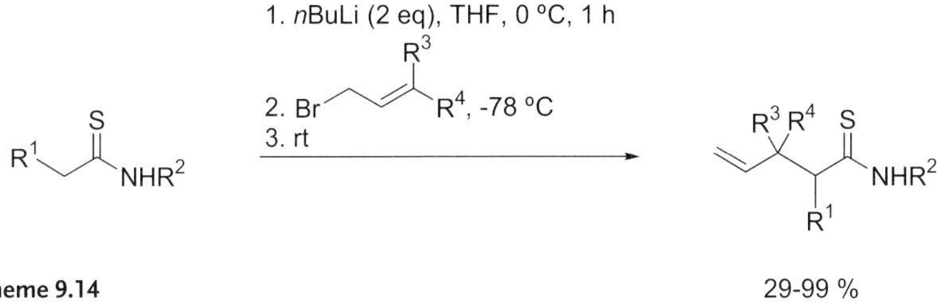

1. *n*BuLi (2 eq), THF, 0 °C, 1 h
2. Br–CH₂CH=C(R^3)R^4, -78 °C
3. rt

29-99 %

Scheme 9.14

An alternative sequence was the addition of allyl bromides to the thioamides in the presence of potassium carbonate, and heating [52, 80]. It involved isomerization of the *S*-allyl thioimidates to the *S*-allyl keteneaminothioacetal, prior to the rearrangement, which explains the drastic conditions required (160–200 °C). Similarly, allylation on nitrogen could be achieved under Pd(II) catalysis [52, 80] (without initial isomerization).

Thioamides with a variety of substituents are accessible by a [3,3] shift. They include diunsaturated derivatives from *α*-unsaturated thioamides [53, 81], 2-allyl thiolactams [52, 54], thioindole derivatives [82, 83], thioimides [82, 83], pyrroles [84], and compounds bearing carbonyl [85, 86] or nitrile groups [87].

9.2.2.6 **Thioketenes**

Schaumann and his group [88–91] disclosed that thioketenes are efficiently produced by rearrangement of (silylethynyl) vinyl sulfides ($R^1 = SiMe_3$) under mild conditions (Scheme 9.15). The products can be trapped by amines, providing *γ*-unsaturated thioamides. In the absence of the silyl group, thioketenes could not be isolated, but were converted in thioamides by reaction with amines [92].

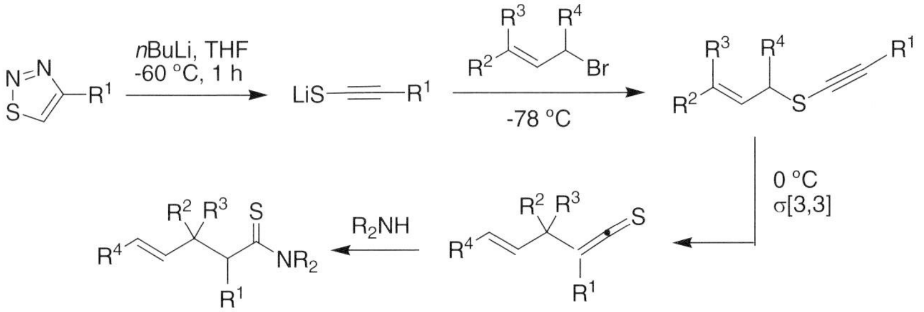

*n*BuLi, THF, -60 °C, 1 h; -78 °C; 0 °C, σ[3,3]; R_2NH

Scheme 9.15

9.2.2.7 Rearrangement of Tricoordinated Sulfur Derivatives: Sulfonium Salts or Sulfoxides

So far, there is only a single report of a strong catalytic effect mediated by metal salts coordination of the allyl double bond and not the heteroatom (see Section 9.2.3), in contrast to the many observations in the oxygen series.

However, a number of examples have demonstrated that decreasing the electron density of the sulfur atom significantly facilitates the thio-Claisen rearrangement.

Two reports [93, 94] have shown that allyl vinyl sulfonium salts rearranged readily to form thionium salts, which were subsequently transformed into a thioacetal (Scheme 9.16). It is noteworthy that the initial investigation was made in the context of a possible path for an indole alkaloid biosynthesis [93]. The rearrangement of an allyl aryl dithiacation was also shown to proceed by an analogous route [95].

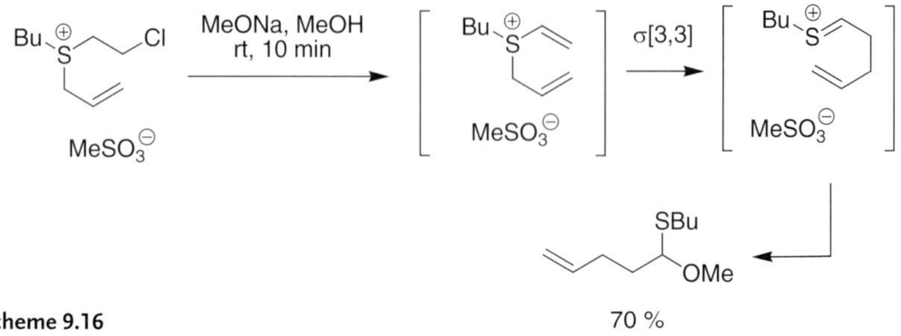

Scheme 9.16

Zwitterionic variants have been reported, first by Bellus and his group [96–101]. The reaction of dichloroketene with allyl sulfides led to 4-unsaturated thiolesters, presumably from the rearrangement of an intermediate bearing an oxyanion and sulfonium salt moieties, as depicted in the example of Scheme 9.23. This variant was later used by other groups [102–104].

The charge acceleration was also recognized early in the sulfoxide series [105] and exploited elegantly for the synthesis of sulfines (Scheme 9.17). The groups of Block [106, 107] and Hwu [108, 109] disclosed that oxidation of allyl vinyl sulfides to the corresponding sulfoxides led to a remarkable effect on the [3,3] process, which can even take place below 0 °C [108, 109]. A comparative study measured a 45-fold acceleration of the kinetic rate [108]. In line with the above considerations (Section 9.1.2), this could be attributed [106] to the weakness of the C–S(O) bond (*ca.* 55 kcal/mol). It was proposed that the (*Z*) selectivity of the formation of the sulfine is under thermodynamic control, the (*E*) isomer being the kinetic product [106]. Investigation of stereoelectronic effects has shown [110] that the formation of the C–C bond is favored in a position antiperiplanar to the more electron-rich vicinal bond.

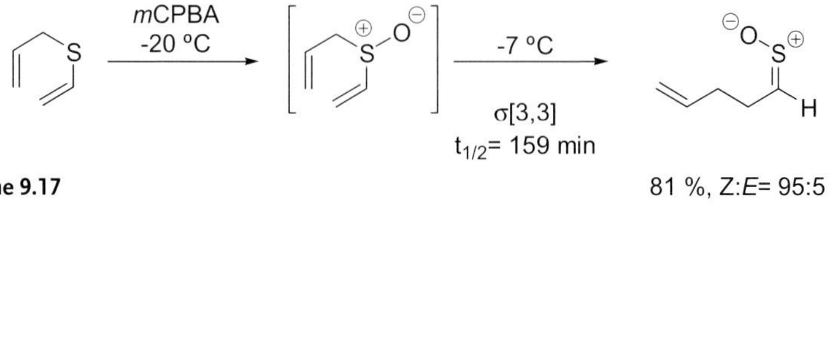

Scheme 9.17

This study was made in the context of the antithrombotic activity of unsaturated compounds isolated from garlic, such as ajoene [107].

Alternative routes to the substrates were explored by Julia and his group [111–113]. An Andersen type reaction of alkyl Grignard reagents with 1-alkenesulfinates presumably led to the expected sulfoxides, which rearranged to (*Z*) γ-unsaturated sulfines in good yields (Scheme 9.18). Alternatively, the source of alkylsulfinyl moiety can be the sulfinate. An allenylsulfinyl moiety led to γ-acetylenic sulfines. The convergence of these methods was then fully reported [113].

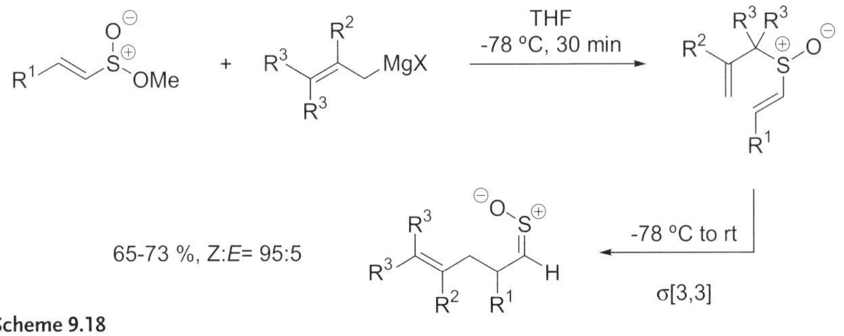

Scheme 9.18

The electron richness of the sulfur atom can be further decreased by introduction of a higher oxidation state. The thermolysis of allyl vinyl sulfone, at rather high temperature, was shown to give the corresponding sulfene, which was trapped by appropriate reagents [113].

9.2.3
Catalysis

Very little information is available about the effect of catalysts towards the thio-Claisen transposition. Meyers and his group had noticed this, and screened a variety of additives. Palladium(II) and nickel(II) complexes (10 mol%) significantly accelerated the rearrangement of hindered keteneaminothioacetals, which could be performed at 25 °C, instead of heating to 140 °C without catalyst [114]. Some variation of the diastereoselectivity was observed. The authors proposed that palladium(II) coordinates to the allylic double bond. The effect of $CeCl_3$ has also been reported [115].

9.3
Rearrangement with Stereochemical Control

The thio-Claisen rearrangement has been successfully used for stereoselective carbon–carbon bond formation. Depending on the substitution of the pericyclic nucleus, two types of stereocontrol can be involved: an *internal control*, closely related to the configurations elements of the heterodiene, or an *external control* by asymmetric induction of a remote stereogenic center. The reported examples will be distinguished by the nature of the inducer, thus covering stereogenic centers on carbon or sulfur atoms or cyclic chiral auxiliaries. In addition, efficient use of axial chirality appeared very recently.

9.3.1
Relative Control Exclusively Through Double-Bond Configurations

Excellent control of the stereochemical outcome of the transposition can be achieved through the C=C bonds, on the condition that structural purities of these unsaturations are excellent. The configuration of the ketene double bond is directly governed by the common *cis*-selectivity of deprotonation of the thiocarbonyl precursors (thioamide [47, 77–79] or dithioester [42, 48]) and that of the allyl chain by the electrophile employed in the *S*-alkylation step. The rearrangement was shown to be stereospecific, with reliable correlations between the configurations of the substrates and products. This was in agreement with the relationships previously observed in other variants of the Claisen rearrangement and interpreted by a six-membered chair-like transition state. As an example, the transposition of (*Z,E*) keteneaminothiocetals afforded *anti* thioamides, whereas *syn* epimers were obtained from (*Z,Z*) compounds (Scheme 9.19) [79]. Although only racemic compounds are thus attainable, the results have been further exploited with more sophisticated systems possessing additional stereochemical elements (see the next section).

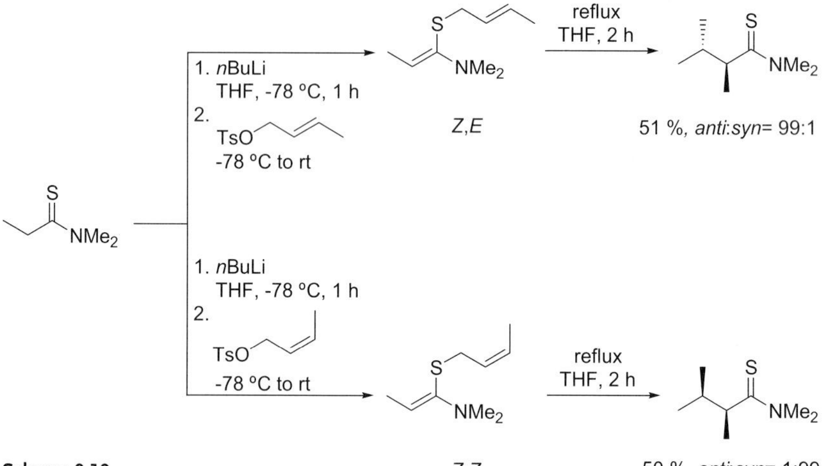

S
NMe_2
1. *n*BuLi
THF, -78 °C, 1 h
2. TsO
-78 °C to rt
S
NMe_2
Z,E
reflux
THF, 2 h
S
NMe_2
51 %, *anti*:*syn*= 99:1
1. *n*BuLi
THF, -78 °C, 1 h
2. TsO
-78 °C to rt
S
NMe_2
Z,Z
reflux
THF, 2 h
S
NMe_2
50 %, *anti*:*syn*= 1:99

Scheme 9.19

Efficient asymmetric induction with 1,3-chirality transfer has also been described with substrates possessing a stereogenic center in the position *alpha* to the sulfur atom and an allylic moiety with a controlled configuration. Despite interesting synthetic features, reported examples are scarce and limited to the zwitterionic Bellus variant of the Claisen rearrangement [98, 100].

9.3.2
Control Through an Asymmetric Carbon Center

Introduction of a stereogenic center outside the diene provides another means of controlling the relative configuration.

Deprotonation of a dithioester bearing a carbon with methyl and *t*-butyl substituents, followed by alkylation, afforded a ketenedithioacetal that rearranged [116] at 101 °C into the anticipated *α*-allylated dithioester in a 95:5 diastereomeric ratio and 89% yield (Scheme 9.20). Although the relative configuration could not be assigned, a model involving exclusively steric factors was proposed. In contrast, a lower diastereoselectivity of 75:25 was observed when the same inducer was attached to the terminal carbon atom of the allyl chain [117].

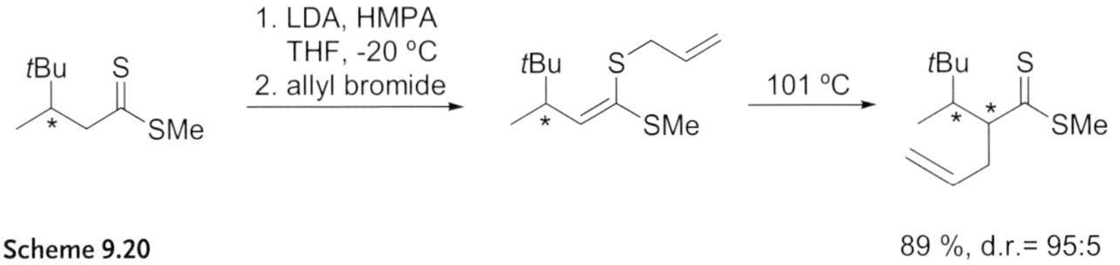

1. LDA, HMPA THF, -20 °C
2. allyl bromide

101 °C

89 %, d.r.= 95:5

Scheme 9.20

A related development with a carbon possessing a hydroxy group was also reported by our group. It allowed the control of the relative configuration of two or three contiguous stereogenic centers, depending on the structural type of the precursor (Scheme 9.21).

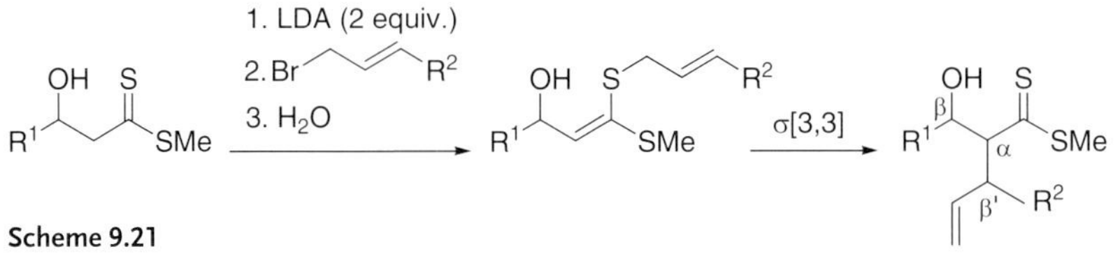

1. LDA (2 equiv.)
2. Br–CH₂CH=CH–R^2
3. H_2O

σ[3,3]

Scheme 9.21

Thus, the rearrangement of *α*-hydroxy *S*-allyl ketenedithioacetals ($R^2 = H$), readily obtained via the dianion of *β*-hydroxy dithioesters, took place (Scheme 9.21) at room temperature and afforded *syn* *α*-allyl *β*-hydroxydithioesters with excellent diastereoselectivities, ranging from 85:15 to 99:1 (seven examples) [44, 118]. This relative configuration, regardless of the double bond geometry, was interpreted by a transition state model based mainly on electronic effects, where the hydroxy group and the hydrogen atom lie in the outside and inside positions, respectively. The attack of the allyl chain, considered the electrophile, takes place on the more electron-rich face of the ketenedithioacetal (nucleophile) *anti* to the alkyl group (Scheme 9.22). A similar *syn* diastereoselectivity was also obtained with the analogous aminothioacetals [119]. Application of the methodology with aldols already

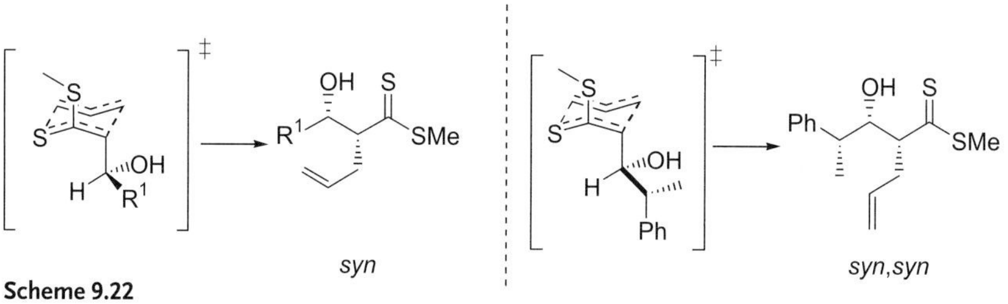

syn *syn,syn*

Scheme 9.22

possessing two vicinal stereocenters in a *syn* relationship gave rise to *syn-syn* dithioesters [120]. The best result with the formation of a single diastereoisomer in 92% yield was obtained for the structure given above.

Another group also investigated the rearrangement of the same substrates in the presence of zeolites. According to the authors, an opposite *anti* configuration was reached [121] and the surprising influence of the zeolite was interpreted by host–guest interactions, supported by molecular simulation data [122].

Starting with *S*-crotyl derivatives ($R^2 = Me$), two stereogenic centers (Scheme 9.21) were created in a single step and both internal and external controls were involved [123]. The relative ($C_\alpha C_{\beta'}$) configuration should correlate in principle to the double bonds geometries and the ($C_\alpha C_\beta$) configuration depends on the asymmetric induction of the stereogenic center. The previously mentioned *cis*-deprotonation of dithioesters allowed the preparation of the four possible ketenedithioacetals with pure geometric integrity. Thermal activation (80 °C) was required for completion within a few hours, without detection of dehydration products (yields: 79–98%). The rearrangement is stereospecific, and each ketenedithioacetal leads predominantly to one of the four possible diastereoisomeric dithioesters (Fig. 9.2). The correlations that are observed are listed in Table 9.1. Interestingly, the sense of the asymmetric induction depends on the geometry of the crotyl chain. With a (*Z*) geometry, a *syn* ($C_\alpha C_\beta$) product is formed (entries 1 and 2) whereas a reverse *anti* configuration is available with an (*E*) one (entries 3 and 4). Moreover, the ($C_\alpha C_{\beta'}$) configurations were consistent with those predicted by a classical pseudocyclic transition state. Models were proposed to account for the observed selectivities.

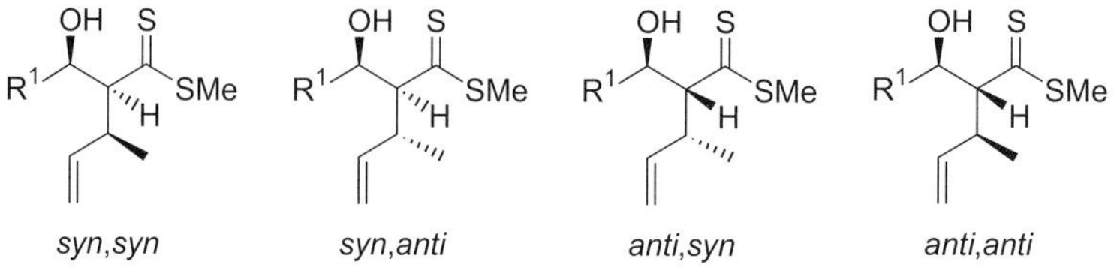

syn,syn *syn,anti* *anti,syn* *anti,anti*

Fig. 9.2 Diastereoisomers available from *S*-crotyl α-hydroxy ketenedithioacetals.

Tab. 9.1 Correlations observed with crotyl substrates.

Entry	Configurations Ketenedithioacetal[a)]	Major dithioester (%)[b)]
1	*ZZ*	*syn-syn* (77–79)
2	*EZ*	*syn-anti* (72–75)
3	*EE*	*anti-syn* (37–42)
4	*ZE*	*anti-anti* (63–73)

a) The first letter refers to the ketenedithioacetal C=C geometry and the second to the crotyl chain.
b) Proportion in the diastereomeric mixture.

Bellus reported another interesting example of 1,2-asymmetric induction by a carbon center possessing an oxygenated group (Scheme 9.23). The reaction of enantiomerically enriched allyl thioethers (ee ≥ 90%) with dichloroketene, generated *in situ* by the reaction of trichloroacetyl chloride with activated zinc, generates transiently a sulfonium salt that rearranged to enantiomerically enriched *syn* thioesters with high diastereoselectivities (de ≥ 90%) [99]. In addition, application of the methodology to the conversion of γ-butyrolactones was described. The change from the silyloxy substituent to an amino group still favors *syn* products but the diastereoselectivity is lowered, not exceeding 80% [101]. Comparison of the energies of the possible transition states, calculated by a semi-empirical AM1 method, is consistent with the observed influence of the heteroatom.

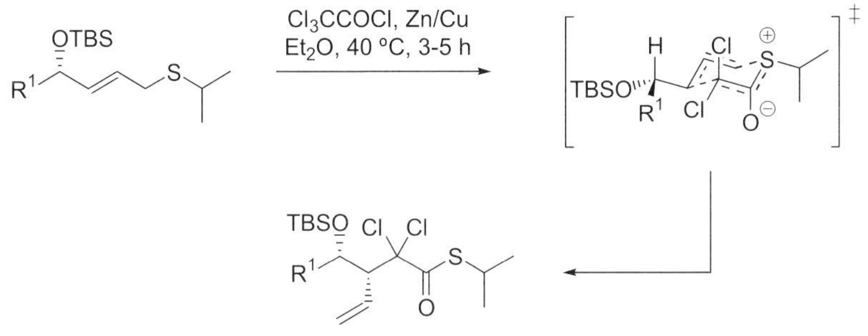

Cl_3CCOCl, Zn/Cu
Et_2O, 40 °C, 3-5 h

Scheme 9.23 76-87 %, >90 % *de*, >95 % *ee*

Finally, reaction of a dihydrothiazine oxide with vinylmagnesium bromide at –60 °C, followed by workup, afforded a sulfine as a single diastereoisomer. The reaction presumably occurred via ring opening of the sulfur Diels–Alder adduct to a vinyl sulfoxide, followed by isomerization [124]. Although the stereochemistry was not identified, the structure depicted below was postulated (Scheme 9.24).

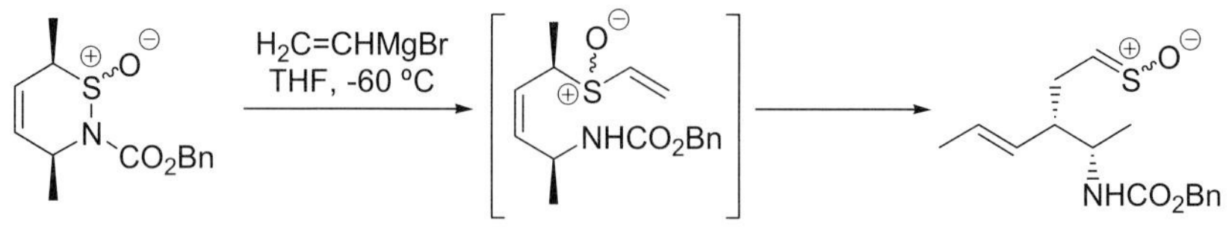

Scheme 9.24

9.3.3 Stereogenic Sulfur Center

The efficiency of sulfoxides as chiral inductors for the thio-Claisen rearrangement has been demonstrated with *N*,*S*-ketene acetals derived from racemic 2-sulfinyl dithioacetates [72]. Moreover, these reactions were successfully extended to the enantiopure series using chiral thioamides as substrates [125]. The source of chirality is the easily available and cheap *D*-glucose [126]. Remarkably, both enantiomers of the starting 2-sulfinyl thioacetamides can be selectively obtained from the same *D*-glucose.

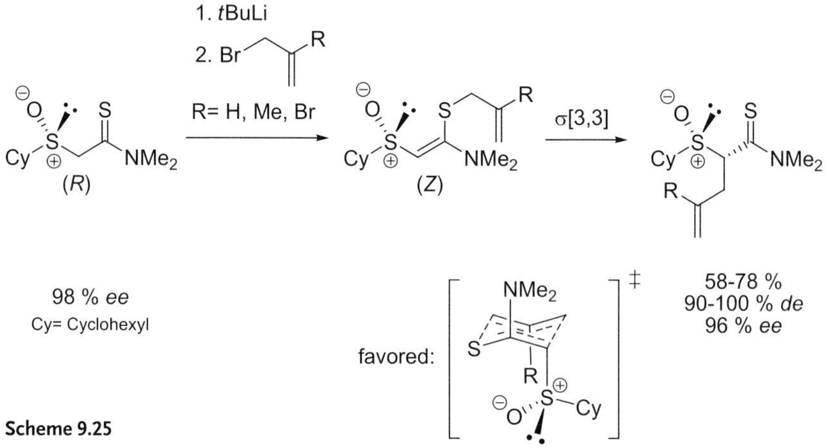

Scheme 9.25

These were readily converted into (*Z*)-keteneaminothioacetals by selective *cis*-deprotonation with *t*-BuLi followed by *S*-allylation, which proceeded with retention of configuration. The [3,3] sigmatropic shifts of the *N*,*S*-ketene acetals occurred smoothly with excellent dia- and enantioselectivities (Scheme 9.25). These highly diastereoselective transformations are assumed to proceed under orbital rather than steric control. The postulated transition state is a pseudocyclic chair, where the best donor substituent of the sulfinyl group – the lone pair – is oriented *anti* to the attack of the allylic chain (electrophile) towards the *N*,*S*-ketene acetal double bond (nucleophile). The oxygen atom occupies the inside position.

Moreover, precursors bearing a crotyl or a cinnamyl chain, with controlled configuration of the double bonds, afforded open-chain products possessing three contiguous asymmetric centers [125]. It was observed that the formation of one diastereomer largely predominates over the three others and, in two examples, even a single diastereomer was formed. The stereochemical course of theses reactions was mostly explained by a pseudocyclic chair transition state, but an unusual switch to a boat transition state was observed in the case of the *trans*-cinnamyl substrate.

9.3.4
Cyclic Chiral Auxiliary

The first examples of an asymmetric thio-Claisen rearrangement induced by a cyclic chiral auxiliary were described by Reddy and Rajappa [127]. *N*-Nitrothioacetyl derivatives of (*L*)-aminoacid esters were used as substrates (Scheme 9.26). After allylation in the presence of DBU, the resulting *S*-allyl derivatives underwent facile thio-Claisen rearrangements upon heating at 50 °C but with modest chiral induction (up to 66% de).

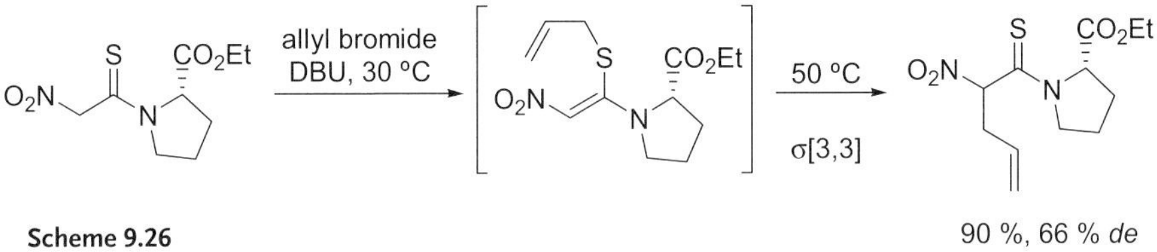

allyl bromide, DBU, 30 °C; 50 °C, σ[3,3]

90 %, 66 % *de*

Scheme 9.26

Analogous selectivities were observed [128] with a bicyclic proline derived auxiliary, which made interesting use of the waste material for a drug synthesis.

Similarly, but in the cyclic series, the thio-Claisen rearrangement of substrates derived from chiral thiolactams was reported to be facile but poorly stereoselective [129]. The low stereocontrol observed in both cases may be explained by the lack of facial selectivity resulting from the free rotation around the C–N bond of the *N,S*-ketene acetals. This critical issue has been solved either by constructing a rigid bicyclic framework or by using C_2-symmetric amines as chiral inductors. The former strategy, developed by Meyers et al. [45], involved bicyclic thiolactams, which were transformed into *N,S*-ketene acetals by deprotonation with LDA, followed by *S*-allylation with various allyl halides (Scheme 9.27).

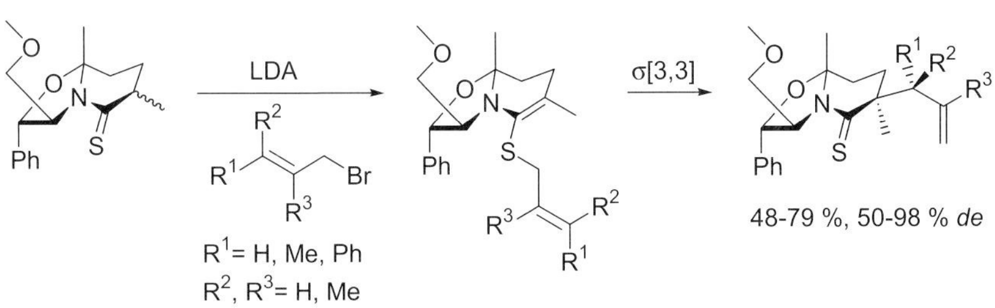

LDA; σ[3,3]

R^1 = H, Me, Ph

R^2, R^3 = H, Me

48-79 %, 50-98 % *de*

Scheme 9.27

The thio-Claisen rearrangement of the *S*-allyl derivatives proceeded in good yields (48–79%) and led preferentially to the *exo*-diastereomers (up to 98% de). The presence of an alkyl R^3 substituent was detrimental to the selectivity, which then dropped to 50%, due to steric interactions with the vinyl methyl group. This high *exo*-facial selectivity stands in contrast to the *endo*-selectivity usually observed in the enolate alkylation reactions of other 5,5- and 5,6-bicyclic lactams. All factors explaining such a preference for the *exo*-product formation have not been determined yet. From a screening of various bicyclic thiolactams [130], it has been demonstrated that the substitution pattern of the oxazolidine ring was determinant for the *exo*/*endo* selectivity of the transposition, although it is quite remote from the sites of rearrangement. Indeed, in the absence of *endo* substituents the selectivity is much lower. Noteworthy is that such transformations cannot be performed in the oxygen series because the corresponding *O*,*N*-ketene acetals cannot be accessed.

Another attractive asymmetric variant of the thio-Claisen rearrangement, which is also well suited for the formation of quaternary centers, has been reported by the Rawal group [131]. The chiral inductor is the C_2-symmetric *trans*-2,5-diphenylpyrrolidine. The [3,3] sigmatropic shifts proceed with excellent yields, high *syn*/*anti* selectivities and good to excellent asymmetric induction (Scheme 9.28). In particular, the rearrangements of substrates bearing a *cis*-substituent R^1 on the allylic moiety give the *anti*-diastereomers with 99% de. The high stereocontrol is rationalized by steric interactions between the phenyl groups of the pyrrolidine ring and the allylic moiety. The level of discrimination will be highest in the presence of the R^1 substituent that is the closest to the pyrrolidine phenyl group. The chiral auxiliary can be cleaved under reductive hydrolysis conditions to furnish chiral substituted pent-4-en-1-ols.

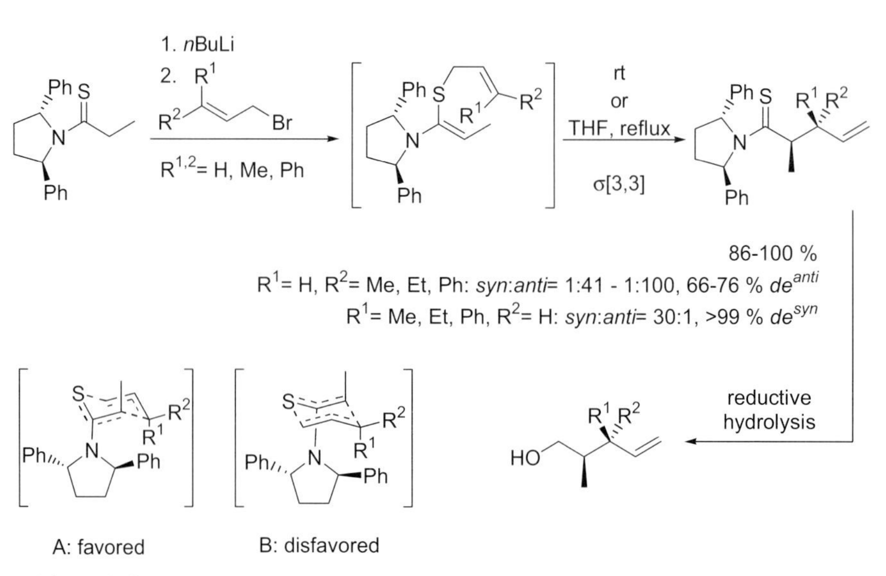

Scheme 9.28

9.3.5 Axial Chirality

The first example of asymmetric thio-Claisen rearrangement based on axial chirality has been described by our group and involves atropisomeric thioanilides bearing an *ortho-tert*-butyl group [132]. These axially chiral thioamides were subjected to LDA deprotonation followed by *S*-allylation with various allyl halides. The resulting *N,S*-ketene acetals were not observed but readily underwent [3,3] sigmatropic shifts (Scheme 9.29). Theses transformations proceed in good to high yields (78–84 %) and diastereoselectivities of up to 88:12. The observed stereochemical course is explained by a back approach of the allylic moiety as the front approach is highly hindered by the *ortho-tert*-butyl group on the perpendicular arene plane.

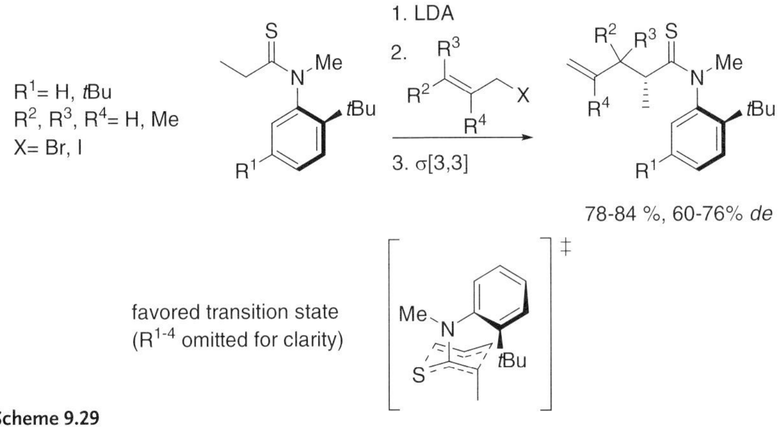

R^1= H, *t*Bu
R^2, R^3, R^4= H, Me
X= Br, I

1. LDA
2. R^2(R^3)C=C(R^4)CH$_2$X
3. σ[3,3]

78-84 %, 60-76% *de*

favored transition state
(R^{1-4} omitted for clarity)

Scheme 9.29

9.4 Applications in Organic Synthesis

9.4.1 Synthesis of Heterocycles

A number of sulfurated heterocycles can be prepared using this transposition. The principal synthetic routes have been already surveyed in previous reviews [38, 40].

It is not our intention to cover all of this large area of research. This subsection will only show some of the structures that have been synthesized according to almost the same strategy. An allylthio- or a propargylthio unsaturated compound underwent [3,3] transposition to afford a thiophenol, which undergoes cyclization (Scheme 9.30).

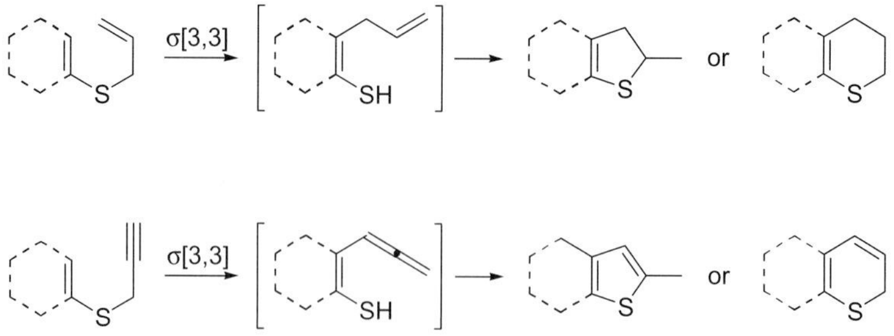
σ[3,3]
SH
or
σ[3,3]
SH
or

Scheme 9.30

Selected examples of five- and six-membered derivatives thus obtained are depicted in Figures 9.3 and 9.4.

In addition, thiophenes can also be prepared [62, 150, 151] by thermal rearrangement of divinyl disulfides followed by H_2S elimination (Scheme 9.31).

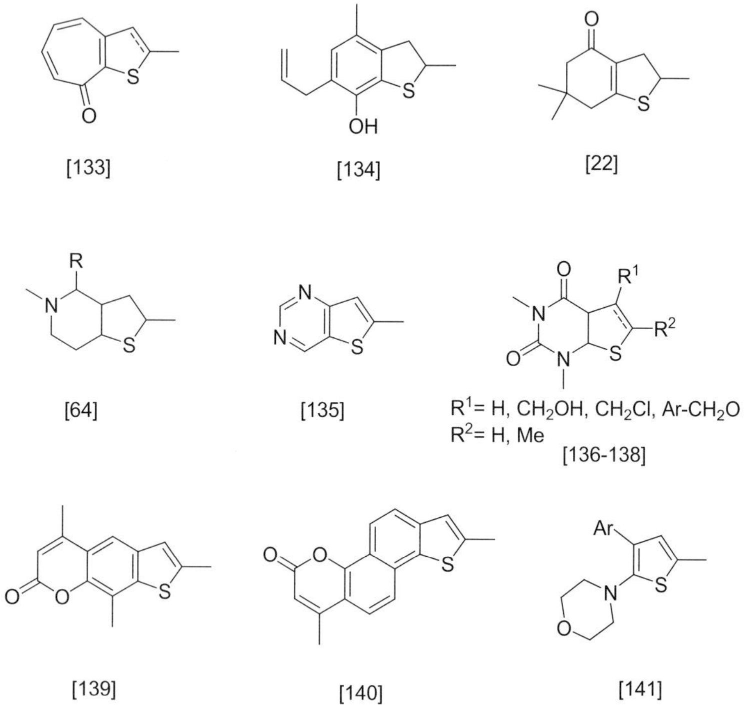
[133]
[134]
[22]
[64]
[135]
R1= H, CH2OH, CH2Cl, Ar-CH2O
R2= H, Me
[136-138]
[139]
[140]
[141]

Fig. 9.3 Examples of 5-membered heterocycles.

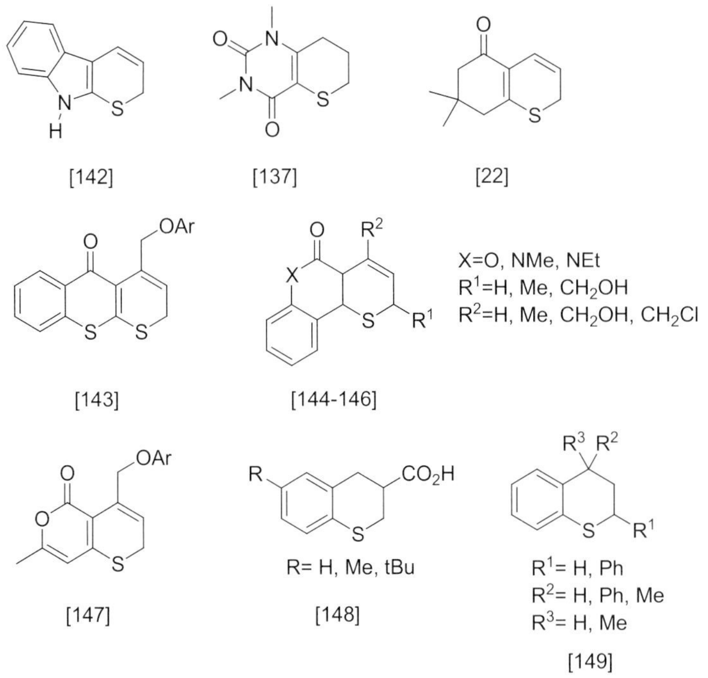

Fig. 9.4 Examples of 6-membered heterocycles.

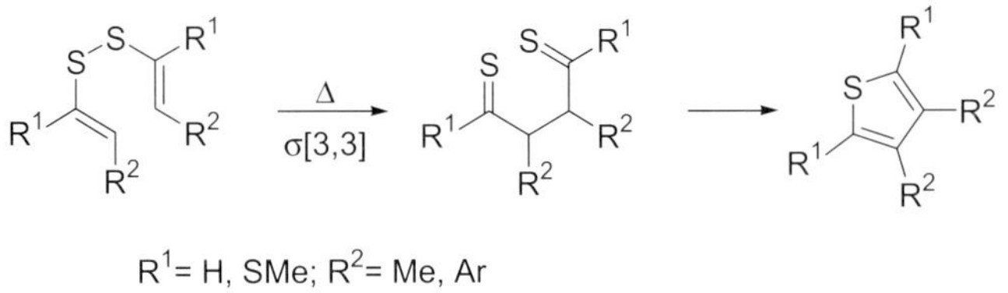

Scheme 9.31

9.4.2
Synthesis of Natural Products and Construction of Building Blocks

Although the thio-Claisen transposition affords various possibilities of synthesis with stereocontrol, few examples of natural product synthesis have been described. Takano's group was a pioneer in this field with the synthesis of nine-membered indole alkaloids (Scheme 9.32) related to cleavamine [152] and velbanamine [153]. The key step involved the transformation of a sulfonium salt into a corresponding ketene amino thioacetal, which underwent the thio-Claisen transposition. A similar strategy was used for the synthesis of the strychnos alkaloid framework [154].

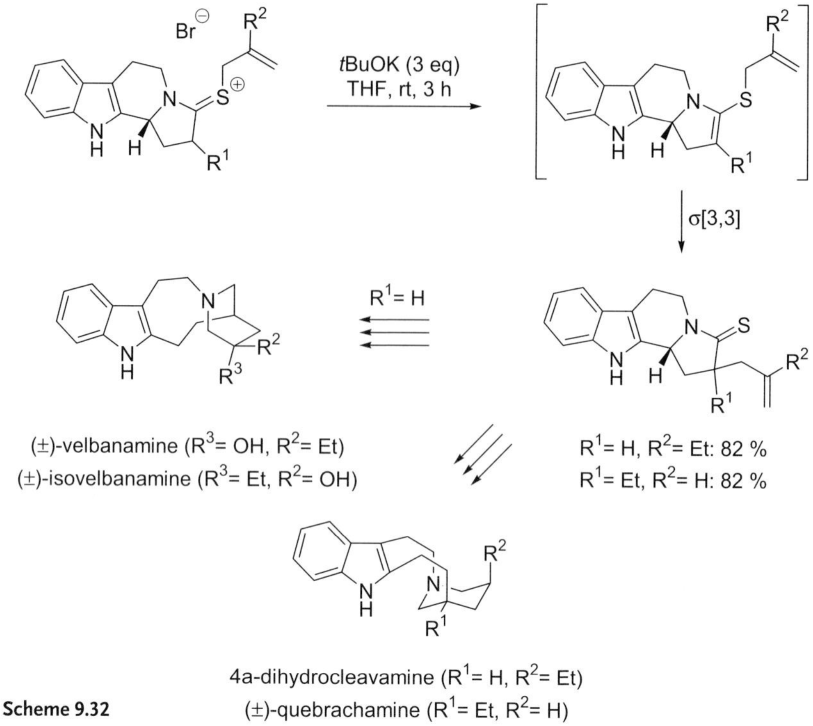

tBuOK (3 eq)
THF, rt, 3 h

σ[3,3]

R^1= H

(±)-velbanamine (R^3= OH, R^2= Et)
(±)-isovelbanamine (R^3= Et, R^2= OH)

R^1= H, R^2= Et: 82 %
R^1= Et, R^2= H: 82 %

4a-dihydrocleavamine (R^1= H, R^2= Et)
(±)-quebrachamine (R^1= Et, R^2= H)

Scheme 9.32

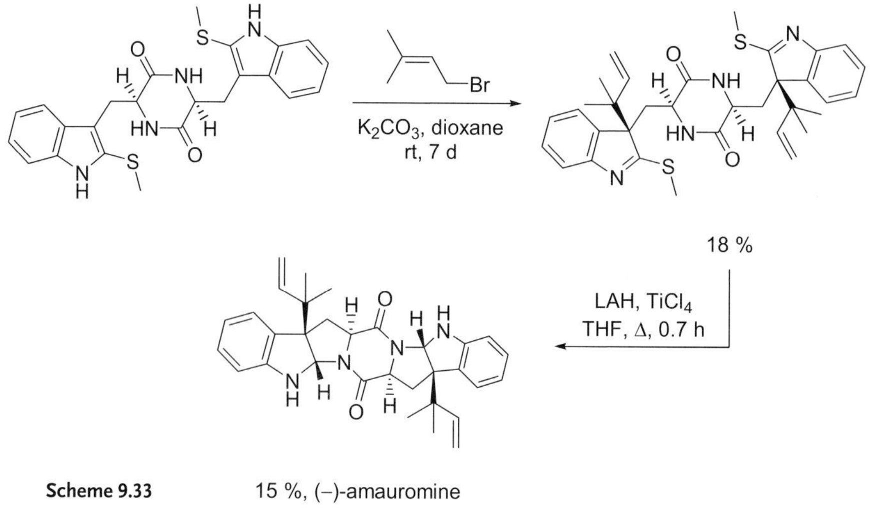

K_2CO_3, dioxane
rt, 7 d

18 %

LAH, $TiCl_4$
THF, Δ, 0.7 h

15 %, (–)-amauromine

Scheme 9.33

The first total synthesis of amauromine [155], another indole alkaloid, has been achieved via a sulfonium salt (not isolated). It involved two simultaneous thio-Claisen rearrangements as key steps (Scheme 9.33). This reaction was also postulated as being involved in the indole alkaloid biosynthesis [93, 156].

Instead of allyl bromide derivative, vinyl diazoacetate in the presence of rhodium acetate also allowed the formation of substituted indolines in excellent yields (Scheme 9.34). *N*-deprotonated allyl thionium was postulated as intermediate [157], no reaction being observed with an *N*-methyl indole analogue.

R^2 ... SEt, N–H; $Rh_2(OAc)_4$, CH_2Cl_2, rt; Cl, CO_2Et, N_2; [R^2, CO_2Et Cl, S⊕, Et, N⊖]; R^2, CO_2Et, Cl, SEt, N; 98 % (3:1 *dr*)

Scheme 9.34

Lemieux and Meyers have used a chiral bicyclic thiolactam to produce vicinal stereogenic quaternary centers, as in the sesquiterpene (–)-trichodiene [158]. A single diastereoisomer was formed during this process (Scheme 9.35). Surpris-

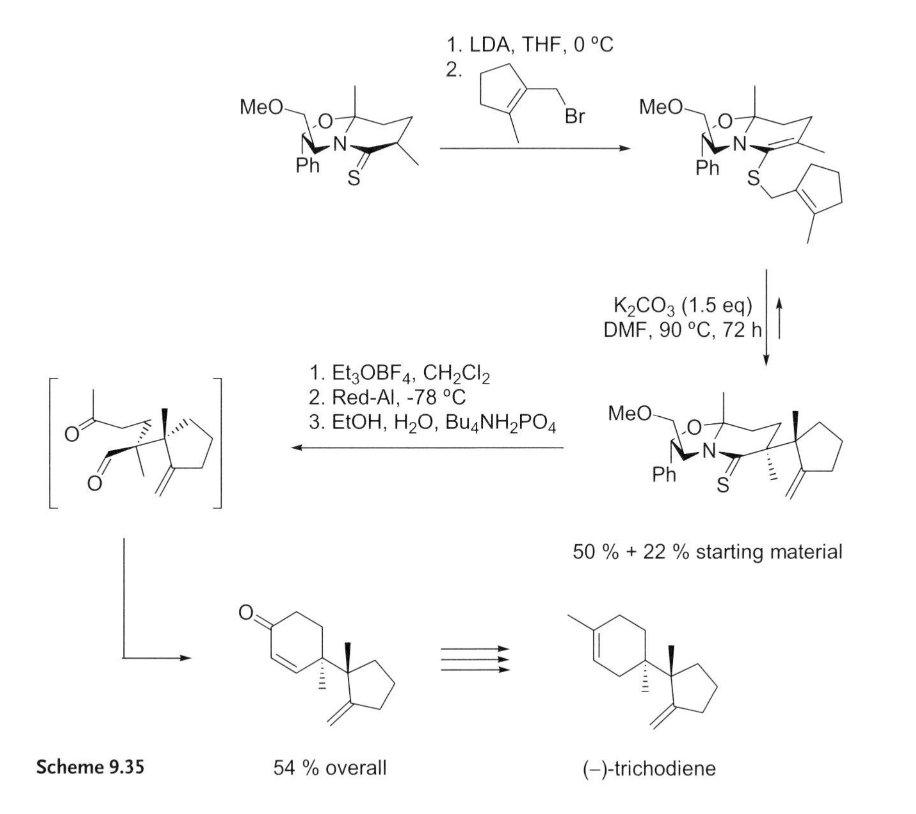

Scheme 9.35

ingly, and in contrast to acyclic thioamide series, the reversibility of the thio-Claisen transposition of the *S*-alkylthioenamine was observed, probably due to the steric hindrance of the product.

The same strategy was used to synthesize enantiopure cyclohexenones with spiro-connected cyclopentenes [159], via two sequential thio-Claisen rearrangements (Scheme 9.36).

first σ[3,3]

second σ[3,3]

or

R^1= Me, CH_2OH
R^2= H, Me

Scheme 9.36

Yoshida et al. have used two different and consecutive sigmatropic [3,3] transpositions (thio-Claisen and Cope) to synthesize the enantiopure (+)-ethyl lanceolate (*E* and *Z*) from (+)-limonene [81].

All of the examples using thio-Claisen transposition for the construction of building blocks consisted of the cyclization of the rearranged adduct. Thus, octahydro[2]pyridinones [160] have been prepared by radical ring closure; differently substituted *γ*-lactams [161, 162] and *γ*-lactones (Scheme 9.37) have been synthesized by electrophilic olefin heterocyclization. The influence of a sulfur substituent on the selectivity of the iodolactonization has been demonstrated by our group [115]. Indeed, whereas the sulfinyl group was not efficient (yield = 60%, dr = 72/28), an excellent 1,3-induction (dr = 96/4) and yield (98%) has been obtained with an *α*-sulfanylamide function.

1. $CeCl_3$ (cat.) THF, rt, 20 h
2. O-O formed in situ

(69 %, d.r. 99:1)

P_4S_{10}
CH_2Cl_2, rt

(90 %)

I_2, THF, H_2O

60 %
trans:cis= 72:28

98 %
trans:cis= 96:4

Scheme 9.37

9.5 Conclusion

The sulfur version of the Claisen rearrangement has emerged as a facile route towards the formation of C–C bonds and as an attractive synthesis of thiocarbonyl compounds. It was performed with successful control of both relative and absolute configurations and provides access to a variety of sulfur heterocycles. Relatively few applications to the synthesis of natural products have been reported so far, but they illustrate the advantages of the sulfur version.

References

1 C. D. Hurd, H. Greengard, *J. Am. Chem. Soc.* **1930**, *52*, 3356–3358.
2 E. N. Karaulova, D. S. Meilanova, G. D. Gal'pern, *Doklady Akad. Nauk. S.S.S.R.* **1957**, *113*, 1280–1282.
3 E. N. Karaulova, D. S. Meilanova, G. D. Gal'pern, *Zhur. Obshchei. Khim.* **1957**, *27*, 3034–3040.
4 C. Y. Meyers, C. Rinaldi, L. Bonoli, *J. Org. Chem.* **1963**, *28*, 2440–2442.
5 H. Kwart, *Phosphorus Sulfur* **1983**, *15*, 293–310.
6 H. Kwart, R. E. Evans, *J. Org. Chem.* **1966**, *31*, 413–419.
7 H. Kwart, M. H. Cohen, *J. Org. Chem.* **1967**, *32*, 3135–3139.
8 H. Kwart, C. M. Hackett, *J. Am. Chem. Soc.* **1962**, *84*, 1754–1755.
9 H. Kwart, J. L. Schwartz, *J. Org. Chem.* **1974**, *39*, 1575–1583.
10 H. Kwart, W. H. Miles, A. G. Horgan, L. D. Kwart, *J. Am. Chem. Soc.* **1981**, *103*, 1757–1760.
11 H. Kwart, J. L. Schwartz, *Chem. Commun.* **1969**, 44.
12 H. Kwart, M. H. Cohen, *Chem. Commun.* **1968**, 319–321.
13 H. Kwart, M. H. Cohen, *Chem. Commun.* **1968**, 1296–1297.
14 H. Kwart, T. J. George, *J. Chem. Soc., Chem. Commun.* **1970**, 433–434.
15 Y. Makisumi, *Tetrahedron Lett.* **1966**, 6399–6403.
16 Y. Makisumi, A. Murabayashi, *Tetrahedron Lett.* **1969**, 1971–1974.
17 Y. Makisumi, A. Murabayashi, *Tetrahedron Lett.* **1969**, 2449–2452.
18 Y. Makisumi, A. Murabayashi, *Tetrahedron Lett.* **1969**, 2453–2456.
19 P. J. W. Schuijl, L. Brandsma, *Recl. Trav. Chim. Pays-Bas* **1968**, *87*, 929–939.
20 F. C. V. Larsson, S. O. Lawesson, *Tetrahedron* **1972**, *28*, 5341–5357.
21 L. Dalgaard, L. Jensen, S.-O. Lawesson, *Tetrahedron* **1974**, *30*, 93–104.
22 L. Dalgaard, S. O. Lawesson, *Acta Chem. Scand. Ser. B* **1974**, *28*, 1077–1090.
23 H. Plieninger, H. P. Kraemer, H. Sirowej, *Chem. Ber.* **1974**, *107*, 3915–3921.
24 J. L. Fourrey, E. Estrabaud, P. Jouin, *J. Chem. Soc., Chem. Commun.* **1975**, 993–994.
25 P. J. W. Schuijl, H. J. T. Bos, L. Brandsma, *Recl. Trav. Chim. Pays-Bas* **1969**, *88*, 597–608.
26 P. Metzner, T. N. Pham, J. Vialle, *Nouv. J. Chim.* **1978**, *2*, 179–182.
27 P. Metzner, T. N. Pham, J. Vialle, *J. Chem. Res.* **1978**, (S) 478–479.
28 S. Yamabe, S. Okumoto, T. Hayashi, *J. Org. Chem.* **1996**, *61*, 6218–6226.
29 B. Gomez, P. K. Chattaraj, E. Chamorro, R. Contreras, P. Fuentealba, *J. Phys. Chem. A* **2002**, *106*, 11227–11233.
30 Y. Makisumi, T. Sasatani, *Tetrahedron Lett.* **1969**, 1975–1978.
31 P. Beslin, D. Lagain, J. Vialle, *J. Org. Chem.* **1980**, *45*, 2517–2518.
32 F. Freeman, M. Y. Lee, H. Lu, X. Wang, *J. Org. Chem.* **1994**, *59*, 3695–3698.
33 R. Arnaud, Y. Vallée, *J. Chem. Soc., Perkin Trans. 2* **1997**, 2737–2743.

34 R. Arnaud, V. Dillet, N. Pelloux-Léon, Y. Vallée, *J. Chem. Soc., Perkin Trans. 2* **1996**, 2065–2071.
35 C. Jenny, H. Heimgartner, *Helv. Chim. Acta* **1987**, *69*, 374–385.
36 L. Brandsma, P. J. W. Schuijl, D. Schuijl-Laros, J. Meijer, H. E. Wijers, *Int. J. Sulfur Chem., Part B* **1971**, *6*, 85–90.
37 L. Morin, J. Lebaud, D. Paquer, R. Chaussin, D. Barillier, *Phosphorus Sulfur* **1979**, *7*, 69–80.
38 A. V. Anisimov, E. A. Viktorova, *Khim. Geterotsikl. Soedin.* **1980**, 435–449.
39 P. Metzner in *Topics in Current Chemistry – Organosulfur Chemistry I*, P. Page (ed.) Springer, Berlin, **1999**, vol. 204, p. 127–184.
40 K. C. Majumdar, S. Ghosh, M. Ghosh, *Tetrahedron* **2003**, *59*, 7251–7271.
41 P. Metzner, *Synthesis* **1992**, 1185–1199.
42 P. Beslin, P. Metzner, Y. Vallée, J. Vialle, *Tetrahedron Lett.* **1983**, *24*, 3617–3620.
43 N. A. Bunina, M. L. Petrov, A. A. Petrov, *Zh. Org. Khim.* **1979**, *15*, 2306–2312.
44 P. Beslin, S. Perrio, *J. Chem. Soc., Chem. Commun.* **1989**, 414–416.
45 P. N. Devine, A. I. Meyers, *J. Am. Chem. Soc.* **1994**, *116*, 2633–2634.
46 T. Fujisawa, K. Umezu, T. Sato, *Chem. Lett.* **1985**, 1453–1456.
47 Y. Tamaru, T. Harada, S. Nishi, M. Mizutani, T. Hioki, Z. Yoshida, *J. Am. Chem. Soc.* **1980**, *102*, 7806–7808.
48 P. Beslin, Y. Vallée, *Tetrahedron* **1985**, *41*, 2691–2705.
49 C. Goasdoué, N. Goasdoué, M. Gaudemar, M. Mladenova, *J. Organomet. Chem.* **1981**, *208*, 279–292.
50 P. Metzner, R. Rakotonirina, *Tetrahedron* **1985**, *41*, 1289–1298.
51 S. Takano, E. Yoshida, M. Hirama, K. Ogasawara, *J. Chem. Soc., Chem. Commun.* **1976**, 776–777.
52 H. Takahata, T. Suzuki, M. Maruyama, K. Moriyama, M. Mozumi, T. Takamatsu, T. Yamazaki, *Tetrahedron* **1988**, *44*, 4777–4786.
53 Y. Tamaru, T. Harada, Z. Yoshida, *Tetrahedron Lett.* **1978**, *25*, 2167–2170.
54 S. Takano, M. Hirama, K. Ogasawara, *Chem. Lett.* **1982**, 529–532.
55 K. Kondo, I. Ojima, *J. Chem. Soc., Chem. Commun.* **1972**, 62–63.
56 K. Oshima, H. Yamamoto, H. Nozaki, *J. Am. Chem. Soc.* **1973**, *95*, 4446–4447.
57 K. Oshima, H. Takahashi, H. Yamamoto, H. Nozaki, *J. Am. Chem. Soc.* **1973**, *95*, 2693–2694.
58 M. Schoufs, J. Meijer, P. Vermeer, L. Brandsma, *Synthesis* **1978**, 439–440.
59 E. J. Corey, J. I. Shulman, *J. Am. Chem. Soc.* **1970**, *92*, 5522–5523.
60 L. Brandsma, H. D. Verkruijsse, *Recl. Trav. Chim. Pays-Bas* **1974**, *93*, 319–320.
61 L. Morin, D. Paquer, *C. R. Acad. Sci., Ser. C* **1976**, *282*, 353–356.
62 D. Barillier, L. Morin, D. Paquer, P. Rioult, M. Vazeux, C. G. Andrieu, *Bull. Soc. Chim. Fr.* **1977**, 688–692.
63 L. Dalgaard, S.-O. Lawesson, *Tetrahedron* **1972**, *28*, 2051–2063.
64 F. Tubéry, D. S. Grierson, H.-P. Husson, *Tetrahedron Lett.* **1987**, *28*, 6461–6464.
65 H.-P. Wu, R. Aumann, S. Venne-Dunker, P. Saarenketo, *Eur. J. Org. Chem.* **2000**, 3463–3473.
66 P. J. W. Schuijl, L. Brandsma, *Recl. Trav. Chim. Pays-Bas* **1969**, *88*, 1201–1204.
67 H. Takahashi, K. Oshima, H. Yamamoto, H. Nozaki, *J. Am. Chem. Soc.* **1973**, *95*, 5803–5804.
68 S. Berrada, S. Désert, P. Metzner, *Tetrahedron* **1988**, *44*, 3575–3586.
69 S. Apparao, S. S. Bhattacharjee, H. Ila, H. Junjappa, *J. Chem. Soc., Perkin Trans. 1* **1985**, 641–645.
70 R. Gompper, B. Kohl, *Tetrahedron Lett.* **1980**, *21*, 917–920.
71 M. Yokoyama, K. Tsuji, M. Hayashi, T. Imamoto, *J. Chem. Soc., Perkin Trans. 1* **1984**, 85–90.
72 C. Alayrac, C. Fromont, P. Metzner, N. T. Anh, *Angew. Chem., Int. Ed. Engl.* **1997**, *36*, 371–374.
73 M. Lemarié, Y. Vallée, M. Worrell, *Tetrahedron Lett.* **1992**, *33*, 6131–6134.
74 K. Hartke, G. Gölz, *Chem. Ber.* **1974**, *107*, 566–569.
75 G. Otani, S.-I. Yamada, *Chem. Pharm. Bull.* **1973**, *21*, 2112–2118.
76 S. Takano, S. Tomita, M. Takahashi, K. Ogasawara, *Chem. Lett.* **1987**, 1379–1380.

77 Y. Tamaru, Y. Amino, Y. Furukawa, M. Kagotani, Z. Yoshida, *J. Am. Chem. Soc.* **1982**, *104*, 4018–4019.

78 Y. Tamaru, M. Mizutani, Y. Furukawa, O. Kitao, Z. Yoshida, *Tetrahedron Lett.* **1982**, *23*, 5319–5322.

79 Y. Tamaru, Y. Furukawa, M. Mizutani, O. Kitao, Z. Yoshida, *J. Org. Chem.* **1983**, *48*, 3631–3639.

80 Y. Tamaru, M. Kagotani, Z. Yoshida, *Tetrahedron Lett.* **1981**, *22*, 4245–4248.

81 Y. Tamaru, T. Harada, Z. Yoshida, *J. Am. Chem. Soc.* **1980**, *102*, 2392–2398.

82 B. W. Bycroft, W. Landon, *J. Chem. Soc., Chem. Commun.* **1970**, 168.

83 T. Nishio, *J. Chem. Res.* **1989**, (S) 204–205; (M) 1567–1580.

84 K.-E. Teo, G. H. Barnett, H. J. Anderson, C. E. Loader, *Can. J. Chem.* **1978**, *56*, 221–225.

85 F. C. V. Larsson, S.-O. Lawesson, *Tetrahedron* **1974**, *30*, 1283–1288.

86 R. Gompper, W.-R. Ulrich, *Angew. Chem., Int. Ed. Engl.* **1976**, *15*, 301–302.

87 T. Sasaki, A. Kojima, M. Ohta, *J. Chem. Soc. (C)* **1971**, 196–200.

88 E. Schaumann, F. F. Grabley, *Tetrahedron Lett.* **1977**, 4307–4710.

89 E. Schaumann, F. F. Grabley, *Liebigs Ann. Chem.* **1979**, 1746–1755.

90 E. Schaumann, F. F. Grabley, *Chem. Ber.* **1980**, *113*, 3024–3038.

91 E. Schaumann, J. Lindstaedt, *Chem. Ber.* **1983**, *116*, 1728–1738.

92 J. Meijer, P. Vermeer, H. J. T. Bos, L. Brandsma, *Recl. Trav. Chim. Pays-Bas* **1974**, *93*, 26–29.

93 B. W. Bycroft, W. Landon, *J. Chem. Soc., Chem. Commun.* **1970**, 967–968.

94 J. N. Harvey, H. G. Viehe, *J. Chem. Soc., Chem. Commun.* **1995**, 2345–2346.

95 N. Furukawa, *Bull. Chem. Soc. Jpn* **1997**, *70*, 2571–2591.

96 R. Malherbe, D. Bellus, *Helv. Chim. Acta* **1978**, *61*, 3096–3099.

97 R. Malherbe, G. Rist, D. Bellus, *J. Org. Chem.* **1983**, *48*, 860–869.

98 R. Öhrlein, R. Jeschke, B. Ernst, D. Bellus, *Tetrahedron Lett.* **1989**, *30*, 3517–3520.

99 U. Nubbemeyer, R. Öhrlein, J. Gonda, B. Ernst, D. Bellus, *Angew. Chem., Int. Ed. Engl.* **1991**, *30*, 1465–1467.

100 B. Ernst, J. Gonda, R. Jeschke, U. Nubbemeyer, R. Öhrlein, D. Bellus, *Helv. Chim. Acta* **1997**, *80*, 876–891.

101 J. Gonda, M. Martinková, B. Ernst, D. Bellus, *Tetrahedron* **2001**, *57*, 5607–5613.

102 G. Rosini, G. G. Spineti, E. Foresti, G. Pradella, *J. Org. Chem.* **1981**, *46*, 2228–2230.

103 E. Vedejs, R. A. Buchanan, *J. Org. Chem.* **1983**, *49*, 1840–1841.

104 B. D. Johnston, E. Czyzewska, A. C. Oehlschlager, *J. Org. Chem.* **1987**, *52*, 3693–3697.

105 Y. Makisumi, S. Takada, Y. Matsukura, *J. Chem. Soc., Chem. Commun.* **1974**, 850.

106 E. Block, S. Ahmad, *J. Am. Chem. Soc.* **1985**, *107*, 6731–6732.

107 E. Block, S. Ahmad, J. L. Catalfamo, M. K. Jain, R. Apitz-Castro, *J. Am. Chem. Soc.* **1986**, *108*, 7045–7055.

108 J. R. Hwu, D. A. Anderson, *Tetrahedron Lett.* **1986**, *27*, 4965–4968.

109 J. R. Hwu, D. A. Anderson, *J. Chem. Soc., Perkin Trans. 1* **1991**, 3199–3206.

110 A. Mukherjee, E. M. Schulman, W. J. le Noble, *J. Org. Chem.* **1992**, *57*, 3120–3126.

111 J.-B. Baudin, M.-G. Commenil, S. Julia, Y. Wang, *Synlett* **1992**, 909–910.

112 J.-B. Baudin, M.-G. Commenil, S. Julia, L. Toupet, Y. Wang, *Synlett* **1993**, 839–840.

113 J.-B. Baudin, M.-G. Commenil, S. Julia, Y. Wang, *Bull. Soc. Chim. Fr.* **1996**, *133*, 515–529.

114 D. J. Watson, P. N. Devine, A. I. Meyers, *Tetrahedron Lett.* **2000**, *41*, 1363–1367.

115 V. Blot, V. Reboul, P. Metzner, *J. Org. Chem.* **2004**, *69*, 1196–1201.

116 S. Désert, P. Metzner, M. Ramdani, *Tetrahedron* **1992**, *48*, 10315–10326.

117 S. Désert, P. Metzner, *Tetrahedron* **1992**, *48*, 10327–10338.

118 P. Beslin, S. Perrio, *Tetrahedron* **1991**, *47*, 6275–6286.

119 P. Beslin, S. Perrio, *Tetrahedron* **1992**, *48*, 4135–4146.
120 P. Beslin, B. Lelong, *Tetrahedron* **1997**, *63*, 17253–17264.
121 R. Sreekumar, R. Padmakumar, *Tetrahedron Lett.* **1997**, *38*, 2413–2416.
122 L. Wang, B. Li, Q. Jin, Z. Guo, S. Tang, D. Ding, *J. Mol. Catal. A: Chemical* **2000**, *160*, 377–381.
123 P. Beslin, S. Perrio, *Tetrahedron* **1993**, *49*, 3131–3142.
124 R. S. Garigipati, R. Cordova, M. Parvez, S. M. Weinreb, *Tetrahedron* **1986**, *42*, 2979–2983.
125 S. Nowaczyk, C. Alayrac, V. Reboul, P. Metzner, M.-T. Averbuch-Pouchot, *J. Org. Chem.* **2001**, *66*, 7841–7848.
126 C. Alayrac, S. Nowaczyk, M. Lemarié, P. Metzner, *Synthesis* **1999**, 669–675.
127 K. V. Reddy, S. Rajappa, *Tetrahedron Lett.* **1992**, *33*, 7957–7960.
128 J. Wilken, S. Wallbaum, W. Saak, D. Haase, S. Pohl, L. N. Patkar, A. N. Dixit, P. Chittari, S. Rajappa, J. Martens, *Liebigs Ann. Chem.* **1996**, 927–934.
129 S. Jain, N. Sinha, D. K. Dikshit, N. Anand, *Tetrahedron Lett.* **1995**, *36*, 8467–8468.
130 D. J. Watson, C. M. Lawrence, A. I. Meyers, *Tetrahedron Lett.* **2000**, *41*, 815–818.
131 S. He, S. A. Kozmin, V. H. Rawal, *J. Am. Chem. Soc.* **2000**, *122*, 190–191.
132 S. Dantale, V. Reboul, P. Metzner, C. Philouze, *Chem. Eur. J.* **2002**, *8*, 632–640.
133 H. Takeshita, K. Uchida, H. Mametsuka, *Heterocycles* **1983**, *20*, 1709–1712.
134 A. V. Anisimov, A. A. Girshkyan, K. A. Gaisina, E. A. Viktorova, *Khim. Geterotsikl. Soedin.* **1994**, 480–482.
135 M. R. Spada, R. S. Klein, B. A. Otter, *J. Heterocyclic Chem.* **1989**, *26*, 1851–1857.
136 H. Inoue, M. Takada, M. Takahashi, T. Ueda, *Heterocycles* **1977**, *8*, 427–432.
137 K. C. Majumdar, N. K. Jana, A. Bandyopadhyay, S. K. Ghosh, *Synth. Commun.* **2001**, *31*, 2979–2985.
138 K. C. Majumdar, N. K. Jana, *Synth. Commun.* **2000**, *30*, 4183–4196.
139 G. R. Wellman, *J. Heterocyclic Chem.* **1980**, *17*, 911–912.
140 Z.-F. Tao, X. Qian, *Phosphorus, Sulfur Silicon Relat. Elem.* **1996**, *114*, 109–113.
141 F. Matloubi Moghaddam, H. Zali-Boinee, *Tetrahedron Lett.* **2003**, *44*, 6253–6255.
142 S. Takada, Y. Makisumi, *Chem. Pharm. Bull.* **1984**, *32*, 872–876.
143 K. C. Majumdar, A. T. Khan, S. Saha, *Synth. Commun.* **1992**, *22*, 901–912.
144 K. C. Majumdar, M. Ghosh, M. Jana, D. Saha, *Tetrahedron Lett.* **2002**, *43*, 2111–2113.
145 K. C. Majumdar, S. K. Ghosh, *Tetrahedron Lett.* **2002**, *43*, 2115–2117.
146 K. C. Majumdar, U. K. Kundu, S. K. Ghosh, *Org. Lett.* **2002**, *4*, 2629–2631.
147 K. C. Majumdar, U. K. Kundu, S. Ghosh, *J. Chem. Soc., Perkin Trans. 1* **2002**, 2139–2140.
148 B. Gopalan, K. Rajagopalan, S. Swaminathan, K. K. Balasubramanian, *Synthesis* **1976**, *6*, 409–411.
149 A. W. Jensen, J. Manczuk, D. Nelson, O. Caswell, S. A. Fleming, *J. Heterocyclic Chem.* **2000**, *37*, 1527–1531.
150 E. Block, S. H. Zhao, *Tetrahedron Lett.* **1990**, *31*, 4999–5002.
151 F. C. V. Larsson, L. Brandsma, S.-O. Lawesson, *Recl. Trav. Chim. Pays-Bas* **1974**, *93*, 258–260.
152 S. Takano, M. Hirama, T. Araki, K. Ogasawara, *J. Am. Chem. Soc.* **1976**, *98*, 7084–7085.
153 S. Takano, M. Hirama, K. Ogasawara, *J. Org. Chem.* **1980**, *45*, 3729–3730.
154 S. Takano, M. Hirama, K. Ogasawara, *Tetrahedron Lett.* **1982**, *23*, 881–884.
155 S. Takase, Y. Itoh, I. Uchida, H. Tanaka, H. Aoki, *Tetrahedron* **1986**, *42*, 5887–5894.
156 R. Plate, A. W. G. Theunisse, H. C. J. Ottenheijm, *J. Org. Chem.* **1987**, *52*, 370–375.
157 A. V. Novikov, A. R. Kennedy, J. D. Rainier, *J. Org. Chem.* **2003**, *68*, 993–996.

158 R. M. Lemieux, A. I. Meyers, *J. Am. Chem. Soc.* **1998**, *120*, 5453–5457.
159 R. M. Lemieux, P. N. Devine, M. F. Mechelke, A. I. Meyers, *J. Org. Chem.* **1999**, *64*, 3585–3591.
160 J. G. Sosnicki, *Synlett* **2003**, 1673–1677.
161 H. Takahata, T. Takamatsu, T. Yamazaki, *J. Org. Chem.* **1989**, *54*, 4812–4822.
162 H. Takahata, T. Takamatsu, Y.-S. Chen, N. Ohkubo, T. Yamazaki, T. Momose, T. Date, *J. Org. Chem.* **1990**, *55*, 3792–3797.

10
Aza-Claisen Rearrangement

Udo Nubbemeyer

10.1
Introduction

The replacement of the basic heteroatom X = O of a Claisen rearrangement by a nitrogen atom defines the so-called aza-Claisen rearrangement and the 3-hetero Cope rearrangement, respectively. For a long period of time, aza-Claisen rearrangements were regarded as a sophisticated variant of the widely used oxygen analog. Because of the much more drastic reaction conditions, sparse applications have been published. Most investigations have been restricted to fundamental research on aza-Claisen rearrangements. About 30 years ago, the perception tended to change, because the nitrogen atom in the central position of the sigmatropic core systems was discerned as an ideal anchor for catalysts such as protons, Lewis acids and for chiral auxiliaries in enantioselective rearrangements. Charge acceleration allowed a significant reduction of the reaction temperature, rendering the process a suitable key step in complex molecule syntheses. Finally, the development of very mild zwitterionic variants enables one to classify the aza-Claisen rearrangement as a powerful method in synthetic organic chemistry. This chapter gives an overview of aza-Claisen rearrangements [1].

10.2
Aromatic Simple Aza-Claisen Rearrangements

Initial efforts to conduct aza-Claisen rearrangements of simple N-allyl aniline **1** failed. No *o* and *p* allyl aniline was isolated upon heating the reactant to > 250 °C, decomposition only occurred to give propene and aniline **2** [2]. The first successful uncatalyzed thermal aromatic aza-Claisen rearrangement was described three decades later with the synthesis of 2-allyl-1-naphthylamine **4-*o*** from α-N-allylnaphthylamine **3** at 260 °C [3]. Marcinkiewicz reasoned that the aza analog of the original Claisen rearrangement required a higher activation energy of about 6 kcal mol^{-1}. Hence, fused aromatic systems only displayed some migration of the ally moiety

The Claisen Rearrangement. Edited by M. Hiersemann and U. Nubbemeyer

ISBN: 978-3-527-30825-5

Scheme 10.1

avoiding the complete rupture of the aromatic system during the course of the reaction (Scheme 10.1).

In addition to the Marcinkiewicz experiment, Inada analyzed several thermal rearrangements of simple N-allyl naphthylamines **3** (PG = H) [3c]. Generally, the *ortho* products **4-*o*** with inverted allyl systems predominated, but in most attempts some deallylated and decomposed side products also formed. In contrast, the corresponding sulfonamido reactants **3** (PG = $MeSO_2$ and Tol-SO_2) gave rise to the formation of the *para* products **4-*p*** minimizing steric repulsion of the allyl side chain and the sulfonamide in the *o* intermediates **4-*o***. However, it could not be excluded that the reaction partly passed through a dissociative reaction path in competition with two consecutive 3,3 sigmatropic rearrangements (aza-Claisen/Cope). Recently, 4-quinolone-fused heterocycles with potential *anti*-tumor activity have been prepared using an aza-Claisen rearrangement/RCM (ring-closing metathesis) strategy [3d]. Starting from quinolone **5**, the amination rearrangement tandem reaction delivered the intermediate **6** with 76% yield. After protection of the aniline as an acetamide, the RCM gave the desired target azepine **7** (85.5% yield, Scheme 10.2).

a) 5 eq. $(allyl)_2NH$, dioxane, 120°C, sealed tube. b) 1. Ac_2O, pyridine, rt; 2. 10 mol% Grubbs (I) catalyst, CH_2Cl_2, rt. PMP = 4-methoxyphenyl

Scheme 10.2

The reduction of ring strain served as a powerful promoting factor in simple aromatic aza-Claisen rearrangement to run the reaction at acceptable temperatures [4]. 4-Phenoxy 1,2 benzoquinone **8** was treated with 2-vinylazetidine **9** (n = 2) and 2-vinyl aziridine **9** (n = 1), respectively. The azetidine adduct **10** (n = 2) could be isolated in 61% yield. Heating in toluene induced the aza-Claisen rearrangement to give the benzoquinone annulated azocine **11** (n = 2) with 100% yield. In contrast, the corresponding aziridine adduct **10** (n = 1) could not be trapped, the azepine **11** (n = 1) was isolated in 55% yield indicating an efficient aza-Claisen rearrangement run at ambient temperature (Scheme 10.3).

n = 1, R = H: **10** not detected, direct reaction to **11** at 23°C in MeCN
n = 2, R = Me: reaction **10** to **11** in PhMe, Δ

Scheme 10.3

Focusing on efforts involving heterocycles, Makisumi reported on early uncatalyzed 3,3 rearrangements starting from pyrazolinones **12** (X = NPh) and isoxazolinones **12** (X = O), respectively [5a,b]. Initial N-allylation of the imine function in **12** gave rise to the formation of the corresponding pyrazolones and isoxazolones **13** (extended -system). The heating of intermediates **13** (X = NPh) induced allyl group rearrangements to give the 4-allylated species **14**. The crotyl derivative suffered from a complete allyl inversion (R^3 = H/Me) pointing out a cyclic transition state. In contrast, the oxazolone reactants **12** (X = O) displayed a less homogeneous behavior. The terminally unsubstituted compounds **13** (R^3 = H/H) rearranged as expected. The crotyl derivative (R^3 = H/Me) resulted a 54:46 mixture of **13** and allyl inverted product **14**, the dimethylallyl derivative **13** (R^3 = Me/Me) gave no rearrangement. The unique sense of the rearrangement was proven by subjecting product **14** to the reaction conditions, which resulted in no retro allyl shift to regenerate the reactant **13** (Scheme 10.4).

X = NPh, R^1 = Me, R^2 = Me, allyl, $R^{3'}$ = H: R^3 = H/H; H/Me
X = O, R^1 = Ph, R^2 = Me: R^3, $R^{3'}$ = H, H, H; H, H, Me; Me, H, H; Me, Me, H

Scheme 10.4

The rearrangement of a 5-allylamino uracil derivative **16** succeeded in refluxing tetraline to give a mixture of starting material **16** and the product 6-allyl-aminouracil **17** with 24% yield (Scheme 10.5) [5c].

Since uncatalyzed simple aromatic aza-Claisen rearrangements required 200 to 350 °C reaction temperature to undergo the 3,3 sigmatropic conversion, such drastic conditions often caused low yields. Charge acceleration promised to run the process at lower temperatures making the reaction more attractive for extended synthetic use. Proton acid-catalyzed rearrangements at lower reaction temperatures had initially been reported by Russian groups [6a]. Intending to generate Uhle's ketone **21** as a precursor for the ergot alkaloid total syntheses, the N cyclo-

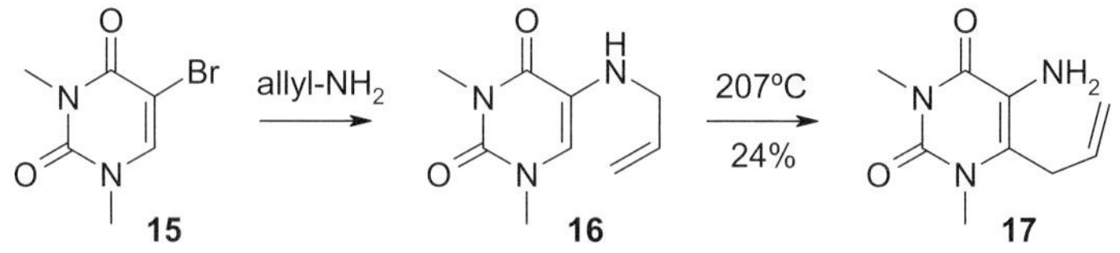

Scheme 10.5

pentenyl aniline **18** (n= 1, R^1 = CN) was heated in the presence of trifluoroacetic acid to give the regioisomers **20** and **19** with 38 and 10% yield, respectively. The replacement of the nitrile against other electron acceptors (n= 1, R^1 = CO_2Me, NO_2, F, CF_3) resulted in the predominant formation of the regioisomer **19** (2:1–10:1, Scheme 10.6).

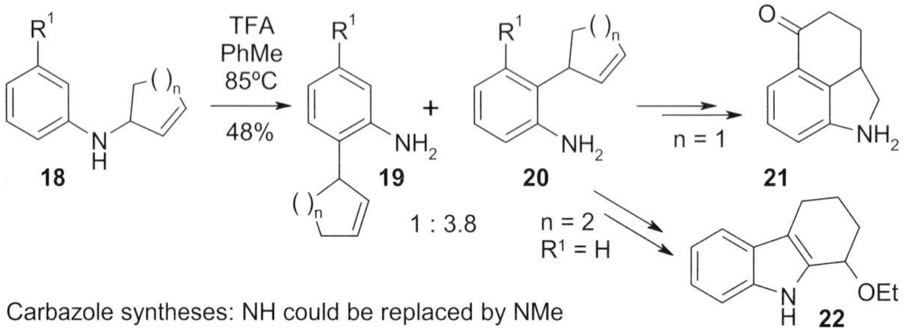

Carbazole syntheses: NH could be replaced by NMe

Scheme 10.6

Majumdar published several aza-Claisen rearrangements of 2-cyclohexenyl-1-anilines **18** (n= 2, R^1 = *o* and *p* H, Me, Br) [6c,d]. The reaction was carried out upon heating the reactant in EtOH/HCl. The corresponding 2-cyclohexenylanilines **19** (n= 2) were obtained with 50 to 90% yield (no regioisomers were obtained because of the reactant substitution pattern). A preceding cyclization to give indole derivatives **22** could be achieved in a separate step. Treatment of the rearrangement products **19** with $Hg(OAc)_2$ in a suitable alcohol in the presence of acetic acid induced tetrahydrocarbazole **22** formation. The tricyclic products **22** were synthesized with 70–85% yield. Finally, carbazoles could be obtained after DDQ dehydrogenation (Scheme 10.6).

Pyrimidine annulated heterocycles fused at positions 5 and 6 to uracil have been synthesized via a three-step sequence starting from uracil **23** [7]. Firstly, the reaction with 3-bromo cyclohexene gave the N-allyl-vinyl core system **24** in 80% yield. Upon heating **24** in EtOH in the presence of HCl, aza-Claisen rearrangement gave rise to the formation of the C-cyclohexenyl uracil **25** in 38% yield. Final bromination (→**26**) and dehydroganation steps (→**27**) allow one to synthesize the desired tricyclic fused uracil systems (Scheme 10.7).

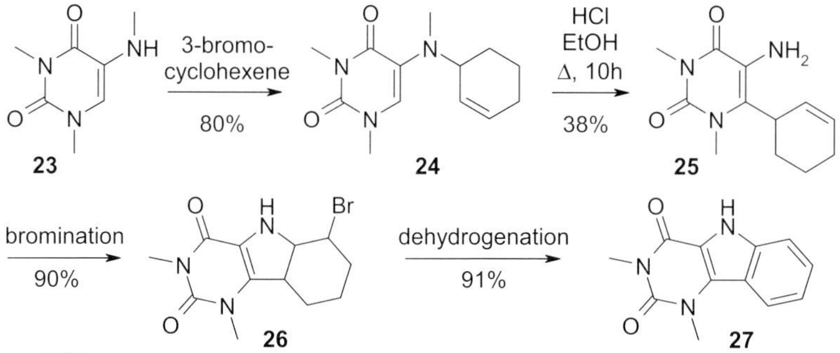

Scheme 10.7

Ward reported on a mild proton catalyzed variant of aromatic aza-Claisen rearrangements. A series of N-(1,1-disubstituted-allyl) anilines **28** have been rearranged in the presence of 10 mol% of *p*TsOH (→**29**) to give **30** with 53 to 95% yield. The reaction proceeded with complete allyl inversion. The formation of side products such as carbinols originating from a H_2O addition to the nascent double bond could be suppressed by running the conversion in aqueous acetonitrile. Electron-withdrawing substituents R^1 in the *p* position accelerated and electron-rich functional groups R^1 retarded the rearrangement (Scheme 10.8) [8a].

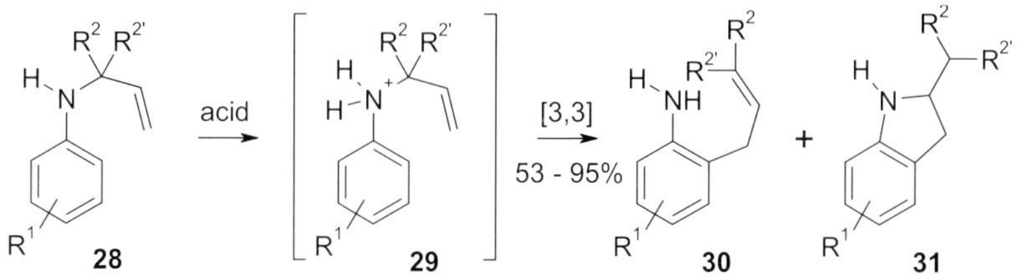

R^1 = H, CO_2Et, OMe. $R^2/R^{2'}$ = Me, Me/Et, Et, -$(CH_2)_5$-, Me/Bn

Scheme 10.8

A successful application of a proton mediated aza-Claisen rearrangement has been described as a key step of the total synthesis of (+)-okaranine J **34** displaying potent insecticidal activity [8b]. Ganesan's reaction, starting from **32**, proceeded at rt to give the desired 7-allylindole derivative **33** with 84% yield (Scheme 10.9).

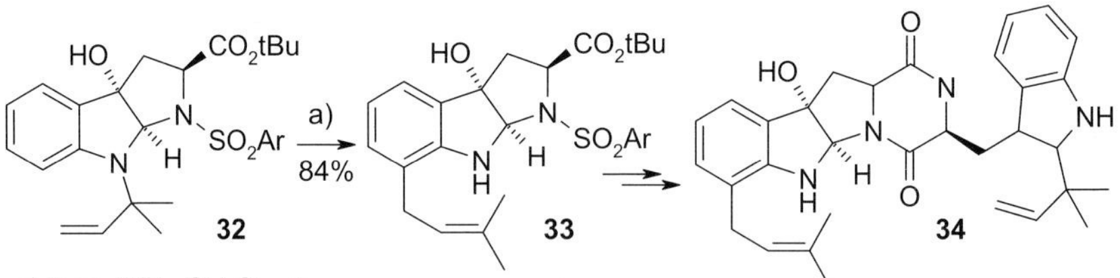

a) 5 eq. TFA, CH_2Cl_2, rt

Scheme 10.9

Lewis acid catalysts also promised to serve as a rate-enhancing factor in aromatic aza-Claisen rearrangements. Early investigations reported on decreased reaction temperatures of < 180 °C but the allyl group migration tended to deliver inhomogeneous products **36** originating from 1,3 and 3,3 processes [9]. Obviously, intermolecular allyl transfer reactions had passed through competing reaction mechanisms. However, Kishi involved such a reaction in an Echinulin **37** total synthesis (Scheme 10.10) [9c].

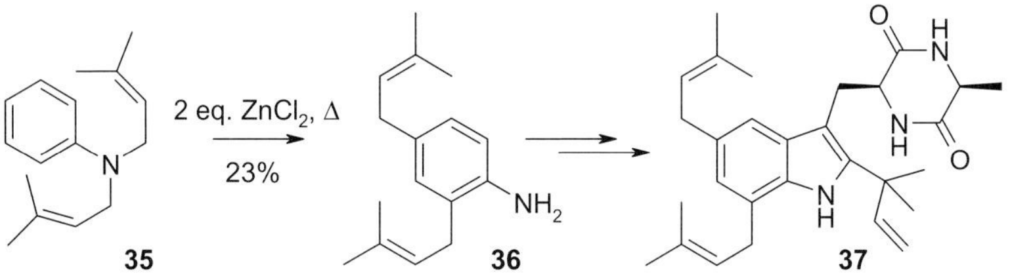

Scheme 10.10

Schmid investigated N-crotyl aniline (type **28**) rearrangements in the presence of $ZnCl_2$. Short reaction times and maximal 1 eq. of the Lewis acid caused a smooth intramolecular aza-Claisen rearrangement to give the *o*-product (type **30**) with allyl inversion of the crotyl chain. In contrast, prolonged reaction times and higher $ZnCl_2$ amounts delivered additional *p*- and cationic cyclization products (type **31**) and isomerized olefins (styrenes) [9d]. In the presence of two *o*-substituents, the *p*-product occurred exclusively (Scheme 10.8, Lewis acid promotion) [9f].

Bicyclic N-allyl anilines **40**, and annulated quinolines **41**, could be prepared by means of hetero Diels–Alder cycloadditions of cyclopentadiene **39** (n = 1) and cyclohexadiene **39** (n = 2), respectively, and imine **38** [9e]. The reaction of cyclopentadiene **39** (reaction as dienophile) in the presence of strong Lewis acids such as BF_3 and SO_2 directly resulted the formation of the quinoline derivative **41** (n = 1, β benzoyl) with 87–88% even at ambient temperature. The use of weaker Lewis acids (CuCl) and low temperatures (–20 °C) enabled to isolate some unstable bicycle **40** (8%, R = NO_2), which could be completely converted into the quinoline **41** upon heating or after treatment with a stronger Lewis acid. In the case of the conversion of cyclohexadiene **39** (n = 2), the acceptor substituted anilines **38** (R = NO_2, Cl) predominantly reacted as dienophiles to give the bicycles **40** with

54–69% yield, the donor derivatives directly led to the quinolines **41** (61–73% yield). However, heating the bicycles **40** to reflux temperature in the presence of BF_3 induced the aza-Claisen rearrangement to give the phenanthridines **41** with almost quantitative yield. The relative arrangement of the benzoyl group played an important role considering the stereochemistry of the products. The *exo*-configured **40** stereospecifically gave the *cis* product **41** (β benzoyl), while the *endo*-configured **40** gave a mixture of *cis* and *trans* phenanthridines **41**. As expected, the ring junction of the new bicyclic system **41** was found to be *cis* (Scheme 10.11).

Scheme 10.11

Stille conducted the first systematic investigations concerning ammonium Claisen rearrangements [10]. Broad variation of suitable acids, solvent, concentration and reaction time showed that $AlCl_3$, $ZnCl_2$ and $BF_3 \cdot Et_2O$, respectively, in refluxing toluene (110 °C) and refluxing mesitylene (140 °C) led to the best results. The treatment of N-allylanilines **42** with acid caused the formation of an intermediate ammonium salt **43**, which underwent 3,3 sigmatropic rearrangement to generate the 2-allylaniline **44**. Bicyclic derivatives **45** were found in some attempts as a minor side product. 4-Alkoxysubstituents decelerated the rate and the corresponding 3-alkoxy-regioisomers enhanced the rearrangement. In this latter case, the regiochemistry could not be influenced, mixtures of 1,2,3 and 1,2,5 trisubstituted products **44** were obtained. Microwave-assisted aza-Claisen rearrangements of N-allylanilines **42** proceeded in very short reaction times (3′–6′) in the presence of Zn^{2+} montmorillonite as a catalyst [10b,c]. Now, the 3,3 sigmatropic rearrangement cationic cyclization tandem process predominated to generate 2-methylindole derivatives **45** with high yield (Scheme 10.12).

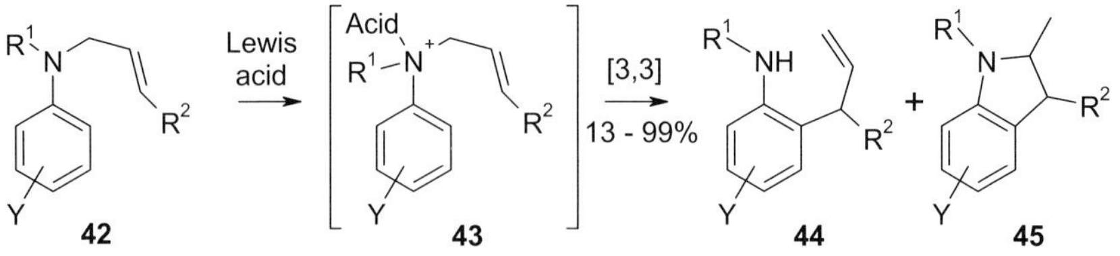

R^1 = Me, Bn. R^2 = H, *n*Pr. Y = H, 3- or 4-OMe. Lewis acid = $AlCl_3$, $ZnCl_2$, BF_3
microwave assistance (3' - 6'):
Y = H: R^1 = H (80%), Me (80%), Et (82%), Bn (78%). β-Naphthyl, R^1 = H (76%)
R^1 = H: Y = 2-Me (78%), 4-F (82%), 4-MeO (85%), 4-NC (79%), 4-O_2N (72%)

Scheme 10.12

Lai used a BF_3-catalyzed rearrangement of N-allylaniline **42** and N-allylindoline **46** as a key step in benzimidazole analogs **49** (X = N) and indole analogs **49** (X = C) of mycophenolic acid (MPA, immunosuppressant and some *anti* psoriasis activity) [11]. The syntheses of the metabolically more stable compounds required a suitably functionalized 7-alkyl indole **48** as a key intermediate. Upon heating of N-allylindoline **46** (R = H) in sulfolane to about 200–210 °C in the presence of $BF_3 \cdot Et_2O$, the aza-Claisen rearrangement, via **47**, delivered 7-allylindoline **48** (R = H) in 47% yield. Generally, such reaction conditions enables one to rearrange a set of aniline **42** and indoline derivatives **46** in 31–57% yield. Further steps allows one to complete the syntheses of three target molecules **49** (X = C) [11b,c] displaying different substitution patterns. One of them (R′ = $CONH_2$) showed significant antitumor activity. One benzimidazole compound **49** (X = N) [11a] was synthesized via the same sequence starting from a nitrogen derivative of 2,3-dehydro-**48** (X = N; Schemes 10.12 and 10.13).

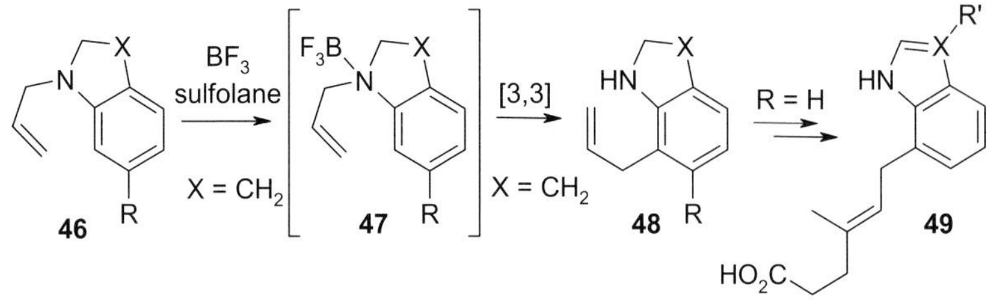

R = H: 200 – 210°C, 2h, 47%
R = OMe: 185 – 190°C, 1h, 31%
R = NO_2: 170 – 175°C, 0.5h, 52%

X = N: no R'
X = C: R' = H, $CONH_2$, CO_2Me

Scheme 10.13

With the intention to investigate the biologically active conformer of the tumor-promoting Teleocidines, Irie and Wender synthesized conformationally restricted analogs of indolactam V **50** (R = Me) [12]. Starting from desmethylindolactam V **50** (R = H), the reaction with allylbromide gave the corresponding N-allyl compound **51**. The heating of the material in the presence of sub-stoichiometric amounts of $ZnCl_2$ and $AlCl_3$ (0.45 mol%) produced the 5-allylated product **52** with

26.4% and 36% yield, respectively. The best results concerning a 3,3 sigmatropic rearrangement/cationic cyclization tandem reaction were achieved upon heating **51** to 140 °C (xylene) in the presence of $AlCl_3$. A mixture of **52**, *β*-Me-**53**, *α*-Me-**53** and **54** was obtained in 29.3% yield and a 43:5.4:9.2:1 ratio. Likewise, cationic cyclization was enforced subjecting 5-allyl indolactam **52** to the $AlCl_3$ promoted reaction conditions resulting **53**/**54** (OH). Alternatively, the OH group in **52** was protected as an acetate (OAc). This compound was then subjected to a Pd(II) catalyzed amination to give OAc-**54** in 67% yield [12d].

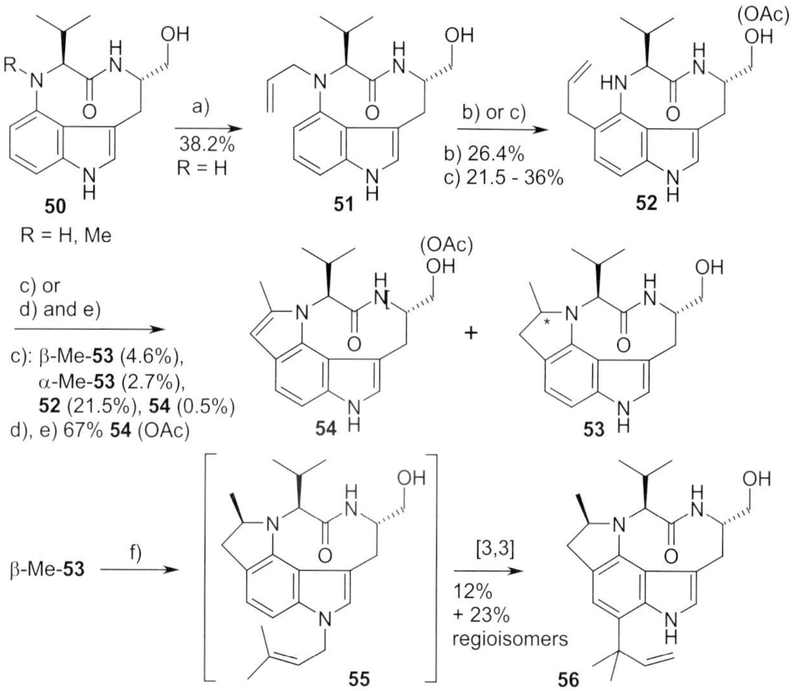

a) allyl-Br,$NaHCO_3$. b) 0.7 eq. $ZnCl_2$,xylene, sealed tube . c) 0.45 eq. $AlCl_3$, xylene 140°C, sealed tube. d) Ac_2O, Py. e) 10% $PdCl_2(MeCN)_2$, 1,4 benzoquinone, 10 eq. LiCl, THF. f) prenyl-Br, AcOH, AcONa

Scheme 10.14

β-Me-**53** and **54** displayed significant tumor promotion activity, which was found to be enhanced after the introduction of a further terpenoid side chain. *β*-Me-**53** was initially treated with prenylbromide to give the intermediate N-adduct **55**. The subsequent aza-Claisen rearrangement gave rise to the formation of 7-prenyl indole **56** as one compound of a mixture of regioisomers (35% yield overall, Scheme 10.14).

Desmethylindolactam G (**50**, R = H, replace *i*Pr with H) was subjected to the same sequence (**50**→**54**). The first allylation step succeeded with 52% yield, but the aza-Claisen rearrangement/cationic cyclization gave only 9.5% of the product mixture **52–54** (replace *i*Pr with H, Scheme 10.14).

Photocyclization of allyl tetrahydroquinoline **60** (n=1) and allyl indoline **60** (n=0) delivered tricyclic compounds **62** with high regioselectivity. The reactant C-allyl systems **60** were produced by means of aza-Claisen rearrangements [13]. Initially, N-allylation succeeded upon treatment of the amine **57** with allyl bromide **58** to give the substrates **59**. Subsequent heating to 140–150 °C in the presence of $ZnCl_2$ caused the 3,3 sigmatropic rearrangements to **60**, the indole derivative **60** (n=0) was obtained in 42% yield along with 9% of the *p*-product **61** (n=0) generated via a final Cope rearrangement (for quinoline: no yield given). It should be pointed out that the reactions incorporating the cinnamyl system (R=Ph) always produced compounds with a 3-phenyl side chain indicating dissociative reaction paths prior to a 3,3 sigmatropic rearrangement (Scheme 10.15).

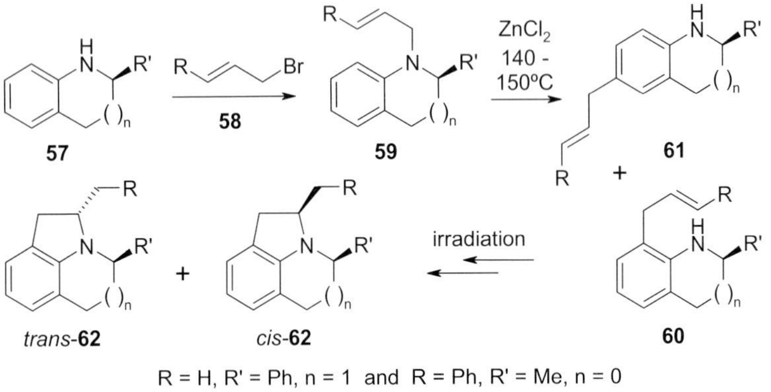

Scheme 10.15

Suitable rearrangement systems could also be generated by N-allylation of a N,N-dialkylaniline **63** to give an ammonium salt **64**. Upon heating **64** in glycerol/H_2O to about 140 °C rearrangement occurred to give the *o* allyl aniline **65** and the *p* product **66** (*o,o* disubstituted material: R′ ≠ H, final Cope rearrangement) with high yield (Scheme 10.16) [14].

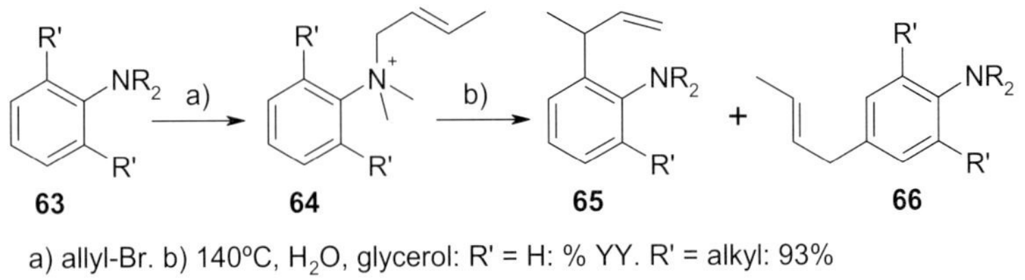

Scheme 10.16

10.3
Aliphatic Simple Aza-Claisen Rearrangements

In addition to the aromatic analogs, simple aliphatic aza-Claisen rearrangements were developed in the early sixties. Generally, the use of a central nitrogen atom offers two advantages. The vinyl double bond of the sigmatropic framework could be prepared with a high *E* selectivity, since a bulky C1 substituent and the chain-branched nitrogen will adopt a maximal distance around the enamine moiety. Furthermore, only two valences of the nitrogen were occupied by the allyl and vinyl substituent of the rearrangement system, the third one could potentially display an optically active sub-unit intending to run the rearrangement under the influence of an external chirality inducing side chain (auxiliary control).

Intending to use the aliphatic simple aza-Claisen rearrangement to generate new C–C bonds, several prerequisites had to be considered. The first problem to be solved was the smooth and selective generation of the allyl vinyl amine backbone. The second challenge is the 3,3 sigmatropic reaction: the rearrangement should pass through a single, highly ordered transition state to give rise to a diastereoselective formation of the product. Since both the reactant and product might suffer from imine enamine equilibration, some difficulties concerning the unique sense of the rearrangement and the stable configuration of a stereogenic α center (with respect to the new imine) have to be taken into account.

In analogy to the experience accumulated with aromatic systems, the induction of the uncatalyzed rearrangements required maximal reaction temperatures of up to 250 °C excluding the presence of a variety of functional groups upon running the processes. Hill investigated some aza-Claisen rearrangements of simple allyl vinyl amines **67** [15]. As expected, the enamine moiety predominantly displayed the *E* configuration right after the condensation. Thus the rearrangement went through a chair-like transition state **68–1** arranging the bulky substituents in a quasi equatorial position leading to the *E* γ,δ unsaturated aldehyde **71** after the final hydrolysis of imine **69**. The corresponding *Z* olefin **72** was found as the minor product because of the sterically more encumbered chair-like transition state **68–2** (axial R^2; Scheme 10.17).

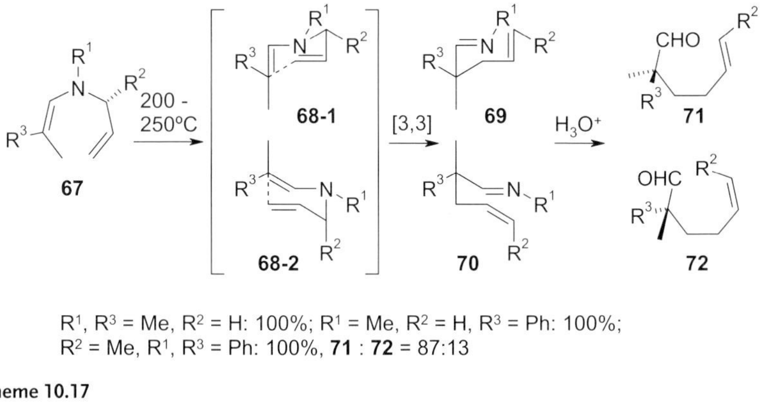

R^1, R^3 = Me, R^2 = H: 100%; R^1 = Me, R^2 = H, R^3 = Ph: 100%;
R^2 = Me, R^1, R^3 = Ph: 100%, **71** : **72** = 87:13

Scheme 10.17

N-allyl/N-vinyl anilines were characterized by a rearrangement of the aliphatic moiety. The aromatic system had not been touched during the course of the process (Winterfeldt: quinolone synthesis [15c]). The rearrangement of a NH allyl vinyl amine failed [15d].

Starting from a mixture of *cis* **75** and *trans* **74** 2,3-divinylaziridines (derived from the thermal decomposition of azide **73**), *cis* material **75** selectively underwent thermal rearrangement with significantly reduced reaction temperature [15e]. The additional loss of ring strain allowed conversion of the reactant aziridine **74** at 140 °C in xylenes to the product **76** even though a boat-like transition state had to be passed. Up to 40% overall yield of **76** was obtained (Scheme 10.18).

Scheme 10.18

N-Allyl difluoro enamines **77** suffered from a rearrangement 2+2 cycloaddition domino process upon heating to about 140 °C. The effectiveness of final cycloaddition (**79**) strongly depended on the bulkiness of R^3. Sterically encumbered substituents R^3 (CPh_2OSiMe_3) supported the rearrangement (**78**). In contrast, a small R^3 ($SiMe_3$, Ph) led to substantial amounts of the cycloadduct **79** (Scheme 10.19) [15f].

Scheme 10.19

Charge acceleration was found to be a promising tool to achieve a significant decrease of the reaction temperature of simple aza-Claisen rearrangements. A temperature of only 80–120 °C was required upon running the sigmatropic rearrangement in the presence of a proton and a Lewis acid, respectively.

The first results were reported by Hill et al. investigating the Lewis-acid-catalyzed formation of imines **82** [16]. In the presence of $TiCl_4$ 3-butenyl-2-N-phenylamine **81**, diverse aldehydes **80** were mixed resulting in a sequence of initial enamine **82** formation and a consecutive aza-Claisen rearrangement to give product **83**. Monosubstituted aldehydes **80** ($R^1 = H$) partly underwent dialkylation side reactions. A final aqueous cleavage gave γ,δ-unsaturated aldehydes **84**. The double bond always suffered from allyl inversion, the major part of all products displayed *E* olefins, indicating that the mechanism passes through a chair-like transition state. Furthermore, the chiral information of the reactant allylamine **81** was shifted into the newly formed C–C bond of the product **83** (1,4 chirality transfer). One product, **84**, served as a key compound in a polyzonimine **85** (millipede pheromone) synthesis (Scheme 10.20).

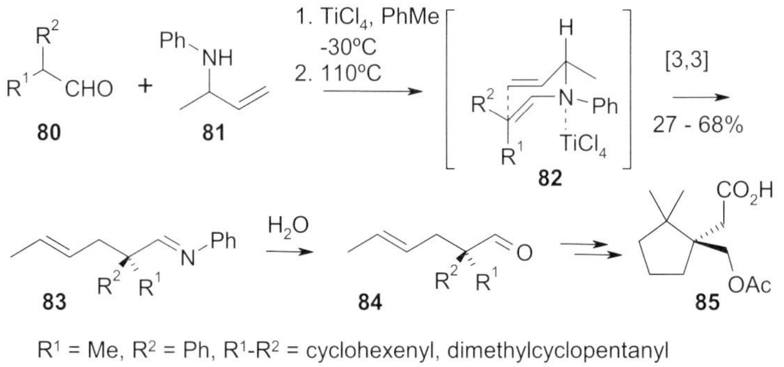

R^1 = Me, R^2 = Ph, R^1-R^2 = cyclohexenyl, dimethylcyclopentanyl

Scheme 10.20

Auxiliary directed Lewis-acid-catalyzed aza-Claisen rearrangements succeeded in employing optically active N-allyl phenylethylamines **87** [16b]. After conducting the sigmatropic reaction, the resulting imine **89** was cleaved and the corresponding aldehyde was immediately reduced to give the carbinol **90**. Mosher analyses allowed the determination of ee values. The stereochemical outcome of the conversions could be rationalized by the predominant chair-like transition state intermediates **88**. Envisaging the synthesis of carbinols **90** bearing a single stereogenic center ($R^2 = H$), the ee remained low (30%). The best results were reported upon generating two adjacent new carbon centers in **90** ($R^1 = Ph$, $R^2 = Me$). Up to 90% ee was obtained by utilizing a quaternary center after running the conversion at lower temperature (55 °C) pointing out an effective influence of the auxiliary center (Scheme 10.21).

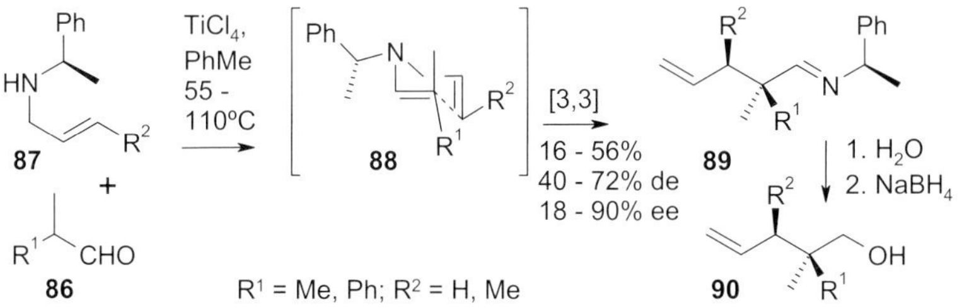

R^1 = Me, Ph; R^2 = H, Me

Scheme 10.21

Mariano investigated proton acid catalyzed aza-Claisen rearrangements of isoquinuclidenes **91** [17]. After an initial O protonation (**92**), the bond reorganization could be rationalized by two sequences. Either a 3,3 sigmatropic rearrangement directly caused the formation of the tricyclic product **95**, or an initial iminium salt fragmentation **93** led to the dienone **94**, which suffered from a consecutive intramolecular Michael addition to give the formal rearrangement product **95**. If R^2 was CH_2CO_2Et, a final cyclization generated lactam **96** exhibiting a lycorane type ring architecture (Scheme 10.22).

*p*TsOH, H_2O, PhH, 80 - 120°C; [3,3]; R^2 = CH_2CO_2Et; R^1 = Ac

R^1 = H, Ac, 1-methyl-2,5-dioxolanyl. R^2 = H, CH_2CO_2Et

Scheme 10.22

A set of systematic investigations has been published by Stille et al. [18]. Starting from allylamine **97** ($R^3 = H$), an optimized three-step sequence of initial imine **99** formation with aldehyde **98** N-acylation to **101** using acid chloride **100** and subsequent $LiAlH_4$ reduction delivered the desired rearrangement systems **102** in 60–96% yield overall. Even though the enamide **101** formation was found to be unselective with respect to the vinyl double bond, the enamine **102** was isolated with substantially higher *E* selectivity pointing out some epimerization during the course of the reduction. The ability to generate defined configured allyl and enamine moieties in the 3,3 sigmatropic rearrangement framework raised the question of achieving internal and external asymmetric induction upon running aza-Claisen reactions [18d]. The first investigations focused on conversions of type **102** reactants. Allylalcohols **105** were converted into the corresponding allylamines **97a** via Overman trichloro acetimidate rearrangement and a consecutive saponification [19]. The three-step sequence led to the allyl vinyl systems **102** with high *E* selectivity concerning the enamine sub-unit and 56 to 89% yield.

The aza-Claisen rearrangement proceeded upon heating the enamines **102** to >100 °C (dioxane, toluene) in the presence of 0.3 to 0.8 eq. of HCl. In every case, the intermediately formed imines **103** were reduced by means of $LiAlH_4$ to give the corresponding amines **104** as stable products. The success of the rearrangement strongly depended on the substitution pattern of the aldehyde **98** involved:

branched starting material **102** ($R^2 \neq$ H) underwent a smooth conversion to give the new amines **104**. In contrast, non-branched derivatives **102** (R^2 or R^3 = H, small rings) suffered from competing reactions such as oligomerization and reductive amination [18a]. The replacement of the originally used HCl by different Lewis acids led to an extension of the original limitations. In the presence of one equivalent of Me_3Al a series of N-allyl enamines **102** could be rearranged to give **104** with high yield (up to 99%, Scheme 10.23) [18b].

The rearrangement of unsymmetrical allylamines **97b** ($R^3 \neq$ H) was investigated to exclude any competing 1,3 rearrangement during the course of the reaction. Allyl vinyl amines **102** were generated via condensation starting from allylamine **97b** and isobutyraldehyde **98** (R^1 = H, R^2 = Me). Subjecting such substrates **102** to the acid-accelerated rearrangement conditions led to smooth conversions to the intermediates **103**, which were finally reduced with $LiAlH_4$ to give the stable amines **104** (displaying *E* double bonds, $R^3 \neq$ H, $R^{3'}$ = H). Careful analyses (GC, ^{1}H NMR) of the products **104** ($R^3 \neq$ H) pointed out that all rearrangements proceeded stereospecifically excluding any crossover products (Scheme 10.23) [18c].

1. Cl_3CCN
2. 140°C
3. 6N NaOH

105

97 (R^3, $R^{3'}$ = H)
97a (R^3 or $R^{3'}$ = H)

98

$-H_2O$

99

RCOCl
100
Et_3N

101

1. $LiAlH_4$
2. aq. NaOH

102

[3,3]
acid, Δ

103

97b

98
$-H_2O$

1. $LiAlH_4$
2. aq. NaOH

104

R = *i*Pr, *i*Bu, Cy. R^1 = H. R^1-R^2 = $(CH_2)_4$, $(CH_2)_3$.
R^2 = H, Me, Et, Ph. R^3 = H, *n*Pr, Ph
Yield rearrangement: acid = HCl: 10 - 98%;
$TiCl_4$: 84 - 92%; Me_3Al: 56 - 99%

Scheme 10.23

Though the rearrangements of the reactants **102** delivered single regioisomers, the products **103** ($R^{3'}$ = *n*Bu/R^3 = H) were isolated as mixtures of diastereomers after the rearrangement reduction sequence. Likewise, the 3,3 sigmatropic process must have passed chair and boat-like transition states, or, alternatively, the products **103** suffered from some epimerization via imine enamine equilibrium. Intending to destabilize boat-like transition states, the aza-Claisen systems **102** ($R^{3'}$ = Me/R^3 = Me) were heated in the presence of an appropriate acid, the rearranged crude imines **103** were immediately reduced by $LiAlH_4$ and DIBALH, respectively. Analyzing the products **104**

($R^{3'}$ = Me/R^3 = Me), revealed that substantially higher diastereoselectivities could be achieved, indicating a nearly exclusive passage through a chair-like transition state during the course of the rearrangement. It was noteworthy that the conversion of the Ph/Me-substituted reactant **102** (R^1 = H, $R^2/R^{2'}$ = Ph/Me) displayed a comparatively high selectivity on generating **104** even though **102** was used as an E/Z vinyl amine mixture. Obviously, the presence of the acid allowed a fast E/Z interconversion prior to the sigmatropic process enabling to rearrange predominantly the isomer with minimized repulsive interactions. The second minor isomer found after running this special sequence was the corresponding *Z* olefin built up via a minor chair transition state (Scheme 10.23).

The first investigations to use a chiral substituent attached to the free valence of the nitrogen for efficient asymmetric induction gave disappointing results [18d]. The rearrangement of N-allyl enamines **106** in the presence of proton and Lewis acids, respectively, produced γ,δ-unsaturated imines as mixtures of diastereomers with acceptable yield and regioselectivity. The de determined after the final $LiAlH_4$ reduction to amine **107** varied between 8 and 20%. Passage through a single transition state conformation failed. Further efforts investigating sterically more-demanding auxiliaries to enable a more effective 1,5 asymmetric induction (1,4 induction including an intermediately built chiral ammonium center) are strongly recommended (Scheme 10.24).

1. acid, Δ, [3,3]
2. $LiAlH_4$
3. aq. NaOH
72% - 86%

106 → *syn*-**107** + *anti*-**107**

R^1 = H, R^2 = Me
R^1 = Me R^2 = H

acid $TiCl_4$: 72%, 20% de, HCl: 86%, 8% de
acid $TiCl_4$: 77%, 15% de

Scheme 10.24

Ring-expansion reactions starting from optically active 1,2-divinyl pyrrolidines **109** promised a smooth generation of nine-membered azonine derivatives **110** bearing defined configured centers and double bonds, respectively [18d]. Firstly, N-BOC prolinal **108** was converted into the divinyl pyrrolidine **109** via Wittig olefination (mixture of *E* and *Z* olefins), protective group removal and a final condensation with phenyl acetaldehyde with 69% (R = Ph) and 79% (R = Me) yield. The rearrangement/$LiAlH_4$ sequence of **109** (R = Me) delivered the azonine **111** (via **110**) as a mixture of diastereomers concerning the stereogenic centers. The olefin was exclusively *Z*, indicating passage through a boat-like transition state. Since the final diastereomer pattern in **111** represented the same ratio as present in the starting material **109**, the result could be interpreted as a sigmatropic rearrangement characterized by a complete 1,3 chirality transfer. However, a thermodynamic reason cannot be excluded, since the reactant **109** and the intermediately formed γ,δ-unsaturated imines **110** might have suffered from some equilibration as observed in several reactions discussed above. In contrast, subjecting the divinyl pyrrolidine **109** (R = Ph) to the rearrangement conditions, no azonine derivative **110** could be isolated. The *E*/*Z* mixture of diastereomers **109** was completely converted into the *E* material **109**. The result was rationalized by a reversible aza-Claisen rearrangement: The initial aza-Claisen reaction led to the nine-membered ring *cis*-**110**, which, after appropriate conformational relaxation suffered from a final 1-aza-Cope rearrangement to regenerate the starting material *E*-**109**, but with inverse absolute configuration (not proven). This special result pointed out that the unique sense of the aza-Caisen rearrangement should be taken in account when planning to use such a process as a key step in total syntheses of complicated compounds (Scheme 10.25).

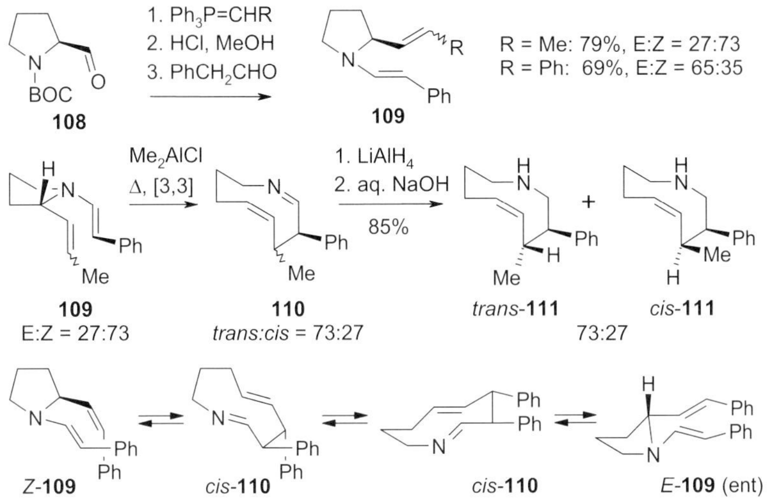

Scheme 10.25

The regioselective enamine formation starting from non-symmetric ketimines **112** has been investigated by Welch [20]. Treatment of **112** with methyl trifluorosulfonate at low temperatures caused N-methylation. The formed iminium salts **113** were then deprotonated by means of proton sponge. Kinetically motivated deprotonation gave rise to the formation of the least substituted enamines **114**, which underwent aza-Claisen rearrangement upon warming up to room temperature. Acid hydrolysis gave the product ketones **115** with 13 to 55% yield (GC, NMR analyzed). The occurrence of a 3,3 sigmatropic process has been proven by deuterium labeling experiments (replacement of marked protons by deuterium, Scheme 10.26).

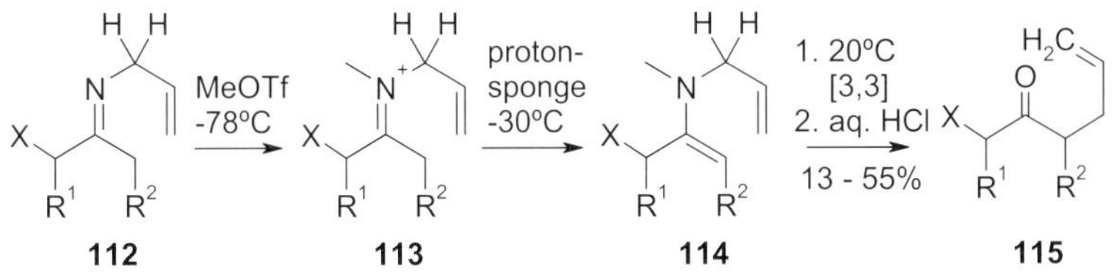

Scheme 10.26

Palladium (0) catalysis promised to be an efficient tool to accelerate aza-Claisen rearrangements. Indeed, the reaction temperatures could be kept between 50 and 100 °C using a Pd(0) catalyst and a strong proton acid. The major drawback reported was that the palladium-assisted aza-Claisen reaction delivered 1,3 as well as 3,3 rearrangement products [21a,b]. The least hindered allyl terminus always formed the new C–C bond restricting the process to symmetric allyl systems.

The synthesis of the tricyclic core of the cytotoxic marine alkaloid Madangamine required an efficient method to generate the central quaternary carbon function. Weinreb employed an aza-Claisen rearrangement in the presence of a palladium catalyst [21c]. After treatment of ketone **116** subsequently with TOSMIC and DIBALH, the formed carbaldehyde **117** was reacted with diallylamine in the presence of $Pd(OCOCF_3)_2/PPh_3$. Initially, the enamine **118** was formed, which underwent diastereoselective aza-Claisen rearrangement. The γ,δ unsaturated imine **119** was cleaved with aqueous HCl and the corresponding aldehyde **120** was isolated in 68% yield. Several further steps allowed one to complete the synthesis of the core fragment **121** of the natural product (Scheme 10.27).

Scheme 10.27

Alternatively, a significant temperature-decreasing effect was achieved after conversion of the central nitrogen into a per-alkylated ammonium salt. Early observations were published by Brannock and Burpitt as well as by Opitz and Mildenberger [22]. In most runs, enamines **122** were obtained via the condensation of secondary amines and aldehydes. Investigating the C alkylation of enamines **122**, the use of crotylhalides and tosylates **123** gave some aldehyde **125** via direct enamine C alkylation along with varying amounts of unsaturated aldehyde **128**. The occurrence of the latter could be rationalized by an initial N-alkylation (**126**) and a subsequent 3,3 sigmatropic aza-Claisen rearrangement to give the product **127** with allyl inversion. The substitution patterns of the nitrogen and the enamine β-C determined the outcome of the reaction: bulky N substituents suppressed any N-allylation. In contrast, small and cyclic N substituents and highly substituted β-C facilitated the rearrangement enabling to generate quaternary carbon centers. The

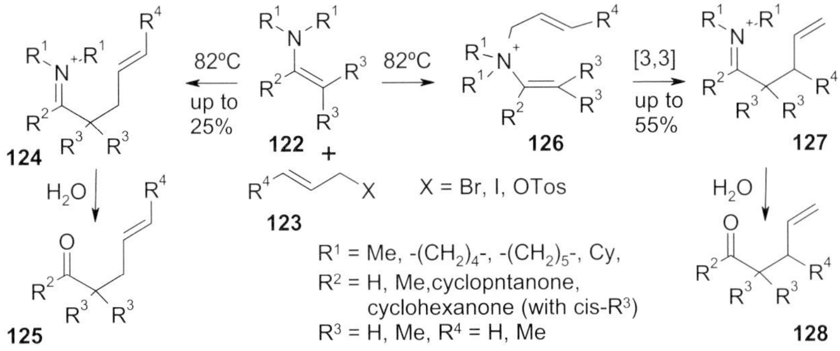

Scheme 10.28

use of propargyl halides allowed one to produce some allene via the N-alkylation rearrangement cascade (Scheme 10.28).

McCurry employed such a cationic aza-Claisen rearrangement as a key step in spiro-sesquiterpene total synthesis [22c]. Initially, aldehyde **129** was converted into the corresponding pyrrolidine enamine **130**. Subsequent treatment with *trans* crotylbromide deliverd the ammonium salt **131**, which suffered from 3,3 sigmatropic rearrangement upon heating to 82 °C. The diastereomers **132** were obtained with 75% overall yield, in a ratio near 1:1 depending on the special reaction conditions pointing out the competing transition states **131–1** and **131–2**. However, the syntheses of racemic *β*-vetivone **134** and *β*-vetispirene **133** succeeded after several further steps (Scheme 10.29).

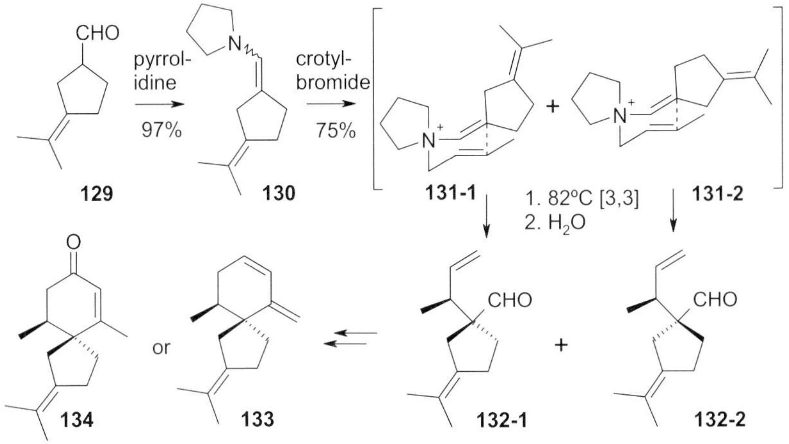

Scheme 10.29

A domino process of enamine **136** formation, N-allylation, aza-Claisen rearrangement and a final Mannich condensation was introduced by Florent [22g]. Aldehyde **135** was subsequently treated with pyrrolidine and allyl iodide **137** to give an *E*/*Z* mixture of the ammonium salts **138**. Heating to 80 °C induced the Claisen rearrangement. The newly formed iminium ions **139** underwent intramolecular Mannich cyclizations. The final amine elimination delivered the spiro ketones **140** with 38% yield as a 2:1 mixture of diastereomers. The formed material should serve as a key compound in diverse cyclopentenone prostaglandine total syntheses (Scheme 10.30).

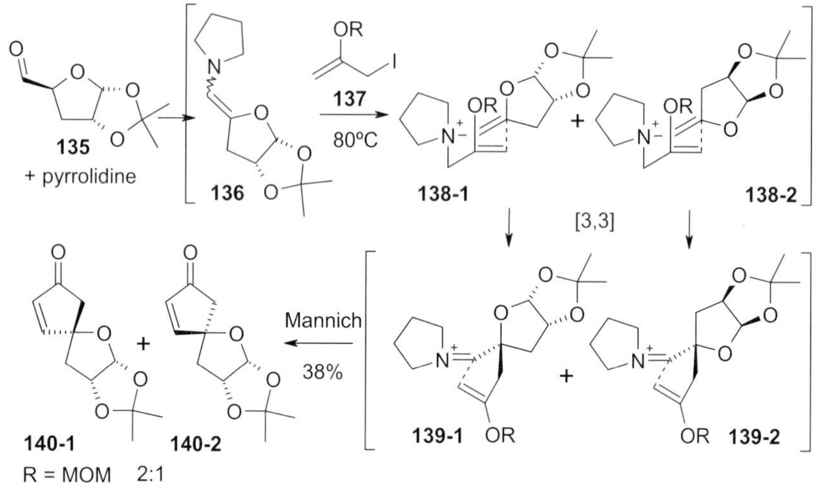

Scheme 10.30

Inoue used a tandem double bond isomerization aza-Claisen rearrangement to generate trisubstituted *E* double bonds with defined configuration [22f,h]. Starting from N,N diallyl ammonium salts **141**, the treatment with base in a protic solvent caused the abstraction of the most acidic/least hindered proton of one allyl system to allow the shift of the double bond into a vinyl situation in **142**. The formed allyl vinyl system suffered from an immediate 3,3 sigmatropic rearrangement at rt or somewhat above to build up the corresponding γ,δ unsaturated aldehyde **143**. The nascent *E* olefin originated from passage through a chair-like transition state with minimized repulsive interactions. Alternatively, the iminium salt formed right after the rearrangement could be reduced by means of a hydride donator. Propargyl amines **144** isomerized to give the allenyl allylamines **145**, which rearranged to generate α,β aldehydes (after hydrolysis) and allylamines **146** (after reduction), respectively (Scheme 10.31).

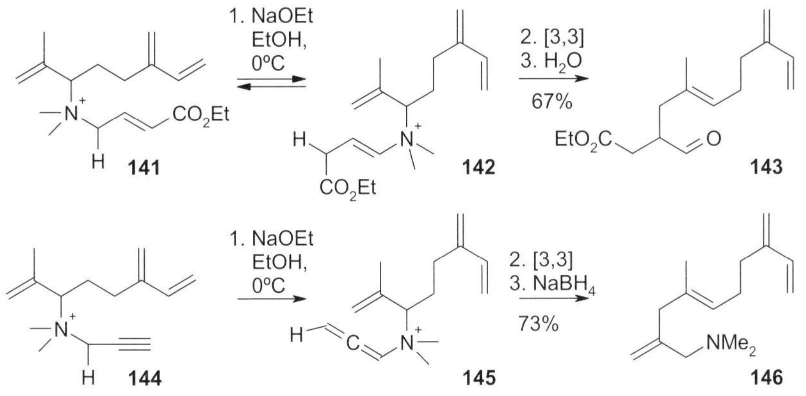

Scheme 10.31

An alternative access to allyl vinyl rearrangement systems was introduced by Gilbert [22e]. The condensation of diethyl(diazomethyl)phosphonate **147**, the ketone **148**, and a secondary allylamine **149** gave the allyl vinylamine **150**, which could be activated (**126**) for rearrangement via N-methylation using dimethylsulfate (Scheme 10.32).

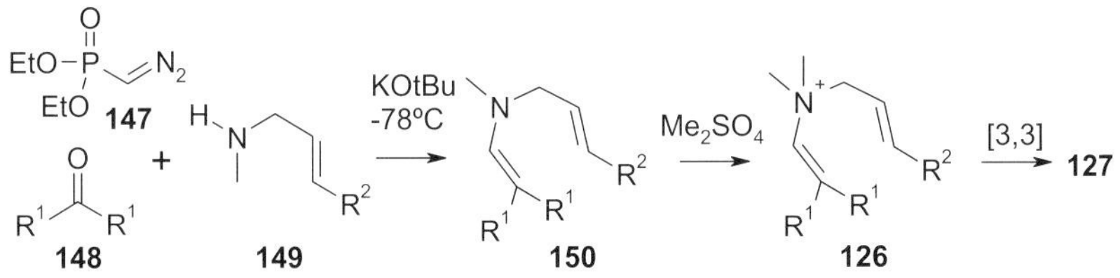

Scheme 10.32

Dieneamines **151** reacted with croylbromide to give the simple aza-Claisen products **152** with about 66% yield along with 17–19% of the conjugated system (isomer-**152**) [22d]. In contrast, the corresponding cinnamyl derivatives suffered from a final Cope rearrangement to give the unsaturated ketone (30% yield). Furthermore, a second alkylation-rearrangement cascade allowed one to introduce two cinnamyl substituents (**153**) with up to 56% yield. The stereochemical outcome of the aza-Claisen rearrangement could always be rationalized by a chair-like transition state (Scheme 10.33).

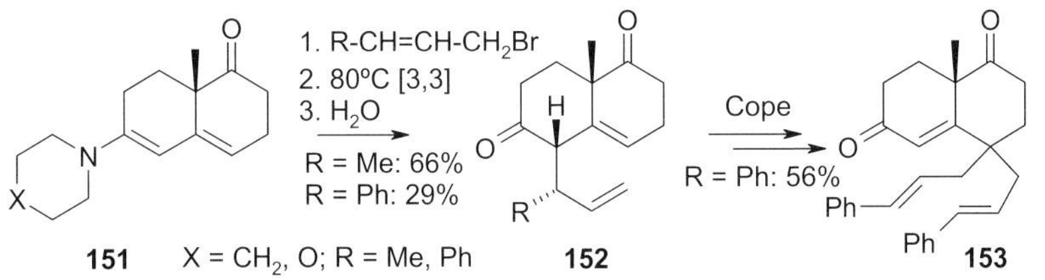

Scheme 10.33

Maryanoff used such a reaction to generate tricyclic tetrahydroisoquinoline derivatives **158** bearing a quaternary center with defined configuration [23]. Initially, indolizidine **154** was allylated. The open-book shape of **154** forced the allylbromide to attack the *exo* face, building up a chiral ammonium center with defined configuration in **155**. TFA-mediated H_2O elimination delivered the vinyl ammonium sub-unit in **156**. The quinolizidinine system **156** (n= 2) rearranged slowly at 23 °C to give a single diastereomer **157** (n= 2), $NaBH_4$ reduction gave the amine **158** (n= 2) with 90% yield. Bridgehead H and allyl groups exclusively displayed *cis* arrangement, pointing out a complete 1,3 transfer of the chiral information from the ammonium center toward the new quaternary carbon position. In contrast, the indolizidinine systems **156** (n= 1) required higher reaction temperatures to induce the rearrangement. After about 2 h at 100 °C, iminium salt **157** (n= 1) was formed as a 95:5 mixture of allyl/bridgehead H *cis* and *trans* relative configuration. The final $NaBH_4$ reduction produced the indolizidinines **158** (n= 1) in a 91:9 ratio of diastereomers and 73% yield. The loss of chiral information has been referred to as a 5 to 10% portion passing through a dissociative, non-concerted reaction path at the elevated temperatures. This hypothesis was supported by deuterium labeling experiments (>90% allyl inversion, Scheme 10.34).

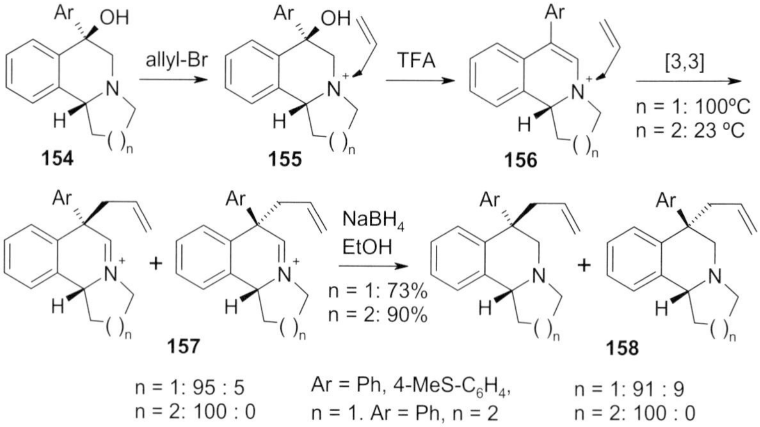

Scheme 10.34

Winter reported on aza-Claisen rearrangements in transition-metal complexes [24]. *Cis* ruthenium dichloride complexes **159** added butadiyne in the presence of non-nucleophilic $NaPF_6$ to generate a cationic butatrienylidene intermediate, which was trapped by a regioselective γ-addition of dialkyl allylamines to produce a vinyl allylammonium salt **160**. This material underwent an immediate aza-Claisen rearrangement. The formation of a resonance-stabilized iminium salt **161** was thought to serve as the driving force upon running this process. While simple allylamines gave smooth reactions at ambient temperature, the rearrangement of sterically more demanding compounds failed. The rearrangement of a propargylamine proceeded at elevated temperature and the product β,γ allenylamine was obtained with 90% yield. Extensive computational studies gave a somewhat reduced activation barrier of the allyl vinyl ammonium system compared to the uncharged analog (Scheme 10.35) [25].

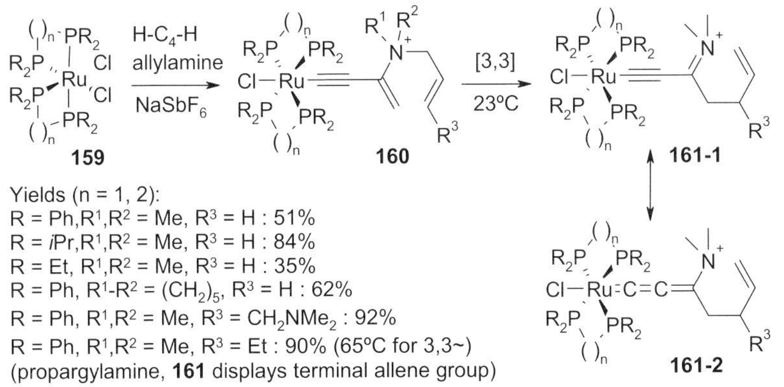
H-C4-H
allylamine
NaSbF6
[3,3]
23°C
159
160
161-1
161-2
Yields (n = 1, 2):
R = Ph,R1,R2 = Me, R3 = H : 51%
R = iPr,R1,R2 = Me, R3 = H : 84%
R = Et, R1,R2 = Me, R3 = H : 35%
R = Ph, R1-R2 = (CH2)5, R3 = H : 62%
R = Ph, R1,R2 = Me, R3 = CH2NMe2 : 92%
R = Ph, R1,R2 = Me, R3 = Et : 90% (65°C for 3,3~)
(propargylamine, 161 displays terminal allene group)

Scheme 10.35

10.4
Amide Acetal and Amide Enolate Claisen Rearrangements

Analogously to the oxy analogs of Eschenmoser and Johnson rearrangement, amide acetal and amidines were tested within 3,3 sigmatropic processes. Because of the electron-rich character of the en diamine **162** the rearrangement to **163** required a roughly 280 °C reaction temperature. In contrast, an acceptor-substituted allyl system **164** rearranged to give **165** at 25 °C (Scheme 10.36) [26].

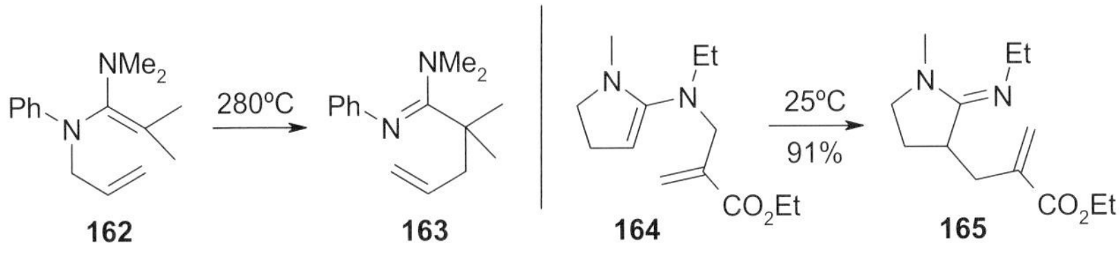
NMe2
Ph
N
280°C
25°C
91%
Et
CO2Et
162
163
164
165

Scheme 10.36

Kurth developed an auxiliary-directed asymmetric aza-Claisen rearrangement [27]. Starting from valinol derived oxazolines **166** an N allylation succeeded using hard allyl tosylates **167** to give the corresponding oxazolinium salts **168**. Deprotonation with a strong base delivered the allyl vinyl system **169**, which underwent rearrangement upon heating to 155 °C in an inert solvent to give the oxazolines **170** with >80% yield and 87:13 dr. The stereochemical outcome could be rationalized by an initial diastereoselective formation of the *E* ketene N,O acetal double bond in **169**. Then, the system preferentially adopted chair-like transition state geometries. The ally moiety could be arranged *anti* (major, **169–1**) and *syn* (minor,

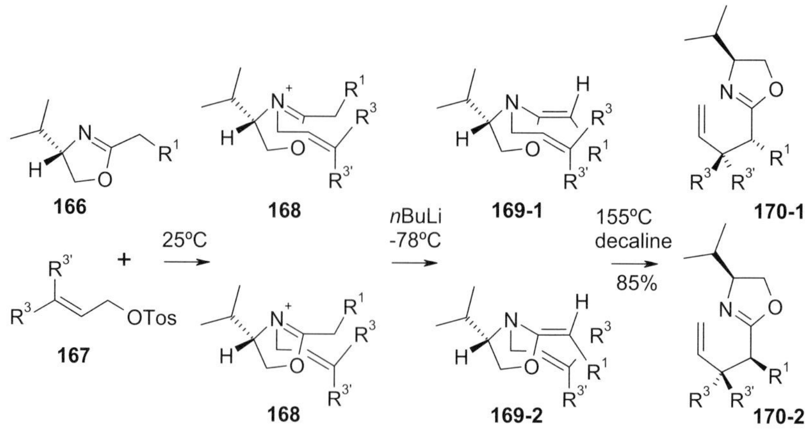

R^1 = Me, R^3, $R^{3'}$ = H: 87:13. R^1, $R^{3'}$ = H, R^3 = Me: 87:13. R^1, R^3 = H, $R^{3'}$ = Me: 14:86

Scheme 10.37

169–2) with respect to the auxiliary isopropyl group. The non-complete auxiliary-directed diastereoselection caused the formation of the mixture of diastereomers **170** (Scheme 10.37).

However, passage through alternative boat-like transition states could not be excluded. Placing a single substituent in position R^1, high diastereoselectivities of up to 98:2 (81% yield) could be achieved. In the presence of $R^1 = H$ and $R^3/R^{3'} \neq H$, the diastereoselectivities varied between 76:24 and 97:3 (21–78% yield) depending on the substitution pattern. If $R^1/R^3/R^{3'} \neq H$, two new stereogenic centers could be built up with 79–92% de [27d]. The generation of a defined quaternary center succeeded in selected cases [27f]. As an application, the ketene aminal aza-Claisen rearrangement was employed as a key step in a (+) dihydropallescensin-2 **176** total synthesis (**171**→**176**, Schemes 10.37 and 10.38) [27e].

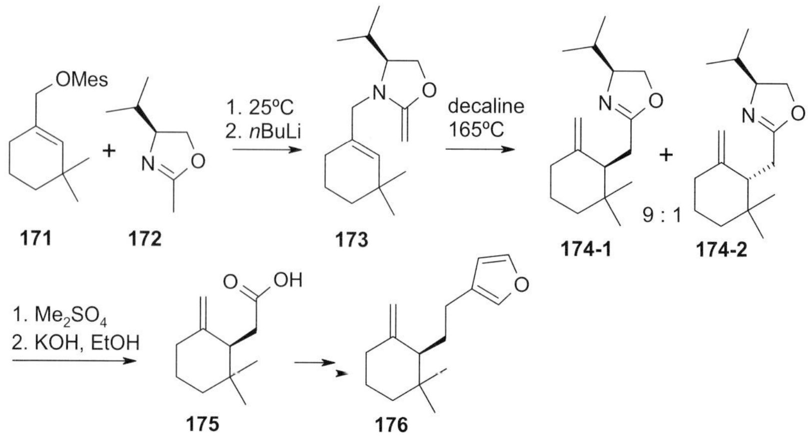

Scheme 10.38

Anionic N-allyl amide enolate rearrangement promised to be a versatile tool for highly effective sigmatropic processes. In analogy to the oxygen systems (Ireland esterenolate and silylketene acetal rearrangement), relatively mild reaction conditions should be applicable. In contrast to the oxygen systems, the fragmentation to generate a ketene and an amide anion was unlikely to occur. Furthermore, the bulky nitrogen forced the nascent amide acetal to adopt the *Z* configuration offering the advantage of achieving high internal asymmetric induction (high simple diastereoselectivity) upon forming the product γ,δ-unsaturated amides.

Basic systematic investigations of amide enolate rearrangements concerning reaction conditions and diastereoselectivity have been published by Tsunoda et al. [28]. *E* and *Z* N-crotyl propionic acid amides **177** were initially deprotonated with LDA at −78 °C in THF to form the corresponding *Z* enolates **178**, respectively [28a,b]. After exchanging THF with a high-boiling non-polar solvent rearrangement was induced upon heating. The corresponding *anti* γ,δ unsaturated amides **179** were generated with high yield. A high diastereoselectivity was achieved starting from the reactant olefin *E*-**177** via a chair-like transition state *E-c*. In contrast, the corresponding *Z* crotyl reactant *Z*-**177** gave the product **179** with high yield, too, but with a low diastereoselectivity of *syn*-**179**:*anti*-**179** = 63:37 pointing out the passage through competing chair- and boat-like transition states *Z-c* and *Z-b* (Scheme 10.39).

E-**178** into **179**: 135°C, 4h, 94% yield, ratio *anti*-**179**:*syn*-**179**: 99.4:0.6
Z-**178** into **179**: 148°C, 14h, 90% yield, ratio *anti*-**179**:*syn*-**179**: 37:63

Scheme 10.39

The high yield and the high internal asymmetric induction obtained via the rearrangement of *E*-**177** to *anti*-**179** raised the question concerning the efficiency of additional external asymmetric induction [28a,b]. The best prerequisite provided amide **180**, placing the chiral center of the phenethyl side chain (attached to the nitrogen) adjacent to the 3,3 sigmatropic rearrangement core system. The chiral auxiliary-directed rearrangement of the N-crotylamide *E*-**180** led to the corresponding γ,δ-unsaturated amides *anti*-**181** with 85% yield, no *syn*-**181** amides were found, indicating the exclusive passage through chair-like transition states *E-c*.

The careful analysis of the product *anti*-**181** gave a composition of *anti*-$\mathbf{181}_a$:*anti*-$\mathbf{181}_b$ of 8.5:91.5. The outcome could be rationalized by the predominant passage through a chair-like transition state *E*-c_b causing the formation of *anti*-$\mathbf{181}_b$ with 83% de. The removal of the auxiliary proceeded by initial acylation of the secondary amide **181** with acid chloride **182**. Then, neighboring-group assisted saponification of **183** gave the optically active γ,δ-unsaturated acid **184** with 63–82% overall yield. The chiral auxiliary phenethyl amine could be recovered in two further steps in 71% yield overall starting from **185**. Investigating a series of chiral auxiliaries attached to amide *E*-**180**, most *S* configured auxiliaries predominantly led to product *anti*-$\mathbf{181}_b$ (yield: 66–90%, de: 70–84%). As an exception, (2*S*)-3,3-dimethylbutyl-2-amine resulted *anti*-$\mathbf{181}_a$ as the major product (38% yield, 80% de, Scheme 10.40) [28c,j].

185: 1. NaH, $(EtO)_2CO$, DMF. 2. 48% aq. NaOH, 120°C → *S*-1-phenethylamine (71%)

Scheme 10.40

Anticipating the employment of the amide enolate Claisen rearrangement as a key step in natural product total synthesis, scope and limitations of the process have been investigated [28d]. N-allyl amino and α-hydroxy acetamides *E*-**186** were initially treated with base to generate the corresponding anionic species. If possible, lithium chelate formation (enolate O and α heteroatom) should fix the *Z* enolate geometry. The glycolic acid amides *E*-**186** (R^1 = OH, OTBS) underwent smooth rearrangement upon heating to give the γ,δ-unsaturated amides **187** with 59–95% yield. The chiral auxiliary-directed rearrangements led to de's of 34 to 73%. The *N*-crotyl glycine derivatives *E*-**186** ($R^1 = NH_2$, NHBOC) displayed different reactivity. Despite varying the conditions, the NHBOC did not give any rearrangement product. In contrast, *E*-**186** ($R^1 = NH_2$) already underwent rearrangement at ambient temperature to build up **187** with 81–89% yield and 78% de after running the auxiliary-directed conversion (Scheme 10.41).

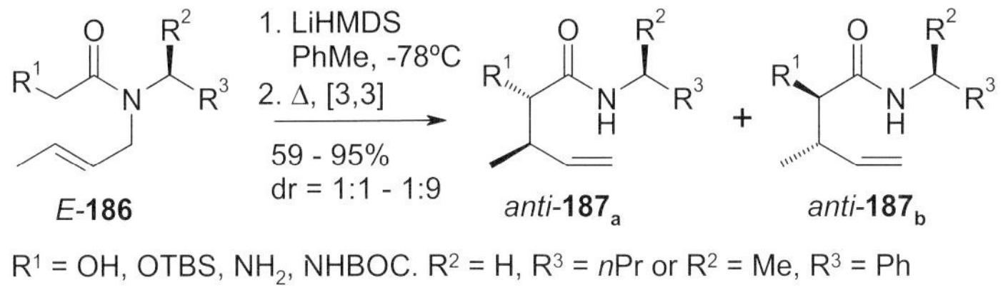

R^1 = OH, OTBS, NH_2, NHBOC. R^2 = H, R^3 = *n*Pr or R^2 = Me, R^3 = Ph

Scheme 10.41

The first applications in total syntheses started from amide **188**. The rearrangement, under standard conditions, led diastereoselectively to amide **189** in 77% yield; no traces of other diastereomers were found. Amide **189** was converted into (–)-isoiridomyrmecin **190** via diastereoselective hydroboration. Oxidative work-up and a final lactonization in the presence of a *p*TsOH (41.4% yield over three steps) [28f]. Alternatively, $LiAlH_4$-reduction of **189** resulted in amine **191** (80%), a final five step sequence enabled one to synthesize (+)-α-skytanthine **192** with 60% yield (Scheme 10.42) [28g].

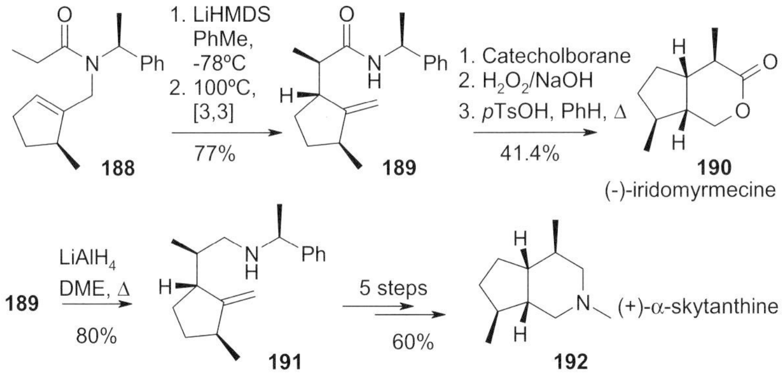

Scheme 10.42

As an extension of the method mentioned above, a successful rearrangement starting from silylenolethers as an electron-rich N-allyl moiety has been published. Allylamide **193** was synthesized in two steps starting from (+)-phenethylamine and acrolein via a condensation acylation sequence. The auxiliary-directed aza-Claisen rearrangement of **193** produced a mixture of two diastereomers, the major compound **194** was isolated by means of column chromatography. Iodocyclization and subsequent reductive removal of the halide resulted γ-butyrolactone **195**, which served as the western half of the bislactone framework of antimycine A_{3b} **196**. Starting from **195** the total synthesis of **196** (component of antimycine A mixture: oxidoreductase inhibitor) was completed in several further steps (Scheme 10.43).

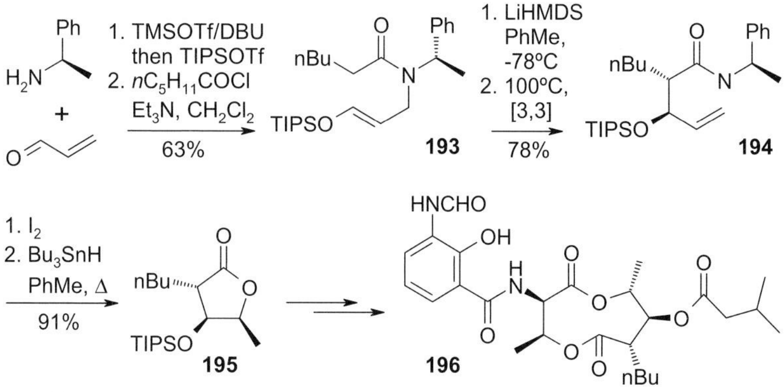

Scheme 10.43

The synthesis of the medium-sized lactam **199** started from 2-vinyl piperidine **197** [29]. The amide enolate aza-Claisen rearrangement led to the corresponding ten-membered ring lactam **199**. Reacting terminally unsubstituted olefins as in **197**, a complete 1,4-chirality transfer was observed pointing out the highly efficient internal asymmetric induction. The stereochemical outcome of the process was rationalized by the chair-like transition state **198** minimizing repulsive interactions. As described above, the amide enolate in **198** should have been *Z* configured. One optically active azecinone **199** (R = Me, Et side chain) served as key intermediate in an asymmetric total synthesis of fluvirucin A_1 (Scheme 10.44).

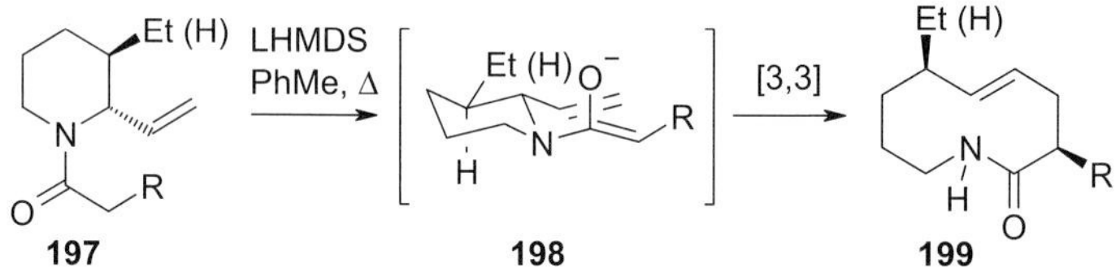

R = H (H): 40%, R = Me (H): 75%, R = OMe (H): 84%, R = Me (Et): 74%

Scheme 10.44

Somfai enhanced the driving force of some amide enolate aza-Claisen rearrangements by choosing vinylaziridines as reactants [30]. The additional loss of ring strain offered the advantage to run most of the reactions at room temperature to synthesize unsaturated chiral azepinones. Various substitution patterns have been tested to investigate the scope and limitations of this process. The stereochemical properties of the 3,3 sigmatropic rearrangement enabled the transfer of stereogenic information of easily formed C–N bonds completely to new C–C bonds by means of the highly ordered cyclic *endo* transition states **201**. The configuration of the allyl double bond was retained throughout the reaction. Furthermore, the defined enolate geometry of the *in situ* formed ketene aminal double bond caused a high internal asymmetric induction leading to one predominant relative configuration of the newly generated stereogenic centers.

The reactant divinylaziridines **200** were synthesized via ex-chiral pool sequences starting from optically active *α*-amino acids. Rearranging divinylaziridines **200** via a boat-like transition state **201** explained the stereochemical outcome of the reactions to give the azepinones **202** in 60 to 85% yield. The rearrangement of an *α* branched amide **200** (replace CH_2R^2 by $CH(R^2)CH_3$) only required a higher reaction temperature of about 65 °C to induce the conversion. A single diastereomer **202** was generated bearing *β*-methyl and *α*-amino function. Surprisingly, the *α*-NHBOC group was found to be replaced by a urea sub-unit, indicating a defined substitution during the course of the reaction (neighbor-group assisted cleavage of the BOC protective group, Scheme 10.45).

LHMDS, THF, -78°C - rt; [3,3], 60 - 85%

200 **201** *endo* **202**

R^1 = H, OBn, Bn, R^2 = H, Me, OBn, R^3 = H, Me, OBn, NHBoc R^4 = H, CH_2OBn

Scheme 10.45

Neier tested a variety of reaction conditions starting from amide keteneacetal **204** [31]. Thermal aza-Claisen rearrangements were induced upon heating reactant **204** in decaline to 135–190 °C. Though some product **205** could be obtained, the best yields achieved were about 40%. Especially the substituted allyl moieties caused poor yields and the formation of diastereomer mixtures. Lewis acid mediation was investigated by treating reactants **204** with 0.7 mol% of $ZnCl_2$ in PhMe with heating to 85 °C. No product **205** was formed, *β*-ketoamides **206** were isolated pointing out the predomination of a competing reaction path. Further testing of various Lewis acids did not result in any rearrangement product **205**. The Overman rearrangement conditions employed the soft electrophile $Pd(PhCN)_2Cl_2$ to accelerate bond reorganization [32]. The reaction mechanism was likewise rationalized by a Pd(II) catalyzed sigmatropic rearrangement or by a cyclization fragmentation sequence. However, rearrangement-type products should be formed. In the present investigations, the reaction of N,O-acetal **204** gave some product **205**; 50 mol% Pd(II)-catalyst were necessary to obtain a maximum of 48% yield. In contrast, the cyclohexenylamine derivative **204** produced no amide **205**. Instead, bicyclic imides **207** were found with low yield indicating passage through a Pd(II)-mediated cyclization *β*-hydride elimination tandem process (Scheme 10.46).

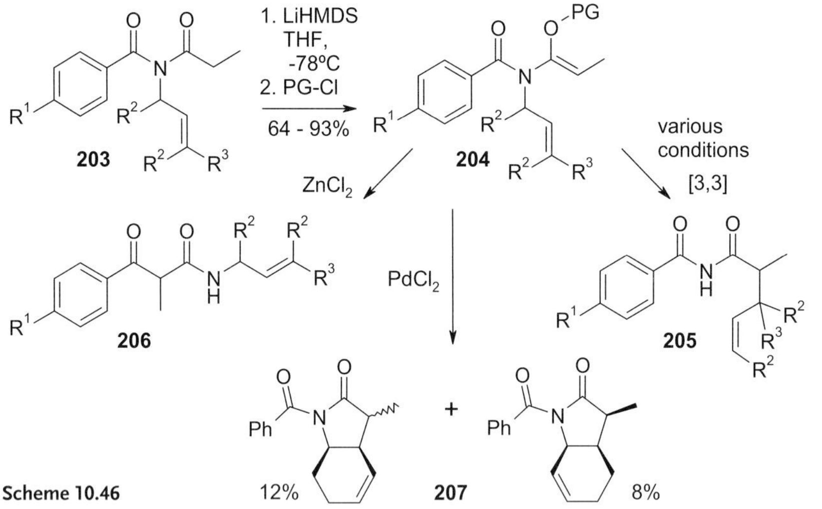

Scheme 10.46

10.5 Zwitterionic Aza-Claisen Rearrangements

Charge-accelerated ammonium and amide enolate aza-Claisen rearrangements allowed one to reduce the originally high reaction temperatures of 200–350 °C to about 80–140 °C. Considering the significant nucleophilicity of some tertiary amines **208**, the addition to neutral electrophiles must cause an initial charge separation. Constructing such a zwitterion using an allylamine **208** as nucleophile and a triple bond **209** or an allenic species **213** as electrophile, the combination gives rise to an aza-Claisen rearrangement framework **210/214**. A consecutive 3,3 sigmatropic bond reorganization to **211/215** should profit from charge neutralization holding out the prospect of a further decrease of the reaction temperature. A significant extension of the limitations in tolerating more complicated substitution patterns and functional groups within the reaction as well as higher stereoselectivities seems to be achievable. Additional base and (Lewis) acid catalysis could be employed to support the reaction. One problem should be pointed out: the addition step of nucleophile **208** and electrophile **209/213** is potentially reversible. Consequentially, the fragmentation of **210/214** to regenerate the reactants does not occur as a negligible process (entropy, charge neutralization). Intending to use the zwitterionic intermediate **210/214** for Claisen rearrangement a sufficient lifetime of the charge-separated species must be taken in account to achieve the highly ordered transition state as a prerequisite for the 3,3 sigmatropic process. Analyzing the systems suitable for the intermolecular zwitterionic aza-Claisen rearrangement (addition/rearrangement tandem process), three combinations were found to be of interest (Scheme 10.47).

The first one can be described as the Michael addition of the N-allylamine **208** to an acceptor-substituted triple bond of **209** to form an intermediate N-allyl enammonium enolate **210** (alkyne carbonester Claisen rearrangement). The anion stabilizing group

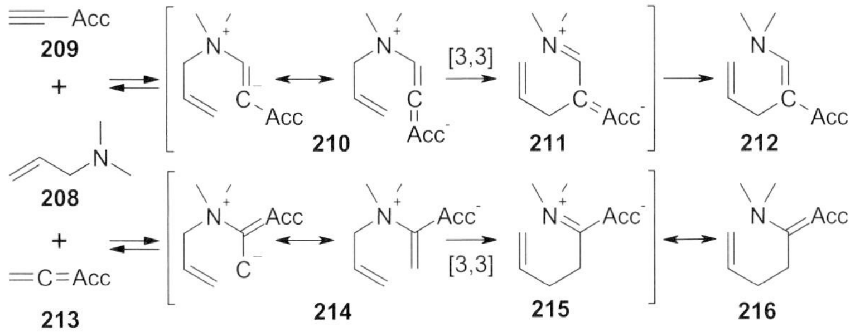

Acc = acceptor, example: in **209**: Acc = CO_2Me, in **213**: Acc = O, $CHCO_2Me$.

Scheme 10.47

is placed in position 1 of the rearrangement core system. 3,3 sigmatropic rearrangement delivers an iminium enolate **211**, which undergoes immediate charge neutralization to form an acceptor-substituted enamine **212** (Scheme 10.47).

The second mechanism starts with an addition of the N-allylamine **208** to the cumulated acceptor system of a ketene **213** (Acc = O) to form an intermediate N-allyl ammonium enolate **214** (ketene Claisen rearrangement) [33]. The anion stabilizing group is predominantly placed in position 2 of the rearrangement core system, resonance stabilization can place the anion in position 1, too. Then, 3,3 sigmatropic rearrangement gives rise to an iminium analog carboxylate (**215**, $Acc^- = O^-$), which represents the zwitterionic mesomer of the charge neutral amide **216** (Acc = O,).

The third mechanism starts with the addition of N-allylamine **208** to the cumulated acceptor system of an allene carbonester **213** (Acc = $CHCO_2Me$) to form an intermediate N-allyl ammonium amide enolate **214** (allene carbonester Claisen rearrangement). The anion-stabilizing group is exclusively placed in position 2 of the rearrangement core system. Then, 3,3 sigmatropic rearrangement generates an iminium enolate (**215**, $Acc^- = CH^-CO_2Me$), which represents the zwitterionic mesomer of the charge neutral vinylogous amide **216** (Acc = $CHCO_2Me$, Scheme 10.47).

All reaction mechanisms presented here should be understood as hypotheses to rationalize the outcome of the processes. Alternative explanations such as stepwise and dissociative mechanisms cannot be excluded. However, as long as constitution and configuration of the product can be described as that of a 3,3 sigmatropic rearrangement, the present hypotheses seem acceptable.

10.5.1
Alkyne Carbonester Aza-Claisen Rearrangements

Mariano et al. developed alkyne carbonester Claisen rearrangements as key steps in alkaloid syntheses. A convincing application of such a process was published in 1990 describing the total synthesis of the Rauwolfia alkaloid Deserpidine **223** [34]. Isoquinuclidene **217** was synthesized in several steps starting from a dihydropyridine. The heating of a mixture of amine **217** and propiolester **218** in MeCN to about 80 °C induced the addition rearrangement sequence to give the isoquinoline derivative **220**. Careful optimization of the reaction conditions improved a crucial role of the diethyl ketal function to obtain a smooth reaction. The addition of the alkynoic acid proceeded *anti* with respect to the ketal building up zwitterion **219** bearing the suitable 3,3 sigmatropic rearrangement framework. Without any indole protecting group (PG = H) only 39% of the desired product was obtained. Rearranging the sulfonamide of **217** (PG = $PhSO_2$) an acceptable 64% yield of ketone **221** was isolated after consecutive cleavage of the ketal. Heating in aqueous acetic acid induced Wenkert cyclization to generate tetracyclus **222** as a mixture of three diastereomers. Several further steps allowed completion of the total synthesis of deserpidine **223** (Scheme 10.48).

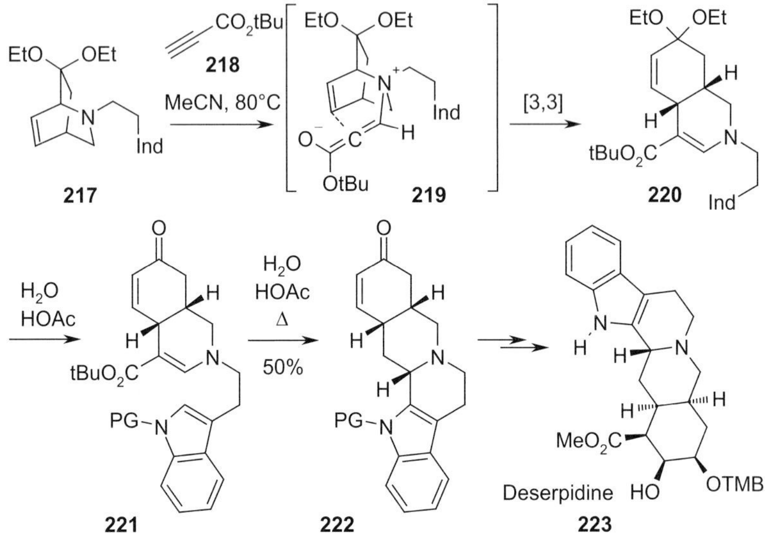

Ind = N-PG-3-Indoyl, yield **217** in **221**: PG = H: 39%, PG = $PhSO_2$: 64%

Scheme 10.48

Vedejs and Gingras investigated intermolecular aza-Claisen rearrangements of acetylene dimethyldicarboxylates **225** and methyl propiolate **234** and various acyclic and cyclic N-allylamines **224**, **230** and **233** [35]. Proton and Lewis acids, respectively, were found to accelerate the Michael addition step of the amine to the triple bond. Hence, the reaction temperature could be lowered to somewhere in range of 0–23 °C in most of the experiments; optimized conversions were allowed to be run at −40 to −60 °C. The reaction of allylamine **224** with DMAD **225** initially formed the Michael adduct **226**. The acid was thought to promote the addition and to stabilize **226**. A subsequent rearrangement enabled one to isolate enamine **227** in up to 99% yield. Acidic cleavage allowed one to remove the enamine and to build up the corresponding ketoesters. In the presence of a sterically demanding allyl system and, likewise, a nucleophilic counter-ion of the acid (e.g., benzoate from benzoic acid), the yield of product **227** decreased. A competing reaction activated the nucleophile to degrade the intermediate **226** by means of an S_N process. Up to 50% yield of enamine **229** and allyl compound **228** could be isolated (Scheme 10.49).

Lewis acid catalysts offered the advantage of using chiral ligands intending to achieve some catalyst-directed asymmetric induction upon generating enamine **227**. The first experiments using (±)-BINOL catalyst (27 mol%) had a significant rate acceleration, but no experiment with enantiopure material was described. An auxiliary-directed asymmetric rearrangement was investigated staring from prolinol derivative **230**. The addition should have given the intermediate **231** bearing a chiral ammonium center. The consecutive 3,3 sigmatropic rearrangement gave the enamine **232** with 62% yield and 83% de as determined after auxiliary cleavage/decarboxylation and final amidation with chiral 1-phenylethylamine. Here, the use of (+)-BINOL led to a slight decrease of the ee to about 80% (mismatched combination?). Ring expansions were studied upon rearranging 2-vinyl pyrrolidine **233** (n= 1) and piperidine **233** (n= 2) with propiolesters **225** and **234**, respectively. Best results were obtained using TsOH as accelerating acid. In most cases, NMR-scale experiments were conducted allowing one to generate medium-sized ring systems **236** via adduct **235** with 50–70% yield (Scheme 10.49).

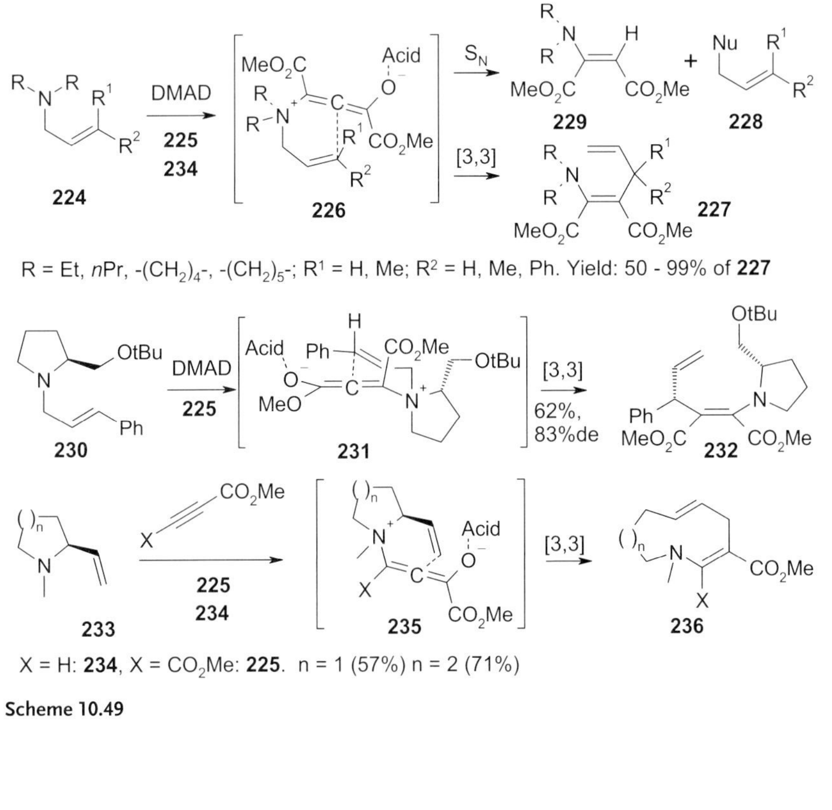

DMAD
225
234
224
MeO2C
Acid
226
SN
229
228
Nu
[3,3]
227
R = Et, nPr, -(CH2)4-, -(CH2)5-; R1 = H, Me; R2 = H, Me, Ph. Yield: 50 - 99% of 227
OtBu
230
231
62%,
83%de
232
CO2Me
233
235
236
X = H: 234, X = CO2Me: 225. n = 1 (57%) n = 2 (71%)

Scheme 10.49

10.5.2
Ketene Aza-Claisen Rearrangements

The N-allyl ammonium enolate Claisen rearrangement requires the addition of a tertiary allylamine to the carbonyl center of a ketene to generate the sigmatropic framework. The first crucial point of this so-called ketene Claisen rearrangement is the formation of the ketene with a lifetime sufficient for the consecutive attack of the nucleophile and avoiding the competing 2+2 cycloadditions. Roberts tested stable diphenylketene **237** [36]. N-benzyl aza-norbonene **238** (R = Ph) was treated with diphenylketene **237** upon heating to reflux in acetonitrile for 6 d. Passage through the addition (→**239**) rearrangement sequence gave the desired bicyclic material **240** with 59% yield. Ultrasonication allowed one to reduce the reaction time to 12 h (61% yield). Already at 0 °C to 23 °C, the sterically less-hindered N-methyl derivative **238** (R = H) suffered from the intermolecular Claisen rearrangement to give **240** in 53% yield. The use of stable ketenes was mandatory, otherwise only 2+2 cycloadducts of ketene and olefin and degradation products were observed (Scheme 10.50).

R = Ph: 80°C, 6d, 59%. R = Ph, ultrasonication, 12h, 61%. R = H: 0°C - 23°C, 13h, 53%

Scheme 10.50

A significant acceleration of the ketene reactivity could be achieved using electron-deficient species such as dichloroketene **241**. An early experiment was published by Ishida and Kato treating a vinyl aziridine **242** with dichloroketene **241** to form a azepinone **243** with 39% yield [37]. Pombo-Villar described a rearrangement of optically pure N-phenethyl azanorbonene **244** to synthesize the α,α-dichloro-δ-valerolactam **247**. At 0 °C in CH_2Cl_2, dichloroketene **241** – generated *in situ* from dichloroacetyl chloride **245** and Hünig's base – was reacted with amine **244**. The addition step led to the hypothetical zwitterion **246**, which underwent immediate 3,3 sigmatropic rearrangement. The bicyclic enantiopure lactam **247** was obtained with 61% yield. Higher reaction temperatures led to tarry side products and the yield was significantly decreased. The α,α dichloro function of **247** could be reduced by means of $Zn/NH_4Cl/MeOH$ to give the dechlorinated material **248**, which served as a starting material in the (–) normethylskytanthine **249** synthesis (Scheme 10.51).

Scheme 10.51

Edstrom used terminally unsubstituted 2-vinyl pyrrolidine and piperidine, respectively, and dichloroketene **241** to achieve the ring expansion to nine and ten-membered lactams [38].

Starting from N-benzyl-2-vinyl pyrrolidine (n=0) and piperidine (n=1) **250**, respectively, the ketene Claisen rearrangement using dichloroketene **241** generated *in situ* led to the corresponding azoninone and azecinone **253** (R^1 = Bn) in 64 and 96% yield. Replacing the N protective group by the more electron rich PMB substituent (R^1 = PMB), the yields of **253** were observed to decrease to 52 and 54%, respectively. Here, Edstrom used trichloroacetyl chloride **251** and activated zinc to generate dichloroketene **241** via a reductive dechlorination at 0 to 62 °C. Simultaneously, Lewis acidic $ZnCl_2$ was formed which might have activated the ketene and stabilized the zwitterionic intermediate **252** to support the rearrangement. Though the double bond included in the medium-sized ring was found to be exclusively *E* configured, both rearrangements suffered from a complete loss of chiral information because of the use of terminally symmetric substituted olefins in **250** (=CH_2) and ketenes **241** (=CCl_2, Acc=O) as reactants. The NMR spectra of the azecinone **253** (n=1) were characterized by the coexistence of two conformers. In contrast, the nine-membered ring **253** (n=0) was revealed as a single species. Both medium-sized lactams had been used in transannular ring contractions to yield the corresponding quinolizidinones **255/256** (n=1) and indolizidinones **255** (n=0), respectively. Finally, the quinolizidinone **256** (n=1, *E*=I) was employed as a key intermediate in a total synthesis of D,L-epilupinine **257** (Scheme 10.52).

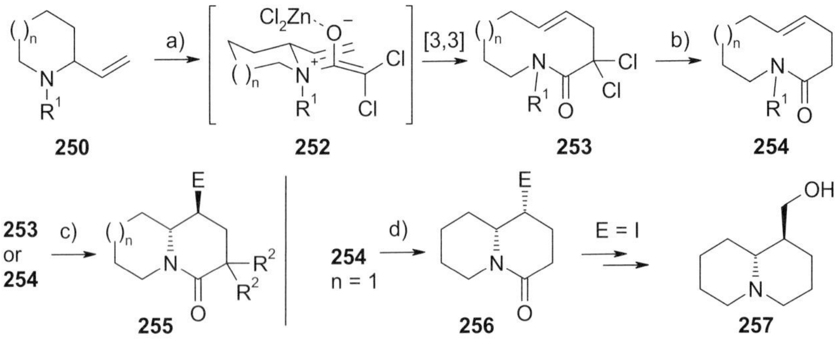

n = 0, 1 R^1 = Bn, PMB, R^2 = H, Cl, E-X = PhSeCl, I_2, TMSI
a) $Cl_3CC(O)Cl$ **251**, Zn/Cu, THF, rt, 62°C, Yield: 52 - 96%. b) Zn/Ag, HOAc, Yield: n = 0 (45%), n = 1 (86%). c) E-X, MeCN, rt, (-R^1-X), yield: 64 - 88%. d) E-X, MeCN, rt, (-R^1-X), yield: 62 - 74%

Scheme 10.52

The mild reaction conditions and the obviously high potential driving force of the ketene Claisen rearrangement suggested the using the process for more complex systems [39]. The first series of this type of reaction suffered from severe limitations. On one hand, predominantly electron-deficient ketenes **213** added to the allylamines **208** and useful yields of the amides were exclusively achieved reacting dichloroketene **241**. On the other hand, the rearrangement was restricted to either monosubstituted olefins in the amino fragment (**250**) or the driving force had to be increased by a loss of ring strain (**238**, **242**, **244**) during the process. The reaction path was rationalized as pointed out in Scheme 10.53: Initially, the ketene **213** was generated from a suitable precursor **258** (i.e., acid halide). Then, a reversible addition of the ketene **213** to allylamine **208** gave the intermediate zwitterion **214**. Here, dichloroketene was found reactive enough to push the equilibrium towards adduct **214** and diphenylketene **237** was stable enough to survive several addition elimination cycles without suffering from competing reactions (diketene formation, etc.). Finally, zwitterion **214** underwent the sigmatropic rearrangement to give amide **216** (Scheme 10.53). The careful analysis of a range of conversions showed that two further competing processes had to be mentioned. (1) The tertiary amines **208** and the acid chlorides **258** (X = Cl) initially formed acylammonium salts **259**, which underwent a von Braun-type degradation by an attack of the nucleophilic chloride ion ($X^- = Cl^-$) at the allyl system to give allyl chlorides **260** and carboxylic acid amide functions **261** [40]. (2) The reaction of acyl chlorides **258** led to the corresponding ketenes **213**, while the allylamines **208** were deactivated as ammonium salts **208**-HCl (Schotten–Baumann conditions, Scheme 10.53).

Best results: **258** (X = F) + $AlMe_3$ ⟶ Me-H + **213**···$AlF(Me)_2$ $\xrightarrow{208}$ **214**···$AlF(Me)_2$ ⟶ **216**

Scheme 10.53

A significant part of the restrictions could be abolished implementing three actions concerning the processing. First of all, the rearrangements were run in the presence of stoichiometric amounts of a Lewis acid (LA), especially trimethyl aluminum. Currently, most α-substituted carboxylic acid halides **258** (X = Cl, F) as precursors of the ketenes **213** could be employed, thus overcoming the restriction concerning the ketene component, but until now, the rearrangement failed using α,α difunctionalized carboxylic acid halides. The Lewis acid might have increased the acidity of the α protons by interacting with the carbonyl group in **213**, **258** and **259**, respectively, facilitating the formation of the intermediate zwitterions **214** and/or the Lewis acid had stabilized the zwitterionic intermediate **214** suppressing the elimination of ketene **213**. Furthermore, allylamines **208** bearing 1,2-disubstituted double bonds could be successfully rearranged overcoming a restriction concerning the carbon framework (Scheme 10.53).

Replacement of the acyl chlorides **258** (X = Cl) by the corresponding acyl fluorides **258** (X = F) as the substitutes of the ketenes **213** efficiently suppressed the von Braun-type degradation as a major competing reaction. The fluoride counterion was known to be less nucleophilic but more basic. In the presence of trimethyl aluminum, the potential formation of a stable Al–F bond (F–$AlMe_2$, methane evolved) should have eliminated the fluoride as a latent nucleophile. The acyl fluorides **258** were found to be less reactive compared to the corresponding acid chlorides causing some difficulties in the rearrangement with n-alkyl carboxylic acid derivatives. Such transformations needed longer reaction times, and the yield of the corresponding rearrangement products were moderate [39d,e].

The use of a second base trapped all proton acids generated during the course of the rearrangement. In most cases, a two-phase system of solid potassium carbonate as a suspension in dichloromethane or chloroform gave the best results even though excessive HX formation during the course of the reaction could be avoided by employing the combination of acid fluoride/Me_3Al (formation of methane and dimethylaluminum fluoride, Scheme 10.53).

Stereochemical Results: 1,3 Chirality Transfer and Internal Asymmetric Induction
Employing the optimized reaction conditions upon running various reactions, the stereochemical advantages of the Claisen rearrangements were combined with an efficient synthesis of the azoninones **265** and **266** bearing defined *E*-configured double bonds in the medium-sized rings [39]. As is known for all Claisen rearrangements, a complete 1,3-chirality transfer was observed by treating *E*-allylamines **262** (R^1, R^4 = H) with acetyl chloride **263** (R^5 = H) [39a]. Both enantiomers of the core framework were constructed starting from the same *L*-(−)-proline derivative choosing either an *E* (R^4 = H) or a *Z* (R^3 = H) allylamine **262** [39f]. Furthermore, a high internal asymmetric induction could be observed involving *α*-substituted

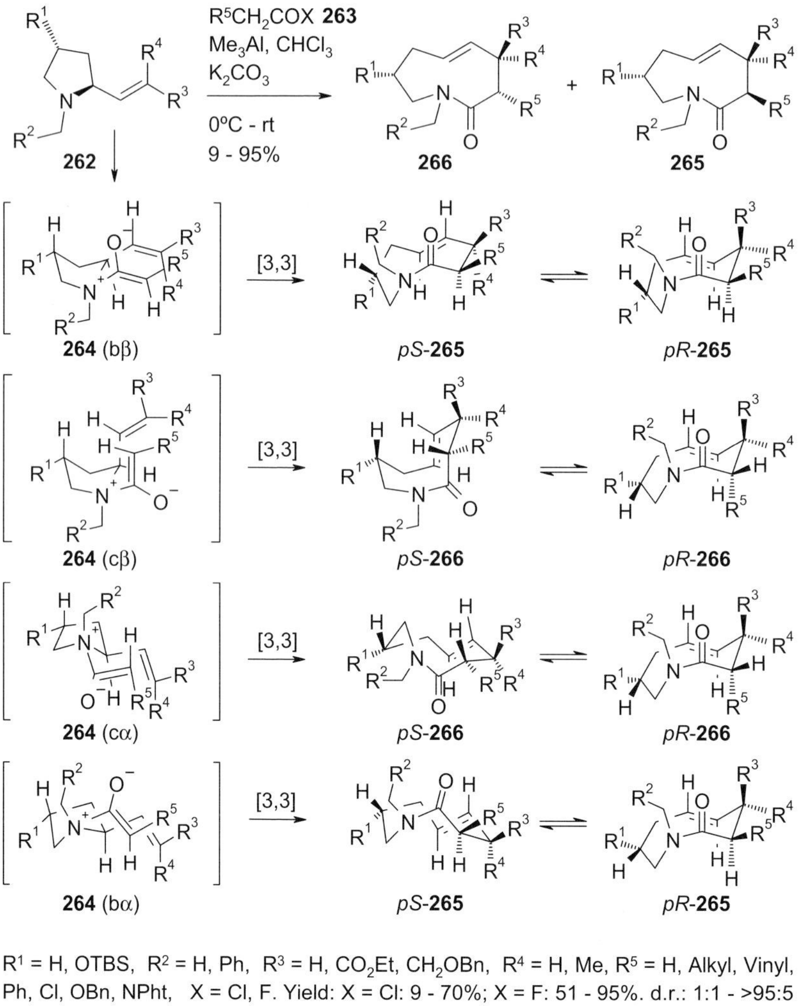

R⁵CH₂COX 263
Me₃Al, CHCl₃
K₂CO₃
0°C - rt
9 - 95%
262
266
265
264 (bβ)
264 (cβ)
264 (cα)
264 (bα)
[3,3]
pS-265
pR-265
pS-266
pR-266
R¹ = H, OTBS, R² = H, Ph, R³ = H, CO₂Et, CH₂OBn, R⁴ = H, Me, R⁵ = H, Alkyl, Vinyl, Ph, Cl, OBn, NPht, X = Cl, F. Yield: X = Cl: 9 - 70%; X = F: 51 - 95%. d.r.: 1:1 - >95:5

Scheme 10.54

acyl halides **263** ($R^5 \neq H$) in the synthesis of the lactams. In most cases the diastereomeric excess was >5:1 in favor of the 3,4-*trans* lactam **266**. The phenylacetyl halide rearrangement ($R^5 = Ph$) gave a nearly equal mixture of *cis* and *trans* azoninones **265** and **266** ($R^5 = Ph$, Scheme 10.54).

The stereochemical outcome of the rearrangement of **262** ($R^1 = H$) was explained by a chair-like transition state **264** (c*α*) with minimized repulsive interactions and a defined *Z* enolate geometry (as expected for all amide enolates) [39b,d]. However the chair-like transition state **254** (c*β*) could not be excluded: both **254** (c*β*) and (c*α*) resulted in the same diastereomer *pS*-**266**!

Surprisingly, the rearrangement of the 4-*t*-butyldimethylsilyloxy-2-vinylpyrrolidines **262** ($R^1 = OTBS$, R^3, $R^4 = H$) took another course. The stereochemical outcome had to be rationalized by a boat-like transition state **264** (b*β*) to give the 3,8-*trans* lactams **265** ($R^1 = OTBS$). The corresponding *cis* product **266** ($R^1 = OTBS$) resulting from the expected chair-like intermediate **264** (c*β*) had only once been isolated as a minor compound. The completeness of the 1,4 chirality transfer should be pointed out [39]. Obviously, the configuration of the intermediately generated stereogenic ammonium center in **264** had to be considered. Rearranging the 2-vinylpyrrolidines **262** ($R^1 = H$), the N-acylation should have been directed by the adjacent side chain to the opposite face of the five membered ring to give **264** (*α*) (1,2 *anti* induction, as found by analyzing appropriate acylammonium salts). Consequentially, the rearrangement proceeded via a chair-like transition state **264** (c*α*), as expected for the acyclic 3,3 sigmatropic reaction, leading predominantly to lactams **266**. In contrast, the N acylation of the 2,4-*trans* disubstituted pyrrolidines **262** ($R^1 = OTBS$) had been directed by the bulky silylether to generate a *syn* arrangement of vinyl and acyl groups in an intermediate ammonium salt **264** (*β*) (1,3 *anti* induction, 1,2 *syn*). Then, an appropriate conformation to undergo a Claisen rearrangement was presumably the boat-like form **264** (b*β*) with minimized 1,3 repulsive interactions resulting in the lactams **265**. However, the 2,4-*cis* disubstituted pyrrolidine **267** ($R^1 = OTBS$, R^3, $R^4 = H$) gave the expected lactam diastereomer **266** via a chair-like transition state conformation **268** (Schemes 10.54 and 10.55).

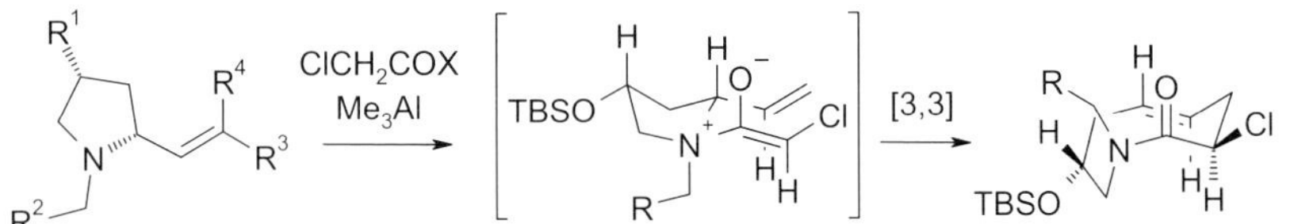

$R^1 = OTBS$, R^2, R^3, $R^4 = H$, $R^5 = Cl$, $X = Cl$

Scheme 10.55

The lactam and olefin unit characterized the heterocyclic cores **265** and **266** as constrained ring systems, the conformations of which were found to be strongly dependent on the substitution pattern and the relative configuration of the stereogenic centers. The planar chiral properties of the medium-sized rings with internal *trans* double bonds had to be taken in account for analyzing the nine-membered rings [41]. The rearrangements of the 2*S* vinylpyrrolidines **262** via a boat-like transition state **264** (b) initially effected the formation of the medium-sized ring with *pS*-arrangement of the *E*-double bond (*pS*-**265**). This planar diastereomer *pS*-**265** was obviously unstable. Finally, the epimerization (flipping of the *E* double bond) to give the *pR* arrangement *pR*-**265** of the olefin with respect to the ring, generated the most stable and rigid conformation. Preliminary force field calculations of the azoninones **265** and molecular mechanics calculations of the related *E*/*Z*-1,5-nonadiene confirmed these observations [42]. Nevertheless, a high activation barrier had to be passed to achieve the change of the planar chiral information (*pS*-**265**→*pR*-**265**). This fact allowed the isolation and characterization of the conformers of the nine-membered rings [39c,e,f]. In contrast, the lactams **266** (R^4 = H), generated via chair-like zwitterions **264**(c), were found to be generated directly in a stable *pS* arrangement of the *E* double bond *pS*-**266** (Schemes 10.54 and 10.55).

The proof of this principle gave the aza-Claisen rearrangement of vinyl pyrrolidine **262** (R^1, R^3, R^4 = H, R^2 = Ph) and chloroacetyl fluoride **263** (R^5 = Cl) under standard conditions. A mixture of two diastereomers *pS*-**265** and *pS*-**266** was obtained in 77% yield and a ratio of 1.4:1, which could be separated by means of column chromatography and preparative HPLC. On handling these compounds, any warming to 30–40 °C was avoided to maintain the planar chiral properties (*pS*) of the diastereomers resulting from the ring expansion. For testing the conformational stability of both compounds, the separated diastereomers *pS*-**265** and *pS*-**266** were heated to about 60 °C [43]. After 3 to 10 h a second diastereomer *pR*-**265** occurred starting from *pS*-**265**, indicating the flipping of the double bond with respect to the ring. All spectral data of the new diastereomer *pR*-**265** were identical with those determined for lactam *pS*-**266** except the specific rotation proving the formation of the enantiomer. Lactam *pS*-**266** suffered the analogous process generating *pR*-**266** (enantiomer of *pS*-**265**), but the conversion was found to be incomplete. The relative arrangement of the double bond and the stereogenic center of the diastereomers were proven via NOE analyses. While the lactam *pS*-**266** was characterized by a single set of peaks (almost rigid conformation), the spectral data of lactam *pS*-**265** indicated the coexistence of two conformers (double set of peaks in the ^{1}H and ^{13}C spectra) potentially originating from some mobility of the lactam function (Scheme 10.56) [39f].

The azoninones **265** and **266** with defined stereochemical properties served as key compounds in natural product syntheses. Firstly, the planar chiral information was used to generate stereospecifically new stereogenic centers depending on the defined conformation of the nine-membered rings [44]. Upon treatment of p*S*-**266** (R^1, R^4, R^5 = H, R^2 = Ph, R^3 = CO_2Et) with PhSeCl in MeCN, the *anti* addition of $[PhSe]^+$ and the lactam lone pair to the double bond gave an intermediate acylammonium ion **269**, which suffered from immediate von Braun degradation to form benzylchloride and the indolizidinone **270** as a single regio- and stereoisomer (ring contraction) with 70% yield. Several further steps allowed the complete total synthesis of (–) 8-epidendroprimine **271** [44a]. A mixture of p*S*-**265**/p*S*-**266**

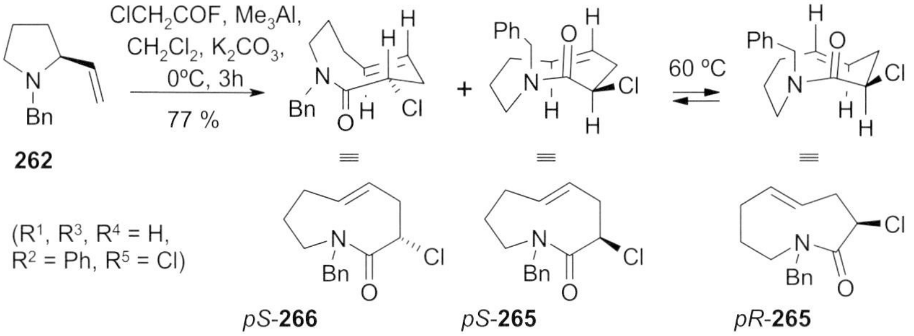

Scheme 10.56

(R^1, R^3, $R^4 = H$, $R^2 = Ph$, $R^5 = Cl$) was epoxidized by means of MCPBA to give the diastereomeric mixture of epoxy azonanones **272** with defined epoxide configuration (stereospecific cycloaddition). Chlorine and benzyl group were removed by hydrogenation using Pearlman's catalyst to give a single hydroxyindolizidinone **273** after regio- and stereoselective intramolecular oxirane opening. Several further transformations enabled one to complete the (+) pumiliotoxin 251D **274** synthesis (Scheme 10.57) [44b,c].

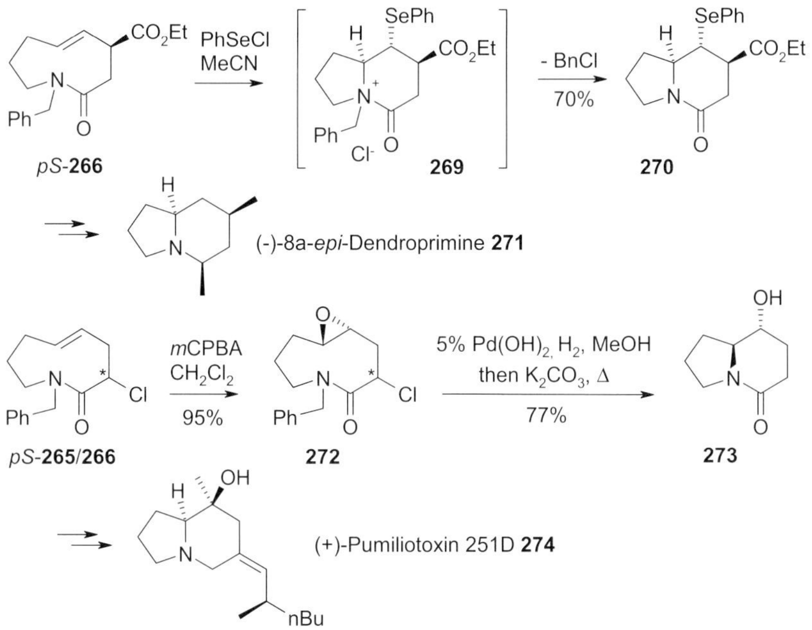

Scheme 10.57

An alternative pathway using a zwitterionic aza-Claisen rearrangement to generate azoninones was described by Hegedus [45]. 2-Vinylpyrrolidines **275** and chromium carbene complexes **276** underwent photochemical reactions in the presence of a Lewis acid to give the corresponding nine-membererd ring lactams **277** bearing *E* double bonds in up to 71% yield. Though reactants and products suffered from some instability against Lewis acids, the presence of the zinc chloride or dimethylaluminum chloride was mandatory to start the rearrangement. In contrast to the classical ketene Claisen process, electron-rich ketene equivalents such as alkoxy or amino ketenes could be used, since the donor substituents stabilized the chromium carbene complex **276**. Furthermore, α,α-disubstituted lactams were synthesized but the stereoselectivity observed was low. The determination of the stereochemical outcome of the reaction proved that the 1,4 chirality transfer was not complete: A Mosher analysis of an appropriate azoninone gave a loss of about 10% of the chiral information. A chiral carbene complex **276** (R^1 = oxazolidinyl) was found to have a negligible influence on the stereoselectivity of the rearrangement. Generally, the present variant of the rearrangement was found to be very sensitive to any sterical hindrance. Additional substituents in any position (e.g., $R^2 \neq$ H) led to a severe decrease of the yield and the stereoselectivity. Additionally, one example of rearranging a 2-vinylpiperidine **278** was given. The corresponding azecinone **279** was formed in about 33% yield (Scheme 10.58).

In analogy to Edstrom's experiments [38], the nine- and ten-membered ring lactams **277** and **279** underwent regio- and stereoselective transannular ring contractions to give the corresponding indolizidinones and quinolizidinones, respectively (Scheme 10.58).

Scheme 10.58

The first systematic efforts to investigate the internal asymmetric induction of ketene Claisen rearrangements were published by Yu [46]. Some simple *E* and *Z* N-crotylpiperidines **280** were treated with propionyl, methoxyacetyl and fluoroacetyl chloride **281**, respectively. Renouncing on Lewis acid support, the reactants were combined in toluene at 0 °C in the presence of solid K_2CO_3 as a proton acceptor (Schotten–Baumann conditions). The piperidides **283** were isolated in 38–61% yield. The formation of side products (von Braun degradation) had not been reported. All reactions were found to be highly diastereoselective. The stereochemical outcome could be rationalized by the initial formation of the zwitterion **282** with a defined double bond and a defined *Z* ammonium enolate geometry due to steric and/or electronic reasons. A chair-like transition state gave rise to the formation of the *anti* product *anti*-**283** starting from *Z*-**280** and the *syn* product *syn*-**283** starting from *E* crotylpiperidine *E*-**280**, respectively. Alternatively, the use of the corresponding crotylpyrrolidines and the employment of *in situ* formed propionyl bromide gave somewhat lower yields and diastereoselectivities, respectively (Scheme 10.59).

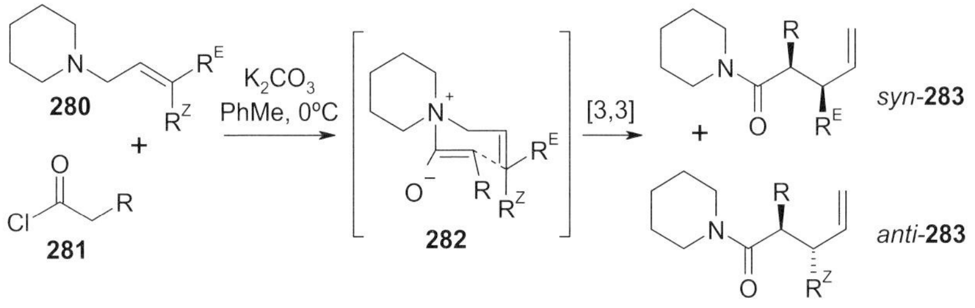

E-**280**: R = Me (41% yield, *syn*/*anti* = 94:6), R = OMe (44%, 92:8), R = F (61%, 95:5)
Z-**280**: R = Me (38% yield, *syn*/*anti* = 3:97), R = OMe (39%, 5:95), R = F (57%, 4:96)

Scheme 10.59

A more recent systematic investigation of the internal asymmetric induction in ketene Claisen rearrangement was contributed by MacMillan [47]. *E* and *Z* N-allylmorpholine derivatives **284** were reacted with acid chlorides **281** in the presence of Hünig's base supported by 5 to 20 mol% of a Lewis acid (Einhorn conditions). The product γ,δ-unsaturated morpholine amides **286** were obtained with 70 to 95% yield. The conditions reported indicate the presence of a high concentration of nucleophilic chloride ions in the reaction medium. Though acid chlorides and tertiary amines tend to a rapid formation of N acylammonium salts, even in the presence of sub-stoichiometric amounts of a Lewis acid, no von Braun degradation generating allyl chlorides and carboxylic acid morpholine amides had been mentioned (Scheme 10.53). The best results were achieved running the rearrangement in the presence of $Yb(OTf)_3$, $AlCl_3$ and $TiCl_4$. As expected, a very high internal asymmetric induction was found in most experiments. Compared to the *Z* amine *Z*-**284** resulting *anti*-**286** as the major compound, the corresponding *E* configured reactants *E*-**284** gave somewhat higher yields and diastereoselectivities forming predominantly the *syn* amides *syn*-**286**. However, the use of α-alkyloxyacetyl chlorides caused decreased diastereoselectivities of 86:14 to 90:10. These findings were in accordance to previous reports [39, 49]. The stereochemical outcome could be rationalized by the intermediate formation of the zwitterion **285** with *Z* enolate geometry, which rearranged via a chair-like transition state to give the desired product **286**. Furthermore, new quaternary carbon centers could be built up by means of the present protocol. The rearrangement of 3-ethyl-3-methyl substituted allylamine *E*-**284** allowed one to generate the corresponding amide *syn*-**286** with 72% yield and 99:1-*syn* selectivity. The cyclohexenylamine **287** could be converted into the amide **288** with 75% yield and 99:1 *syn*-diastereoselectivity, too (Scheme 10.60).

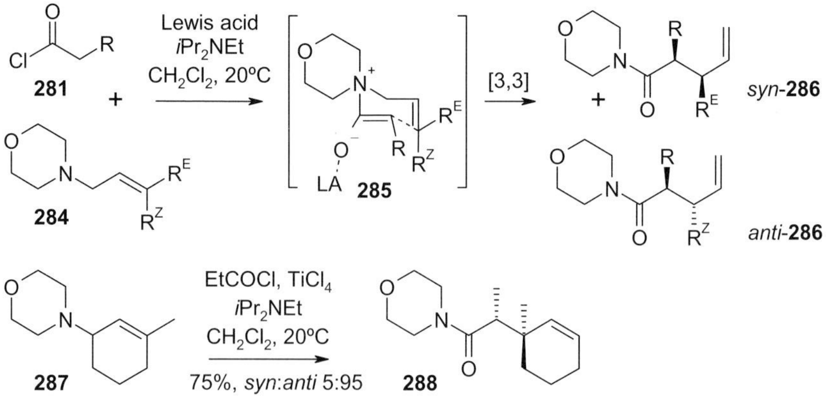

R = Me, NPht, SPh, OBn. R^E = H, Me, Ph, Et, Cl. R^Z = H, Me, Cl
Lewis acid = $Yb(OTf)_3$, $AlCl_3$, $TiCl_4{\cdot}2THF$. Yield: 70 - 95%, d.r. = >86 : 14

Scheme 10.60

The present protocol enabled one to run tandem ketene aza-Claisen rearrangements [47b]. The reaction of allyl systems **289** bearing two allylamine fragments gave the corresponding diamides **293** with high yield and high simple diastereoselectivity as well as an excellent 1,3 asymmetric induction. The amount of Lewis acid had to be increased to about 200 to 400 mol%. The process was explained by two consecutive sigmatropic rearrangements. In the first step, the *E*-allylmorpholine moiety of **289** reacted with the ketene fragment from **281** passing through the zwitterionic intermediate **290** displaying the well-known *Z* enolate geometry. The

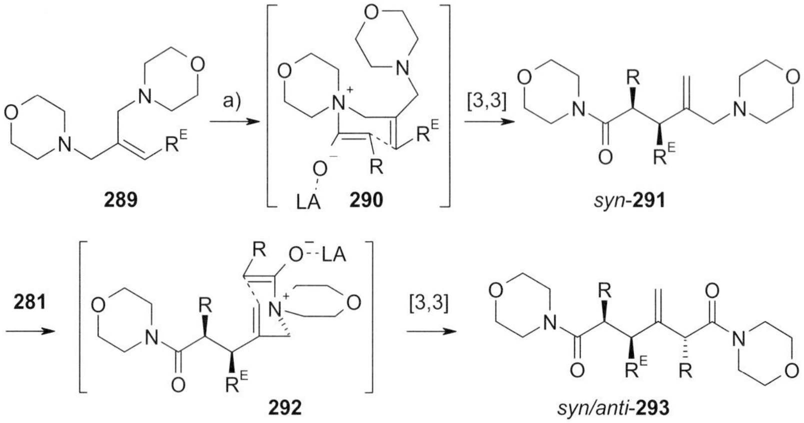

a) $RCH_2C(O)Cl$ **281**, Lewis acid, *i*Pr_2NEt, CH_2Cl_2, 20°C. Yield: 70 - 99%, d.r. > 9:1
R = Me, Bn, NPht, OPiv. R^E = Me, CN, Cl, OBz, SPh. Lewis acid = $Yb(OTf)_3$, $AlCl_3$, MgI_2, $TiCl_4{\cdot}2THF$

Scheme 10.61

chair-like transition state gave rise to the formation of the *syn* intermediate *syn*-**291**. Then, the newly generated allylmorpholine in **291** suffered from a second rearrangement. The addition of another acid chloride **281** led to the zwitterion **292**, which was immediately transformed into the diamide *syn/anti*-**293** via a chair-like conformation with minimized repulsive interactions. High yields of 71% to 99% have been reported. In most runs a major diastereomer could be detected with >92:8 diastereoselectivity (Scheme 10.61).

Stereochemical Results: External Asymmetric Induction/Remote Stereocontrol

The low reaction temperatures of the zwitterionic ketene aza-Claisen rearrangements suggested testing further stereodirecting properties. An efficient external chiral induction within the addition rearrangement sequence requires not only a highly ordered transition state of the six core atoms, but also a defined arrangement of the external chiral sub-unit with respect to the rearrangement framework. A single defined transition state conformation always causes a highly selective reaction. With the intention to achieve a maximal asymmetric induction via remote stereocontrol, the chiral information should predominantly be placed next to the nascent sigma bond formed during the course of the rearrangement and carrying the new chiral centers. A stereogenic center in the allylamine moiety adjacent to C3 fulfilled such prerequisite. Particularly, the defined C atom–heteroatom bonds offer the advantage of acting as potential stereodirecting sub-units. Since the electron rich 1,2-vinyl double bond must attack the allyl system at position 6, an adjacent C–X (nucleophilic X) bond should adopt an *anti* arrangement with respect to the incoming donor. In other words, the extended C–X–σ^* orbital is *syn* coplanar positioned with respect to the attacking vinyl double bond. The transition state is stabilized by an additional delocalization of some electron density in the empty *anti*-binding orbital. Such weak electronic effects have been successfully used in ketene thia-Claisen rearrangements [48].

Allylamines **294** bearing an additional chiral center adjacent to C6 (R^1 = **a**–**d**) have been efficiently synthesized via short ex-chiral pool sequences starting from *D*-mannitol (→ **c**), *L*-malic (→ **b**) and *L*-lactic acid (→ **a**) and *L*-proline (→ **d**), respectively [49]. Subjecting the allylamines **294** to α-monosubstituted acid halides **295** in the presence of Me_3Al at 0 °C in a two phase system of CH_2Cl_2 and K_2CO_3, the rearrangement delivered the unsaturated amides. Analyzing the stereoselection properties of the conversion to **296**–**299**, the 1,2-asymmetric induction was found to be mostly >90:10 in favor of the *syn* product–even in the presence of the nitrogen as directing function. The minor *anti* diastereomer only occurred in appreciable amounts, if acetyl chloride **295** (R_2 = H) was used as the C2 source. Furthermore, the simple diastereoselectivity (internal asymmetric induction) was high, allowing the diastereoselective generation of two new stereogenic centers in a single step bearing a variety of functional groups. The formed γ,δ-unsaturated amides **296**–**299** represented useful intermediates for natural product total syntheses as demonstrated completing the synthesis of (+)-dihydrocanadensolide **300** [49b] and the formal synthesis of (–)-petasinecin **302** (**301** = known precursor, Scheme 10.62) [39d].

K_2CO_3, 0°C or rt, CH_2Cl_2 or $CHCl_3$,

$AlMe_3$ X = F, Cl [3,3]

R^1 = **c** R^1 = **d** Lit.

R^2 = H, Me, Cl-CH_2-CH_2, *i*Pr, vinyl, butadienyl, Ph, Cl, OBn, NPht
Yield: 45 - 85%
d.r. (simple): 7:4 - >95:5
d.r. (induced): 1:1 - >95:5

Scheme 10.62

The stereochemical outcome of the reaction could be explained by a clearly preferred transition state. Generally, the ketene equivalent (from **295**) and the allylamine **294** (R^1: **c**) combined to form a hypothetical intermediate acyl ammonium enolate with a defined *Z* enolate geometry (in *b*-**296**, *c*-**297**, *c*-**298**), as expected for amide and acyl ammonium enolates. Adopting the chair-like conformations *c*, the *anti* arrangement of the attacking enolate and the guiding heteroatom (N and O at C6a) favored *c*-**297** facing *c*-**298**, the *anti*/*syn* product **297** was isolated as the major compound. Surprisingly, R^2 substituents characterized by extended *Π*-systems led to the *syn–syn* products **296** selectively. Here, the high remote stereocontrol must involve an alternative boat-like transition state conformation *b*-**296**. In Scheme 10.63 the hypotheses concerning **294** (R^1: **c**) are lined out. With respect to the inverted configuration of the directing center adjacent to C6 in **294** (R^1: **a**, **b**, **d**) and **296–299** (R^1: **a**, **b**, **d**), the enantiomer stereotriades were formed (Scheme 10.63).

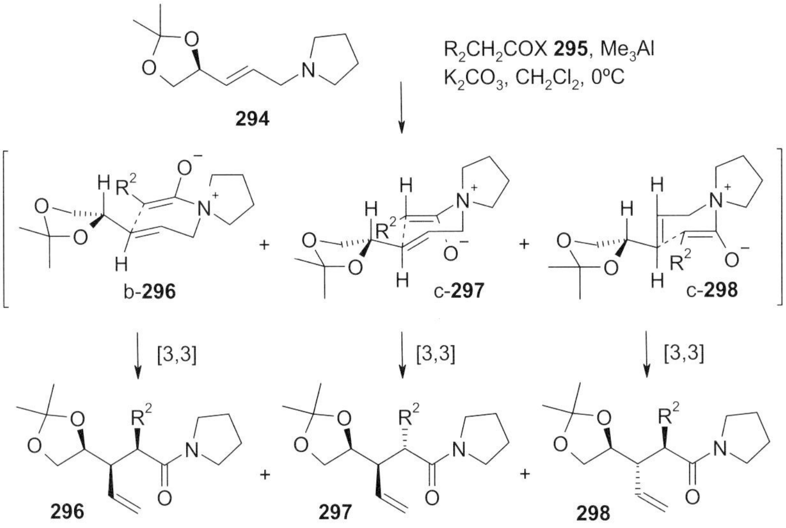

Scheme 10.63

Stereochemical Results: External Asymmetric Induction/Auxiliary Control

Auxiliary-controlled 3,3 sigmatropic rearrangements represent an almost classical approach to introduce chiral information into a more complicated rearrangement system. The major advantage is the reliable control of the stereochemical outcome of the reaction. All products are diastereomers, e.g., the separation of minor compounds should be more or less easy and the well-known spectroscopic analyses will always be characterized by defined and reproducible differences. However, the auxiliary strategy requires two additional chemical transformations: the attachment and the removal of the auxiliary must be carried out in two steps. The efficiency of these steps influences the whole sequence. High yields and avoiding stereochemical problems are the prerequisite of each step. Furthermore, the synthesis and recycling of an auxiliary have to be taken into account. Overall, a set of advantages and problems have to be considered before deciding on an auxiliary strategy.

An aza-Claisen rearrangement enables one to attach a chiral auxiliary to the central nitrogen via the third binding valence next to the 3,3 sigmatropic framework. In analogy to the remote stereocontrol prerequisites mentioned above, a low reaction temperature was crucial to guarantee a restrained conformational mobility of the potential transition state. The system should be forced to take a single reaction path for obtaining a high diastereoselectivity.

The ketene aza-Claisen rearrangement seemed to fulfill these prerequisites using *L*-(–) proline derivatives **303** as chiral auxiliaries [50]. Several N-allyl pyrrolidines **306** were synthesized via a Pd(0)-catalyzed amination of the corresponding allylmesylates **304** and **305**, respectively. The double bond was always *E* configured. The treatment with chloro- and suitably protected α-amino acetyl fluorides **307**, respectively, in the presence of solid K_2CO_3 and trimethyl aluminum in

$CHCl_3$/0 °C, led to the formation of the corresponding γ,δ unsaturated amides **308** and **309**. Again, the charge neutralization served as an efficient driving force allowing one to conduct the reactions at such low temperatures. In particular, the use of azido acetyl fluoride **307** ($R^3 = N_3$) enabled a subsequent reductive cyclization to generate *D*-proline *L*-proline dipeptides **313** allowing the introduction of varying substituents in the new *D*-proline moiety (Scheme 10.64).

R^1 = H, CO_2Me, CO_2tBu, CH_2OTBS, CH_2OBn. R^3 = Cl, N_3, NPht, (alkyl)NBOC, (alkyl)NCBZ.
Yield: 13 - 91%, 1:1 ->95:5

$[Pd_2(dba)_3]$; PPh_3, Et_3N; THF, rt

Me_3Al, K_2CO_3, $CHCl_3$

R^2 = H, Ph,

Scheme 10.64

The removal strategy of the auxiliary should be chosen depending on the auxiliary and the substitution pattern of the amide **308**/**309**. Generally, the iodo lactonization (**310**) reduction sequence (**311**) as described by Tsunoda [28] and Metz [51] led to smooth cleavages of all types of amides **309**. Particularly the prolinol auxiliaries **309** ($R^1 = CH_2OTBS$) offered further advantages. In the presence of acid-stable substituents R^2 and R^3, a neighboring group-assisted transesterification ($R^1 = CH_2OTBS$) with HCl/MeOH allowed one to convert the amides **309** into the corresponding esters **312**. Alternatively, the auxiliary can be used as a leaving group in an intramolecular metal organic reaction of **309** ($R^3 = CH_2$–Ar–Br) to generate a cyclic ketone **314** without any loss of the chiral information (Scheme 10.65).

Discussing the stereochemical outcome of the Claisen rearrangements, two aspects had to be considered. On one hand, the relative configuration of the new stereogenic centers was found to be exclusively *syn* in **308** and **309** suggesting a chair-like transition state *c-α* and *c-β*, respectively, including a *Z* acyl ammonium enolate structure (complete simple diastereoselectivity/internal asymmetric induction, Scheme 10.64).

On the other hand the external asymmetric induction strongly depended on the chiral auxiliary. The careful analysis of the hypothetical zwitterionic intermediates *c-α* and *c-β* indicated the formation of a stereogenic ammonium center. In terms of the well-known 1,3 chirality transfer of 3,3 sigmatropic rearrangements, the

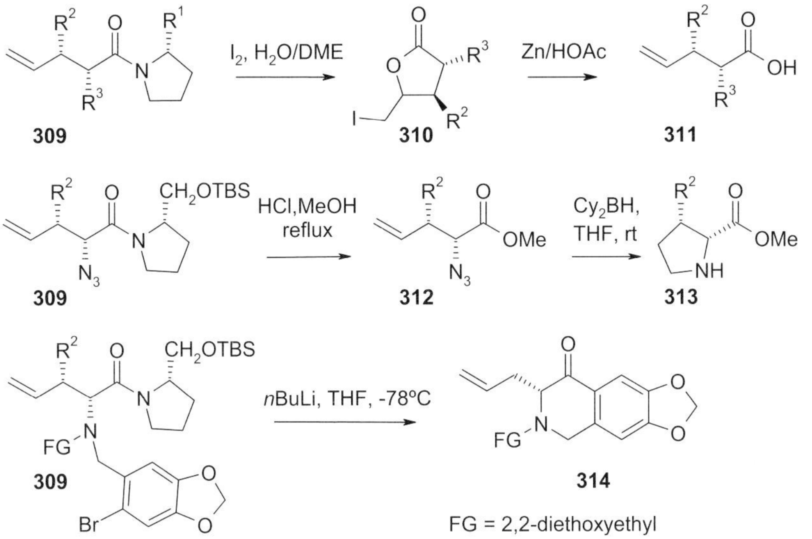

Scheme 10.65

present reaction allowed one to shift the chiral information from the ammonium center (1) to the enolate C (3). The amide **308/309** α-carbon atom was built up with a defined configuration after passing through the above-mentioned chair-like transition state *c-β/c-α* including the defined olefin geometry and the equatorial arrangement of the bulky (chain branch) part of the auxiliary. Consequentially, the crucial step of the whole process must have been the diastereoselective addition of the ketene equivalent from **307** on generating the zwitterionic intermediates. Thus, by employing auxiliaries bearing the small proline methyl ester substituent ($R^1 = CO_2Me$) in **306**, the reaction with non-hindered acid fluorides **307** gave the corresponding amides **308/309** with low or moderate diastereoselectivity indicating unselective N-acylation. In contrast, conversions at lower temperatures or with bulky-substituted acid fluorides **307** resulted in significantly higher selectivities (more selective acylation). The use of reactant allylamines **306** bearing the bulky proline *tert*-butylester and the OTBS prolinol auxiliaries as R^3, respectively, were characterized by a high auxiliary-directed diastereoselectivity indicating passage through a defined acylation rearrangement path via *c-α*. Presently, the OTBS prolinol ($R^1 = CH_2OTBS$) is the auxiliary of choice because of the easy introduction, the high auxiliary directed induction of chirality, the stability against a set of consecutive processes and the simple cleavage by the neighbor group-assisted amide **309**-ester **312** conversion (Schemes 10.64 and 10.65).

Presently, the zwitterionic aza-Claisen rearrangement has been developed as a reliable method to synthesize suitably protected non-natural α-amino acid derivatives, e.g., C-allyl glycines of type **312** and 3-arylprolines of type **313**.

The major disadvantage of the classical auxiliary-controlled 3,3 sigmatropic rearrangements is still the requirement of two additional chemical transformations:

the attachment and the removal of the auxiliary must always be considered. The efficiency of these steps influences the usability of the whole sequence. Since coordination of the Lewis acid metal salt at the core heteroatoms of the 3,3 sigmatropic system was found to accelerate the process, the proximity of the Lewis acid ligands should allow to influence the stereochemical outcome of the rearrangement. Hence, the use of chiral ligands should cause an external chiral induction. In conclusion, a Lewis acid carrying chiral ligands should serve as a chiral auxiliary. The separate attachment and the final removal of the auxiliary could be saved, the enantioselective Claisen rearrangement arose as a more straightforward process. Generally, such a reaction should be run in a catalytic sense, but the increased complexation ability of the product in comparison to the reactants mostly inhibited the release of the Lewis acid right after a rearrangement step until the aqueous cleavage. It is understood that the stereochemical properties of the products had to be carefully analyzed using chiral GC, HPLC and derivatization techniques, respectively.

MacMillan [52] investigated the intermolecular aza-Claisen rearrangements treating *N*-allyl morpholines **315** with glycolic acid chlorides **316** in the presence of a chiral-chelated Lewis acid. So termed "magnesium BOX" systems **319** and **320** gave the best results concerning yield (up to 95%) and chirality transfer (up to 97% ee) generating the amides **318**. Usually, two to three mol equivalents of the chiral metal complex **319** and **320** had to be employed to achieve a satisfactory ee. The glycolic acid framework seemed to play a crucial role in terms of asymmetric induction: a non-chelating α-oxygen substituent R^3 produced amides **318** with only moderate enantiomeric excess. In contrast, the use of benzyloxy acetyl chloride allowed one to achieve a very high ee. It seemed reasonable, that this α-oxygen substituent enabled an efficient chelation of the chiral modified Lewis acid in **317** causing the high level of external chirality transfer. The substitution pattern of the allyl morpholine remained variable, the use of *E* and *Z* olefin, respectively, led to the defined formation of enantiomer amides **318** with comparable asymmetric induction. Until now, the catalytic enantioselective aza-Claisen rearrangement involving sub-stoichiometric amounts of the chiral information remained undiscovered (Scheme 10.66).

2-3 eq. M^C, iPr_2NEt, -20°C; [3,3]

315 + **316** → [**317**] → **318**

M^C-1: **319**, 2 [I⁻]

M^C-2: X^1 = Ph, X^2 = H
M^C-3: X^1 = Ph, X^2 = Cl
M^C-4: X^1 = 4-MeO-C_6H_4, X^2 = Cl

320, 2 [I⁻]

R^1 = H, Me, Ph. R^2 = H, CH_2OBz, 4-(NO_2)-Ph, CO_2Et, Cl. $R^{2'}$ = H, Me, Cl.
R^3 = Me, Bn, Ac, TBS, Ph, 4-Cl-Ph. Yield: 42 - 95%, d.r.: >84:16. ee: 38 - 97%

Scheme 10.66

10.5.3
Allene Carbonester Aza-Claisen Rearrangements

The third type of the zwitterionic aza-Claisen rearrangement is called the N-allyl ammonium enolate Claisen rearrangement [53]. The first step of this tandem process was a Lewis acid catalyzed Michael addition of a tertiary allylamine **321** to the β carbon center of an allene carbonester **322**. The formed hypothetical zwitterion **323** must be characterized by a highly resonance-stabilized anion. Additional support should have given the O-coordinated Lewis acid. Then, the allyl vinyl ammonium moiety of **323** underwent a 3,3 sigmatropic rearrangement to give the unsaturated ester **325** bearing two new stereogenic centers in the γ and δ position. The formation of the vinylogous carbamate in **325** and the charge neutralization served as potential driving forces allowing one to run such a reaction at 23 °C. The best results were obtained using 5 to 10 mol% of $Zn(OTf)_2$. The products **325** were isolated with 75–97% yield and excellent diastereoselectivity of up to > 98:2. Variation of the allylamine **321** and the allene substitution pattern in **322** gave the first insight into the scope and limitation of the transformation. It should be pointed out that the present protocol enabled one to generate defined quaternary centers. Furthermore, the double bond geometry of the allylamine **321** moiety allowed one to predict the stereochemical outcome of the reaction. The geranyl derivative E-**321** rearranged upon treatment with benzyl pentadienoate **322** to give the *syn* product *syn*-**324** (methyl groups) with 94% yield and 98:2 dr. In contrast, the analogous reaction of nerylamine Z-**321** delivered the corresponding *anti* derivative *anti*-**324** (methyl groups) with 93% yield and 98:2 dr, too. The stereochemical outcome could be rationalized by the favored chair-like transition state in **323**, the vinyl double bond should have been *E* configured because of the arrangement of R^3 and the bulky ammonium center with maximal distance. Enantioselectively-catalyzed experiments will be reported in the future (Scheme 10.67).

10 mol% $Zn(OTf)_2$, CH_2Cl_2, 23°C; **321**; **322**; **323** LA; [3,3]; **324**

E-**321** → (5 mol% $Zn(OTf)_2$, **322**, CH_2Cl_2; 94%) → *syn*-**324**

Z-**321** → (93%) → *anti*-**324**

R^1 = H, Me. R^2 = H, Me, *i*Pr, Ph. $R^{2'}$ = H, Me. R^3 = Me, *i*Pr, vinyl, Ph, Cl, NPht
R' = Me, $-(CH_2)_4-$, $-(CH_2)_5-$. Yield: 75 - 97%, d.r.: >91:9

Scheme 10.67

10.6
Alkyne Aza-Claisen Rearrangements

Propargylamines could serve as suitable allyl moiety in aza-Claisen rearrangements. The 3,3 sigmatropic bond reorganization led to allenes, which easily underwent consecutive processes like nucleophile addition and cyclization in a tandem process.

In analogy to the allyl naphthalenes the thermal behavior of N-propargyl naphthylamines **325**/**332** was investigated [54]. Upon heating and β propargyl systems **325**/**332** to about 240–260 °C, ring closure was found yielding 13/16% of piperidino naphthalene **330**/**334** and 57/41% pyridino naphthalene **329**/**333**, respectively. The process (α series) could be rationalized by starting an initial aza-Claisen rearrangement to generate the allene **326**. Then, a 1,5 H-shift (**327**) and a final electrocyclization led to the tricycle **328**, which finally suffered from disproportionation or, likewise, dehydrogenation to form the products **329**/**330**. In contrast, the corresponding N-methyl derivatives (R′ = Me) gave the indole systems **331**/**335** with 28–51% yield. Obviously, the intermediate allene **326** suffered from a fast addition of the amine to the terminal double bond (the 1,5 H-shift was suppressed for steric reasons, Scheme 10.68).

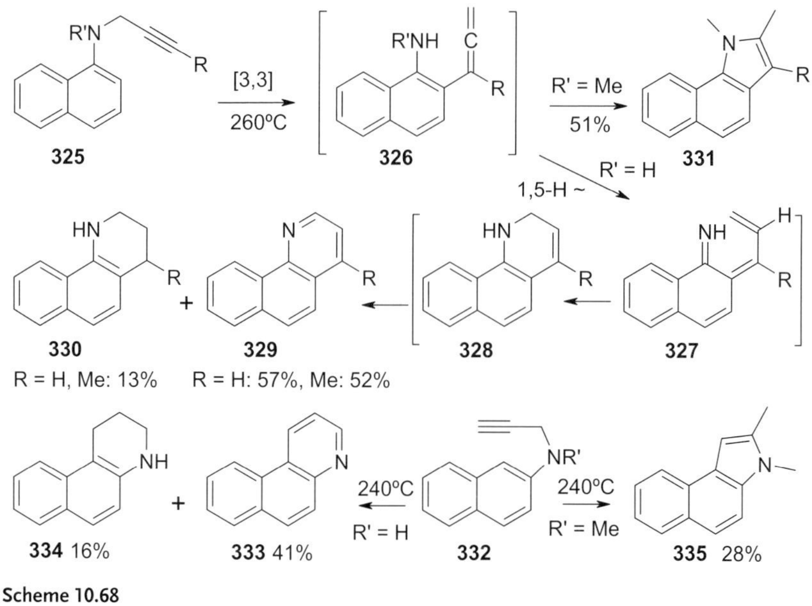

Scheme 10.68

Frey developed a pyrrole **338** synthesis starting from vinyldibromides **336** and enamines **337** [55]. In the presence of a strong base (KO*t*Bu) an initial dehydrobromination of **336** led to an alkynylbromide **339**. A consecutive equilibration was found to be crucial. Involving activating aryl substituents Ar (Ph, Naphthyl), a reversible base-induced H-shift, should have formed the corresponding allene **340**.

Without such a substituent, no cyclization took place. Then, the nucleopilic attack of the enamine **337** nitrogen proceeded to give the propargyl vinyl framework in **341** ready for the sigmatropic reaction. At 23–65 °C, the aza-Claisen rearrangement generated the β,γ-allenylimine **342**, which underwent a final 5-exo-trig cyclization to produce the pyrrole **338**. The present procedure allowed one to build 1,2,3,5 tetrasubstituted pyrroles with 32 to 50% yield overall including annulated bicyclic structures (Scheme 10.69).

R^1 = Me, OMe, OEt. R^2 = Me. R^1-R^2 = -$(CH_2)_3$-, -$[CH_2C(Me)_2CH_2]$-. R' = H, Me, Ph, Bn

Scheme 10.69

A related amination rearrangement cyclization tandem sequence was introduced by Cossy [56]. Starting from cyclic epoxyketones **343** the reaction with propargylamines **344** caused an oxirane opening condensation process to generate the enaminoketones **345**. Upon heating in toluene to reflux, aza-Claisen rearrangement delivered the intermediate allenyl imines **346**, which suffered keto-enol tautomerism to **347** and a final cyclization to give the annulated pyrroles **348** with 33 to 80% yield overall. In contrast to all other reactions, the conversion of the N-benzyl propargylamine **343** (R = Bn) proceeded at ambient temperature; the best yield obtained 80%. The whole process could be run as a one-pot reaction without any heating. Furthermore cycloketones **349** and propargylamines **344** gave rise to the formation of simple N-propargyl enamines **350**. The aza-Claisen rearrangements of these systems required significantly prolonged reaction times to achieve about 60 to 70% conversion of **350**. However, the corresponding pyrroles **351** were isolated in 50 to 60% yield recommending the procedure for further investigation (Scheme 10.70).

An uncatalyzed amination/aza-Claisen rearrangement/cyclization cascade described by Majumdar et al. was terminated by a final six-membered ring formation [57]. N-propargyl enamines **354** had been generated from vinylogous acid chlorides **352** and propargylamines **353** by means of a Michael addition elimination process with 68–77% yield (X = O, coumarines) and 80–90% yield (X = S), respectively. Upon heating the propargyl vinyl amines **354** in *o*-dichlorobenzene to about 180 °C, a cascade of rearrangement and cyclization steps allowed one to

a) MeOH, H_2O. b)PhMe, 110°C
R = Me: n = 1 (66%), n = 0 (33%). R = Bn, n = 1 (80%, all steps in MeOH at 23°C)

PhH, 80°C
X = CH_2 :60%, C(ethylenedioxy): 60%, NBOC: 50%

PhMe 110°C 50 - 60%

343 344 345 346 347 348 349 350 351

Scheme 10.70

generate the tricyclic products **355** and **356** with 56–72% yield (X = O) and 60–90% yield (X = S). Generally, the exocyclic olefin **355** was obtained as the major compound, if any, the material characterized by an endocyclic double bond **356** was isolated as a side product, which could be converted into **355** by prolonged heating (1,3 H-shift). Though propargylamine **354** displayed a propargyl vinyl amine as well as a propargyl vinyl (aryl) ether sub-unit, the aza-Claisen rearrangement proceeded. The ether system remained untouched despite a broad variation of the aryl system. The reaction path was rationalized by starting with an aza-Claisen rearrangement to produce allene imine **357**. Imine-enamide tautomerism led to the vinylogous amide **358**, which suffered from a 1,5 H-shift to build up 1-azahexatriene **359**. Then, an electrocyclic ring closure formed the dehydropiperidine **356**, which finally underwent double bond migration to give the coumarin derivative **355** as the major compound (Scheme 10.71).

Additionally, dimedone derivative **360** and propargylamine **353** could be combined to give the alkynyl vinyl amine **361**. The rearrangement cyclization cascade could be induced upon heating until reflux in chlorobenzene. The annulated piperidines **362** and **363** were isolated with 75 to 80% yield. In analogy to the coumarin series, the product **362** displaying the exocyclic double bond was formed as the major product. The endocyclic olefin **363** was obtained as the side product (up to about 20%, Scheme 10.72).

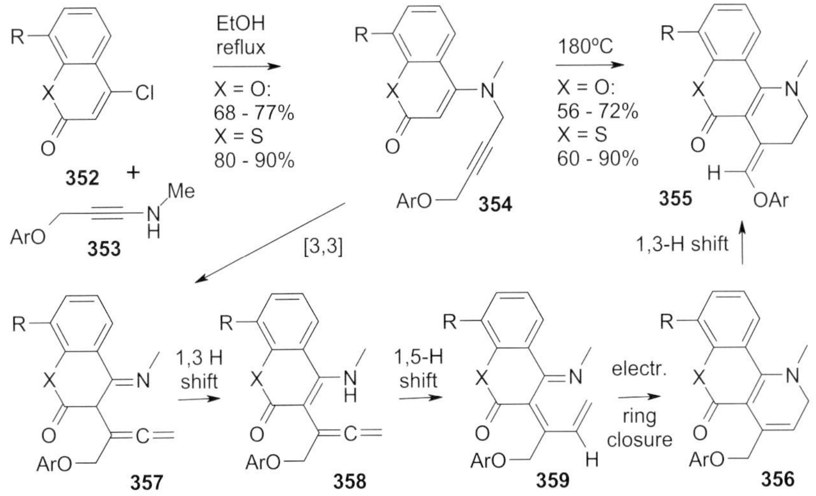

X = O: R = H, Me. Ar = Ph, 2-Tol, 4-Tol, 3,5-dimethylphenyl, 4-Cl-phenyl, 4-nitrophenyl,
X = S: R = H. Ar = Ph, 2-Tol, 4-Tol, 2-Cl-phenyl, 4-Cl-phenyl, 2,4-dichlorophenyl

Scheme 10.71

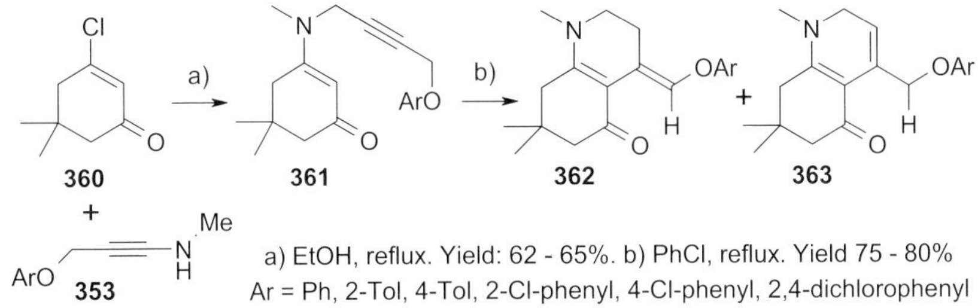

a) EtOH, reflux. Yield: 62 - 65%. b) PhCl, reflux. Yield 75 - 80%
Ar = Ph, 2-Tol, 4-Tol, 2-Cl-phenyl, 4-Cl-phenyl, 2,4-dichlorophenyl

Scheme 10.72

10.7 Iminoketene Claisen Rearrangements

The iminoketene Claisen rearrangement was investigated by Walters et al. [58]. Motivated by the early publication from Brannock and Burpitt in 1965 [22a], N-allyl amides **364** were activated by means of a strong water removing reagent like Ph_3PBr_2 and PPh_3/CCl_4. The dehydratization at 20 °C led to an highly active intermediate, the hypothetical N-allyl iminoketene **365**, which underwent immediate aza-Claisen rearrangement to generate the product γ,δ-unsaturated nitrile **366**. The low reaction temperature of the present protocol recommended the process for further investigation. Extensive variation of the water-removing reagent and the conditions showed that the originally introduced activated triphenylphosphine produced best results. Additionally, the combination of trimethylphosphite/iodine and Et_3N was found to be useful for reacting α-heteroatom-substituted amides (R^1 = OBn, NPht, etc.). Quaternary centers in α-position to the nascent nitrile functions (R^1, $R^{1'} \neq$ H) were generated smoothly. In most cases oxygen substituents placed anywhere on the reactant resulted in low yields because of side reactions presumably caused by an oxygen–phosphorous interaction (Scheme 10.73).

a) Ph_3P, CCl_4, Et_3N, MeCN, 23°C. b) Ph_3PBr_2, Et_3N, CH_2Cl_2, 23°C. c) $(MeO)_3P$, I_2, Et_3N, CH_2Cl_2, 23°C. R^1 = Ph, Bn, p-MeO-C_6H_4-, p-CF_3-C_6H_4-, MeO_2CCH_2, Et, BnO, Br-$(CH_2)_3$-, Br-$(CH_2)_4$-, MOMO, NPht, o-$HOCH_2$-C_6H_4-. $R^{1'}$ = H, Me, Ph. R^2 = H, R^2-R^{3Z} = -$(CH_2)_3$-, R^{3Z} = H, Me. R^{3E} = H, Me. Yield: 36 - 94%

Scheme 10.73

The rearrangement of *E* and Z-N-crotylamines **364** (R^3, *E* or *Z* = Me) gave the corresponding nitriles **366** with 82 and 68% yield, respectively. Disappointingly, the product was obtained as an inseparable mixture of *syn/anti* diastereomers **366** indicating a low simple diastereoselectivity. Obviously, the intermediate ketene imine fitted neither a chair- nor a boat-like conformation. Hence, a low axis-to-center chirality induction was operative, and the *E* and *Z* reactants gave a 1.1:1 and a 1.6:1 ratio in favor of the major compound (isomer not determined, Scheme 10.73).

Further information concerning the stereochemical properties of the rearrangement was evaluated by submitting rigid cyclohexane derivatives **373/374** to the reaction conditions. In 1975, House described the allylation of a cyclohexyl cyanide **367** [59]. The initial deprotonation with LDA led to a ketene imine anion **368**, which was then treated with allylbromide. Two potential paths rationalized the outcome. An N-allylation generated the intermediate ketene imines **369/370**, which underwent aza-Claisen rearrangement to deliver the nitriles **371/372**. Alternatively, the direct C-allylation of **368** produced the nitriles. The ratio was 88:12 in favor of the axial nitrile **371**. Walters presumed that the aza-Claisen rearrangement of the allyl amides **373** and **374** should have given the same nitriles **371/372** with comparable dr after passing through the ketene imines **369/370**. In fact, the reaction of the axial amide **373** led to the corresponding nitriles with 41 to 46% yield and 75:25 ratio in favor of the axial nitrile **371**. In contrast, the equatorial amide **374** was converted into the imidate **375** and the azadiene **376**, only traces of the nitriles **371/372** were found. It seemed reasonable, that the iminoketen Claisen rearrangement was sensitive concerning sterically encumbered situations. The formation of the ketene imines **369/370**, starting from axial amide **373**, appeared to represent a sterically favored process leading to the nitriles **371/372**. In contrast, the formation of the ketene imines **369/370** starting from equatorial amide **374** must have been disfavored and the system gave rise to the formation of the competing products **375/376** (Scheme 10.74).

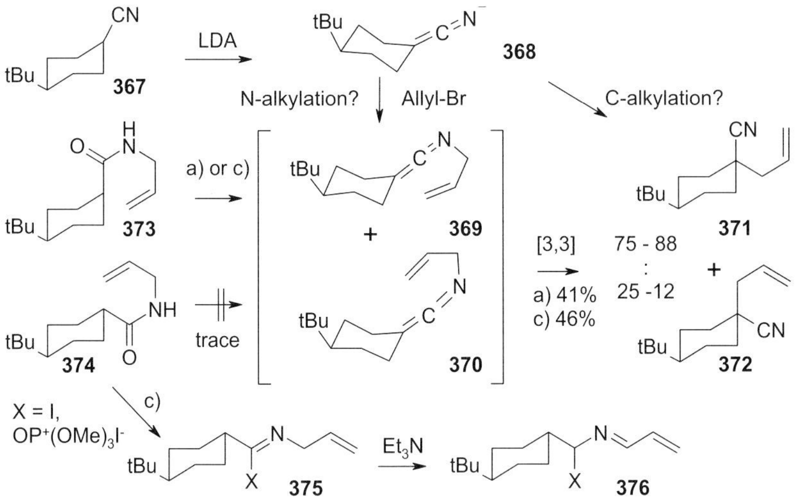

Scheme 10.74

Preliminary investigations were undertaken rearranging propargylamides **377**. In the presence of an alkyl substituent R^1 (R^1 = Bn, Et), the use of standard reaction conditions caused the dehydratization to give intermediate **378**. The final aza-Claisen rearrangement delivered allenylnitrile **379** with moderate yield. The reaction cascade of the phenyl derivative **377** (R^1 = Ph) suffered from a final double bond migration to give the $\alpha,\beta,\gamma,\delta$-unsaturated nitrile **380** (56% yield, Scheme 10.75).

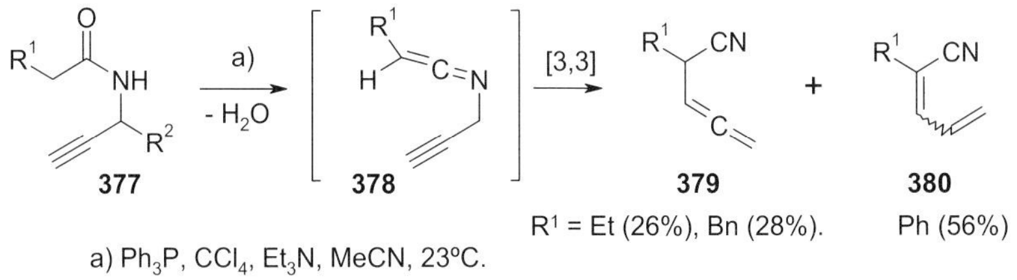

Scheme 10.75

In 1993, the first application of the Walters protocol in natural product syntheses was reported [60]. N-allylamide **381** could be converted into a 1:1 mixture of the diastereomer nitriles **382** with 56% yield. Despite the mild reaction conditions, no external 1,2 asymmetric induction (remote stereocontrol) was operative conducting such a rearrangement. The diastereomers were separated and the cleavage of the nitriles allowed one to build up the lactones **383**. The *syn* lactone *syn*-**383** was involved as a key intermediate in a (+)-canadensolide **384** total synthesis. The *anti* lactone *anti*-**383** enabled the completion of a total syntheses of (+)-santolinolide A **385** and (–)-santolinolide B **386** via several further steps (Scheme 10.76).

Ph$_3$P, CCl$_4$, Et$_3$N, MeCN, 23°C. 56%

381 → *syn*-382 + *anti*-382 (1:1)

66% — 1. KOH. 2. TFA — 60%

β-Me: **385**. α-Me: **386** ⇐ *anti*-**383**; *syn*-**383** ⇒ **384**

Scheme 10.76

Molina reported on a related process [61]. N-allyl azides **387** had been subjected to a Staudinger reaction to generate phosphine imines **388**. Then, the addition of stable ketenes **389** (synthesized separately) caused an aza-Wittig reaction to give iminoketenes **390**, which underwent immediate aza-Claisen rearrangement to produce the γ,δ-unsaturated nitriles **391** (method A). The driving force of the present cascade was high enough to generate two adjacent quaternary carbon centers. The diastereoselectivities observed on generating the nitriles varied between 1:1 and about 4:1, the configuration of the major compound had not been determined (Scheme 10.77).

Alternatively, phosphine imines **388** had been treated with various phenylacetyl chlorides **392** (method B). Surprisingly, phosphonium salts **393** were isolated with 25 to 97% yield, which could be deprotonated by means of a base to build up the corresponding phosphoranes **394** (66–89% yield). Upon heating to 90–130 °C nitriles **391** were formed in 25 to 60% yield. This outcome was explained by an initial extrusion of Ph_3P=O to generate ynamines **395**. The consecutive isomerization delivered iminoketenes **390**, which underwent usual iminoketene Claisen rearrangement to produce the nitriles **391** (Scheme 10.77).

R^{1E} = Me, vinyl, Ph, R^{1Z} = H, Me. R^2 = Ph, *p*-tol, *p*-Cl-C_6H_4-, *p*-F-C_6H_4-. $R^{2'}$ = H, Et, Ph

Scheme 10.77

References

1 Claisen rearrangement recent reviews: a) H. Frauenrath in Houben–Weyl (Methods of Organic Synthesis), Stereoselective Synthesis E21d, G. Helmchen, R. W. Hoffmann, J. Mulzer, E. Schaumann Eds, Thieme Stuttgart **1995**, 3301–3756. b) D. Enders, M. Knopp, R. Schiffers, *Tetrahedron Asymmetry* **1996**, *7*, 1847–1882. c) H. Ito, T. Taguchi, *Chem. Soc. Rev.* **1999**, *28*, 43–50. d) S. M. Allin, R. D. Baird, *Curr. Org. Chem.* **2001**, 395–415. e) M. Hiersemann, L. Abraham, *Eur. J. Org. Chem.* **2002**, 1461–1471. f) Y. Chai, S. P. Hong, H. A. Lindsay, C. McFarland, M. C. McIntosh, *Tetrahedron* **2002**, *58*, 2905–2928. g) K. C. Majumdar, T. Bhattacharyya, *Ind. J. Chem.* **2002**, *79*, 112–121. h) U. Nubbemeyer, *Synthesis* **2003**, 961–1008.

2 a) F. B. Dains, R. Brewster, J. Blair, W. Thompson, *J. Am. Chem. Soc.* **1922**, *44*, 2637–2643. b) F. L. Carnahan, C. D. Hurd, *J. Am. Chem. Soc.* **1930**, *52*, 4586–4595.

3 a) S. Marcinkiewicz, J. Green, P. Mamalis, *Tetrahedron* **1961**, *14*, 208–222. b) S. Marcinkiewicz, *Bull. Acad. Pol. Sci., Ser. Sci. Chim.* **1971**, *19*, 603–609. c) S. Inada, R. Kurata, *Bull. Chem. Soc. Jpn.* **1981**, *54*, 1581 –1582 and references cited therein. d) C. Pain, S. Célanire, G. Guillaumet, B. Joseph, *Synlett* **2003**, 2089–2091.

4 L. Viallon, O. Reinaud, P. Capdevielle, P. Maumy, *Tetrahedron Lett.* **1995**, *36*, 4787–4790.

5 a) Y. Makisumi, *Tetrahedron Lett.* **1966**, 6413–6417. b) Y. Makisumi, T. Sasatani, *Tetrahedron Lett.* **1969**, 543–546. c) B. A. Otter, A. Taube, J. J. Fox, *J. Org. Chem.* **1971**, *36*, 1251–1255.

6 a) R. N. Khusnitdinov, R. N. Fakhretdinov, I. B. Abdrakhmanov, A. A. Panasenko, L. M. Khalilov, U. M. Dzhemilev, *J. Org. Chem. USSR (Engl. Transl.)* **1983**, *19*, 920–924 and references cited therein. b) S. J. Danishefski, G. B. Phillips, *Tetrahedron Lett.* **1984**, *25*, 3159–3162. c) K. C. Majumdar, R. N. De, S. Saha, *Tetrahedron Lett.* **1990**, *31*, 1207–1208. d) K. C. Majumdar, U. Das, *Can. J. Chem.* **1996**, *74*, 1592–1596.

7 K. C. Majumdar, N. K. Jana, *Monatsh. Chem.* **2001**, *132*, 633–638.

8 a) M. A. Cooper, M. A. Lucas, J. M. Taylor, D. Ward, N. M. Williamson, *Synthesis* **2001**, 621–625. b) J. M. Roe, R. A. B. Webster, A. Ganesan, *Org. Lett.* **2003**, *5*, 2825–2827.

9 a) C. D. Hurd, W. W. Jenkins, *J. Org. Chem.* **1957**, *22*, 1418–1421. b) M. Elliot, N. F. James, *J. Chem. Soc. (C)* **1967**, 1780–1782. c) N. Takamatsu, S. Inoue, Y. Kishi, *Tetrahedron Lett.* **1971**, 4661–4664. d) M. Schmid, H.-J. Hansen, H. Schmid, *Helv. Chim. Acta* **1973**, *56*, 105–124. e) V. Lucchini, M. Prato, G. Scorrano, P. Tecillat *J. Org. Chem.* **1988**, *53*, 2251–2258. f) K. Yu, C. W. Jones, *Organometallics* **2003**, *22*, 2571–2580.

10 a) L. G. Beholz, J. R. Stille, *J. Org. Chem.* **1993**, *58*, 5095–5100. b) R. Sreekumar, R. Padmakumar, *Tetrahedron Lett.* **1996**, *37*, 5281–5282. c) J. S. Yadav, B. V. Subba Reddy, M. Abdul Rasheed, H. M. Sampath Kumar, *Synlett* **2000**, *487–488.*

11 a) G. Lai, W. K. Anderson, *Tetrahedron Lett.* **1993**, *34*, 6849–6852. b) W. K. Anderson, G. Lai, *Synthesis* **1995**, 1287–1290. c) G. Lai, W. K. Anderson, *Tetrahedron* **2000**, *56, 2583–2590.*

12 a) K. Irie, F. Koizumi, Y. Iwata, T. Ishii, Y. Yanai, Y. Nakamura, H. Ohigashi, P. A. Wender, *Bioorg. Med. Chem. Lett.* **1995**, *5*, 453–458. b) K. Irie, T. Isaka, Y. Iwata, Y. Yanai, Y. Nakamura, F. Koizumi, H. Ohigashi, P. A. Wender, Y. Satomi, H. Nishino, *J. Am. Chem. Soc.* **1996**, *118*, 10733–10743. c) K. Irie, Y. Yanai, K. Oie, J. Ishizawa, Y. Nakagawa, H. Ohigashi, P. A. Wender, U. Kikkawa, *Bioorg. Med. Chem.* **1997**, *5*, 1725–1737. d) L. S. Hegedus, G. F. Allen, J. J. Bozell, E. L. Waterman, *J. Am. Chem. Soc.* **1978**, *100*, 5800–5805.

13 O. Benali, M. A. Miranda, R. Tormos, S. Gil *J. Org. Chem.* **2002**, *67*, 7915–7918.

13 a) H. Katayama *Chem. Pharm. Bull.* **1978**, *26*, 2027–2033. b) H. Katayama, *J. Chem. Soc., Chem. Commun.* **1980**, 1009–1011. c) K. Kaneko, H. Katayama, Y. Saito, N. Fujita, A. Kato, *J. Chem. Soc., Chem. Commun.* **1986**, 1308–1311.

15 a) R. K. Hill, N. W. Gilman, *Tetrahedron Lett.* **1967**, 1421–1423. b) R. K. Hill, G. R. Newkome, *Tetrahedron Lett.* **1968**, 5058–5062. c) G. Schmidt, E. Winterfeldt, *Chem. Ber.* **1971**, *104*, 2483–2488. Failure of aza-Claisen rearrangement: d) T. Sasaki, A. Kojima, M. Otha, *J. Chem. Soc., Section C, Organ. Chem.* **1971**, 196–200. e) C. Vogel, P. Delavier, *Tetrahedron Lett.* **1989**, *30*, 1789–1792. f) M. S. J. Gomes, L. Sharma, S. Prabhakar, A. M. Lobo, P. M. C. Glória, *J. Chem. Soc., Chem. Commun.* **2002**, 746–747. g) H. Amii, Y. Ichihara, T. Nakagawa, T. Kobayashi, K. Uneyama, *J. Chem. Soc., Chem. Commun.* **2003**, 2902–2903.

16 a) R. K. Hill, H. N. Khatri, *Tetrahedron Lett.* **1978**, *19*, 4337–4340. b) P. D. Bailey, M. J. Harrison, *Tetrahedron Lett.* **1989**, *30*, 5341–5344.

17 a) P. S. Mariano, D. Dunaway-Mariano, P. L. Huesmann, R. Beamer, *Tetrahedron Lett.* **1977**, 4299–4302. b) P. S. Mariano, D. Dunaway-Mariano, P. L. Huesmann, *J. Org. Chem.* **1979**, *44*, 124–133. c) Y. Chen, P. L. Huesmann, P. S. Mariano, *Tetrahedron Lett.* **1983**, *24*, 1021–1024.

18 a) G. R. Cook, J. R. Stille, *J. Org. Chem.* **1991**, *56*, 5578–5583. b) G. R. Cook, N. S. Barta, J. R. Stille, *J. Org. Chem.* **1992**, *57*, 461–467. c) N. S. Barta, G. R. Cook, M. S. Landis, J. R. Stille *J. Org. Chem.* **1992**, *57*, 7188–7194 d) G. R. Cook, J. R. Stille, *Tetrahedron* **1994**, *50*, 4105–4124.

19 L. E. Overman, *Angew. Chem. Int. Ed. Engl.* **1984**, *23*, 579–586, *Angew. Chem.* **1984**, *96*, 565–573.

20 J. T. Welch, B. De Korte, N. De Kimpe, *J. Org. Chem.* **1990**, *55*, 4981–4983.

21 a) S.-I. Murahashi, Y. Makabe, *Tetrahedron Lett.* **1985**, *26*, 5563–5566. b) S.-I. Murahashi, Y. Makabe, K. Kunita, *J. Org. Chem.* **1988**, *53*, 4489–4495. c) N. Matzanke, R. J. Gregg, S. M. Weinreb, *J. Org. Chem.* **1997**, *62*, 1920–1921. For photo-aza-Claisen rearrangements see: c) B. Vogler, R. Bayer, M. Meller, W. Kraus, *J. Org. Chem.* **1989**, *54*, 4165–4168

22 a) K. C. Brannock, R. D. Burpitt, *J. Org. Chem.* **1961**, *24*, 3576–3577. b) G. Opitz, *Liebigs Ann. Chem.* **1961**, *650*, 122–134. c) P. M. McCurry, R. K. Singh, *Tetrahedron Lett.* **1973**, 3325–3328. d) P. Houdewind, U. K. Pandit, *Tetrahedron Lett.* **1974**, 2359–2362. e) J. C. Gilbert, K. P. A. Senaratne, *Tetrahedron Lett.* **1984**, *25*, 2303–2306. f) K. Honda, S. Inoue, *Synlett* **1994**, 739–741. g) C. Kuhn, L. Skaltsounis, C. Monneret, J.-C. Florent, *Eur. J. Org. Chem.* **2003**, 2585–2595. h) K. Honda, H. Yasui, S. Inoue, *Synlett* **2003**, 2380–2382.

23 D. F. McComsey, B. E. Maryanoff *J. Org. Chem.* **2000**, *65*, 4938 –4943.

24 a) R. F. Winter, F. M. Hornung, *Organometallics* **1997**, *16*, 4248 –4250 b) R. F. Winter, K. W. Klinkhammer, *Organometallics* **2001**, *20*, 1317–1333.

25 R. F. Winter, G. Rauhut, *Chem. Eur. J.* **2002**, *8*, 641–649.

26 R. Gompper, B. Kohl, *Angew. Chem.* **1982**, *94*, 202; *Angew. Chem.* Int. Ed. Engl. **1982**, *21*, 198.

27 a) M. J. Kurth, O. H. W. Decker, *Tetrahedron Lett.* **1983**, *24*, 4535–4538. b) M. J. Kurth, O. H. W. Decker, H. Hope, M. D. Yanuck, *J. Am. Chem. Soc.* **1985**, *107*, 443–448. c) M. J. Kurth, O. H. W. Decker, *J. Org. Chem.* **1985**, *50*, 5769–5775. d) M. J. Kurth, O. H. W. Decker, *J. Org. Chem.* **1986**, *51*, 1377–1383. e) M. J. Kurth, C. J. Soares, *Tetrahedron Lett.* **1987**, *28*, 1031–1034. f) M. J. Kurth, E. G. Brown, *Synthesis* **1988**, *362–366.*

28 a) T. Tsunoda, O. Sasaki, S. Itô, *Tetrahedron Lett.* **1990**, *31*, 727–730. b) S. Itô, T. Tsunoda, *Pure & Appl. Chem.* **1990**, *62*, 1405–1408. c) T. Tsunoda, M. Sasaki, O. Sasaki, Y. Sako, Y. Hondo, S. Itô, *Tetrahedron Lett.* **1992**, *33*, 1651–1654. d) T. Tsunoda, S. Tatsuki, Y. Shiraishi, M. Akasaka, S. Itô *Tetrahedron Lett.* **1993**, *34*, 3297–3300. e) T. Tsunoda, S. Tatsuki, K. Kataoka, S. Itô, *Chem. Lett.* **1994**, 543–546. f) S. Itô, T. Tsunoda, *Pure & Appl. Chem.* **1994**, *62*, 2071–2074. g) T. Tsunoda, F. Ozaki, N. Shirakata, Y. Tamaoka, H. Yamamoto, S. Itô, *Tetrahedron Lett.* **1996**, *37*, 2463–2466. h) T. Tsunoda, T. Nishii, T. Suzuki, S. Itô, *Tetrahedron Lett.* **2000**, *41*, 7667–7670. i) T. Nishii, S. Suzuki, K. Yoshida, K. Arakaki, T. Tsunoda, *Tetrahedron Lett.* **2003**, *44*, 7829–7832. j) S. G. Davies, A. C. Garner, R. L. Nicholson, J. Osborne, E. D. Savory, A. D. Smith, *J. Chem. Soc., Chem. Commun.* **2003**, 2134–2135.

29 a) Y. G. Suh, J. Y. Lee, S. A. Kim, J. K. Jung, *Synth. Commun.* **1996**, *26*, 1675–1680. b) Y. G. Suh, S. A. Kim, J. K. Jung, D. Y. Shin, K. H. Min, B. A. Koo, H. S. Kim, *Angew. Chem.* **1999**, *111*, 3753–3755, Angew. Chem. Int. Ed. **1999**, *38*, 3545–3547.

30 a) U. M. Lindstroem, P. Somfai, *J. Am. Chem. Soc.* **1997**, *119*, 8385–8386. b) U. M. Lindstroem, P. Somfai, *Synthesis* **1998**, 109–117. c) U. M. Lindstroem, P. Somfai, *Chem. Eur. J.* **2001**, *7*, 94–98.

31 K. Neuschütz, J.-M. Simone, T. Thyrann, R. Neier, *Helv. Chim. Acta* **2000**, *83*, 2712–2737.

32 L. E. Overman, L. A. Clizbe, R. L. Freerks, C. K. Marlowe, *J. Am. Chem. Soc.* **1981**, *103*, 2807–2815.

33 Originally, ketene Claisen rearrangements were described by Bellus and Malherbe in the reaction of allyl sulfides and dichloro ketene: a) R. Malherbe, D. Bellus, *Helv. Chim. Acta.* **1978**, *61*, 3096–3099. b) R. Malherbe, G. Rist, D. Bellus, *J. Org. Chem.* **1983**, *48*, 860–869.

34 a) P. S. Mariano, D. Dunaway-Mariano, P. L. Huesmann, R. L. Beamer, *Tetrahedron Lett.* **1977**, 4299–4302. b) P. S. Mariano, D. Dunaway-Mariano, P. L. Huesmann, *J. Org. Chem.* **1979**, *44*, 124–133. c) F. A. Kunng, J. M. Gu, S. Chao, Y. Chen, P. S. Mariano, *J. Org. Chem.* **1983**, *48*, 4262–4266. d) S. Chao, F. A. Kunng, J. M. Gu, L. Ammon, P. S. Mariano *J. Org. Chem.* **1984**, *49*, 2708–2711. e) A. W. Baxter, D. Labaree, S. Chao, P. S. Mariano, *J. Org. Chem.* **1989**, *54*, 2893–2904. f) A. W. Baxter, D. Labaree, H. L. Ammon, P. S. Mariano, *J. Am. Chem. Soc.* **1990**, *112*, 7682–7692.

35 E. Vedejs, M. Gingras, *J. Am. Chem. Soc.* **1994**, *116*, 579–588.

36 a) M. Ishida, H. Muramaru, S. Kato, *Synthesis* **1989**, 562–563. b) S. M. Roberts, C. Smith, R. J. Thomas, *J. Chem. Soc. Perkin. Trans. I*, **1990**, 1493–1495.

37 a) M. Ishida, H. Muramaru, S. Kato, *Synthesis* **1989**, 562–563. b) M. M. Cid, U. Eggnauer, H. P. Weber, E. Pombo-Villar, *Tetrahedron Lett.* **1991**, *32*, 7233–7236 c) M. M. Cid, E. Pombo-Villar, *Helv. Chim. Acta* **1993**, *76, 1591–1607.*

38 a) E. D. Edstrom, *J. Am. Chem. Soc.* **1991**, *113*, 6690–6692. b) E. D. Edstrom, *Tetrahedron Lett.* **1991**, *32*, 5709–5712.

39 a) M. Diederich, U. Nubbemeyer, *Angew. Chem. Int. Ed. Engl.* **1995**, *34*, 1026–1028. b) M. Diederich, U. Nubbemeyer, *Chem. Eur. J.* **1996**, *2*, 894–900. c) A. Sudau, U. Nubbemeyer, *Angew. Chem. Int. Ed. Engl.* **1998**, *37*, 1140–1142. d) S. Laabs, A. Scherrmann, A. Sudau, M. Diederich, C. Kierig, U. Nubbemeyer, *Synlett* **1999**, 25–28. e) A. Sudau, W. Münch, U. Nubbemeyer, J.-W. Bats, *J. Org. Chem.* **2000**, *65*, 1710–1720. f) A. Sudau, W. Münch, U. Nubbemeyer, J.-W. Bats, *Eur. J. Org. Chem.* **2002**, 3304–3314.

40 a) J. von Braun, *Chem. Ber.* **1907**, *40*, 3914–3933. b) J. H. Cooley, E. J. Evain, *Synthesis* **1989**, 1–7.

41 a) The Schlögl nomenclature is used for terming the planar chiral properties: K. Schlögl, *Top. Curr. Chem.* **1984**, *125*, 27–62. Alternatively, with the use of *P* (*pR*) and *M* (*pS*) b) V. Prelog, G. Helmchen, *Angew. Chem.* **1982**, *94*, 614–631; *Angew. Chem. Int. Ed. Engl.* **1984**, *21, 567–583.*

42 a) D. N. J. White, M. J. Bovill, *J. Chem. Soc. Perkin Trans. II* **1977**, 1610–1623. b) G. Guella, G. Chiasera, I. N'Diaye, F. Pietra, *Helv. Chim. Acta* **1994**, *77*, 1203–1221. c) A. Deiters, C. Mück-Lichtenfeld, R. Fröhlich, D. Hoppe, *Org. Lett.* **2000**, *2*, 2415–2418.

43 Preliminary kinetic investigations of the conversion of *pS*-**157** into *pR*-**157**: half-life period at 43 °C: 347 min, half-life period at 55 °C: 84 min, half-life period at 65 °C: 26.4 min, half-life period at 7 °C: 10.5 min, activation energy (ΔG*) of the epimerization: about 24 kcalmol^{-1} (±3).

44 a) M. Diederich, U. Nubbemeyer, *Synthesis* **1999**, 286–289. b) A. Sudau, W. Münch, U. Nubbemeyer, J.-W. Bats, *Chem. Eur. J.* **2001**, *7*, 611–621. c) A. Sudau, W. Münch, U. Nubbemeyer, J.-W. Bats, *Eur. J. Org. Chem.* **2002**, 3315–3325.

45 C. J. Deur, M. W. Miller, L. S. Hegedus, *J. Org. Chem.* **1996**, *61*, 2871–2876.

46 C.-M. Yu, H.-S. Choi, J. Lee, W.-H. Jung, H.-J. Kim, *J. Chem. Soc. Perkin Trans. I* **1996**, 115–116.

47 a) T. P. Yoon, V. M. Dong, D. W. C. MacMillan, *J. Am. Chem. Soc.* **1999**, *121*, 9726–9627. b) V. M. Dong, D. W. C. MacMillan, *J. Am. Chem. Soc.* **2001**, *123*, 2448–2449.

48 a) U. Nubbemeyer, R. Öhrlein, J. Gonda, B. Ernst, D. Bellus, *Angew. Chem.* **1991**, *103*, 1533–1535; *Angew. Chem. Int. Ed. Engl.* **1991**, *30*, 1465–1467. b) B. Ernst, J. Gonda, R. Jeschke, U. Nubbemeyer, R. Öhrlein, D. Bellus, *Helv. Chim. Acta* **1997**, *80, 876–891.*

49 a) U. Nubbemeyer, *J. Org. Chem.* **1995**, *60*, 3773–3780. b) U. Nubbemeyer, *J. Org. Chem.* **1996**, *61*, 3677–3686.

50 a) S. Laabs, W. Münch, J.-W. Bats, U. Nubbemeyer, *Tetrahedron* **2002**, *58*, 1317–1334. b) N. Zhang, U. Nubbemeyer, *Synthesis* **2002**, 242–252. For a preliminary result see ref. 39d.

51 P. Metz, *Tetrahedron* **1993**, *49*, 6367–6374.

52 T. P. Yoon, D. W. C. MacMillan, *J. Am. Chem. Soc.* **2001**, *123*, 2911–2912.

53 T. H. Lambert, D. W. C. MacMillan, *J. Am. Chem. Soc.* **2002**, *124*, 13646 –13647.

54 a) H. Scheurer, J. Zsindely, H. Schmid, *Helv. Chim. Acta* **1973**, *56*, 478–489.

55 a) H. Frey, *Synlett* **1994**, 1007–1008. b) for an early work see 15c

56 J. Cossy , C. Poitevina, L. Sallé, D. Gomez Pardo, *Tetrahedron Lett.* **1996**, *37*, 6709–6710.

57 a) K. C. Majumdar, S. Gosh, S *Tetrahedron* **2001**, *57*, 1589–1592. b) K. C. Majumdar, S. K. Samanta, *Tetrahedron Lett.* **2001**, *42*, 4231–4233. c) K. C. Majumdar, S. K. Samanta, *Tetrahedron* **2001**, *57*, 4955–4958. d) K. C. Majumdar, S. K. Samanta, *Synthesis* **2002**, 121–125.

58 a) M. A. Walters, C. S. McDonough, P. S. Brown, A. B. Hoem, *Tetrahedron Lett.* **1991**, *32*, 179–182. b) M. A. Walters, A. B. Hoem, H. R. Arcand, A. D. Hegeman, C. S. McDonough, *Tetrahedron Lett.* **1993**, *34*, 1453–1456. c) M. A. Walters, A. B. Hoem, *J. Org. Chem.* **1994**, *59*, 2645–2647. d) M. A. Walters, *J. Am. Chem. Soc.* **1994**, *116*, 11618–11619. e) M. A. Walters, A. B. Hoem, C. S. McDonough, *J. Org. Chem.* **1996**, *61*, 55–62.

59 H. O. House, J. Lubinkowski, J. J. Good, *J. Org. Chem.* **1975**, *40*, 86–92.

60 U. Nubbemeyer, *Synthesis* **1993**, 1120–1128.

61 a) P. Molina, M. Alajarin, C. Lopez-Leonardo, *Tetrahedron Lett.* **1991**, *32*, 4041–4044. b) P. Molina, M. Alajarin, C. Lopez-Leonardo, J. Alcantara, *Tetrahedron* **1993**, *49*, 5153–5168.

11
Mechanistic Aspects of the Aliphatic Claisen Rearrangement

Julia Rehbein and Martin Hiersemann

The synthetic application of a reaction and the physico-chemical approach of unravelling its mechanism are often strictly separated. However, a thorough knowledge of mechanistic aspects is desirable in order to influence the course and outcome of a reaction in a systematic way and to insure optimal results with respect to rate and selectivity. On the other hand, unexpected and/or unexplainable experimental observations may guide the chemist to interesting mechanistic aspects.

The Claisen rearrangement [1] is a prototypic reaction for which the chemist profits from the combination of theoretical understanding and synthetic application. The synthetic organic chemist certainly considers the Claisen rearrangement as one of the most powerful C–C-connecting transformations and the Claisen retron can be found in various potential target molecules. The challenge to the physical organic or computational chemist was, and still is, to unravel the exact nature of the transition state of the Claisen rearrangement because the rearrangement is characterized by a certain "flexibility" with respect to the extent of the bond-making and bond-breaking step in the transition state. It turns out that the nature (and thereby the energy and geometry) of the transition state is significantly dependent on the substituents and solvents and that the energy and geometry of the transition state is pivotal for the rate and stereoselectivity of the rearrangement. It is therefore desirable to develop a general predictive model to foresee whether a certain substituent or solvent will be beneficial for the rate and the selectivity of the Claisen rearrangement.

Catalyst development for carbon–carbon bond-forming sigmatropic rearrangements may be one of the primary goals for future work in the field of method development. Rational design of a catalyst that unfolds its activity by way of transition-state stabilization requires a detailed understanding of the nature of the transition state. In order to stabilize the transition state, the catalyst has to interact with the transition state. This interaction will be initiated by binding to the substrate, which, in the case of the Claisen rearrangement, is an allyl vinyl ether. Therefore, the catalyst can be considered as a new substituent on the allyl vinyl ether framework. By studying the consequences of the nature and position of sub-

The Claisen Rearrangement. Edited by M. Hiersemann and U. Nubbemeyer

ISBN: 978-3-527-30825-5

stituents on the transition state, one can gain basic information on how the catalyst has to be designed in order to stabilize the transition state most effectively. Our engagement in the development of chiral catalysts for the Claisen rearrangement sparked our interest in the theoretical aspects of the Claisen rearrangement and we are very thankful that the specialist originally assigned to this chapter suddenly and unexpectedly gave us the opportunity to write this review. Through the eyes of a synthetic organic chemist, this chapter was written to summarize the available kinetic data and the results from theoretical investigations that are concerned with the aliphatic Claisen rearrangement of allyl vinyl ethers [2]. In order to enable the reader to draw his or her own conclusions, we will summarize the relevant literature chronologically and with minimal subjective interpretation.

> *"It is assumed that the initial effect of heat on the system C=C-O-C-C=C is to alter the position of the pair of electrons which bind the allyl group to the oxygen so that semi-ionization occurs ... Actual separation into ions does not occur, but the semi-ionization promotes other ionic disturbances at the double bonds. This effect, combined with the spatial proximity of the atoms at the end of the systems, brings about temporary ring closure and readjustment of electrons as shown in the following sequence of steps."*
> *(Charles D. Hurd and Maxwell A. Pollack, 1938 [3], Scheme 11.1)*

O heat

Scheme 11.1 An early attempt to devise a mechanism for the Claisen rearrangement according to Hurd and Pollack (1938) [3].

The first kinetic study on thermal rearrangement of aliphatic allyl vinyl ether at reduced pressure was reported by Schuler and Murphy in 1950 [4]. The gas-phase study established that pressure did not influence the rate of the rearrangement. An activation energy of 30.6 kcal mol^{-1} was determined from the Arrhenius equation ($k\,[s^{-1}] = 5.0 \times 10^{11} \exp[-30600/RT]$) at 0.51 bar. At 80 °C, the Arrhenius equation affords a first-order rate constant of $0.0055 \times 10^{-5}\,s^{-1}$. A large negative entropy of activation was determined (–7.7 cal $mol^{-1}\,K^{-1}$ at 180 °C) from the Eyring equation, indicating a cyclic transition state in accordance with the currently accepted pericyclic mechanism of the Claisen rearrangement.

The gas-phase rearrangement of the 2-methyl-substituted allyl vinyl ether was investigated in 1952 by Stein and Murphy [5]. Again, it was found that the rearrangement follows first-order kinetics and that the rate constant was independent of pressure. The Arrhenius plot delivered the equation $k\,[s^{-1}] = 5.4 \times 10^{11} \exp(-29300/RT)$ for the rearrangement of allyl isopropenyl ether. The activation energy for the Claisen rearrangement of allyl isopropenyl ether was found to be 1.3 kcal mol^{-1} lower than for allyl vinyl ether. This result was the first quantification of a substituent effect on the rate of the Claisen rearrangement. The entropy of activation for allyl isopropenyl ether was identical to the value determined for allyl vinyl ether (–7.7 cal $mol^{-1}\,K^{-1}$). Calculation and comparison of the gas-phase rate constants for the rearrangement of these two simplest allyl vinyl ethers at 80 °C demonstrate the rate-accelerating effect of the 2-methyl group (Scheme 11.2).

80 °C
gas phase

R	k × 10^{-5} [s^{-1}]	k_{rel}
H	0.0055	1
Me	0.038	6.9

Scheme 11.2 Gas-phase rate constants calculated from the Arrhenius equation according to Stein and Murphy (1952) [5]. $R = 1.985792$ cal $mol^{-1}K^{-1}$, $T = 353$ K.

In 1960, a study by Burgstahler showed experimentally that the thermal Claisen rearrangement (170–180 °C, 15–20 min, sealed tube) of acyclic allyl vinyl ethers that were derived from secondary allylic alcohols proceeded stereoselectively to provide rearrangement products containing stereogenic *E*-configured double bonds. The double-bond configuration was assigned on the basis of the infrared absorption of the double-bond isomers [6].

An early kinetic study on the influence of substituents on the rate of the rearrangement in a condensed phase was performed by Ralls et al. (Scheme 11.3) [7]. A significant influence of the nature of the substituent R on the rate of the rearrangement was determined. Crossover experiments supported the assumption of a concerted mechanism of the rearrangement. The study also verified a small but significant catalytic effect of solid NH_4Cl on the rate of the reaction, an effect that was initially reported by Claisen [1] and later confirmed by Lauer and Kilburn [8].

115 °C, neat

R	k × 10^{-5} [s^{-1}]	k_{rel}
H	4.18	1
Me	9.35	2.2

Scheme 11.3 Influence of substituents on the rate of the rearrangement according to Ralls et al. (1963) [7].

It was first demonstrated by Hill and Edwards in 1964 that 1,3-chirality transfer can be accomplished by the Claisen rearrangement of an aliphatic allyl vinyl ether (Scheme 11.4) [9]. This remarkable piece of work clearly supports the suprafacial nature of the thermal intramolecular rearrangement.

180-185 °C
sealed tube
(81 %)

(+) (−)

Scheme 11.4 The first asymmetric aliphatic Claisen rearrangement by 1,3-chirality transfer according to Hill and Edwards (1964) [9].

The results of an investigation concerned with the stereochemical course of the thermal Claisen rearrangement of all four double-bond isomers of crotyl propenyl ether were published by Vittorelli et al. (1968) [10a] and Hansen and Schmid (1974) [10b]. Based on the relationship between the double-bond configuration and the relative configuration of the rearrangement product, they concluded that the Claisen rearrangement proceeded preferentially (ca. 95:5) through a chair-like transition state.

The Claisen rearrangement of 1-methallyl vinyl ether was investigated in the gas phase and in solution by Frey and Montague in 1968 (Scheme 11.5) [11]. The rearrangement was first order in both the gas phase and in solution and no pressure dependence was observed when the rearrangement was performed between 4 and 200 mbar. 4-Pentenal was formed as an E/Z=25–30/1 mixture of double-bond isomers. Arrhenius plots afforded the rate constants for the rearrangement in the gas phase $\{k\,[s^{-1}] = 2.09 \times 10^{11} \exp(-27870/RT)\}$, in 1-decene $\{k\,[s^{-1}] = 6.03 \times 10^{11} \exp(-25740/RT)\}$ and in benzonitrile $\{k\,[s^{-1}] = 8.13 \times 10^{11} \exp(-25300/RT)\}$. The 4-methyl group accelerates the rearrangement at 80 °C in the gas phase by a factor of 18 compared to the parent allyl vinyl ether. The author concluded that this rate acceleration is "... undoubtedly due to the inductive effect of the methyl group and suggests that there is a δ+ charge on the carbon atoms adjacent to the oxygen atom in the transition state." The solvent effect on the rate and the entropy of activation was explained by postulating "... increased solvation of the transition complex in highly polar media ..."

A study by Perrin and Faulkner in 1969 demonstrated that the E/Z ratio of the product from the Claisen rearrangement may be predicted from the relative thermodynamic stabilities of the chair and boat conformation of cyclohexanes that

160 °C

solvent	$k \times 10^{-5}$ [s^{-1}]	k_{rel}	$\Delta S^{\#}_{430K}$ [cal mol^{-1} K^{-1}]	$\Delta G^{\#}_{430K}$ [kcal mol^{-1}]	E_a [kcal mol^{-1}]
gas phase	139.7	1	−9.45	31.08	27.87
1-decene	488.0	3	−11.9	30.01	25.74
benzonitrile	1102.1	8	−111.3	29.32	25.30

Scheme 11.5 Claisen rearrangement of 1-methallyl vinyl ether according to Frey and Montague (1968) [11]. Rate constants at 160 °C calculated with R = 1.985792 cal mol^{-1} K^{-1}, T = 430 K.

carry the same substituent as the allyl vinyl ether. Their data also support the commonly accepted preference of the Claisen rearrangement for a chair-like transition state [12].

The ring-expanding Claisen rearrangement of 2-methyl-2-vinyl-5-methylene-tetrahydrofurane was investigated by Rhoads and Watson in 1971 (Scheme 11.6) [13]. As a consequence of the strained bicyclic transition state, a rather high reaction temperature was required to observe a comparable rate constant.

170 °C, neat

$k \times 10^{-5}$ [s^{-1}] = 4.13 ± 0.07
E_a = 33.6 ± 0.6 kcal/mol
$\Delta H^{\#}$ = 32.7 ± 0.6 kcal/mol
$\Delta S^{\#}$ = –5.6 ± 1.2 cal/mol

Scheme 11.6 Ring-expanding Claisen rearrangement according to Rhoads and Watson (1971) [13].

Due to the exothermicity of the rearrangement and under consideration of the Hammond postulate, one would expect that the aliphatic Claisen rearrangement proceeds through an earlier, reactant-like transition state. By determining the secondary kinetic deuterium isotope effect, Gajewski and Conrad verified this proposal in 1979 [14]. Furthermore, they suggested a transition state with more bond-breaking than bond-making character and that radical stabilizing groups at C-1, C-4 and C-6 will stabilize the transition state to a greater extent than radical stabilizing substituents at C-2 and C-5 (Scheme 11.7). The relative reaction rates, however, are dependent on the relative stability of the transition state *and* the ground state.

early dissociative transition state with

character

radical stabilizing substituents at C-1, C-4 and C-6
should stabilize the transition state

Scheme 11.7 Schematic presentation of the nature of the transition state according to Gajewski and Conrad (1979) [14].

Burrows and Carpenter launched the first systematic study concerning substituent effects on the rate of the aliphatic Claisen rearrangement (Table 11.1) [15]. For this purpose, the kinetic data of the cyano-substituted allyl vinyl ether **2a–e** were determined in di-*n*-butyl ether at temperatures between 55 and 185 °C depending on the substrate structure. The turnover was determined by analytical HPLC with UV-detection. Compared to the gas-phase study of Schuler and Murphy in 1950, a significant lower activation energy for the rearrangement of allyl vinyl ether **1** was determined in solution (25.4 versus 30.6 kcal mol^{-1}). Furthermore, a decreased entropy of activation was determined for the rearrangement of **1** in *n*-Bu_2O compared to the value for the gas phase experiment (–15.9 versus –7.7 cal mol^{-1} K^{-1}). Compared to the parent allyl vinyl ether **1**, the rate-accelerating effect of the relatively small cyano group as a π-acceptor substituent is particularly pronounced if present at the 2- or 4-position of the allyl vinyl ether skeleton (**2b**, **2c** compared to **1**, Table 11.1). The 5-cyano substituent exerts a moderate rate-accelerating effect whereas the presence of a cyano substituent at the 1- or 6-position leads to a rather small but significant rate retardation (**2d**, **2a**, **2e** compared to **1**, Table 11.1).

In order to explain the observed rate effects of substituents, Burrows and Carpenter used simple Hückel molecular orbital theory to evaluate the influence of a substituent on the transition state and the reactants (Scheme 11.8) [16]. Their

Table 11.1 Substituent effect on the rate of the Claisen rearrangement according to Burrows and Carpenter (1981) [15].

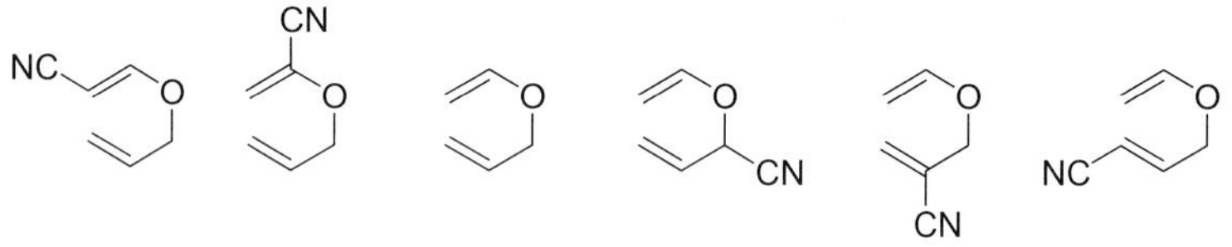

ave[a)]	T_{exp} [°C]	$\Delta H^{\#}$ [kcal mol^{-1}]	$\Delta S^{\#}$ [cal mol^{-1} K^{-1}]	k (100 °C) $\times 10^{-5}$ [s^{-1}][b)]	$k_{relative}$ (100 °C)
2a	124–174	27.08 ± 0.09	–11.6 ± 0.2	0.036	9
2b	66–115	22.84 ± 0.19	–13.4 ± 0.5	6.190	1547.5
1	113–173	25.40 ± 0.65	–15.9 ± 1.5	0.046	11.5
2c	55–101	22.33 ± 0.20	–13.0 ± 0.6	15.668	3917
2d	90–140	22.58 ± 0.26	–18.0 ± 1.3	0.884	221
2e	135–185	28.76 ± 0.54	–11.2 ± 1.2	0.0040	1

a) ave = allyl vinyl ether; solvent: *n*-Bu_2O.
b) Calculated from the Eyring equation with $k_B = 1.38066 \times 10^{-23}$ J K^{-1}, $h = 6.62618 \times 10^{-34}$ J s, $R = 8.31451$ J K^{-1} mol^{-1}, 1 cal = 4.187 J.

model rests on the basic assumption that substituent effects on reactions with quasi-aromatic transition states could be evaluated from the ability of the substituents to participate in the delocalization process [17]. The model for the transition state is dependent on the number of participating atoms. Completely conjugated hydrocarbons are chosen for even-numbered rings, whereas in the case of odd numbers the corresponding Möbius orbital array is used. In this model, electron-donating substituents are represented by carbanions and electron-withdrawing groups as carbenium ions, respectively.

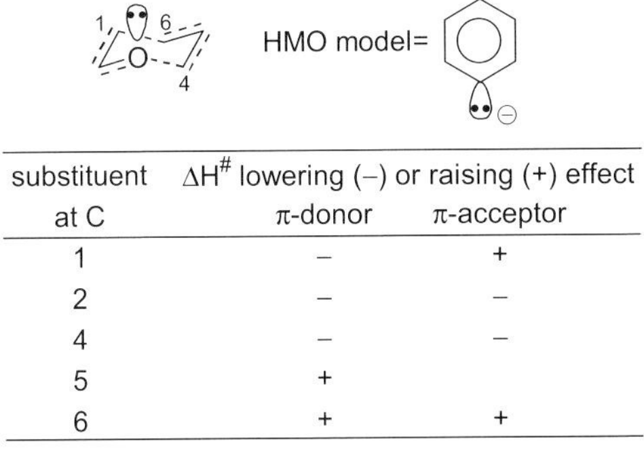

substituent at C	$\Delta H^{\#}$ lowering (–) or raising (+) effect π-donor	π-acceptor
1	–	+
2	–	–
4	–	–
5	+	
6	+	+

Scheme 11.8 Predicting substituent effects on the rate of the Claisen rearrangement based on simple HMO theory (Burrows and Carpenter, 1981, [16]).

The arbitrary prediction that in even-numbered rings electron donors influence the reaction rate in the same way as electron acceptors is a consequence of this treatment. The π-energies of ground and transition state are calculated by means of Hückel molecular orbital theory and compared to the π-energies of the unsubstituted reference systems. The model has been tested on different pericyclic reactions such as electrocyclic ring opening reactions, cycloadditions and sigmatropic shifts. The authors concluded their study on the Claisen rearrangement by noting that "... qualitative predictions about substituent effects on the aliphatic Claisen rearrangement can apparently be made by focusing on the delocalized model for the transition state and that explicit inclusion of such a structure is probably necessary for a proper description of substituent effects on [3,3]-sigmatropic migrations ..."

The thermal Claisen rearrangement of so-called diosphenol allyl ether has been investigated by Ponaras [18]. In 1983, he reported half-life times for the rearrangement in different solvent systems. Selected data are given in Schemes 11.9 to 11.11. Unfortunately, the half-life times have been determined at varying temperatures, which limits the quantitative comparability of many of the available data.

THF, 66 °C

X	$t_{1/2}$ [h]	$k \times 10^{-5}$ [s^{-1}]	k_{rel}
O	340	0.057	1
N–NH–C(=O)–OMe	22	0.875	15.4
N–N(Na)–C(=O)–OMe	1.5[a)]	12.836	225

a) 85 % yield.

Scheme 11.9 Influence of the nature of the electronwithdrawing group at C-2 on the rate of the Claisen rearrangement according to Ponaras (1983) [18].
k calculated from reported $t_{1/2}$ according to $k = \ln 2/t_{1/2}$.

solvent, T

solvent	T [°C]	$t_{1/2}$ [h]	$k \times 10^{-5}$ [s^{-1}]	k_{rel}
THF	66	22	0.875	1
MeOH	66[a)]	15	1.284	1.5
$MeOH/H_2O$ 1/1	79[b)]	0.8	24.077	27.5
toluene	111	0.8	24.077	27.5

a) 91 % yield. b) 84 % yield.

Scheme 11.10 Influence of the solvent on the rate of the Claisen rearrangement according to Ponaras (1983) [18].
k calculated from reported $t_{1/2}$ according to $k = \ln 2/t_{1/2}$.

solvent, 66 °C

solvent	$t_{1/2}$ [h]	$k \times 10^{-5}$ [s^{-1}]	k_{rel}
THF	30	0.642	1
HMPA	20	0.963	1.5
MeOH	4[a)]	4.814	7.5

a) 95 % yield.

Scheme 11.11 Influence of the solvent on the rate of the Claisen rearrangement according to Ponaras (1983) [18].
k calculated from reported $t_{1/2}$ according to $k = \ln 2/t_{1/2}$.

A quantification of substituent effects on the rate of the Ireland–Claisen rearrangement was undertaken by Ireland et al. [19] and Curran and Suh [20]. The comparison of the rate constants for three acyclic allyl vinyl ether containing different substituents at C-6 illustrated a significant accelerating effect of the methoxy substituent (Scheme 11.12). The comparison of the rate constant for the parent allyl vinyl ether **1** (Table 1) and **3a** (Scheme 11.12) indicates the immense accelerating effect of the siloxy group at the 2-position of an allyl vinyl ether.

ave	R	$k \times 10^{-5}$ [s^{-1}]	k_{rel}
3a	H	1.4 ± 0.1	1
3b	Me	1.7 ± 0.1	1.2
3c	OMe	4.8 ± 0.1	3.4

Scheme 11.12 Kinetic data for the Ireland–Claisen rearrangement of acyclic allyl vinyl ether according to Curran and Suh (1984) [20] (ave=allyl vinyl ether).

The same study provided kinetic data for the Ireland–Claisen rearrangement of cyclic allyl vinyl ethers (Scheme 11.13). "Cyclic" is indicative of the fact that either the allyl or the vinyl ether double bond is part of a ring.

X	$k \times 10^{-5}$ [s^{-1}]	k_{rel}
O	44 ± 0.1	1
CH_2	49 ± 0.1	1.1

Scheme 11.13 Rate constants for the Ireland–Claisen rearrangement of cyclic allyl vinyl ether according to Curran and Suh (1984) [20].

In order to elucidate the origin of the rate-accelerating effect of the 2-trimethylsiloxy group, Gajewski and Emrani studied the relevant secondary deuterium kinetic isotope effects for the Claisen rearrangement of 2-(trimethylsiloxy)-3-oxa-1,5-hexadiene [21]. The results of this study led them to the conclusion that the presence of the Me_3SiO group induces a transition state with much more bond-breaking character compared to the transition state of the parent allyl vinyl ether. They proposed that this alteration of the transition-state structure is a conse-

Scheme 11.14 Using the resonance formalism in order to visualize the stabilizing effect of a 2-siloxy-substituent on the transition state.

quence of the enhanced stability of the 2-(trimethylsiloxy)-1-oxallyl segment compared to the corresponding 2-hydrogen-1-oxallyl moiety in the transition state of the parent allyl vinyl ether.

An empirical approach to the substituent effects in [3,3]-sigmatropic rearrangements was reported by Gajewski and Gilbert in 1984 [22]. In order to explain the influence of substituents on the rate of the rearrangement, two correlation equations were applied to calculate the values of $\Delta G^{\#}$ for a concerted process based on the $\Delta G^{\#}$ values for the non-concerted alternatives (Scheme 11.15). Comparison of the calculated and experimental data led to the conclusion that, "there is, in fact, a fair quantitative correlation between calculated and experimental values ..."

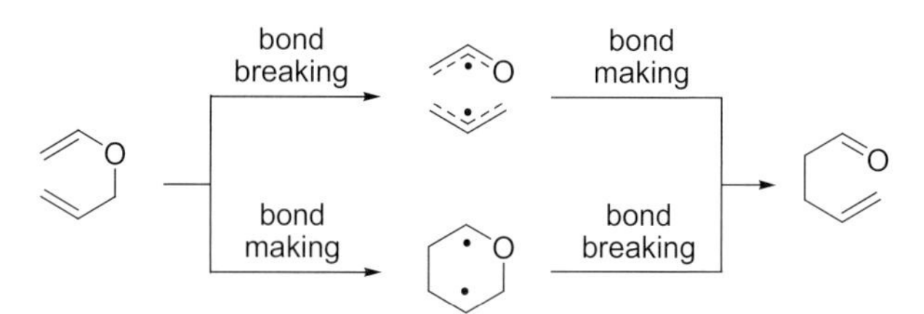

Scheme 11.15 Non-concerted alternatives to a concerted mechanism [23].

The first computational study on the Claisen rearrangement was published by Dewar and Healy in 1984 [23]. The semiempirical MNDO (modified neglect of differential overlap) calculation characterized the transition state as "... an early one, resembling the reactant in structure ..." and the rearrangement as concerted but not synchronous with a transition state of highly polarizable biradicaloid character. Typically, for the semi-empirical approach, bond formation exceeds bond breaking and the transition state is best described by a 1,4-diyl-species. Despite the fact that semi-empirical approaches, due to their experimental parametrization for ground states, do not perform very well in modeling transition-state structures, Dewar and Healy [23] used the MNDO method for calculating activation parameters. They found that the Claisen rearrangement of allyl vinyl ether and its various methoxy- and cyano-substituted derivatives proceeds via a concerted but non-synchronous transition state. They considered radical stabilizing effects of certain substituents to predict the rate-acceleration by electron-donor and electron-acceptor substituents at C-2 and C-5. Except for the electron-donating group at C-5 the predictions matched the experimental results. It was proposed that the failure was due to solvent effects.

For the Ireland–Claisen variant, Wilcox and Babston [25] reported kinetic data in order to quantify the influence of alkyl substituents of different size at the 1- and 5-position on the rate of the rearrangement (Schemes 11.16 to 11.19). This study also considered the effect of the vinyl ether double-bond configuration on the rate of the rearrangement.

OTBS O R **4** — $CDCl_3$, 35 °C → OTBS O R

ave	R	k × 10^{-5} [s^{-1}]	k_{rel}
4a	H	10.8	2.4
4b	Me	4.6	1
4c	Et	5.8	1.3
4d	*n*-Pr	6.38	1.4
4e	*i*-Pr	7.65	1.7
4f	*neo*-Pent	34.4	7.5
4g	CH_2SiMe_3	5.25	1.2

Scheme 11.16 Kinetic data for the Ireland–Claisen rearrangement according to Wilcox and Babston (1986) [24].

Following conventional wisdom, it was not surprising that the increased steric size of the methyl group in **4b** compared to the hydrogen-substituted **4a** led to a decreased rate for the rearrangement (Scheme 11.16). However, a further increase in steric bulk of the C-5 substituent R resulted in an accelerated rearrangement. The *neo*-pentyl-substituted substrate **4f** rearranged 7.5 times as fast as the methyl-substituted allyl vinyl ether **4b**. Wilcox suggested a ground-state destabilization based on a combination of enthalpic and entropic effects of the C-5 substituent. The study was further extended to 1-methyl-substituted *E*- and *Z*-configured allyl vinyl ethers **5** and **6** (Scheme 11.17) [26]. Remarkably, for R=H, the *Z*-configured allyl vinyl ether **6a** rearranged 1.2 times faster than the *E*-configured **5a**. A qualitative analysis of the putative six-membered cyclic chair-like transition state would predict that the rearrangement of **6a** via the transition state **8** is less favorable due to the axial methyl group compared to the transition state **7** for the rearrangement of the *E*-configured allyl vinyl ether **5a** (Scheme 11.18). This apparent discrepancy between qualitative transition state analysis and experimental result may be due to significantly different ground-state stabilities for **5a** and **6a**. According to

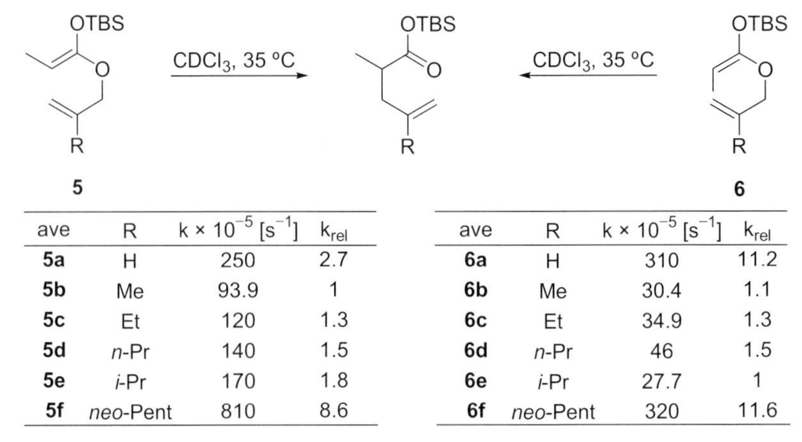

ave	R	k × 10^{-5} [s^{-1}]	k_{rel}
5a	H	250	2.7
5b	Me	93.9	1
5c	Et	120	1.3
5d	*n*-Pr	140	1.5
5e	*i*-Pr	170	1.8
5f	*neo*-Pent	810	8.6

ave	R	k × 10^{-5} [s^{-1}]	k_{rel}
6a	H	310	11.2
6b	Me	30.4	1.1
6c	Et	34.9	1.3
6d	*n*-Pr	46	1.5
6e	*i*-Pr	27.7	1
6f	*neo*-Pent	320	11.6

Scheme 11.17 Influence of the vinyl ether double-bond configuration on the rate of the Ireland–Claisen rearrangement [26].

Wilcox and Babston the *E*-configured trisubstituted vinyl ether double bond is significantly less stable than the *Z*-configured [26]. The destabilizing interaction of the vinyl ether double-bond substituents are less important in the transition states **7** and **8** based on a qualitative analysis that calls for bond elongation and re-hybridization toward the sp^3-carbon atoms. With the exception of **5a**, an increased size of the substituent R leads to rate acceleration for the rearrangement of **5b–f** as discussed for the rearrangement of **4**. The viability of the concept of destabilizing 1,3-diaxial interaction in the transition state **8** can be demonstrated by comparison of the individual rates for the rearrangement of the *E*-configured **5b–f** and the *Z*-configured allyl vinyl ethers **6b–f** (Scheme 11.18). The rearrangement of **5b–f**, via the transition state **7** that lacks the 1,3-diaxial interaction between Me and R, proceeds consistently faster as indicated in Scheme 11.18.

ave	R	k_{rel}
5b/6b	Me	3.1
5c/6b	Et	3.4
5d/6d	*n*-Pr	3.1
5e/6e	*i*-Pr	6.1
5f/6f	*neo*-Pent	2.5

TBSO O R ‡ **7**

TBSO O R ‡ **8**

Scheme 11.18 Schematic presentation of the putative cyclic six-membered chair-like transition state of the Ireland–Claisen rearrangement. **8** should be destabilized by 1,3-diaxial interactions between the methyl group and R.

Wilcox and Babston (1984) also studied the influence of different solvents on the rate of the rearrangement (Scheme 11.19) [27]. Although the effect was classified as "only small," we consider the solvent effect as notable because its existence raises the question of why solvent polarity has an influence on the rate of a sigmatropic rearrangement.

OTBS — solvent, 35 °C → OTBS

5b

solvent	$t_{1/2}$ [min]	k × 10^{-5} [s^{-1}]	k_{rel}
$CDCl_3$	251	4.6	1.7
CD_2Cl_2	315	3.7	1.4
acetone-d_6	333	3.5	1.3
DMF-d_7	335	3.5	1.3
CCl_4	350	3.3	1.2
pyridine-d_5	384	3.0	1.1
benzene-d_6	431	2.7	1
D_3CCN	435	2.7	1

Scheme 11.19 Influence of different solvents on the rate of the Claisen rearrangement indicated by half-life times according to Wilcox and Babston (1984) [27]. Rate constants were calculated according to $t_{1/2} = \ln 2/k$.

The effect of the methoxy group at the 4-, 5- or 6-position of the allyl vinyl ether on the rate of the Claisen rearrangement was investigated by Coates et al. in 1987 using ^{1}H NMR data for their kinetic analysis [28]. Compared to the unsubstituted allyl vinyl ether, the methoxy substituent can either decrease or dramatically increase the rate of the rearrangement (Scheme 11.20). A very impressive 3800-fold rate acceleration was observed for the 4-methoxy-substituted allyl vinyl ether compared to 5-methoxy-substituted allyl vinyl ether. The rate-accelerating effect of the 6-methoxy substituent was less pronounced but still significant.

R^6, R^5, R^4 — benzene-d_6, 80 °C → R^6, R^5, R^4

OMe	k × 10^{-5} [s^{-1}]	k_{rel}
R^4	6.21	3857
R^5	0.00161	1
R^6	0.612	380
-*	0.0649	40

*in nBu$_2$O

Scheme 11.20 Kinetic data for methoxy-substituted allyl vinyl ether according to Coates et al. (1987) [28].

A solvent effect on the rate of the rearrangement was also determined (Scheme 11.21). The significance of the solvent effect depended on the position of the methoxy substituent. For the 6-methoxy-substituted allyl vinyl ether **9c**, the rearrangement proceeded 68 times faster in methanol-d_4 than in benzene-d_6.

ave	T [°C]	solvent	k_{rel}
9a	65	benzene-d_6	1
		acetonitrile-d_3	2.1
		methanol-d_4	18
9b	139	benzene-d_6	1
		acetonitrile-d_3	1.5
		methanol-d_4	-
9c	80	benzene-d_6	1
		acetonitrile-d_3	3.2
		methanol-d_4	68

Scheme 11.21 Relative rate of the Claisen rearrangement depending on the position of a methoxy substituent and the solvent [28].

In order to explain the observed substituent and solvent effects on the rate of the Claisen rearrangement, Coates et al. [28] suggested a dipolar character for the transition state: "... partial delocalization of a non-bonded electron pair from the donor substituent generates a significant degree of enolate–oxonium ion-pair character that stabilizes the TS more than the ground state, ..." (Scheme 11.22).

Scheme 11.22 The dipolar character of the transition state [28].

The influence of a hydrogen-bond-donating solvent was also studied. The rate of the rearrangement of the cyclic allyl vinyl ether **11** increased by changing the solvent from acetonitrile-d_3 to ethanol-d_6 to 80% aqueous ethanol-d_6 (Scheme 11.23). At the same time, more negative entropies of activation were determined. Coates et al. [28] suggested that hydrogen bonding to the allylic ether oxygen atom may be responsible for the rate-accelerating effect of protic solvents. The more negative entropies of activation could then be interpreted assuming an "... increased ordering of solvent molecules arising from stronger hydrogen bonding in the TS".

solvent	k_{rel} (96.9 °C)	$\Delta S^{\#}$ (cal K^{-1} mol^{-1})
acetonitrile-d_3	1	-18.9 ± 2.6
ethanol-d_6	1.9	-28.0 ± 3.4
80 % aq ethanol-d_6	3	-31.8 ± 1.5

Scheme 11.23 Influence of protic solvents on the rate of the Claisen rearrangement [28].

The ongoing quest for an explanation of the rate-accelerating effect of the chorismate mutase has captivated researchers for decades. In this context, independent results concerning the substituent and solvent effects on the rate of the Claisen rearrangement from two research groups have been jointly published [29]. Gajewski et al. (1987) investigated several acyclic and cyclic 2-methoxycarbonyl-substituted allyl vinyl ethers. The selected kinetic data depicted in Scheme 11.24 were determined by integration of ^1H NMR spectra and verify the rate effect of the phenyl substituent at the allylic ether segment of the allyl vinyl ether. A particularly pronounced rate-accelerating effect was detected if the phenyl group was positioned at C-4 of the allyl vinyl ether (**10e**). The corresponding half-life time (1.1 h) indicates a significant rate of the Claisen rearrangement even at room temperature.

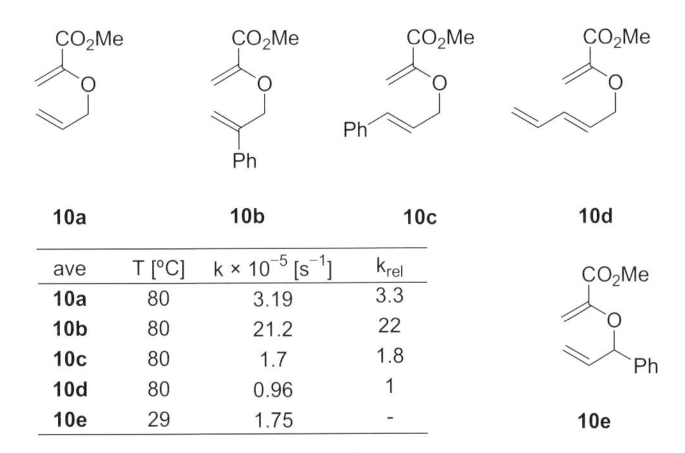

ave	T [°C]	k × 10^{-5} [s^{-1}]	k_{rel}
10a	80	3.19	3.3
10b	80	21.2	22
10c	80	1.7	1.8
10d	80	0.96	1
10e	29	1.75	-

Scheme 11.24 Substituent effect on rate of the Claisen rearrangement in CCl_4 [29]. The data were calculated from the reported half-life times according to $t_{1/2} = \ln 2/k$.

solvent	$k \times 10^{-5}$ [s^{-1}]	k_{rel}
cyclohexane	0.299	1
CCl_4	0.42	1.4
CH_3OH	0.58	1.9
CH_3OH, H_2O (3/1)	4.18	14

Scheme 11.25 Solvent effect on the rate of the Claisen rearrangement according to Gajewski et al. (1987) [29].

As exemplified in Scheme 11.25, Gajewski also observed a small but significant solvent effect on the rate of the rearrangement which led him to the conclusion that, "There must be some polar character in the transition state judging by the response to polar solvents."

In the second part of the study, Ganem and Carpenter also detected substituent (at C-4) and solvent effects on the rate of the Claisen rearrangement [29]. In accordance with the data given in Scheme 11.26, a mixture of water and methanol provided the most significant accelerating effect on the rate of the rearrangement.

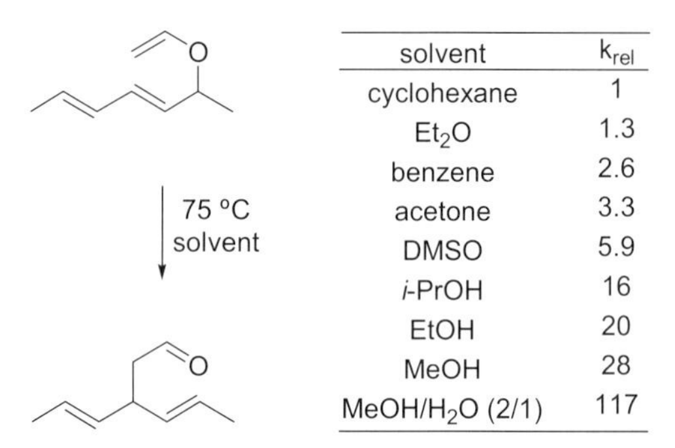

solvent	k_{rel}
cyclohexane	1
Et_2O	1.3
benzene	2.6
acetone	3.3
DMSO	5.9
i-PrOH	16
EtOH	20
MeOH	28
$MeOH/H_2O$ (2/1)	117

Scheme 11.26 Solvent effect on the rate of the Claisen rearrangement according to Gajewski et al. (1987) [29].

In order to explain the experimental results, Gajewski argued with the radical stabilizing ability of the substituents, whereas Ganem and Carpenter gave heavier emphasis to the dipolar nature of the transition state.

The results of a computational study of the transition structures of the Claisen rearrangement were published by Vance et al. in 1988 [30]. The authors provided an analysis of the geometrical and electronic properties of the boat and chair transition structures for the Claisen rearrangement of the parent (unsubstituted) allyl vinyl ether. The results were based on calculations on the RHF level of theory with correlation-corrected energies obtained by MP2/6–31G* frequency analysis. It was found that larger basis sets in combination with polarization functions and electron correlation are vital for predicting energies close to experiment. In contrast to the semi-empirical calculations, the transition state is more dissociative, i.e., shows more bond-breaking than bond-forming character. Geometrical properties (bond lengths) showed certain parallels to the Cope rearrangement and analogous allylic radicals. The charge distribution does not change significantly going from ground state to transition state. Exceptions are C-4 and C-6, which become more negative and more positive, respectively. A comparison of the boat and the chair transition structures revealed that the chair is more stable than the boat by 6.6 kcal mol^{-1} at the RMP2/6–31G* level of theory.

The influence of the nature of the solvent on the rate of the rearrangement was further investigated by Brandes et al. (1989) [31]. The kinetic data were obtained by integration of the corresponding ^1H NMR spectra. As outlined in Scheme 11.27, the rate of rearrangement of sodium carboxylate was significantly faster in pure water than in any other solvent combination. Pure trifluoroethanol was found to be superior compared to methanol in accelerating the rearrangement.

O $(CH_2)_6CO_2Na$ —(60 °C, solvent)→ O $(CH_2)_6CO_2Na$

solvent	solvent ratio	$k \times 10^{-5}$ [s^{-1}]	k_{rel}
H_2O	-	18	66.7
H_2O/CF_3CH_2OH	3/1	6.9	25.6
H_2O/CF_3CH_2OH	1/1	3.7	13.7
H_2O/CF_3CH_2OH	1/3	2.8	10.4
CF_3CH_2OH	-	2.6	9.6
H_2O/MeOH	3/1	11	40.7
H_2O/MeOH	1/1	4.6	17.0
H_2O/MeOH	1/3	1.6	5.9
MeOH	-	0.79	2.9
H_2O/DMSO	3/1	3	11.1
H_2O/DMSO	1/1	1.6	5.9
H_2O/DMSO	1/3	0.93	3.4
H_2O/DMSO	1/9	0.27	1

Scheme 11.27 Solvent effect on the rate of the Claisen rearrangement according to Brandes et al. (1989) [31].

A very similar observation was made for the corresponding ester (Scheme 11.28) [31]. The rearrangement proceeded faster in trifluoroethanol compared to any other solvent combination investigated. However, with respect to the nature of the transition state of the Claisen rearrangement, Brandes et al. concluded that, "... even in polar, hydroxylic solvents, the Claisen rearrangement has substantially less polar character than *bona fide* solvolyses." [31].

Semi-empirical calculations on the AM1 level were performed to study the Claisen rearrangement of various substituted allyl vinyl ethers by Dewar and Jie

(1989) [32]. The results of open- and closed-shell calculations implied that, "... the TSs should be regarded as closed-shell species with relatively little biradical character...". It was furthermore predicted that the rearrangement proceeds through a chair-like

$(CH_2)_6CO_2Me$ —(60 °C, solvent)→ $(CH_2)_6CO_2Me$

solvent	solvent ratio	$k \times 10^{-5}$ [s^{-1}]	k_{rel}
CF_3CH_2OH	-	4.7	56.0
H_2O/MeOH	1/1	3.6	42.9
H_2O/MeOH	1/3	1.4	16.7
MeOH	-	0.72	8.6
H_2O/DMSO	1/3	0.82	9.8
H_2O/DMSO	1/9	0.50	6.0
EtOH	-	0.51	6.1
i-PrOH	-	0.42	5.0
MeCN	-	0.25	2.98
acetone	-	0.18	2.1
benzene	-	0.17	2.0
cyclohexane	-	0.084	1

Scheme 11.28 Solvent effect on the rate of the Claisen rearrangement [31].

transition state and affords preferentially *E*-configured rearrangement products. Various substituent effects were analyzed computationally. Due to the large exothermicity of the Claisen rearrangement, the authors predicted a single transition state with an intermediate character between biradicaloid (concerted asynchronous) and aromatic synchronous. The calculated substituent effects on the activation parameters were discussed in terms of MO theory, polarizability of the transition state and steric effects.

The substituent and solvent effects on the rate of the Claisen rearrangement may be explained by assuming that the nature of the transition state of the Claisen rearrangement is located in a continuum between two mechanistic extremes (Scheme 11.29). The two extremes reflect the possibility of a rearrangement that proceeds through a transition state with either an early bond-making or an early bond-breaking character. In the case of an early bond-making process, the mechanistic extreme may be visualized by the zwitterion **11** or the 1,5-diyl **12**. On the other hand, the mechanistic extreme that results from an early bond-breaking process may be expressed by the ion pair **13** or the radical pair **14**.

As pointed out by Gajewski, "The substituent rate effects can be explained by alteration in transition-state structure between the two extremes ... in response to the nature and position of substituents" [33]. In other words, one would expect that, for instance, the presence of radical stabilizing group at C-5 (R^5 = Ph) would alter the nature of the transition state of the Claisen rearrangement toward an

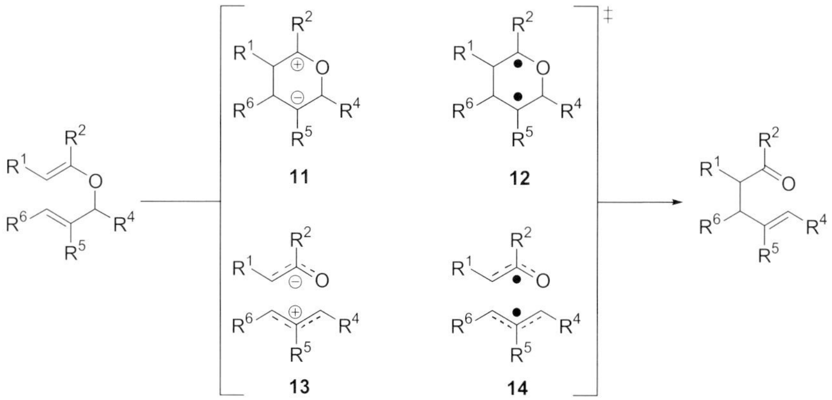

Scheme 11.29 The nature of the transition state is located in the continuum between the mechanistic extremes of an early bond-braking or bond-making process. However, the aromatic nature of the transition state should be considered.

early bond-making character with radical character at C-2 and C-5 due to its ability to stabilize the putative radical at C-5. In order to differentiate between the radical and the ionic effects of substituents, Gajewski compared the rate effects of different substituents from which he expected that they either stabilize charges or radicals [34]. For this purpose, the trifluoromethyl group was chosen as a substituent that has the ability to act as an electron-withdrawing substituent but cannot stabilize radicals. Introducing the F_3C group at the C-2 position of the allyl vinyl ether led to no significant solvent effect on the rate of the rearrangement (Scheme 11.30). According to Gajewski et al. [34], this observation indicates that the corresponding transition state is not polarized.

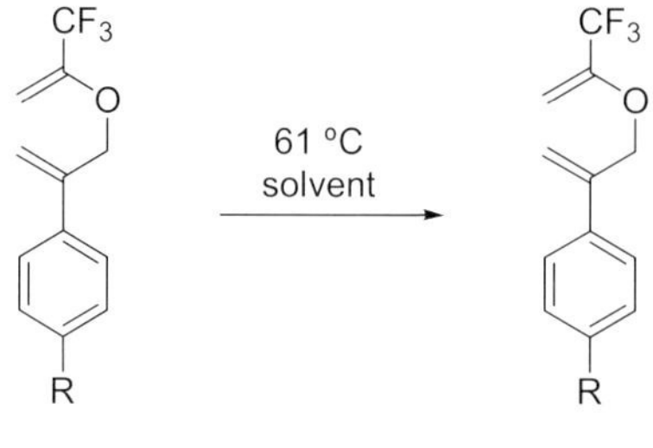

61 °C
solvent

R	solvent	$k \times 10^{-5}$ [s^{-1}]	k_{rel}
H	cyclohexane-d_{12}	3.09	1
H	methanol-d_4	5.47	1.8
OMe	cyclohexane-d_{12}	3.33	1.1

Scheme 11.30 Kinetic data determined from ^{1}H NMR data [34].

A very instructive part of the study compared the rate effect of several substituents at the 2-position of the allyl vinyl ether (Scheme 11.31). Although the rate constants were determined in different solvents, a general trend is quite obvious. The presence of an electron-withdrawing group clearly accelerates the rearrangement compared to the parent allyl vinyl ether or the 2-methyl-substituted derivative. From the three electron-withdrawing groups studied, the cyano group provided the greatest accelerating effect whereas the methyoxycarbonyl and the trifluoromethyl group showed an equally beneficial effect on the rate of the Claisen rearrangement. Gajewski et al. proposed that ground-state destabilization might be responsible for the observed rate-accelerating effect of the trifluormethyl group compared to the parent allyl vinyl ether [34].

R^2 — 80 °C, solvent → R^2

R^2	solvent	$k \times 10^{-5}$ [s^{-1}]	k_{rel}
H	benzene-d_6	0.0711	1
Me	benzene	0.3316	4.7
CO_2Me	CCl_4	2.9715	41.8
CF_3	cyclohexane	2.7622	38.8
CN	n-Bu_2O	5.077	71.4

Scheme 11.31 Rate constants for 2-substituted allyl vinyl ethers calculated from the Arrhenius parameters [34]. $R = 1.985792$ cal mol^{-1} K^{-1}, 1 cal = 4.187 J.

The kinetic data presented so far clearly indicate the susceptibility of the C-4 position of the parent allyl vinyl ether to rate-accelerating substituent effects. Consequently, the presence of a C-4 *n*-pentyl substituent led to a 100 times faster rearrangement (Scheme 11.32). Replacement of the remaining hydrogen atom on the C-4 by a methyl or trifluoromethyl group further increased the rate of the rearrangement. Compared to the trifluoromethyl group, the methyl group exerted a slightly increased rate-accelerating effect.

R^4 n-C_5H_{11} — 80 °C, C_6D_6 → R^4 n-C_5H_{11}

R^4	$k \times 10^{-5}$ [s^{-1}]	k_{rel}
H	7.0761	1
Me	217,17	30,7
CF_3	69.8232	9.9

Scheme 11.32 Rate constants for 4-substituted allyl vinyl ethers calculated from the Arrhenius parameters [34]. $R = 1.985792$ cal mol^{-1} K^{-1}.

Finally, the lack of a rate-accelerating effect of a single trifluoromethyl group if present as the substituent at the 4-position compared to the parent allyl vinyl ether was interpreted by Gajewski et al. [34] as an indication for the importance of the radical stabilizing capability of a substituent at C-4 (Scheme 11.33). A polarized transition state should have been stabilized by the presence of a C-4 trifluoromethyl group. Instead, according to Gajewski et al. [34], the radical stabilizing capability of the 4-cyano substituent was more significant leading to a 200-fold rate acceleration.

80 °C
solvent

O
R^4
O
R^4

R^4	solvent	$k \times 10^{-5}$ [s^{-1}]	k_{rel}
H	benzene-d_6	0.0711	1
CF_3	benzene-d_6	0.0968	1.4
CN	*n*-Bu_2O	13.6269	192

Scheme 11.33 Rate constants for 4-substituted allyl vinyl ethers calculated from the Arrhenius parameters [34]. $R = 1.985792$ cal mol^{-1} K^{-1}.

With their solvation model AM1-SM2 Cramer and Truhlar examined the influence of water on the reaction rate of the Claisen rearrangement [35]. They compared different methoxy-substituted allyl vinyl ethers with respect to solvation energy changes caused on one hand by interactions with the first hydration shell, which is expressed quantitatively in the $G°_{CDS}$ term and, on the other hand, by electronic effects (ΔG_{ENP}). The results of the study indicate that polarization and hydrophilic effects are responsible for the observed aqueous acceleration. Particularly, the polarization of the oxygen atom appears to dominate the significance of the ΔG_{ENP} term. The frequently discussed hydrophobic effect, however, participated with only a small and substrate-independent contribution to the solvation energy. Despite the correct prediction of trends for the rate acceleration as a function of the substitution pattern, the absolute values are underestimated. This is due to the semi-empirical approach of calculating the Mulliken charges, which provided the basis of further calculations of the electric polarization effects.

In order to attribute observed solvent effects on rates and equilibrium constants of different reactions to physical properties of the solvent, Gajewski evaluated different multiparameter approaches [36]. He compared four equations, namely the KOPMH, Taft, Beak and Swain equations, to explain the solvent effects on the Claisen rearrangement. Each equation consisted of at least two parameters that represent measurable properties of the solvent. By comparison of the correlation coefficients, he found that Taft's four-parameter approach most accurately described the experiment ($r = 0.994$) followed by the KOPMH and Beak equation (r=0.967 and 0.968, respectively). The significant parameters therein are the ced and hydrogen-bonding terms, whereas the Kirkwood–Onsager term, which describes dipolar interactions, is of minor importance. On the molecular level, these results can be explained by the negative activation volume, accounting for the importance of the ced term and partial formation of the carbonyl-function in the transition state, which operates as a possible hydrogen-bond acceptor.

Severance et al. studied the consequences of hydration on the Claisen rearrangement of the parent allyl vinyl ether computationally on the RHF/6–31G(d) level of theory [37]. A profile for the change in free energy of hydration (ΔG_{hyd}) along the minimum energy reaction path (MEP) was calculated using Monte Carlo (MC) simulations. The calculation predicted that the transition state has the lowest Gibbs free energy of hydration. The computed rate acceleration was not due to partial ionization in the transition state. Instead, the MC simulations predicted for the transition state an enhanced polarization of the oxallylic fragment, an increased solvent accessibility of the oxygen atom and an increased dipole moment. It was suggested that, "... the results promote catalyst designs that incorporate two or more hydrogen-bond-donating groups positioned to interact with the oxygen in the TS."

Kupczyk-Subotkowska et al. studied heavy-atom kinetic isotope effects (KIE) of the thermal Claisen rearrangement of the parent [2-^{14}C]-, [4-^{14}C]-, [6-^{14}C]- and [^{18}O]-allyl vinyl ether [38]. From the experimental KIEs they concluded, "It is concerted; a two-step process in which one bond is made (or broken) before the other is broken (or formed) is ruled out." From isotope effect calculations (BEBOVIB program) it was deduced that, "... the C_4–O bond is 50–70% broken and the C_1–C_6 bond 10–30% formed in the transition structure."

For the parent allyl vinyl ether, solvent effects on the rate of the Claisen rearrangement were studied computationally by Davidson and Hillier applying a number of variants of the continuum model [39]. The authors found that on the RHF/6–31G* level, the polarizable continuum model (PCM) was most successful in predicting the barrier lowering in di-*n*-butyl ether and water. Based on calculated atomic charges and the dipole moment, the authors concluded that, "... there is considerably more electron polarization in the transition state than in the ground state ..."

By studying the secondary kinetic isotope effect at the C-4 and C-6 of the parent allyl vinyl ether in *m*-xylene and aqueous methanol, Gajewski and Brichford gained evidence against an ionic transition state for the Claisen rearrangement [40]. They concluded that, "... bond breaking has not proceeded to a larger extent in the aqueous media than in *m*-xylene". They explicitly stated that the rate acceleration in aqueous media due to increased hydrogen bonding to the transition state and/or due to an increased solvent cohesive energy density is not inconsistent with the experimental KIEs.

Geometries, charge separation and kinetic isotope effects for the transition structure of the Claisen rearrangement of the parent allyl vinyl ether were predicted on different levels of theory (RHF/6–31G*, MCSCF/6–31G*, CASSCF/6–31G*) by Yoo and Houk in 1994 [41]. The isotope effects calculated on the CASSCF/6–31G* level were closest to the experimental values. The optimized MP2/6–31G* structure of the transition structure was 1,4-diyl-like whereas the CASSCF/6–31G* calculation predicted more oxallyl–allyl radical-pair character.

In a computational study published in 1994, Wiest et al. reported results from density functional theory (DFT) calculations using large basis sets [42]. Because of the relative flatness of the potential hypersurface of the Cope and Claisen rearrangements, computational approaches had been equivocal in predicting a realistic picture of the transition state. Therefore, local and non-local density functional theory was examined for their ability to distinguish between the possible 1,4-diyl and aromatic transition states of the [3,3]-sigmatropic rearrangements. Geometric parameters of the chair and boat conformers, energies and kinetic isotope effects (KIEs) were calculated and compared to experimental data. For both types of [3,3]-rearrangement, generalized-gradient-approximation approaches like BLYP or the hybrid functional B3LYP were necessary to produce good agreement with experimental results. For example, using non-local DFT in combination with 6–31G* and 6–31+G** basis sets led to estimations of activation energies that were within 1 kcal mol^{-1} of the experimental value. In comparison to MP2 and RHF methods, B3LYP gave a better description of the actual transition structure, which is best described as having aromatic character.

$d_{C\text{-}C}$ 2.312 Å

[O]‡

$E^{\#}$ (160 °C, gas phase)= 26.8 kcal/mol (B3LYP/6-31G*)
$E^{\#}$ (80 °C, gas phase)= 30.6 kcal/mol (Schuler, Murphy)

$d_{C\text{-}O}$ 1.902 Å

Scheme 11.34 Results from a computational study by Wiest et al. [42].

Sehgal and co-workers found an important correlation between solvent effects and transition-state characteristics by modeling the Claisen rearrangement in aqueous solution by a combined quantum mechanical and molecular mechanical approach [43]. Key geometrical parameters of the optimized transition structures for substituted allyl vinyl ethers were calculated on the RHF/6–31G* level of theory. First they showed that the calculated change of dipole moment and the free energy of solvation between the transition state and the reactant in water is correlated in a free energy relationship, e.g., with increasing dipole moment the solvent effect is also enlarged. The authors pointed out that it is crucial to use an appropriate transition-state geometry in condensed phase simulations for good predictions because the dipole moment is a function of charge separation and computational methods that predict a transition state that is too close and will certainly underestimate solvent effects. Furthermore, it was found that solvent effects are a function of the substitution pattern, as substituents can influence the charge distribution and bond lengths in the transition state. By applying a dipolar model to the transition state, it was predicted that electron-donating groups at the allylic fragment and electron-withdrawing groups at the oxallylic fragment will lead to an increase of ionic character and dipole moment. A large synergistic effect was calculated when a disubstituted allyl vinyl ether with R^2= cyano and R^6= methoxy was studied.

A computational study concerning the aqueous acceleration of the Claisen rearrangement of the parent allyl vinyl ether was published in 1995 by Davidson and Hillier [44]. Two explicit water molecules and an implicit description of the remaining solvent using PCM were considered for the calculation of the reactant and the transition state on the RHF/6–31G* level of theory. For the reactant, it was predicted that only one water molecule is associated with the ether oxygen atom by hydrogen bonding. However, in the transition state, the two water molecules are hydrogen bonded to ether oxygen atom with bond lengths close to 2 Å. The predicted energies for the reactant and the transition state indicate a "substantial lowering of the free energy of activation upon solvation by the two water molecules". It was furthermore predicted that "... upon solvation the transition state is clearly more dissociative" and that "... bond breaking has occurred to a larger extent in aqueous solution than in the gas phase." However, the predicted kinetic isotope effects at C-4 were not in agreement with the reported experimental values.

In the context of the chorismate mutase problem, Wiest and Houk studied substituted acyclic allyl vinyl ethers as model systems for chorismate [45]. Three different levels of theory were used (RHF/6–31G*, RHF/6–31+G*, BLYP/6–31G*) and all three predicted a "... loose, aromatic-type transition state." It was found that the influence of substituents on the geometry and energy of the calculated transition structure can be dependent on the computational method. The authors concluded, however, that "... relative activation energies for all three methods follow the same pattern and give reasonable estimates of substituents effects on the energies of activation." Calculated relative rate constants were found to be in good agreement with experimental data. Remarkably, a very large transition-state stabilization was calculated in the presence of a carboxylic group at either C-2 or C-6. This result was interpreted as either a consequence of a stabilizing interaction between the negative charge on the carboxylate and a partial positive charge in the allylic segment of the transition state or as the result of the higher polarizability of the aromatic-type transition state compared to the substrate. The presence of two carboxylate groups destabilized the transition state by electrostatic repulsion. A comparison of calculated and experimental kinetic isotope effects revealed that the RHF calculations afforded KIEs that were systemically too high whereas the BLYP/6–31G* calculation provided values in much better agreement with the experimental data. Several cyclic allyl vinyl ethers as models for chorismate were also investigated using the same computational methods. In order to gain insight into the molecular mode of catalysis of the chorismate mutase, the consequences of the interaction of hydrogen bond donors with substituted and non-substituted allyl vinyl ethers were calculated on the RHF/6–31G** level of theory. The most significant effect was predicted for the interaction of an ammonium ion or a proton with the ether oxygen atom of the parent allyl vinyl ether. The ammonium ion binds much more tightly to the allylic ether oxygen atom in the transition state than in the allyl vinyl ether. The forming and breaking bond are elongated and the transition structure is considerably polarized. The corresponding activation energy is lowered by 11.4 kcal mol^{-1} compared to the Claisen rearrangement in the absence of ammonium ion [46]. These trends are much more pronounced in the protonated allyl vinyl ether: "Although the activation energy is lowered by 39.3 kcal mol^{-1}, the calculated transition structure has to be described as a loose complex between an enol and an allyl cation."

level of theory	$\Delta H^{\#}_{298K}$ [kcal mol^{-1}]	$\Delta S^{\#}_{298K}$ [cal mol^{-1} K^{-1}]	ΔH_{298K} [kcal mol^{-1}]
HF/6-31G*	39.26	−7.9	−4.47
HP2/6-31G*	16.03	-	−6.27
MP4(SDTQ)/6-31G*	20.57	-	−6.40
QCISD(T)/6-31G*	23.18	-	−5.56
B3LYP/6-31G*	20.16	−8.3	−4.00

Scheme 11.35 Activation enthalpy, activation entropy and reaction enthalpy for the thio-Claisen rearrangement of allyl vinyl sulfide predicted on different levels of theory (1996).

The thermal thio-Claisen rearrangement of allyl vinyl sulfide was studied theoretically for the first time in 1996 [47]. Energies were predicted with various methods using the 6–31G* basis set (Scheme 11.35). In this regard, the authors stated that "... B3LYP hybrid functional which is less expensive in computational time gives $\Delta E^{\#}$ values close to those computed at our highest level of theory." Comparison of the predicted activation barrier for the Claisen rearrangement of allyl vinyl ether [41] and allyl vinyl sulfide indicated a lower activation barrier for the thio-Claisen rearrangement. The calculated reaction enthalpy was rather low and is in agreement with the experimentally observed reversibility of the thio-Claisen rearrangement of allyl vinyl sulfide. Intrinsic reaction coordinate calculations on the HF/6–31G* and B3LYP/6–31G* level of theory provided the MEP. A natural bond orbital analysis was utilized to discuss the nature of the transition state. Solvent effects on the rate of the rearrangement were simulated using self-consistent reaction-field (SCRF) calculations associated to a continuum model. An activation barrier lowering in all solvents was predicted: "... weak (0.3 kcal mol^{-1}) in alkanes, substantial (2.5 kcal mol^{-1}) in methanol or water."

In 1996, a leading review concerned with the mechanistic aspects of the Claisen rearrangement was published by Ganem [48].

X	d_{X-C4} [Å]	d_{C1-C6} [Å]
O	1.902	2.312
S	2.324	2.127

Scheme 11.36 Comparison of the predicted bond lengths of the forming and the breaking bond at the B3LYP/6–31G* level of theory. The values for X=O have been taken from [42].

The Claisen rearrangement of hydroxy-substituted allyl vinyl ethers was studied computationally on the RHF/6–31G* level of theory by Yoo and Houk [49]. Geometries and energies were reported for the substrates, transition structures and products. Since hydroxy-substituted allyl vinyl ethers cannot and/or have not been investigated experimentally, the computed values were compared to the experimental data reported for the methoxy or siloxy substituted derivatives. The Mullikan charge differences between substrate and transition structure was calculated (RHF/6–31G*) and the authors concluded that, "There is enolate/allyl cation character in the transition state." However, no relationship between the extent of charge separation and magnitude of rate acceleration was calculated and the authors state that, "Accounting for solvation via SCRF calculations did not significantly change the charge separation in the transition state of any of the cases studied ..." A Marcus equation was applied to separate the intrinsic and thermodynamic contribution of substituent effects on the activation energy. For instance, it was calculated that the decreased activation energy for the 2-hydroxy substituted allyl vinyl ether is mainly due to the thermodynamic stability of the rearrangement product, which stabilizes the transition state by an amount proportional to the position of the transition state along the reaction coordinate. On the other hand, the transition state for the rearrangement of the 5-hydroxy substituted allyl vinyl ether is destabilized mainly due to an intrinsic factor. Frontier molecular orbital theory was then used to rationalize the influence of substituents on the intrinsic and thermodynamic contribution to the transition-state stability. The FMO analysis led to the conclusion that the effect of hydroxy substituents on the oxallylic fragment of the allyl vinyl ether is mainly thermodynamic and on the allylic fragment of mainly intrinsic nature. The insightful study closed with the notion that, "The optimized transition structures and the charge separation in the transition state for the substituted allyl vinyl ethers show weak dipolar character in the transition state."

In continuation of the aforementioned study, Aviyente et al. studied computationally the rate effects of a cyano, an amino and a methyl group at C-2 or C-6 of the parent allyl vinyl ether [50]. Reactant, transition and product structures were calculated on the RHF/6–31G* level of theory. Activation energies were determined at the B3LYP/6–31G* and RHF/6–31G* level. It was found that the B3LYP/6–31G* calculation (within 0.5–2.4 kcal mol^{-1}) predicted the experimental values with greater accuracy than the RHF/6–31G* calculation (within 1.0–2.3 kcal mol^{-1}). The calculations predicted that a 2-cyano and a 2-amino function lower the activation energy significantly. A 6-cyano group increased whereas the 6-amino group slightly decreased the activation energy. A 2- or 6-methyl group showed only small effects on the activation energy. The calculated activation energies of the gas-phase rearrangement were compared to the rearrangement in benzene and acetonitrile on the RHF/6–31G* level using the SCRF approach. The authors conclude that, "Except for 6-CN, SCRF calculations are consistent with transition states which are more polar than ground states ... therefore, the activation energies decrease in solution." A Marcus theory formalism was used to dissect the activation energy into a thermodynamic and an intrinsic component. This approach led the authors to the conclusion that the presence of a 2-cyano group mainly lowers the intrinsic barrier and that this lowering arises "from the stabilization of the developing partial negative charge on the enolate group in the transition state." This explanation was further specified by a frontier molecular orbital (FMO) analysis, which revealed that the 2-cyano group shows a stabilizing interaction with the large HOMO coefficient at the C-2 of the transition state. The authors continued their analysis and proposed that the lowering of the activation energy by a 2-amino function is a thermodynamic factor due to the formation of the thermodynamically favorable amide function. If present at C-6, a methyl, a cyano and an amino function should stabilize the reactant by conjugation and, therefore, make the reaction less exothermic. Nevertheless, "The positive charge on the allyl moiety in the transition structure is better stabilized by an electron-donor group at C-6; hence, the intrinsic barrier lowering is substantial for NH_2." In agreement with this explanation, the FMO analysis predicted a stabilizing interaction between the C-6 amino function and the LUMO of the transition state. The substituent rate effects were also analyzed based on the calculated charge distribution in the transition states of the parent and the substituted allyl vinyl ethers. The extent of charge separation was found to be dependent on the nature and position of the substituent. Generally, the oxallylic fragment carries a partially negative and the allylic fragment a partially positive charge. The intrinsic contribution to the activation energy is then a consequence of the ability of a substituent to stabilize partial charges. Consequently, "... the electron acceptor group, CN, can stabilize the partially negative oxallyl moiety of the transition state the most ..."

In 1997, Arnaud and Vallée published a theoretical study concerned with substituent and solvent effects on the rate of thio-Claisen rearrangement of allyl vinyl sulfide [51]. Geometries and energies of the transition structures were predicted on different levels of theory (Scheme 11.37). The authors concluded from the results of their calculation that, "All TSs present the same features, i.e., chair-like, aromatic type transition structures. However, substituents induced sizeable changes in the geometry." It was predicted that a 2-amino-5-cyano disubstitution will lower the activation energy significantly, increase the exothermicity of the rearrangement and that the corresponding transition structure will be highly polarized. The results from SCRF computations associated with a continuum model predicted that "... all TSs are stabilized by solvation." Calculations indicated only a weak effect of solvation on the lengths of the breaking and forming bonds.

NH2 S → [NH2 S]‡ → NH2 S

$\Delta H^{\#}$= 15.8 kcal/mol, ΔH= −19.4 kcal/mol (298 K, B3LYP/6-31G*)

Scheme 11.37 The thio-Claisen rearrangement was studied computationally by Arnaud and Vallée in 1997 [51]. Compared to the only slightly exothermic Claisen rearrangement of allyl vinyl sulfide, the presence of the 2-amino function increases the exothermicity of the rearrangement.

The leading review article concerned with mechanistic aspects of the Claisen rearrangement was published in 1997 by Gajewski [52]. He emphasized solvent effects and provided a useful summary to each paragraph. With respect to the experimental solvent effects on the rate of the Claisen rearrangement his summary was: "The Claisen rearrangement is more than 200 times faster in water than in cyclohexane; solvent studies suggest that both solvent self-association (hydrophobicity in the case of water) and hydrogen bond donation are important factors."

Wood and Moniz reported kinetic data for the Claisen rearrangement of allyl vinyl ethers that contain an enol moiety [53]. Their data underlines the rate-accelerating effect of a substituent in the 4-position, which was nullified when the enol was converted into an enol trifluoroacetate (Scheme 11.38).

40 °C
benzene-d_6

X	R^1	R^4	R^5	$t_{1/2}$ [min]	$k \times 10^{-5}$ [s^{-1}]	k_{rel}
H	Ph	Me	H	8.8	131.28	122.7
H	Me	Me	H	9.2	125.57	117.4
H	Ph	H	H	118	9.79	9.2
CF_3CO	Ph	Me	H	860	1.34	1.3
H	Ph	H	Me	1080	1.07	1

Scheme 11.38 Kinetic data according to Wood and Moniz (1999) [53]. k calculated from $t_{1/2}$ according to $k = \ln 2/t_{1/2}$.

In the context of the chorismate mutase problem, Menger and coworkers reported a detailed computational analysis of the Claisen rearrangement of the parent allyl vinyl ether on the B3LYP/6–31G* level of theory [54]. The Claisen rearrangement of a collection of allyl vinyl ethers that are structurally related to chorismate were also investigated. The results from this part of the study will not be discussed and the reader is referred to the chapter that is concerned with the chorismate mutase catalyzed Claisen rearrangement. For the parent allyl vinyl ether, intrinsic reaction coordinate (IRC) calculations were used to predict the MEP from the transition structure to the nearest local minima, allyl vinyl ether and 4-pentenal. Geometric and energetic data for the transition structure predicted by various levels of theory for the Claisen rearrangement of allyl vinyl ether were compared and the authors concluded that, "Clearly, the Becke3LYP method is superior to other DFT methods (BLYP and SVWN) and MP2, all of which include contributions to electron correlation effects," (Scheme 11.39).

Analysis of the MEP from allyl vinyl ether to 4-pentenal derived from individual IRC calculations led the authors to dissect the reaction path into three segments. The first segment (3 kcal mol^{-1} energy increase) is "characterized by conformational reorganization around the three σ-bonds in the allyl vinyl ether." Within the

$d_{C\text{-}C}$ 2.314 Å (33 % formed)

δ= +0.211

δ= −0.211

$d_{C\text{-}O}$ 1.904 Å (51 % broken)

$\Delta H^{\#}$ (gas phase)= 27.2 kcal/mol (B3LYP/6-31G*)
$\Delta S^{\#}$ (gas phase)= −5.3 eu (B3LYP/6-31G*)
$\Delta H^{\#}$ (gas phase)= 29.4 kcal/mol (Schuler, Murphy)
$\Delta S^{\#}$ (gas phase)= −8.4 eu (Schuler, Murphy)

Scheme 11.39 Results from the computational study by Menger (1999) [54]. Menger suggested a "delocalized aromatic transition state with low sensitivity to polar effects."

second segment (15 kcal mol^{-1} energy increase) "the calculated structures maintain non-stationary chair-like shapes with gradual compression of the reacting termini from 4.0 to 2.6 Å." The third and last segment (10 kcal mol^{-1} energy increase) "is characterized by substantial bond reorganization." From the calculated data for the transition structure and the reactant, the authors suggested three potential options to accelerate the Claisen rearrangement: compression of reacting termini of the allyl vinyl ether to shorter contact distance, stabilization of the increased dipole moment of the transition structure by ion-dipole interaction and interaction of a catalyst with the non-bonding electrons of the oxygen atom of the allyl vinyl ether.

The comparison of experimental and theoretically predicted kinetic isotope effects (KIEs) can be used to probe the accuracy of the computationally predicted transition structure provided that the experimental KIEs have been determined accurately. A comparison of predicted and experimental KIEs by Meyer et al. [55] led to the conclusion that, "There is a firm disagreement in about half the cases between predicted and literature experimental heavy atom KIEs for both the aliphatic and aromatic Claisen rearrangements". Therefore, they reinvestigated the experimental KIEs for the Claisen rearrangement of allyl phenyl and allyl vinyl ether and compared the determined values (solution) with the calculated data (gas phase). For the Claisen rearrangement of allyl vinyl ether, the transition structure was calculated on the MP4(SDQ)/6–31G* level of theory and the predicted KIEs were in excellent agreement with the new experimental KIEs. The authors collected and compared previously reported data for the calculated distance between C-1/C-6 and O/C-4 and added their own predicted data (Scheme 11.40).

The publication of Singleton et al. [55] concluded with a discussion of the nature of the transition state of the Claisen rearrangement of allyl vinyl ethers. In the context of the 1,4-diyl versus aromatic versus diallyl controversy, the authors pointed out that the "Claisen transition state cannot be described as a tradeoff between the 1,4-diyl and allyl/oxallyl extremes – it has the structural properties of both" and the importance of the aromatic character of the transition state was emphasized.

A computational study aimed at the investigation of the dynamical solvent effects for the Claisen rearrangement of the parent allyl vinyl ether was published by Hu et al. in 2000 [56]. The reactant, product and transition structure were calculated in the presence and absence of two water molecules on the B3LYP/6–31G**

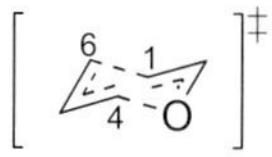

level	d_{C4-O} [Å]	d_{C1-C6} [Å]
AM1	1.58	1.84
MP2/6-31G*	1.80	2.20
MP4(SDQ)/6-31G*	1.85	2.20
RHF/6-31G*	1.92	2.27
B3LYP/6-31G*	1.90	2.31
B3LYP/6-311+G**	1.95	2.38
CASSCF/6-31G*–6e/6o	2.10	2.56

Scheme 11.40 Summary of predicted geometric data of the transition structure calculated on various levels of theory (gas phase) [55].

level of theory. Considering a transition state in which two water molecules are hydrogen-bonded to the oxygen atom of the allyl vinyl ether they concluded that "the transition state is stabilized by enhanced hydrogen bonding of the water molecules to the solute oxygen." Based on calculated bond lengths and partial charges they predicted that "... solvation causes the transition state to be looser and more dissociative" and that "... the presence of the water molecules increases the polarity of the transition state more than that of the reactant."

The geometry and energy of the reactants, transition structures and products of the Claisen rearrangement of cyano-, amino- and trifluoromethyl-substitued allyl vinyl ethers were predicted by Aviyente and Houk on the B3LYP/6–31G* level of theory [57]. A Marcus theory approach was used to separate the thermodynamic and intrinsic contributions of the substituents to the activation energy. Concerning the nature of the transition state it was predicted that "the transition structures have polarized diradical character, because C–O bond breaking exceeds C–C bond making in the transition state." Data that predict the intrinsic and thermodynamic contributions to the activation energy for CN, NH_2 and CF_3 at any position of the allyl vinyl ether have been tabulated. From the available data, an example is depicted in Scheme 11.41 that provides the calculated data for the synthetically interesting 2-substituted allyl vinyl ethers. The data clearly demonstrate that substituents influence the predicted nature of the transition state with respect to "looseness," dipole moment and charge separation. Following the Marcus formalism, the transition state stabilization by the 2-amino substituent is due to the thermodynamic stability of the rearrangement product (amide function). The 2-cyano and the 2-trifluoromethyl substituents are predicted to stabilize the transition state due to an intrinsic factor. The reason for the intrinsic stabilization was discussed in terms of charge and/or radical stabilizing abilities of the corresponding substituents.

The preference for a chair-like transition for the thermal Claisen rearrangement of acyclic allyl vinyl ethers is usually very dependable. However, the Claisen rearrangement of cyclic allyl vinyl ethers is sometimes accompanied by a deteriorated

X	transition structure d_{O-C4} [Å]	d_{C1-C6} [Å]	μ [D]	Δδ	ΔE [kcal/mol]	$\Delta E^{\#}$ [kcal/mol]	$\Delta\Delta E_{intrinsic}$ [kcal/mol]	$\Delta\Delta E_{thermodyn}$ [kcal/mol]
NH_2	1.83	2.34	1.78	0.20	−33.3	21.0	−0.6	−6.1
CN	1.90	2.27	4.60	0.28	−16.5	23.9	−4.7	+0.9
CF_3	1.93	2.31	3.10	0.25	−17.0	23.9	−4.5	+0.7
H	1.90	2.31	2.10	0.21	−18.3	27.7	0	0

Scheme 11.41 Predicted structural and energetic data for the Claisen rearrangement of 2-substitued allyl vinyl ethers according to Aviyente and Houk (2001) [57]. Gas-phase calculations were performed on the B3LYP/6–31G* level of theory. μ = dipole moment, $\Delta\delta$ = charge separation in the transition state, ΔE = reaction energy, $\Delta E^{\#}$ = activation energy, $\Delta\Delta E_{intrinsic}$ = relative intrinsic contribution to the activation energy; $\Delta\Delta E_{thermodyn}$ = relative thermodynamic contribution to the activation energy.

simple diastereoselectivity, which can be explained by a competition between boat- and chair-like transition states. In order to address this problem theoretically, Houk et al. [58] investigated the Claisen rearrangement of various 2-methoxy-substituted cyclohexenyl vinyl ethers on the B3LYP/6–31G* level of theory.

Scheme 11.42 Activation energies for the Claisen rearrangement of cyclohexenyl vinyl ethers via either a chair- or a boat-like rearrangement have been predicted on the B3LYP/6–31G* level of theory [58].

The calculations indicated that the preference for either the chair- and/or the boat-like transition state is a consequence of the attempt to minimize non-bonding interactions between R^1 and OMe with the cyclohexenyl ring substituents in the transition state.

Since the initial kinetic study of Hurd and Pollack 55 years ago, the combined application of physical organic and computational methods have provided some instructive details of the mechanism of the Claisen rearrangement. Nowadays, it is safe to say that the aliphatic Claisen rearrangement in the absence of strong acids proceeds through an early transition state, which is characterized by a concerted but asynchronous bond-breaking and bond-making process. However, the extent of bond-breaking and bond-making is highly dependent on substituents and solvents, as is the geometry and energy of the transition state. Most likely, there will never be a single unique description for the "real" nature of the transition state for the Claisen rearrangement of every allyl vinyl ether under all reaction conditions. Whether the nature of the transition state of the Claisen rearrangement can be best described as ionic or biradicaloid and to what extent aromaticity is relevant depends very much on the utilized level of theory in computational studies and subjective interpretations of experimental results. With respect to catalysis, it is quite obvious what the interested chemist can do to develop a catalyst for the Claisen rearrangement. A "simple" hydrogen bond donor that provides two hydrogen-bond-donating sites to the allylic ether oxygen atom should suffice! At least in theory ...

References

1 L. Claisen *Chem. Ber.* **1912**, *45*, 3157–3167.

2 For an excellent review that covers mechanistic aspects of the Claisen rearrangement, see: F.E. Ziegler *Chem. Rev.* **1988**, *88*, 1423–1452.

3 C.D. Hurd, M.A. Pollack *J. Am. Chem. Soc.* **1938**, *60*, 1905–1911.

4 F.W. Schuler, G.W. Murphy *J. Am. Chem. Soc.* **1950**, *72*, 3155–3159.

5 L. Stein, G.W. Murphy *J. Am. Chem. Soc.* **1952**, *74*, 1041–1043.

6 (a) A.W. Burgstahler *J. Am. Chem. Soc.* **1960**, *82*, 4681–4685. See also: (b) E. Marvell, J. Stephenson *J. Org. Chem.* **1960**, *25*, 676–676.

7 J.W. Ralls, R.E. Lundin, G.F. Bailey *J. Org. Chem.* **1963**, *28*, 3521–3526.

8 W.M. Lauer, E.I. Kilburn *J. Am. Chem. Soc.* **1937**, *59*, 2586–2588.

9 R.K. Hill, A.G. Edwards *Tetrahedron Lett.* **1964**, *5*, 3239–3243.

10 (a) P. Vittorelli, T. Winkler, H.-J. Hansen, H. Schmid *Helv. Chim. Acta* **1968**, *51*, 1457–1461. (b) H.-J. Hansen, H. Schmid *Tetrahedron* **1974**, *30*, 1959–1969.

11 H.M. Frey, D.C. Montague *Trans. Faraday Soc.* **1968**, *64*, 2369–2374.

12 C.L. Perrin, D.J. Faulkner *Tetrahedron Lett.* **1969**, *10*, 2783–2786.

13 S.J. Rhoads, J.M. Watson *J. Am. Chem. Soc.* **1971**, *93*, 5813–5815.

14 (a) J.J. Gajewski, N.D. Conrad *J. Am. Chem. Soc.* **1979**, *101*, 2747–2748. (b) J.J. Gajewski, N.D. Conrad *J. Am. Chem. Soc.* **1979**, *101*, 6693–6704.

15 C.J. Burrows, B.K. Carpenter *J. Am. Chem. Soc.* **1981**, *103*, 6983–6984.

16 C.J. Burrows, B.K. Carpenter *J. Am. Chem. Soc.* **1981**, *103*, 6984–6986.

17 B.K. Carpenter *Tetrahedron* **1978**, *34*, 1877–1884.

18 A.A. Ponaras *J. Org. Chem.* **1983**, *48*, 3866–3868.

19 R.E. Ireland, R.H. Mueller, A.K. Willard *J. Am. Chem. Soc.* **1976**, *98*, 2868–2877.

20 D.P. Curran, Y.G. Suh *J. Am. Chem. Soc.* **1984**, *106*, 5002–5004.

21 J.J. Gajewski, J. Emrani *J. Am. Chem. Soc.* **1984**, *106*, 5733–5734.

22 J.J. Gajewski, K.E. Gilbert *J. Org. Chem.* **1984**, *49*, 11–17.

23 M. S. Dewar, E. F. Healy *J. Am. Chem. Soc.* **1984**, *106*, 7127–7131.

24 J.J. Gajewski *Acc. Chem. Res.* **1980**, *13*, 142–148.

25 C.S. Wilcox, R.E. Babston *J. Am. Chem. Soc.* **1986**, *108*, 6636–6642.

26 We are aware that the consequent application of the IUPAC nomenclature would lead the opposite *E*/*Z*-assignment due to the presence of the silicone. However, for the sake of comparability with non-Ireland-type allyl vinyl ethers, we prefer to use the *E*/*Z* descriptors under disregard of the priority of the silicon.

27 C.S. Wilcox, R.E. Babston *J. Org. Chem.* **1984**, *49*, 1451–1453.

28 R.M. Coates, B.D. Rogers, S.J. Hobbs, D.P. Curran, D.R. Peck *J. Am. Chem. Soc.* **1987**, *109*, 1160–1170.

29 J.J. Gajewski, J. Jurayj, D.R. Kimbrough, M.E. Gande, B. Ganem, B.K. Carpenter *J. Am. Chem. Soc.* **1987**, *109*, 1170–1186.
30 R.L. Vance, N.G. Rondan, K.N. Houk, F. Jensen, W.T. Borden, A. Komornicki, E. Wimmer *J. Am. Chem. Soc.* **1988**, *110*, 2314–2315.
31 E.B. Brandes, P.A. Grieco, J. J. Gajewski *J. Org. Chem.* **1989**, *54*, 515–516.
32 M.J.S. Dewar, C. Jie *J. Am. Chem. Soc.* **1989**, *111*, 511–519.
33 J.J. Gajewski *Acc. Chem. Res.* **1997**, *30*, 219–225.
34 J.J. Gajewski, K.R. Gee, J. Jurayj *J. Org. Chem.* **1990**, *55*, 1813–1822.
35 C.J. Cramer, D.G. Truhlar *J. Am. Chem. Soc.* **1992**, *114*, 8794–8799.
36 J.J. Gajewski *J. Org. Chem.* **1992**, *57*, 5500–5506.
37 (a) D.L. Severance, W.L. Jorgensen *J. Am. Chem. Soc.* **1992**, *114*, 10966–10968. (b) W.L. Jorgensen, J.F. Blake, D. Lim, D.L. Severance *J. Chem. Soc., Faraday Trans.* **1994**, *90*, 1727–1732.
38 L. Kupczyk-Subotkowska, W.H. Saunders, H.J. Shine, W. Subotkowski *J. Am. Chem. Soc.* **1993**, *115*, 5957–5961.
39 M.M. Davidson, I.H. Hillier *J. Chem. Soc., Perkin Trans. 2* **1994**, 1415–1417.
40 J.J. Gajewski, N.L. Brichford *J. Am. Chem. Soc.* **1994**, *116*, 3165–3166.
41 H.Y. Yoo, K.N. Houk *J. Am. Chem. Soc.* **1994**, *116*, 12047–12048.
42 O. Wiest, K.A. Black, K.N. Houk *J. Am. Chem. Soc.* **1994**, *116*, 10336–10337.
43 A. Sehgal, L. Shao, J. Gao *J. Am. Chem. Soc.* **1995**, *117*, 11337–11340.
44 M.M. Davidson, I.H. Hillier *J. Phys. Chem.* **1995**, *99*, 6748–6751.
45 O. Wiest, K.N. Houk *J. Am. Chem. Soc.* **1995**, *117*, 11628–11639.
46 Menger later (1999) confirmed this value on the B3LYP/6–31G* level of theory, see Ref. [54]
47 R. Arnaud, V. Dillet, N. Pelloux-Léon, Y. Vallée *J. Chem. Soc., Perkin Trans. 2* **1996**, 2065–2071.
48 B. Ganem *Angew. Chem.* **1996**, *108*, 1014–1023; *Angew. Chem. Internat. Ed. Engl.* **1996**, *35*, 936–945.
49 H.Y. Yoo, K.N. Houk *J. Am. Chem. Soc.* **1997**, *119*, 2877–2884.
50 V. Aviyente, H.Y. Yoo, K.N. Houk *J. Org. Chem.* **1997**, *62*, 6121–6128.
51 R. Arnaud, Y. Vallée *J. Chem. Soc., Perkin Trans. 2* **1997**, 2737–2743.
52 J.J. Gajewski *Acc. Chem. Res.* **1997**, *30*, 219–225.
53 J.L. Wood, G.A. Moniz *Org. Lett.* **1999**, *1*, 371–374.
54 N.A. Khanjin, J.P. Snyder, F.M. Menger *J. Am. Chem. Soc.* **1999**, *121*, 11831–11846.
55 M.P. Meyer, A.J. DelMonte, D.A. Singleton *J. Am. Chem. Soc.* **1999**, *121*, 10865–10874.
56 H. Hu, M.N. Kobrak, C. Xu, S. Hammes-Schiffer *J. Phys. Chem. A* **2000**, *104*, 8058–8066.
57 V. Aviyente, K.N. Houk *J. Phys. Chem. A* **2001**, *105*, 383–391.
58 M.M. Khaledy, M.Y.S. Kalani, K.S. Khuong, K.N. Houk, V. Aviyente, R. Neier, N. Soldermann, J. Velker *J. Org. Chem.* **2003**, *68*, 572–577.

Subject Index

The Claisen Rearrangement. Edited by M. Hiersemann and U. Nubbemeyer

ISBN: 978-3-527-30825-5

b

d

o

p

u

v

Related Titles

A. Berkessel, H. Gröger

Asymmetric Organocatalysis

From Biomimetic Concepts to Applications in Asymmetric Synthesis

2005. ISBN 978-3-527-30517-9

J. Zhu, H. Bienaymé (Eds.)

Multicomponent Reactions

2005. ISBN 978-3-527-30806-4

M. A. Sierra, M. C. de la Torre

Dead Ends and Detours

Direct Ways to Successful Total Synthesis

2005. ISBN 978-3-527-30644-2

F. Zaragoza Dörwald

Side Reactions in Organic Synthesis

A Guide to Successful Synthesis Design

2005. ISBN 978-3-527-31021-0

R. Mahrwald (Ed.)

Modern Aldol Reactions

2004. ISBN 978-3-527-30714-2

K. C. Nicolaou, S. A. Snyder

Classics in Total Synthesis II

More Targets, Strategies, Methods

2003. ISBN 978-3-527-30685-5